AF686934

GLOBAL STRENGTH OF SHIPS

For student and professional alike this book provides an all-encompassing overview of the modern theory of global ship strength. Novices will find clear descriptions of the well-established methods, both mathematical and numerical, used worldwide currently. Researchers will find detailed descriptions of the ideas underlying the theoretical basis of modern techniques, whereas professionals will benefit from the fundamentals of research results that have found application in recent rules and design practice. Covering both state-of-practice and state-of-the-art of the subject in a modern and up-to-date manner, this volume offers readers a deeper understanding. It includes many examples of the application of the theory to problems providing the foundation to developing software. One chapter is dedicated to tracing the development of ship structural design from prehistory to today, allowing the reader to comprehend how design and construction practice have evolved and the pivotal turning points in a long and diverse pattern of development.

DR P. A. CARIDIS taught courses in ship strength and inspection, repair and maintenance of ship structures at the National Technical University of Athens for 30 years. He was a member of the International Ship and Offshore Structures Congress committees (1991–2022). Dr Caridis is the author of the recently published companion text *Local Strength of Ship Structures* (Taylor & Francis, 2024) as well as the highly respected publication *Inspection, Repair and Maintenance of Ship Structures, Second Edition* (Witherby Seamanship International, 2009).

'This is a very useful book for engineering and naval architecture students, as well as professionals interested in this topic. It reflects the long teaching experience of Piero A. Caridis and will certainly be used as a reference book across many countries.

It includes extensive coverage of the most important topics required for the analysis and design of global ship structures, all conveniently condensed into one book.

The book deals with the core topics of loads, structural response and longitudinal strength, then expands into various chapters on more recent topics of engineering practice that derive directly from research, which is not often found in other books.

With coverage on uncertainty modelling, probabilistic models of loading and hull girder response, and structural reliability, this book includes the topics on which practising engineers, in particular, may want an update.'

Professor Emeritus Carlos Guedes Soares, University of Lisbon

'This book compiles foundational theories for assessing the global strength of ships during the preliminary design stage into a single volume, offering students a valuable starting point for understanding structural fundamentals in ship design.'

Professor Jeom Kee Paik FREng, University College London

GLOBAL STRENGTH OF SHIPS

Analysis and Design Using Mathematical Methods

P. A. CARIDIS

National Technical University of Athens

CAMBRIDGE
UNIVERSITY PRESS

CAMBRIDGE
UNIVERSITY PRESS

Shaftesbury Road, Cambridge CB2 8EA, United Kingdom

One Liberty Plaza, 20th Floor, New York, NY 10006, USA

477 Williamstown Road, Port Melbourne, VIC 3207, Australia

314–321, 3rd Floor, Plot 3, Splendor Forum, Jasola District Centre, New Delhi – 110025, India

103 Penang Road, #05-06/07, Visioncrest Commercial, Singapore 238467

Cambridge University Press is part of Cambridge University Press & Assessment, a department of the University of Cambridge.

We share the University's mission to contribute to society through the pursuit of education, learning and research at the highest international levels of excellence.

www.cambridge.org
Information on this title: www.cambridge.org/9781009465908

DOI: 10.1017/9781009465892

© P. A. Caridis 2025

This publication is in copyright. Subject to statutory exception and to the provisions of relevant collective licensing agreements, no reproduction of any part may take place without the written permission of Cambridge University Press & Assessment.

When citing this work, please include a reference to the DOI 10.1017/9781009465892

First published 2025

A catalogue record for this publication is available from the British Library.

A Cataloging-in-Publication data record for this book is available from the Library of Congress.

ISBN 978-1-009-46590-8 Hardback

Cambridge University Press & Assessment has no responsibility for the persistence or accuracy of URLs for external or third-party internet websites referred to in this publication and does not guarantee that any content on such websites is, or will remain, accurate or appropriate.

Dedicated to the memory of
Douglas Faulkner, Ronald Cameron and Neil Miller
as well as
The Department of Naval Architecture and Ocean Engineering,
University of Glasgow

Contents

Preface

In recent years a number of texts have appeared on the subject of marine structures, several of which focus on the application of computer-based methods and in particular the finite element method. This has proved its worth in the analysis and design of ship structures and indeed of all types of engineering structures. As implied in the title of this text, the approach followed is to describe basic aspects of ship structural analysis and design using purely mathematical tools; at the same time an emphasis is placed on developing an understanding of the physics of the behaviour of these structures. It is felt that such an understanding is essential, firstly for its own sake, and secondly in order to be able to use computer-based tools effectively. In addition to mathematical models, reference is made to test results as well as full-scale measurements that have been used to gain a deeper insight into ship structural behaviour.

The text is aimed at engineers both within and outside academia, because, to paraphrase the historian Eric Hobsbawm, 'if academic engineers are to justify the resources society devotes to their subject, modest though these are, they should not write exclusively for other academic engineers'. To this end an effort has been made to present the 'state-of-practice' as well as the 'state-of-the-art'. By state-of-practice we understand to be the well-established methods in use by industry, whereas state-of-the-art denotes the approaches that are described in some of the recent research literature and are gradually finding application at the present time or will do so in the near future. The state-of-practice is covered in the earlier chapters of the book, whereas the state-of-the-art is discussed in the later chapters. An understanding of where we stand today may be aided further by tracing developments through time and by discovering the reasons for which particular decisions were taken at various stages in the past. These may have been of a practical nature, reflecting conditions that were different to those of today, or else were dictated by the lack of scientific knowledge that led to the application of different design and construction practices. One chapter of the book is devoted to providing a historical background to the evolution of ship structural practice from very early on up to the present day.

Merchant ships are the subject of this book, and due account is taken of the IACS Common Structural Rules (CSR) for bulk carriers and oil tankers, which draw upon the extensive research work that has been carried out in recent years. An effort is made to provide a background to some of the most important theoretical results upon which the Rules are based. The CSR are applicable to oil tankers and bulk carriers, which at present account for approximately 70 per cent of the world fleet. The introduction of these rules represents an important step forward in ensuring the safety and adequacy of ship structures, since they deal with the subject at a deeper level and in a much more thorough manner than those of individual societies in the past. At the same time, it is necessary to acknowledge the fact that rule development will continue in the future as new applications and new research results will be forthcoming. In principle, classification societies are free to diverge from past practice, provided they adhere to the Goal-Based Standards stipulated by the IMO.

This book deals exclusively with global strength issues, whereas a companion volume, entitled *Local Strength of Ship Structures*, published in 2024, deals with the response of major structural components such as superstructures and bulkheads as well as local strength problems. Inevitably, gaps in understanding may arise when following this top-down approach so that in principle it is necessary to consult both texts. The decision to follow this approach has been influenced primarily by the need to gain a grasp of the characteristic ship-related issues such as longitudinal strength first, before dealing with the more specialised topic of local strength. The volume of material that needs to be presented has necessitated the publication of two separate texts. Global strength issues arise in design and analysis when a ship is considered in its entirety, whereas local strength issues pertain to the behaviour of particular types of structural members.

The majority of ocean-going ships are large vessels that are generally made of some type of steel alloy. It is thus necessary to focus on such types of ships, although reference is made to the use of other materials that are gradually being used more widely in ship construction.

Chapter 1 provides an introduction to ship structures, their particularities of form and their working environment. An overview of current practice in ship structural design of both merchant ships and warships is also included.

Chapter 2 deals with the evolution of ship structures from prehistoric times up to the present. Ancient practices in structural arrangements and fabrication are described and changes in these are traced up to the period of the industrial revolution. The changes that followed as a consequence of scientific progress and the introduction of iron and, subsequently, steel are then described. The last part of the chapter deals with the use of mathematical methods in the nineteenth and twentieth centuries and information technology in the 20th century. This continues to the present day.

The types of loads that act are discussed in the three chapters that follow. Chapter 3 is an introductory chapter on sea loads whereas Chapter 4 deals with the primary (longitudinal) bending loads of the hull girder. Chapter 5 focuses on hull structure–related, mechanical and cargo-related loads.

Hull girder response is the subject of the next four chapters. Chapters 6, 7 and 8 treat linear elastic response, nonlinear response and hull girder vibration, respectively. In Chapter 9 the probabilistic modelling of sea loads as well as structural response are described.

In Chapter 10 the structural design of the hull girder for strength is discussed. Limit states and their application to ship structural design are discussed next, whereas in the last part of the chapter the methodology followed by IACS to determine design loads for use with the Common Structural Rules for bulk carriers and oil tankers is presented.

This brings to a conclusion the presentation of the state-of-practice in ship structural analysis and design. In the chapters that follow the more recent ideas and techniques that have been developed are described. These include reliability theory, the use of the finite element method and optimum ship structural design.

Chapter 11 deals with the concept of uncertainty and the techniques used to allow for it. Uncertainty is a feature of random loading and structural response, and basic ideas of probability and reliability theory are described. An altogether different aspect of uncertainty arises in the decision-making process in ship structural design. In order to resolve this, a different approach is required, and to this end fuzzy logic has been used and is described in the latter part of the chapter.

Chapter 12 provides an account of ship structural reliability theory with a number of examples to illustrate its use. Chapter 13 deals with the application of the finite element method in ship structural analysis and design, followed by example applications. Chapter 14 discusses the development and application of optimisation theory in ship structural design as well as some of the techniques that have been developed in recent years that make use of specialised computer software and operations research methods. Examples are included to illustrate the manner in which these techniques may be applied.

The appendices that are included at the end of the book deal with a number of subjects that supplement the main text. These include certain topics from mathematical theory as well as extracts of relevant rules and standards.

This book and its companion volume are the outcome of the author's involvement with ship structures that started during his undergraduate studies in the

Department of Naval Architecture and Ocean Engineering (NAOE) of Glasgow University under the guidance of D. Faulkner. The Department of NAOE was the oldest university department of naval architecture in the world, founded in 1883, the year in which the John Elder Chair of Naval Architecture was endowed. In 2001, the Department joined the Department of Ship and Marine Technology of the University of Strathclyde to form the Department of Naval Architecture, Ocean and Marine Engineering of both universities. Professor Faulkner was an inspiring teacher, and a serious study of his writings gave one the opportunity to gauge the complexity of the various problems that a naval architect who has to deal with structures faces. It became possible to see how complex problems can be resolved using deceptively simple ideas of mechanics once a deep understanding of the physical aspects of the problem has been acquired. This is an invaluable lesson for any engineer involved in the design of marine structures of any kind.

At this point the author wishes to express his gratitude to Professor Jerolim Andrić of the University of Zagreb for reviewing the example of the structural design of a RoPax ship included in Chapter 14.

Abbreviations

2D	two-dimensional
3D	three-dimensional
ABS	American Bureau of Shipping
ABS GN	American Bureau of Shipping Guidance Note
AD (CE)	Anno Domini (Common Era)
ADH	advanced double hull
AEW	aft end of way
AH	Grade A high strength (steel)
AHL	alternate hold loading
AHP	analytic hierarchy process
ALARP	as low as reasonably practicable
ALS	accidental limit state
AP	aft perpendicular
ASME	American Society of Mechanical Engineers
BCE	Before Common Era
BEM	boundary element method
BL	ballast load
BM	bending moment
BSP	hydrodynamic pressure (beam sea)
BSR	roll motion (Beam Sea)
BV	Bureau Veritas (Class Society)
CAD	computer-aided design
CAM	computer-aided manufacturing
CATS	Cost (to) Avert (a) Ton (of oil) Spilt
CB	centre of buoyancy
CBR	case-based reasoning
CDP	concept design phase
CG	centre of gravity
COV	coefficient of variance
CS	centre of shear
DMF	design modification factor

DNV GL	Det Norske Veritas Germanischer Lloyd (Class Society)
DOF	degrees-of-freedom
DSS	design sea state
DSS	double side skin
DWT	deadweight
EDW	equivalent design wave
EDWF	equivalent design wave (following sea)
EDWH	equivalent design wave (head sea)
ETA	expected time of arrival
ETA	event tree analysis
FBC	Ferry Borges-Castanheta
FEA	finite element analysis
FEM	finite element method
FI	frequency index
FIS	fuzzy inference system
FL	full load
FLS	fatigue limit state
FMEA	failure modes and effects analysis
FMECA	failure modes effects and criticality analysis
FN	frequency number of fatalities
FOSM	first-order second moment
FP	fore perpendicular
FPSO	floating production storage and offloading
FRP	fibre-reinforced plastic
FSA	formal safety assessment
FSM	vertical bending moment in following sea
FSO	floating storage and offloading
FTA	fault tree analysis
GA	general arrangement
GA	genetic algorithm
GBS	goal-based standard
GDP	gross domestic product
GL	Germanischer Lloyd
GT	gross tonnage
GWS	global wave statistics
HAZOP	hazard and operability studies
HSLA	high strength low alloy
HSM	vertical bending moment in head sea
HY	high yield
IACS	International Association of Classification Societies
IACS CSR	IACS Common Structural Rules
ICAF	implied cost to avert a fatality

IMO	International Maritime Organisation
INCOSE	International Council on Systems Engineering
IAHR	International Association for Hydraulic Research
IPDD	integrated product and process development
IRCN	Institut Recherche Centre Navale
ISFEM	intelligent super-sized finite element method
ISO	International Standards Organisation
ISSC	International Ship and Offshore Structures Congress
ISUM	idealised structural unit method
JBP	Joint Bulker Project
JPDF	joint probability distribution function
JTP	Joint Tanker Project
LCB	longitudinal centre of buoyancy
LCG	longitudinal centre of gravity
LNG	liquid natural gas
LP	linear programming
LPG	liquid petroleum gas
LQI	Life Quality Index
LR	Lloyd's Register
LRFD	load resistance factor design
LS	limit state
LSD	limit state design
LTCD	long-term cumulative distribution method
MARPOL	marine pollution
MCDA	multi-criteria decision attribute
MCDM	multi-criteria decision-making
MCR	maximum continuous rating
MCT	moment to change trim
MLE	maximum likelihood estimator
MOGA	multi-objective genetic algorithm
MOPSO	multi-objective particle swarm optimisation
MPP	multi-purpose carrier
MSC	Maritime Safety Committee
MTTF	mean time to failure
MVFOSM	mean-value first-order second moment
NA	neutral axis
NKK	Nippon Kaiji Kyokai
NLFEA	nonlinear finite element analysis
NLP	nonlinear programming
OBO	ore-bulk-oil
OECD	Organisation for Economic Co-operation and Development
OSA	pitching acceleration in oblique sea

OST	wave torsional moment in oblique (stern quartering) sea
PCA	progressive collapse analysis
PDP	preliminary design phase
PL	partial load
PM	Pierson Moskowitz (Spectrum)
PNA	plastic neutral axis
PRHA	preliminary hazard analysis
PSF	partial safety factor
RANS	Reynolds-Averaged Navier Stokes
RAO	response amplitude operator
RCM	risk control measure
RCO	risk control option
RI	redundancy index
RI	reliability index
RINA	Royal Institution of Naval Architects
RMS	root-mean-square
ROPAX	roll-on roll-off passenger
RORO	roll-on roll-off
RPM	revolutions per minute
RSI	reserve strength index
SDOF	single degree of freedom
SF	shear force
SL-7	Sea Land Services
SI	severity index
SLP	sequential linear programming
SLS	serviceability limit state
SM	section modulus
SOLAS	safety of life at sea
SORM	second-order reliability method
SPL	sound pressure level
SRSS	square root of sums
SS	superstructure
STF	Salvesen Tuck Faltinsen
SWATH	small waterplane area twin hull
SWBM	still-water bending moment
TEU	twenty-foot equivalent unit
TMPC	thermo-mechanically processed controlled
TPCM	tonnes per centimetre immersion
TS	Takagi-Sugeno
UK	United Kingdom
ULCC	ultra-large crude carrier
ULS	ultimate limit state

UNCTAD	United Nations Committee on Trade and Development
UR	unified requirement
US	United States
USD	United States Dollar
USS	United States ship
VCG	vertical centre of gravity
VWBM	vertical wave bending moment
VBM	vertical bending moment
VEGA	vector evaluated genetic algorithm
VLCC	very large crude carrier
VLOC	very large ore carrier
VPD	virtual process development
WBM	wave-induced bending moment
WLLN	weak law of large numbers
WSD	working stress design

1
Ship Structures and Structural Design Practice

> The task of assessing the adequacy of a ship's structure is perhaps the most complex structural engineering problem there is.
>
> *E. C. Tupper*

In this book we shall describe concepts and techniques that have been used in the analysis and design of ship structures, the intention being to provide an understanding of their behaviour using the principles of engineering mechanics. At the outset it is important to underline that the form and layout of ships are unique amongst engineering structures because of the purpose for which they are built and the environment within which they operate.

It is possible to employ sophisticated computer-based tools in ship structural analysis and design, but in order to make responsible and effective use of these tools, it is necessary to have a firm understanding of the characteristics and behaviour of ship structural members. This can only be achieved by developing a sound physical insight of the mechanics involved, in parallel with a good command of the corresponding mathematical tools. Difficulties arise from the fact that it is not easy to develop an intuitive understanding of ship structural behaviour because the environment within which they operate, the loadings to which they are subjected to, the materials used and their form lie outside most people's everyday experience.

We shall begin by providing a brief description the types of structures we shall be dealing with, starting from an overview of the world fleet of merchant ships.

1.1 The World Fleet

An idea of the present-day makeup of the world fleet can be gained from Figures 1.1 to 1.3, which contain information on ship types and distribution by deadweight and ship type. Figure 1.1 includes practically all known non-naval ship types, classified according to their purpose.

Clearly the most important group is that of cargo ships, which are used to carry dry and wet cargoes. Passenger ships still play an important role, although the role of ships in passenger transportation nowadays is limited. More important is that of leisure, which is catered for by cruiseships. There exists also a large, diversified fleet of workships that

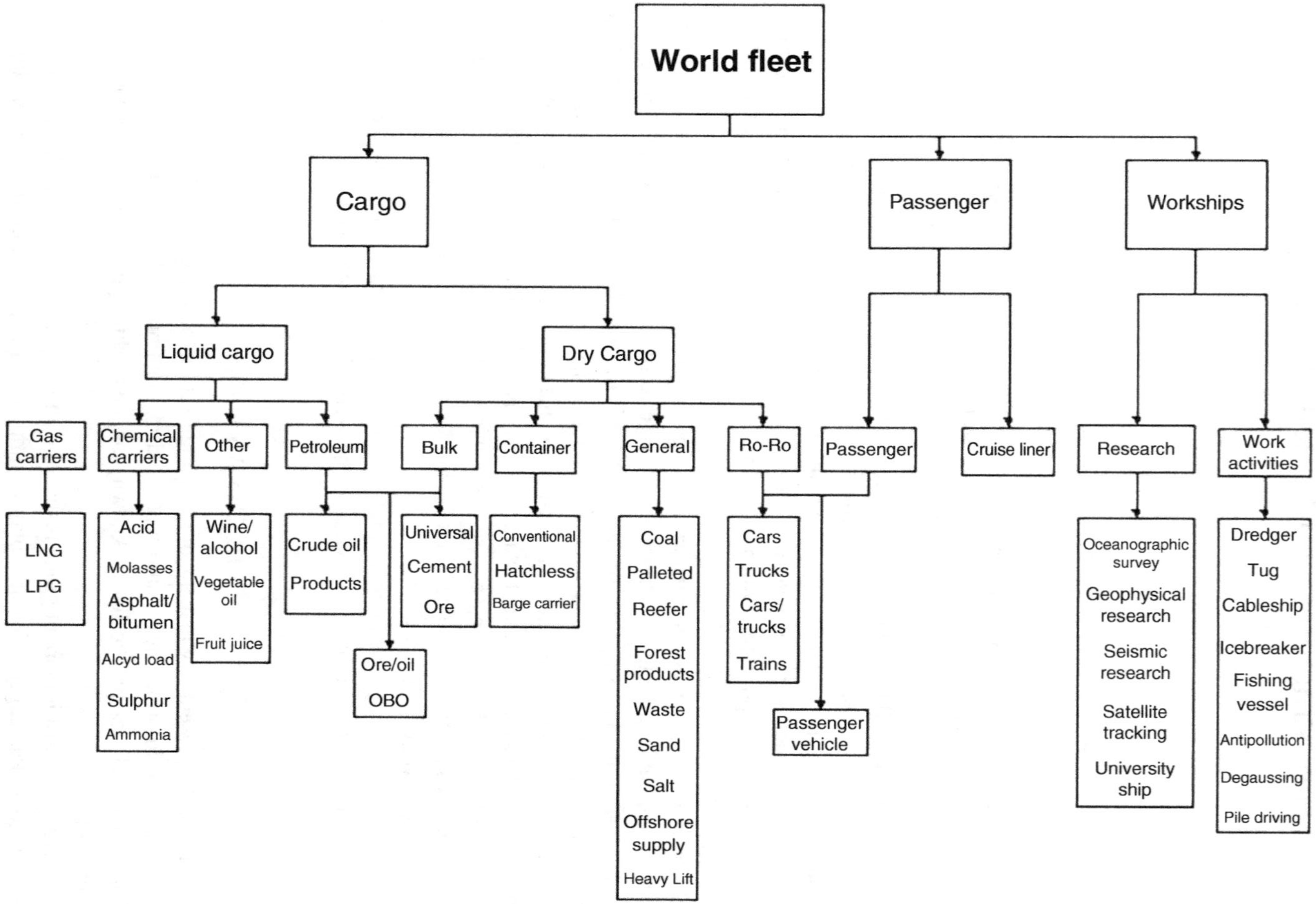

Figure 1.1 Classification of world fleet of merchant ships.

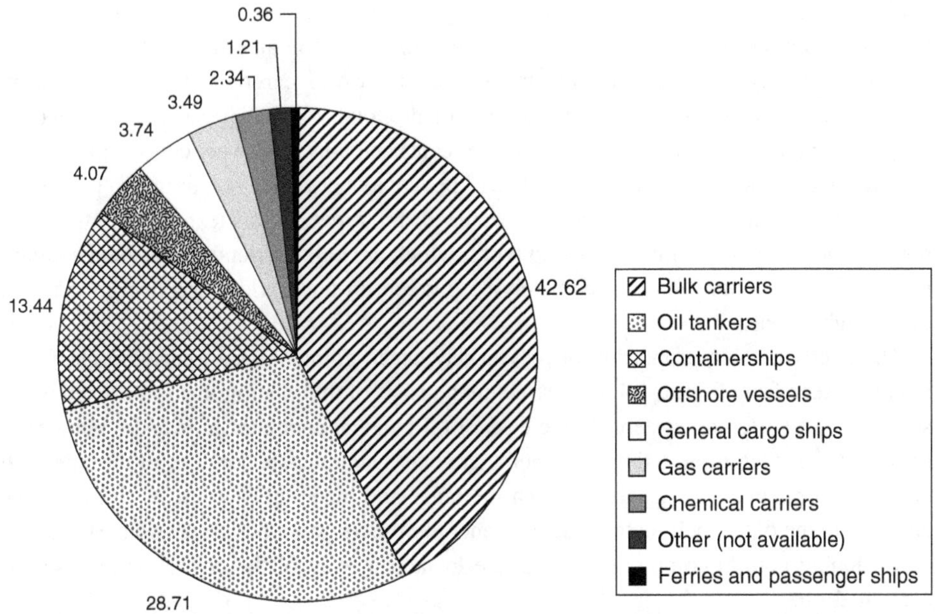

Figure 1.2 World cargo fleet breakdown [28]. (Percentage by deadweight tonnage.)

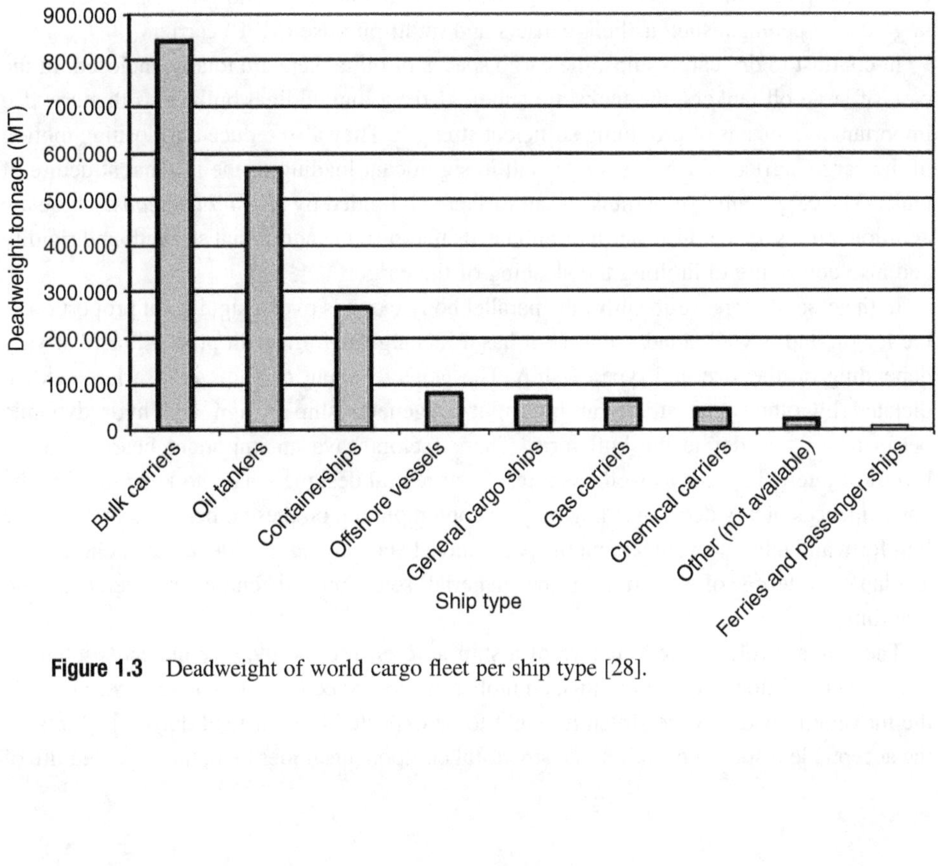

Figure 1.3 Deadweight of world cargo fleet per ship type [28].

perform specialised duties and that cannot be grouped together with the other types. Their structural features are such that they need to be considered on an individual basis.

Figures 1.2 and 1.3 contain information on the world cargo fleet. It is seen from Figure 1.2 that the overwhelming proportion of the total volume of cargo transported by sea is currently carried in bulk carriers (43 per cent) and tankers (29 per cent). At this point in time container ships account for another 14 per cent whereas the remaining 15 per cent is transported in a large number of other specialized ship types. In this text we shall therefore concentrate on the more important ship types, which are bulk carriers, tankers and containerships. These account for approximately 85 per cent of the world cargo fleet in terms of deadweight capacity.

The overwhelming proportion of present-day ocean-going ships are made of steel, and as can be seen from Figure 1.4, their structure consists of thin plating that is stiffened using some type of stiffening system. In the case of oil tankers, containerships and a number of bulk carriers, in addition to the outer shell there exists an inner shell in way of the side shell and bottom plating. The spaces between the outer and inner shells are used to isolate the cargo from the outer shell plating and are generally used to store ballast seawater. In certain cases double bottom compartments that are located in way of the centreline are used to carry fuel oil.

One important role of the double skin is to enhance the pollution prevention capability of the ship. If it is constructed in a cellular form, it furthermore enhances the torsional rigidity of the structure. This is necessary in the case of containerships and other ships that have large deck openings, such as bulk carriers and multi-purpose (MPP) carriers

In contrast to dry cargo ships, the cargo spaces of oil tankers are totally enclosed. In the case of large oil tankers, the tanks are subdivided by longitudinal bulkheads that are also important as a means of providing sufficient strength. They also reduce the sloshing motion of the cargo carried, since this may result in significant loading of the internal structure of tanks. The cargo tanks of oil tankers are further subdivided by *swash bulkheads*. These are additional transverse non-watertight bulkheads that provide additional strength and rigidity and also contribute to limiting the sloshing of the cargo.

In the case of large cargo ships, the parallel body extends over a significant proportion of the length and is wall-sided – that is, it has a rectangular form with proportions that vary depending on the size and type of ship. The general layout of ships is to a large extent dictated by other than structural issues; it is general ship design and hydrodynamic performance that define the hull form. These factors have an important bearing on the layout of internal spaces as well. As a result, structural designers have to keep in mind the consequences of any decisions they may be contemplating on these other aspects. Once the hull form and internal arrangement of spaces have been defined, a wide range of choices are available in terms of structural form, material selection, stiffening arrangements and scantlings.

The primary role of the structure of a ship is to ensure its integrity under its intended operating conditions. Related to this, a number of more specific issues arise. These include the maximum load that the structure ought to be expected to withstand during its lifetime, the acceptable response of individual structural components under load, the expected life of

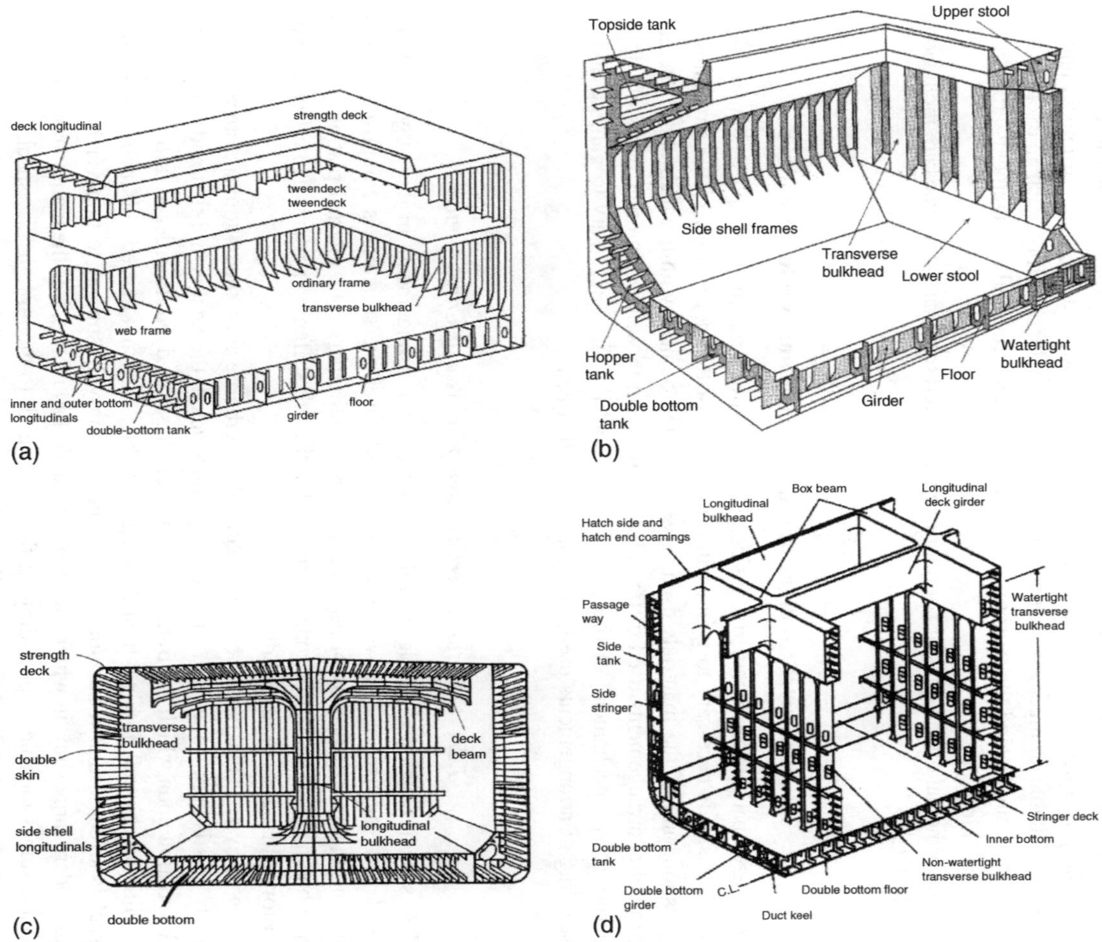

Figure 1.4 Cargo spaces of various merchant ship types. (a) General cargo (MPP) ship; (b) bulk carrier; (c) oil tanker; (d) containership.

the hull structure, the costs of construction, maintenance and repair and, last but not least, the preservation of the environment. These issues will be discussed throughout the book.

In order to appreciate the range of questions that arise, an overview of the different types of configurations and structural layouts that are encountered in present-day ships will be provided. The range in size, function and layout is wide, and it is important to understand how the principles of mechanics can be applied to study existing ship structures and to propose suitable arrangements for a new design. To be able to perform such an exercise it is essential to understand the reasons for which particular structural arrangements are used in different ship types.

The principal dimensions of commonly encountered merchant ship types are included in Table 1.1.[1]

1.2 Structural Configurations, Materials and Modes of Failure

Before embarking on detailed considerations of structural analysis and design of ship structures, it is necessary to gain an understanding of the morphology of the structures in question, and of the materials used, given the environment within which ships operate. Structural arrangements used differ substantially from those used in other fields of engineering (civil, mechanical, offshore, aerospace), and thus it is necessary to gain an understanding of arrangements used, given the loads that act.

1.2.1 Structural Arrangements in Way of Cargo Spaces

The sketches included in Figure 1.4 show structural arrangements in the cargo-carrying spaces of some representative ship types. In all cases the structure is composed of thin plating that is stiffened by orthogonal systems of stiffeners. The internal plating of decks and bulkheads is also orthogonally stiffened, although stiffening systems can differ widely in form, materials used and dimensions.

In many texts that describe the structure of waterborne craft, the bottom structure is cited first. This is because when a ship is being constructed, the keel is laid first and acts as the foundation upon which the frames and shell plating are added. In order to fulfil this role in traditional shipbuilding in which the length of ships does not exceed 100 metres or so, the keel is sized so that it has sufficient strength to provide support to transverse members. In modern large ships in which the length exceeds 100 metres, it is the main (strength) deck that assumes the role of providing sufficient strength to the hull girder, especially when it bends in the longitudinal direction. The bottom structure and the strength deck complement each other in terms of their strength capability and work in tandem in that they act as flanges of the hollow girder that resembles a ship's hull when it bends in the longitudinal

[1] Data for LNG, LPG, containerships and passenger ships from [27].

Table 1.1 *Main dimensions of various ship types [27].*

Ship type and size category	*Cargo capacity*	*Length BP (m)*	*Breadth (m)*	*Draught (m)*
Bulk carriers	(metric tonnes)			
Handysize	10,000–40,000	110–200	20–30	8.0–12.5
Handymax/Supramax	40,000–60,000	200–220	30–32	11.5–12.5
Panamax	60,000–80,000	220–225	32–37	12.5–14.0
Capesize	>80,000	250–350	40–65	15.0–23.0
Containerships	(TEU)			
300–500	5,000	101	17.9	6.3
630–850	10,000	129	22	7.9
1,300–1,500	20,000	165	27	10.0
2,000–2,200	30,000	191	30.4	11.4
2,600–2,900	40,000	211	30.6	12.5
2,100–4,600	30,000–60,000	187–285	32.3	11.3–13.4
4,300–5,600	60,000–70,000	260–268	37.2–40	12.7–14.0
6,300–6,700	80,000–100,000	285–292	40.0–42.8	13.5–14.5
7,200–7,900	100,000–105,000	332–336	42.8	14.0–14.5
8,000–8,500	99,000–102,000	308–312	42.8	14.0–14.5
12,500	140,000	380	56	15
13,500–21,300	170,000–200,000	400	56–59	16
Oil tankers	(metric tonnes)			
Small	5,000–10,000	100–120	16–19	3.0–7.5
Handysize	20,000–30,000	130–160	22–26	8.5–9.0
Handymax	30,000–47,000	138–174	23–32	9.0–11.0
Panamax	70,000–80,000	228.6	32.2	11.0–12.6
Aframax	85,000–115,000	233–239	42–44	11.0–13.6
Suezmax	120,000–180,000	256	45–52	13.5–15.5
VLCC	200,000–300,000	333	59–60	17.7–20.5
ULCC	>300,000	327–440	65–70	21.5–22.8
Gas carriers	(gross tons)			
LNG carriers	20,000	166	27.8	8.4
	30,000	188	31.4	9.2
	50,000	223	36.7	10.4
	80,000	260	42.4	11.5
	100,000	281	45.4	12.1
LPG carriers	3,000–5,000	92–109	16.1–18.6	6.3–7.3
	5,000–10,000	109–136	18.6–22.7	7.3–8.9
	10,000–30,000	136–193	22.7–31.1	8.9–12.1
	30,000–50,000	193–228	31.1–36.0	12.1–14.0
Passenger and cruiseships	(gross tons)			
	3,000	88	16.5	4.3
	5,000	104	18.6	5.0
	10,000	131	21.8	6.4
	20,000	165	25.7	7.8
	30,000	189	28.2	7.8

Table 1.1 (*cont.*)

Ship type and size category	*Cargo capacity*	*Length BP (m)*	*Breadth (m)*	*Draught (m)*
	50,000	224	32.3	7.8
	70,000	250	32.3	8.1
	100,000	281	32.3	8.1

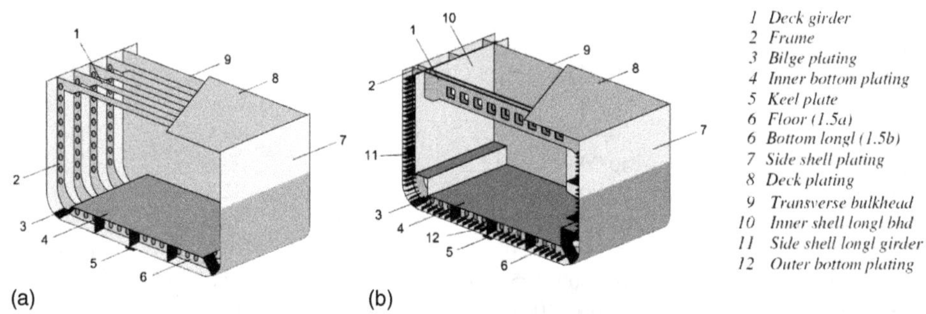

Figure 1.5 Stiffening systems. (a) Transverse stiffening system; (b) longitudinal stiffening system.

direction. Both of these components consist of thin plating that in many cases is longitudinally stiffened.

Since the structure is additionally subjected to transverse loads such as hydrostatic pressure and the weight of the cargo, these components have to be able to carry these loads as well. To support the side shell structure in carrying lateral loads, transverse frames are fitted (item 2 in Figure 1.5a). Frames form part of a stiffening arrangement that generally extends over the whole internal perimeter of any transverse section (frame-floor-frame-deck girder), as seen in Figures 1.4 and 1.5a. Continuity in the transverse direction ensures stiffness and strength. The presence of transverse members such as frames also increases the compressive strength of longitudinal members where these are fitted by acting as points of support lying between stiff members such as bulkheads.

The side shell structure that comprises the outer shell plating, its stiffening and the inner longitudinal bulkhead if fitted, as in the case of the oil tanker and containership (Figures 1.4c, 1.4d), ensures watertightness of the cargo compartment. It is subjected to both lateral and in-plane loading in the longitudinal direction. The critical region of the side shell structure is the strake (a strake is a longitudinal series of shell plates) that is joined to the deck plating (*sheer strake*), which carries the highest in-plane loads. The lateral loading consists of the hydrostatic pressure that acts below the waterline and wave impact above the waterline.

The most important major internal structural components are the *transverse bulkheads* (item 9 in Figures 1.5a–b, also shown in Figure 1.4). These serve a number of functions. Firstly, they subdivide the internal spaces, thereby enabling a variety of cargo types to be transported. Secondly, they provide support to the deck, bottom and side shell structures, since they are much more rigid than transverse frames. Thirdly, they contribute to the torsional strength of the hull girder. Last but not least, they ensure survivability of the ship when in a damaged condition by preventing the flooding of adjacent compartments.

In large ships and especially in oil tankers, the cargo tanks are subdivided by *longitudinal bulkheads* as well (Figure 1.4c). These enhance the longitudinal strength and also provide the number of compartments that are necessary to transport different types of cargoes. Transverse *swash bulkheads* may also be fitted in order to provide sufficient strength to tank structures. These are not watertight and permit the motion of the liquid cargo within the tank, although their presence dampens the fluid motion and thus reduces the magnitude of *sloshing* loads that may otherwise act on the internal structure.

In addition to cargo spaces, the other major components of the hull girder are the fore and aft ends (Section 1.2.3) as well as the superstructures. The loads that act on the fore and aft ends arise from the action of the sea environment and include wave impacts of all types. In the case of the aft end, the external plating is additionally loaded as a result of propeller action. A large part of the volume enclosed within the fore and aft peaks is used for the carriage of ballast seawater and, these tanks are used to adjust trim, since they are furthest from the longitudinal centre of gravity of the ship. Since the most important external loads that act here are lateral loads, in order to withstand these, the fore and aft structures are transversely stiffened. The structure needs to be sufficiently rigid, especially in way of the propulsion equipment in order to withstand the vibration loads that arise as a result of main engine and propeller shaft operation.

In cargo ships, superstructures house the crew and navigation equipment. They are located at the aft end and thus do not contribute to longitudinal strength. In the case of large passenger ships, however, superstructures are particularly important structural items, as they extend over a large proportion of the length of the ship.

1.2.2 Stiffening Systems

In theory, flat plating can carry both lateral and in-plane loads without the need for any type of stiffening. However, it quickly becomes apparent that the structure becomes very inefficient from the points of view of weight, cost and ease of fabrication; in addition, limitations to ship size are imposed. Stiffening systems are thus used and, for ease of fabrication, are arranged in the form of orthogonal grids.

In vessels whose length does not exceed 100 metres, the lateral loads that act, such as hydrostatic pressure and cargo weight, are critical. The most efficient system of stiffening to withstand these loads is the *transverse* stiffening system, which consists of closely spaced frames that subdivide the spacing between *web frames* (Figure 1.4a). In an efficient design the frame extends around the perimeter of the hull envelope and, by a system of brackets that connect the ends of each member, ensures an efficient transfer of loads. Thus, along the bottom structure, *floors* are aligned with side shell frames. These provide continuity and act as supports to the frames by being loaded in compression. They also carry lateral loads, by supporting the inner bottom plating upon which rests the cargo within each hold, this being their primary function. In way of the under-deck structure, *beams* provide continuity to frames and, as in the case of floors, provide support to the strength deck by withstanding lateral loads such as the weight of deck equipment, deck cargo and deck wetting loads. In the case of cargo ships with large deck openings, beams also contribute to the rigidity of the deck. In this way a rectangular *ring frame* is formed, which is an efficient structural arrangement.

In the case of ships whose length exceeds 100 metres, it is longitudinal strength that is critical, and the most efficient stiffening system is the longitudinal one. This consists of closely spaced longitudinal stiffeners whose cross-section is much smaller than that of frames. A large variety of forms is used depending on the magnitude, direction of loads as well as the size and type of vessel. Shipyards around the world use different kinds of sections according to national industry standards and availability, but in every case these are required to fulfil minimum strength requirements that are expressed in classification society rules in terms of the respective section moduli.

In certain ship types both transverse and longitudinal stiffening systems are used (*mixed stiffening system*). Examples are the MPP cargo ship and the bulk carrier shown in Figures 1.4a and 1.4b, respectively. The MPP vessel is seen to be transversely stiffened throughout except in way of the double bottom structure. The bulk carrier, on the other hand, is transversely stiffened only in way of the side shell plating between the top side and lower hopper tanks. Elsewhere it is longitudinally stiffened. In bulk carriers the transverse bulkheads are vertically stiffened with corrugations instead of the more usual flat, vertically stiffened plating. This is done in order to reduce weight and fabrication costs.

One other system that has been proposed in order to simplify fabrication and reduce weight is the *unidirectional* stiffening system. In this system the number of transverse members is substantially reduced when compared to conventional designs. Strength and stiffness are provided by a double hull system in which the inner and outer plating are connected by longitudinal members (Figure 1.6). It has been used by the US Navy in the design of a 43,000

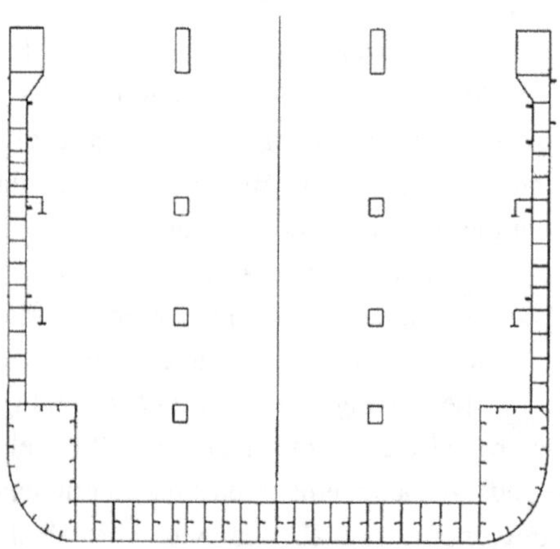

Figure 1.6 Unidirectional system in advanced double hull [2].

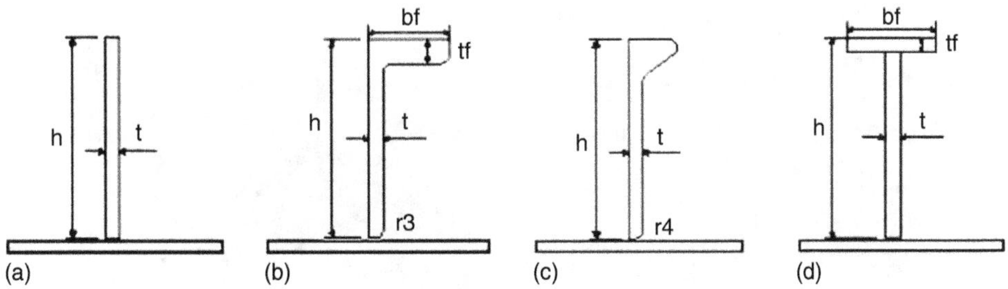

Figure 1.7 Stiffener sections used in ship structures [7]. (a) Flatbar; (b) angle bar; (c) bulb section; (d) T-bar. Reprinted courtesy of Det Norske Veritas, Høvik.

dwt tonne containership as well as by certain Japanese shipyards and has its origin in the cellular type of structure used in containerships. Such an arrangement provides the necessary torsional strength that is required when there are large openings in the strength deck.

Stiffeners can be subdivided into rolled and built-up sections. For primary (longitudinal) strength purposes the types used are shown in Figure 1.7.

Flatbars: Flatbar stiffeners are used as strength deck and longitudinal bulkhead longitudinals. They have lower torsional strength than other types of stiffeners. However, they are easier to produce and hence are competitively priced.

Angle bars: Angle bars have a higher section modulus than flatbars but are prone to twisting when exposed to lateral loading due to their asymmetry about the web axis. They are more prone to fatigue cracking as well as grooving (necking) corrosion.

Bulb profiles: These are widely used because they have a higher section modulus than flatbars. As in the case of angle bars, they are prone to twisting and bending about the line of attachment to the plating because of their asymmetry about the web axis.

T-bars: These have a higher section modulus than other sections and, being symmetrical, are not prone to twisting. As a result, they have good fatigue strength characteristics. Care is needed when selecting the slenderness of the web and the flange in order to avoid local buckling.

1.2.3 Structural Arrangements in Way of the Fore and Aft Ends

The use of transverse stiffening systems is evident in both the fore and aft ends of the vessel (Figure 1.8).

As mentioned previously, the most important loading in these regions of the hull girder are lateral static and time-varying external loads as well as vibration excitation in the case of the machinery spaces aft.

1.2.4 Materials Used in Ship Structures

Figure 1.9 gives an overview of the use of materials in merchant ships. Mild steel is generally used for smaller ships that operate at speeds below 20 knots. For larger ships that

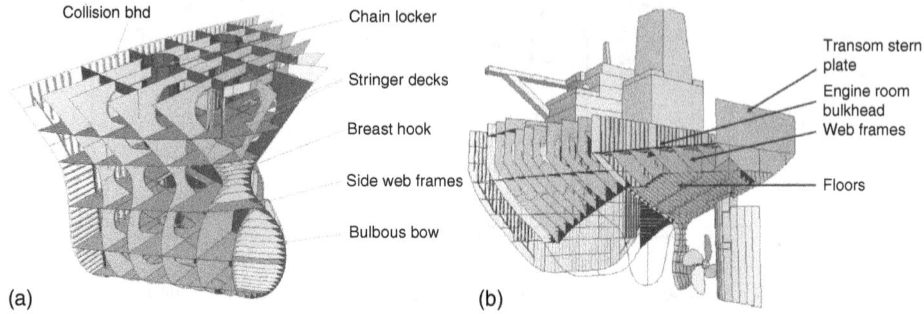

Figure 1.8 Structural arrangements in way of fore and aft ends of a cargo ship [7]. (a) Fore end structural arrangement; (b) aft end structural arrangement. Reprinted courtesy of Det Norske Veritas, Høvik.

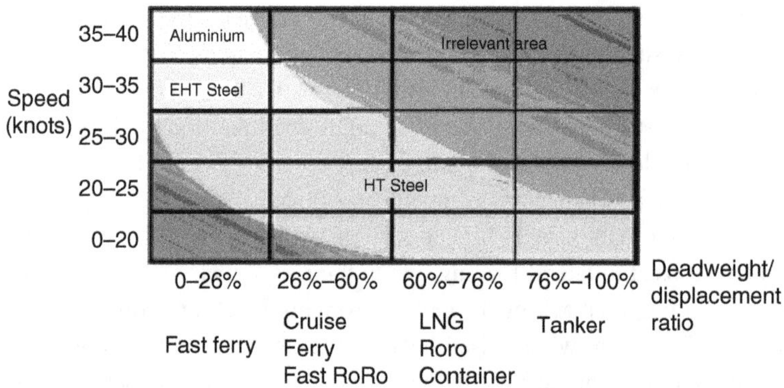

Figure 1.9 Suitability of materials according to the deadweight to displacement ratio and speed [16].

have a high deadweight-to-displacement ratio, such as oil tankers and bulk carriers, in addition to mild steel, high tensile steel is used in order to reduce weight and facilitate fabrication. If, however, the speed requirements exceed 30 knots, as in the case of fast passenger craft, extra high tensile steel or aluminium is used to reduce the weight even further, thereby reducing powering requirements for the desired payload.

The most important criteria upon which material selection is based are strength, stiffness and cost. Other criteria are notch toughness and ease of fabrication, ship construction being a competitive commercial activity. In more specialised applications materials with enhanced properties such as low flammability and low susceptibility to corrosion are required.

Mechanical and Other Properties

Conventional ocean-going ships have been built using steel since the second half of the 19th century. The use of steel in conjunction with improvements in fabrication techniques has permitted the multifold increase of ship size. Its use has continued to the present day

Table 1.2 *Major iron ore producers [9].*

Company	*Country*	*2017 one quarter production, In million tonnes (mt)*
Vale	Brazil	91.8
Rio Tinto	Brazil	85.0
BHP Billiton	Australia, Brazil	55.6
Fortescue Metals	Australia	50.1
Anglo-American	South Africa et al.	15.7
Arcelor Mittal	several countries	14.2

Table 1.3 *Comparison of manufactured material costs.*

Material	*Cost (US$/mt)*
Shipbuilding steel	600–2,000
Aluminium	Steel price × 5
FRP	2,000
Composites	15,000

and will most likely continue in the foreseeable future, given its availability and comparatively low cost (Tables 1.2 and 1.3).

Iron ore is abundantly available in the earth's crust and is mined around the world. Major producers and exporters are Brazil and Australia, represented by the top four companies included in Table 1.2.

The cost of shipbuilding steel, as that of other potential candidates for ship construction, depends on supply and demand and thus fluctuates. For purposes of comparison, Table 1.3 gives an idea of relative costs as they were in 2020.

The success of steel as a prime candidate for ship construction depends not only on its low cost but also on its mechanical properties, that is, its strength, stiffness, hardness, ductility, fatigue and fracture toughness.

Strength is a measure of the ability of a material to transmit a specified load without suffering failure, generally by yielding. For structural applications it is expressed in terms of a critical *yield stress* whose units are N/mm^2 (MPa) (Table 1.4).

Stiffness is a measure of the amount of distortion that a structural member suffers when subjected to a particular load and should be distinguished from elastic modulus which characterizes a material rather than a member. It is measured in units of force/elongation (N/m).

Hardness is a measure of the extent to which a structural member is indented when struck by an object [10]. Thus, the impression of a striker on a soft material is much greater than the one left on a hard material. There are many types of tests that are based either on the measurement of depth (Rockwell, Brinell) or the size of the indentation (Vickers and others) made by a striker under known conditions. For relatively soft metals and for steel,

Table 1.4 *Properties of metals used in shipbuilding [26].*

	Low carbon steels		*Aluminium (5000 series)*		*Titanium (α alloy)*	
Properties	*Range*	*Mean*	*Range*	*Mean*	*Range*	*Mean*
Physical properties						
Density (kg/m^3)	7,640–8,080	7,860	2,640–2,720	2,670	4.42–4.84	4.54
Mechanical properties						
Hardness, Brinell	86.0–562	247	28.0–120	68.5	290–411	348
Hardness, Knoop	103–682	281	68.0–150	98.9	315–450	379
Hardness, Rockwell B	30.0–105	89.5	49.0–75.0	57.1		
Hardness, Vickers	22.0–661	268	68.0–137	89.9	304–480	368
Tensile Strength, Ult. (MPa)	250–2,450	788	110–450	268	825–1,580	1,070
Tensile Strength, Yld (MPa)	140–2,400	593	40–435	194	759–1,410	981
Elongation at Break (%)	1.00–48.0	20.8	3.50–35.0	13.8	4.00–18.0	11.2
Modulus of elasticity (GPa)	183–213	202	68.9–72	70.0	105–123	114
Compressive Yield Strength (MPa)	152–1,800	1,480	160–325	261	860–1,280	1,030
Bulk Modulus (GPa)	148–163	160				
Compressive Modulus (GPa)			70.0–72.4	71.7		
Poisson's Ratio	0.23–0.30	0.29	0.33	0.33	0.31–0.34	0.322
Charpy Impact (J)	8.13–339	60.1			9.50–25.0	18.7
Fatigue Strength (MPa)	207–772	664	82.7–159	127	140–1,160	497
Fracture Toughness (MPa√m)	66.0–82.0	72.4			24.0–105	55.5
Shear Modulus (GPa)	62.1–82.7	79.5	25.9–26.4	26.0	41.0–46.0	43.5
Shear Strength (MPa)			75–270	159	550–760	672
Thermal properties						
CTE, linear (μm/m-°C)	10.1–16.6	12.6	23.8–26.6	25.6	8.30–11.7	9.26
Specific Heat Capacity (J/g-°C)	0.448–0.615	0.478	0.880–0.904	0.897	0.368–0.670	0.542
Thermal Conductivity (W/m-K)	12.0–93.0	43.6	108–205	147	6.10–10.9	7.19
Melting Point (°C)	1,230–1,530	1,440	568–657	623	1,300–1,680	1,650

respectively, there is a very simple relation between the *Vickers Hardness* H_V and the yield stress σ_0 or the tensile strength σ_t, expressed in kg/mm^2:

$$H_V \approx \begin{cases} 3\,\sigma_0 & \text{for soft metals} \\ 3.2\,\sigma_t & \text{for steel} \end{cases}$$

Ductility of a material is a measure of the extent of distortion it can undergo without fracturing when subjected to external loading. Thus low-ductility materials will fracture when subjected to comparatively low loads.

Ductility can be quantified in terms of the fracture strain, which is the engineering strain at which a test specimen fractures during a uniaxial tensile test. Another commonly used measure is the reduction of area at fracture.

Fracture toughness denotes the ability of a material to resist fracture.[2] It is denoted K_{Ic} and has units of MPa$\sqrt{m}$. The subscript '*Ic*' denotes mode *I* crack opening under a normal tensile stress perpendicular to the crack, since the material can be made thick enough to resist shearing (mode *II*) and tearing (mode *III*).

Table 1.4 gives the numerical values for selected properties of metals used in shipbuilding. A study of this table reveals the comparative advantages of each material. For hull structural use, *rolled steel* is used, and this may be either of ordinary or of higher strength. Cast and forged steel is also used for items such as rudder components, the stem and stern, anchors and anchor chains. Ongoing efforts to improve the properties of shipbuilding steels and processes are aimed at reducing the cost of assembly and fabrication (*TMCP* steels, laser welding) and also to improve performance under particular conditions, in particular container ships, oil and chemical tankers (corrosion resistance).

Aluminium is used in superstructures and for navigation instrument components. It is also used in vessels up to 80 metres in length especially when weight reduction is a high priority. Its stiffness (65 GPa) is one-third of that of steel (207 GPa), and thus the resulting deflections render it impracticable for use in the hull structure of large vessels.

Titanium has been used in the construction of a small number of yachts in Japan because of its corrosion resistance and its low density. In the shipbuilding industry it is more commonly used for mechanical components (heat exchangers, piping and valves), although efforts are underway to make its use more widespread by reducing production costs [17]. It is also used in the offshore industry for risers and ballast systems. In the past it was used by the Soviet navy in submarine construction.

Fibre-reinforced plastic (*FRP*) materials are used for specialist applications such as the hull construction of mine-hunters, patrol craft, yachts, etc. A major technical drawback of composite materials is their low fire resistance.

Table 1.5 lists the elements found in low-carbon steels and their presence whereas Table 1.6 lists the effects of the most important of these elements when used in steel alloys.

Classification Society Requirements for Shipbuilding Steels

Table 1.7 includes the most important mechanical properties that classification societies normally require of steels used in hull construction. These are yield strength and fracture toughness. Shipbuilding steels are classed as of ordinary or of high strength, and within each class they are allocated grades that are used to distinguish their fracture toughness, which increases from *A* through to *E*. Below a certain temperature, termed the *transition temperature*, steels that normally exhibit ductile behaviour behave in a brittle manner and can fracture under low loads. The transition temperature is influenced by the chemical composition of the alloy, grain size and heat treatment. By modifying one or more of these parameters it is possible to reduce the transition temperature, thereby increasing fracture toughness.

[2] If the material contains a defect or crack the property is *notch toughness*.

Table 1.5 *Presence of elements in low-carbon steels [19].*

Element	*Range (%)*	*Average value (%)*
Aluminium, Al	0.010–1.5	0.412
Boron, B	0.00050–0.0060	0.00273
Carbon, C	0.0030–0.80	0.182
Chromium, Cr	0.20–10	0.891
Cobalt, Co	7.5–12	8.84
Copper, Cu	0.020–1.5	0.372
Iron, Fe	63–100	96.0
Lead, Pb	0.15–0.36	0.251
Manganese, Mn	0.10–2.2	0.783
Molybdenum, Mo	0.080–4.8	0.570
Nickel, Ni	0.030–18.5	3.13
Niobium, Nb (Columbium, Cb)	0.0050–0.15	0.0485
Nitrogen, N	0.0010–0.070	0.0200
Phosphorus, P	0.0010–0.40	0.0362
Silicon, Si	0.010–1.0	0.283
Sulphur, S	0.0010–0.50	0.0432
Titanium, Ti	0.010–1.4	0.416
Vanadium, V	0.0050–0.95	0.0912
Zirconium, Zr	0.010–0.15	0.0211

Table 1.6 *Effects of the most important elements on steel alloys.*

Element	*Effect on steel alloy*
Carbon	Increases toughness, yield stress, tensile strength Decreases ductility, max. elongation, welding properties
Manganese	Increases tensile strength and improves weldability Deoxidises steel and forms manganese sulphide (MnS)
Silicon	Increases toughness, tensile strength and yield strength Reduces maximum strain and cold-forming ductility
Phosphorus	Decreases ductility and notch toughness
Sulphur	Decreases ductility, weldability and notch toughness
Nitrogen	Reduces resistance against brittle fracture
Hydrogen	Reduces ductility and weldability
Copper	Improves corrosion properties

For hull construction, the International Association of Classification Societies (IACS) Common Structural Rules (CSR) give detailed guidance as to the class and grade of steel to be used, depending on the thickness of the individual member, its exact location and the nature of the load it is subjected to (Table 1.8). More precisely, whether the member is within $0.4L$ from the midship section, and also whether it forms part of the *primary*

Table 1.7 *Mechanical properties of shipbuilding steels [12].*

Class	*Grade*	*Yield stress (N/mm^2)*	*Tensile strength (N/mm^2)*	*Elongation $5.65\sqrt{S_o}$ A_5 (%) min*	*Charpy V-notch tests* Test temp (°C)	*Average energy min. (J)* Long.	Trans.
Ordinary strength	A	235	400–490	22	–	–	–
	B				0	27	20
	D				–10	27	20
	E				–40	27	20
Higher strength	AH 32[3]				0	31	22
	DH 32	315	440–590	22	–20	31	22
	EH 32				–40	31	22
	AH 36				0	34	24
	DH 36	355	490–620	21	–20	34	24
	EH 36				–40	34	24

Table 1.8 *Material grades [12].*

Thickness t (mm)	*Material class* I	II	III
$t \leq 15$	A, AH	A, AH	A, AH
$15 \leq t \leq 20$	A, AH	A, AH	B, AH
$20 \leq t \leq 25$	A, AH	B, AH	D, DH
$25 \leq t \leq 30$	A, AH	D, DH	D, DH
$30 \leq t \leq 35$	B, AH	D, DH	E, EH
$35 \leq t \leq 40$	B, AH	D, DH	E, EH
$40 \leq t \leq 51$	D, DH	E, EH	E, EH

structure, the *secondary structure*[4] or is considered to be a special item, subjected to more severe loading. From Table 1.7 it is evident that as the thickness is increased, so do fracture toughness requirements.

Indeed, for a specimen under mode *I* loading with a tensile stress that develops normal to the crack, tests have shown that fracture toughness decreases as the thickness is increased.

1.2.5 Modes of Failure

Failures in ship structures can be classified as *major*, involving the entire hull girder, thereby leading to loss of the vessel; or *minor*, involving a single structural member or a

[3] The letter *H* designates higher strength, and the number is the yield stress expressed in kgf/mm^2 (1kgf ≡ 9.81 N).

[4] These terms are discussed in more detail in Section 1.3 and in Chapter 10. The rules give details as to the appropriateness of material class. 'Special' items are considered to be the sheer strake and the stringer plate of the strength deck, deck strakes i.w.o. longitudinal bulkheads, strength deck plating in way of hatch openings, the bilge strake and longitudinal hatch coamings [12].

Figure 1.10 Major failures involving hull girder collapse.

limited number of members. In this case overall loss or incapacitation of the vessel does not take place.

Major failures are generally complex events that involve more than one failure mode, such as extensive elasto-plastic buckling, deep plastic collapse, fatigue, fracture and weld tearing. In the majority of cases when a ship is lost at sea, it is not possible to verify events in order to understand how collapse took place. There exist, however, a number of cases such as those illustrated in Figure 1.10 that have enabled investigators to understand what happened and what the prime cause was. Of most interest to the structural designer are failures that are due to inadequate design, from which useful lessons can be learnt. It has to be said that these are rare; at least they are rarely important enough to be documented in public. Two occasions that involve inadequate structural design are:

a. The design of strength members in the cargo tanks of the first-generation very large crude carriers (VLCCs) (early 1970s). Shipyard design offices and classification societies at the time relied on extrapolating the empirical rules that were used up to that time for existing ships, which were much smaller. The result was inadequate strength of the transverse members, which led to local buckling collapse. This could have had even more serious consequences.
b. The use of high-strength steel in second-generation VLCCs. High-strength steel permits the use of lighter scantlings. However, as a result, the stress range increases accordingly. This leads to a reduced fatigue life, and a large number of cracks during the early stages in the life of second-generation VLCCs were discovered.

The important lesson that the maritime industry has learnt from failures such as these is that when innovation or departure from well-established practice is contemplated, it is necessary to resort to engineering mechanics methods that have a firm theoretical foundation. These can be either purely mathematical such as those described in the present text, measurements in situ, computational methods or a combination of these.

Table 1.9 includes a list of major problems that have been faced and solved in the design and operation of ultra large ships that were built during the second half of the 20th century. They include structural as well as non-structural problems. We note that expertise from a

Table 1.9 *Technical challenges related to the design of ultra-large ships [22].*

Oil tankers

a) Failure of transverse frames in cargo tanks solved by measurements and stress computation
b) 300,000 tonnes dwt with high-strength (HS) steel: early fatigue cracking solved by fatigue tests, detail new designs
c) 550,000 tonnes dwt: springing phenomenon solved by ship behaviour and structural response computation verified by hull stress monitoring

Bulk carriers

a) First generation: lamellar tearing at bulkhead foot solved by steel plate testing requirement (Z-grade plating)
b) Hatch coaming and hatch cover stiffness: hold tightness solved by hatch coaming deformation computations
c) Ballast in cargo holds: sloshing and damages solved by model tests and upper tank shape design
d) Fatigue cracks and domino effect: shell brittle fracture solved by stress computations, fatigue and crack propagation studies
e) Fast loading by gravity (16,000 metric tonnes/hr): bottom pressure determined by theoretical calculations and impact structural response computations

Containerships

a) Large open deck: torsion and fatigue problems arising in hatch corners solved by measurements at sea, then stress computations
b) Wave impacts: slamming, slapping, green water solved by hydrodynamic computations, hull monitoring
c) Dynamic wave loads: hull girder springing, whipping solved by stress computation and fatigue verification
d) Motion accelerations: on deck container lashing solved by load and stress computations, lashing design
e) Propulsion power: shaft alignment, rudder erosion solved by elastic alignment, rudder shape and material

Cruiseships

a) Superstructures: hull girder deformation loads solved by structural stress computations
b) Large wide spaces: structural discontinuities solved by structural computations
c) Large shell opening: structural discontinuities solved by structural computations
d) Passenger comfort: reduced vibration, noise solved by structural computations (finite element studies)
e) Propulsion: increased manoeuvrability requirements provided by pods

number of fields has been drawn upon to provide permanent solutions to these problems that in certain cases are quite complex.

Minor failures, on the other hand, arise from any one of a large number of causes and result in plastic bending, elasto-plastic buckling, fatigue and/or fracture of one or more members. In ship structures minor failures do not cause overall collapse of the hull girder. This is because the structure is highly redundant, since loads are redistributed amongst neighbouring members and are thus able to be sustained. Figure 1.11 shows two cases of buckling that involve individual structural members. In both cases the cause was overloading. In the first case dynamic loading of the hull girder was transmitted by the vertical

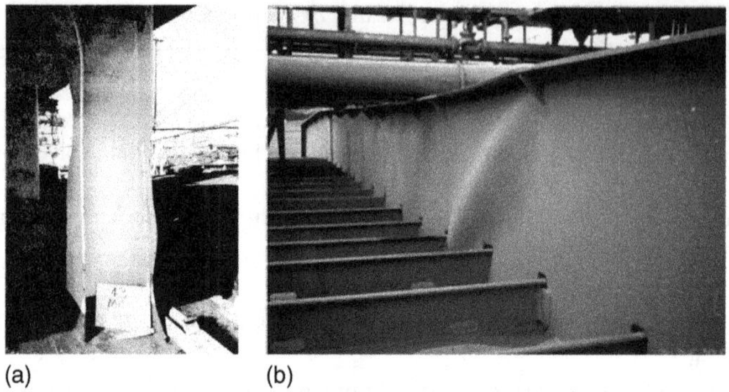

Figure 1.11 Buckling collapse of structural members (minor failures). (a) Compressive buckling of stanchion. (b) Shear buckling of deck web.

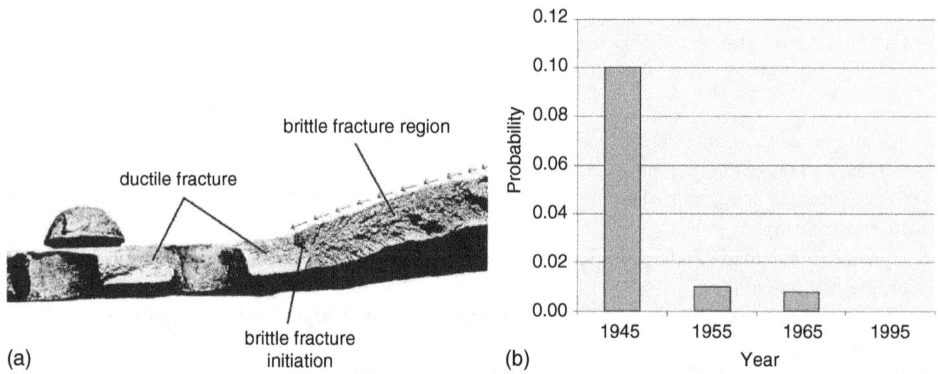

Figure 1.12 Ductile and brittle fracture. (a) Liberty ship plating [5]. (b) Probability of occurrence of brittle fracture [18].

stanchion from the deck below to that above. In the second case it is likely that excess vertical load was applied to the transverse deck beam by the piping as a result of inertial accelerations during heavy seas. This led to shear failure of the web that, as noted, is not stiffened.

In both cases the members were not designed to withstand the loads that led to their failure. We note that buckling is elasto-plastic, so that the members have collapsed and cannot sustain any further load. The repairs required thus involve cropping and renewal. Furthermore, since the excess load is attributable to environmental factors and is not the result of normal wear and tear, to avoid repetition of failure, the structure will have to be redesigned.

An altogether different type of failure is illustrated in Figure 1.12a, which shows a portion of the shell plating of a Liberty ship. On the left-hand side of the section two rivet holes are visible. Between the two holes and also on their right-hand side the material has failed in a ductile manner. From the location to which the arrows point the material has failed in a brittle manner, as evidenced by the herringbone pattern and the rough surface. This particular type of brittle fracture is now of historical interest, given material quality has improved drastically since the time that these fractures occurred. Figure 1.12b shows the dramatic reduction in the probability of occurrence of brittle fracture during the second half of the 20th century. The figure for 1995 is approximately 0.0012, which is one order of magnitude lower than that for 1965. It has to be said that tearing may still occur, mainly as a result of excess dynamic loading in severe weather conditions combined with a weakened structure as a result of corrosion.

1.3 A Systematic Approach for Ship Structural Problems

A first impression of the various types of ship structural arrangements was gained from Figure 1.4. Our objective is to be able to perform strength analyses of the entire hull girder as well as of each and every individual structural component. However, in the case of individual components, it is not possible to do this directly because of the complex geometry and the nature of the applied loads. The task has to be broken down into a number of stages, or levels of analysis. This makes it possible to understand the role and contribution of individual members as well as major sections of the hull girder such as transverse frames, bulkheads, decks and so forth.

1.3.1 A Top-Down Approach in Ship Structural Analysis and Design

Modern ship structures are the result of progressive evolution, with past experience acting as a guide. It is only relatively recently that an in-depth understanding of structural behaviour and how individual members interact with each other has become possible. When ships grew in size on an unprecedented scale after the Second World War, strength problems arose, and it became apparent that a proper understanding of structural behaviour is needed, based on engineering principles. Structural arrangements in use today have in many cases been chosen on the basis of experience rather than analysis. Nevertheless, the means available to study structures today, which include analytical as well as numerical techniques, have proven invaluable in both analysis and design. In many cases they have reaffirmed past choices, whereas in others it has become apparent that new solutions are required.

It is possible to perform a *multi-level assessment* of the structure of a ship in more than one way, depending on the aim of a particular study. If the study is limited to a prediction of overall response, a simplified structural model of the ship, or part of it, is sufficient. In such a model, only essential features are retained. In some other cases, information may be needed for detail design and a *top-down* approach is followed.

Alternatively, the study can be initiated at the level of a component and then built up in a sequential manner, eventually encompassing the entire vessel. This is termed the *bottom-up* approach. The two approaches have different aims and are complementary in that results obtained from one assist in the other. The essential information needed when a particular member is designed concerns the nature of the loads that act on it and the manner in which it is supported by the adjoining structure. Once answers to these questions are available,

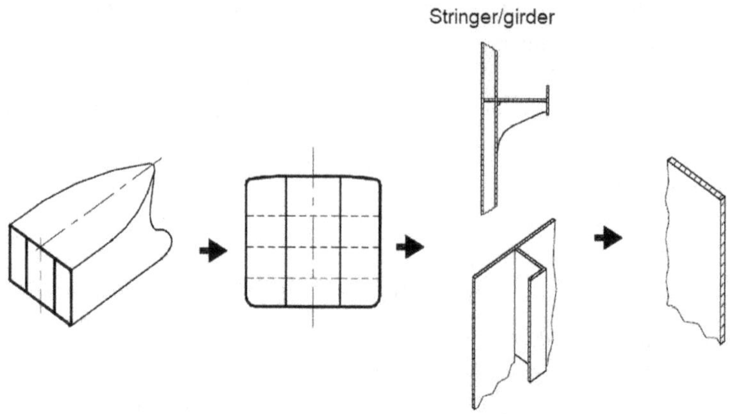

Figure 1.13 The top-down approach in ship structural analysis and corresponding levels [7]. Reprinted courtesy of Det Norske Veritas, Høvik.

form, materials and scantlings can be determined. This requires an understanding of loading at every level as well as that of load transmission mechanisms.

In the design of ships for longitudinal strength, the traditional approach involves analysing a typical *transverse section*. This achieves economy of effort, as the structural arrangement is duplicated to a significant extent along the length of the ship. Therefore, if the effort is concentrated on a critically loaded section whose scantlings are then extended sufficiently in the longitudinal direction, an inherently safe design will, in theory at least, result. This has to be supplemented by strength calculations for the fore and aft ends, as their layout and loading are different from those of the wall-sided, cargo-carrying compartments.

Figure 1.13 shows typical components of the structure that are studied at different levels. Before proceeding to identify different levels, it is necessary to identify the different types of response that may arise. For example, when considered in its entirety, the hull girder behaves as a beam. However, at a local level, the behaviour can be modelled as a grillage, as an assembly of panels, as a stiffened plate or else as an individual member. A successful subdivision is thus possible once the characteristic behaviour of particular subunits has been identified. In Figure 1.13 the hull girder is subdivided into major components such as the transverse bulkhead shown here, and following this into the typical members of which it consists, such as stiffened plate members. At the most detailed level, the behaviour of an individual plate that is supported by the orthogonal stiffening system is studied.

Other ways of subdividing the structure are possible and indeed are used. Major subsections or modules such as a portion of the midship region that extends fore and aft of an entire cargo compartment can be studied, as is done in finite element analysis. Another example is a complete superstructure, which consists of several decks and related appendages. These are subdivided into their major components such as decks, side shell structure and bulkheads.

In order to gain a complete picture of the response at one particular level, it is necessary to sum the effects at all higher levels. For example, when studying stiffened plate behaviour, we may need to determine beforehand the behaviour at the hull girder level and at the major (gross panel or grillage) component level, inasmuch as these affect local behaviour. When studying a stiffened plate, it is assumed that its boundaries are rigid, although in reality this is not the case. The effect of finite boundary stiffness is taken into account when the analysis is performed at the immediate higher level, such as that of the deck transverses, which in turn are attached to longitudinal girders and the side shell structure. Therefore, as we consider the structure in greater detail, it is necessary to have results for stresses and deflections at all higher levels. This is the underlying justification for a top-down approach in the study of ship structures.

1.3.2 Structural Modelling and Models for Global Strength Assessment

Structural engineers make use of *structural models* that are selected in part on the basis of past practice, although it is always necessary to have an understanding of the suitability of the model and its limitations. Needless to say, the accuracy of the results will depend in each case on the suitability of the chosen model. Therefore, the need for an understanding of the mechanics of the response and the selection of a suitable model cannot be overemphasised.

A structural model is an abstraction created in order to focus on a particular part or aspect of the behaviour of a structure whilst at the same time enabling a mathematically tractable solution of the problem to be obtained. The development and use of structural models must be based on an understanding of the behaviour of the particular structure or structural component under the specified loading conditions.

Many different structural models have been developed and are in use. In the case of ship structures, it is necessary to differentiate between global and local strength considerations.

When considering global strength of ships, the most important structural model in use in the case of longitudinal bending is the *linear elastic beam* (Figure 1.14). The analogy with beam behaviour is based on the form of ship hulls and the distribution of the primary loads.

In linear elastic beam theory, due account is taken of the longitudinal (axial) stresses along the length of hull girder and within each transverse cross-section, but the effect of shear stresses on the longitudinal stresses is not included (Chapter 6). The contribution of shear stresses varies along the length of the hull girder, and maximum values are expected at the quarter-length positions. Shear stresses are also important in ships that are fitted with longitudinal bulkheads. It has to be added that the linear elastic (Euler–Bernoulli) beam model has been validated from full-scale measurements and has widespread use in the marine industry.

A more accurate structural model, which does allow for the effect of shear stress, is the *Timoshenko beam*. Timoshenko beam theory is presented in Chapter 8, which deals with hull girder vibration. This is a type of response that necessitates the use of a more refined beam theory because linear beam theory is unable to predict the higher mode frequencies of vibration with sufficient accuracy.

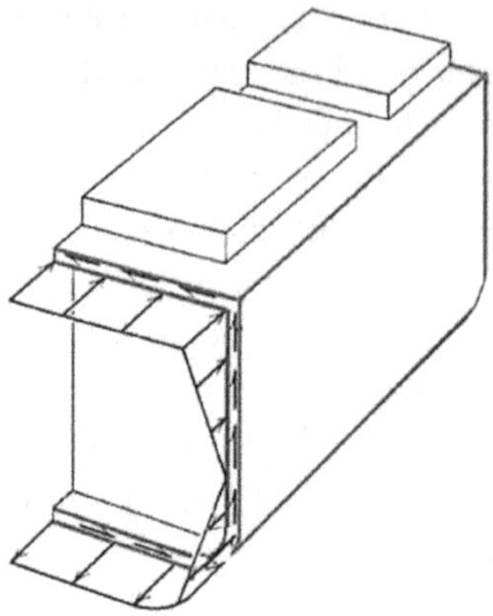

Figure 1.14 The hull girder in longitudinal bending can be modelled as a beam [11]. Reprinted courtesy of the Society of Naval Architects and Marine Engineers.

Other than longitudinal bending the other important distortion mode of the hull girder is due to *torsion*. This is important in the case of open-deck ships such as containerships (Chapters 6 and 7). In this case also the hull girder is treated as a beam, although a number of complications arise that need to be accounted for in order to obtain meaningful results. The first point to make is that it is necessary to consider the entire hull girder and not rely on the analysis of a typical cross-section, as is done in longitudinal bending analysis. The reason for this is that the geometry of the hull girder varies continually along the length of the ship, as does the distribution of the loading (torque). In the torsion theory of beams, the important basic assumption is that the transverse cross-sections along the length of the beam do not distort in their plane, nor do they distort in the direction normal to their plane. On this basis a *linear theory of torsion* is developed (Saint-Venant theory), and this provides a first-order analysis of torsion behaviour (Chapter 6).

In real structures, however, the assumptions upon which linear torsion theory is based do not hold, and in the case of ships it is necessary to consider the distortions of transverse sections. These out-of-plane distortions are termed *torsional warping*, and to this end *warping torsion* theory has been developed, with various degrees of sophistication (*Vlasov beam* and *beam-shell theories*). Warping arises in general in prismatic sections that are subjected to torque, and if the ends of the structure are free, *free warping* will arise. In the case of ship hull girders, however, the ends of the structure are attached to torsionally stiff regions such as the fore and aft ends and the engine-room spaces, all of which are enclosed. As a result, *constrained warping* arises, and to this end several theories have been advanced (Pedersen beam model, Senjanović model). This question is discussed in Chapter 7.

All of the above beam theories that are used to represent the global behaviour of ships neglect local effects. The stresses predicted by beam models are thus *nominal stresses*, to which local stresses need to be added.

In addition to the beam theories discussed above, the hull girder can be modelled using numerical techniques, the most important of which is the *finite element method* and its

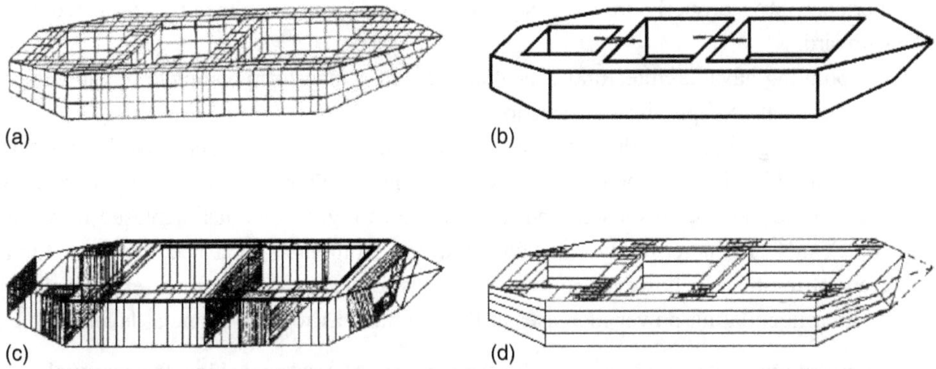

Figure 1.15 Structural models for hull girder analysis [29]. (a) Finite element model. (b) Discontinuous beam model. (c) Module-element model. (d) Finite strip/finite element model.

variants. Because finite element modelling is laborious and time consuming, a variety of simplified models have been developed. Examples are presented in Figure 1.15 and discussed in Chapter 13.

1.4 The Practice of Ship Structural Design

We shall now turn our attention to the question of how ship structures are designed in practice. An overview of industrial practice will be given in the remainder of this chapter whereas detailed ship structural design procedures are described in Chapters 10 and 14.

Ships are designed by entities that vary in size, complexity and degree of sophistication. First there is the merchant shipbuilding industry that is served by yard design offices, consultants and owners' design teams. Second there are national bodies such as navies that have at their disposal dedicated teams of experts and research teams that may or may not be directly employed by them. For organisations such as these, ship design, including structural design, is a major task that has to be conducted with a view to ensuring that defence requirements are successfully fulfilled – an issue that carries especial weight. The financial outlay for any government in such cases is considerable, and decisions have to be correct from the beginning, particularly since the lifetime of a new class of ship may extend to several decades.

1.4.1 Merchant Ship Structural Design

In merchant shipping the owner may range from a large multinational corporation that operates a large fleet to a small private owner who operates a few second-hand vessels and who may at one point in the life of his company decide to place an order with a yard for a new ship. The needs of owners are served in various ways:

- By forming or by using existing design teams from within their own organisations. These will have a deep understanding of owners' requirements and can transfer available

experience into a properly drafted design specification, which can then be negotiated with a shipyard.
- By appointing naval architectural consultancies that can perform complete or part design studies as well as supervise construction.
- By negotiating directly with a shipyard on the basis of one of the standard designs that may be on offer. This is a practice that is becoming more common, especially as designs have become more sophisticated and the volume of regulations has increased in recent years. Yards are increasingly reluctant to divert from their own specifications for a variety of reasons. If, however, an owner intends to place an order for a series of ships, then their negotiating power increases at the technical as well as financial levels.

Stripped down to its bare essentials, for the majority of merchant ships, the essential task is to produce a design of the hull structure that will satisfy the current classification society rules and other statutory requirements given:

- Ship type
- Main dimensions
- Operating range
- Carrying capability (cargoes, passengers)

As the design proceeds, the structural design group will interact with the other members of the design team so that it becomes necessary to introduce corrections. The tools that are available to the structural designer vary in accuracy and complexity, so a judicious use of methods at every stage of the design is necessary in order to devote available resources efficiently. Figure 1.16 shows how information flows in ship structural design in the design office of a major shipyard [18].

To illustrate how this question is faced by practitioners we shall give a brief account of the design procedure followed by a ship design bureau experienced in passenger and cruiseship design [23]. Structural design represents one component of the design process. In order to initiate structural calculations for a new design project, it is necessary to have an idea of the general arrangement, functions and capacities of the proposed ship. The design team relies on

- A reference ship design
- Available open literature and design studies
- In-house similar designs and studies

in order to initiate proposals.

1. The first task is to decide on space allocations, and this is done by preparing a rough, first-stage general arrangement. An important question is then to decide on the type of stiffening system to be used for each part of the section, following which the frame spacing is selected. These decisions are based on the reference vessel. The general arrangement plan can then be adjusted accordingly. For passenger ships and cruiseships, which have a multiple deck arrangement with complicated piping and other systems, it is convenient to use a transverse stiffening system for the side shell structure whereas the bottom structure and decks are longitudinally stiffened.

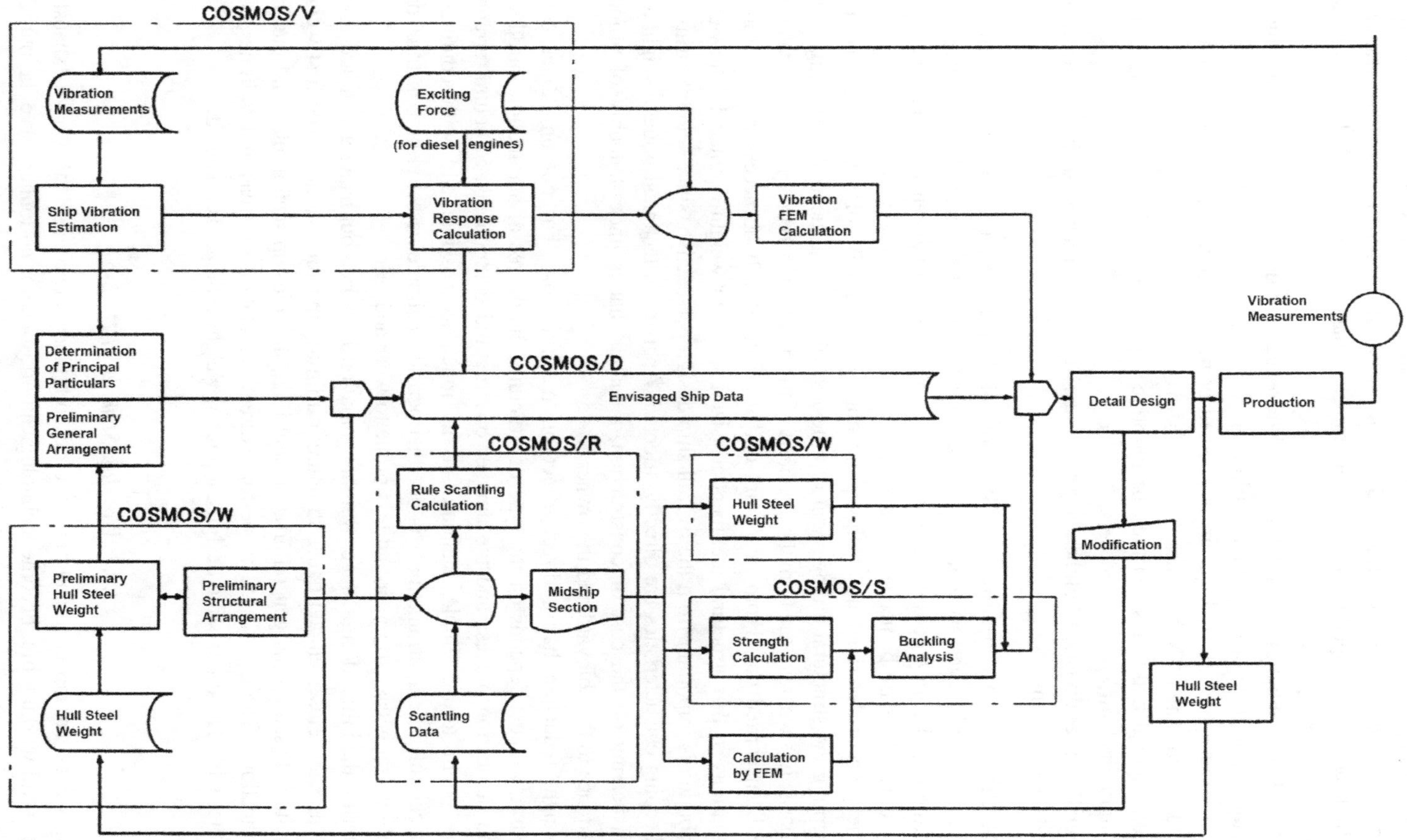

Figure 1.16 Data flow between structural design modules of the COSMOS software procedure [18].

2. The spacing between web frames is chosen and is a multiple of the ordinary frame spacing. The positions of pillars and supporting bulkheads are then selected. Transverse bulkheads are arranged in place of web frames. Pillars are positioned below beams and/ or longitudinal girders.
3. Deck longitudinal girders are introduced to reduce the unsupported deck area and are arranged in place of longitudinal stiffeners.
4. The watertight bulkhead system is selected bearing in mind subdivision requirements and the spacing of web frames.
5. Longitudinal bulkheads can then be introduced bearing in mind the spacing of deck longitudinal stiffeners and girders.
6. Finally, fire division bulkheads are arranged in accordance with the watertight bulkhead system.

Once this stage has been reached, it is possible to prepare a general arrangement plan. As is evident, it is necessary to introduce matters relating to the structural arrangement early on in the ship design process. In this way serious revisions of the general arrangement plan that would otherwise be needed are avoided.

The process described up to this point consists essentially of a series of decisions concerning the structural arrangement of a new layout using a successful existing design as a basis. It is what we can term the *synthesis* of the design. Up to this point no calculations have been carried out. From this point on, however, it will be necessary to introduce calculations so that structural members are sized and their weight estimated. The weight distribution is important from the overall ship design perspective, since stability, trim and sea-keeping characteristics are directly affected. Apart from technical issues, weight also has a bearing on the cost of construction, a matter that is understandably of primary importance to the shipyard and the owner.

Weight reduction has a direct effect on material costs but not on labour costs. Construction cost depends on the type of structure, the degree of automation introduced in the fabrication process and the number of structural components. Structural arrangements ought to be simple, standardised as much as possible and with continuity. A successful general arrangement is one that permits a simple vessel to be constructed; a simple vessel is easy to repair and less prone to wear and tear.

During the initial design stage, weight estimation is carried out using a standard list of systems and items included (steel structure, outfitting, machinery, etc.). The first weight estimate is based on the general arrangement plan, the midship section plan and the ship specifications. These are used to perform direct calculation of areas and volumes, and together with data on existing designs, an initial weight estimate is obtained.

1.4.2 Warship Structural Design

Naval vessels are designed and built in order to replace older vessels that have exhausted their useful working life, become technologically obsolete or no longer serve the purpose

Figure 1.17 Warship designs. (a) Belh@rra frigate (France). (b) Los Angeles class submarine (USA).

for which they were originally conceived. The opportunity is taken to incorporate in a new design lessons learnt from past practice as well as new technologies. However, a very important aspect of naval design is that it offers the opportunity to innovate by considering competing proposals that may differ radically from one another (Figure 1.17). This is not common practice in merchant shipbuilding. Innovation can touch upon all aspects of a new design, including structural arrangements and materials used.

This, however, has resulted in a much more complicated design and production process that has brought with it increased cost and time. Efforts have been made to manage design and production in more efficient ways. The US Navy has introduced project management techniques that have proven successful in a commercial environment, in an effort to reduce development time and cost. These are grouped under *systems engineering*, which, according to the International Council on Systems Engineering (*INCOSE*), is defined as follows:

> Systems engineering is an interdisciplinary approach that focuses on defining customer needs and required functionality early in the development cycle. It involves documenting requirements, then proceeding with design synthesis and system validation while considering the complete problem. By integrating all disciplines and specialty groups into a team effort, a structured development process is formed that proceeds from concept to production to operation [13].

Systems engineering is thus not a specific technique but a general approach that may be implemented using a number of methods. Of these the most relevant to ship design has proven to be *Concurrent Engineering* and *Integrated Product and Process Development (IPPD)* [6]. Systems engineering techniques have proven successful when the particular characteristics of the case to which they are to be applied have been correctly identified. Examples of successful application of concurrent engineering are the car industry, whereas IPPD has been used in the aerospace industry. Both of these industries are involved in the design and production of complex engineering products.

In order to implement a systems engineering approach in the maritime sector it is necessary to identify *stakeholders* that may be bodies or groups of individuals that will be directly or indirectly involved with the ship and thereby in its operation and performance throughout its working life. In the case of merchant ships, stakeholders are owners, designers, shipyards, crews, charterers, passengers and insurers. Bodies that can also be considered as stakeholders are classification societies, flag states and port authorities. In the case of naval ships, stakeholders are the naval command that extends from the vessel's commanding officer to the Joint Chiefs of Staff, crew, designers, shipbuilding yard, subcontractors, equipment manufacturers and suppliers, and so forth.

The functional requirements of stakeholders are expressed in non-technical terms, such as needs and expectations. Systems engineers then transform stakeholders' requirements into technical requirements. The most important issues that differentiate a systems engineering approach are that all stakeholders' requirements are addressed early on – that is, from the initial design stage – which is most amenable to radical changes and also that the entire life cycle of the product is addressed.

Systems engineering approaches thus offer advantages in the case of novel designs and can be effective in an environment in which where there is structured interaction between stakeholders. The traditional (design spiral) approach is suitable in the case of parametric-based design studies.

A comparison of the naval with the merchant ship structural design procedure shows that the techniques used are similar but not the same (Figures 1.18 and 1.22). One important difference in the design specification concerns the load profiles of these two categories of ships. Warships have to be able to operate under severe weather conditions without any reduction in operational ability. Other differences are:

- Their speed profile is significantly more demanding than that of merchant ships.
- Hydrodynamic loading is different due to differences in hull form.
- They need to have an acceptable standard of survivability under military threats.
- Warships have to be able to be replenished at sea from other ships and also rely on vertical replenishment using helicopters.

It should be added, however, that warship designers are now drawing upon the expertise of the merchant shipbuilding industry. Until recently, merchant ship and naval ship design standards developed in parallel. An important difference between them is that in the first case rules focus on standard ship types whereas in the second they focus on applications to specific cases, especially with regard to loading, as mentioned earlier. Navies are now collaborating with classification societies and have drafted rules for the construction and maintenance of warships. This is in contrast to past practice, in which they retained in-house technical expertise that relied on their own research, the most advanced ship design methods available as well as their own operating experience. The improvements in the theoretical basis of the rules as well as merchant ship design practice means that a large part of the design effort can be carried out using commercially available expertise. This has enabled governments to cut costs and at the same time to expand the activities of entities such as classification societies, designers, commercial shipyards and so forth. Needless to say, there remain many technical and operational aspects that are of exclusive interest to naval designers.

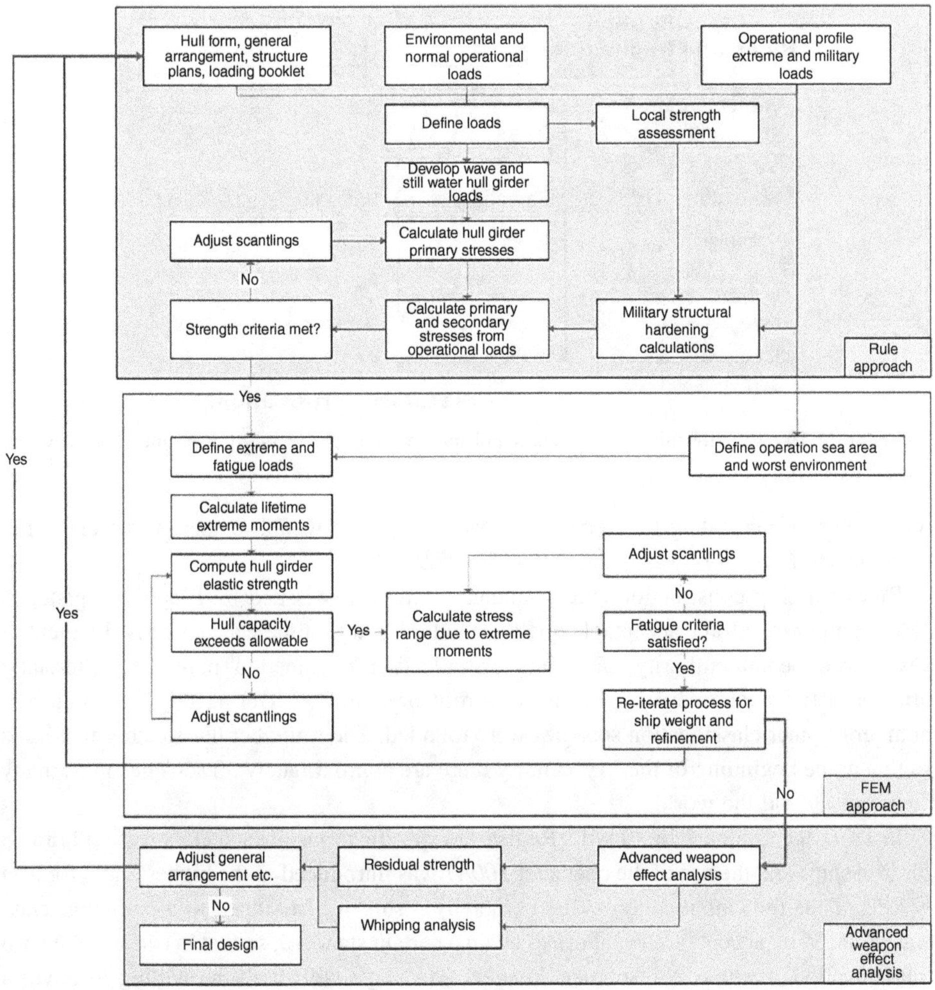

Figure 1.18 Warship structural design procedure [1].

1.4.3 Classification Societies and Their Evolving Role

For the overwhelming majority of merchant ships the structure is designed and constructed in accordance with the rules of classification societies. It is useful to reflect at this point on the role of the class society today, which is not as clear as it may appear at first glance. To appreciate the changes that have occurred over time it is necessary to go back to the beginning. How Lloyd's Register started out is well known; in the late 18th century, insurers who wished to provide cover needed to know the quality and seaworthiness of individual ships in order to assess the risk involved in each case [20]. The Register Society was thus formed (later Lloyd's Register of Shipping) and was allocated the task of grading the condition of ships and of granting them a class. Originally five classes were available: *A, E, I, O* and *U*

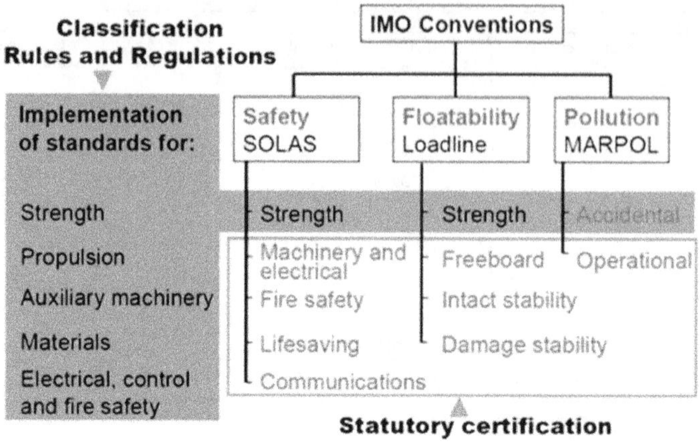

Figure 1.19 Relationship between class and statutory requirements [8]. Reprinted courtesy of Elsevier.

with a numeral indicating the number of years that a ship would remain in that class. For example, the *Cutty Sark* was granted *16A1* in 1869 [24].

Rules for ship construction and maintenance were first issued by Lloyd's Register in 1834 in order to enforce acceptable quality standards. At no time was there involvement of any governmental authority, and thus classification remained a purely commercially oriented activity, carried out by a not-for-profit organisation. During the 19th century a number of other classification societies were founded. Their number has steadily increased so that at the beginning of the 21st century there are approximately 50 societies in as many countries around the world.

In 1870, the reference by Lloyd's Register to specific terms in each class was withdrawn for iron ships and the use of the character *100A1* was introduced as the highest class for iron vessels. Thus the classification system whereby a ship was assigned to a particular class was replaced by acceptance or failure to attain a certain standard, so that in a sense, the term 'classification' became a misnomer. Insurers nowadays will rarely be willing to cover a ship that does not have valid class certificates. Risk (and hence premium) is quantified on the basis of other, non-technical factors such as insured value, type of vessel, owner's track record, fleet size and age profile.

In 1835, Lloyd's Register introduced rules concerning the minimum freeboard of ships, the so-called Lloyd's Rule. In 1876, this became a statutory requirement for UK-flagged merchant ships. Over time, national administrations began to delegate survey and certification tasks to classification societies, and this number has continued to grow. In this way the role of class societies has expanded a great deal. At the same time statutory requirements have also increased, making it necessary for both parties to ensure compatibility between rules and statutory requirements (Figure 1.19). The SOLAS regulations require that all ships be classed, so that classification is now also a statutory requirement. Class, in turn, also requires ships to have valid statutory certificates.

Since the middle of the 19th century, technological advances have changed the face of shipping. Classification societies have had to be at the forefront of developments in order to

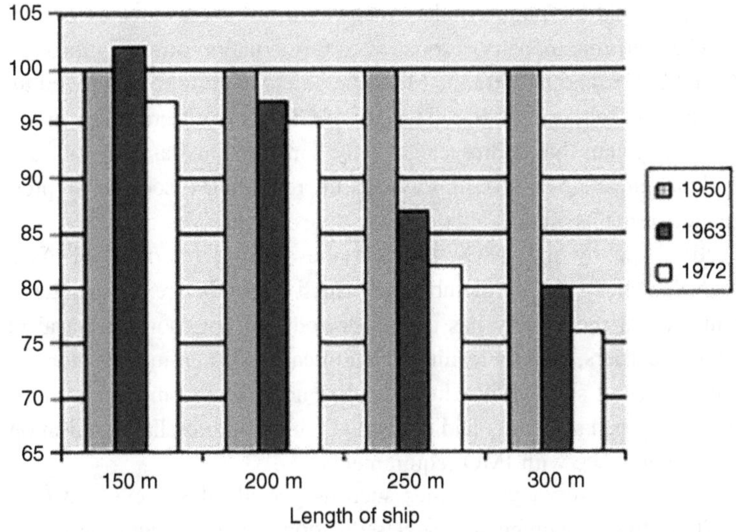

Figure 1.20 Percentage reduction in oil tanker section modulus requirement over time [18].

be in a position to assess the quality of construction and thus of inherent safety. To succeed in this they have expanded their research and development activities to such an extent that they nowadays retain a large concentration of technological expertise, with which the majority of owners and individual yards cannot compete. As a result, they play an increasingly important role in technological innovation in the maritime industry. There have been a number of fields in which even the largest shipowning companies do not possess suitable resources or expertise.

It is also necessary to note that the original purpose of construction and classification rules was to ensure safety of the ship, and as such they were not intended to play the role of 'design manuals', for use by yards and consultancies. However, the scale and complexity of the design and construction process in conjunction with the rapid technological developments that have taken place over the past two centuries have made it difficult for builders and owners to maintain a fully independent viewpoint with respect to design and construction practice. As a result, the influence that rules have had has extended beyond ensuring the safety of seafarers and cargoes.

In recent years, there has been intense competition between classification societies. Throughout this period, advances in structural analysis and ship design permitted reductions in scantlings (Figure 1.20). This fact, in conjunction with a number of ship losses, led owners to question whether their ships were being built as robustly as before. The reduction in scantlings meant that yards could build ships at a lower cost. At the same time the fees paid by yards to class societies reduced as a result of market competition. The end result was that class societies did not have the resources to devote to plan approval and new construction supervision as before. Transfers of class also took place in the case of aged ships which were in need of costly repair work, since the standards of acceptance varied, and continue to do so, between different class societies. A number of environmental disasters occurred, and the publicity they received in the media created political pressure for change. The maritime industry responded to this situation primarily through initiatives of the IMO and IACS, and in 2003 the Maritime Safety Committee approved the introduction of *Goal-Based Standards (GBS)*. The aim of GBS is to ensure safety at sea through a multi-level (tier) system that addresses not only construction standards but also safety of seafarers and passengers (*SOLAS*). In principle, the role of class societies in preparing rules for construction remains intact, although the rules themselves are assessed against the safety objectives set by the IMO. The activities of the International Association of Classification Societies (IACS), members of which currently are 10 societies, have been all-important, in that the society has already introduced common structural rules for oil tankers and bulk carriers, thereby leading to a unification of minimum standards at a global level for these types of ships. Individual societies now have to introduce rules that are in compliance with agreed standards laid down by their own umbrella organisation; they also have to be in compliance with IMO requirements.

When it comes to everyday activities such as periodical surveys and other types of inspections, it has to be remembered that classification societies assume their historical role, which is to limit themselves to assessing the condition of the ship. The responsibility for any mishap or incident lies solely with the owner, as the classification society does not assume any liability. We thus see that, on the one hand, class has an overwhelming influence on ship design, construction and the standard of maintenance, and on the other it is the owner who is responsible in the case of any untoward event.

1.4.4 IMO Goal-Based Standards

Safety at sea, in a general sense, is the most important concern of the International Maritime Organisation (IMO). In 2004, a proposal was submitted to the IMO in which a radically new approach to ensuring safety at sea was advocated. According to this proposal, all activities related to maritime transport are integrated in such a way that safety at sea is served. The fundamental shift in the approach to safety from the status quo consisted of a process of integration of rules and regulations through the establishment of GBS (Figure 1.21). These do not specify the means of achieving compliance but set goals that allow alternative ways of achieving compliance.

An example will serve to illustrate this approach. Consider the requirements for bulk carrier deck plate thickness. In a prescriptive regulation an expression of the type

$$t = f(L, s, z_s, \ldots) + t_k \quad (mm) \tag{1.1}$$

is usually given, whereas in a goal-based regulatory environment the requirement would be given as:

"Deck plating should not fail during ship's life of X years if operated in a Y (e.g. global) environment"

A number of equally acceptable rule formulations could be demonstrated to fulfil the specified objective, such as, for example:

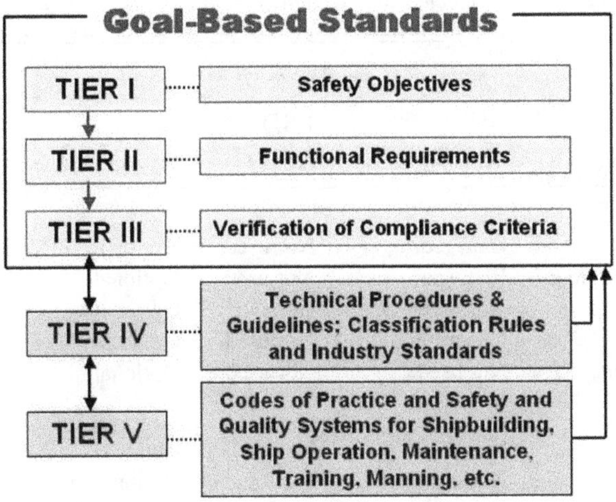

Figure 1.21 Goal-Based Standards according to IMO MSC 78/6/2. Reprinted courtesy of IACS, London.

- Minimum thickness with specified stiffener spacing, corrosion allowance and with defined inspection/repair intervals.
- A different minimum thickness without a corrosion allowance but with more frequent inspection intervals.
- A different minimum thickness with specified coating requirement, cathodic protection requirement and inspection intervals.

It is hoped, therefore, that classification societies will be able to explore in a much more free and creative manner the alternative paths to fulfil the goals that have been set, hopefully on the basis of rational analysis. This would be an improvement to current practice, which has been to compete by setting essentially different structural weight requirements, which correspond to different construction costs. These, however, rely on essentially similar strength formulations, which in the absence of transparency do not have a demonstrable scientific basis.

The implementation of GBS entails the fulfilment of the requirements of a series of levels (Tiers), as indicated in Figure 1.21. It is envisaged that such an approach will eventually encompass and integrate all activities that relate to the safety of ocean-going vessels. The upper Tiers (I, II and III) are the responsibility of the IMO whereas Tiers IV and V are implemented by classification societies, governmental authorities, etc. A compendium of relevant IMO texts is included in Appendix B.

1.4.5 An Assessment of Rules and Standards

As mentioned in Section 1.4.3, there have been significant developments in the rules for ship construction. Table 1.10 summarises the methodologies used in rules published by

Table 1.10 *Approaches used in selected ship structural rules for classification [15].*

	Approach used				
Rule	Prescriptive	Working stress	LSD (LRFD)	First principles	Simulation-based design
Merchant ships					
CSR	Min req.[1]	Gen. req.[2]	Not used	Optional	Optional
ABS	Min & gen. req.	Req. spec vsls[3]	Not used	Optional	Optional
BV	Min req.	Not used	Gen. req.	Optional	Optional
DNV	Min req.	General req.	Not used	Optional	Req. spec vsls
LR	Min & gen. req.	Not used	Not used	Optional	Optional
Naval ships					
ABS	Min req.	Not used	Not used	Gen req.	Optional
GL	Min req.	Not used	Gen. req.	Optional	Optional
LR	Min req.	Gen req.	Not used	Optional	Optional

[1] Minimum requirements: Scantlings (and strength) are to be at least equal to those required.
[2] General requirements: Requirements are to be satisfied as stipulated.
[3] Required for specific vessels: Tankers, bulk carriers and containerships with $L > 150$ m.

several classification societies. It has to be said that changes are continuous and that there is a gradual shift away from prescriptive to first principles and simulation-based design.

Some brief comments regarding the approaches mentioned in the table are pertinent.

Prescriptive requirements stipulate that scantlings and strength requirements, generally expressed in terms of member thickness and section moduli, have to exceed specified minimum values. As such, a strength standard is guaranteed for each component of the structure. In fact, all rules are based on relatively simple concepts of structural engineering (beam theory, frame behaviour). In the *working stress* approach, scantlings are selected so that under normal operating conditions the stress that arises in each component does not exceed a specified value. Both of these approaches have a history of application in ship structures. *Limit state design* (*LSD*) or *load resistance factor design (LRFD*), on the other hand, has been used in civil engineering applications and has only recently been introduced in ship structural design, although it has been in use in the offshore industry. It has a probabilistic basis and makes use of partial safety factors so that a consistent failure probability of all structural components is assured. A *first principles approach* aims at modelling structural response using rational criteria on the basis of certain simplifying assumptions with regard to load application (quasi-static wave loading etc). The stresses that result in a structural member are obtained using mechanics of materials theory.

Simulation-based design essentially consists of the development of a finite element model of a large part of the structure or of the entire structure. A series of specified load conditions are then applied to it, which permits the response to be assessed at a global, intermediate or local level for each load condition. It may involve interaction with environmental loads, in which case computational fluid dynamics software may also be used.

In this text we shall consider the response of the entire hull girder and in doing so will apply structural engineering concepts to analysis and design. Reference will also be made to probabilistic approaches (Chapters 9, 11 and 12). This will enable us to carry out a first-principles approach and also develop an understanding of the theoretical basis of classification society rules.

Rules and standards for the construction of engineering structures in general and ships in particular need to include, either explicitly or implicitly:

- An idealisation approach (structural model)
- A definition of loads (model or idealisation of loads)
- A definition of response
- Suitable safety factors

In [15] the comparison of rules included an assessment of the following:

1. Wave pressure loading, since this is the most important component of external loading and is subject to different interpretations and treatment. In general, the load on a structural member is based on adding hydrostatic pressure to wave pressure
2. Required minimum scantlings
3. Corrosion additions
4. Ultimate strength capacity of the hull girder

With regard to wave pressure loading, it was found that the design wave pressure:

- Varies between class societies
- Varies or may be constant over the flat of bottom
- Varies in way of the design draught waterline

Furthermore it was found that the safety factor is low throughout: that is, the applied pressure can be exceeded for deeper draught conditions. Results from this comparison study are summarised in Table 1.11.

The required minimum scantlings are prescriptive and based on load-stress-based rules. They are not related to load calculations and they overrule other requirements if necessary. An example comparison showed that for the thickness of the bottom plating of a 150-m ship with 800-mm frame spacing, requirements vary from 11 mm to 13 mm.

Corrosion additions for double bottom ballast tank plating vary from 3 mm down to 1 mm for merchant ships and 0.5 mm for naval ships. Hull girder longitudinal strength requirements were unified under IACS UR S11, and rules for bulk carriers and oil tankers have a common basis, following the publication by IACS of the Harmonised Common Structural Rules.

Of interest to the designer and the owner is the effect that the choice of rule has on the structure of a particular ship type, particularly its weight. Table 1.10 summarises findings for the application of different rules to the design of various ship types. Differences between these were found to be negligible. This is understandable, since these rules have been based on a large database of ships of these types and configurations. The Common

Table 1.11 *Comparison study for the structural design of common ship types [15].*

Type of ship	*Rules and software packages used*	*Findings*
Small general cargo	DNV – NAUTICUS GL – POSEIDON BV – MARS LR Naval – EXCEL	Small differences (<20%), BV requirements lower than others. Corrosion addition requirements reduce differences
General cargo $L > 90$ m	DNV – NAUTICUS GL – POSEIDON LR Naval – EXCEL	Very similar results
Bulk carriers $L > 150$ m	JBR DNV – NAUTICUS LR Naval – EXCEL	Very similar results
Oil tankers $L > 150$ m	JTR – NAUTICUS GL – POSEIDON BV – MARS	Very similar results

Structural Rules result in an increase in weight only through corrosion allowances – that is, scantlings requirements are essentially the same as in earlier rules. As a result, owners who wish to introduce a greater degree of *robustness*[5] in the structure will need to request that scantlings exceed minimum requirements or else introduce innovative arrangements and/or materials. As a final comment, it has to be said that the CSR, being particularly thorough and far-reaching, add complexity, and their interpretation requires a good understanding of structural mechanics, in contrast to the prescription-based rules of the past. The added complexity is a source of possible misinterpretations and errors, and it is up to the classification societies to publish rules that, on the one hand, comply with the CSR and, on the other, are clear and easy to apply.

Ship structural design has thus been transformed in many cases into a process of rule compliance. However, it has to be said that the use of optimisation software in conjunction with operations research techniques can lead to structural designs that are competitive in terms of payload and operating (fuel) expenses. This is particular true in the case of multi-deck ships.[6] The fact is that with the development of clearly identifiable vessel types such as oil tankers and containerships that have more or less standardised configurations, rules can provide very detailed design guidance which, in the majority of cases, facilitates the work of the designer. Improvements in vessel structural designs will thus have to follow improvements in standards. This situation has been acknowledged as such by the maritime community and is an important reason for which GBS have been introduced.

Figure 1.22 shows the flow of calculations that a structural design team will generally need to follow. This, as discussed in Chapter 10, forms part of the ship design process.

[5] *Robustness* (η) is a measure of insensitivity (or stability) with respect to parameters that cannot be controlled. It is defined quantitatively to be equal to the ratio of the mean of the attribute value μ to the standard deviation σ resulting from uncertain parameter values. It is thus the ratio of predictability to unpredictability.

[6] The use of an advanced optimisation procedure in the structural design of a RoPax ship is described in Chapter 14.

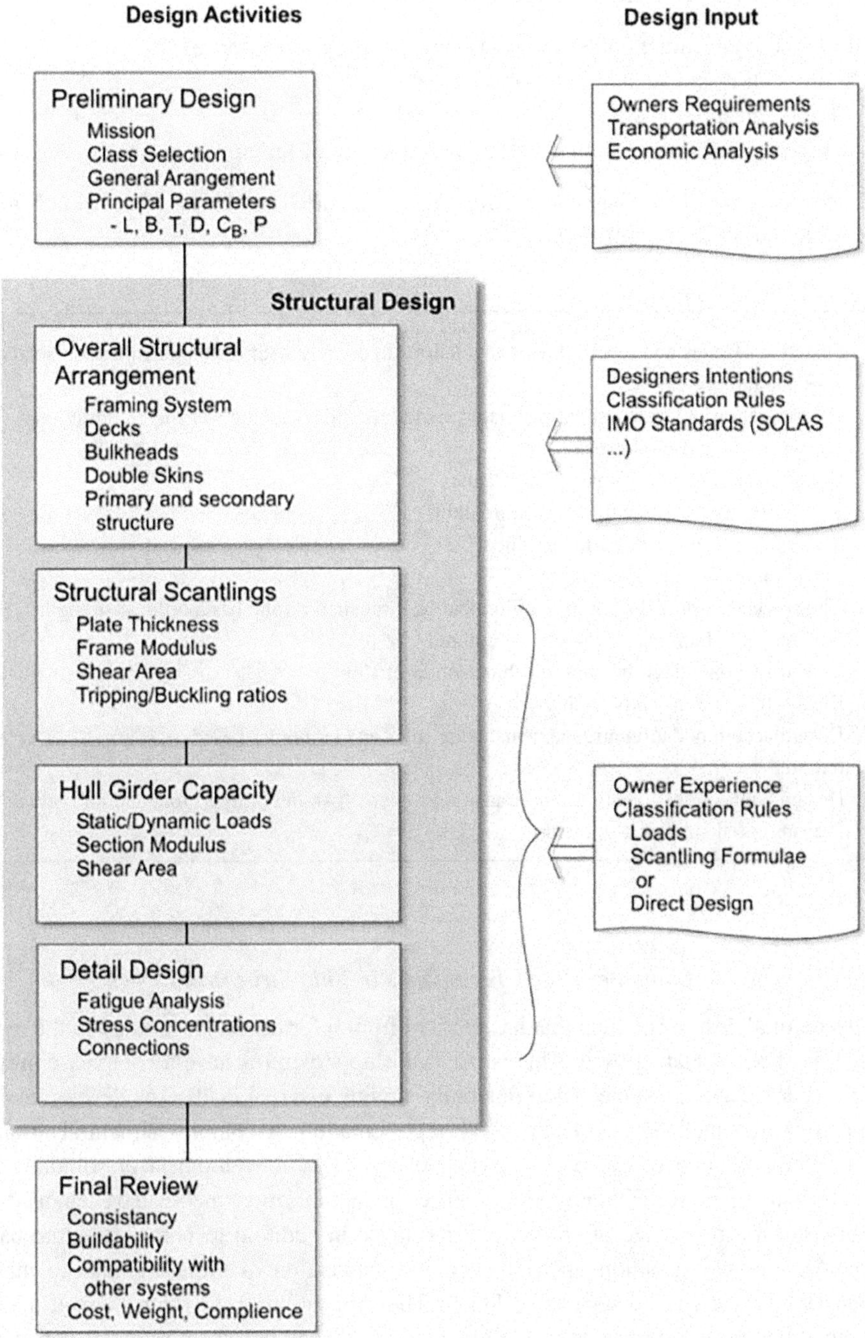

Figure 1.22 Rule-based design of merchant ships [15]. Reprinted courtesy of the Ship Structure Committee, Washington, DC.

To conclude, a structural standard in principle ought to:

1. Be based on accurate models of in-service loading and response
2. Address all response and failure mechanisms
3. Address uncertainties in all aspects of models
4. Incorporate safety factors that reflect consequences of failure

The basic steps of rule-based ship structural design are listed below. This list is complemented by the flow chart of Figure 1.22.

1. Selection of rules and other relevant regulations to comply with (IMO, classification society, national authorities)
2. Determination of structural arrangements, materials and stiffening systems (longitudinal, transverse, mixed) (synthesis)
3. Determination of scantlings by use of rules
   3.1 Longitudinal strength (section modulus)
   3.2 Application of rules to modified basis design to select spacing and stiffener section moduli for midship section
4. Check whether global strength requirements are satisfied on the basis of local strength requirements. Increase scantlings as required.
5. Extend section design beyond midship section of ship
6. Design fore and aft ends of ship
7. Design accommodation arrangements using rules and computer-based tools as required
8. Design connections, openings and transition regions (parallel body to fore and aft end)
9. Prepare initial structural drawings

1.4.6 Computer-Based Techniques in Ship Structural Design

Ship structural design is a field that has profited from information technology to the extent that it has been reshaped by it. This is because ship structures have a complex configuration, are large and are subjected to randomly varying external loads. The development of computer-based methods has enabled the representation of complex structural arrangements, the modelling of randomly varying loads and the development of solutions that involve both of these. Furthermore, advances in optimisation theory have enabled the adoption of a broad range of criteria of acceptance in addition to purely structure-based ones. A rigorous exposition of the theory and application of these methods to marine structures would require a separate volume.[7] The application of the finite element method in ship structural design has now become well established and is an essential part of the

[7] A text that covers these subjects is [11].

most basic activity, which is plan approval. It is now impossible to seek approval without including results from finite element procedures for a variety of ship types and cases. The application of the finite element method with examples from ship structural analysis and design is discussed in Chapter 13.

1.4.7 Good Practice in Ship Structural Design

In closing this chapter it is necessary to mention some points of guidance in ship structural design that experience has shown to be of value. Clearly, computer-based methods will have built-in checks so that the issues raised are taken into account, but it is important that the designer is able to have a first-hand idea of the quality of the design by referring to a brief (mental) check list.

Requirements that are demanded of the structure of a sea-going vessel in general are:

- Sufficient working strength. This is ensured by adherence to classification society rules but may need to be assessed in the case of an unconventional hull structure or arrangement.
- Requisite fatigue strength. This is also satisfied by adherence to classification society rules. Again, should extraordinary requirements be imposed, first-principles methods will have to be resorted to.
- Ultimate longitudinal strength. The designer ought to have an idea of the ultimate longitudinal strength of the hull girder and of the effect that alternative structural arrangements may have on its value, hence the true safety margin against overall collapse.
- Producibility. This is of importance to the shipyard. Shipyards are well aware of production techniques and particularities of hull girder fabrication methods.
- Inspectability and repairability. These qualities are of most importance to the owner and the classification society.

Beyond the above-listed requirements there are additional features that a well-designed structure will exhibit. *Robustness* is one such quality. We desire that the consequences of structural failure are not disproportionate to the effect that causes failure. In a robust structure, failure of a single member does not lead to the collapse of the whole structure. In order to design a structure so that it proves to be robust, it needs to have built-in *redundancy*. A redundant structure is one with additional internal load paths or external supports in excess of the minimum required for stability. An example of a structure with insufficient redundancy would be a three-legged oil platform which, if one leg collapses (as may occur should a ship collide with it), will collapse as a whole. Additional legs and braces increase redundancy, thereby leading to a more robust structure.[8]

[8] The *Alexander L. Kjelland*, a semi-submersible oil rig operating in the North Sea, collapsed after fractures developed in one of the braces and foundered with a severe loss of life. Robustness as an explicit design aim is discussed in Chapter 12.

In ship structures, failure occurs at a local level (buckling, yielding and fracture), so that overall failure occurs when the consequences of one of these failure modes propagate in an uncontrolled manner or lead to a different failure mode. For example, a heavily corroded structure that is subjected to continuous pounding in severe seas may suffer fracture or tearing. The structure will then not be able to withstand further loading, and the fracture may propagate to the extent that one or more compartments become flooded or, in a worst-case scenario, the ship founders. Such a sequence of events is the result of one failure mode leading to a different one, thereby causing a reduction in structural redundancy.

In general, because of their configuration, ship structures have a greater degree of redundancy than do civil engineering structures. This is because they consist of thin stiffened plating, in contrast to civil engineering structures that consist of beams and frames. When failure occurs in a plated structure, loads are borne by adjoining members. In the case of beam-type structures, the degree of redundancy depends on the spatial configuration of the members, the direction of loading and the support conditions. It has to be said that redundancy is not considered explicitly in the design of marine structures in the same way that subdivision is used to ensure adequate damage stability. For example, when designed properly, frames can exhibit sufficient bending strength but also redundancy, due to secondary load paths created by axial stresses in the plate and frame. As a second example we can consider the vertical support provided to a tweendeck inside a cargo hold. In general, pillars are used to support the deck. Clearly a larger number of pillars, each bearing a smaller proportion of the total load, will lead to a more robust structure. By adopting such a standpoint a *ductile* structure will result – that is, one for which the consequences of local failure will lead to a gradual loss of strength and not to sudden unloading, which may occur in an uncontrolled manner and have much more serious consequences. The issue of redundancy is discussed further in Chapter 7 and that of robustness in Chapter 14.

We shall end this chapter by referring to the issues of *alignment* and *continuity*. These are closely linked, and it is worth to consider continuity first. Structural continuity at a global level is not under absolute control of the structural designer, as there will always be structures that end abruptly, such as superstructures. In such cases it is important to ensure that stress concentrations are avoided by suitable detail design at a local level. Past practice in which longitudinal stiffeners were ended in way of transverse bulkheads and were attached to them using various arrangements has been superseded by having them pass through the bulkhead and by being suitably attached to it (Figure 1.23). The problem that arises when a longitudinal stiffener is attached to bulkhead plating is that the bending stiffness of the two members is very different. If a stiff member is attached to a more flexible one, the latter bends about the former.[9] In the case of stiffener–bulkhead connections, this may give rise to cracks in

[9] This is termed a *hard spot*.

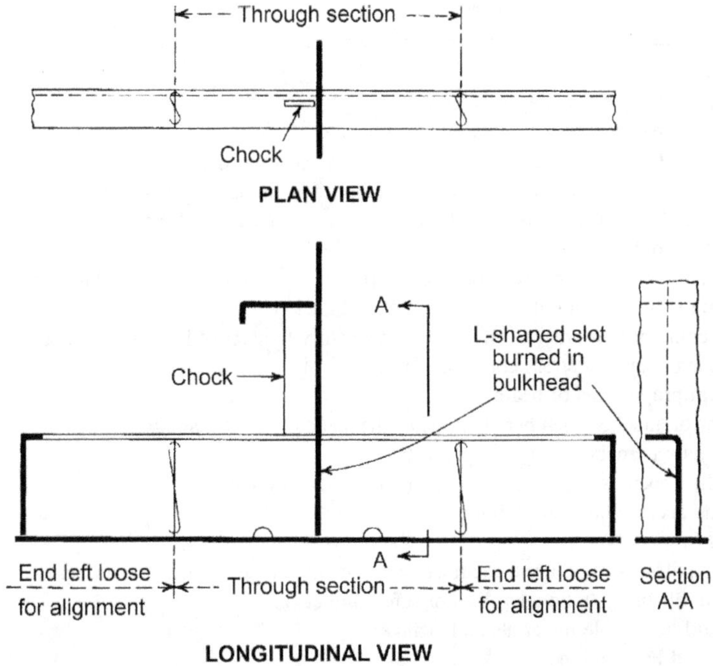

Figure 1.23 Connection of a continuous longitudinal stiffener with a transverse bulkhead [3]. Reprinted courtesy of the Society of Naval Architects and Marine Engineers.

the bulkhead plating. The aim therefore is to ensure a connection in which both members will distort by approximately the same extent. One way in which this can be is achieved is by increasing the bulkhead plate bending stiffness by fitting a collar plate. Another way is to align vertical stiffeners on the bulkhead with the longitudinal stiffeners of both the deck and the bottom structure, thereby forming a ring frame within each compartment. This, however, is not effective in the case of the longitudinal stiffeners that are attached to the side shell plating, and it is for this reason that many fatigue cracks arise in way of the connections of longitudinal stiffeners with web frames and transverse bulkheads.

The question of *misalignment*, on the other hand, is one in which the designer can exert much greater control. Misalignment may arise as a result of design decisions or else during fabrication. Alignment is a necessary prerequisite to proper load transfer and, when used effectively, can reduce the loads to which members may otherwise be unnecessarily subjected. Thus it pays to have supporting members such as pillars located beneath each other in successive decks.

Table 1.12 summarises some of the points raised and gives an idea of issues that are under the control of the structural designer.

Table 1.12 *Recommendations for good practice.*

a) General
- Keep it simple – everything in general.
- Incorporate feedback from in-service experience.
- Ensure that loads are readily transferred, thereby ensuring that the structure is highly redundant (for example, by making use of multiple rather than single supports, the use of brackets at stiffener ends, the use of foot and head plates and chocks at pillar ends).

b) Structural Layout
- Ensure alignment of the structure (such as pillars supporting successive layers of decks, stiffening members on either of bulkheads and decks).
- Ensure continuity in order to enable the satisfactory transfer of loads between adjacent members and the avoidance of stress concentrations.
- Optimum proportions of beam.
- Stiffness of attached members is to be of the same order of magnitude.
- Avoid sharp corners in way of openings.
- Strengthen openings and select positions of welds with care.
- Facilitate access and inspectability.

c) Structural Properties
- Bending stiffness of adjacent members is to be similar (avoidance of hard spots).
- Uniform reliability (rational values of safety indices).
- Use readily available materials and sizings.

d) Materials and Fabrication
- Judicious use of high-tensile steel.
- Reduce lengths of welds to reduce the number of defects such as stress concentrations (for example, by the use of larger plate sizes or alternative structural arrangements or materials).
- Simplify stiffening systems (for example, consider the use of the unidirectional system) as well as the connections between structural members.

References

1. Ashe G. et al. (2009) Naval Ship Design. Committee V.5 Rept. in: Jang C. D. Hong S. Y. (Eds) *Proc. 17th Int. Ship & Offsh. Struct. Cong.* Vol. 2. Seoul: Seoul Nat. Univ. 289–307.
2. Beach J. E. Bruchman D. D. Sikora J. P. (2003) Advanced Double Hull Concept Reduces Costs Increases Capabilities. *AMPTIAC Quart.* Vol. 7 5–11.
3. Buermann T. M. (1969) Structural Components. Chapter V in: d' Arcangelo A. M. (Ed) *Ship Design and Construction.* Jersey City, NJ: SNAME 191–250.
4. Chalmers D. W. (1993) *Design of Ships' Structures.* London: HMSO Publications.
5. d' Arcangelo A. M. (1964) *A Guide to Sound Ship Structures.* Centreville, MD: Cornell Maritime Press.
6. Department of Defense (1998) *Integrated Product and Process Development Handbook.* Washington, DC: DOD.
7. Det Norske Veritas (2005) *Hull Structure Course.*
8. Dimitracopoulos J. (2008) *Lloyds Register Presentation.* Piraeus.
9. "EconStats : Iron ore prices and iron ore producers" *Econstats.com.* [Online]. Available: www.econstats.com/rt_ironore.htm. [Accessed: 19 December 2020].
10. "Hardness" *Uni-kiel.de.* [Online]. Available: www.tf.uni-kiel.de/matwis/amat/mw1_ge/kap_8/advanced/t8_4_2.html. [Accessed: 19 December 2020].

11. Hughes O. F. Paik J. K. (2009) *Ship Structural Analysis and Design*. Jersey City, NJ: SNAME.
12. International Association of Classification Societies (2008) *Common Structural Rules for Double Hull Oil Tankers*.
13. "International Council on Systems Engineering Website" *Incose.org*. [Online]. Available: www.incose.org/. [Accessed: 19 December 2020].
14. International Maritime Organisation (2010) Adoption of the International Goal-based Ship Construction Standards for Bulk Carriers and Oil Tankers Resolution MSC.287(87).
15. Kendrick A. Daley D. C. (2006) Comparative Study of Naval and Commercial Ship Structure Design Standards. Ship Struct. Comm. Rept. SSC-446 Washington.
16. Moan T. (2006) Marine Structures for the Future. Keppel Inaugural Lecture Singapore.
17. Nippon Steel (2002) Application of Titanium to Construction Civil Engineering and Ocean Development *Nippon Steel Technical Rept. No 85*.
18. Okumoto Y. Takeda Y. Mano M. Okada T. (2008) *Design of Ship Hull Structures*. Berlin: Springer Verlag.
19. "Online Materials Information Resource – MatWeb" *Matweb.com*. [Online]. Available: www.matweb.com. [Accessed: 19 December 2020].
20. Pomeroy V. (2010) Lloyd's Register marks 250 years. Lloyd's Register of Shipping London.
21. Pradillon J.-Y. et al. (2012) Design Methods. Committee IV.2. in: Fricke W. Bronsart R. (Eds) *Proc. 18th Int. Ship & Offsh. Struct. Cong.* Vol. 1. Hamburg: Schiffahrts-Verlag Hansa GmbH & Co KG 507–576.
22. Segretain J.-F. (2009) Gigantism in Shipbuilding and Correlative Safety Requirements. *Le Rendez-vous de l' Assurance des Transports* Bureau Veritas Presentation, Cannes.
23. "Ship design process" *Deltamarin.com* 30-Sep-2014. [Online]. Available: https://deltamarin.com/shipdesign/ship-design-process/. [Accessed: 19 December 2020].
24. Society for the Registry of Shipping (1818) *The Register of Shipping for the Year 1818*. London.
25. *Steelguru.com*. [Online]. Available: https://steelguru.com/. [Accessed: 19 December 2020].
26. Suzuki S. Muraoka R. Obinata T. Endo S. Horita T. Omata K. (2004) Steel Products for Shipbuilding JFE Technical Rept. No. 2.
27. Takahashi H. Goto A. Abe M. (2006) Study on Standards for Main Dimensions of the Design Ship. Technical Note of Nat. Institute for Land and Infrastructure Management Japan Vol. 309.
28. UNCTAD (2019) Review of Maritime Transport.
29. Ziliotto F. et al. (1991) Quasi-static Load Effects. Committee II.1 Rept. in: Hsu P. H. Wu Y. S. (Eds) *Proc. 11th Int. Ship & Offsh. Struct. Cong.* Vol. 1. Amsterdam: Elsevier Applied Science 163–238.

2

The Evolution of Ship Structures from Antiquity to the Present Day

εἰπὲ δέ μοι γαῖάν τε: τεὴν δῆμόν τε πόλιν τε,
ὄφρα σε τῇ πέμπωσι τιτυσκόμεναι φρεσὶ νῆες:
οὐ γὰρ Φαιήκεσσι κυβερνητῆρες ἔασιν,
οὐδέ τι πηδάλι' ἔστι, τά τ' ἄλλαι νῆες ἔχουσιν:
ἀλλ' αὐταὶ σασι νοήματα καὶ φρένας ἀνδρῶν,
καὶ πάντων σασι πόλιας καὶ πίονας ἀγροὺς
ἀνθρώπων, καὶ λαῖτμα τάχισθ' ἁλὸς ἐκπερόωσιν
ἠέρι καὶ νεφέλῃ κεκαλυμμέναι: οὐδέ ποτέ σφιν
οὔτε τι πημανθῆναι ἔπι δέος οὔτ' ἀπολέσθαι.
Οδύσσεια Θ' φνε'-φξγ'

Name your country, your people and your city,
so that our ships that have a mind and can think will take you there directly.
Because the ships of the Phaeacians have neither helmsman
nor steering oar like all other ships,
but know themselves the thoughts and desires of humans
as well as all cities, even the fertile fields
of men, and are so fast that speedily they cross the seas
that may be covered in thick mists and fog; nor are they
afraid of losing their way or of suffering any evil.
Odyssey, 8th Rhapsody, 555–563

To understand and appreciate the origins of the problems that the naval architect has to face, it is useful to trace the evolution of ship structures through time. This chapter is dedicated to this end, and changes in needs, fundamental scientific understanding as well as the means available for ship construction from the prehistoric period until the present day are described. The description is, of necessity, a brief outline of a long and diverse pattern of development. Recent advances in technology have enabled important discoveries of finds that have remained on the seabed for centuries to come to light. The underwater world is now beginning to reveal its secrets, and many more discoveries are expected. In parallel, the history of science and technology is a subject that has grown significantly as texts are being studied and interpreted more intensely than ever before. In this chapter an attempt is made to bring together the findings of two hitherto distinct fields of research, for a specific purpose – those of underwater archaeology and the histories of science and technology.

The use of iron and steel in shipbuilding dates back to the mid-19th century; prior to that, wood was the material used, since the very early period in the history of waterborne vessels. In this chapter we shall briefly consider the various stages in the development of the construction of sea-going ships, with an emphasis on matters that relate to structural integrity and strength.

2.1 Prehistoric Period

The first vessels in which humans crossed the water were as follows [19]:

- *Rafts*, made of materials such as logs and reeds, which have a density less than that of water.
- *Skin boats,* consisting of skins sewn together over a skeleton frame of wood or bone.
- *Bark boats*, made by stripping bark from a tree to form a continuous cylinder in which the ends are stitched together and the boat is reinforced internally by a light frame.
- *Dugouts*, made of hollowed-out tree trunks (Figure 2.1a).

Of all primitive craft, dugouts were the only type that could be developed into a more sophisticated form. This was possible because ribs and frames could be securely attached to the log, thereby providing support to any shell structure added at a later stage. Two different methods were employed to achieve this. In the first of these, termed the *expanded dugout*, the log was hollowed out through a narrow slot to form a cylinder. The sides were then forced apart, thus giving greater breadth and freeboard. Finally, the sides were cut to the thickness of planks and then reinforced by frames (Figure 2.1b). In the case of the

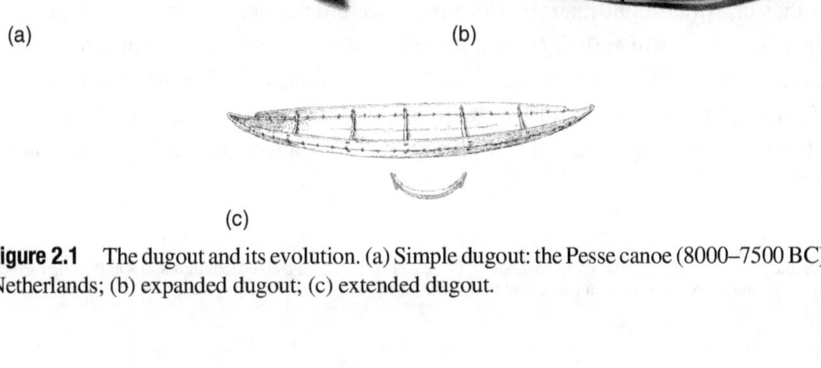

(a) (b) (c)

Figure 2.1 The dugout and its evolution. (a) Simple dugout: the Pesse canoe (8000–7500 BC), Netherlands; (b) expanded dugout; (c) extended dugout.

extended dugout, the freeboard was increased by attaching planks to the dugout, which thus assumed the role of a keel (Figure 2.1c). Such vessels were primarily used on rivers and lakes.

Vessels that traversed inland waterways were the precursors to sea-going ships, since the conditions they encountered were not as hazardous as the ones on the open sea.

Amongst the most important developments that took place during this early period in the history of ship construction is the manner in which the components of the structure were attached to each other. In the earliest vessels the planking was lashed together using reeds, cords or thongs; indeed, this manner of joining was in use in the Indian Ocean until the advent of the Portuguese. The evidence from Egypt indicates that the use of wood in ship construction was highly developed early on, and techniques such as the use of *mortise-and-tenon* joints, *butt* joints, *scarph* joints and *rabbets* were used. These techniques are still in use today.

Developments took place on other fronts as well; the need for the provision of sufficient longitudinal strength led to the development of the *keel,* which henceforth provided the backbone for the strength of the hull and at the same time a point of reference during the erection process. New rigging techniques using square rather than rectangular sails necessitated the use of stronger supports for the masts. Superstructures were added, as were provisions for multiple arrangements of rowers.

The primary structural problem throughout this early period was not so much the action of external loads as the need to keep the hull components securely attached to each other while providing watertight integrity. The seriousness of this problem becomes clear when we consider the increases that took place in vessel size, from the dugout to the Nile riverboats,[1] and also the severity of conditions encountered when attempting a voyage across the open sea.

Other developments related to the *fabrication sequence* as well as to the manner in which the planks are attached to each other (Figure 2.2). Plank-built ships originated from the most developed form of the dugout, which by its very nature provides sufficient longitudinal and transverse strength and does not require any internal stiffening. The evolution of plank-built ships enabled larger vessels to be built while simultaneously reducing the wall thickness. The question then arose as to how the planks ought to be attached to one another. Continuity of the hull form led to adjacent planks to be *edge-joined (carvel).* Two techniques were developed to achieve this in prehistoric times, the *mortise-and-tenon* joint and *sewing/lacing*, the aim being to ensure that planks were attached to each other in a watertight manner. Both these techniques were based on the presumption that the planking was of sufficient strength to withstand external as well as internal loads, as in the case of the dugout. We shall discuss these techniques in more detail in a later section. In order to build ships using either of these methods, it was necessary to have available the right tools, and it was in fact the invention of the appropriate tools that enabled these

[1] The Dashur riverboats had a length of approximately 10 m, a beam of 2.3 m and a depth of 0.8 m. On the other hand, the the Cheops papyriform boat was 40 m in length and 8 m in the beam.

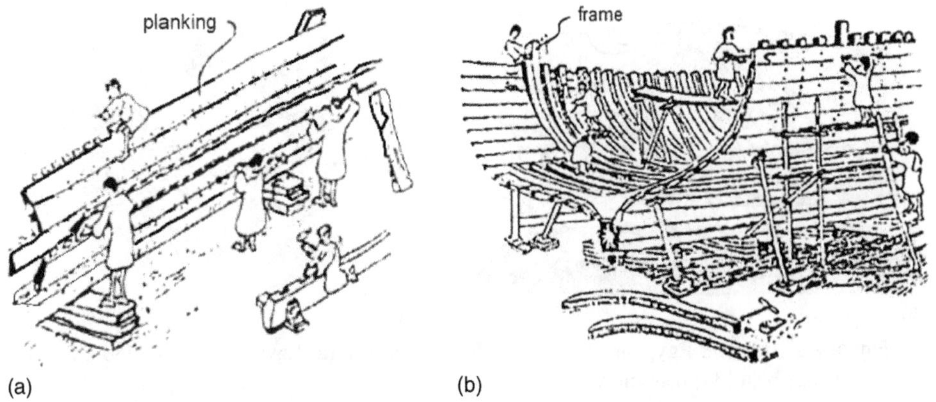

Figure 2.2 Fabrication sequence. The shell-first method. (a) Laying of keel and fabrication of shell. (b) Attachment of internal frames to hull. Drawings by J.-M. Gassend.

techniques to reach a high level of craftsmanship along the shores of the Mediterranean throughout the period of antiquity.[2]

From the point of view of fabrication technology, the distinguishing feature of the different structural types is whether the hull planks are supported by one another or whether they are held in position by being attached to a rigid system of frames. They can thus be seen to be either *edge joined* or *non-edge joined*. The two important stages in the shell-first method are shown in Figure 2.2.

From the point of view of structural strength, the most important development that took place many centuries later is in the shift in reliance on the hull planking to the use of *frames*, in order to provide sufficient transverse strength and stiffness to the hull. This practice has continued to the present day. In antiquity, strength and structural integrity were provided by the shell alone, which consisted of thick planks attached to each other in a variety of ways. With the passage of time, internal stiffening was gradually introduced, thus assuming a more important role. This enabled a reduction in the thickness of the shell planking, which simplified the fabrication procedure. At the same time, the spacing of the joints increased.

2.1.1 Egypt

The earliest source of information on the construction of sea-going ships in Egypt is a relief dating from 2450 BCE, which depicts vessels returning from an overseas military expedition (Figure 2.3a). The form of the hull was long and slender with overhanging fore and aft ends, similar to that of inland waterways craft that date from that era. The shell planking was edge-joined with a minimum of internal stiffening. To be fit to sail on open water, and

[2] In northern Europe where the saw was not yet known, vessels were primarily *lapstrake* (*clinker*-built) until very much later. In lapstrake construction planks overlap each other. This requires less precision in cutting timber, and vessels were thus limited in size. Later on all ships were carvel-built.

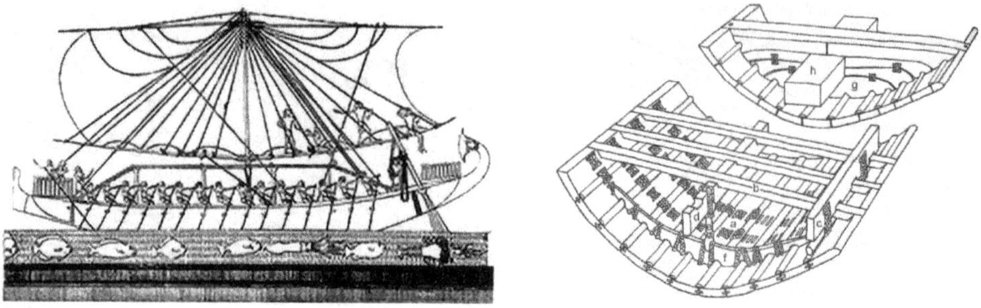

(a) (b)

Figure 2.3 Ancient Egyptian ships and hull construction. (a) Egyptian vessel 2450 BCE; (b) Cheops boat [34] (assembly using transverse lashings).

to withstand wave loading, these ships were equipped with a rope truss in the form of a heavy cable that extended from the fore to the aft end and which was carried on upright forked sticks. A lever permitted the crew to twist the cable, thus increasing the tension and preventing the sagging of the fore and aft ends of the vessel. In addition, the narrow rope netting was attached to the upper part of the hull at deck level, in way of what we now term the *sheer strake*. It is believed that this rope provided support to the side planking upon which the deck structure rested and to which beams were attached.

The next available evidence dates from one thousand years later (1500 BCE). There are significant improvements in the hull form which, although of the same general shape as that of the earlier period, is more graceful and with cleaner lines. An important innovation is the extension of the deck beams through the planking in the side shell, thereby increasing the strength and removing the necessity for the rope netting. The heavy cable that provided stiffness in the longitudinal direction was still in use. Indeed, this practice was followed much later (fifth century BCE) by the Athenians who used it to provide sufficient longitudinal stiffness for the hull of the *trieres*.[3]

Towards the end of the second millennium BCE (circa 1200 BCE) the Egyptians came into contact with other maritime nations whose ships had superior design features.[4] Of these the most important was the fact that the vessels were both sail and oar powered, thus improving manoeuvrability and speed. Several additional features are evident in the ships depicted in the Medinet Habu relief in Egypt. These concern the existence of a type of ram, modifications to the rigging arrangements and the existence of raised platforms to facilitate military operations. The rope truss necessary for longitudinal strength of the vessel, seen on earlier designs, is not evident in these depictions. We may therefore conclude that these vessels were much more strongly built, with a keel that acted as a support for transverse frames, to which the side shell planking was attached.

We have seen that developments in ship construction that, to our knowledge, were not recorded took place over a period of time and in regions that we cannot pinpoint with any

[3] *Trieres* (Greek: τριήρης). Also called *trireme*, after the Latin *triremis*.

[4] The enemies that the Egyptians had to face were the Sea Peoples who were probably Phoenicians and Achaean Greeks.

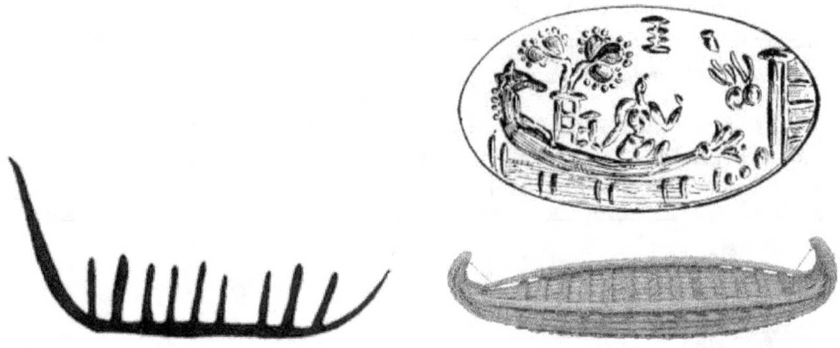

(a) (b)

Figure 2.4 Prehistoric craft from the Aegean and Ionian Seas. (a) Neolithic rock carving of ship (4000 BCE) (Strophylas, Andros) [53]. (b) Reed craft from Mochlos, Crete, and a reconstruction of the papyrella.[5]

degree of accuracy. We do know of developments in the Levant, with the rise and expansion of the Phoenicians, who traded throughout the Mediterranean. Depictions that have been discovered show vessels, both merchantmen and naval, with a characteristically Egyptian profile, manned by men who appear to be Syrian. The vessels are, importantly, not fitted with the rope truss needed to ensure longitudinal strength. It is concluded that structural strength is provided otherwise, by the hull itself. That can only have been a more developed system of internal framing than that used in Egyptian ships.

2.1.2 Greece

The Aegean Sea is an expanse of water filled with literally hundreds of islands of various sizes. A large proportion of these couldn't, and still cannot, sustain a human community beyond the bare essentials of existence. It would therefore appear natural for the inhabitants of these islands to turn to the sea to reach other coastal communities and to trade, in order to survive and improve their living conditions. This is in fact what happened, and led to the development of a maritime tradition that extends to the present day but whose origins are lost in prehistory. The inhabitants of the islands of the Aegean developed sea-going craft to suit their needs.

To understand the development of sea-going vessels during this early period it is necessary to turn to the evidence available, both physical and written. The physical evidence consists in large part of ship models, depictions on household ware, seals and rock carvings (Figure 2.4a). The remains of a wreck dating from the Minoan period were located off the coast of Crete in 2010 and reported in [4]. Another important find is the Late Bronze Age shipwreck at Point Iria, along the north coast of the Argolid discovered by

[5] The image of the bundle (reed) craft appears on a gold ring discovered by Evans. The reconstruction of the papyrella was directed by H. Tzalas following the discovery of vessels being built in this manner on the island of Corfu during the mid-20th century by the paleontologist A. Sordinas.

Table 2.1 *Chronology of Bronze Age Aegean Greece [12]*

	Cycladic civilization (Cyclades)	*Minoan civilization (Crete)*
1	Early Cycladic (EC): 3300–2000 BCE	Early Minoan (EM): 3650–2160 BCE
2	Kastri (EH II–EH III): ca. 2000–1500 BCE	Middle Minoan (MM): 2160–1600 BCE
3	Convergence with MM from ca. 1500 BCE	Late Minoan (LM): 1600–1170 BCE

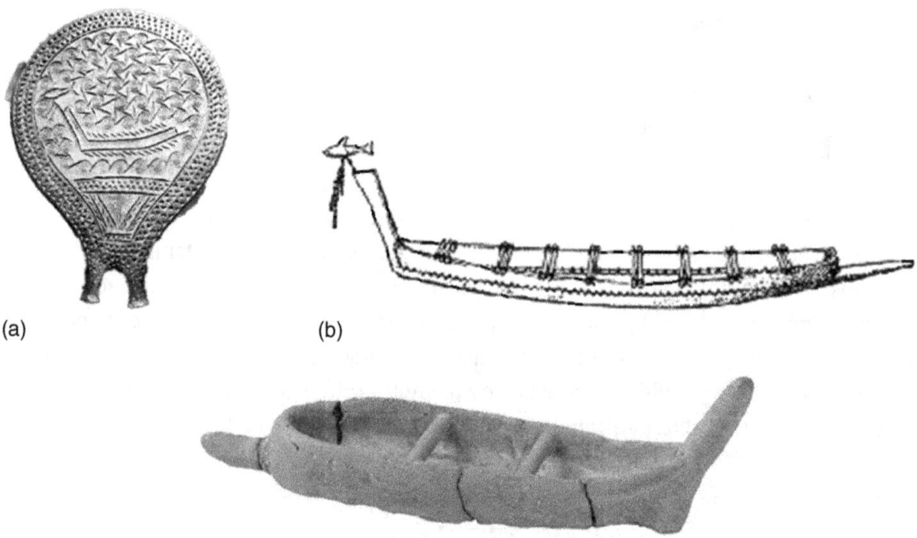

(a) (b) (c)

Figure 2.5 The early Cycladic longboat (2800–2300 BCE) (Type I vessel) [57]. (a) Ship incised on a clay vessel (Halandriani, Syros). (b) Sketch of longboat depicted in (a).[6] (c) Ship models (2800–2400 BCE) (Heraklion Museum, Crete).

Tsouchlos in 1962 [42]. With the passage of time more evidence becomes available: written evidence is available from the eighth century BCE (Homeric epics, although these refer to ships of the 11th century BCE). The first difficulty that arises is that of chronology, because developments took place at different times in different geographical regions.

Our interest lies in the evolution of sea-going craft in the Aegean Sea (Cyclades and Crete) during the Bronze Age up to and including the beginning of the Iron Age (4000–1000 BCE).

Table 2.1 indicates the generally accepted chronological breakdown for the islands of the Cycladic archipelago in the Aegean Sea and Crete.

[6] During this period boats were paddle-powered [58].

During the Bronze Age, which ranges from 4000 BCE to 1000 BCE, the available information permits the classification of vessels to be based on (a) vessel type (conceptual basis) and (b) technological criteria.[7]

Vessel type (conceptual basis)	Technological criteria[8]
1. Dugout-based craft	A. Fabrication (assembly) method
2. Plank-built oared sailing vessel	B. Nature of primary structure
3. Oared galley	C. Type of oared propulsion

The three major phases in the evolution of sea-going craft during this period are as follows:

1. The extension of the dugout, probably before 4000 BCE.
2. The extended dugout becomes the first plank-built ship ca. 2300/2100 in the Aegean.
3. The introduction of the oared ship, by Mycenaean shipwrights, ca. 1350 BCE.

The process of evolution from the simple dugout to the expanded dugout (Figures 2.1a–b) and thence to the extended dugout (Figure 2.1c) probably took place before 4000 BCE. In the Aegean, the development of the first plank-built vessel from the expanded dugout (Figure 2.5a) took place during the 2300–2100 BCE period.

The earliest vessels (Type I) are the *Cycladic longboats* (Figure 2.5) that are considered to have been paddle driven. It is seen from the figure that the stem projects beyond the stempost, the hull has a slender form, and the stern rises substantially.

Towards the end of the Early Bronze age, which coincides with that of the Early Minoan age, the vessels depicted on seals and earthenware (Figures 2.6a, 2.6b) are equipped with a mast as well as oars. The number of oars, averaging out at five, indicates that the vessel was shorter in the length and broader in the beam than in earlier periods.

Towards the end of the Middle Minoan era, the best-known representations of ships are the frescoes discovered at Akrotiri (Santorini) (Figure 2.6c). These include depictions of sea-going craft of different sizes and types. The frescoes show a procession of luxuriously outfitted large vessels which appear to be paddle and oar powered. These are accompanied by smaller escort vessels that are sail powered.

The ships shown in Figure 2.7 have a single bank of oars and are also sail powered.

The timeline depicted in Figure 2.8 includes three dating systems, each of which has a specific geographic connotation. The Minoan concerns Crete, the Cycladic concerns the islands of the Aegean Sea, and the Helladic concerns mainland Greece. These dating systems are based on the archaeological findings in the respective regions.

[7] Wedde [57] proposes a typology that is based on vessel design, means of propulsion and hull form. It is this typology that is followed here.

[8] 1. Fabrication (assembly) method: laced/stitched, intermediate hybrid, mortise-and-tenon.
2. Nature of primary structure: Dugout core, hog, T-hog, keel.
3. Type of oared propulsion: Moneres (single row of oars), dieres (two rows of oars), trieres (three rows of oars).

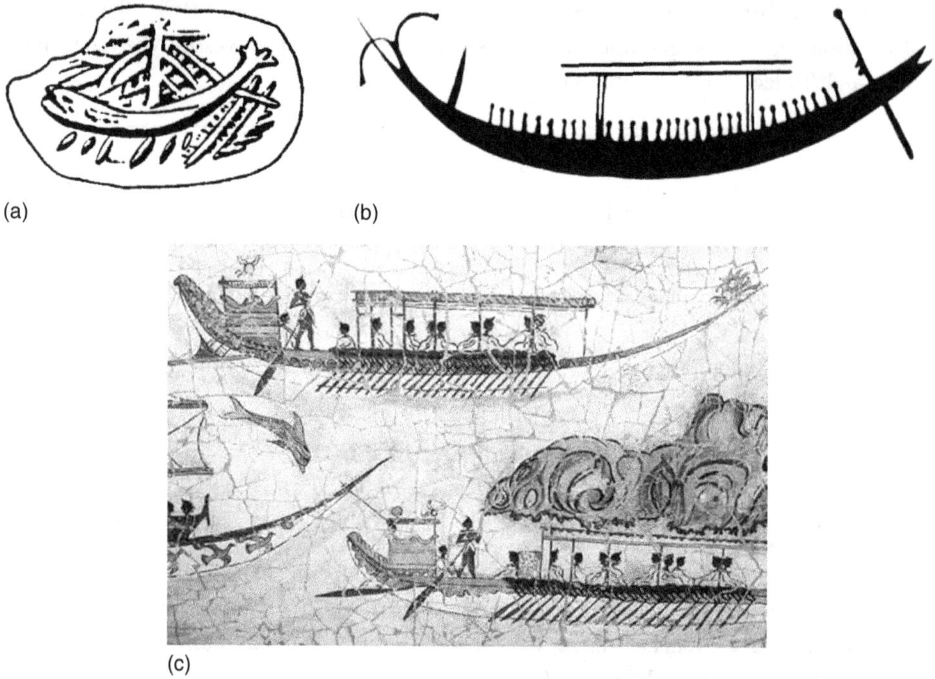

Figure 2.6 Evolution of vessels in Crete and the Aegean [57]. (a) Type II vessel (Malia). (b) Protopalatial vessel (Type III) (Middle Minoan II). (c) Middle Minoan III era vessels (Type IV) (Akrotiri fresco).

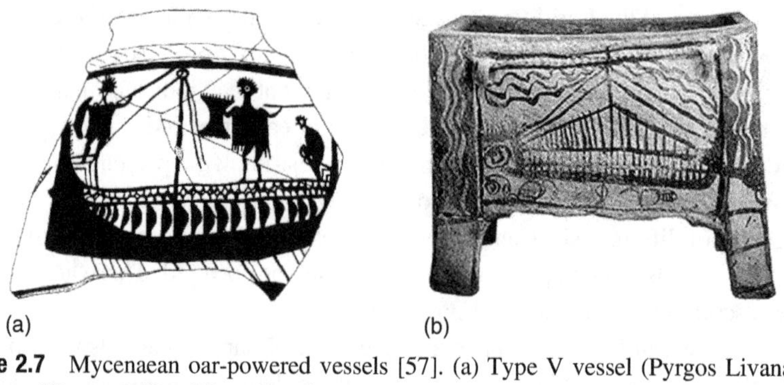

Figure 2.7 Mycenaean oar-powered vessels [57]. (a) Type V vessel (Pyrgos Livanaton). (b) Type VI vessel (Heraklion, Crete).

2.1.3 Homer's Descriptions of Ships

It is generally accepted that Homer, the poet who wrote the *Iliad* and the *Odyssey*, lived during the eighth century BCE. The events that he described have been dated to around 1200 BCE.[9] A careful study of his texts as well as those of later writers of antiquity (Herodotos,

[9] There has been a lot of effort expended in dating the events described by Homer. A recent attempt at dating the return of Odysseus proposed April 1207 BCE, based on astronomical data as well as seasonal, geographic and environmental observations [40]. In another recent attempt a much earlier date has been proposed for the siege of Troy (1312 BCE) [21].

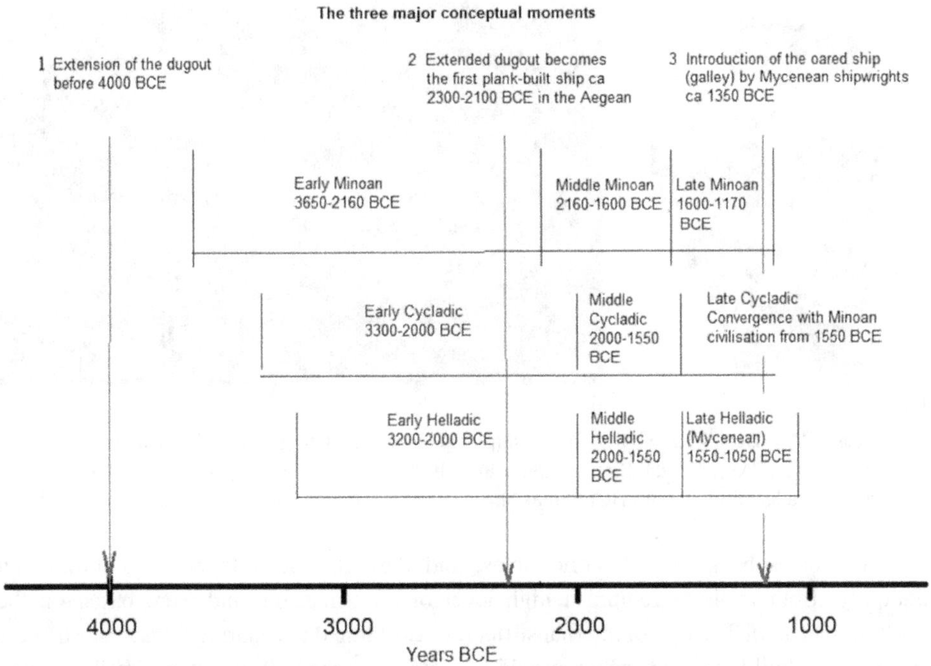

Figure 2.8 Timeline of the development of sea-going craft in prehistoric Greece.

Thucydides, Apollonios of Rhodes, Diodoros of Sicily and the poet Alkaios) reveals that his descriptions refer to vessels of a period that preceded the eighth century BCE.

There are a large number of mentions of ships in the two epics, the most important of which is the detailed description of the construction of a boat by Odysseus in the *Odyssey* (5th Rhapsody 244–257) [22]. It is apparent that the assembly method was that of sewing/lashing rather than the mortise-and-tenon joint.

The smallest vessels were equipped with a single mast and 20 oars, whereas the larger had 50 oars (*pentekontors*). There existed vessels with larger crews, up to 100 oarsmen, that were used as merchant ships. A pentekontor had a slender hull, and its overall length ranged up to 40 metres, to accommodate the 25 oarsmen on either side as well as the fore and aft deck structures. It would have a beam of 3–4 metres, and thus $L/B \approx 10$. The ships at the time of the Trojan war were light and were beached at night by the crew. They had a low freeboard, raised bow and stern and a curved bulwark (Figure 2.9a). There is mention of a deck in way of the fore and aft ends of a vessel. Figure 2.9b illustrates the wreck of a ship recently discovered at the bottom of the Black Sea and preserved until today due to the inactive (*anoxic*, or oxygen-free) environment. Research continues, and further results are expected.

2.2 Fabrication Methods

In this section we will provide a brief description of the two fabrication methods used widely in the Mediterranean world and further (Indian Ocean) in antiquity. Evidence of

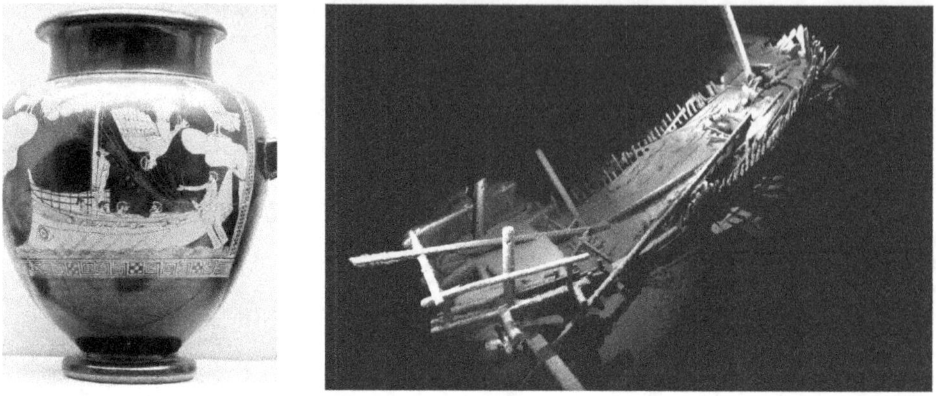

(a) (b)

Figure 2.9 Depiction of Odysseus' ship and a wreck dating from classical antiquity. (a) Odysseus' ship and the Sirens (Athenian fifth-century BCE krater). (b) Photogrammetric model of a late fifth-century BCE Greek ship lying in the Black Sea.[10]

their use dates back to prehistoric times, and they continued to be used throughout antiquity. Both methods require a high level of craftsmanship and were based on the *shell-first method*. The role of internal stiffening was limited to ensuring transverse stiffness (retention of hull form) rather than providing strength to the hull. It was the shell planking in conjunction with its method of fabrication that ensured sufficient hull strength.

2.2.1 The Mortise-and-Tenon Joint

The aim of the *mortise-and-tenon* joint was to ensure a smooth, continuous and watertight exterior hull surface of sufficient strength. This was achieved by accurate machining of the planking and by joining of the planks to each other by inserting closely spaced rectangular pegs along their longitudinal edges (Figure 2.10). These pegs were kept in place by cylindrical *dowels*. Local strength was ensured by use of planking of sufficient thickness whereas accurate craftsmanship ensured overall strength and a faired hull. Additional structural members contributed to hull form retention, and these were fitted both internally (transverse frames) and externally (longitudinal *wales*). Of primary importance was the keel, which was the first member to be laid and which provided sufficient longitudinal strength and stiffness to the entire hull girder. It also acted as a means of resistance against grounding and impact with submerged objects.

This type of fabrication has been found in ships dating from prehistoric times (Dashur boat, 18th century BCE; Uluburun ship, 14th century BCE) right through to the Middle Ages (Byzantium, 10th century AD). It reached its heyday with the construction of the Athenian trieris (fifth century BCE), at which point shipbuilding practice resembled more cabinet-making than carpentry [8].

[10] University of Southampton Black Sea Maritime Archaeology Project.

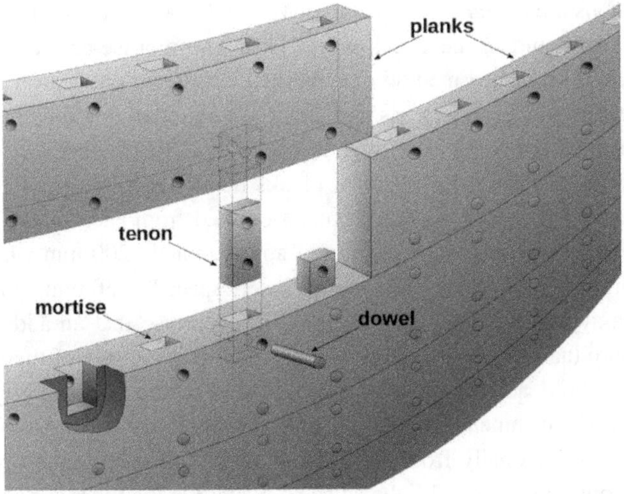

Figure 2.10 The mortise-and-tenon joint.

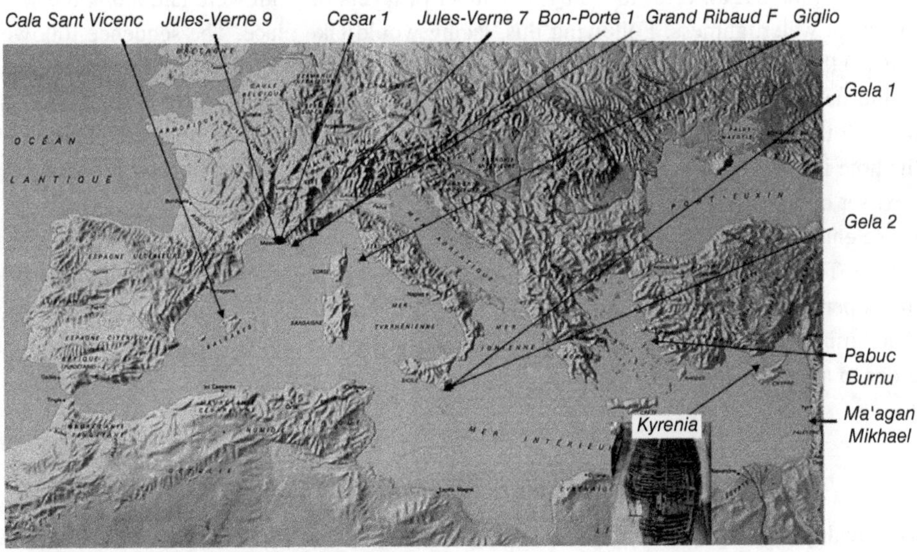

Figure 2.11 Wrecks of laced ships [45].

2.2.2 Laced Ships

In addition to the mortise-and-tenon joint, another method of assembly was used in antiquity, and in some parts of the world has continued to be used until modern times. *Laced* or *sewn* ships were built by the Greeks from the Bronze Age until approximately the fourth century BCE. A small number of wrecks dating from the sixth century BCE (Figure 2.11) provide evidence of the technique used. In some of these ships both assembly techniques have been used. The transition from sewn/laced to mortise-and-tenon occurs over a fairly long period (sixth–fourth century BCE) as evidenced from shipwrecks found in various locations (*Jules Verne 7→Ma'agan Mikhael→ Kyrenia*) so that sewn ships from later dates are not found in the Greek world [45]. The mortise-and-tenon joint is much stronger than the sewn/lashed method of construction, and for this reason it prevailed, and much larger vessels such as the trieris and later the polyremes could be built.

The sewn/laced method of assembly is actually a hybrid method because it uses cylindrical dowels to ensure that adjacent planks do not translate parallel to each other, and lacing to ensure that they do not become detached from each other. The dowels are inserted into circular openings at a spacing of approximately 200 mm whereas the lacing threads through openings that are much more closely spaced (≈50 mm) (Figure 2.12). The dowels act as a system of unpegged mortise and tenons, so that an additional system is required to ensure that the planks are held together; this is the role of the lacing system.

For a cord to hold adjacent planks together in a secure and watertight manner, an efficient system of attachment is required. In Greek ships the system consisted of cylindrical holes driven diagonally from the internal surface of each plank towards the outer surface, at the point of contact with the adjacent plank. On the inner surface of the plank a triangular notch was formed, and it is through this notch that the hole passes through the plank (Figure 2.12a). Prior to lacing, a number of layers of cloth were laid along the seam to ensure watertightness. Following this, lacing would take place. The sequence followed was (1) through the hole from the inner to the outer hull surface, (2) back through the hole in the next plank from the outer to the inner surface, (3) directly across the seam, (4) through the first hole from the inner side to the outer hull surface once again, (5) through the hole in the adjacent plank from the outer to the inner surface, (6) diagonally across the next set of holes (IIXIIXII).

Retention of hull form (transverse stiffness) is provided by the frames (Figures 2.12a, 2.12b). These are trapezoidal in section and are narrower at the bottom and rounded along the upper surface. The frames are lashed into place and are widely spaced (approximately 1,000 mm apart). The frames are recessed where they come into contact with the seams, in order to avoid chafing of the lacing.

2.3 Greece during the Historical Period

The quality of ship fabrication and assembly techniques in the Greek world improved considerably since the Trojan War. As mentioned in the previous section, up to the sixth century BCE, structural members were attached to each other either by use of the mortise-and-tenon joint or else were laced.[11] At this juncture archaeological evidence points to a complete transition to the mortise-and-tenon type joint. This ensured a close fitting that was of high quality. Following the assembly of the shell planking, the internal framing was added. This was the method of construction followed in Greece and later on in the Roman world and was employed for all types of sea-going craft until the Byzantine period.

[11] See, for example, the wrecks included in Figure 2.11 [45].

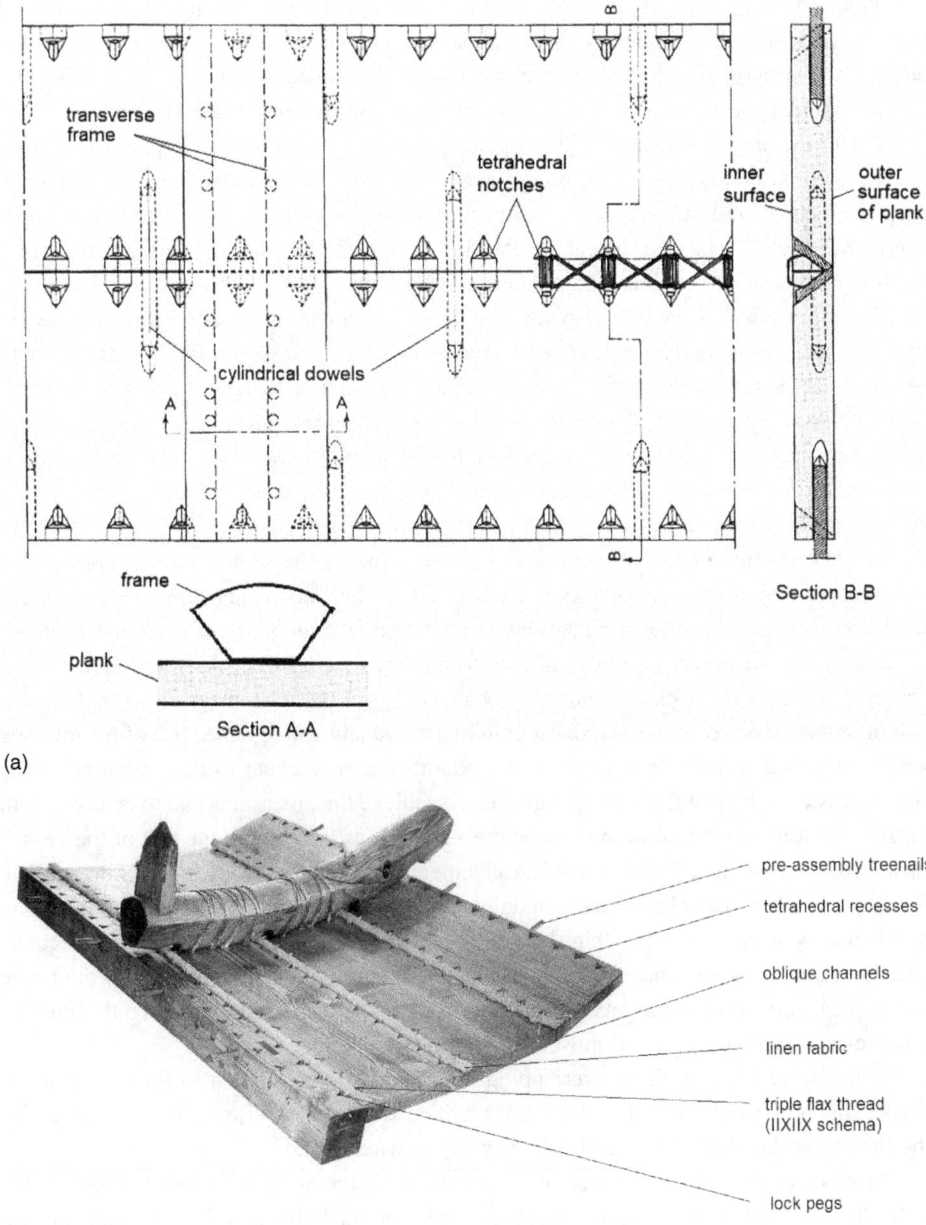

Figure 2.12 Lacing of outer shell planks and frames (sixth century BCE) [45, 50]. (a) Plan view of laced plank (*Jules Verne 9* wreck); (b) laced section.

To summarise, the fabrication sequence consisted of (1) laying the keel, (2) joining the stem and sternposts using scarph joints, and (3) installing a series of temporary frames to act as guides for the construction of the shell planking. The shell planking was then built up, being faired by eye, and finally the temporary frames were replaced by permanent ones.

The plank thickness was in the 35–100 mm range, depending on the particular section of the hull and available timber quality. Frame spacing varied from 100mm to 300 mm, although a variety of fabrication systems have been noted in discovered wrecks [8]. An analogous approach was employed during the reconstruction of an Athenian *trieris*.[12]

Prior to the seventh century BCE, warships essentially acted as troop platforms. Once the ram was introduced, however, its use demanded increased manoeuvrability, speed and proper structural design. The number of oars with which warships were fitted prior to the *trieris* was found to be insufficient for the full exploitation of their ramming capabilities. During that time, an increasing number of Greek city states were in the process of establishing navies. This provided an impetus for innovation, and as a result, design improvements were rapid. In order to increase oar power, vessels with increased length beyond that of the pentekontor were built but were found to have poor manoeuvring characteristics. Shipwrights then experimented with additional levels and arrangements of oars, ultimately reaching a satisfactory solution. Ameinoklis of Corinth is mentioned by Thucydides as having designed and built the first trieris, at least in Greece, around 705 BCE. Triereis served as the standard ship for navies for several centuries and formed the basis of several subsequent warship designs. The additional level of oars was provided in Greek triereis by adding an outrigger, in contrast to the Phoenician design in which the third level was positioned above the second level, with a consequent increase in freeboard.

Initially the trieris was designed as a swift, light hull with full decking installed only in way of the fore and aft ends (*aphract*[13]). Later on, decking was installed throughout the length of the vessel (*cataphract*). In order to make efficient and repeated use of the ram, the vessel itself had to be able to absorb an amount of energy equal to that imparted to the struck vessel, without suffering any damage. To achieve this, measures had to be taken both locally – that is, in way of the bow structure – as well as throughout the rest of the vessel. In way of the bow, a sufficiently stiff arrangement was installed [8], with the ram actually forming part of the vessel's structure and not added as an appendage to it. Additionally, the hull surface was fabricated as a thin shell of high quality, by minimising local imperfections and defects such as gaps between planks, thickness variations and so on, that would give rise to local stress concentrations following impact, and consequent failure. To this end the mortise-and-tenon joint proved most successful.

Frames were fitted in three overlapping tiers and were not joined to form continuous rings. This prevented failures in the frames, which would otherwise have occurred due to the flexing of the shell planks and their expansion when wet.

The design calculations that were carried out for the reconstruction trieris (Figure 2.13) indicated that the length-to-depth ratio ($L/D = 13$) was such that high longitudinal bending stresses in the hull were induced.[14] The trieris was therefore a ship designed to the limit.

[12] *trieris*: denoting 'in threes.' A long-standing debate of whether this signified the number of levels of oarsmen or the number of oarsmen per oar has been resolved in recent years in favour of the former.

[13] *aphract*: without decking (open galley) (Gk: άφρακτος). In contrast, a cataphract was a ship with a raised deck.

[14] Olympias' main dimensions: Length = 36.9 m, Breadth = 5.5 m, Draught = 1.25 m, L/D = 13 (hence freeboard = 1.59 m). The trieris built by Dupuy de Lôme and Jal in 1840 had the following main dimensions: Length = 39.25 m, Breadth = 5.5 m, Depth = 2.18 m, giving L/D = 18. Alexandris [1] meanwhile gives as acceptable ranges for the above: $40 > L > 35$ m, $5.5 > B > 5$ m, $D = 2.6$ m (freeboard = 1.6 m, $T \approx 1$ m). The mean value for L/D is thus approximately 14.4.

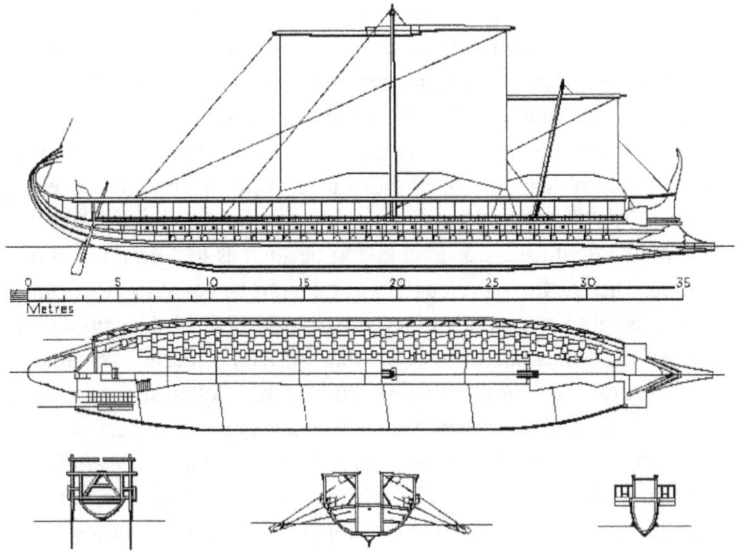

(a)

(b)

Figure 2.13 The Athenian trieris [14, 36]. (a) Profile, plan view and sections in way of aft end, amidships and bow of vessel. (b) The *Olympias*: a trieris reconstruction at sea under sail and oar power. Reprinted courtesy of Cambridge University Press.

During the fifth century BCE, two ropes,[15] each 47 mm in diameter, were used to provide additional longitudinal stiffness to the hull. These ran the length of the vessel below the main deck and were securely attached at the fore and aft ends. They had a low enough

[15] The Greek term is *hypozoma* (pl. *hypozomata*) (Gk: ὑποζώματα).

modulus of elasticity to extend sufficiently and to contract under tensile loading when the vessel was in a hogging condition. The presence of these ropes reduced the maximum tensile stress at deck amidships when in a hogging condition from 5.7 N/mm^2 to 4.6 N/mm^2, a figure considered acceptable for wooden ships during the 19th century.

Shear loading, on the other hand, was carried by mortise-and-tenon joints, which, for this reason in addition to that mentioned previously, were much more closely spaced than in smaller vessels. The number of joints required was of the order of 20,000.

The design wave-height used was assumed to be such as to permit the upper two levels of oars to be used whereas the wavelength was taken to be equal to the length of the ship. This design wave-height corresponds to roughly one-fortieth of the ship's length[16].

Subsequent to the Athenian trieris, warship structural design remained essentially unchanged for several centuries, although during the Hellenistic and Roman period, *polyremes*[17] were built, and these had many more levels and numbers of oarsmen. Other types of vessels such as the Roman *liburnian*[18] and the Greek *hemiolia*[19] made their appearance, but they did not represent an advance on past practice in terms of structural design or fabrication technique. With the development of weapons during this period such as the catapult, military power came to rely less and less on the performance of the ship itself.

Information on the structural design of merchant ships during the classical age and the later Hellenistic period has been obtained in part from the remains of a small number of ships, notably the Porticello wreck that dates from the fifth century and the Kyrenia ship dated to the late fourth century. The Porticello wreck was discovered in the straits of Messina, off the Italian mainland, and is by inference estimated to have had a length of approximately 20 metres. It was built in the same manner as all vessels of that period: shell first, with the frames fixed to the hull planking using copper nails. Holes were drilled on the outside of the hull planking, *treenails* were inserted and then copper nails were driven into the treenails to ensure a tight fit. The ends of the nails were clenched and driven into the frame surface. Where the fit between the frame and the hull planking was not perfect, shims were used. To prevent leaks, the outer surface of the planking was covered by small patches of lead sheeting. Other objects that were found and which testify to their use at that time are *cleats* and *wooden toggles*. The latter were made on a lathe and were used as a quick-release anti-luffing device to assist with the handling of sails when the ship sailed into the wind. Mortise-and-tenon joints were used to assemble the side shell planking.

The Kyrenia ship was found to have been built in a similar manner and was about 80 years old when she sank. She had a length of 14 metres and operated as a small freighter (Figure 2.14a). Her construction reveals the quality of workmanship and understanding of the behaviour of wooden structures at that time; apart from the absence of the keelson, all

[16] Compare with the L/20 wave height used until recently.

[17] For a description of the different types of polyremes (Gk: πολυήρεις), see [8] and [18].

[18] The *liburnian*, named after the Liburni, an Illyrian tribe, had two levels of oars, having been developed by pirates from the simple oared warship that was originally fitted with one level of oars.

[19] Equipped with one and a half files of oars on each side.

Figure 2.14 The Kyrenia ship reconstruction. (a) At sea under sail power. (b) Hold side structure.

other structural elements to be found in ships for the next 2,000 years were present. Her keel was of a composite structure, consisting of the keel proper, a false keel, a two-piece stem and a kneed sternpost. Mortise-and-tenon joints were used to assemble the hull planking, spaced at 120 mm intervals. The framing system consisted of floors and *futtocks*,[20] alternating with each other. The frames were nailed to the hull planks but not to the keel, which was *rockered*[21] and *rabbeted*.[22]

The hull planking was strengthened by a set of *wales*[23] that extended the full length of the vessel. Her hold was fitted with internal decking up to the required height in order to facilitate the storage of cargo (Figure 2.14b). Lead sheeting was found attached to the outer surface of the hull indicative of repair work during her lifetime to prevent ingress of water.

It was the battle of Aktion in 31 BCE, which took place between the fleet of the emperor Octavian and those of Mark Anthony and Cleopatra in which the Roman ships defeated the ram-equipped Egyptian and Greek combined fleet, that signified that fundamental changes in the naval architecture of oared warships were taking place. The Roman fleet relied less on the agility and speed of their ships and more on the military strength of the soldiers that were on board.[24] This represented a return to the tactics that preceded the development of the light and swift ships of the Greeks. From this point onwards the ram was relegated to

[20] *Futtock:* a frame that extends over only part of the section, and normally originates in way of the keel. Ordinary frames, on the other hand, extend up to the connection of the side shell to the main deck.

[21] The longitudinal profile of a *rockered* keel follows the sheer of the vessel – that is, it is curved upwards towards the fore and aft ends. Rockered keels contribute to the manoeuvrability of a vessel but reduce its speed. Longitudinal hull bending would be induced in a long, modern hull by a rockered keel, although in a small wooden vessel it aids by providing resistance to hogging.

[22] *Rabbet* (or *rebate*): a groove incised along a timber to receive the edge of a plank or strake. The most common rabbets are the *keel rabbet*, which receives the *garboard*; and the rabbet of the stem and stern post, which receives the strake ends.

[23] *Wale*: a plank of increased thickness extending over the full length of the vessel (strake). Its role is to provide additional strength and stiffness.

[24] A fifth-century BCE Athenian trieris carried 14 soldiers (hoplites) on board whereas a Roman trireme could have up to 120 soldiers, in addition to the oarsmen.

the role of an auxiliary weapon of attack and ceased to dominate ship design and structural arrangements.

The three-pronged ram was replaced by the Romans by an upcurving tusk-like ram, similar to that used in early Phoenician biremes [56]. It thus became a weaker weapon that could be used only once. The other modifications that were introduced were of a functional and operational nature and did not affect the structural arrangement in any significant way. The most important of these was the positioning of the oars in openings in the hull planking, thereby dispensing with the rowing frames that had been in use until then. At the fore end of the vessel a forecastle deck was added which incorporated the stempost into its structure, whereas at the aft end a cabin for the commanding officer was installed. These changes were introduced after the first century CE.

Contemporary with the battle of Aktion was a vessel discovered at Madrague de Giens, in the south of France, which was a large merchant vessel with a length of 40 metres. This particular wreck provides direct evidence of a *laminated hull* that was assembled using two layers of planking, with thicknesses of 60 mm and 40 mm, respectively. This double hull was fabricated using closely spaced mortise-and-tenon joints. The internal stiffening system consisted of floor timbers and half-frames each with futtocks.

2.4 Byzantium and Venice

Evidence of ship construction during the ensuing period comes from a variety of sources. With regard to warships, details of early Byzantine vessels are given by Dain [15] who discusses the contents of several contemporary texts, including that of the Byzantine emperor Leo the Wise.[25] The ships of the Byzantine navy evolved from the designs of the imperial Roman fleet, and by the time of the emperor Justinian in the fifth century CE, the term *dromon* (*δρόμων*) had been introduced. Dromon denotes 'swift' (racer) and was applied to more than one variation of a basic design.[26]

The standard bireme dromon had two levels of oars (Figure 2.15); the structural arrangement essentially followed that of past practice but, in addition to the structural members used in the past, keelsons were fitted. It is believed that dromons were fitted with a full deck (*katastroma – κατάστρωμα*), supported by beams, upon which deckhouses (*castles*) were installed, in way of the stern and also midway between the masts. Following Roman practice, the hypozoma was replaced by a stiffener (solid) keel which was located above the inner bottom and which also supported crutches in which the yards rested when lowered. A degree of uncertainty exists with regard to the general arrangement of this class of vessel, as there is no pictorial or other evidence analogous to that of ships of the classical period.

[25] Leo wrote extensively on military matters and in particular on naval tactics (see his *Naumachica* and *Taktika*).

[26] Three classes of dromon were in use, in increasing vessel (and crew) size: *ousiakos*, *pamphylos* and *chelandion* (dromon proper).

Figure 2.15 A bireme dromon of the 10th century.[27]

On the other hand, details of the construction of merchant ships during the Byzantine era come from, amongst other sources,[28] two wrecks that were discovered a short distance from Yassi Ada, a small island in the southeastern Aegean, off the coast of Asia Minor (Figure 2.17). Both vessels were merchantmen. The first of these dates from the fourth century CE, and it is apparent from the findings that half-frames were used to assist the shipwright in shaping the hull. This provides evidence of the gradual transition from the shell-first to the frame-first type of construction. The second wreck dates from the seventh century AD, and it is evident that in this case, economy of construction was a factor that had been considered: that is, the ship was built for a finite lifetime, in contrast to the ships of antiquity that were built to last as long as possible. The ancient method of joining hull planking, the mortise-and-tenon joint, played a less important role, judging by the size of the tenons and the increase in their spacing with the passage of time (Figure 2.16).

To the south of Yassi Ada, off the coast at Bozburun, another wreck dating from the ninth century has been discovered (Figure 2.17). The ship at Bozburun has enabled the piecing together of evidence of the use of geometrical design methods when constructing the ship's structure. This is the earliest evidence of the use of such methods and links the earlier Yassi Ada ships to the Serçe Limani vessel discussed following this. Furthermore, it provides a basis for the understanding of the origin of the geometrical sketches and drawings of Venetian shipwrights of the 15th century onwards (documents of this later period are also discussed later in this section). In the Bozburun ship a standard measure of

[27] Scale model exhibited in the Hellenic Maritime Museum, Piraeus, Greece.
[28] Two wrecks found in the Tantura Lagoon, Israel, from around 500 AD were built frame-first [10].

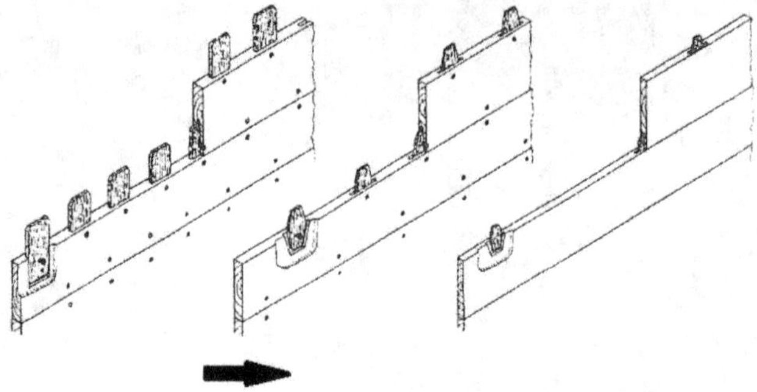

Figure 2.16 Evolution of the mortise-and-tenon joint during the Byzantine era.

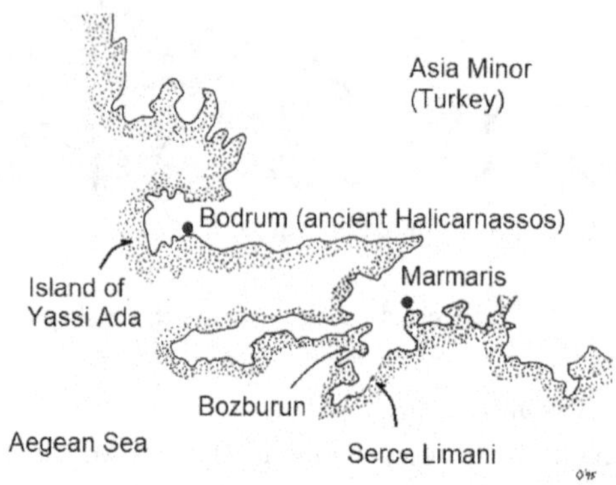

Figure 2.17 Location of Byzantine wrecks along the south-western coast of Asia Minor (Turkey).

34.5 cm and multiples of this have been used to select the spacing of important members of the ship's structure. The use of the *proportional rule* indicates a gradual evolution, as well as that repeatability, efficiency and predictability were inherent in the new methods. The proportional rule enabled the shipbuilder to predetermine the locations of key framing timbers in the hull's structure and, by transposing this framework to those key stations, predict the span and inclination of particular floor timbers prior to construction. Figure 2.18 includes a sketch indicating the use of the proportional rule, similar to that which could have been used by the shipwright at the time, before construction of the ship.

The use of geometrical design techniques and the implementation of the proportional rule are described in the following quote:

> The shipwright most likely began with a simple sketch to build the conceptual framework that governed the cross-sectional shape of the ship at various stations. Such a drawing may have comprised a rectangle representing a framework approximately 3.45 m long and 2.02 m high, divided into three sections with two additional vertical lines, 69 cm to either side of the centerline. The lower half of the rectangle was divided into three sections as the shipwright added two more horizontal lines across its width, at heights of 11.5 and 23 cm above the baseline. To finish, he marked two

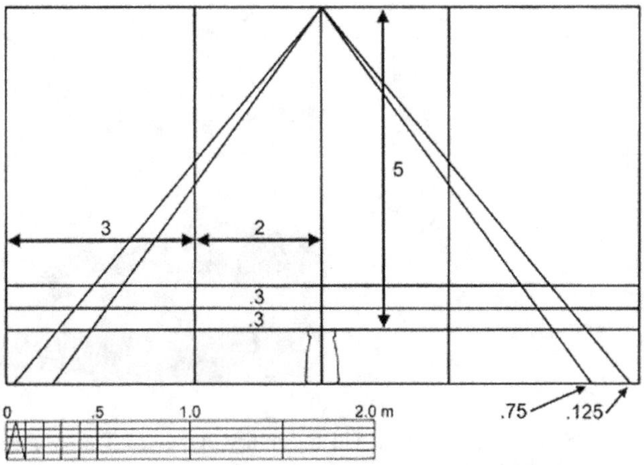

Figure 2.18 Use of proportional rule in the structural design of the Bozburun ship [20].

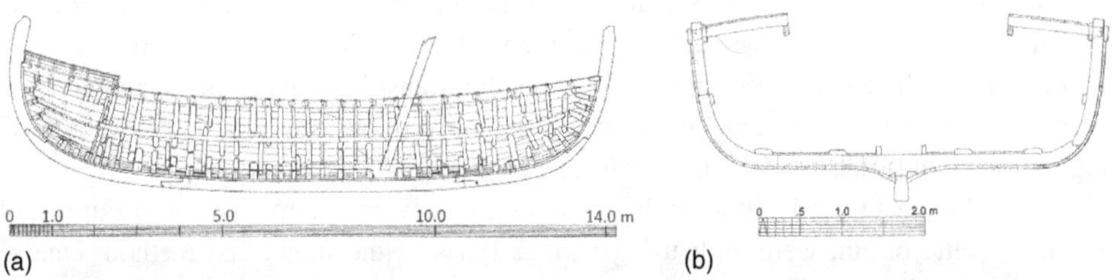

Figure 2.19 The Bozburun ship (ninth century AD) [20]. (a) Longitudinal profile; (b) midship section.

intermediate widths on the baseline, approximately 4.3 and 25.8 cm away from either vertical edge, and joined these marks to the top of the centerline with four diagonal lines. [20]

When complete, the sketch most likely looked very similar to the framework in Figure 2.19.

A fourth wreck, that of a small Byzantine merchantman dating from around 1024 to 1025 CE, which was found further south at Serçe Limani (Figures 2.17 and 2.20), provides further evidence of the continuing process towards the frames-first type of construction. Mortise-and-tenon joints are not used anymore. The fabrication sequence is described in some detail in [51]. Following the assembly of the keel and posts, two frames and a number of floor timbers in way of the ends of the vessel were erected into position. The planking was then added, and there is evidence that this procedure was assisted by cleats. Before completing the assembly of the side shell planking, the frames were placed in position. Finally the upper strakes of the side shell were added. What is of importance to the naval architect, however, is the evidence of the use of simple geometric projections and ratios to design the hull form (transverse sections). It appears that mould lofting and batten projections of hull shapes were used. A simple measure of length close to the Byzantine foot was applied, and multiples of it were used in spacing the frames and the sizing of the timber

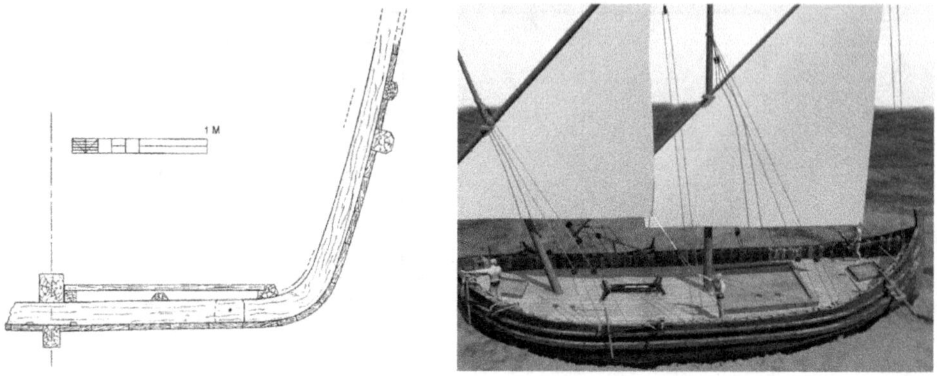

(a) (b)

Figure 2.20 The Serçe Limani wreck (11th century AD). (a) Transverse section [51]. (b) Reconstruction model [61].

sections. The Serçe Limani wreck marks the turning point between the shell-first method and the modern approach, based on the frames-first method of construction for the Mediterranean. The exact date and location of this transition cannot be accurately pin-pointed because it was the outcome of changes that took place in different geographical regions and concerned various ship types and their shipwrights.

It should be mentioned that a number of vessels discovered in north-western Europe, of Romano-Celtic origin, were built using various types of the frame-first method. One of these, the Port Berteau II ship which was found in the river Charente near the Atlantic coast, is dated to around 600 CE [46]. However, this method of construction disappears after the end of the Roman era, to reappear at a much later point in time in northern Europe.

Vessels similar to those used by the Byzantines were used by the Muslims in the east. Little information is available on these types of ships, let alone their structural design. Similarly, in the west, there is a dearth of information on ship design during the period CE 1000–1250, during which a variety of ships came into existence. Of these, the *galea* (galley, Greek *γαλέα*) is the type which rose to prominence during the centuries to follow. The term *galea* is first encountered in the writings of the emperor of Byzantium Leo the Wise, in which it is used to designate a small dromon. Eventually, the term *galley* came to denote a type of vessel that was principally oar-powered and sail-assisted and which became a standard ship-of-the-line in many navies, over a long period of time. The first recorded description of a galley dates from the late 13th century and concerns the fleet of Charles I of Anjou. The structure of these particular ships was fabricated in the conventional manner; one particular innovation noted is the use of *forked frames* (*furcate*) in way of the extremities of the vessel, instead of floors and futtocks. These forked frames were single timbers curved to form the frame.

It was Venice, however, that achieved mastery of the seas and dominated trade in the Eastern Mediterranean over several centuries. The Venetian galleys were constructed in a manner superior to those of any other contemporary fleet, since they established methods of good practice that were followed by shipwrights rather than relying on the individual expertise of each man.

Documents that have survived[29] provide an understanding of the processes involved and the end product and are probably the earliest surviving on ship construction. The Venetian galley is considered to be the successor to the Byzantine dromon since some of the master shipwrights that were recruited to work in the Arsenale in the early 15th century originated and brought their expertise from the East (Rhodes[30] and the Peloponnese[31]).

The steps followed in the design process were:

1. Determination of the form and size of the main frame (*corba maistra*).
2. Determination of position, form and size of reference frames (*cavi di sesto*) at the fore and aft ends of the ship. The form of the reference frames was based on that of the main frame.
3. Determination of curvature of the frame top timbers.[32]

The most important stage in the design of a new ship was the establishment of the corba maistra, which together with the keel, stem and stern defined the hull form (Figure 2.22). From the corba maistra and the fore and aft reference frames, intermediate frames were defined (*partisoni*). A mathematical-graphical procedure was used to generate the transverse sections at the partisoni along the length of the ship. This procedure is described in the 15th-century manuscript of Giorgio Trombetta da Modone [55] and more recently in [18] and [30]. A number of methods (*mezzaluna, brusca*) were implemented to determine the spacing of a sequence of points that were used to define the rise of floor and that of the keel in the transverse and longitudinal directions respectively. These points were used to interpolate between the offsets of the corba maistra and those of the cavi di sesto in such a way as to generate a faired surface. The job of the master shipwright was to prepare the sketches for the individual frames which were then fabricated by the carpenters of the Arsenale. The extremities of the hull beyond the cavi di sesto where the hull surface had of necessity a complex double curvature were probably fabricated in the ancient manner – that is, hull planking first – to be followed by the installation of the internal stiffening members. In Portuguese shipyards, during the 16th century, the hull form in way of the fore and aft ends was defined by the use of *ribbands*, which extended the full length of the vessel and were tangential to the already erected frames.

This approach, although not rigorous by modern standards, could be used to build ships on the basis of a successful existing design. This in itself was an important step forward, especially when compared with common practice at the time of the one-off empirically based designs that were generally produced. It could not be used for innovative ship forms, much the same as today's classification society rules do not allow for unconventional ship types without resorting to a first-principles method. The Venetian approach ensured a

[29] The *Libro di Marineria* (mid-16th century) is based on earlier manuscripts (early 15th century) and describes the method of construction of the galley. It contains drawings and a series of ship designs, including those of small craft. This text is now better known as *Fabrica di galere* (as renamed by Jal). See also the manuscript of Michael of Rhodes (internet website and book). Another document, dated to 1445, was authored by Giorgio Trombetta and contains valuable information on hull form definition.

[30] Teodoro Baxon and his successors; also Michael of Rhodes.

[31] Giorgio Trombetta da Modone.

[32] This step is mentioned by Pre Theodoro and other contemporary writers.

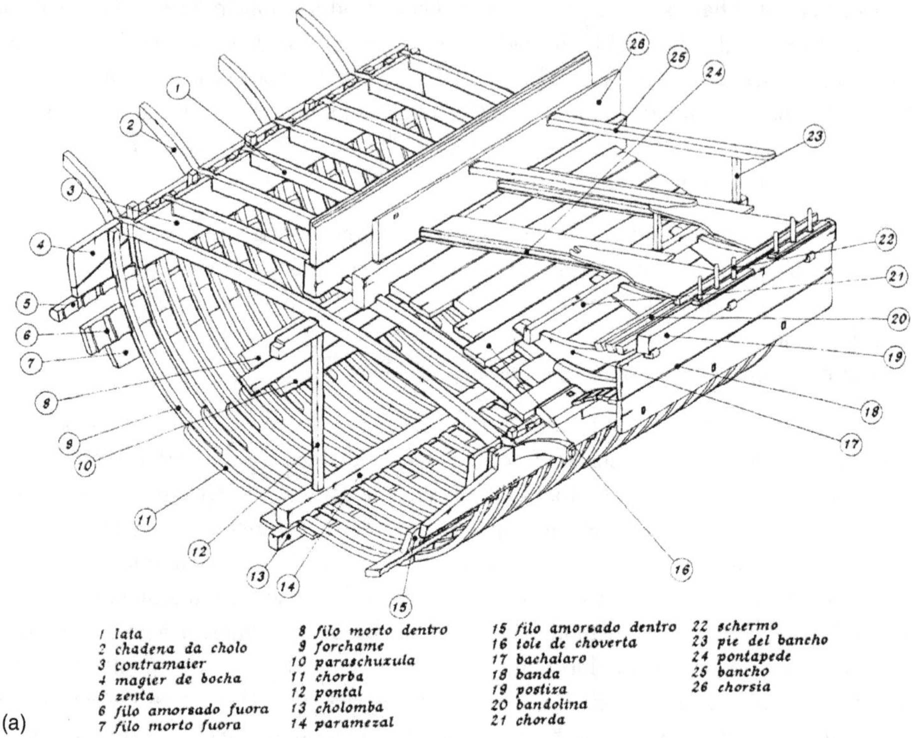

(a)

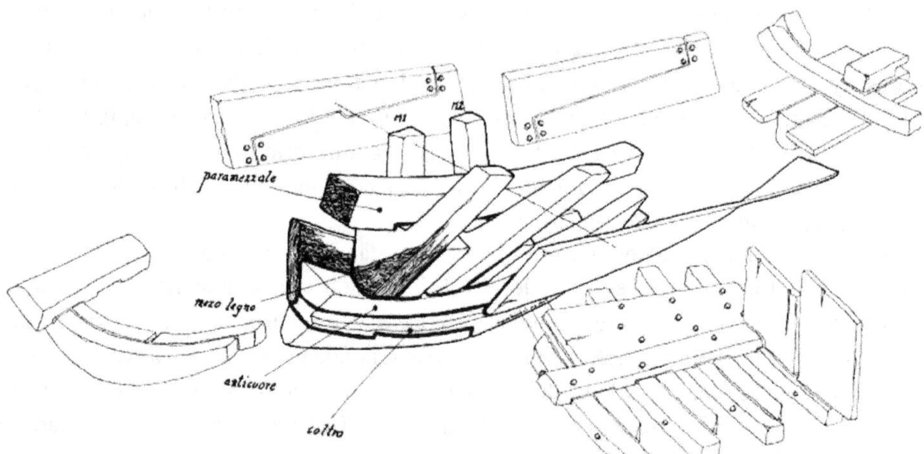

(b)

Figure 2.21 Galley structural details[33] [3]. (a) Midship section; (b) bottom structure.

[33] The terms in Figures 2.21 and 2.22 are in the Venetian dialect of 16th-century Italian.

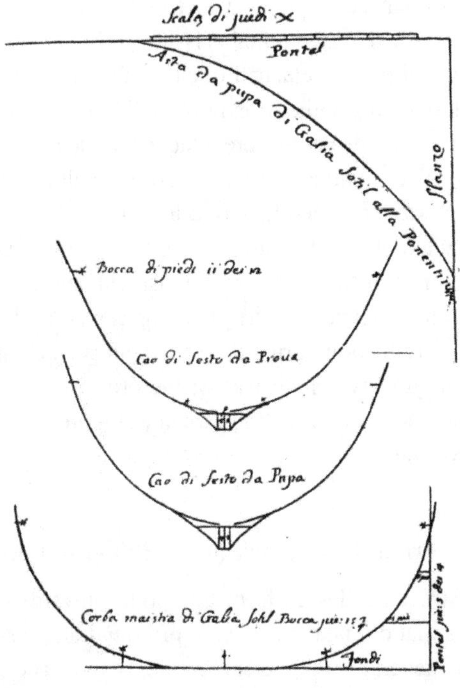

Figure 2.22 The *corba maistra* and *cavi di sesto* for a *galia sotil* (light galley) [55].

uniform standard of construction and performance that was based on best practice. This was implemented for all galleys in the fleet, so that vessels differed little from one another.

2.5 China

Independently of developments in the West, there was much going on in shipbuilding in China. Evolution in ship construction in the East followed a parallel but not identical path to that outlined earlier. The availability of bamboo, which is hollow and floats easily, had a profound influence on the construction and structural integrity of waterborne craft from very early on. It was used as early as 5500 BCE to build rafts.

There exists evidence that these rafts led to the development of the junk from around the third century CE. Curved bamboos whose diameter reached 30 cm were split to form semi-cylindrical structures that were then attached to each other. It is believed that the septa in the bamboo gave rise to the concept of the watertight compartment. Rigidity of the vessel was ensured by means of transverse bulkheads rather than by the use of additional internal stiffening in the form of frames.

Vessel size increased using planks in a carvel-built hull which was supported internally by watertight bulkheads. Thus, there was no need for a keel, stem and stern posts. The stern was built in the form of a transom, and longitudinal rigidity was provided by the use of wales. The *shachuan*, which was a shallow-draught vessel used for trading in the Yellow Sea, was thus developed. This led to the 11th-century warship of the Song dynasty.

The construction and connection of the watertight bulkhead to the shell planking as evidenced from a number of excavated remains is shown in Figures 2.23a–c. The bulkhead itself consisted of horizontal planks attached to each other using tenons. In order to attach the bulkhead to the shell planking, either iron hook-nail joints (Figures 2.23a–b) or wooden tongues (Figure 2.23c) were used. These were attached to the bulkhead planking using iron nails (Figures 2.23a–c). Watertightness was ensured by caulking with jute or hemp fibres and then sealing with a mixture of sifted lime or burnt oyster shell with boiled tung oil. This type of construction resulted in a particularly strong hull and necessitated a high degree of craftsmanship. It is of interest to compare and contrast with ancient Mediterranean practice, especially with regard to the thickness of the planking as well as the joining methods. Ship construction reached its culmination during the 15th century in China with the construction of a fleet of ships whose greatest length was of the order of 120 metres [34] (*Zhen Zhe* "treasure" ships, a reconstruction of which is shown in Figure 2.23e). Clearly these are the largest wooden ships ever built.

2.6 From the Carrack to the Ship-of-the-Line

Developments in ship structural design from this point onwards did not take place until much later. The changes that did take place were profound and far-reaching and related to the use of sail and the general arrangement of the vessel. The interaction of northern European with Mediterranean shipbuilding brought about developments in both regions. The lateen sail, believed to have been introduced by the Arabs in the sixth century and used for many centuries in the Mediterranean Sea, was gradually replaced by the square sail, commonly used in northern Europe (*cogs*). One other feature of northern European vessels adopted in the Mediterranean area was the use of the stern rudder, which replaced the twin side rudders used until then. In the north, on the other hand, ships ceased to be clinker built and began to follow the Mediterranean tradition of carvel-built ships. The use of the carvel technique enabled the use of thicker planking and thus the construction of stronger hulls. This made it possible to build larger ships. The use of more than one mast also spread from the Mediterranean northwards around the beginning of the 15th century. By today's standards, ships continued to be small, so longitudinal strength was not an issue.

The 15th century saw the development of ship types such as the *carrack*[34] and the *caravel*[35]; these were later succeeded by the *galleon* (Figure 2.24b). The carrack was the result of the merging of elements found in northern European ship types (cog, hulk) with the Mediterranean *roundship* (merchant ship).[36] It sailed in both the western Mediterranean and northern European waters and was built throughout Europe.

What affected the layout of the structure was the installation of multiple decks and castles (superstructures) in way of the fore and aft ends of the ship. Castles could be fabricated flush with the vessel sides rather than being positioned inboard, as in the case of

[34] The term *carrack* has its origins in antiquity (Gk. *kerkouros* (*κέρκουρος*) and Akkadian *kalakku*). In England, carracks were known as *great ships*.

[35] The term *caravel* is thought to originate from the Dutch *kraweel*, a large nail used in ship construction.

[36] Also known as nef (Gk. *nafs* (*ναύς*), ship).

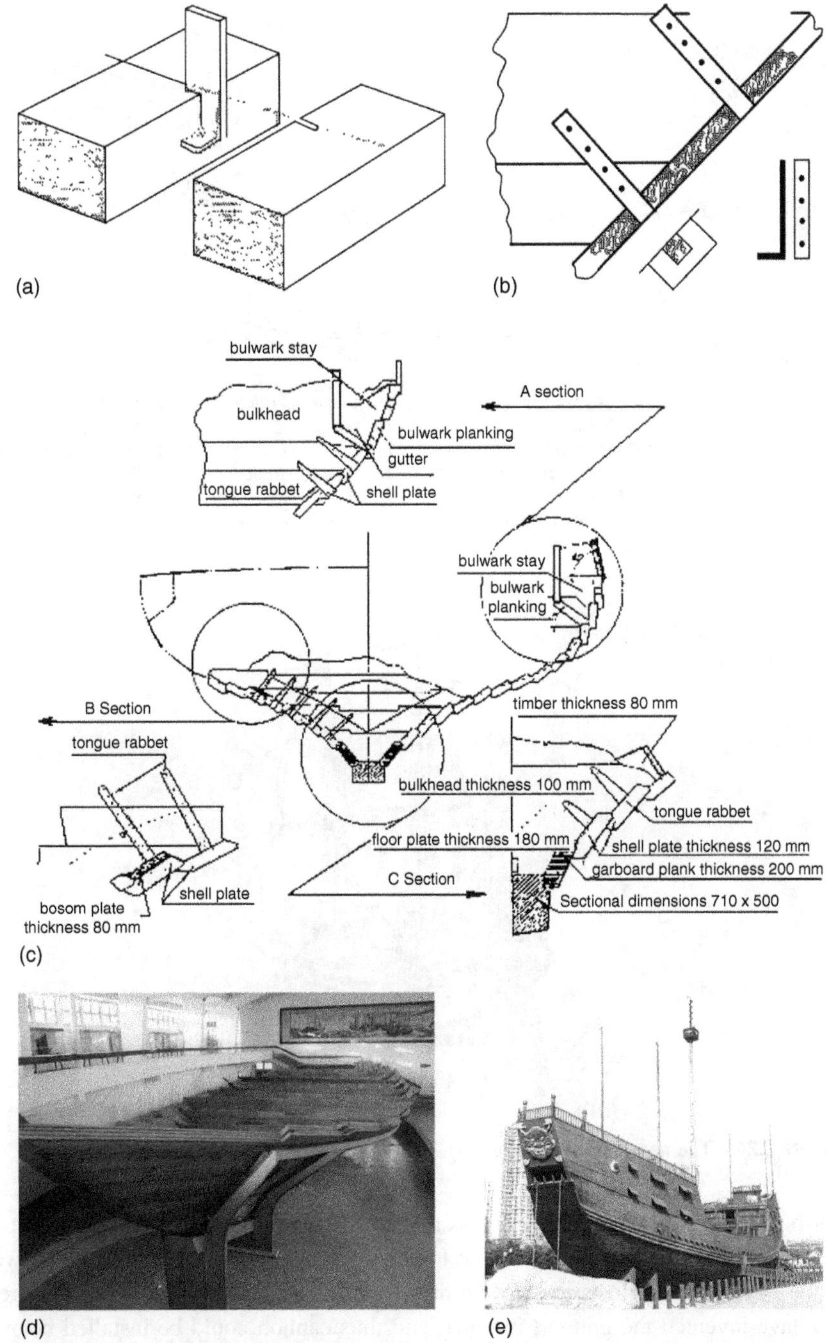

Figure 2.23 Chinese ship construction during the Middle Ages [59]. (a) Hook-nail joint. (b) Bulkhead connection to shell planking. (c) Section showing shell planking and bulkhead connections. (d) Quangzhou boat (reconstruction). (e) Zhen Zhe treasure ship (reconstruction). Reprinted courtesy of the Royal Institution of Naval Architects, London.

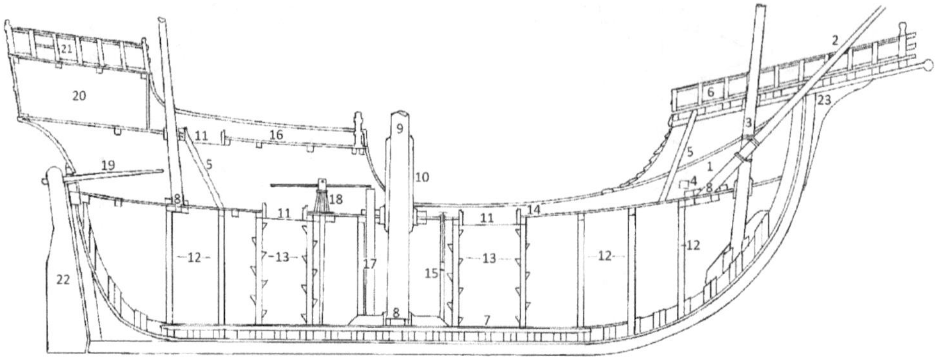

1. *Tilla*
2. *Bowsprit*
3. *Foremast*
4. *Horizontal bitt*
5. *Ladder*
6. *Forecastle bulwark*
7. *Keelson*
8. *Mast step*
9. *Mainmast*
10. *Mainmast wedges*
11. *Hatchway*
12. *Hold pillars*
13. *Stepped pillar*
14. *Waist*
15. *Pumps*
16. *Quarterdeck*
17. *Bitt*
18. *Capstan*
19. *Tiller*
20. *Admiral's cabin*
21. *Quarterdeck bulwark*
22. *Rudder*
23. *Knee*

(a)

(b)

Figure 2.24 The nao and the galleon. (a) The *Santa Maria* (nao)[37] [41]. (b) Galleon.

clinker-built ships. Changes also arose as a result of developments in the guns that were used; initially these were installed on the fore and aft castles, but as their size and weight grew, they were moved closer to the waterline. Descharges, a French shipwright in Brest, is said to have invented the gunport in 1501, and thus cannon could be installed below the

[37] This drawing depicts the structure of the 1992 reconstruction model of the *Santa Maria* that is believed to have been a nao and not a caravel, as previously thought [18]. The nao was a much larger vessel than the caravel, and in the case of the *Santa Maria* was slightly in excess of 100 tons.

weather deck. This improved firing accuracy as well as the ship's transverse stability. To improve these qualities further, a pronounced tumblehome was used, and a transom stern replaced the rounded stern. The continuing improvements in naval gunnery meant that ships could engage each other from a distance. In order to provide a smaller target for the enemy, it was essential to reduce the height of the fore and aft castles. There were other disadvantages to the use of high castles: they reduced sail-carrying capacity and increased wind resistance, thus rendering the ship inherently unstable. Since vessels had to be able to sail in a straight line in order to fire a broadside, gun accuracy also suffered. Carracks nevertheless served their purpose well as both merchant ships and warships until the early 16th century and were used in the voyages of exploration.

The caravels, in comparison, were more efficiently designed with a larger overall length-to-beam ratio and with lower superstructures than those used in carracks. They were much smaller and thus much more manoeuvrable and were generally preferred to carracks by explorers who sailed close to shore in order to survey newly discovered lands.

The developments in weaponry led to the evolution of the *galleon* during the middle of the 16th century, in both England and the Netherlands. Galleons were purpose-built warships with a much lower forecastle and a larger L_{OA}/B ratio, which made them much faster and more manoeuvrable.

Apart from the changes that resulted from weaponry requirements, there were other developments that took place in ship design during the 16th century that were of a different nature. The new ideas stemming from post-Renaissance mathematics and natural sciences were spreading and began to be felt in the realm of shipbuilding. In England, Baker initiated the use of fine paper drawings and used arithmetic and geometry to develop new hull forms.[38]

The most important problem a shipwright had to tackle at that time was how to design a new hull form (and consequently a new ship structure) on the basis of an existing one. This problem was related to a number of other questions, such as the manner in which the hull form changes in the longitudinal direction and the selection of linear dimensions in order to obtain a required volume given an existing design whose linear dimensions and volume are known. Baker communicated with English mathematicians and Venetian shipwrights and used Albrecht Dürer's work on geometry to develop geometrically based techniques (proportional scales) that could be used to design the midship frame and other sections along the length of the ship. The process of *ship design* thus became distinct from that of construction. The importance of shipbuilding was being reassessed by the Portuguese João Batista Lavanha who, in revising Vitruvius' classification system, classed it as a branch of architecture, thereby introducing the term *naval architecture*.[39]

One of the persistent problems of wooden ships, especially the larger ones, was hogging in the longitudinal direction. Hogging in wooden sailing ships was not a longitudinal strength problem but one of fabrication technique, and had been acknowledged as such

[38] See his *Fragments of Ancient English Shipwrightry*. A review of Baker's approach is included in [24].
[39] *Livro Primeiro da Architectura Naval*, dated to the early 17th century.

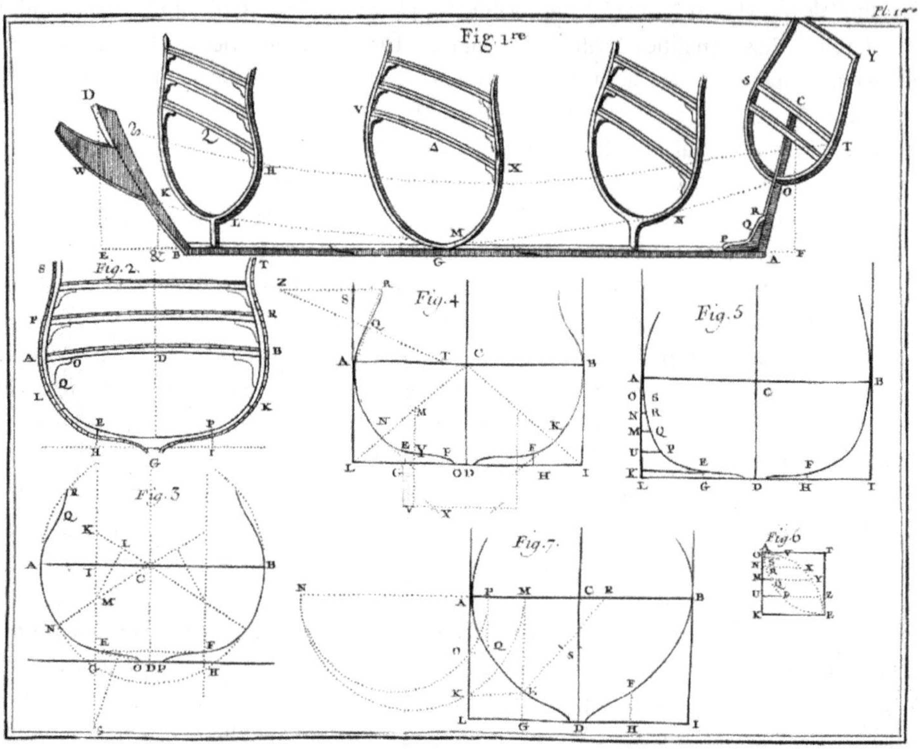

Figure 2.25 Hull construction according to Bouguer.

by Gobert, Bouguer and Young [39]. In wooden ships, the shell planks are not secured to each other between adjoining strakes in as effective a manner as in welded plating. Caulking material was the only substance to be found between individual strakes in the longitudinal direction, and when caulking friction was overcome, the planks were able to slip longitudinally against each other. Since at the fore and aft ends there was insufficient buoyancy to carry the overhang of the structure, the planking of the upper strakes and decking moved and a slack developed. Evidence of early attempts to remedy this problem was the use of diagonal braces that were fitted inside hold frames, as found in the wreck of the *Mary Rose*. However, this solution fell into disuse for a long period, to be mentioned much later in Bouguer's *Traité du Navire* that was published in 1743. The inherent weakness of the types of joints used meant that the strength capability of the hull planking could not be exploited to its full capacity, thereby imposing a limit on ship size. It is for this reason that in the era of wooden ships, lengths did not exceed 50–60 metres despite the advances made in joining methods. As a result, during the period that lasted from the age of the galleon until the ship-of-the-line of the late 18th century, little changed in ship structural design. One other important problem with the use of wood in ship construction was rot, which meant that the structure had a finite life and required extensive maintenance.

2.7 The Industrial Revolution and Beyond

The situation was to change, however, because of developments on several fronts. The *industrial revolution* brought about new, more efficient methods of iron production. Previously, the fuel used to fire furnaces was organic (wood), but this was replaced by fossil fuels (coal). In addition, the invention of the steam engine opened up new opportunities for ship propulsion. At a theoretical level, important developments had already taken place in strength of materials theory and were continuing. Finally, the theory of elasticity was being formulated as a new field of physical science as a consequence of the scientific revolution initiated during the 17th century. We shall outline developments in each of these.

Iron was first used in ship construction during the early part of the 19th century. The use of the steam engine as a means of ship propulsion, which took up a significant amount of space, necessitated the lengthening of ships in order to retain cargo-carrying capacity. This, as we have seen, imposed additional strength problems which traditional methods of wooden ship construction were not capable of resolving. The solution was to use iron for internal stiffening members that were used to support a conventional wooden shell. With the passage of time, a greater proportion of the hull structure was fabricated in iron, so that after a few decades, decks and cargo battens were the only remaining elements made of wood (Figure 2.26).

Figure 2.26 The *Glenlee* (1896). (Riverside Museum, Glasgow.)

In early iron ships, the structural layout resembled that of wooden ships; the scantlings were massive by today's standards, as it was not possible to select dimensions in a rational manner. Lloyds Register of Shipping classed its first iron ship in 1832 and acknowledged the fact that at that time there was insufficient experience with iron ships. Rules for iron ships were issued in 1855 in tabular form, following the layout used in wooden ship construction rules. Figure 2.27 shows that, in a manner analogous to the wooden ship shown in Figure 2.25, the bottom structure was heavily built whereas the deck structure was lightly built. This is in direct contrast to ships built today whose length exceeds 100 metres, for which longitudinal strength is critical. In this case the deck is the critical region of the hull girder.

The use of steel in shipbuilding dates from around 1870. Initially the Bessemer process was used, and production costs were high. When the open hearth process was introduced, however, quality improved and costs fell. As a result, the use of steel became widespread in the shipbuilding industry towards the end of the 19th century.

There exist several differences between iron or steel ship structures and those made of wood. In structural terms, the important differences are that iron and steel structures have significantly greater strength; secondly, members can be attached to each other much more efficiently. As a result, a greater proportion of the strength of the material can be exploited. Iron and steel ship structures can thus be fabricated using significantly smaller scantlings, so that they can be classed as *thin-walled.* Appropriate theories pertaining to thin-walled constructions are therefore required. However, at the time at which iron and steel ships were first built, the theoretical basis for the analysis of such structures had not been developed. Furthermore, in the absence of prior experience and accurate guidance from rules, it was only possible for constructors to select scantlings for iron and steel ships using wooden ship structures as a basis, which resulted in excessive factors of safety. As we have seen, until this point in time, questions relating to such issues as strength, stiffness and local failure of structural members in ships had never been considered using rigorous scientific techniques.

The foundations of the theory of strength of materials had been laid during the 17th and 18th centuries, but it had not been developed enough to enable it to be applied in a routine manner to ship structural problems. In 1759, Euler [17] described the use of beam theory to calculate the maximum bending moment of a ship amidships, so that by the end of the 18th century, in principle, it would have been possible to estimate the longitudinal bending moments and stresses [54]. The calculation of deflections would have been much more difficult because of uncertainty regarding the bending stiffness of the hull girder, not least because of the anisotropy of wood.

In 1814, Young submitted a report to the British Board of the Admiralty recommending the use of *oblique riders* in order to overcome the problem of hogging due to slack between wooden planking, as described previously. A longitudinal bending analysis of a 74-gun ship was also included in this report, which was submitted to the Royal Society as a paper [60].

A classic example of an early iron ship is the *SS Great Eastern* (1858–1889), which at the time of construction was significantly larger than any other ship built, having a length in excess of 200 metres. It was one of the first ships designed using beam theory and it proved

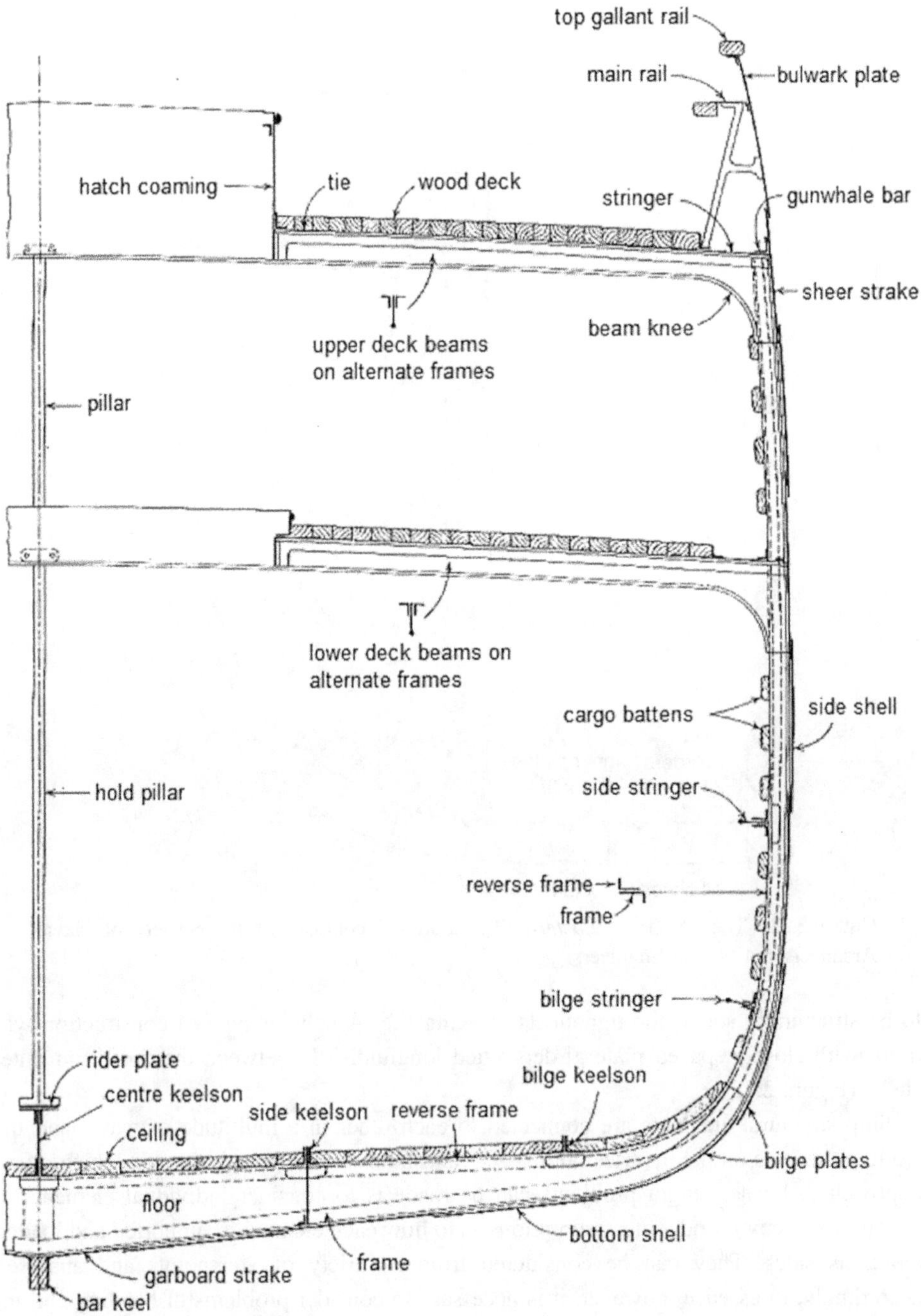

Figure 2.27 Iron sailing ship (19th century) [2]. Reprinted courtesy of the Society of Naval Architects and Marine Engineers.

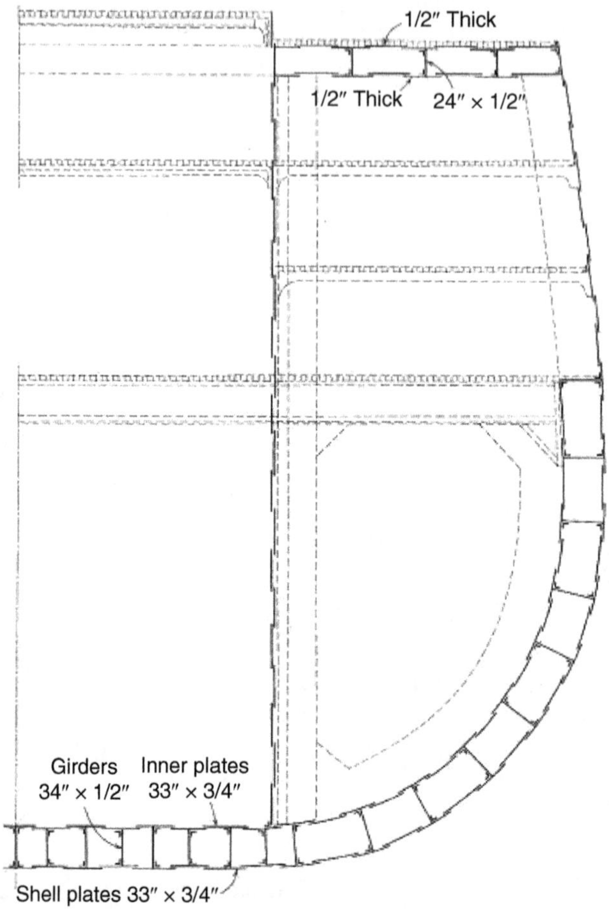

Figure 2.28 The *SS Great Eastern* [2]. Reprinted courtesy of the Society of Naval Architects and Marine Engineers.

to be structurally sound throughout its working life. A cellular type of construction was used, with closely spaced plate girders fitted longitudinally between the inner and outer shells (Figure 2.28).

Ship structural elements are connected to each other in a multitude of ways, and the loading that acts on them varies both space- and time-wise. The only way in which a first approach to local strength problems can be made is to consider individual elements in isolation, thereby introducing assumptions as to how each element is supported and loaded along its sides. They can be considered from a variety of viewpoints and analysed accordingly. In essence, however, it is necessary to consider problems of bending and in-plane loading of stiffened plating. A first-order approximation to the bending problem can be obtained using strength of materials approaches[40] for assemblies of beam-type structures

[40] In the theory of the strength of materials, it is assumed that when calculating stresses and deflections of beams, plane sections remain plane and the material is elastic.

which, using suitable assumptions, can be extended to the bending of plated structures. However, in-plane loading requires the analysis of the *stability* of both column and plate-type structures. The 18th century saw developments in stability theory for column-type structures, and Euler presented his formula for the elastic behaviour of columns in compression [16]. However, the influence of material plasticity was not understood, and for a long time Euler's formula was not used, because as the column length was reduced, it increasingly overestimated the strength of the column. Experimental work during the 19th century led to an understanding of material behaviour, and thus approximate solutions were obtained. The importance of Euler's formula has since been appreciated, and it has become established as the cornerstone of compression strength analysis of prismatic structures.

On the other hand, to obtain solutions to the equations that describe the behaviour of thin plates, it is necessary to make use of *elasticity theory*. The 18th century had seen important advances in mathematical theory as well as in the study of structures, exclusively for civil engineering applications. Contributions by Bernoulli, Coulomb, Euler and Lagrange laid the basis of the mathematical treatment of mechanics and strength of materials.

During the first half of the 19th century, the *equations of elasticity* were developed, thus initiating the mathematical study of the behaviour of solids. Navier considered the behaviour of elastic bodies on the basis of molecular theory and presented the differential equations of equilibrium for isotropic elastic bodies [37]. The development and acceptance of these equations, however, met with a fundamental conceptual difficulty that arose from the fact that mathematical analysis is built on the concept of continuous space in which infinitesimal segments can be considered and thus differential and integral calculus can be applied. On the other hand, Newton's molecular theory represented media as consisting of discrete particles, so that matter is essentially discontinuous. The application of mathematical analysis could only be justified by invoking the law of large numbers, since even a small quantity of matter contains a very great number of particles ('molecules')[41]. In other words, a statistical approach was adopted. Navier also underscored the importance of calculating the *limit to the elastic behaviour* of solids rather than the *ultimate load*, as was done previously.

His work was advanced by Cauchy who introduced the concepts of *stress* and *strain* in the theory of elasticity [11], based on that of pressure in fluids. Cauchy's formulation of the linear theory of elasticity remains unchanged to this day. He did not start from a specific law of molecular interaction and showed that the state of stress at an internal point in a body is determined by a set of nine functions. At equilibrium, these functions satisfy three partial differential equations. For small deflections of an elastic body, the set of functions describing the state of stress can be related linearly to the corresponding deformations. The concepts of *principal stress* and *principal direction* were also introduced by Cauchy as well as the differential equations of equilibrium of an elemental rectangular parallelepiped. Green introduced the principle of *conservation of elastic energy* and derived the equations of elasticity following Lagrange's approach in *Mécanique Analytique*, having first introduced the concept of *strain energy*.

[41] The proof of the existence of molecules came with Einstein's 1905 paper on Brownian motion.

The foundations of the theory of elasticity were thus laid by Navier, Cauchy, Poisson and Green. One of the most important contributions to the solution of a specific problem was that of Saint Venant, who presented the *semi-inverse method* to solve the equation of equilibrium of a solid bar of arbitrary cross-section under torsion [47]. In the semi-inverse method a series of assumptions were made concerning the nature of the deformations of the body, which was then shown to satisfy the equations of equilibrium and the boundary conditions. The equations of equilibrium were applied to several problems by Lamé and Clapeyron, and later on, the first book on elasticity theory was published [29]. Elasticity theory continued to be developed during the 20th century, with an emphasis on the existence of solutions and the study of specific problems.

The first step towards the static analysis of thin plates was taken by Jacques Bernoulli, who in 1789 obtained an approximate form of the equation of equilibrium of a thin membrane under lateral loading. In 1811, Sophie Germain used a variational approach to obtain the plate equation but omitted a warping term that was later introduced by Lagrange. Poisson also obtained the equations describing the lateral deflection of a plate and considered certain boundary conditions [43]. In 1823, Navier presented the first exact theory of bending of rectangular plates and solved the equation of equilibrium using Fourier series [38]. Important advances in plate theory were subsequently made by Kirchhoff [26], whose two assumptions regarding the behaviour of thin plates in bending [27] are the basis of small-deflection theory. He also extended plate bending theory to the case of large deflections [27].

The first exact solution to the stability problem of a rectangular thin plate, which concerned uniaxial *edge compression* for simply-supported edges, was given by Bryan in 1891 [5]. However, ship-type structural elements are supported in ways that are seldom simple, and the applied loading is also complex, having components in more than one direction [35]. Bubnov [6] made significant contributions and at the beginning of the 20th century used plate bending theory in warship design. He showed that plate deflections under hydrostatic pressure are usually not small, so that both bending and stretching of the middle surface arise. This makes it necessary to consider *large deflections*, which requires a new set of equations of equilibrium, originally obtained by Kirchhoff. The final form of these equations was given by von Kármán in 1923 [25]; the equations were recast in order to include initial distortions terms by Marguerre [33]. The equations presented by Marguerre were in a form from which it was not easy to obtain solutions. This presented a significant hindrance to their direct application in everyday situations. One other important reason for their limited use is the fact that although plate theory is an essential tool in ship structural analysis, it cannot provide answers to all ship structural problems, since ships' plating is generally supported by an orthogonal stiffening system. It is thus necessary to consider the behaviour of the stiffening system as well, along with its interaction with the plating.

In the mid-20th century, a number of structural models and techniques of varying complexity and sophistication were developed,[42] all of which have been used to provide answers when considering the structure at a *gross panel level* (bulkhead, deck section, double bottom, etc.). These models were generally based on beam theory, and solutions have been

[42] Simple grillage, plate beam, orthotropic plate, moment relaxation.

presented in the form of tables and graphs. They continue to evolve, and apart from their direct use, they also serve as benchmark solutions for computer-based procedures (grillage theory [13], orthotropic bending theory [48], orthotropic in-plane behaviour [32]).

Even though solutions at a gross panel level were obtained with sufficient accuracy for design purposes, given the lack of data in related fields (loading), this still left open the question of *global hull strength* and, more specifically, *longitudinal strength*. This, as already discussed, had been a matter of concern to designers at a much earlier period for different reasons. Early in the 19th century, Young had developed an approach to longitudinal strength. This, however, was not used for a long time because, as mentioned previously, the problem relating to longitudinal strength that existed was not one of overall strength but of fabrication (leakages arising from hogging of the wooden structure).

The underlying assumptions of this approach were fundamentally sound, and it has formed the basis of the semi-empirical methods in use well into the 20th century.

After the Second World War, with the increase in the size of ships, several new problems arose. Longitudinal strength of the hull girder had already emerged as the dominant strength issue in the structural design of ships. The importance of longitudinal strength arises from the fact that the stresses in the hull girder due to longitudinal bending exceed those due to other types of loading (such as the hydrostatic pressure exerted on the outer shell plating and the inner structure) when the length of the ship exceeds 90–100 metres. This is the case for a large proportion of merchant ships built since the second half of the 19th century. The method used was to apply a standard wave of length equal to that of the ship and of a height proportional to the length. The trochoidal waveform was used, and a maximum allowable stress in the hull girder was selected. The midship section was then designed by selecting a section modulus such that the maximum stress was nowhere exceeded for the design loading conditions.[43]

One other challenge which presented itself with the increase in ship size was that of *transverse strength*. Members that contribute to the transverse strength of the structure also support those members that carry loads in the longitudinal direction. As a result, when excessive loads are imposed, not only the transverse strength capability is impaired but also that in the longitudinal direction. Transverse strength is particularly important in configurations with non-uniform distributions of weight athwartships (oil tankers, bulk carriers). Insufficient transverse strength has led to fatigue cracks arising in the web portion of deep frames.

Ships designed with large openings in the main deck (containerships, general cargo carriers) have suffered from insufficient *torsional strength*, which manifested itself through the appearance of cracks around the hatch corners and coamings, loss of hatch cover watertightness and shear buckling of the deck plating between openings. The torsional loads that arise are primarily the result of sea loads, but they are also due to asymmetrical loading of cargo.

The development of a plethora of techniques to study the local behaviour of the structure made it clear that it was possible, depending on the nature of the problem being considered,

[43] See Chapter 5.

to make use of or even develop a new conceptual model. For example, when studying the linear elastic bending response of a laterally loaded stiffened plate, it makes sense to use a model that is accurate for the range of bending loads considered whilst not necessarily being capable of rendering other types of response (e.g. large deflections, plasticity, torsional bending, warping) with the same degree of accuracy. This also holds true when considering the global response of the hull girder. Thus, a box girder analogy provides a more accurate representation of longitudinal bending than a simple beam.

2.8 Toward Rational Ship Structural Design

In parallel with the changing face of ships and shipbuilding, important developments were taking place in other fields that were to bring about profound changes in the thinking and execution of ship structural analysis and design. From the outset, the *information technology revolution* contributed to the solution of complex engineering problems. These problems required new theoretical tools as well as new methods of solution. Engineering structures are generally large, complex and loaded in a multitude of ways, making it impossible in a large number of cases to approach the real situation with any degree of accuracy using conventional mathematical tools. The development of numerical analysis and particularly the finite difference and the finite element method enabled the accurate mathematical representation of such structures and the loads that act on them. On the other hand, developments in computer hardware have enabled the solution of mathematical problems that would not otherwise have been possible. From an early stage, it was the solution of engineering problems that has provided targets for efforts in formulating and advancing the finite element method.[44]

The size and complexity of ship structures render them prime candidates for treatment by computer-based tools,[45] although several decades were to pass before these could be used on a routine basis in design work. Today, classification societies require finite element (FE) studies to form part of the calculations submitted for the approval of a new design for several ship types.

Until recently, structural design was dominated by the need to satisfy strength requirements – that is, to ensure that the structure possesses sufficient strength so that it will not collapse under the maximum load that may arise throughout its operating life. However, progress in several unrelated fields has led to the realisation that the situation is much more complicated. In order to design a ship's structure, it is necessary to consider other aspects as well. These include the effect of time, the structure's flexibility and the effect this has on the performance of vital equipment on board and, more recently, on the environment. These additional requirements have led to the concept of the *limit state*, a term that has its origin in

[44] Triangular finite elements were conceived by J. H. Argyris when studying the design of delta-wing jet fighters for the UK Royal Air Force at the end of the Second World War.

[45] An important class of problems that arises concern complex three-dimensional configurations such as superstructures of varying breadth and length, openings, etc. The results obtained showed that in many cases, the necessary simplifying assumptions made concerning boundary conditions when considering individual components of the structure were incorrect. It was found that supports are generally elastic and involve displacements and rotation of the boundaries of individual elements.

probability theory. Probability theory has played a major role in ship structural design, primarily through its use in modelling the marine environment and consequently of the loads that act on the hull structure. The stochastic nature of the marine environment as well as the variability in the properties and response of the structural assembly itself have meant that strength capability can be expressed in a rational manner only in statistical terms. Significant progress in this direction has been made over the past half-century, and the results of the work done have contributed to important changes in approach which lie behind the IACS rules for bulk carriers and oil tankers [23].

Thus, during the second half of the 20th century, we have witnessed a transition from an *evolutionary approach*, in which a new design is based on past practice whilst at the same time adhering to elementary mechanics of materials theory, to a *partially deterministic approach* using strength equations from beam and plate theory, etc., to a *partially probabilistic and reliability-based approach* [28].

In the latter, the environment and the structural response are described in statistical terms. Acceptability of the structural design is then assessed by comparing the probability of failure with a target failure probability. The development of numerical techniques to achieve an accurate representation of the structure has meant that it is now possible to perform elaborate parametric studies, thus optimising a design against a number of constraints.

In this text the deterministic approach is implemented in Chapters 3–8, 10, 13 and 14. The probabilistic approach is discussed in Chapters 9, 11 and 12.

References

1. Alexandris K. A. (1950) *Η Θαλασσία Δύναμις είς την Ιστορίαν της Αρχαίας Ελλάδος. (Maritime Power in the History of Ancient Greece).* Athens: Royal Hellenic Navy Historical Service.
2. Arnott D. (1955) Developments of Ship Structures. In Arnott D. (Ed), *Design and Construction of Steel Merchant Ships*. Jersey City, NJ: SNAME, 62–94.
3. Bondioli M. (2006) Private communication.
4. Bonn-Muller E. (2009) First Minoan shipwreck. *Archaeology* Vol. 63 No 1.
5. Bryan G. H. (1891) On the stability of a plane plate under thrusts in its own plane, with application to the buckling of the sides of ships. *Proceedings of the London Mathematical Society* Vol. 22, 54–67.
6. Bubnov I. G. (1912) *Theory of Structure of Ships*. St. Petersburg.
7. Caridis P. A. (2012) The evolution of ship structures from antiquity to the present day. In *Proceedings of the RINA Conference on Historic Ships III*, 5–6 December, London, 81–100.
8. Casson L. (1971) *Ships and Seamanship in the Ancient World*. Princeton, NJ: Princeton University Press.
9. Castro F. (2007) Rising and narrowing: 16th century geometric algorithms used to design the bottom of ships in Portugal. *International Journal of Nautical Archaeology* Vol. 36, 148–154.
10. Catsambis A., Ford B., Hamilton D. L. (2011) *The Oxford Handbook of Maritime Archaeology*. Oxford: Oxford University Press.

11. Cauchy A. (1823) Recherches sur l' équilibre et le movement intérieur des corps solides ou fluides, élastiques ou non élastiques. (Investigations on the equilibrium and motion of solids or fluids, elastic and inelastic). *Bulletin de la Société Philomatique* 9–13.
12. Chronology Overview. *Dartmouth.edu*. Available: www.dartmouth.edu/~prehistory/aegean/?page_id = 67 (accessed 19 December 2020).
13. Clarkson J. (1965) *The Elastic Analysis of Flat Grillages*. Cambridge: Cambridge University Press.
14. Coates J. (1991) The ancient Greek trireme: An early pre-stressed shell. In Smith C. S., Dow R. S. (Eds). *Advances in Marine Structures Vol. 2*. London: Elsevier Applied Science, 452–468.
15. Dain A. (1943) *Naumachica*. Paris: Editions Les Belles Lettres.
16. Euler L. (1759) Sur la force des colonnes. *Memoires de l' Académie de Berlin*, Tom. XIII, 252–282.
17. Euler L. (1771) *Sur le roulis et le tangage: Examen des efforts qu' ont a soutenir toutes les parties d' un vaisseau dans le roulis et dans le tangage. (On Roll and Pitch: Review of the Supporting Forces of All Parts of a Vessel in Roll and Pitch)*. Paris: Académie Royale des Sciences.
18. Gardiner R. (Ed). (1995) *The Age of the Galley: Mediterranean Oared Vessels Since Pre-classical Times*. London: Conway Maritime Press.
19. Greenhill B., Morrison J. (1996) *The Archaeology of Boats and Ships: An Introduction*. Annapolis, MD: Naval Institute Press.
20. Harpster M. J. (2005) A Re-Assembly and Reconstruction of the 9th-Century AD Vessel Wrecked off the Coast of Bozburun, Turkey. Doctoral Thesis, Texas A&M University, Austin, TX.
21. Henriksson G. (2012) The Trojan war dated by two solar eclipses. *Mediterranean Archaeology and Archaeometry* Vol. 12, 63–76.
22. Homer *Odyssey*.
23. International Association of Classification Societies (2020) *Common Structural Rules for Bulk Carriers and Oil Tankers*. London: IACS.
24. Johnston S. (1994) Making Mathematical Practice: Gentlemen, Practitioners and Artisans in Elizabethan England. PhD Thesis, Cambridge University.
25. von Kármán T. (1923) *Beitrage zür Technischen Physik und Technischen Mechanik. (Contributions to Technical Physics and Engineering Mechanics.)*. Berlin: Springer.
26. Kirchhoff G. R. (1850) Über das gleichgewicht und die bewegung einer elastischen scheibe. (On the equilibrium and motion of an elastic disc.) J. r. ang. Math. (Crelle), 51–88.
27. Kirchhoff G. R., Wien W. (1877) *Mechanik*. 2nd ed. Leipzig: B.G. Teubner.
28. Kline R. G. (1993) Hull structural analysis. In Benford H. (Ed), *A Half-century of Maritime Technology 1943-1993* Jersey City, NJ: SNAME, 54–64.
29. Lamé G. (1852) *Leçons sur la Théorie Mathématique de l' Elasticité des Corps Solides. (Lectures on the Mathematical Theory of Elasticity of Solid Bodies)*. Paris: Editions Bachelier.
30. Lane F. C. (1934) *Venetian Ships and Shipbuilders of the Renaissance*. Baltimore, MD: Johns Hopkins University Press.
31. Landström B. (1961) *The Ship: An Illustrated History*. New York: Doubleday & Co.
32. Mansour A. (1971) On the nonlinear theory of orthotropic plates. *Journal of Ship Research* Vol. 15, 266–276.
33. Marguerre K. (1938) Zür Theorie der Gekrümmter Plätte Grosser Formänderung (On the Theory of Bent Plates with Large Deformations.) *Proceedings of the Fifth International Congress on Applied Mechanics*, Cambridge, 93–101.

34. McGrail S. (2001) *Boats of the World: From the Stone Age to Medieval Times*. Oxford: Oxford University Press.
35. Meleshko V. V. (2003) Selected topics in the history of the two-dimensional biharmonic problem. *Applied Mechanics Reviews* Vol. 56, 33–85.
36. Morrison J. S., Coates J. F., Rankov N. B. (2000) *The Athenian Trireme: The History and Reconstruction of an Ancient Greek Warship*. 2nd ed. Cambridge: Cambridge University Press.
37. Navier C. L. (1821) *Mémoire sur les lois de l' équilibre et du movement des corps solides élastiques. (Dissertation on the Laws of Equilibrium and Motion of Elastic Solid Bodies.)* Paris: Académie des Sciences.
38. Navier C. L. (1823) *Bull. Soc. Philom*. Paris.
39. Nowacki H. (2006) *Developments in Fluid Mechanics Theory and Ship Design Before Trafalgar*. Berlin: Max Planck-Institüt für Wissen-schaftsgesichte.
40. Papamarinopoulos St. P., Preka-Papadema P., Antonopoulos P., Mitropetrou H., Tsironi A., Mitropetros P. (2012) A new astronomical dating of Odysseus' return to Ithaca. *Mediterranean Archaeology and Archaeometry* Vol. 12, 117–128.
41. Pastor X. (2010) *Ships of Christopher Columbus* (Anatomy of the Ship series). London: Conway Maritime Press.
42. Phelps W., Lolos Y., Vichos Y. (Eds.). (1999) The Point Iria Wreck: Interconnections in the Mediterranean ca. 1200 BC. 2nd ed. *Proceedings of the International Conference on the Island of Spetses*, 19 September 1998. Athens: Hellenic Institute of Marine Archaeology.
43. Poisson S. D. (1829) Mémoire sur l' équilibre et le mouvement des corps élastiques. (Memoire on the equilibrium and motion of elastic bodies). Mém acad., Vol. 8, 1829 & Mémoire sur les équations générales de l' équilibre et du movement des corps élastiques et des fluides *J. éc. pol.* (Paris), cahier 20, 1831.
44. Polzer M. (2011) Early shipbuilding in the Eastern Mediterranean. In Catsambis A., Ford B., Hamilton D. L. (Eds), *The Oxford Handbook of Maritime Archaeology*. Oxford: Oxford University Press 349–378.
45. Pomey P. (2015) The origins of ancient Greek shipbuilding: Sewn boats and their evolution. *First International Seminar on Ancient and Traditional Shipbuilding*. Athens: National Technical University of Athens.
46. Rieth E. (2003) From wreck to shipyard: The example of the Port Berteau II wreck France (VIIth c. AD). In Nowacki H., Valleriani M. (Eds), *Shipbuilding Practice and Ship Design Methods from the Renaissance to the 18th Century: A Workshop Report*. Berlin: Max Planck-Institüt für Wissenschaftsgesichte.
47. de Saint-Venant A. J. C. B. (1856) Mémoire sur la Torsion des Prismes, avec des considérations sur leur flexion, ainsi que sur l' équilibre intérieur des solides élastiques en général, et des formules pratiques pour le calcul de leur résistance à divers efforts s' exerçant simultanément. (Memoire on torsion of prismatic sections, with considerations on their bending, as well as the internal equilibrium of elastic solids in general and practical formulas for calculating their resistance to various forces applied simultaneously.) *Mem. Acad. Sc. Sav. Étr.* Vol. 14 (1855) 233–560 and *J. math. Liouv*. ser. 2 Vol. 1 (1856) 89–189.
48. Schade H. A. (1941) Design curves for cross-stiffened plating under uniform bending load. *Trans SNAME* Vol. 49, 154–182.
49. Sokolnikoff I. S., Specht R. D. (1946) *Mathematical Theory of Elasticity*. New York: McGraw-Hill Book Co Inc.
50. Steffy J. R. (1998) The development of ancient and medieval shipbuilding techniques. Proceedings of the International Symposium on Archaeology of Medieval and Modern Ships of Iberian-Atlantic Tradition, Lisbon, 49–60.

51. Steffy J. R. (1982) The reconstruction of the 11th century Serçe Liman vessel. *International Journal of Nautical Archaeology and Underwater Exploration* Vol. 11, 13–34.
52. Tartaron T. F., Rothaus R. M., Pullen D. J. (2003) Searching for prehistoric Aegean harbors with GIS geomorphology and archaeology *Athena Review* Vol. 3, 27–36.
53. Televantou C. A. (2013) The Neolithic settlement at Strofilas Andros: Expanding the horizons of Cycladic prehistory and iconography. *Cycladic Seminar*. Athens: The Archaeological Society at Athens.
54. Timoshenko S. P. (1983) *History of Strength of Materials*. New York: Dover Publications Inc.
55. Trombetta manuscript, British Library, Cotton Titus A XXVI No.3 27 b [-60].
56. University of Southampton (2006) *History and archaeology of the ship*. Department of Archaeology Lecture Notes Southampton.
57. Wedde M. (2015) Typology and beyond: Defining moments in the history of early Aegean watercraft. *First International Seminar on Ancient and Traditional Shipbuilding*. Athens: National Technical University of Athens.
58. *Wikipedia.org*. Available: https://en.wikipedia.org/wiki/Aegean_civilizations (accessed 19 December 2020).
59. Xi L., Chalmers D. W. (2004) The rise and decline of Chinese shipbuilding in the Middle Ages. *Trans RINA* Vol. 146 Part A2, 59–69.
60. Young T. (1814) Remarks on the Employment of Oblique Riders, and on Other Alterations in the Construction of Ships. Being the Substance of a Report Presented to the Board of Admiralty, with Additional Demonstrations and Illustrations. *Phil. Trans. Roy. Soc.*, 303–336.
61. *Amazonaws.com*. Available: https://s3.amazonaws.com/classconnection/830/flashcards/9312830/jpg/sl_reconstruction_with_rig-1518872D893590B2A6F-thumb400.jpg (accessed 9 June 2021).

3

Sea Loads on Ship Structures

Venez à mon secours, Grand Dieu: ma Barque est si petite et la mer est si grande...

Come to my help good Lord: my boat is so small and the sea is so vast...

Old Breton fisherman's prayer

One of the most important questions that arise when considering the behaviour of ship structures is that of load identification and estimation. This topic has been laden with difficulties, and until recently, major simplifications have had to be made when selecting values for use in design.

A closely related issue is that of load classification. Loads have been classified on the basis of a variety of criteria, such as the nature of the load [18],[1] the type of response, both of these, or the level of structure at which they act. In [21] a classification that is based partly on the nature of the load and partly on the type of response has been followed,[2] whereas in [14] and [32] the classification is based on the time-domain variation (frequency) of the applied load.[3]

From the perspective of structural design, load classification assists in determining the nature of the response of the structure, which in turn leads to an appropriate method of analysis. In this sense load classification is but one, albeit an important, step in assessing structural behaviour. Generally speaking, it is specific features of the load, the structure and of the load in conjunction with the structure that results in a particular type of response. For this reason it is necessary to consider each case on an individual basis in order to decide on an appropriate line of attack. To do this it is necessary to identify load features as well as the response of the particular structural component.

In order to be able to develop a framework that will enable this to be achieved, we shall consider the problem in stages. To start with, it will be necessary to describe the nature of the loads, propose a rational basis for their classification and identify their properties. At the same time it is necessary to identify in general terms the expected types of response of ship structural components and to describe the pertinent features of each of these. Following

[1] (a) Static loads; (b) thermal gradients; (c) wave bending loads; (d) dynamic loads (extreme and cyclic).

[2] (a) Static loads; (b) low-frequency dynamic loads; (c) high-frequency dynamic loads; (d) impact loads.

[3] (a) Static loads; (b) gradually applied loads; (c) pulsating or alternating loads; (d) dynamic loads. Dynamic loads are furthermore classed as low frequency (e.g. wave action), slam or impulsive, and vibratory.

this, we shall have to consider the loads in conjunction with structural response in order to decide on a suitable approach. In a number of cases the choice of method is self-evident. In others it becomes necessary to examine matters much more closely.

3.1 The Operating Environment: A Hypothetical Scenario

Before discussing the loads that act on surface ships, we will outline a fictitious but nonetheless realistic operating scenario for a dry cargo ship. We have selected a bulk carrier that performs a single voyage in a loaded condition, an example that is representative of an important class of ships.

This exercise will help us pinpoint the most important loads that act on the hull structure and bring to our attention others that may otherwise slip by unnoticed. Following this, we will approach the question of load classification in a much more meaningful way.

Some comments of a general nature that are applicable to all types of ships are necessary before we focus on the voyage of the bulk carrier in question. Perhaps we ought to begin with the construction of the vessel on the building berth. The fabrication process induces stresses in the plating and stiffeners as well as distortions. These stresses are higher than any metal-forming stresses that may already exist, especially in way of welds and plate–stiffener connections. If fabrication or material defects are present, these may become a cause for concern at a later stage. Following completion of construction, the vessel is launched; this is a procedure during which the bending moment amidships increases substantially, a condition that is taken into account by the shipyard. Following launching and whilst afloat, outfitting is carried out, during which the addition of further loads results in changes in the longitudinal shearing force and bending moment distributions.

We now turn our attention to the various stages of the voyage of a bulk carrier that is employed in the transportation of iron ore from South America to the Far East. The loading port is assumed to be located along the east coast of South America. The temperature of the seawater varies diurnally from 23°C to 27°C on average, with the air temperature varying from 26°C to 33°C.

Before entering the loading port, ballast seawater is discharged and transferred to trim the vessel in order to achieve the correct draught for her to be ready for loading. The vessel is then berthed alongside the loading terminal (Figure 3.1). Permissible loading sequences are stipulated in the loading manual used by the chief mate, who is responsible for the loading operation. The most important role of the loading manual is to ensure that hull girder stresses do not exceed permitted values at any stage of the loading operation. In the case of bulk carriers, excess loading may arise in way of bulkheads that separate loaded from empty holds, as the vertical shearing force there may increase substantially.

The ship is loaded using conveyor belts capable of transferring cargo at high rates (up to 16,000 metric tonnes/hour), with the cargo being dropped into the holds from a height of approximately 20 metres. When the holds are empty, the mineral ore strikes the inner bottom directly, so to prevent impact damage, low loading rates are used initially. In the case of large bulk carriers, the loading rates may result in problems due to excessive global

Figure 3.1 Bulk carrier being loaded with iron ore at Porto Sudeste, Rio de Janeiro, Brazil [5].

loading of the hull girder, excessive local loading as well as in the synchronisation of the de-ballasting operation, which is performed at the same time as loading (de-ballasting pumps have a total capacity of 4,000–5,000 m^3/hour). De-ballasting continues during loading in order to maintain correct trim.

In the case of a smaller vessel, its cargo-handling equipment could possibly be used for loading, in which case the main deck would be subjected to local normal loads and bending moments, distributed over an area limited by adjacent hatch openings. In order to operate the ship's cranes, the ship's generators are set in operation. During loading, de-ballasting and/or shifting of ballast also take place to ensure proper trim and draught. It is also possible that the ship takes on bunkers during this stage. As loading proceeds, the longitudinal bending moment and shearing force gradually change in tandem with local stress distributions, especially in way of load discontinuities.

Upon completion of loading, the ship sets sail under its own power, with the assistance of tugs. The operation of the main engine produces vibration that is borne by the engine seats and may possibly propagate throughout the accommodation spaces. The tugs operate either by pulling or pushing at permitted positions above the waterline. As the ship leaves port, it picks up speed, and the propeller begins to impose a fluctuating load on the hull surface and the shaft. Within the engine room, the number of diesel generators in operation is now reduced, since electrical power requirements are not as high as in port. For a diesel engine-propelled ship, an auxiliary boiler is used to produce steam for engine operation and service requirements. Both the main engine and boilers radiate heat within the engine room, thus increasing the temperature in relation to external conditions.

During the voyage, liquids such as bunkers and diesel oil will gradually be consumed. For a slow-speed main diesel engine this will in many cases exceed 30 metric tonnes of bunkers per day, which will bring about load changes at a local level. An operation carried out routinely is ballast seawater exchange, which involves the simultaneous de-ballasting of certain tanks and the ballasting of others. This is done in order to reduce stresses. Ballast seawater exchange[4] is also carried out to prevent the spread of harmful species of micro-

[4] The two methods used are the *sequential exchange method* and the *flow-through exchange method.*

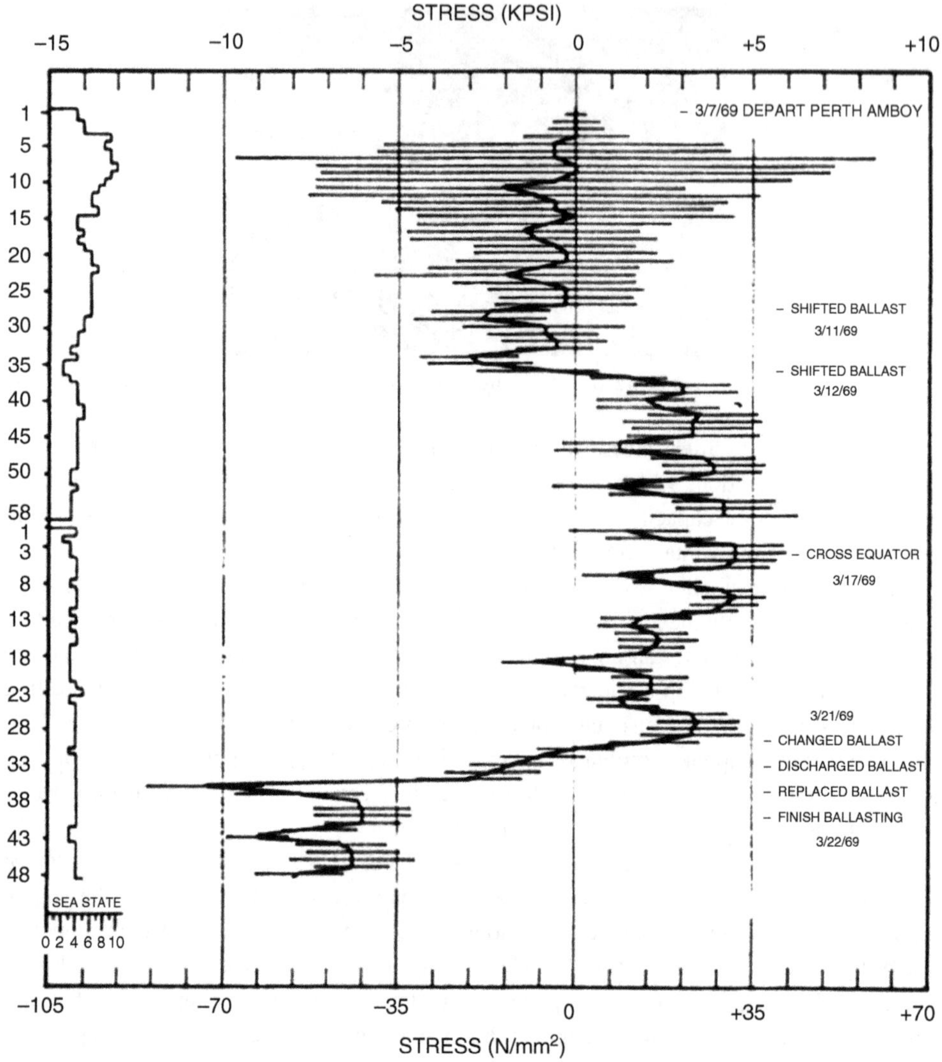

Figure 3.2 Effect of ballast water transfers on deck stress (adapted from [19][5]). Reprinted courtesy of the Society of Naval Architects and Marine Engineers.

organisms. The effect of ballast water transfers is seen clearly in Figure 3.2 that shows stress variations during the ballast voyage of a small oil tanker. Meanwhile, air and seawater conditions (temperature and humidity) change. Since the ship maintains a steady course, the temperature of the hull girder will vary according to whether or not it is exposed to direct sunlight. An additional source of differential heating is the difference in temperature between air and seawater.

[5] These measurements were made onboard the *MT R.G. Follis,* a 66,500 dwt tonne oil tanker. The dark lines represent stillwater bending stresses.

On board the ship there are a number of tanks used for the storage of liquids (ballast seawater, bunkers, marine diesel oil and fresh water). When the ship is at sea, these liquids are subjected to inertial accelerations, depending on the location of the respective tanks with respect to the ship's centre of gravity. Usually a number of tanks are located within the vicinity of the engine room. Others such as ballast tanks are located in the forepeak, the aft-peak, the double bottom and the topside tanks (in the case of bulk carriers). If the accelerations are sufficiently high, sloshing within the tanks may occur, thereby imposing loads on the vertical bulkheads of tanks.

It is possible that during the voyage the ship encounters heavy seas. In such a case, the longitudinal bending moment will increase and will fluctuate as the ship meets with irregular waves. If severe conditions are encountered, the ship's sides in way of the bow and the bow flare will be subjected to wave impact. As waves break against the sides, spray and water will strike the main deck forward. Furthermore, if emergence of the bottom forward takes place, this will possibly be followed by *slamming*. Slamming is followed by *whipping*, which is in effect a high-frequency vibration that propagates from the region of the slam throughout the ship. In this way energy that is transferred to the hull girder during the slam is absorbed by a large part of the structure. Under such circumstances, the ship's speed will have to be reduced and the course altered so as to avoid further bottom slamming.

In addition to bottom slamming, when a ship encounters heavy seas, the quantity of cargo within each hold is subjected to inertial accelerations that impose loads, in addition to those that result from gravity, on the inner bottom structure. The inertial acceleration loads also act on the vertically aligned structural members of the hold.

By the time the ship reaches the discharge port, the consumption of liquids such as bunkers and diesel oil will have brought about non-negligible changes to the longitudinal shearing force and bending moment distributions. The ship may berth upon arrival or else may be required to lie at anchor awaiting its turn. Atmospheric conditions will be different from those at the loading port, thereby imposing a different thermal stress field on the hull girder. Following berthing, discharging can begin. Shore equipment is used, and grabs are lowered into the holds in accordance with the master's approval. At the same time the ship takes on additional ballast in order to maintain proper trim and draught, as well as to maintain stresses at acceptable levels. When the discharging of individual holds nears completion, mechanical equipment such as bulldozers is used to collect the last traces of cargo. At this stage the grabs and the bulldozers used can strike the inner bottom and the lower parts of the hold, resulting in local dents and damage.

If the next voyage is in ballasted condition, the ship will take on ballast water that can reach up to 50–60 per cent of her deadweight capacity. Usually two ballast conditions are specified: light ballast, which is suitable for fair weather – this represents 40–50 per cent of deadweight capacity; and heavy ballast which corresponds to 50–65 per cent of deadweight capacity. The latter is necessary when rough weather is expected.

3.2 Classification of Loads

For the case described in the previous section, we will now consider the loads that act and group them according to their origin. In this particular case the ship has performed a loaded voyage whereas on another occasion it could have been a ballast voyage. Therefore, loads relating to the presence of cargo would then have been irrelevant. Furthermore, if we consider the vessel in port or at anchor, we see that loads arising from the operation of mechanical equipment such as the main engine and propeller would not be relevant. It becomes evident that these two load sources ought to be grouped separately so that they can be included when appropriate.

The loads that act on the hull girder on a continuous basis are those that are due to the environment and those that relate to the hull structure itself. On this basis we propose the following classification:

1. *Environment-related loads*. These loads have their origin in the natural environment, and their magnitudes depend on various physical factors. They are ever-present in the absence of a ship.
   1. Hydrostatic pressure
   2. Hydrodynamic pressure resulting from motion of the sea
   3. Wave action
   4. Thermal gradients
   5. Wind pressure gradients
2. *Hull girder-related loads*. These loads arise as a consequence of the presence of the ship in the marine environment, its form and its motion.
   1. Hydrostatic pressure loading of the entire hull girder
   2. Hydrostatic pressure loading of individual structural members
   3. The combined action of hydrostatic and hydrodynamic pressures on the hull girder
   4. The combined action of hydrostatic pressure, hydrodynamic pressure and wave action on the hull girder
   5. Wave loading on individual components of the hull girder
   6. Wave loading of the entire hull girder
   7. Local loads intrinsic to the structure arising as a result of material defects, the fabrication process and aging
3. *Mechanical equipment loads.* These are the loads, resulting from the presence and operation of mechanical equipment, that do not form an integral part of the hull girder.
   1. Propeller
   2. Main engine operation
   3. Other machinery operation
   4. Thermal gradients arising from equipment operation
   5. Local loads (thrust stool loads, anchor and chain loads, tug operations)
4. *Cargo loads*. These are the loads that arise solely as a result of the presence of cargo/ballast.
   1. Cargo weight
   2. Cargo inertial loads
   3. Cargo loading and discharging
   4. Wet cargo impact loads on the internal structure

5. Loads corresponding to (1–4) for ballast water and other liquids (bunkers, diesel oil, freshwater)
6. Thermal gradients (natural as well as imposed)

During the lifetime of a ship, in addition to the conditions described previously, a number of other circumstances arise, some of which have to be allowed for. These are:

1. Launching
2. Dry-docking
3. Collision and grounding

Of these, launching is a once-in-a-lifetime load, whereas dry-docking loads arise at specified intervals throughout a ship's life. As regards collisions and grounding, these are unwelcome events that may or may not occur during a ship's lifetime. In any case, ships are not designed to be able to withstand the loads that arise under such circumstances without suffering some damage to the hull structure. These conditions then need to be added to the previous list, and the most appropriate group is that of hull-structure related loads, since they arise as a result of human activities.

Lastly we need to be aware of the effect that wear and tear have on the condition of the hull girder. These can result in (1) thickness reductions due to corrosion, (2) fatigue life reduction leading possibly to cracks and (3) local buckling. All of these result in a redistribution of loads, generally at a local level.

Clearly the net load that acts on an individual member will be the result of the interaction of loads that belong to more than one of the above categories. It is then necessary to specify particular circumstances and select the appropriate loads. For example, the loads that act on the inner bottom and vertical plating of the cargo tank of an oil tanker will be the outcome of (1) longitudinal bending of the hull girder, (2) ship motions, (3) the presence of cargo, (4) residual stresses due to welding and (5) thermal fields.

In the remaining sections of this chapter we will discuss environment-related loads and ship-related loads in an introductory manner. The next chapter deals with the longitudinal bending loads of the hull girder (primary bending) whereas loads related to the operation of mechanical equipment, cargo loads and the additional loads listed above are considered in Chapter 5.

3.3 Sources of Environment-Related Loads

Ships operate in what is probably the most destructive environment on the planet and are expected to fulfil their mission for a period that may exceed 25 years. In general, the conditions ship structures are expected to confront are the ocean environment at a global level. Let us now turn our attention to the marine environment and its effect on the performance of marine structures.

3.3.1 Properties of Seawater

Seawater is a dilute solution of sodium chloride ($\approx$3.5%) that acts as an electrolyte that brings about the electrochemical corrosion of metal structures. After some years of

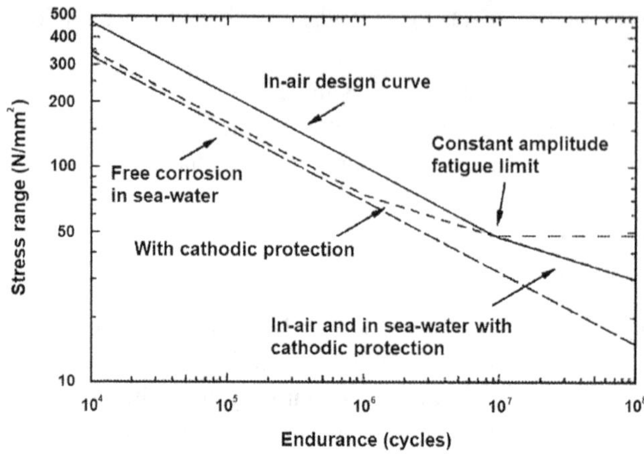

Figure 3.3 Effect of dilute sodium chloride solution on the fatigue strength of steel [6]. Reprinted courtesy of the Ship Structure Committee, Washington, DC.

operation of a ship, corrosion results in thickness reductions of plating and stiffeners and a loss of strength as well as a reduction in fatigue life. Extensive studies have been made of the effects of corrosion on different ship types which led to the development of specific requirements with respect to the ship structures' protection and maintenance.

Classification societies have imposed strict survey guidelines, especially for aged ships, so that problematic areas are identified and dealt with. However, it is not possible to resolve the question of fatigue life reduction because, in the marine environment, fatigue life will reduce regardless of how low the stress range may become. This is illustrated in Figure 3.3 and explains why ships have a finite life, in contrast to land-based structures that may also be subjected to cyclic loading.

Particular circumstances arise when ships sail in waters that contain pollutants and/or are rich in marine life, especially in the tropics. When ships are moored in such waters, the organic substances that are present contribute to the deterioration of the condition of the outer hull surface, thereby causing damage to the protective coating and an increase in surface roughness.

Figure 3.4 illustrates the variation of mean annual temperature on the surface of the earth's oceans. It is seen that in the tropics, temperatures reach 30°C whereas in the regions of the poles the temperature drops to around 0°C. Furthermore, there are diurnal temperature variations (Figure 3.5a) as well as variations due to the position of the sun with respect to a particular ship. In general, temperature increases have a detrimental effect on ship structures due to the increased action of both corrosion and fatigue. Very low temperatures, on the other hand, can cause steel to behave in a brittle manner so that cracks may occur even when the actual loading is low. This problem was identified following the large number of failures that occurred in early welded ships. Extensive investigations followed, and a number of measures were implemented, including improvements to the fracture toughness of shipbuilding steels.

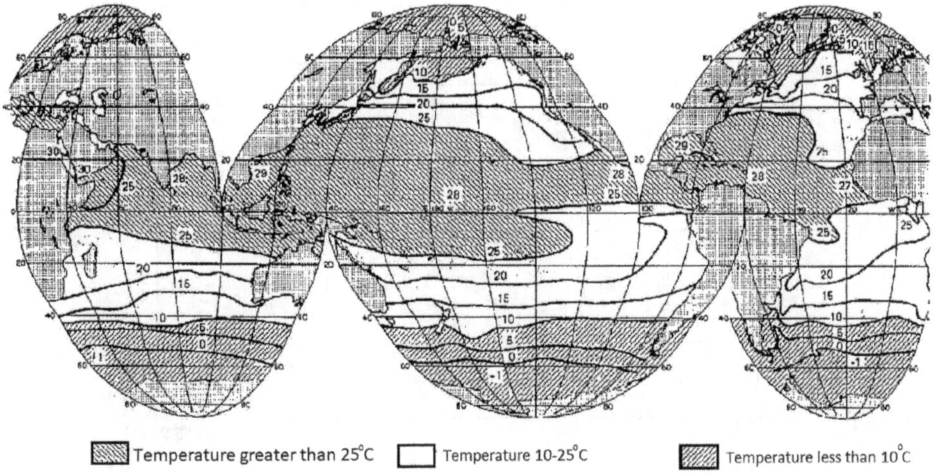

Figure 3.4 Mean sea surface temperatures of the earth's oceans in August [1].[6] Reprinted courtesy of the Society of Naval Architects and Marine Engineers.

More important are the temperature differences between air, seawater and specific regions of the hull structure (such as heated cargo tanks) at any one point in time (Figure 3.5b). These give rise to stress fields that need to be taken into consideration.

When temperatures are low enough, ice is present on the sea surface. This normally occurs in the Arctic Ocean and in the Southern Ocean, in the regions of the poles. In Antarctica, the ice forms seasonally, and thus the depth of the ice pack does not normally exceed 1 metre. In the Arctic, the thickness extends to 3–4 metres over large areas. Seasonal variations in pack thickness and area covered are important, especially in Antarctica. In recent years, as a result of global warming, the extent of ice has reduced and is continuing to do so, so that it is now possible for ships to sail north of the American continent via the Northwest Passage and also north of Eurasia via the Northern Sea Route (Figure 3.6). From the point of view of maritime trade, the latter route represents a reduction in voyage distance of 8,000 nautical miles for the voyage from the Far East to northern Europe. The loads that act on the outer shell plating during this type of voyage can be significant, depending on the depth of the ice pack (Figure 3.7). Ships that perform voyages under such conditions need to be appropriately designed.

3.3.2 Wind Pressure

In the case of sea-going ships, *wind pressure* is not a critical factor, since structures are normally designed to withstand much higher loads, such as deck wetting. The variation of wind speed with altitude above sea level has a parabolic form, and a rapid increase from

[6] Originally from: Sverdrup H.U. Johnson M. W. Fleming R. M. (1942) *The Oceans, Their Physics, Chemistry and General Biology*. Englewood Cliffs, NJ: Prentice Hall.

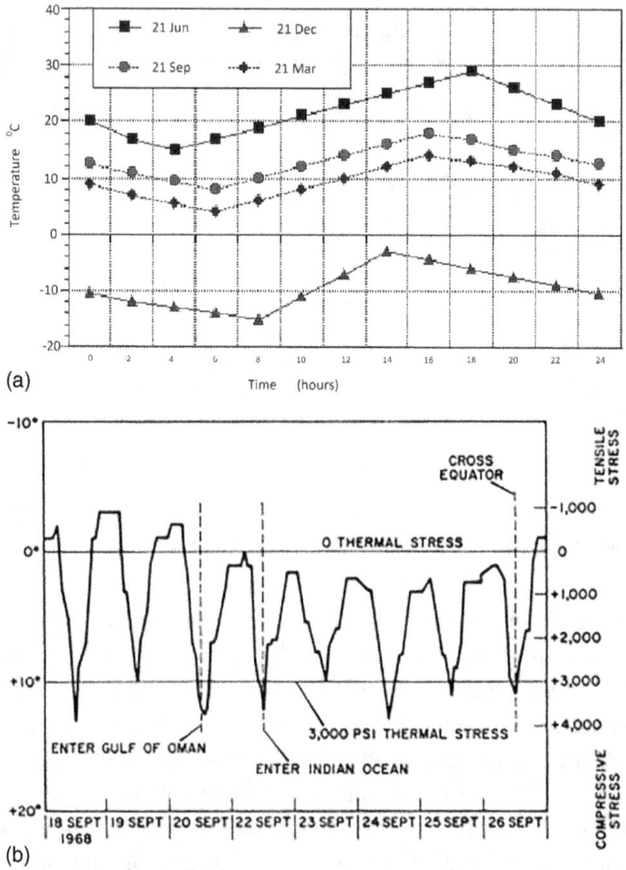

Figure 3.5 Thermal gradients and resulting fields. (a) Hourly variations in surface temperature over a 24-hour period for a location at 45° latitude [28]. (b) Air-seawater temperature differences and resulting thermal stresses (*Esso Malaysia*) [18]. Figure 3.5b reprinted courtesy of the Society of Naval Architects and Marine Engineers.

zero is observed over the first few metres (boundary layer effect). Thus any effect wind pressure might have would concern parts of the structure that are high above the waterline such as superstructures, navigation masts and equipment fitted to them. Detailed considerations are made in the case of such items. In general, in the absence of green-sea loading, it is recommended that a pressure of 0.5 kPa be used. There are several cases in which wind pressure is important. The most important of these is that of ships with extensive superstructures such as cruiseships, as well as ships with extensive deck loading arrangements such as containerships.[7] Wind pressure may impose significant loads on fixed offshore

[7] A method to determine wind loads on ships using data on ship type and length is described in Ueno M. Kitamura F. Sogihnara N. Fujiwara T (2012) A Simple Method to Estimate Wind Loads on Ships. *World Congress on Advances in Civil, Environmental and Materials Research (ACEM'12)*. Seoul, 26–30 August, 2314–2322.

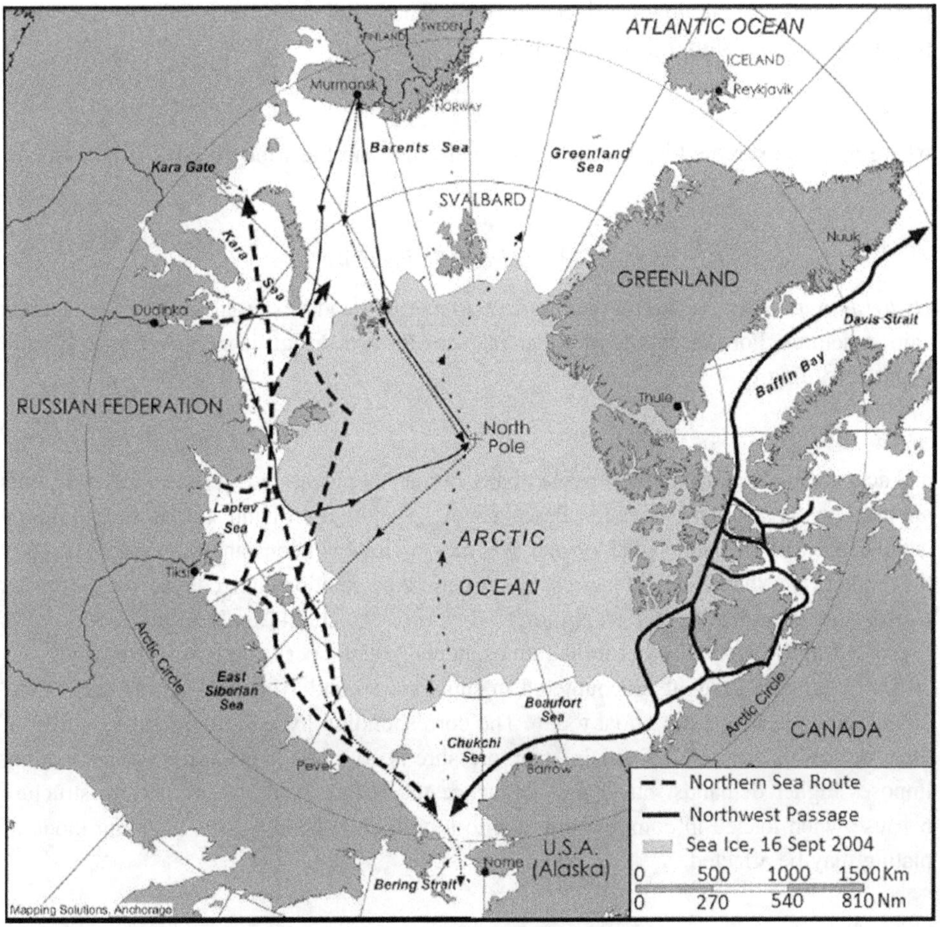

Figure 3.6 The Northern Sea Route and the Northwest Passage [12].

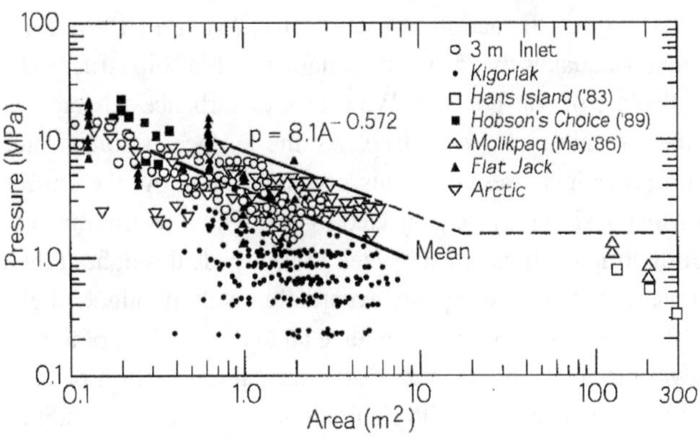

Figure 3.7 Effective ice pressure versus contact area [22]. Reprinted courtesy of Elsevier.

platforms and loaded floating docks as well. The wind pressure acting on the lateral projected area of a ship, p_{wind}, is given by:

$$p_{wind} = 0.004V^2 \tag{3.1}$$

where p_{wind} is expressed in poundals per square foot and the wind speed V is in knots.

3.3.3 Hydrostatic Pressure

Hydrostatic pressure is the most important source of steady external loading that acts on ship structures. For the depths of water relevant to ship design it increases linearly with depth according to

$$p = \rho gh \tag{3.2}$$

and acts perpendicularly to the surface of any immersed object. In Equation (3.2) if SI units are used, the pressure p is given in Pascal (N/m^2). On the right-hand side of the equation ρ is the density of seawater (1025 kg/m^3), g is the gravitational constant (9.81 ms^{-2}) and h is the depth below the surface in metres. For structural calculations it is convenient to express hydrostatic pressure in MPa (N/mm^2).[8] For non-structural applications, pressure is expressed in a variety of other units (atmospheres, pounds per square inch, bars, etc.).

The maximum draughts encountered in ships are around 20 metres, in the case of the largest oil tankers that are afloat today. The corresponding hydrostatic pressure is thus of the order of 0.2 N/mm^2. This hydrostatic pressure is not critical, because in-plane loading imposes higher demands, but it may influence the failure mode of the bottom structure because when loaded in compression, due to longitudinal bending, the buckling mode of plating may be affected.

3.3.4 The Surface of the Sea

The surface of the sea is in direct contact with the atmosphere, and when wind develops, it gives rise to waves through the action of friction (*wind-generated waves*). Other sources of waves exist, such as seaquakes that produce tsunamis, but in ship structural design it is wind-generated waves that are of sole concern. Waves are disturbances characterised by a continuous change in the elevation of the surface. As the surface becomes more disturbed, the loading on a hull girder increases. It is thus necessary to know the expected maximum or *extreme load* to which a ship may be subjected to throughout its lifetime, in order to design a safe structure. Furthermore, ships are subjected to sea loads throughout their working life so that the continuous, repetitive loading may eventually result in fatigue of elements of the hull girder. For both of these reasons we need to have an understanding of the nature of sea loads. *Linear wave theory* provides an adequate mathematical description of the motion of undisturbed ocean waves. The interaction with a ship up to moderate sea conditions is discussed in

[8] 1 MPa $\equiv 10^6$ Pa.

Chapter 4. In recent years nonlinear theories have been developed and used to estimate the loading of ships in rough seas and under extreme conditions.

When a ship is advancing in a seaway, the fluid environment interacts with the solid body in a number of ways. As a result of this interaction, both the motion of the body and the underlying fluid flow are affected. Wave motion gives rise to phenomena that affect the ship in several ways and at different levels. Within the vicinity of the immersed body wave action and the motion of the fluid bring about changes to the pressure on the outer hull surface as the elevation of the sea surface varies continuously and also due to the hydrodynamic forces that act on the immersed body. This is discussed in Chapter 4, in which both the traditional quasi-static theory and linear strip theory are presented. In the following section we shall describe in qualitative terms the nature of certain additional types of loads that act on the hull girder under specific conditions. Their action may affect, to a greater or lesser extent, both the ultimate limit state and the fatigue limit state of certain types of ships. Their study generally requires the use of specialised techniques and test methods. These techniques are a subject that lies beyond the scope of this text.

3.4 Transient Wave Loads

When a ship navigates in rough seas, it may experience a variety of different kinds of impacts with the waves. These impacts result in a rapid exchange of momentum between the ship and the surrounding sea environment. The additional loads that act on the hull girder or specific regions of it are the result of different kinds of *hydrodynamic wave impact* (Figure 3.8).

These occur mostly in way of the fore end and, to a lesser extent, the aft end and the side shell plating of a ship. They have been classified on the basis of the particular region of the ship affected, and thus we have:

Figure 3.8 Different types of hydrodynamic wave impact (bow flare impact, tunnelling, deck wetting).

Figure 3.9 Effects of wave impact. (a) Profile of pressure pulse[9] (*MV Derbyshire*). (b) Damage to bow plating [8]. (c) Damage to side shell plating [9]. Figure 3.9b reprinted courtesy of Det Norske Veritas, Høvik.

- *Slamming* of the bottom structure, in way of the forward and aft ends (Figure 3.11).
- *Wave slap,* in way of side shell plating (Figures 3.8, 3.9c).
- *Bow flare impact* (Figure 3.9b).
- *Deck wetting* (*green seas*) in way of the main strength deck, generally in the forward region (Figures 3.8, 3.9a, 3.22a).

As seen in Figure 3.8, these can arise simultaneously as a ship is advancing in heavy seas. Clearly the analysis of ship structural response under such conditions is a tall order. We shall consider specific aspects of the problem and describe the results available today from the point of view of structural design.

Wave impacts on a solid surface are generally characterized by a rapid increase in pressure, which, after peaking, reduces much more gradually, following a quasi-exponential decline path (Figure 3.9a). Measurements have shown that the maximum pressure can exceed 0.2 N/mm^2, which corresponds to a static pressure of 20 metres of water. The duration may vary in the 25–250 msec range. Wave impact and its effects have been investigated for many years using experimental and theoretical methods in order to improve physical understanding and design practice.

[9] Tarman D. Heitmann E. Derbyshire: Loss of a Bulk Carrier, *Ship Struct. Comm. Case Study II* [*gifle* (French) = slap].

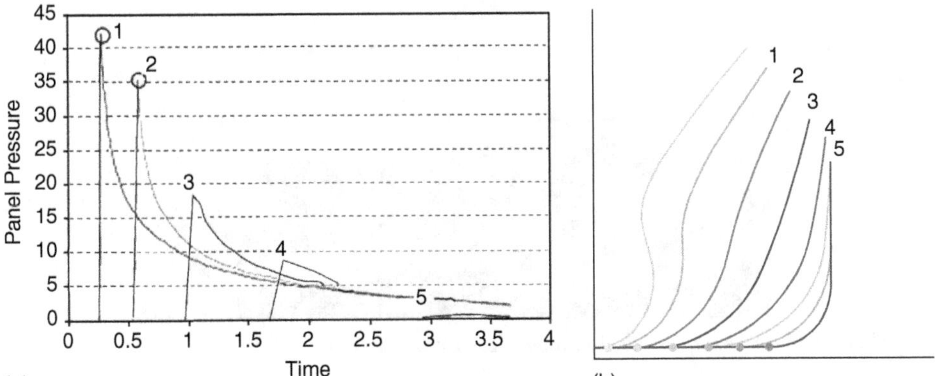

Figure 3.10 Slamming profiles for assessment of structural response [2]. (a) Time-domain variation of panel pressure. (b) Location of panels.

In aged ships it is frequently possible to observe the side shell plating to have set in as a result of wave impacts that the ship has encountered during its working life, so that the positions of the internals are evident from the outside (Figure 3.9b). It is possible that a contributory factor to the serious damage illustrated in Figure 3.9c was weakening of the shell plating by corrosion and fatigue.

3.4.1 Slamming

Slamming occurs when a ship that is travelling at speed in seas that can cause significant heaving and pitching motion, the bottom structure in the region of the fore-end emerges from the surface and re-enters at a high velocity relative to that of the wave surface. It is thus associated with a sudden change in acceleration of the ship. The terms *impact slamming* and *momentum slamming* have been suggested in order to refer to the two major characteristics of the phenomenon [4]. Impact slamming is concerned with the description of response at the point of impact, given that this is highly localised. Momentum slamming refers to the effect of pressure variations around the solid surface as it penetrates the moving fluid after entry. Slamming is thus seen to have a local and a global effect on a ship's structure. The global effect, which is termed *whipping*, consists of flexural vibrations in the two-node and higher modes that propagate along the length of the hull girder and are eventually damped out. In order to allow for slamming effects in design, classification societies provide detailed guidance, an example of which are the slamming profiles included in Figure 3.10.

In order for slamming to occur, the following conditions have to be met [30]:

- Bow emergence.
- The relative velocity of the ship's bow (or stern) to that of the wave has to exceed a threshold value.[10]

[10] It was found from measurements recorded on board *M/S Wolverine State* that slamming did not occur for relative velocities below 12 ft/s (3.66 m/s).

Figure 3.11 Warship at speed in rough seas (bow emergence) [24].

- The phase difference between the ship's motion and the wave has to exceed a critical value.
- The angle between the wave at the region of impact and the bottom surface is small (deep-vee craft are less likely to experience slamming).

There are several other parameters that have an effect on the probability of occurrence of slamming, the most important being ship's length, sea severity, ship's speed, course of the ship relative to the direction of the predominant wave system, ship's loading condition, hull form and, last but not least, the fullness and flatness of the bottom forward. The angular velocity of the bow, and consequently the relative velocity of the bow to the sea surface, is a function of the ship's length. Forward speed affects the severity of slamming as it changes the period of wave encounter and therefore the motions and bow emergence. For a given ship, there are particular speed ranges that are associated with slamming with waves of a given profile. The most severe condition occurs when the predominant direction of the wave system lies within $\pm 30^\circ$ off the bow. This is because pitch and heave, which have a direct effect on bow motion, are at their maxima within this range of values. Bow emergence is much more frequent when a ship is sailing in a light-ballast condition. For this reason, when sailing in rough seas, ships need to be in a deep-ballast condition. Design changes to the transverse sections, whilst maintaining the design waterline intact, have little effect on pitch amplitude and bow acceleration. What does affect slamming severity is the roundness of the bottom structure forward. Blunter forms are more likely to suffer slamming damage (Figure 3.12).

It is apparent then that there are particular types of ships that are more likely than others to suffer from slamming. These are high-speed vessels that include both commercial and military craft. Navy ships in particular are highly weight-critical and need to maintain speed regardless of sea conditions (Figure 3.11). High-speed commercial vessels such as catamarans may suffer from slamming because of their hull form, in particular because of the existence of the horizontal deck structure joining the two hulls. Ocean-going merchant ships of conventional monohull form that suffer slamming damage are large oil tankers, as these have a high block coefficient and a full form in way of the bow region (Figure 3.12a). Large containerships have a pronounced overhanging stern (Figure 3.12b) that is also a potential source of slamming damage.

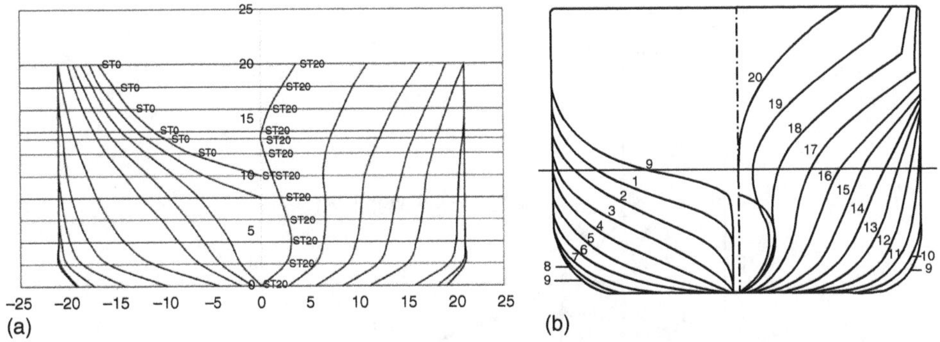

Figure 3.12 Hull forms prone to slamming loads. (a) Bow slamming – VLCC (301 kdwt). (b) Stern slamming – Panamax containership.

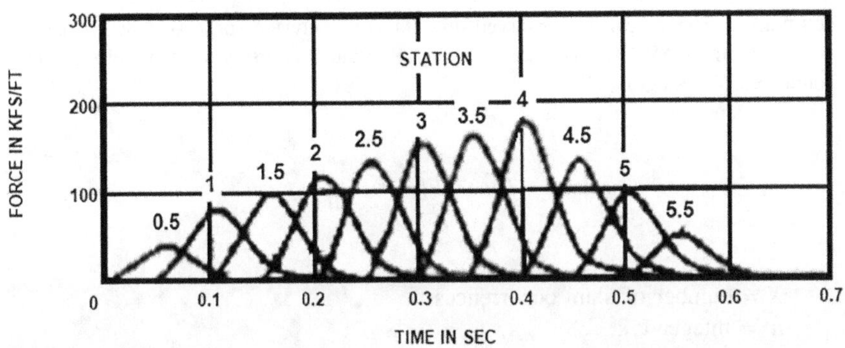

Figure 3.13 Time variation of slamming force along the ship's length [27]. Mariner in light-draught condition sailing in sea state 7: significant wave height = 7.5 m speed = 7 knots. Reprinted courtesy of the Society of Naval Architects and Marine Engineers.

Since the duration of impact is very short (0.1–1 sec) and therefore much shorter than the fundamental period of vibration, *hydroelastic* effects become important. This will be discussed later.

Slamming impacts are highly localised, and it has been found that the position of maximum pressure lies between 12 per cent and 15 per cent abaft the fore perpendicular (Figure 3.13). Its magnitude is directly proportional to the square of the relative velocity at impact. The re-entry location moves, and as a result the pressure pulse also moves, although the peak intensity is maintained. The space-time behaviour of the slamming force is plotted in Figure 3.13.

The time interval between successive slams can be obtained using the Poisson process [27].[11] For the Poisson process, the probability that a slam will occur exactly n times in a specified time interval t_0 is given by

[11] The validity of the Poisson process has been based on model tests and full-scale measurements at sea.

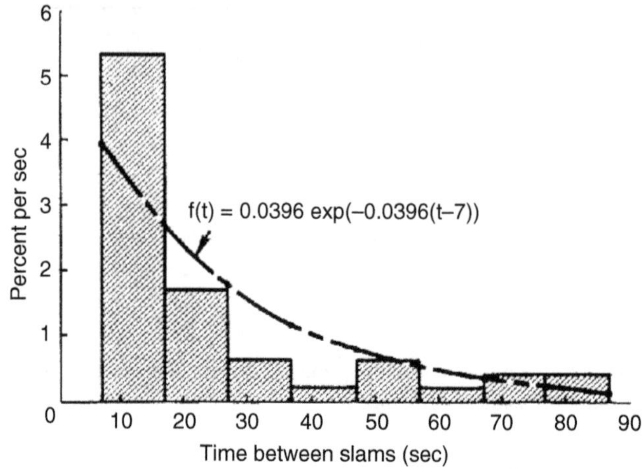

Figure 3.14 Theoretical and observed time intervals between successive slam (measurements on board *SS Wolverine State*) [33]. Reprinted courtesy of the Ship Structure Committee, Washington, DC.

$$P[X = n] = e^{-\nu} \frac{\nu^n}{n!} \tag{3.3}$$

where

X = number of slam occurrences
n = integer $1, 2, \ldots$
ν = average number of slams in a specified time interval $t_0 (= N_s t_0)$.

From probability theory, the time interval between successive slams can be found from the following probability density function:

$$f(t) = N_s e^{-N_s(t - t_*)} \tag{3.4}$$

where N_s is the number of impacts per unit time and t_* is the natural pitching period. Figure 3.14 shows a comparison of the results obtained using (3.4) and the time interval between successive slams observed on the *SS Wolverine State*.

3.4.2 A Phenomenological Description of Slamming, Including Hydroelastic Effects

At the outset it has to be noted that slamming is a phenomenon that is difficult to analyse because of its dependence on a number of factors that are influenced by the environment as well as the particular conditions under which it occurs. Factors such as entrapped air in way of the liquid–solid interface; air bubbles below the sea surface, which are particularly prevalent in storm conditions; the compressibility of water; as well as possible cavitation phenomena all contribute to difficulties in the analysis of slamming and in performing accurate tests on scale models.

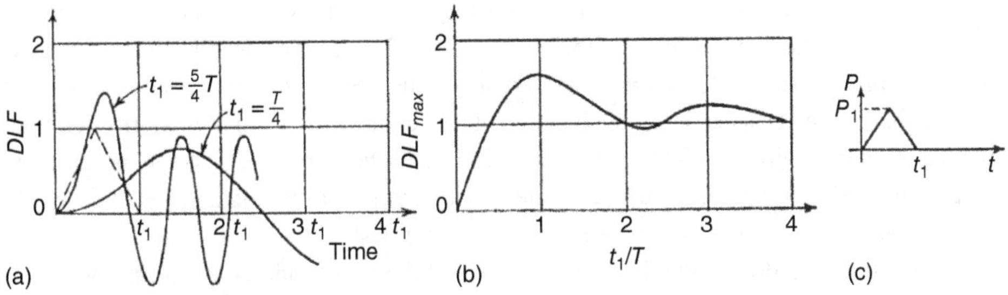

Figure 3.15 Response of SDOF system to triangular pulse excitation [25]. (a) Response for $t_1 = \mathrm{T}/4$, $t_1 = 5\mathrm{T}_1/4$. (b) DLA for $4 \geq t_1/T > 0$. (c) Excitation profile.

Slamming involves the sudden application by a mass of fluid of a force of significant magnitude for a short duration to a solid body. The impact produces a sudden change of acceleration of the fluid that causes a rapid increase in the pressure that is applied to the body. The magnitude of the force is strongly dependent on the angle between the surface of the body and that of the wave (*deadrise angle*) (Figure 3.16b). If this angle is very small, the contact region between the fluid and the solid surface expands rapidly, even if the speed of motion of the body is moderate. Furthermore, the pressure increases significantly, and if the angle is below 5°, the impact phenomenon becomes much more complex. In these circumstances it is possible that just before impact, air is entrapped and compressed between the body and the surface of the liquid. This creates a zone of high pressure that causes a depression of the surface of the liquid. As a result, the liquid first comes in direct contact with the body in way of the outer edges of the air pocket. The compressibility of air then affects the pressure exerted on the body by the entrapped air. The air cushion collapses into smaller bubbles and the air escapes. The liquid then comes into direct contact with the solid surface.[12] What is more important than the effects of air compressibility is that the response of the body itself can affect the magnitude of the pressure. That is, if the impact duration is short enough, *hydroelastic* effects come into play, and the interaction of the fluid with the body can result in significant reductions in the applied pressure loading. To understand this we need to digress and consider what is meant by *hydro-elasticity*.

An early description of the term, put forward by Heller and Abramson, (1959), is that it 'deals with phenomena involving interactions among inertial, hydrodynamic, and elastic forces'. Having in mind the behaviour of an entire ship, Bishop and Price (1979) described the term as being 'that branch of science which is concerned with the motions of deformable bodies through liquids'. Faltinsen (2005) on the other hand, stated that 'hydroelasticity means that the fluid flow and the structural elastic reaction are considered simultaneously and that we have mutual interaction, that is, the elastic vibrations cause a fluid flow with a pressure field and the hydrodynamic loading affects the structural elastic vibrations'.

We shall try to clarify the meaning of the term and thence the significance of these statements by considering in qualitative terms what happens when a fluid in motion comes

[12] It is not always the case that an air cushion forms, as impact may occur in one location with contact spreading rapidly over the surface. It is more likely to occur if the radius of curvature of the wave crest is large and there are no forward speed effects.

into contact with a solid body, and the interaction involves an impact of significant magnitude and short duration. In doing so we shall refer to some basic concepts of hydrodynamics, structural dynamics and linear elastic mechanics. We need to point out that hydroelastic effects can arise at a local level as well as at the level of an immersed body such as an entire ship. Our interest lies with the local phenomenon, as our concern is the effect bottom slamming has on the structure in the vicinity of the impact. The hydroelastic response of the entire ship is a problem of ship dynamics.

In ship structural design the dynamic response of the entire ship is not normally studied out, nor is it done for individual structural members. Instead, in the latter case, use is made of *equivalent static pressures* for each region of the outer hull surface.[13] The equivalent static pressure can be either the maximum value of the average pressure found for an individual panel at any stage during loading or the pressure that would produce the same strains or stresses that result from the various loads that act.

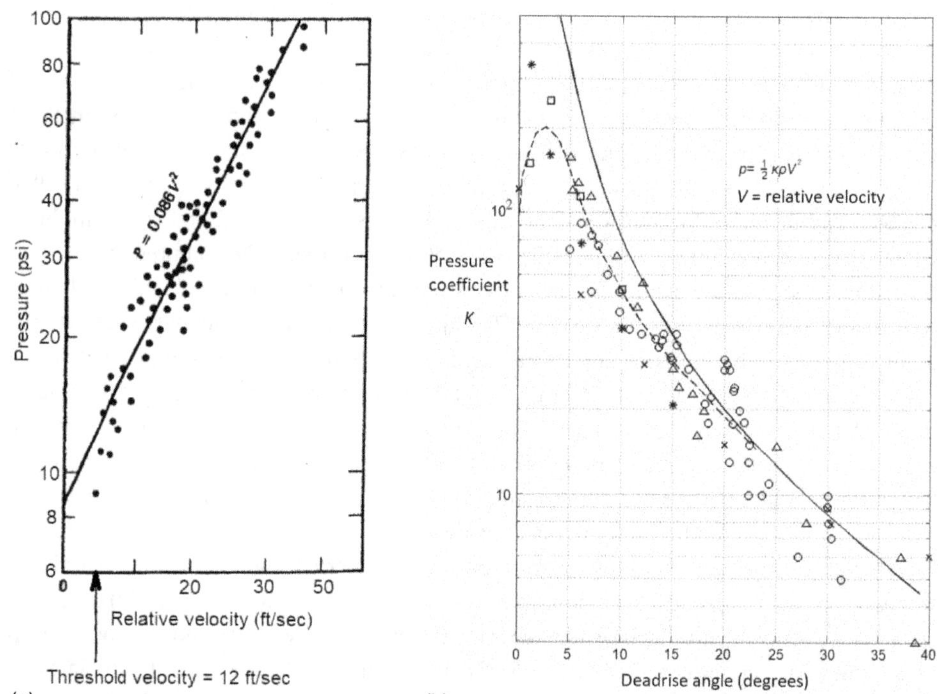

Figure 3.16 Effects of relative velocity and deadrise angle[14] on slamming pressure. (a) Relative velocity [3]. (b) Deadrise angle [15]. Figure 3.16b reprinted courtesy of the Royal Society, London.

[13] See, for example, the IACS Harmonised Common Structural Rules.

[14] If the relative angle between the fluid and the solid surface is greater than 5°, it is accepted that Froude's law of similitude holds, in which case scaling of test results is possible. If the angle is less than 5°, the compressed air affects the slamming pressure, and it is therefore necessary to reduce the ambient pressure in scaled tests.

However, the magnitude of the *effective pressure*, being the pressure experienced by the structure at a local level, depends on fluid motion as well as the mechanical properties of the structure being impacted. This is an important feature of hydroelastic behaviour. Consider two panels that have the same overall dimensions and are subjected to a nominally identical dynamic load. The first panel is light and stiff whereas the second is heavy and flexible, so that $(EI)_1 \gg (EI)_2$, $m_1 \ll m_2$. The natural periods of vibration in air of the two panels are

$$T_1 = \frac{2}{\pi}\sqrt{\frac{m_1 L^4}{(EI)_1}} \qquad \text{and} \qquad T_2 = \frac{2}{\pi}\sqrt{\frac{m_2 L^4}{(EI)_2}}$$

Hence $T_2 \gg T_1$. Figure 3.15 shows the response of a single degree-of-freedom system to a triangular pulse load of maximum value P_1 and duration t_1. The diagram includes the response when the impulse duration $t_1 = T/4$ and $t_1 = 5T_1/4$. It is seen from Figure 3.15a that when $t_1 < T$, the maximum response amplitude is less than the static response, for which the *dynamic load amplification* (DLA) factor equals 1.

Figure 3.15b shows that for very short impact durations $(t_1/T \ll 1)$ the deflection amplitude reduces considerably, so that in the case of a ship's bottom structure, if the duration of a slam is less than the natural period of vibration, the deflection amplitude will be overpredicted by a static analysis. This conclusion is reached on the basis of structural dynamics considerations alone.

Let us now turn to the typical profile of a slam, shown in Figure 3.9a. It is a transient form of loading that is characterised by a near-instantaneous increase in pressure with time. This is followed by a quasi-exponential decrease to zero.

The build-up of the *gifle peak*, or peak pressure, is very short, and during this early phase, the structure responds in a particular way. Prior to impact, it is assumed to be at relative rest. At the instant of impact, the hydrodynamic force imposes a specific relative acceleration profile on it. This is balanced by an inertial force that produces in time a deflection and velocity profile on the body. Within the impacted body, the resulting acceleration profile results in the transmission of *stress waves* at a finite speed. As a result, the increases in deflection and stress are not instantaneous. We can therefore consider this as an initial stage of structural response, and call it the *inertia phase*. The relationship of strain to peak pressure was studied during the inertia phase, and it was found that there is little relationship between the two. That is, maximum pressure is not a suitable determinant of the response of steel and aluminium structures [11]. This was attributed to the fact that the peak pressure is highly localised in spatial terms and its duration is very short.

The expressions quoted previously for the natural period of vibration correspond to dry conditions. When the vibrating body is immersed in water, it is necessary to allow for the *added mass*[15] effect of water. This brings about a reduction in the natural frequency of vibration of the impacted structure.

[15] Section 8.3.2.

The situation is not, however, fully described by the preceding comments. This is because when slamming occurs, the loading is hydrodynamic, and therefore its magnitude depends at any instant in time on the relative velocity of the fluid to that of the body. Therefore, if the impacted body possesses a finite velocity prior to or as a result of impact, the magnitude of this velocity will have a direct influence on the magnitude and distribution of the hydrodynamic load.

Clearly then, in the case of two identical bodies subjected to the same fluid flow and moving at different relative velocities to the flow, the hydrodynamic loads acting on them will not be the same. In the case of the body whose relative velocity to that of the fluid is lower, the effective pressure exerted on it will be lower. As seen from Figure 3.16a, which is a log-normal chart, the effect of relative velocity on maximum pressure is very marked.

The time it takes for the maximum stresses to be reached depends on the natural period of vibration of the structure in water (*highest wet natural period*). From this stage onward the plating is in contact with the liquid throughout. The plate then vibrates in the elastic mode similarly to a wet beam that has an initial space-averaged vibration velocity and zero initial deflection. This phase is termed the *free vibration phase*. The equation of motion of a beam is [11]:

$$M_B \frac{\partial^2 w}{\partial t^2} + EI \frac{\partial^4 w}{\partial x^4} = p(x, w, t) \tag{3.5}$$

where M_B is the structural mass per unit length and breadth of the beam, $w = w(x,t)$ is the deflection in the direction of the z-axis and $p(x, w, t)$ is the pressure that is a function of beam deflections, in addition to x and t. The solution of this equation leads to the following expression for the elastic bending stress in a fibre located a distance z_a from the neutral axis [11]:

$$\sigma_b = E z_a \frac{4V}{\pi \omega_w} \left(\frac{\pi}{L}\right)^2 \cos\left(\frac{\pi x}{L}\right) \sin \omega_w t \tag{3.6}$$

where z_a is the distance from the neutral axis to the stress point. This expression shows that when hydroelastic effects are important, the bending stresses are proportional to the impact velocity V. They reach a peak value at the mid-length position of the beam for a given distance z_a when the beam is simply supported at its ends. In Equation (3.6) V is the *effective local drop velocity* V_e given by:

$$V_e = V + U\alpha + u_* \tag{3.7}$$

where

V = vertical rigid drop velocity of the beam
U = forward speed of beam
α = rigid body rotation about y-axis
u_* = vertical velocity of incident waves ($u_* \ll V$)

The maximum bending stresses occur after approximately one-quarter of the highest wet natural period. During the second half of the natural period the pressure becomes negative, causing cavitation and ventilation.

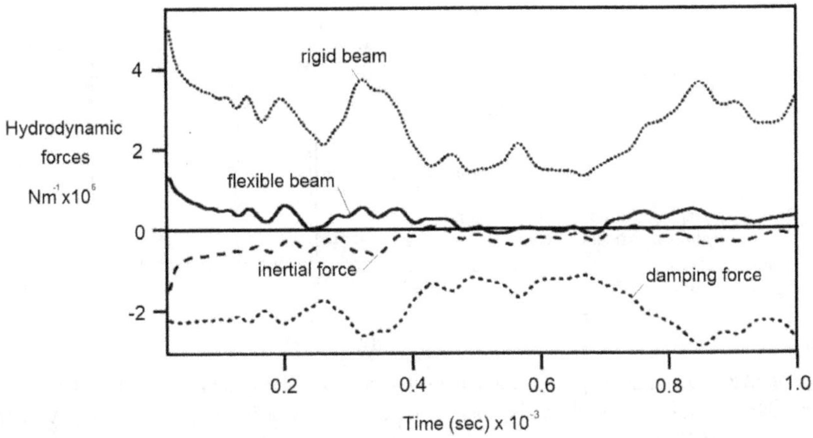

Figure 3.17 Effect of hydroelastic behaviour of a beam subjected to hydrodynamic loading [17].

The effect of hydroelasticity on hydrodynamic loading is illustrated in Figure 3.17 that shows the hydrodynamic force acting on a beam supported at its ends assuming (a) that it is totally rigid and (b) that it has a finite flexibility. It is seen that if the stiffness of a beam is allowed for, the beam is subjected to a much lower hydrodynamic load than if it is assumed to be rigid.

A means of assessing the extent to which hydroelastic effects are important is to calculate the *slamming parameter,* ξ, introduced by Faltinsen [11]:

$$\xi = \frac{\tan\beta}{V}\sqrt{\frac{EI}{\rho L^3}} \tag{3.8}$$

In this expression V is the relative velocity and β is the deadrise angle. Since ξ is inversely proportional to V, it is proportional to the *wetting time* of the beam, which is a measure of the *slamming impulse duration* [20].[16] Furthermore, ξ is inversely proportional to the dry natural period of vibration of the stiffener. We therefore expect a structure to behave hydroelastically for small values of ξ; as ξ increases, hydroelastic effects diminish.

In Figure 3.18 the non-dimensional maximum strain at a critical location of a stiffened panel subjected to slam impact has been plotted against ξ, for different values of non-dimensional constant impact velocity V_{ND}, given by Equation (3.9). The full line corresponds to results obtained allowing for hydroelastic effects, whereas the dotted line represents results from quasi-static orthotropic plate theory.

$$V_{ND} = V\sqrt{\frac{\rho L^3}{EI}} \tag{3.9}$$

[16] The wetting time is, according to Wagner's early theory, equal to tan$\beta(B/V)$.

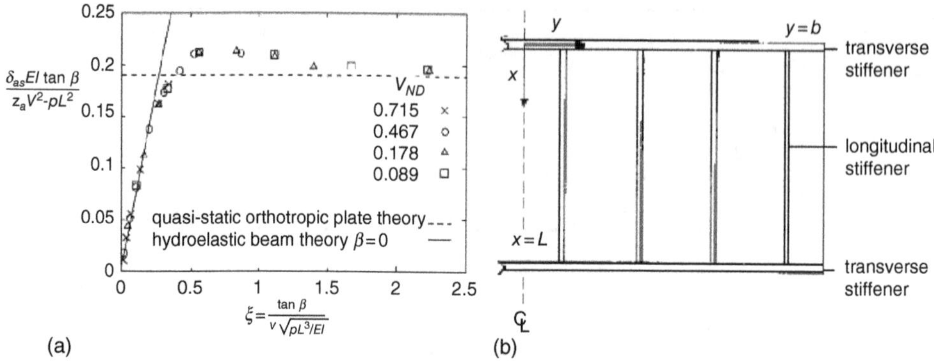

Figure 3.18 Dynamic response of a panel subjected to hydrodynamic impact [11]. (a) Non-dimensional strain versus slamming parameter ξ. (b) Test panel. Reprinted courtesy of the Royal Society, London.

Figure 3.18 shows that for $\xi < 0.40$, the response is hydroelastic, a transition area arises for $0.40 < \xi < 1.50$, and for $\xi > 1.50$ the response can be predicted using quasi-static orthotropic plate theory, in which the strains and stresses are proportional to the square of the relative velocity rather than being directly proportional to it.

3.4.3 Bow Flare Impact

If significant pitching motion takes place, the bow flare may become rapidly immersed and as a consequence will deflect large amounts of water. The resulting fluid pressure will act over a comparatively large area, and this excitation may also produce a whipping response of considerable magnitude. These forces are significant and possibly of the same order of magnitude as slamming forces if there is pronounced flare, such as in the case of an aircraft carrier or a large containership (Figure 3.19). The phenomenon is sketched in Figure 3.20, which shows an incoming wave striking the bow flare. The most important differences with bottom slamming are that bow flare impact is spread over a larger area, has a longer duration and has a lower lateral pressure load. To fully describe the impact force on the bow flare and the resulting structural response, a number of variables (entrapped air, hydroelastic interaction and nonlinear surface mechanics) must be taken into account. This type of response was identified during the 1950s when measurements were made on board ships of the US and Dutch navies in order to understand what caused buckling of the hangar deck and the double bottom structure amidships of several aircraft carriers (Chapter 8). The location of damage was at a significant distance from the point of impact, which was the bow region. A plot of whipping stresses obtained on board the *USS Essex* is reproduced in Figure 8.33.

These measurements indicate the presence of a *whipping stress* whose frequency of 0.83 Hz is approximately equal to the natural frequency of vertical vibration of the hull girder (0.87 Hz) (Figure 8.33). Prior to these measurements it was thought that whipping stresses are related exclusively to bottom slamming. The records obtained demonstrated that bow emergence is not required for whipping. The magnitude of the whipping stress was found

(a) (b)

Figure 3.19 Bow flare in a large cargo ship. (a) Bow flare in large containerships. (b) Bow flare angle.

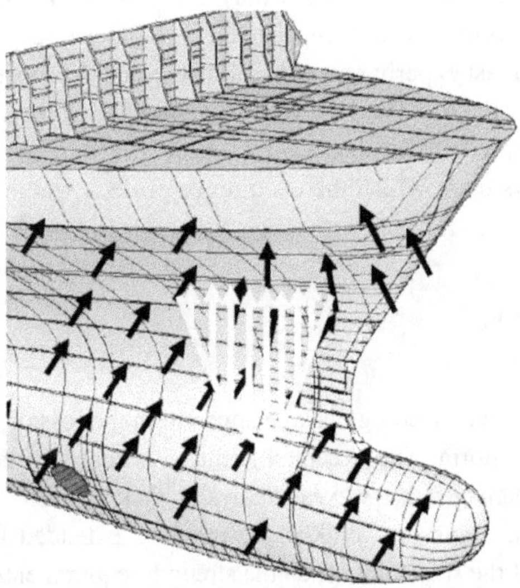

Figure 3.20 Bow flare impact [8]. Reprinted courtesy of Det Norske Veritas, Høvik.

to be significantly higher than the wave-induced stress, so that the validity of quasi-static wave bending moment theory (Chapter 4) was brought into question. It was concluded that since the induced vibration is in the fundamental mode of the hull girder, this can be treated as a single-degree-of-freedom system. Therefore, if the duration of impact is greater than half the natural period of vibration, the maximum compressive stress in the deck will be equal to the sum of the wave-induced stress and the total peak-to-peak variation in whipping stress. The total effect may be found by adding the two maximum values, since the phase angle between them is negligible. The effect of bow flare immersion is thus a nonlinear addition to the sagging wave-bending moment.

The estimation of bow flare slamming loads has been the subject of investigation for many years, starting with the application of the theories of von Kármán and later of Wagner. Numerical approaches involving linear and nonlinear strip theories in which the impact phenomenon is modelled with ever-increasing accuracy are used. This is a nonlinear phenomenon that requires the hydrodynamic load to be predicted with precision as well as the structural response. The latter is also nonlinear because of the effect of high strain rates and deep plasticity. Figure 3.21 summarises the results obtained for a ship model using several different theoretical approaches for hydrodynamic loading.

3.4.4 Deck Wetting

This can arise either as a result of waves breaking over the topsides and striking the main strength deck, mostly in way of the bow region (Figure 3.22a), or when a ship digs into waves when travelling too fast in heavy seas (Figure 3.22b). The former is more common in the case of cargo ships whereas the latter may occur in the case of warships that need to maintain speed. Deck wetting also occurs in fixed offshore structures (FPSOs and jacket platforms). Ship model test experiments [16] indicated that the most serious of these is that of waves breaking over the topsides, known as *swelled waves*. In the worst-case scenarios impact pressures were predicted to reach 50 ton/m^2 [7].

The main parameters upon which the occurrence of deck wetting depends are:

- Freeboard
- Forward speed of vessel
- Significant wave height
- Characteristic wave period

The consequences of deck wetting became apparent in the loss of the *MV Derbyshire* that foundered in a tropical storm. The accepted scenario is that the forward hatch cover was struck and seriously damaged by waves. Seawater then entered cargo hold No. 1, and following failure of the transverse bulkheads, flooding extended to more holds, leading eventually to the loss of the ship. Since then the strength requirements for the forward region of the hull girder of bulk carriers, which have large openings with hatch covers, have been revised and the scantlings increased. The additional loads imposed on the hull girder during deck wetting have consequences only at a local level and do not affect longitudinal strength. These, however, may be of particular importance in the case of offshore vessels such as FPSOs that are permanently stationed and swivel about a point of attachment. They are also important in the design of deck equipment such as mooring and cargo-handling gear.

The probability of occurrence is not easily determined, as bow submergence does not necessarily lead to deck wetting, since large bow flare may prevent this from occurring. As a result, it is necessary to use three-dimensional representations of the hull form when performing theoretical calculations.

Currently there is a significant amount of research being carried out to determine hydrodynamic loading associated with deck wetting, as reported in the ISSC

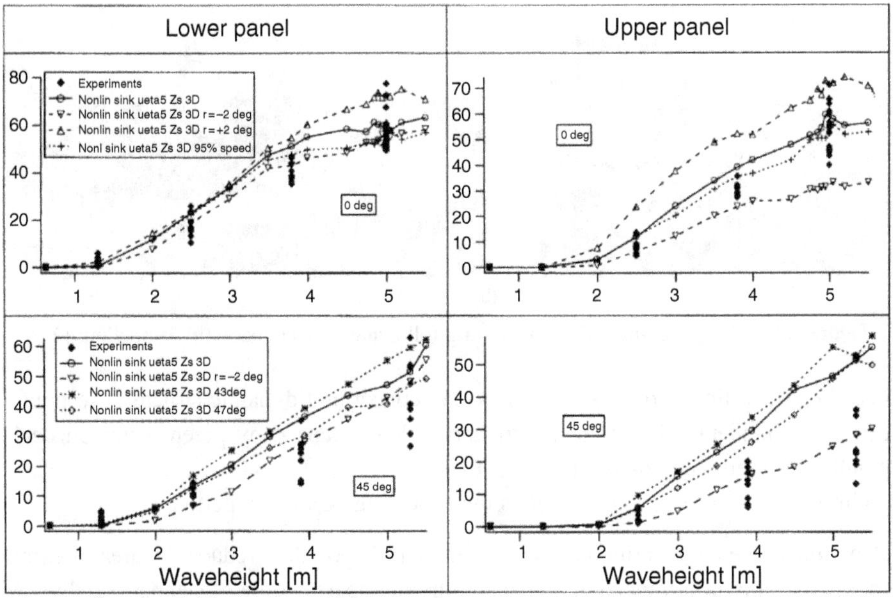

(a)

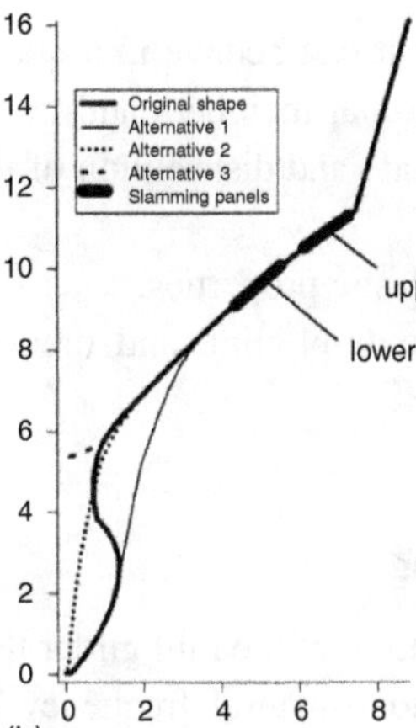

(b)

Description of method	*Linear*	*Nonlinear*				
			Sinkage	*Sinkage $U\eta5$*	*Sinkage $U\eta5$, Zs*	*Sinkage $U\eta5$, Zs, 3D*
Linear ship motions with forward speed	X					
Nonlinear ship motions with forward speed		X	X	X	X	X
Sinkage included in calculating ship motions			X	X	X	X
$U\eta5$ term included in relative velocity				X	X	X
Steady-state elevation included in relative velocity					X	X
Slamming pressures reduced by 30% to allow for 3D effects						X

(c)

Figure 3.21 Comparison of results for hydrodynamic loading due to bow flare slamming using linear and nonlinear theories [13]. (a) Model test results and theoretical predictions for head seas (0°) and quartering seas (45°). (b) Location of panels. (c) Methods used in load prediction. Reprinted courtesy of the Journal of Marine Science and Technology.

(a) (b)

Figure 3.22 Deck wetting. (a) Deck wetting following wave impact. (b) Tunnelling [23].

Proceedings, but little work has been reported concerning dynamic structural response to green-water incidents. The response to impact loads caused by green water on deck is normally analysed using quasi-static methods.

With regard to design considerations the following points are pertinent:

1. Commonsense engineering needs to be applied. If possible, reduce the area of exposed surface to forefoot immersion and slamming (compare profile of slender warships with that of VLCCs).
2. Structural design of forepeak. Design the forepeak such that it can behave in a quasi-hydroelastic manner. In this the relative stiffness of longitudinal members and transverse frames has to be considered with respect to the magnitude and distribution of the impact loading.
3. Consider the development of materials with required hydroelastic properties.
4. Provide practical guidance to masters concerning speed, route-plotting and draught requirements for each particular vessel for each sea state.

3.5 Hydroelastic Resonance (Springing)

Hydroelastic resonance (*springing*) refers to wave-induced vibration of the hull girder that occurs if the wave encounter frequency is close to the two-node natural frequency of vertical vibration of the hull girder (Figure 3.23). It is not the result of transient wave loading, since the excitation forces are active along the entire length of the hull girder. In conventional hull forms it is the vertical two-node vibration mode that is excited by waves. The resulting resonant vibration continues until it is eventually damped out by the structure. Structural damping is generally low, and for this reason loads rapidly increase at resonance; springing can therefore be classed as a steady-state phenomenon.

Ships that have a low fundamental natural frequency of vertical vibration, and are thus more flexible, are large oil tankers and long bulk carriers (Great Lakes bulkers). In large oil tankers, an allowance of 10 per cent has to be made when estimating longitudinal bending strength. Other vessel types that may suffer springing are catamarans, whose lowest fundamental frequency is in the torsional mode, as is that of cargo ships with large deck openings, such as containerships.

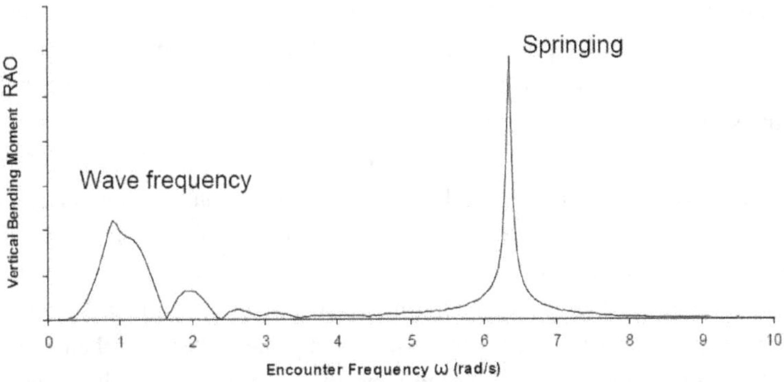

Figure 3.23 Response amplitude operator for vertical bending moment.

Furthermore, ships that may suffer are those with high normal operating speeds, and hence wave encounter frequencies, such as warships, as well as containerships. In one exceptional case springing was observed in a large conventional bulk carrier. This is not generally observed in this type of ship and was attributed to the presence of second-order (sum-frequency) terms involving the cross-coupling of the combined wind and swell-driven wave system.

Springing is a phenomenon that generally occurs in quartering seas when sailing in moderate sea states (waves with short period have a limited height). Springing vibration may arise in the vertical bending, horizontal bending and torsional modes.

Springing is generally considered to be excited by both *linear* and *nonlinear excitation mechanisms*. Linear exciting forces are associated with waves of small wavelength relative to the ship's length, while nonlinear or *sum-frequency* springing arises when the periodic excitation forces act with a higher order of wave encounter frequency [31]. The former can be predicted by linear hydrodynamic theories, while the latter depend on second-order hydrodynamic effects such as the superposition of different wave systems. In spite of the periodic sum-frequency excitation forces having a smaller intensity per unit wave height than the linear forces, they may become of greater importance because they originate from considerably longer waves with higher energy content.

The factors that affect springing are:

- *Hull girder stiffness*. From elementary vibration theory, the natural frequency of vibration is proportional to the square root of the appropriate stiffness. Thus, a lower stiffness results in a reduction in natural frequency rendering it more likely to be close to that of the wave encounter frequency, thereby resulting in more acute springing. In energy terms, resonance occurs for longer waves. These contain more energy and also give higher excitation per unit amplitude (higher RAO) due to less cancellation (ship–wave matching).
- *Bow (and stern) geometry*. Blunt bows give more springing than do slender bows. Bow stem slamming and wave reflection can contribute to springing excitation forces.
- *Structural damping*. Since springing occurs at comparatively high encounter frequencies (Figure 3.8), damping caused by wave generation is low. For high-speed ships, however,

hydrodynamic damping due to lift effects from the transom stern becomes important. For other ships, structural damping is important. Structural damping is very difficult to predict by theoretical methods, making it necessary to rely on measured data from the ship in question or from similar ships.

- *Loading condition.* More cargo means an increase in mass, which results in a lower natural frequency of vibration of the hull girder. A lower natural frequency of vibration will generally produce more springing. However, more cargo also results in a deeper draught. Since most hydrodynamic pressures decay rapidly with draught, this means that the pressure variation (that is, the springing excitation forces) under the forward part of the ship bottom becomes lower. This effect is normally stronger than the effect of a reduced natural frequency, and springing is normally aggravated in the ballast condition. A wide ship may also have a low draught even in loaded condition.
- *Ship's speed.* A higher speed will give resonance for longer waves that have a higher energy content, which in turn will give more springing. Some studies indicate that the springing vibration amplitude increases with the square of the forward speed.
- *Ship's heading angle.* In principle, springing is expected to be most acute in head seas and at its minimum in following seas, since these conditions maximise and minimise the encounter frequency, respectively.
- *Wave condition.* Since relatively high encounter frequencies are needed to give springing, the phenomenon is most pronounced in moderate sea states, where most of the wave energy is contained in relatively short waves. *Linear* springing forces normally come from very short waves, and they are therefore very sensitive to the high-frequency tail of the wave spectrum.
- *Nonlinear* springing forces get a contribution from the interaction between waves with different frequencies and different directions. This gives rise to forces that oscillate with the sum of the frequencies of the different waves. The total springing forces are therefore dependent not only on the frequency distribution of the wave energy but also on the directional distribution.

If the springing response at the two-node vertical frequency is ω_s, this will be excited by waves with encounter frequency ω_s, as predicted by linear theory, but also by waves with encounter frequencies ω_i, ω_j whose sum equals ω_s (wave system S_1). If another wave system, S_2, whose direction differs from that of S_1 is present, the same would apply to two components ω_m, ω_p of S_2. The excitation thus assumes a double-peaked form (Figure 3.24). Second-order interaction (cross-coupling) arises between these two wave systems. If this is accounted for, bidirectional sea states are modelled much more accurately than if linear springing theory is applied.

3.6 Nonlinear Analysis for Springing and Whipping Loads

Linear strip theory generally gives good predictions of vertical ship motions and wave-induced loads in low and moderate regular waves. For the past few decades, this theory has been combined with the work of St. Denis and Pierson [29] as a direct calculation means to

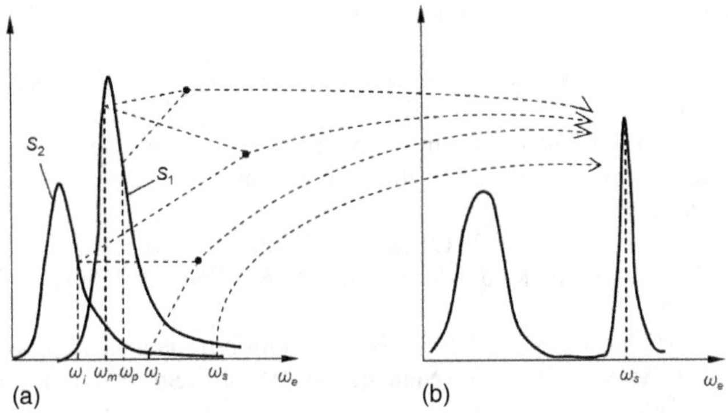

Figure 3.24 Wave excitation due to two wave systems leading to springing [31]. (a) Wave spectral density.[17] (b) Response spectrum.

estimate the probability of exceeding different levels of ship motions and wave-induced loads caused by irregular waves in a given period of time. However, the nonlinear behaviour of those responses is completely lost in this process.

Nonlinearity is obviously existent, particularly in the vertical bending moments and shear forces for ships running in severe or even moderate sea states, which are of most interest from the viewpoint of structural design. This is mainly due to the fact that the linearity assumption of small ship motions relative to the wave surface is violated at the fore quarter of most ships. This violation, together with the usually V-shaped cross-section of the ship hull in that region, introduces significant nonlinear effects in the vertical hydrodynamic forces, though heave and pitch may be predicted accurately by linear theory (Chapter 4). An increase in the bending moments and shear forces in both magnitude and frequency is one of the consequences. As discussed previously, it can cause severe structural damage and reduce fatigue life. An accurate prediction of wave-induced non-linear loads is therefore vital to the integrity and durability of ship structures.

Both springing and whipping are wave-induced loads that are particularly prone to nonlinear effects for both loading and response. In recent research, the finite element method, hydroelasticity theory and nonlinear hydrodynamics theory have been applied to both springing and whipping as well as to a wide range of vibration-related problems that affect hull girder response. A detailed appraisal of the methods used is included in the 2015 ISSC Committee II.2 Report (Dynamic Response).

References

1. Allmendinger E. E. (1990) *Submersible Vehicle Systems Design*. Jersey City, NJ: SNAME.
2. American Bureau of Shipping (2020) *Guide for slamming loads and strength assessment for vessels*. Houston, TX: ABS.

[17] Wave spectra are discussed in Chapter 9.

3. Bhattacharyya R. (1978) *Dynamics of Marine Vehicles*. New York: John Wiley & Sons.
4. Bishop R. E. D. Price W. G. (1979) *Hydroelasticity of Ships*. Cambridge: Cambridge University Press.
5. "Bulk Carrier Being Loaded with Iron Ore at Porto Sudeste, Rio de Janeiro, Brazil." n. d. *Flickr.Com*. Available: www.flickr.com/photos/trafigura_images/26583650343. Accessed 5 January 2021.
6. Burnside O. H. Hudak Jr. S. J. Oelkers E. Chan K. Dexter R. J. (1984) Long-term Corrosion Fatigue of Welded Marine Steels. Ship Struct. Comm. Rept. SSC-326 Washington, DC.
7. Daidola J. C. Mishkevitch V. (1995) Hydrodynamic Impact Loading of Displacement Ship Hulls: An Assessment of the State of the Art. Ship Struct. Comm. Rept. SSC-385 Washington, DC.
8. Det Norske Veritas (2005) *Hull Structure Course*.
9. "Energy Endurance – IMO 6726278." n.d. *Shipspotting.com*. Available: www.shipspotting.com/gallery/photo.php?lid=2384001. Accessed 5 January 2021.
10. Faltinsen O. F. (2005) *Hydrodynamics of High-Speed Marine Vehicles*. Cambridge: Cambridge University Press.
11. Faltinsen O. F. (1997) The effect of hydroelasticity on ship slamming. *Phil Trans. R. Soc. Lond. A* Vol. 355 575–591.
12. "GA Homepage." *Grida.no*. Accessed 19 December 2020. Available: http://maps.grida.no/go/graphic/arctic-sea-routes-northern-sea-route-and-northwest-passage. Accessed 5 January 2021.
13. Hermundstad O. E. Moan T. (2005) Numerical and experimental analysis of bow flare slamming on a Ro–Ro vessel in regular oblique waves. *J Mar Sci Technol* Vol. 10 105–122.
14. Hughes O. (1983) *Ship Structural Design: A Rationally-Based Computer Aided Optimization Approach*. New York: John Wiley & Sons.
15. Kapsenberg G. K. (2011) Slamming of ships: Where are we now? *Phil Trans. R. Soc. Lond. A*. Vol. 369 2892–2919.
16. Kawakami M. (1971) On the impact strength of ships due to shipping green seas-towing experiments of a ship model in regular waves. *Selected Papers from the J. of the Society of Naval Architects of Japan*. Vol. 8 1–16.
17. Kvålsvold J. Faltinsen O. M. (1993) Hydroelastic modelling of slamming against the wet deck of a catamaran. *Proc 2nd Int. Conf. on Fast Sea Transportation (FAST)* Yokohama 681–697.
18. Lewis E. V. Hoffman D. Maclean W. M. van Hooff R. Zubaly R. B. (1973) Load Criteria for Ship Structural Design. Ship Struct. Comm. Rept. SSC-240 Washington, DC.
19. Little R. S. Lewis E. V. (1981) A statistical study of wave-induced bending moments on large oceangoing tankers and bulk carriers. *Trans SNAME* Vol. 79 117–168.
20. Manganelli P. F. Hobbs M. A. (2006) An alternative approach to the design of structures exposed to slamming loads. *2nd Int. Symposium on design and production of motor and sail yachts*. Madrid 11 p.
21. Mansour A. Liu D. (2008) *Strength of Ships and Ocean Structures*. Paulling J. R. (Ed), *Principles of Naval Architecture Series*. Jersey City, NJ: SNAME.
22. Masterson D. M. Frederking R. M. W. (1993) Local contact pressures in ship/ice and structure/ice interactions. *Cold Regions Science and Technology* Vol. 21 169–185.
23. Miller P. H. Stettler J. W. (2009) EN358 Ship Structures. Notes for an Undergraduate Course. Naval Architecture Program, US Naval Academy, Annapolis, MD.

24. "Navi russe nella tempesta." 12 February 2017. Available: www.youtube.com/watch?v=h147Afg_5AU. Acessed 19 December 2020.
25. Norris C. H. Hansen R. J. Holley M. J. Biggs J. M. Namyet S. Minami J. K. (1959) *Structural Design for Dynamic Loads*. New York: McGraw-Hill.
26. Ochi M. K. Motter L. E. (1973) Prediction of Slamming Characteristics and Hull Responses for Ship Design. *SNAME Annual Meeting* 144–176.
27. Paulling J. R. Strength of Ships. In Lewis E. V. (Ed), (1988) *Principles of Naval Architecture Second Revision. Vol. 1 Stability and Strength*. Jersey City, NJ: SNAME 205–299.
28. *Physicalgeography.net*. Available: www.physicalgeography.net/fundamentals/7l.html. Accessed 19 December 2020].
29. St. Denis M. Pierson W. J. (1955) On the Motion of Ships in Confused Seas. *SNAME Annual Meeting* 281–357.
30. Szebehely V. G. Todd M. A. (1950) Ship Slamming in Head Seas. David Taylor Model Basin Rept. No 913.
31. Vidić-Perunović J. Juncher Jensen J. (2005) Non-linear springing excitation due to a bidirectional wave field. *Mar. Struct.* Vol. 18, 332–358.
32. Wah T. (Ed) (1960) *A Guide for the Analysis of Ship Structures*. Washington, DC: US Dept of Commerce.
33. Wheeler T.W. (1970) Analysis of the Slamming Data of the SS Wolverine State. Ship Struct. Comm. Rept. SSC-210 Washington, DC.

4

Primary Loading of Ship Structures

Smooth seas do not make skilful sailors.

African proverb

4.1 Introduction: A Conceptual Model for Primary Hull Girder Bending

Ship structures are large, complex and subjected to external loads that differ in fundamental ways from those that act on land-based and airborne structures. As a result, it is not easy to develop an intuitive understanding of their response. It becomes necessary to develop a systematic approach that will assist in understanding the way in which the hull and its various parts respond to the environment within which it operates, and the effect that the characteristics of the vessel itself may have upon its behaviour.

Before we embark on this path, let us first take another brief look at the structures of certain common ship types. This will provide us with an idea of the types of structures that we will encounter and will also provide us with a lead to more detailed considerations. Figure 1.4 gives an idea of the form and geometry of the structure of the cargo-carrying regions of merchant ships. From Table 1.1 we can get an idea of ships' proportions by considering in particular the L/B and L/D ratios.

At a basic level and crudely put, the hulls of modern cargo ships consist of the cargo-carrying region, which is made up of a number of tanks or holds aligned in the direction of motion of the vessel, to whose extremities are attached hydrodynamically shaped ends. These enclose spaces within which vital functions for the operation of the vessel are performed. Clearly, from the structural viewpoint, the most important region is the cargo-carrying region, and this has a generally rectangular (wall-sided) cross-section that may be subdivided in the longitudinal and/or the transverse direction by bulkheads. All major members of the cross-section, such as decks, sides, bottom and bulkheads, are stiffened using some kind of orthogonal system.

The first impression gained is that of a large and complicated arrangement the exact analysis of which is not easy to accomplish at a global level. Ways of simplifying the task in question have to be obtained. In doing so, approaches that will ensure sufficient accuracy and that can be followed through within acceptable time limits have to be used.

Turning to Table 1.1, the overall dimensions, proportions ($7 > L/B > 5$, $L/D \approx 14$) and the form of the cross-section of modern ships lead directly to the *beam analogy*, which,

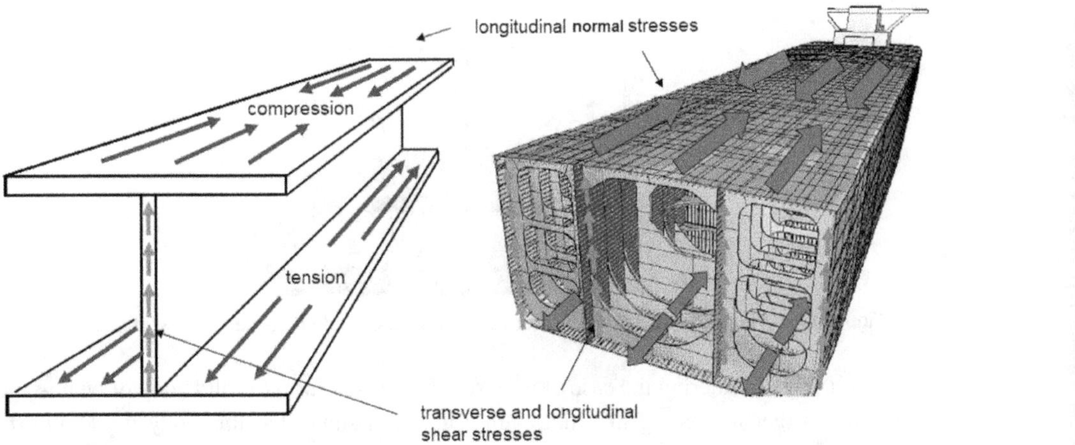

Figure 4.1 The beam analogy for longitudinal bending of the hull girder [8]. Reprinted courtesy of Det Norske Veritas, Høvik.

following Euler's pioneering work during the 18th century [9], was first used at a practical level at around the beginning of the 20th century to represent ship structures at a global level [5]. The model makes use of a flanged beam whose web represents the vertically aligned longitudinally effective material (side shell and longitudinal bulkheads) (Figure 4.1). The flanges correspond to the horizontally aligned material (decks and bottom shell structure). The beam analogy permits many different types of models to be used, depending on the form of the actual structure and the accuracy required of the model. This is a conceptual model found to give satisfactory results when modelling the most important loads that act on conventional mono-hull ships. Details concerning the experimental validation of the beam analogy are included in Section 6.2.5. The important thing to realise from this process is that sufficient insight has been gained of the nature of *global structural response*. It is thus possible to develop a structural model that can be used to quantify these effects.

4.2 Floating Bodies in Still Water

If we now turn our attention to the environment within which ships operate, we note that, in contrast to land-based structures that are normally attached to motionless supports, the sea is in constant motion unless of course a ship is in port. Furthermore, ships translate within this environment. Clearly, we have to resort to a simplification of these conditions in order to be able to understand the effects that each of the external loads may have and the corresponding response of the structure. The simplest situation we can consider is that of a sea devoid of waves, termed a *still-water condition.*[1] We will subsequently consider situations that are closer to the condition a ship encounters at sea.

[1] In classification society rules the term *harbour condition* is used to indicate the absence of waves.

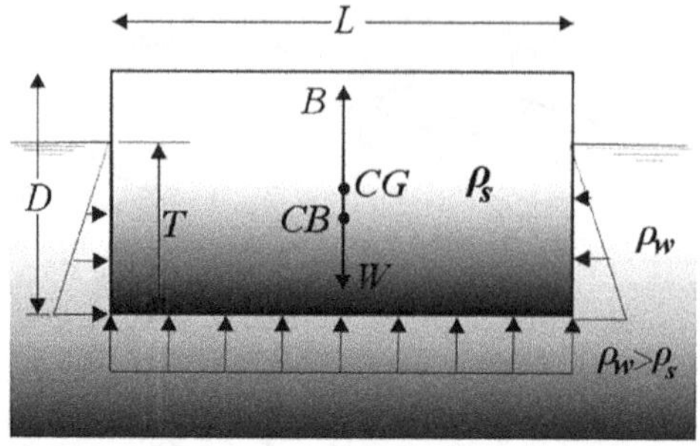

Figure 4.2 Floating body in still water (uniform density and constant cross-section).

We start by considering the case of a vessel of idealised form (a solid body of beam-type proportions) that is floating in a flat calm sea. We assume that this body is of uniform density ρ_s, so that the weight per unit length in the longitudinal direction is constant and also that $\rho_s < \rho_w$, thus ensuring that it floats. The situation is depicted in Figure 4.2.

If we consider the body in its entirety – that is, at a global level – we see that the forces that act on it are its weight, which acts at its centre of gravity, and the buoyancy, which acts at the centre of buoyancy. According to Archimedes' principle, at equilibrium, the buoyancy force is equal to the weight of the displaced liquid. This force acts at the centroid of the immersed volume. Since the body has a constant section, the vertical axes of the centres of buoyancy and gravity coincide and the two forces cancel each other out. As a result, no other forces are necessary to achieve equilibrium.[2]

We consider now a ship whose weight and buoyancy distributions are shown in Figure 4.3. In this case both distributions vary along the length. In order for the ship to float at equilibrium, the draught adjusts itself so that the buoyancy force equals the weight (displacement), and the trim adjusts itself so that the centres of gravity and buoyancy lie on the same vertical axis.

In this way vertical and rotational equilibrium conditions are satisfied at a global level. At equilibrium, the areas under the weight and buoyancy distributions are equal.

For any transverse section located a distance x from the aft perpendicular (AP), the net load $p(x)$ acting is given by:

$$p(x) = w(x) - b(x) \tag{4.1}$$

Clearly then the load acting at a randomly chosen transverse section of the hull girder will not be zero. We can draw the *load curve*, $p(x)$ (Figure 4.3). The sign of the net load at any particular section depends on the relative magnitudes of the weight and the buoyancy force. For the ship to be at equilibrium in the vertical direction, the area under the load curve above the horizontal axis has to be equal to that below the axis. In other words, the sum of the forces acting vertically downwards has to be equal to the sum of forces acting vertically upwards.

In the case shown in Figure 4.3, at the fore and aft regions of the ship the weight exceeds the buoyancy force whereas the opposite occurs amidships. In the aft region, there is excess

[2] That is, the reaction forces and moments that are generally required at the ends in order to restore equilibrium are zero.

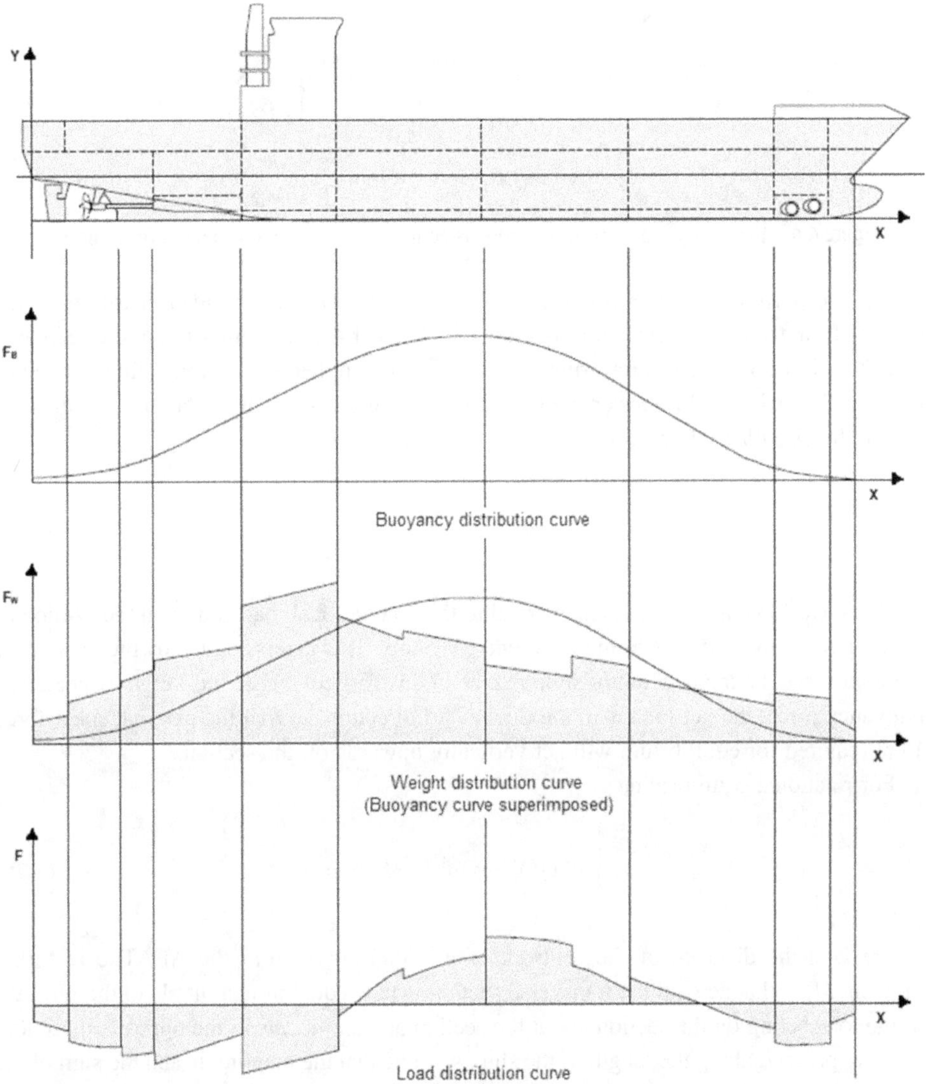

Figure 4.3 Longitudinal weight, buoyancy and load distributions [45].

weight due to the presence of machinery and at the same time the buoyancy force reduces there because of the nature of the hull form. As far as the hull form is concerned, a similar situation arises at the fore end. At the fore and aft extremities the load is zero, since these are not restrained in any way. The nature of the longitudinal vertical bending problem of a ship thus becomes apparent: The ship behaves as a *free-free beam* that is loaded in the vertical direction; it bends in the longitudinal direction in a globally self-equilibrating manner, and is supported by distributed buoyancy forces.

We now consider a transverse section at a distance x from the aft end (Figure 4.4). This section acts as a boundary of the portion of the ship that extends from the aft end up to this

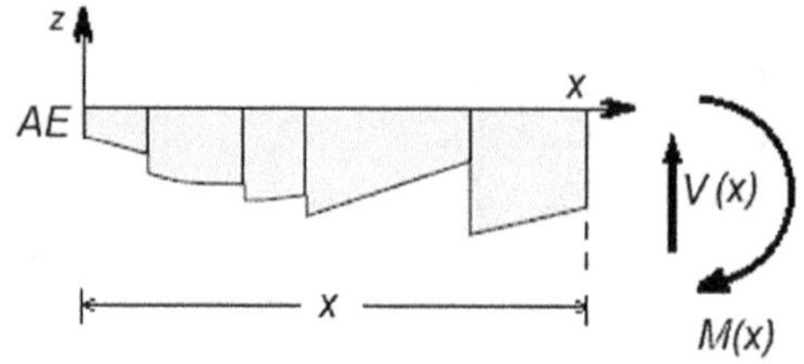

Figure 4.4 Free body diagram showing loads acting on section at a distance x from the AP.

point. The portion of the ship considered is then subjected to the load distribution that extends from the aft end (AE) to x, as shown in Figure 4.4. For equilibrium of the section at x, a shear force $V(x)$ and a restoring couple $M(x)$ are applied by the remaining portion of the ship. These internal forces or *stress resultants* ensure vertical and rotational equilibrium at x in the Oxz plane. For equilibrium in the vertical direction,

$$\int_{AE}^{x} p(x)dx + V(x) = 0 \tag{4.2a}$$

The shear force at x, $V(x)$, is then equal to the total vertical load acting on the portion of the hull girder that extends from the aft end (AE) to x. This is represented by the area under the load curve, taking account of signs. Since in this particular case the weight exceeds the buoyancy force, the net load is in the downward direction, so that the external shear force $V(x)$ required for equilibrium will act vertically upwards on the section.

For rotational equilibrium,

$$\int_{AE}^{x} p(x)(X - x)dx + M(x) = 0 \tag{4.2b}$$

where X is the distance of the centroid of the load curve from the AE. The restoring moment $M(x)$ that acts on the transverse section is thus equal in magnitude to the moment of the load acting on the section about the section at x and it acts in the opposite direction.

If we proceed along the length of the ship, we note that the magnitude and the sign of the stress resultants change in accordance with the load distribution that acts on the portion of the ship up to that point so that shear force and bending moment distributions are required.

4.3 Quasi-Static Wave Loading of the Hull Girder

We now assume that the ship floats in a sea in which there are waves present, as shown in Figure 4.5. The two sketches correspond to the presence of a wave with crests amidships or near the ends. In both of these cases the buoyancy distribution differs from that of the still-water condition considered previously.

In Figure 4.5a, there is an excess of buoyancy amidships and a deficiency of buoyancy in way of the ends of the ship; the situation is reversed in Figure 4.5b. For equilibrium, the mean

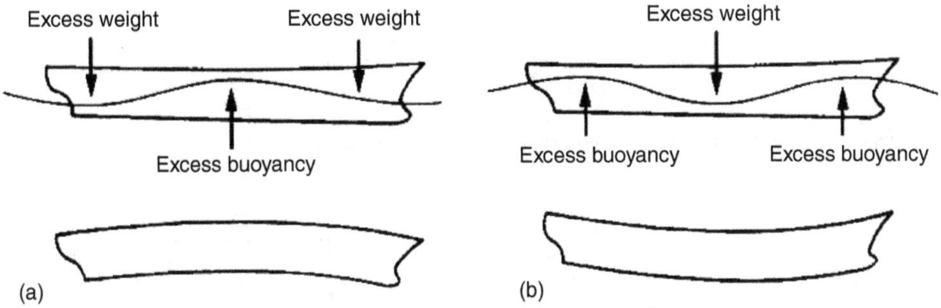

Figure 4.5 Hogging and sagging of the hull girder due to the presence of a wave. (a) Hogging. (b) Sagging.

draught of the vessel has to change for the total buoyancy force to remain equal to the weight. This is because the hull form changes along its length and in particular in way of the fore and aft regions. Furthermore, since the forms of the fore and aft regions are not identical, the ship will also need to trim. In this artificial situation, termed *quasi-static wave loading*, we consider the instant at which the wave has the form shown in Figure 4.5a (or Figure 4.5b). In doing so, we ignore the time history of the sea surface experienced by the ship as well as the motions of the ship itself. The excitation force that acts on a hull girder due to wave motion is dynamic, in the sense that it is time-varying. However, the frequency of the wave excitation force is significantly lower than that of the frequency of oscillation of the hull girder itself. For this reason resonance does not occur, and there is thus no amplification in the response. The dynamic component of the response is therefore negligible, and the response can be treated as static. This is the reason for which the term "quasi-static" is used. Therefore, in quasi-static wave loading analysis, important simplifications are made.

Returning to Figure 4.5, we note in the lower sketches the form the hull girder assumes in each case. When the trough of the wave coincides with the fore and aft ends, the ship is said to be in a *hogging* condition (Figure 4.5a). If, on the other hand, there were to be excess buoyancy at the ends of the ship, the hull girder is said to be in a *sagging* condition (Figure 4.5b).

A variety of forms of quasi-static waves have been used, the most useful being those that best resemble the surface of the sea. It has been found that the critical form of a wave is that whose trough coincides with either the mid-length or the extremities of the ship and whose length is approximately equal to the length of the ship.[3]

Once again we consider a portion of the ship that extends from the aft end to a transverse section at a distance x. For equilibrium, stress resultants act on the transverse section, but their magnitude and direction will differ from those required previously, since the load curve will have changed due to the change in buoyancy distribution.

We will now use beam theory to obtain the shear force and bending moment distributions. From Equation (4.2a) we have that at section x (Figure 4.6):

$$\frac{dV(x)}{dx} = -p(x) \tag{4.3a}$$

[3] This idea was proposed in [39].

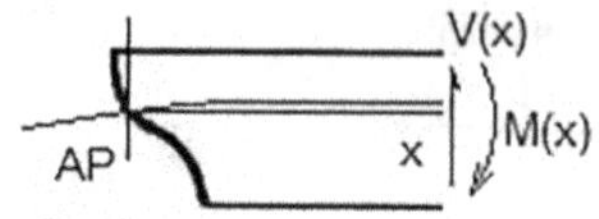

Figure 4.6 Loading on aft portion of ship on hogging wave.

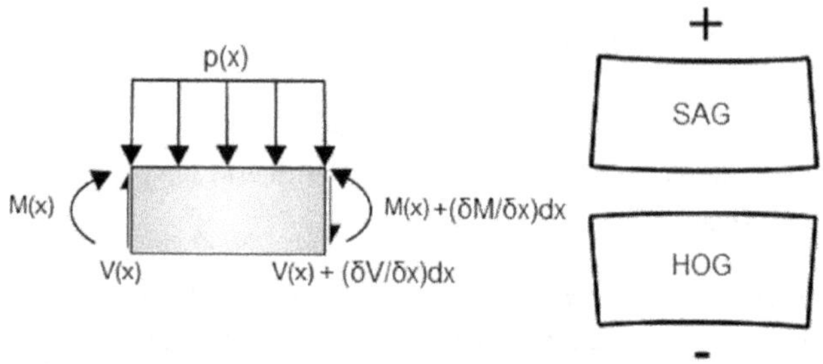

Figure 4.7 Sign Convention.

also, given the sign convention shown in Figure 4.7 we have:

$$\frac{dM(x)}{dx} = V(x) \tag{4.3b}$$

Integrating, at any distance x from the aft perpendicular,

$$V(x) = -\int_{AP}^{x} pdx + C_1 = \int_{AP}^{x} [w - b]dx + C_1 \tag{4.4a}$$

and

$$M(x) = \int_{AP}^{x} Vdx + C_2 \tag{4.4b}$$

Since the ends are free, the shear force and bending moment are zero there and $C_1 = C_2 = 0$. Thus,

$$V(x) = \int_{AP}^{x} [w - b]dx \tag{4.5a}$$

$$M(x) = \int_{AP}^{x} V(x)dx \tag{4.5b}$$

Since the ship is at equilibrium, the weight is equal to the buoyancy force and the *LCG* and *LCB* are equal and lie on the same vertical line. Integrations carried out in Equations (4.4a) and (4.4b) are performed using a suitable numerical integration rule and if a minor discrepancy arises such that there exist residual values V_r and M_r where $V_r \leq 0.03\,V_{max}$ and $M_r \leq 0.06M_{max}$, the shear force and bending moment distributions can be adjusted using a linear approximation along the ship's length, which in effect is equivalent to a minor change in trim. Following the introduction of the trim correction,

$$V_1(x) = V(x) - V_r(x/L) \tag{4.6a}$$

$$M_1(x) = M(x) - M_r(x/L) \tag{4.6b}$$

The following are features of the shear force and bending moment distributions and are the result of Equations (4.5a–b) (Figure 4.8):

1. The net area under the shear force curve is zero.
2. The shear force curve maxima correspond to positions of zero load.
3. The bending moment curve maximum corresponds to zero shear force.

We now need to discuss how the weight and buoyancy distributions are obtained (Figure 4.3). The weight distribution consists of a large number of individual components of varying magnitude and longitudinal extent and is therefore discontinuous. For a particular ship it will change according to the loading condition.

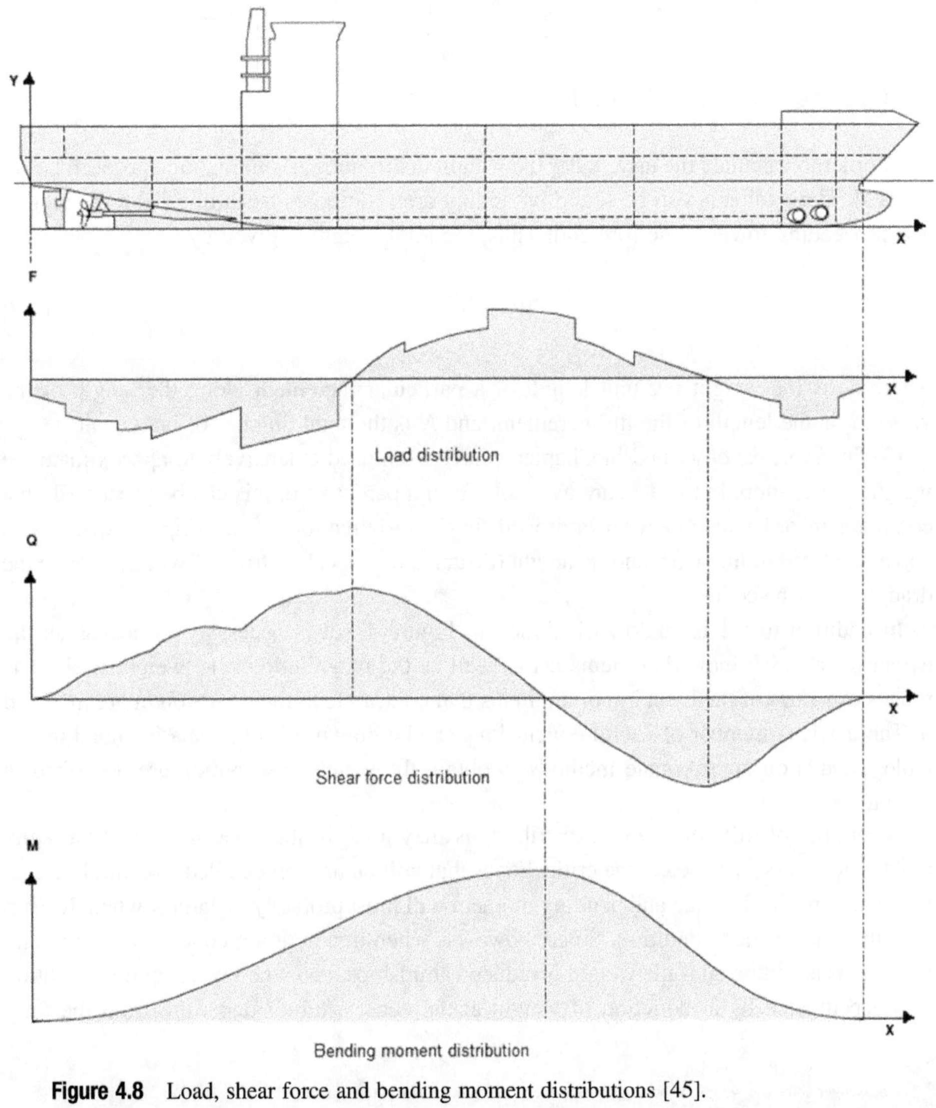

Figure 4.8 Load, shear force and bending moment distributions [45].

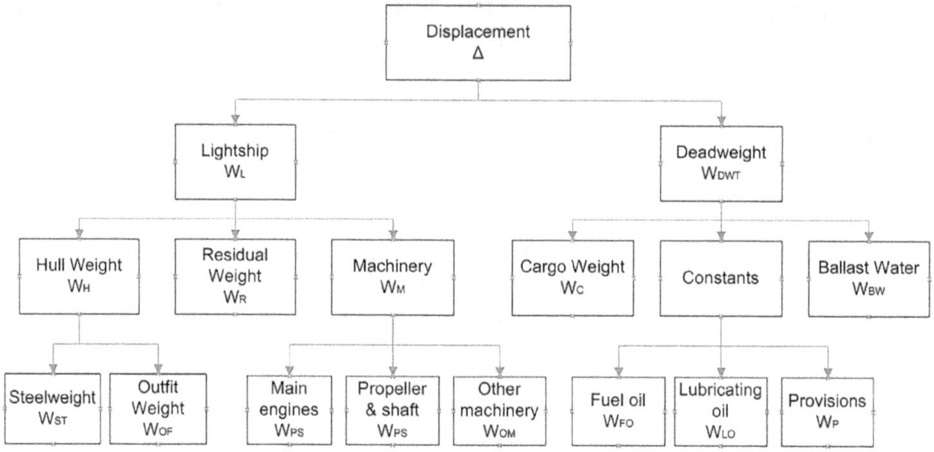

Figure 4.9 Breakdown of weights.[4]

In order to calculate the area under the weight distribution, a summation approach has to be used. The total area can be found by adding area components starting from the aft end and proceeding towards the fore end. Thus, the total weight is given by

$$W = \sum_{i=1}^{N} w_i l_i \tag{4.7}$$

where w_i is the weight per unit length of a particular increment along the length of the vessel, l_i is the length of the ith increment, and N is the total number of increments.

Coffin diagrams described in Chapter 5 have been used extensively to approximate the weight distribution, but if data are available from a parent ship, this can be processed on a computer in order to obtain an improved first approximation. If the ship is poised on a wave, the form of the wave and its height (distance from peak to trough) will determine the draught at each section.

In addition to the breakdown included in Figure 4.9 it is necessary to decide on the extent over which individual items act as well as the magnitude of the weight itself. The ranges of values of the most important items that contribute to the total weight are included in Table 4.1. A number of useful conclusions can be drawn from the data included in this table. Details on approximate methods to obtain the weight distribution are described in Chapter 5.

A number of different weight distributions may arise in the case of a particular ship, making it necessary to select the critical ones that will be used in detailed structural design. In general, the shear force and bending moment will most probably be largest when the ship is in the fully loaded condition. Since, however, when in a hogging condition the bending moment is accentuated if the weight is reduced amidships, and when in a sagging condition it is accentuated by a reduction of weight at the ends, a minor departure from the fully

[4] For passenger ships the weight of passengers and their personal items has to be added.

Table 4.1 *Contribution of major weight components for various ship types*[5]

Vessel type	*Deadweight range*		*DWT/Δ*	W_{ST}/W_L	W_{OT}/W_L	W_{ML}/W_L
	lower	upper	|%|	|%|	|%|	|%|
General cargo (open)[6]	5,000	15,000	65–80	55–64	19–33	11–22
General cargo (closed)	5,000	15,000	60–70	55–63	19–33	11–22
Bulk carriers[7]	20,000	50,000	74–80	68–79	10–17	12–16
	50,000	150,000	80–87	78–85	6–13	8–14
Oil tankers	25,000	120,000	68–83	73–83	5–12	11–16
	200,000	above	83–88	75–88	9–13	9–16
Containerships	10,000	15,000	60–76	58–71	15–20	9–22
	15,000	50,000	60–70	62–72	14–20	15–18
Ro-Ro	$L > 80$ m	16,000	50–60	68–78	12–19	10–20
Reefers	300k ft^3	500k ft^3	45–55	51–62	21–28	15–26
Ferries	$L = 85$ m	$L = 120$ m	16–33	56–66	23–28	11–18
Small passenger		$L = 120$ m	15–25	50–52	28–31	20–29
Large passenger	$L = 200$ m	above	23–34	52–56	30–34	15–20
Fishing vessels	$L = 44$ m	L = 82 m	30–58	42–46	36–40	15–20
Tugs	500 kW	3,000 kW	20–40	42–56	17–21	38–43
River vessels[8]	$L = 80$ m	$L = 110$m	78–79	69–75	11–13	13–19
Coastal vessels	500	1,000	70–75	57–62	30–33	9–12

loaded condition will give the maximum shear force and bending moment acting on the hull girder. For each vessel type, suitable weight distributions can then be specified so that the most severe loading conditions are considered.

The buoyancy distribution depends on the wetted hull form and the weight distribution, and is therefore continuous. Details concerning the hull form are obtained from the lines plan. The loading manuals that are supplied by the shipyard include the most important loading conditions and the resulting draught and trim. For a given draught, the total area under the buoyancy distribution, which represents the buoyancy force, B, is given by

$$B = \int_{AP}^{FP} b(x)dx \tag{4.8}$$

where $b(x)$ is the buoyancy force $(= \rho g A(x)dx)$ acting on a transverse section of the vessel of area $A(x)$ and length dx at a distance x from the aft perpendicular. This is directly proportional to the volume of the immersed portion of the transverse hull section, $A(x)dx$, at that particular location. The form of the buoyancy distribution will depend on the wave profile selected, on its position in relation to the ship and on the hull form. In selecting wave position, it is again necessary to consider critical loading conditions for the hull girder. The purpose of critical conditions is to be able to approximate what are expected to

[5] Based on table 2.1 in [33]. [6] General cargo ships of open and closed tonnage types. [7] Gearless bulk carriers.
[8] Self-propelled river vessels.

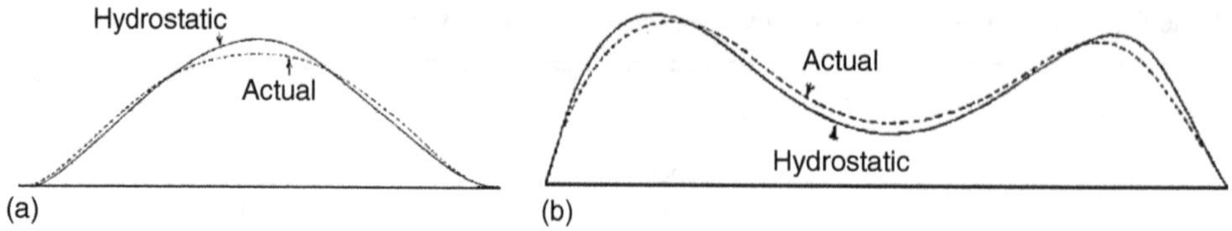

Figure 4.10 Effect of wave profile on buoyancy distribution showing Smith effect[9] [40]. (a) Hogging condition. (b) Sagging condition.

be the most severe loading conditions that the ship is expected to experience during its lifetime.

For a ship supported by quasi-static buoyancy forces, maximum stresses will arise amidships when it is subjected to three-point bending. The closest to this arises when a ship is supported by a wave of length equal to its own length with a crest or trough amidships (Figures 4.5, 4.10).

When in a hogging condition, the fibres above the neutral axis will be in tension whereas those below will be in compression. The opposite will occur when in a sagging condition. For intermediate positions, the stresses will assume intermediate values. It therefore makes sense to consider at least these two conditions in addition to the still-water (harbour) condition.

The primary (longitudinal) bending moment will vary accordingly; we distinguish between the *still-water bending moment*, M_{SW}, and the *wave-induced bending moment*, M_{WV}, which is the additional bending moment that arises in the presence of a wave. The wave-induced bending moment is then the difference between the *total bending moment* M_{tot} and the still-water bending moment. Approximations to its value exist,[10] but it cannot be measured explicitly. Thus,

$$M_{tot} = M_{SW} + M_{WV} \tag{4.9}$$

As seen from Figure 4.11, the wave-induced bending moment varies cyclically about a mean position that is the still-water condition.

4.4 Balancing of the Ship on a Quasi-Static Wave

Whichever case is considered, the buoyancy curve has to enclose an area equal to that enclosed by the weight curve in order to ensure that equilibrium is satisfied. To achieve this, a *balancing procedure* has to be carried out, since the transverse area of each section varies along the length as does the wave profile. First of all, the wave profile and its exact position with respect to the ship have to be selected. A convenient method is to sketch a longitudinal profile of the ship upon which the sectional area (Bonjean) curves

[9] The Smith effect is the result of dynamic pressure in the fluid below a wave train (see Section 4.4). This results in a pressure that is lower than the corresponding hydrostatic one below a wave crest and an increased pressure below a wave trough (Figure 4.10) [42].

[10] The IACS expressions for M_{WV} in hogging and sagging are included in Section 4.9

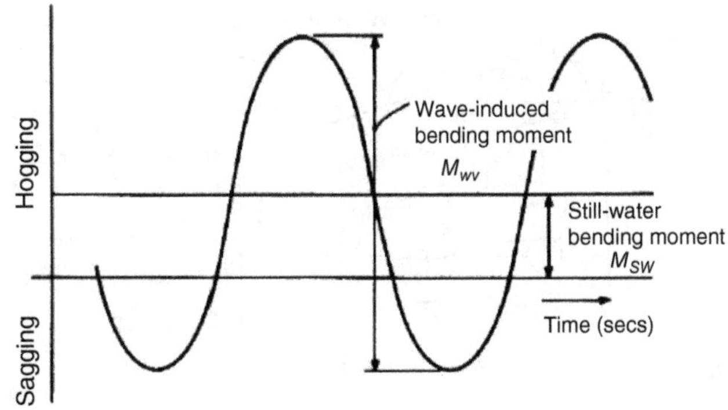

Figure 4.11 Components of the total bending moment [10].

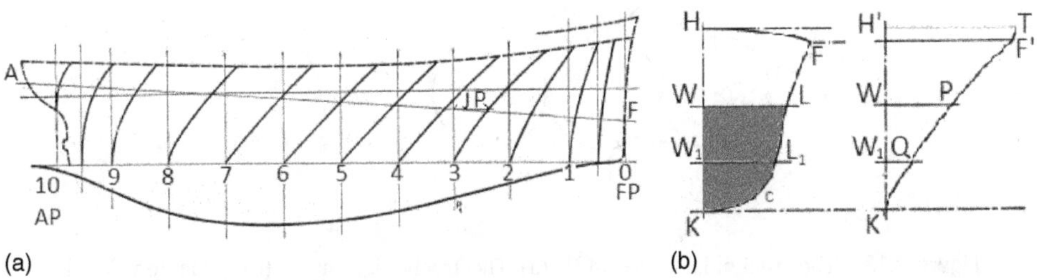

Figure 4.12 Bonjean curves (sectional area curves). (a) Longitudinal profile with Bonjean curves. (b) Section and Bonjean curve.

(Figure 4.12)[11] have been traced out at a number of stations, suitable for the use of a numerical integration rule. The selected waveform can then be drawn on this and the corresponding sectional areas read from the Bonjean curves.

For small displacements or rotations of the waterline, the volume changes and other geometric parameters vary linearly with draught. Furthermore, since the angle of tilt of the initial estimate of the waveform will be small, the geometrical properties of the waterline area (area, moment of inertia, centre of area) will not differ significantly from the true horizontal projection. The balancing procedure requires a trial-and-error approach because not only do the areas under the two curves have to be equal (vertical equilibrium), but also the *LCG* and the *LCB* have to lie on the same vertical line (rotational equilibrium). The procedure is outlined later on, but it has to be added that use of computer software ensures that the correct buoyancy curve is obtained without difficulty.

Let us now consider the wave profile and related issues since it is necessary to select in addition to the wave profile, the length and height of the wave. As in the case of the weight distribution, it is necessary to select a wave profile that corresponds to the most severe loading condition the ship is likely to encounter. Clearly, it is impossible to forecast the

[11] A Bonjean curve is a plot of the area of a transverse section (sectional area curve), as the draught is increased from zero to a maximum value. Curves are obtained for a selected number of stations along the length of the ship.

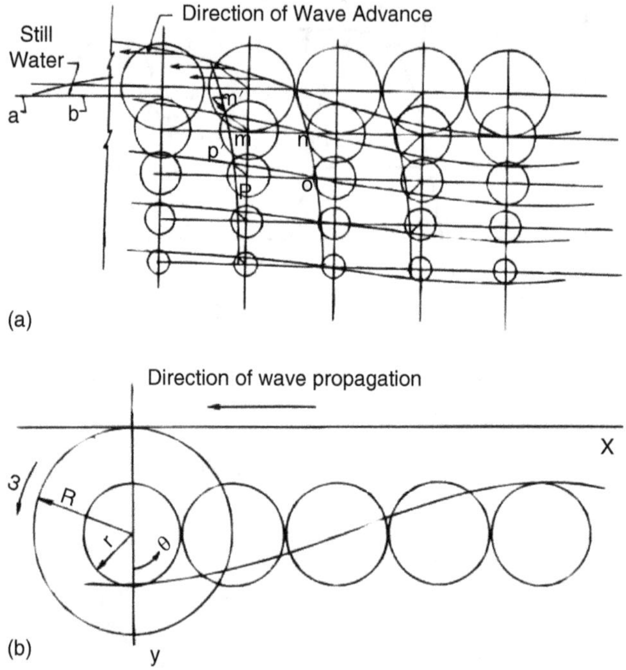

Figure 4.13 The trochoidal curve [48]. (a) The trochoidal curve. (b) Generation of the trochoid.

nature and magnitude of the loads that the ship will be subjected to at a future date, so we have to rely on past experience and engineering judgement in order to develop a reliable basis for a design criterion.[12]

The most satisfactory wave profile, originally proposed by Froude and used until today, is the *trochoid* [15]. Of all mathematically tractable functions the trochoid approximates the condition on the sea surface in the best possible manner (Figure 4.13a). An important geometric property of the *trochoidal curve* is that it is not symmetrical about its axis, as the area below the axis is larger than that above it and the radius of curvature at the crest is less than that in the trough.

The trochoidal curve is traced out by a point lying on a circle of radius r that is included in another circle of radius R, where $R > r$. The outer circle moves on a horizontal line, and the curve is traced out in this manner (Figure 4.13b). The relevant expressions are [38]:

$$x = R\theta - r\sin\theta \tag{4.10a}$$

$$z = r(1 - \cos\theta) \tag{4.10b}$$

[12] To this end, the methodology developed by IACS involves the loading on ship hulls that takes into account conditions encountered in the oceans, their distribution and frequency of occurrence. The approach used is essentially a statistical one and is discussed in Chapters 9 and 12.

A wave whose length differs significantly from the length of the ship will result in loading that is close to the still-water condition. The critical wavelength that will result in maximum longitudinal bending is approximately equal to the ship's length. Data from wave observations performed in the 19 century by Paris, an officer in the French navy, led to the conclusion that length and wave height are statistically correlated. Biles conducted a statistical analysis of Paris's wave observations and other data and proposed that the wave have a height H equal to $L/20$. Since then it has been acknowledged that as length increases, the wave height-to-length ratio increases but not proportionately. Many slightly nonlinear formulations have been proposed and used. Simple proportional relations used in the past give excessively large values for ships whose length exceeds 150 metres and low values for ships with lengths up to 150 metres so that nonlinear relations are nowadays used, such as the following:

$$H = 0.607\sqrt{L} \tag{4.11a}$$

$$H = 24.38Le^{-L/274}(C_B + 0.3)(10^{-3}) \tag{4.11b}$$

where L is the ship's length in metres and C_B is the block coefficient. For Equation (4.11a), the wave profile can be obtained by substituting in Equations (4.10a) and (4.10b). Thus, if $L = 2\pi R$ and $r = H/2 = 0.3045\sqrt{L}$, we obtain [38]:

$$x = \left[\frac{L}{2\pi}\right]\theta - \left[0.3045\sqrt{L}\right]\sin\theta \tag{4.12a}$$

$$z = \left[0.3045\sqrt{L}\right](1 - \cos\theta) \tag{4.12b}$$

Having chosen a suitable wave profile and related geometrical parameters, we can return to the balancing procedure. The procedure is implemented in the following stages [34]:

1. Select and trace out the wave profile on a plot of Bonjean curves so that its mean height line coincides with the still-water draught for which the bending moment is required.
2. Estimate the buoyancy distribution and the geometrical properties of the waterline area by reading off the sectional areas (Bonjean curves) at each station.
3. Trace out the shear force curve, $V(x)$, using Equation (4.5a). If there is surplus buoyancy amidships (hogging condition) or deficiency in buoyancy amidships (sagging condition), the shear force at the *FE* position will assume a non-zero value. To achieve equilibrium, the axis of the wave profile has to be translated in the vertical direction through a distance v. This translation is in effect a change in draught that is positive if the draught is increased and negative if the draught is decreased:

$$v = (-V_r)/(\rho g A_w) \tag{4.13}$$

4. Calculate the residual bending moment M_r at the fore end, $M(FE)$, from Equation (4.5b). If $M_r = -V_r z_{LCF}$, the bending moment is zero at the new draught (z_{LCF} = distance of *LCF* from *FP*). The negative sign is required by the sign convention (Figure 4.7). If this is the case, the balancing procedure is terminated. In general,

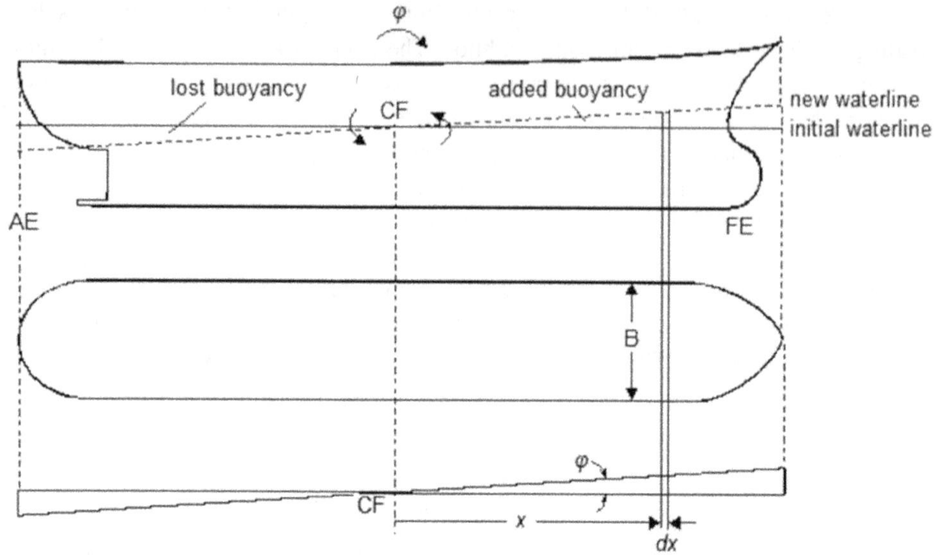

Figure 4.14 Correction to the BM distribution in the balancing procedure [34].

however, the bending moment is not equal to zero at the *FP* and it is necessary to trim the ship by rotating it about the centre of flotation in order to avoid any change in displacement following rotation (Figure 4.14). The angle of rotation is positive in the clockwise direction. The bending moment resulting from the change in trim is given by:

$$\int_{AP}^{FP} x\rho B(x)\varphi x dx = \rho\varphi\int_{AP}^{FP} B(x)x^2 dx = \rho\varphi I_L \tag{4.14}$$

where I_L is the longitudinal moment of inertia of the waterplane area about the transverse axis through the centre of flotation. The angle φ is calculated from

$$M_r + V_r z_{LCF} = \rho g \varphi I_L$$

Hence,

$$\varphi = (M_r + V_r z_{LCF})/\rho g I_L \tag{4.15}$$

5. The shear force and bending moment diagrams are redrawn for the new trim using the latest estimated values. If discrepancies arise at the fore end the procedure is repeated from step 2 onwards. If the differences are within the tolerances specified earlier, the correction is introduced using Equations (4.6a–b).

The geometrical particulars of the waterline area are given by:

$$A_w = \int_{AE}^{FE} B(x)dx \qquad I_L = \int_{AE}^{FE} B(x)x^2 dx \qquad LCF = \int_{AE}^{FE} xB(x)dx \tag{4.16a–c}$$

4.5 Factors That Affect Vertical Bending Moment and Shear Force

It is useful for the designer to be aware of the changes that geometrical and other parameters have on the bending moment and shear force distributions, especially at the early stages of a design that is based on that of an existing ship.

Since the still-water bending moment depends on the form of the load curve, the factors that affect it are the:

- Hull form
- Lightship distribution
- Cargo distribution

Lightship and cargo are the two major components of the weight distribution. For a given ship the lightship distribution remains constant whereas the cargo distribution varies. The hull form, on the other hand, does not change, so the still-water bending moment varies depending on the quantity of cargo and its distribution in the longitudinal direction. Since there are significant differences between the various loading arrangements, it is not possible to make a general statement as to whether the resulting still-water bending moment will be hogging or sagging. However, when the buoyancy force is approximately equal to the weight at each point along the length, shear force and bending moment are at their minimum values. This arises in the homogeneous loading condition (Figure 5.38).

If cargo is added to a ship in a ballast condition, since there will be no weight increase at the ends but only a buoyancy increase, we can expect the stresses in way of the main deck amidships to become increasingly compressive. If, therefore, the ship is initially in a sagging condition, an increase in displacement will bring about an increase in still-water bending moment. In general, when ballasted, a cargo ship is in a hogging condition because of the weight of machinery near the aft end and empty cargo spaces along the parallel middle body. Also, the buoyancy force per metre length is greater amidships, since the ship has a parallel middle body. In such a case, when loading commences, the degree of hogging will reduce and when loading ends sagging may arise.

As regards the effect of length, from beam theory we have that the magnitude of the bending moment will increase when the length is increased.

As regards fullness of form, a measure of which is the block coefficient, for ships of the same principal dimensions the fuller ship will have a larger displacement and thus a larger wave bending moment.

The above reasoning was enshrined in a formula that was proposed in 1874 by John. This formula relates the maximum bending moment to displacement and length ($M = \Delta L/C$ where $C \approx 35$, depending on ship's characteristics[13]). The corresponding value for the maximum shear force based on a simplified distribution is given by $V \approx \Delta/8.75$.

As regards the wave-induced bending moment, we can make use of beam theory to examine the effect of geometrical parameters. We consider a vessel that has a uniform weight distribution and a rectangular cross-section along its length. We assume that it is balanced on a sinusoidal wave of length equal to its own length and that the wave trough

[13] Coefficients for various ship types and dimensions are included in [11].

coincides with the midship section. At any distance x along the length, the wave amplitude, $a(x)$, is given by:

$$a(x) = \left(\frac{H_e}{2}\right)\cos^2\left(\frac{2\pi x}{L}\right) \tag{4.17}$$

where x is the distance from the aft perpendicular and H_e is the effective wave height. For this vessel the shear force at any position x is given by:

$$\begin{aligned} V(x) &= -\int_{AE}^{x} \rho g B a(x) dx \\ &= -\rho g B \frac{H_e}{2} \int_{AE}^{x} \cos^2\left(\frac{2\pi x}{L}\right) dx \\ &= -\rho g B \frac{H_e}{2}\left(\frac{L}{4\pi}\right) \sin\left(\frac{2\pi x}{L}\right)\left[1 + \cos\left(\frac{2\pi x}{L}\right)\right] \end{aligned} \tag{4.18}$$

Integrating, the wave-induced bending moment at any position x is obtained:

$$M_{WV}(x) = \int_{AE}^{x} V_{WV}(x)dx = \rho g B H_e\left(\frac{L^2}{8\pi^2}\right)\left[\frac{1}{2}\sin^2\left(\frac{2\pi x}{L}\right) - \cos\left[\frac{\pi x}{L}\right]\right] \tag{4.19}$$

and at $x = L/2$, the maximum wave-induced bending moment is

$$M_{WVmax} = \rho g H_e B\left(\frac{L}{4\pi}\right)^2 \tag{4.20}$$

We see that M_{wvmax} is proportional to B, L^2, and $\rho g H_e$. Analogous calculations can be carried out for hulls of different geometries supported by waves of different forms [7]. In this particular study hull geometries ranging from a rectangular wall-sided vessel to a V-section and a diamond-shaped vessel in plan (no parallel body) were considered and the following conclusions reached:

1. Wave bending moments increase rapidly with block coefficient (40 per cent greater for $C_B = 0.80$ than for $C_B = 0.60$) (increase in displacement).
2. Sagging wave bending moments are greater than hogging wave bending moments.
3. Wave bending moment increases with wave height.
4. The length of wave to cause maximum bending moment is different for hogging and sagging.
5. The bending moment on a trochoidal wave is larger than on a sinusoidal wave (by 15 per cent).
6. The magnitude of the Smith correction is such that at a constant wave height the wave-induced bending moment reduces by 20–30 per cent. This reduction increases for shorter waves. A formula for this reduction is given below.
7. At a fixed wave height, the wave bending moment is almost constant for lengths of the order of the ship's length.
8. For full-form vessels there is a reduction in bending moments with an increase in draught.

The following formula can be used to allow for the reduction in wave-induced bending moment due to the Smith effect [38]:

$$Corrected\ M_{WV} = M_{WV}\{\ exp(-nT/L)\} \tag{4.21}$$

where

	n	
C_B	*Sagging*	*Hogging*
0.80	5.5	6.0
0.60	5.0	5.3

For a ship on a hogging wave with $L = 200$ m, $T = 10$ m, $C_B = 0.80$ the table gives $n = 6.0$. The correction factor is then equal to 0.74, which amounts to a 26 per cent reduction in wave-induced bending moment. If the same ship is balanced on a sagging wave, the reduction is approximately 22 per cent.

The model tests that have been performed have essentially validated the above conclusions [11]. The effects of draught and ship speed were investigated and were found to be minor. It was found that the maximum total bending moment arises when the wavelength is 1–1.25 times the ship's length and the ship encounters head seas. Theoretical studies have also been performed to determine the effect of speed on the longitudinal distribution of the vertical wave bending moment. In these studies long-crested irregular seas corresponding to the average sea states of the North Atlantic were assumed in order to calculate the average values of the one-tenth highest wave bending moments induced in large oil tankers. It was found that at low speeds in regular head waves the amplitude of the wave bending moment is distributed approximately symmetrically about the midship section, about which the maximum amplitude arises. As the speed increases, the maximum amplitude shifts forward. In irregular seas the effect was not found to be as pronounced. It was concluded that in irregular seas the longitudinal distribution can be considered to be symmetric about amidships [14].

The effect of longitudinal weight distribution is assessed in terms of the effect it has on the radius of gyration about the transverse axis amidships. This depends on the longitudinal disposition of weight, and thus if cargo is transferred from spaces near the ship's ends to spaces near the midship section, the moment of inertia reduces so that the radius of gyration also reduces. This results in an increase in wave bending moment. Hence sagging moments are subject to higher dynamic wave load increases.

4.6 Torsional Loading of the Hull Girder

Torsional loading of the hull girder arises in a variety of circumstances that can be classed as *still-water* (*static*) or *sea-going* (*dynamic*). It may be necessary to consider more than

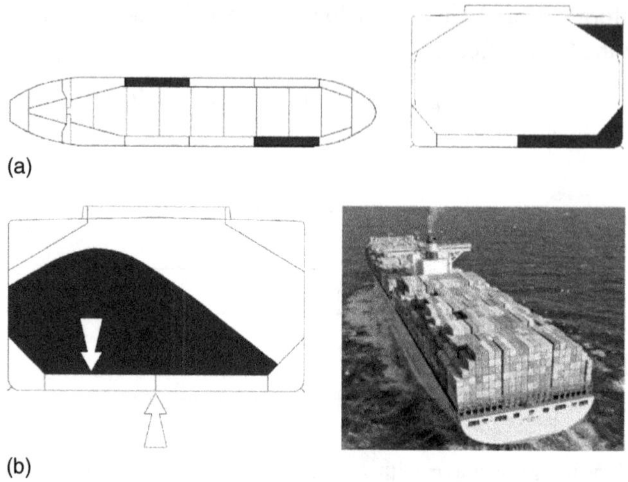

Figure 4.15 Still-water torsional loading of dry cargo ships [20]. (a) Longitudinally and transversely asymmetric distributions of ballast water in bulk carrier. (b) Transversely asymmetric cargo distributions in bulk carrier and containership. Reprinted courtesy of IACS, London.

one condition simultaneously in which case the total torque is the sum of the static and the dynamic component.

In still-water conditions the (static) torque is the result of the asymmetrical distributionof the ballast water and/or dry cargo with respect to the ship's centreline plane (Figure 4.15). In the case of containerships cargo asymmetry is not expected to be a serious condition because any discrepancies in the weight of individual containers will tend to cancel each other out, especially when the number of containers is large.

In the case of bulk carriers the weight of cargo transported within any one hold is significantly greater than that of a number of containers so that the resulting static torque is expected to be much greater.

Dynamic torque arises when a ship is travelling at sea. In this case it is primarily the result of the asymmetric distribution of the hydrostatic and hydrodynamic forces that act on the wetted surface. Vedeler, who was the first to study the torsional loading of ships, considered that the critical condition arises when a ship encounters a regular wave train whose length exceeds that of the ship by a small amount, at an angle of 45°. In later work it was found that the peak torque arises when the angle of incidence is 60° [11]. In addition to external loading, inertial forces resulting from the motion of the ship, and in particular roll, may contribute to dynamic torque.

Figure 4.16a shows a ship proceeding in a regular seaway at an angle of 45° to the direction of the waves. It is assumed that the wave crests coincide with the fore and aft ends of the ship. The differences in draught on either side of each section induce a rolling moment, except in the case of the midship section, which is balanced.

However, the rolling moments in the fore and aft parts of the hull girder are of opposite sign so that the hull girder is subjected to a net torsional moment.

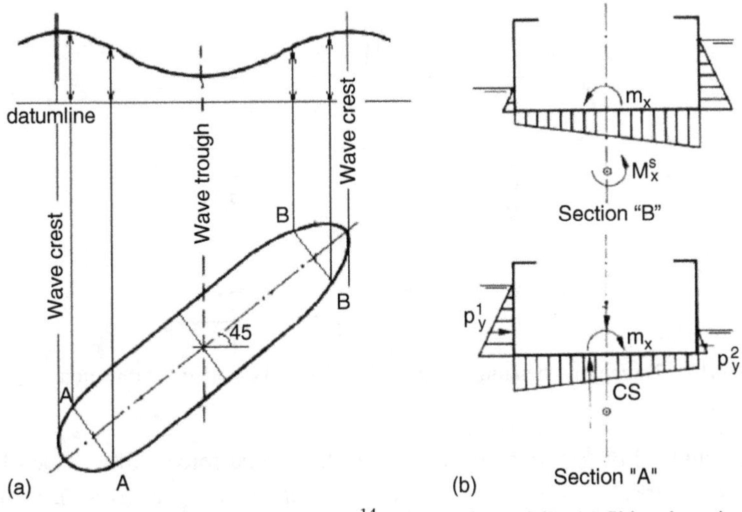

Figure 4.16 Torque loading of a hull girder[14] (adapted from [36]). (a) Ship advancing at an angle to a regular seaway. (b) Hydrostatic components of torque. Reprinted courtesy of the Society of Naval Architects and Marine Engineers.

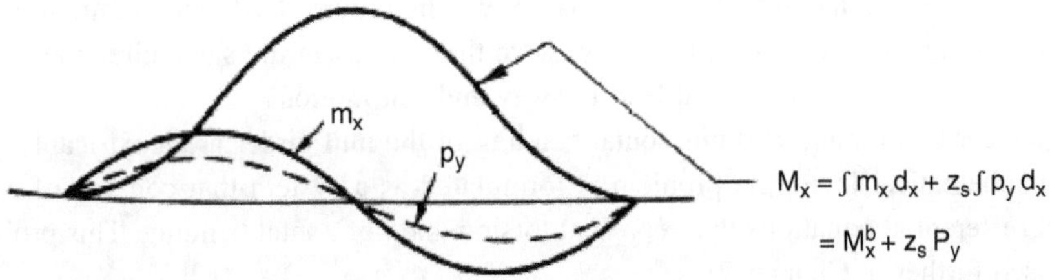

Figure 4.17 Longitudinal distribution of torque acting on hull girder (z_s = lever of the net horizontal force about the centre of shear at any section).

Figure 4.16b shows the distributions of the hydrostatic pressure loads that act at sections AA and BB along the length of the ship. It is seen that these distributions differ. By considering vertical and rotational equilibrium of the resultant of these distributions about the centre of shear, it is seen that they result in a moment about the centre of shear and a horizontal force in the transverse direction. The magnitudes and directions of action of these depend on the respective distributions. In section AA, the moment acts in the clockwise direction whereas in the section BB it acts in the counter-clockwise direction. If the direction of action of these moments were to be the same, the outcome would have been roll motion of the ship. However, since they act in opposite directions, as in Figure 4.16b, the outcome is a twisting moment that is distributed along the length of the ship (Figure 4.17). In closed ship sections the centre of shear is located some distance below the keel, leading to significantly larger moments of the horizontal forces.

[14] p_y = hydrostatic force acting on side 1 (port) or 2 (starboard). M_x = torque arising as a result of the hydrostatic pressure gradient on bottom plating. m_x = net bending moment acting on bottom structure as a result of change in hydrostatic pressure load in transverse direction. CS = centre of shear.

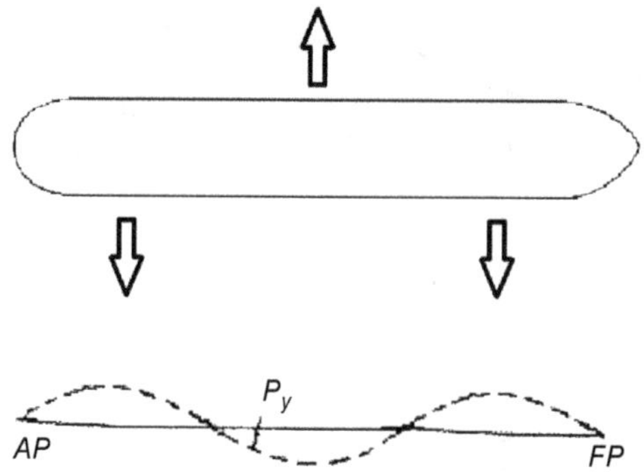

Figure 4.18 Three-point bending of the hull girder in the horizontal direction.

The longitudinal distributions of torque and horizontal force will then be of the form indicated in Figure 4.17 where M_{wt} is the *torsional* (or *twisting*) *moment* acting on the hull girder.

Figure 4.17 shows that the direction of action of the horizontal force P_y changes from starboard to portside forward of the midships region, and in this particular case yawing will arise. If, however, the distribution is as shown in Figure 4.18, three-point horizontal bending of the hull girder arises. Finally, were these to act in the same direction all along the length of the ship, this would lead to sway and roll motion.

It is clear that torsion and horizontal bending of the hull girder are inextricably linked. At a mathematical level, the problem is formulated as a system that consists of coupled partial differential equations that represent torsion and horizontal bending. This problem is discussed further in Chapter 7.

With regard to the longitudinal distribution of the torsional moments, it has been found that

- Torsional moments increase as the heading angle increases from 0° to approximately 60°.
- The torsional moment depends on the type of section, that is, whether it is open or closed.
- The torsional moment is a function of principal dimensions *L, B, D, T* and also depends on the waterplane area coefficient C_{WL}.
- Torsional moments increase with wave height.
- Torsional moments increase with ship's speed ("synchronous rolling").

These findings have been verified by model tests.

Of importance is the maximum value of torque, and a number of formulas have been proposed. For closed sections, assuming a wave height of $L/10$, Abrahamsen [1, 2], proposed the following formula early on [11]:

$$M_{wt} = k\rho LB^3 \qquad \text{(ton.m)} \tag{4.22}$$

where ρ is the density of seawater and k is a constant obtained as a function of the waterplane area coefficient, C_{WL} (Figure 4.19).

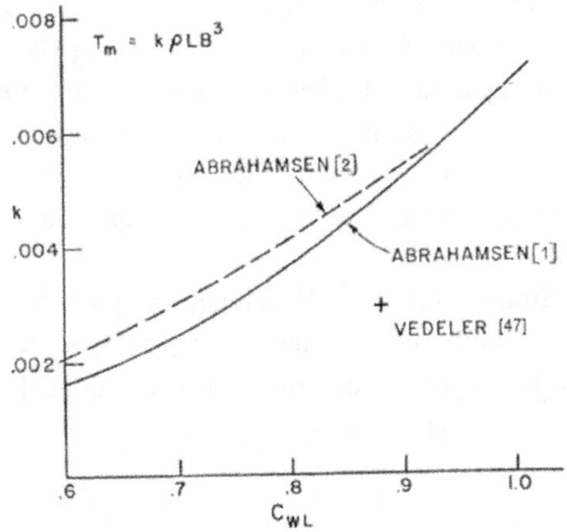

Figure 4.19 Torque coefficient k [11].

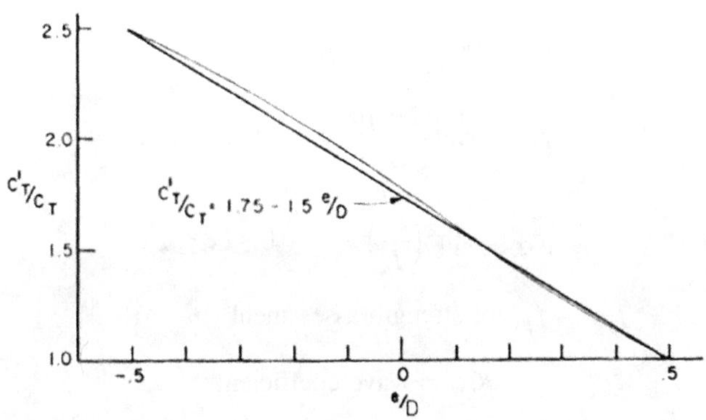

Figure 4.20 Open section torque coefficient C_T' [11].

For open sections, the following formula gives the torque coefficient, C_T', in terms of the torque coefficient for a closed section with the same waterplane area coefficient, C_{WL}, for which k is obtained from Figure 4.20.

$$C_T' = k\left(1.75 - 1.5\frac{e}{D}\right) \tag{4.23}$$

where e and D are the distance of the centre of shear from the baseline (positive within and negative outside the section) and the hull depth, respectively. It is seen that the maximum torque for the open section ranges from that of the equivalent closed section to up to 2.5 times that value, depending on e/D.

As regards the magnitude of the torsional moment, it has been found that in the case of open-section ships (containerships) its maximum value is of the order of 20 per cent of the hogging wave bending moment, although large variations are noted.

The longitudinal distribution of the torsional moment is affected by the geometry of the cross-section along the length of the ship and in particular by the position of the centre of shear with respect to the centroid of the section. One other factor is the degree of torsional restraint offered by the fore and aft end structures of a ship. For ship-type structures the ends are closed sections that have higher torsional rigidity than the parallel section, particularly so for open section ships such as containerships. As a result, warping is restrained and significant warping stresses arise, especially when the parallel section is torsionally flexible. Warping stresses can be reduced by gradually increasing the torsional rigidity towards the ends of the ship.

The IACS Common Structural Rules [22] include a design formula for the wave-induced torsional moment acting on bulk carriers, which are open-deck type ships. This is given as a function of length L, and is expressed in terms of two components M_{wt1}, M_{wt2}:

$$M_{wt} = f_p(|M_{wt1}| + |M_{wt2}|) \tag{4.24}$$

where

$$M_{WT1} = 0.4 f_{t1} C_w \sqrt{\frac{L}{T_{LC}}} B^2 D C_B$$

$$M_{wt2} = 0.22 f_{t2} C_w L B^2 C_B$$

$$f_{t1} = \sin\left|\frac{(2\pi x)}{L}\right| \qquad 0 \leq x \leq L$$

$$f_{t2} = \sin^2\left(\frac{\pi x}{L}\right) \qquad 0 \leq x \leq L$$

$$f_p = f_{ps} \text{ for strength assessment } \left(p_f = 10^{-8}\right)$$

$$C_w = \text{wave coefficient}$$

L, B, D, T, C_b are the ship's principal dimensions and block coefficient.

Figure 4.21 shows the longitudinal distribution of the wave torsional moment for a bulk carrier calculated using Equation (4.24) that has the following particulars:

$L = 185$ m $\quad B = 45$ m $\quad D = 15$ m
$T = 10$ m $\quad C_b = 0.70$

ABS provides the following expression for the still-water torque, T_S, acting on a transverse section amidships:

$$T_S = \pm kBW_T \qquad \text{(kNm)} \tag{4.25}$$

where

$$k = 0.004$$
$$B = \text{breadth of ship (m)}$$
$$W_T = \text{maximum total container weight (kN)}$$

The maximum value of T_S at all other positions along the length of the ship is obtained by multiplying with the distribution factor m_T, shown in Figure 4.22 [3]. The purpose of the distribution factor is to indicate, for each position along the length of the ship, the

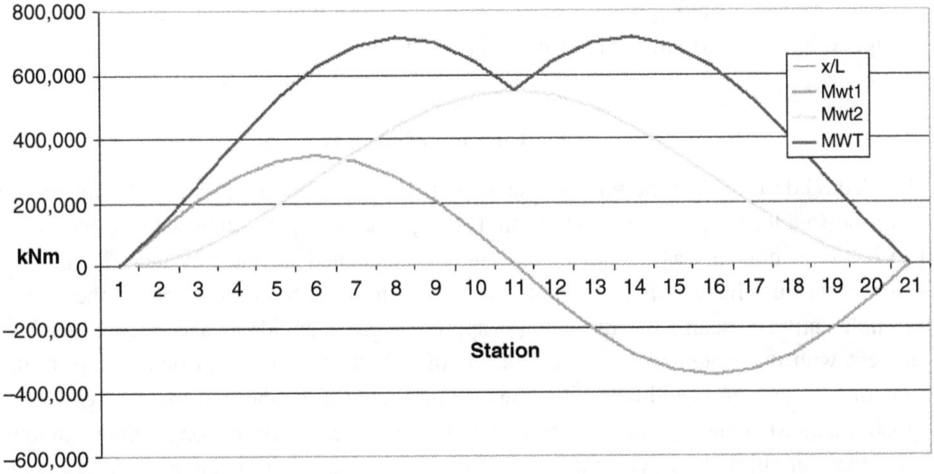

Figure 4.21 Wave torsional moment distribution according to IACS CSR for bulk carriers.

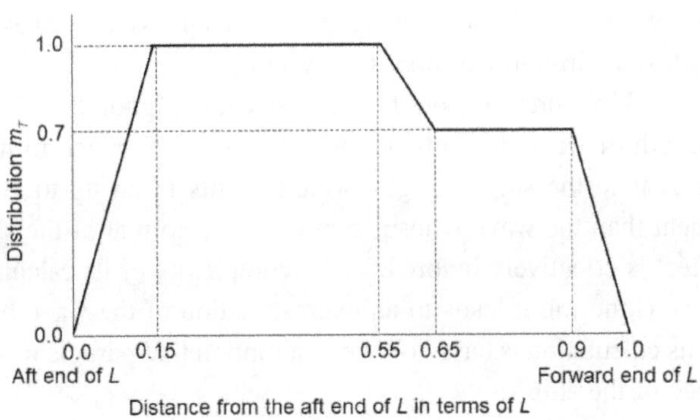

Figure 4.22 Distribution factor, m_T, for the still-water torque, T_S [3].

maximum (design) value for that position, and for this reason the diagram is termed an *envelope curve*; it does not represent a single critical condition for an entire ship.

For bulk carriers, the wave-induced torque amidships is (in kNm) according to ABS, given by the following formula:

$$T_M = kLB^2 d_1 \left[(C_w - 0.5)^2 + 0.1\right]\left[0.13 - (e/d)(c_0/d_1)^{1/2}\right] \tag{4.26}$$

where

$k = 2.7$
$c_0 = 0.14$
$d_1 =$ draught (m) ($d_I > 12.5$ m)
$e =$ vertical distance (m) of the effective centre of shear of the hull measured from the baseline of the vessel, positive upward
$C_w =$ waterplane area coefficient for the scantling draught

At other positions along the length of a ship, the value obtained from Equation (4.25) is multiplied by the distribution factor included in Figure 4.22.

4.7 Wave Loading of Ship Hulls

The method described in the previous sections is based on the premise that the maximum wave-induced loading arises when the ship is in a prescribed quasi-static loading condition. The bending moments and shear forces that are calculated in this way have in the past ensured that ships have sufficient longitudinal strength. It has been found that, in the case of a simple uniform beam, the quasi-static approach gives an identical value of bending moment with that obtained for the amplitude of a fluctuating moment obtained from the dynamic analysis of a rigid beam [6]. This is considered to be the fundamental reason for which the quasi-static approach has proven to be so successful in the longitudinal strength assessment of ships. However, the quasi-static loading condition is an artificial construct that has been used mainly because more accurate approaches were not developed for many years. Important objections to it are that [37, 38]:

1. The trochoidal wave shape does not match the steepness of waves at the crests, as predicted by Stokes's irrotational flow theory [43].
2. The effect of the ship's presence on the wave system is ignored.
3. The critical length of wave depends on the ship's form. For fine-form ships it may be 20 per cent less than the ship's length, which results in an up to 10 per cent higher bending moment than the wave whose length is equal to that of the ship.
4. The Smith effect is effectively ignored, as the complexity of its calculation necessitates approximations. (Ignoring it leads to an overestimation of the wave bending moment.) Furthermore, its calculation is based on the assumption that particle motion is unaffected by the presence of the ship.

Methods have since been developed in which the sea environment and its interaction with the ship are much more accurately accounted for. When a body is floating on the surface of the sea, regardless of whether it is moving or not and also whether the surface is disturbed or not, the interaction that takes place results in both the body and its immediate environment being affected (Figure 4.23). In the case of a ship, the forces that act at each position on the hull surface change continuously as it encounters waves of varying size and direction as it advances. On the one hand, the hull girder is subjected to rigid body motions about all three coordinate axes as well as elastic deflections and rotations; on the other, the environment is affected by the passage of the ship; this leaves behind it a wave system that is markedly different to that ahead of it. This is the result of its interaction with the fluid and also because of propeller action (Figure 4.23).

Clearly the real situation differs substantially from the artificial one considered earlier and is a particularly difficult problem to solve.[15] Nevertheless, there exist solutions that

[15] A closed-form solution for the complete ship–wave problem is not possible for several reasons. Until today, only numerical solutions of formulations in which simplifying assumptions have been made have been obtained.

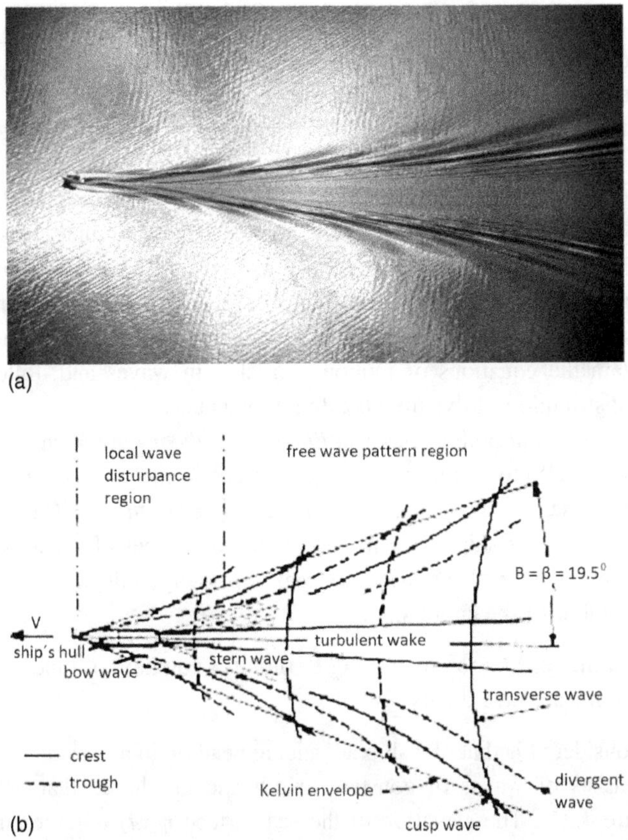

Figure 4.23 Wave pattern created by the passage of a ship. (a) Wave pattern created by a small boat [13]. (b) Nomenclature.

capture the most important parameters, and these methods are used in design procedures. The quasi-static buoyancy and weight distributions considered previously must be replaced by the forces in the fluid that surrounds the floating body and by the inertial reaction forces that act upon it. The fluid forces that arise are the result of the interaction of the wave-induced motions of the water with the body (*Froude–Krylov* force). In other words, the fluid forces that pre-exist within the wave system in the absence of the body are affected by the passage of the floating body. The wave pattern that is created is then the outcome of this interaction (Figure 4.23).

As regards the motion of the body itself, the inertial reaction load acting on it is equal to the product of its mass and its absolute acceleration. The net load acting on the body is then the difference between the fluid force and the inertial reaction load. By performing successive integrations of the longitudinal distribution of the load acting over segments of a small but finite length, the shear force and bending moment can be obtained as in the still-water condition [35].

4.8 Classical Linear Strip Theory

In order to render the ship–wave problem tractable, it is necessary to introduce a number of simplifying assumptions that concern both the environment and the ship. The solution of the equations that describe the behaviour of a volume of water of infinite extent that is bounded by a free surface and a solid surface (seabed) is possible only in certain cases. Available solutions have been based on a *linearization*[16] of the boundary conditions. As regards the ship itself, the hull is initially taken to have a circular cross-section whose radius is equal to the draught, end-effects are ignored, and its mass is considered to be uniformly distributed about the centreline [26]. The method was subsequently adapted to allow for realistic ship hull forms. In the following we will outline the method presented in [25–27] to obtain the equations of motion of a ship in waves and, following that, the corresponding distribution of dynamic bending moments.

In this section we shall outline classical *linear strip theory* in which the solution to the problem is subdivided into a number of stages, described in Table 4.2. Before continuing however, it is necessary to explain the use of the term *strip theory*. The principle of strip theory is to subdivide the ship along its length into a number of sections (strips) and to calculate the loading on each of these sections separately. Integrating along the length renders the behaviour of the entire ship.

1 Description of the sea environment using Bernoulli's equation of fluid flow for a regular sea, assuming linear wave theory.

The problem considered is that of a ship advancing head on in a *regular sea*. A regular sea is one that consists of waves of constant length and amplitude that propagate in one direction (Figure 4.24). The elevation of the sea surface $\eta(x,t)$ is represented by a wave whose amplitude is small in relation to its wavelength. Small-amplitude (Airy) waves are two-dimensional and travel in one direction (Figure 4.24a). The free surface elevation $\eta(x, t)$ of each wave component is assumed to be sinusoidal and is expressed as a function of horizontal position x and time t:

$$\eta(x,t) = a \sin \frac{2\pi}{\lambda}(x - ct) \tag{4.27}$$

where a is the *wave amplitude*, λ is the *wavelength* and c is the *celerity*.

Whilst the wave propagates in one direction, the water particles perform an orbital motion that is circular but whose diameter decreases exponentially with depth. At the crest of the wave the particles move in the direction of propagation of the wave, whereas at the trough they move in the opposite direction (Figure 4.24b). At intermediate positions the velocity component that is tangential to the surface changes continuously between these two extremes. The particles thus do not translate in the direction of wave propagation, although energy (potential and kinetic) is transferred in this direction by their motion. In shallow water, as depth increases, the orbits change from a circle to an ellipse that

[16] *Linearization* of the boundary conditions involves using dimensional analysis to compare the relative magnitudes of the terms that are included in the expressions representing them and then discarding the higher-order terms.

Table 4.2 *Classical linear strip theory*

No	*Stage*
1	Description of the sea environment. Regular sea and linear wave theory. Bernoulli's equation of fluid flow.
2	Ship motions and establishment of equations of motion.
3	Idealised representation of a ship travelling in a regular sea. Component velocities resulting from ship–environment interaction for a representative 'strip'.
4	Introduction of component velocities into Bernoulli's equation using potential flow theory and determination of pressures acting on a strip of the immersed body.
5	Integration of pressures over the length of the body to calculate the hydrodynamic loads (exciting) forces acting on the entire immersed body (dynamic loading of hull girder).
6	Solution of coupled equations of motion for heaving and pitching.
7	Dynamic shear force and bending moment distributions.

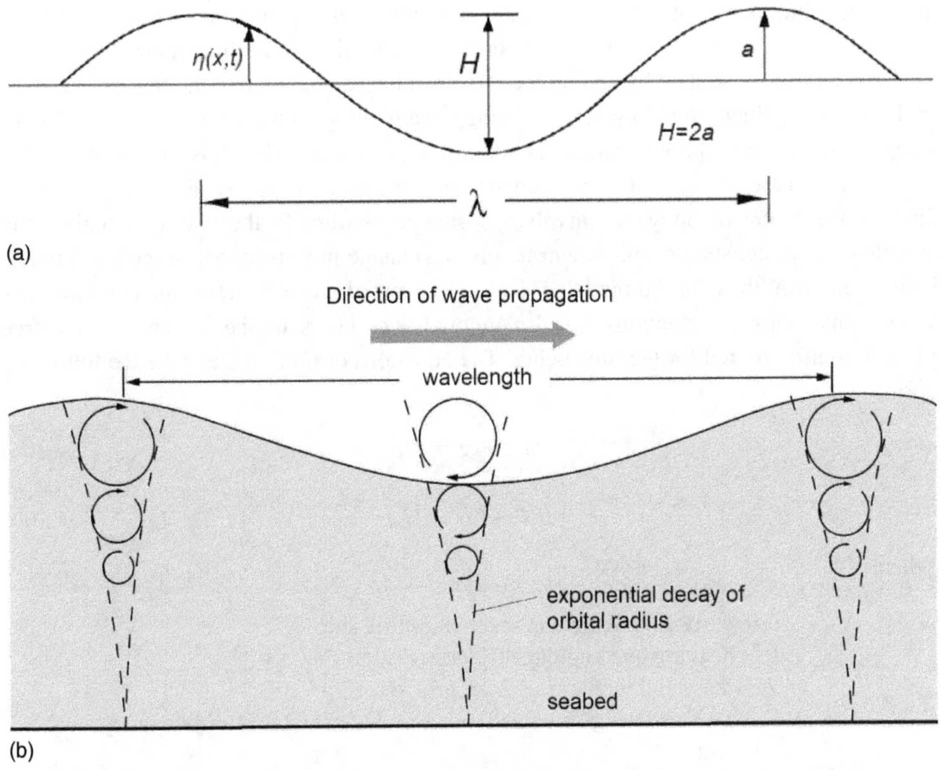

Figure 4.24 Linear wave. (a) Geometrical particulars of linear, free surface gravity wave. (b) Orbital motion and propagation mechanism of surface wave.

becomes increasingly elongated and in way of the seabed the particles perform a simple harmonic motion (Figure 4.24b). As a result of particle motion, a *dynamic pressure* develops within the wave that also oscillates with depth, and thus the total pressure at any fixed point below the surface is the sum of the hydrostatic pressure and the continuously varying dynamic pressure. The resulting pressure distribution is given by Bernoulli's equation for *irrotational* flow[17]:

$$\frac{p}{\rho}+g\eta+\frac{\partial\varphi}{\partial t}+\frac{1}{2}\left[u^2+v^2+w^2\right]=C \tag{4.28}$$

where u, v, w are component perturbation velocities, φ is the velocity potential and $C =$ constant along a streamline. The second term in Equation (4.28) represents the hydrostatic pressure and the third and fourth terms add up to the dynamic pressure. If we let $C = p_{atm}/\rho$, at the free surface

$$\frac{p}{\rho}+g\eta+\frac{\partial\varphi}{\partial t}+\frac{1}{2}\left[u^2+v^2+w^2\right]=0 \tag{4.29}$$

2. Ship motions and corresponding equations of motion

Let us now turn our attention to the ship. As a result of the assumptions mentioned earlier, the motions and loads have components only in the vertical centreline plane. The motion can thus be fully described by adding the effects of heave and pitch to the forward velocity V. Heaving is a linear motion whereas pitching is a rotational motion. The two motions are coupled, since they cannot occur independently of each other. The physical reason for this is that in the case of heave, the maintenance of the centre of gravity in the same vertical line, as the centre of buoyancy involves a change of trim. In the case of pitching, the maintenance of constant displacement involves a change in immersion. Surge has a minor effect and can thus be ignored, so that the forward velocity remains constant. All other translations and rotations are also ignored. The effect of the presence of the free surface is also ignored for the time being. The equations of motion then take the following form:

$$m\ddot{z}+cz=H_f \tag{4.30a}$$

$$J\ddot{\theta}+C\theta=H_m \tag{4.30b}$$

where

$\ddot{z},\ddot{\theta} =$ vertical and angular acceleration of ship
$z,\theta =$ heaving and angular displacement of ship
$m,J =$ mass and moment of inertia of the ship
$c,C =$ unit restoring force and moment due to changes in buoyancy force
$H_f,H_m =$ heaving force and pitching moment about the centre of gravity of the ship due to water pressures generated in flow and due to the ship's own oscillations

[17] If a flow is irrotational, the complete state of velocity at any one point is described in terms of a single function φ of position. For steady motion, the velocity potential is independent of time.

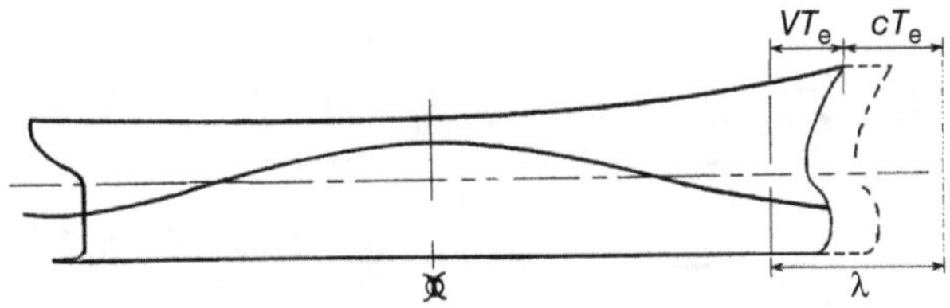

Figure 4.25 Effect of forward speed on wave encounter frequency [12]. Reprinted courtesy of Cambridge University Press.

Since, as mentioned earlier, heaving and pitching are coupled, Equations (4.30a) and (4.30b) form a coupled system so that terms in the excitation forces H_f, H_m that will affect both translational and rotational motion will arise. Experimental data have shown that a complete set of coupling terms has to be included in the two equations of motion so that they have to be written down as follows:

$$a\ddot{z} + b\dot{z} + cz + d\ddot{\theta} + e\dot{\theta} + g\theta = \overline{F}e^{i\omega t} \tag{4.31a}$$

$$A\ddot{\theta} + B\dot{\theta} + C\theta + D\ddot{z} + E\dot{z} + Gz = \overline{M}e^{i\omega t} \tag{4.31b}$$

Our aim is therefore to determine the values of $\overline{F}$ and $\overline{M}$, the *excitation forces*. In order to do so it is necessary to calculate the coefficients *a*, *b*, ..., *g* and *A*, *B*, ..., *G*. Once these have been determined, the hydrodynamic forces acting over the entire ship can be found and the equations of motion solved. The *hydrodynamic coefficients* have to be determined once the behaviour of the ship is adequately described. In the next step we therefore consider the behaviour of the ship.

3. Idealised representation of a ship travelling in a regular sea. Component velocities on a representative strip resulting from ship–environment interaction

We consider a ship advancing at a speed V_s into a system of regular waves propagating in the opposite direction at a speed V_w ($= g/\omega_0$ where ω_o is the propagation frequency). The frequency of encounter ω_e of each wave crest at any fixed point on the hull is then greater than that of the wave system. We let T_e be the period of wave encounter ($T_e < T_o$), and from Figure 4.25 we deduce that $VT_e + V_w T_e = \lambda$. Since the wavelength $\lambda = 2\pi V_w/\omega$, the wave encounter frequency, ω_e, can be written as:

$$\omega_e = \omega_o + \omega_o{}^2 V/g \tag{4.32}$$

We assume that the motion of the ship consists of small translations and rotations and consider an element of length *dx* in the longitudinal direction at a distance *x* from the origin *O* (Figure 4.26).

We disregard the effect that the flow may have on neighbouring regions of the immersed body. If we assume for the moment that the flow about the body is identical to that of the lower half of a fully submerged body, the free surface effects are effectively ignored. By making these assumptions we can replace the three-dimensional problem of a body whose form varies along its length and is floating on a free surface with a two-dimensional

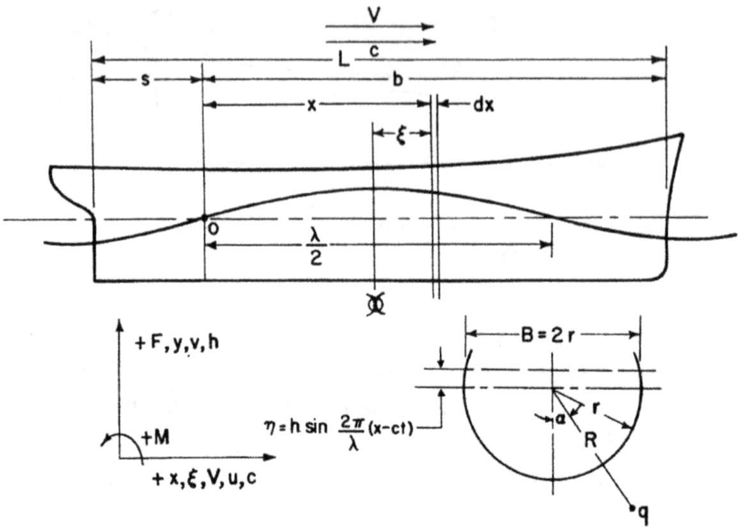

Figure 4.26 Idealised ship section in a regular wave [26]. Reprinted courtesy of the Society of Naval Architects and Marine Engineers.

problem of a regular cross-section. Under such conditions, the flow pattern between sections at x and $x + dx$ results from the following imposed motions:

1. The vertical velocity of the centre of the circle, v.

$$v = \dot{z} + \xi\,\dot{\theta} - V\theta \tag{4.33}$$

2. The vertical component of the wave orbital velocity, v_w.

$$v_w = \dot{\eta}\; e^{2\pi y/\lambda} = -\left[\frac{2\pi ac}{\lambda}\right] e^{2\pi y/\lambda} \cos\frac{2\pi}{\lambda}(x - ct) \tag{4.34}$$

   where a is the wave amplitude, λ is the wavelength, y $(= -Rcos\alpha)$ is the depth below the still-water level in the fluid and $\eta = asin2\pi(x - ct)/\lambda$.

3. The apparent variation of the immersed body radius with time, due to the passage of the wave, $r = r(t)$.

The hydrodynamic force acting on the element of length dx is then found by integrating the vertical component of the pressure over the area:

$$\frac{dF}{dx} = 2r \int_0^{\pi/2} p \cos\alpha d\alpha \tag{4.35}$$

where F is the vertical force and α is the polar angle as defined in Figure 4.26. On the sea surface, the pressure is equal to the atmospheric pressure. The dynamic pressure term accounts for the vertical component of acceleration that results from the orbital particle motion described earlier. The last term represents the second-order dynamic pressure effects that can be ignored, since we are using a linear theory. Furthermore, the hydrostatic pressure does not enter the formulation. Hence Bernoulli's equation for the flow about the two-dimensional hull element becomes:

$$p = \rho \frac{\partial \varphi}{\partial t} \tag{4.36}$$

p is now the first-order dynamic pressure.

4. Introduction of component velocities into Bernoulli's equation using potential flow theory and determination of pressures acting on a strip of the immersed body

Two-dimensional potential flow theory is applied to find the velocity potential φ of the flow about the cylinder for a relative velocity $v - v_w$. The total flow potential is found by adding this to the potential of the wave motion. The resultant has thus components relating to (a) body motion, (b) wave motion and (c) body–wave interaction. Henceforth the forces that result from wave motion and wave–body interaction are called *exciting forces*.[18] The total potential of the flow is then the sum of the potential due to the vertical velocity of the body in calm water φ_{bm}, the potential due to body–wave interaction φ_{bw} and the potential due to wave motion φ_w:

$$\varphi = \varphi_{bm} + \varphi_{bw} + \varphi_w \tag{4.37}$$

where

$$\varphi_{bm} = -v\frac{r^2}{R}\cos\alpha \tag{4.38a}$$

$$\varphi_{bw} = v_w\frac{r^2}{R}\cos\alpha = -\frac{2\pi ac}{\lambda}\frac{r^2}{R}e^{2\pi y/\lambda}\cos\alpha\cos\frac{2\pi}{\lambda}(x - ct) \tag{4.38b}$$

$$\varphi_w = ace^{2\pi y/\lambda}\cos\frac{2\pi}{\lambda}(x - ct) \tag{4.38c}$$

The above approach, which involves the determination and addition of individual components of the total potential, is possible because of the linearity assumptions introduced with regard to the wave profile and the magnitude of the ship's motions. The substitution of the individual terms of Equation (4.38) into the Bernoulli equation (Equation (4.29)) gives the component pressures that are introduced into Equation (4.35) to obtain the component hydrodynamic forces on the element. These are then added to obtain the total force on the element. In order to allow for the discrepancy due to the use of the semicircular cross-section in lieu of the correct hull form, a correction factor has to be introduced; also, the radius r has to be replaced by the local half-breadth $(B^*/2)$. After allowing for these, the total hydrodynamic force on the hull element is found to be [25]:

[18] The exciting force acting at any instant on the floating body is identical in magnitude to that which would act on supports which, if in place, would ensure that the floating body did not translate or rotate about any axis. It is thus the resultant force acting on the body after taking into account body–wave interaction.

$$\frac{dF}{dx} = \left[\overbrace{\rho(Sk_2k_4)\,\ddot{\eta}}^{heaving} + \overbrace{\left[N(\xi) - V_s\rho\frac{d(Sk_2k_4)}{d\xi} \right]\dot{\eta}}^{damping} + \overbrace{\rho g B^* \eta}^{restoring} \right] e^{2\pi y/\lambda} \tag{4.39}$$

The first term of the above equation represents the heaving force, the second term represents the damping force that is involved in the generation of surface waves, and the third represents the restoring force to hydrostatic equilibrium. The terms in this equation are:

S = area of circular hull cross-section
Sk_2k_4 = area taking into account the correct hull form and free surface effects
k_2 = *coefficient of accession to inertia* in vertical flow [49]. This is the ratio of the added mass to the mass of water of given volume. This coefficient allows for the correct hull form and is a function of the section coefficient and the beam–draught ratio (the immersed sectional area, $\pi r^2/2$, is thus replaced by Sk_2).
k_4 = free surface effects correction factor for k_2, using data provided by Ursell[46]
B^* = local breadth
V_s = ship's speed
$N(\xi)$ = damping force per unit vertical velocity $v\left(=\rho g^2\overline{A}^2/\omega_e^2\right)$, where
$\overline{A}$ = ratio of amplitude of waves radiated by oscillating body to the amplitude of heaving motion as calculated by Grim [16]

In this manner a distribution of hydrodynamic forces along the length of the ship is obtained.

5. Integration of pressures over the length of the body to calculate the hydrodynamic (exciting) forces acting on the entire immersed body

This distribution has to be integrated in order to obtain the total *heaving force F* and *pitching moment M* that act on the hull at any instant, according to the following [27]:

$$F = \int_{AP}^{FP} \frac{dF}{dx} dx \tag{4.40a}$$

$$M = \int_{AP}^{FP} \frac{dF}{dx}\left(\xi + \frac{1}{\pi}\frac{dS}{d\xi}\right) dx \tag{4.40b}$$

6. Solution of coupled equations of motion for heaving and pitching

Having evaluated the exciting forces that act on the ship's hull, we are in a position to solve the equations of motion (4.31a), (4.31b) and hence determine the longitudinal distribution of the forces acting on the hull girder. Following that we can determine the dynamic shear force and bending moment distributions.

For each degree of freedom, the equation of motion includes mass, damping and spring terms. We note at this point that since static equilibrium is not considered, the weight distribution and the static buoyancy of the ship are not introduced into the equations. It is

Table 4.3 *Coefficients in equations of motion for coupled heaving and pitching*

Coeff.	*Expression*	*Description*
A	$m + \rho \int (Sk_2k_4) d\xi$	Inertia coefficients that include mass, moment of inertia and sectional inertial masses (Lewis integrals) and their moments. The integrations are performed over the range $\pm L/2$ and ξ is the distance from the midship section.
A	$J + \rho \int (Sk_2k_4) \xi^2 d\xi$	
D	$\rho \int (Sk_2k_4) \xi d\xi$	
D	$\rho \int (Sk_2k_4) \xi d\xi$	
B	$\int N(\xi) d\xi$	Damping coefficients that represent dissipation of energy by radiation of surface waves as well as dynamic coupling that brings about a reduction in heave damping and an increase in pitch damping. Note the dependence of B, E and D on ship speed V_s.
B	$\int N(\xi) \xi^2 d\xi - 2VD - V_s \rho \int \xi^2 d(Sk_2k_4)$	
E	$\int N(\xi) \xi d\xi - 2V\rho \int (Sk_2k_4) d\xi - V_s \rho \int \xi d(Sk_2k_4)$	
E	$\int N(\xi) \xi d\xi - V_s \rho \int \xi d(Sk_2k_4)$	
C	$\rho g \int B^* d\xi$	Restoring coefficients that depend on displacement changes when heaving and pitching in still-water as well as fluid flow when the ship has an instantaneous angle of trim.
C	$\rho g \int B^* d\xi - V_s E$	
G	$\rho g \int B^* d\xi - V_s b$	
G	$\rho g \int B^* \xi d\xi$	

for this reason that the hydrostatic term is deleted in Bernoulli's equation (Equation (4.29)). For the equation of heave, the mass term is then proportional only to the instantaneous vertical acceleration of the element. This is called the *added mass force*,[19] because the energy imparted to the fluid by the ship, which changes the wave pattern, brings about a reduction in the acceleration of the ship; this can be viewed as being equivalent to an increase of the mass of the ship itself, and is not a negligible quantity. The damping term represents the decay of the vertical harmonic motion with time and is proportional to the instantaneous velocity. Finally, the spring term corresponds to the *hydrostatic force* that tends to restore the ship to the undisturbed position in the still-water condition. The latter force is proportional to the instantaneous distance from the still-water position and is equal to the change in mean static buoyancy of the element.

The force and moment excitation forces perform a harmonic oscillation that is described by the term $e^{i\omega t}$, where ω is assumed to be the encounter frequency. *Cross-coupling* is represented by coefficients d, e, g and D, E, G [26]. The role of each type of coefficient included in the equations of motion is described in Table 4.3.

It is important to note that the coefficients as well as the exciting forces are frequency dependent. This means that each solution obtained is valid only for an individual frequency, so that in order to design for the most severe structural loading, calculations have

[19] The added mass concept is discussed in Chapter 8.

to be carried out for the full range of frequencies, wave heights and headings at selected speeds and loading conditions.

In order to facilitate the solution of the system of Equations (4.26a–b), the excitation force and moment on the r.h.s. are written in complex form

$$\overline{F} = F_0 e^{i\sigma} \quad \text{and} \quad \Re e\left[\overline{F}e^{i\omega t}\right] = F_0 \cos(\omega t + \sigma) \tag{4.41a}$$

$$\overline{M} = M_0 e^{i\tau} \quad \text{and} \quad \Re e\left[\overline{M}e^{i\omega t}\right] = M_0 \cos(\omega t + \tau) \tag{4.41b}$$

where F_0, M_0 are the amplitudes of the heaving force and pitching moment given by Equations (4.40a) and (4.40b) and σ, τ are the phase angles at which maxima occur. If a steady state is reached in which a ship is advancing in a regular sea and only pitching and heaving arise, transient behaviour will have been damped out. Thus, only a particular solution of Equations (4.31a–b) is required.

Since the excitation force and moment are harmonic and we are dealing with a linear problem, the response will also take the form of harmonic functions. Therefore,

$$z = \overline{Z}e^{i\omega t} \qquad \text{where} \qquad \overline{Z} = Z_0 e^{i\delta} \tag{4.42a}$$

$$\theta = \overline{\theta}e^{i\omega t} \qquad \text{where} \qquad \overline{\theta} = \theta_0 e^{i\varepsilon} \tag{4.42b}$$

where Z_0, θ_0 are the absolute values of the heaving and pitching amplitudes and δ, ε are the corresponding phase lags. In these, the real part corresponds to the magnitude whereas the imaginary part corresponds to the phase angle of the response.

Substituting the assumed solutions (Equations (4.37a–b)) into Equations (4.26a–b) gives the complex amplitudes as follows:

$$\overline{Z} = \frac{\overline{M}Q - \overline{F}S}{QR - PS} \tag{4.43a}$$

$$\overline{\theta} = \frac{\overline{F}R - \overline{M}P}{QR - PS} \tag{4.43b}$$

where P, Q, R and S are given by

$$\left.\begin{aligned} P &= -a\omega^2 + ib\omega + c \\ Q &= -d\omega^2 + ie\omega + g \\ R &= -D\omega^2 + iE\omega + G \\ S &= -A\omega^2 + iB\omega + C \end{aligned}\right\} \tag{4.44}$$

The solution to the coupled Equations (4.31a) and (4.31b) is then obtained in the following form:

$$\left(-\omega^2 a + i\omega b + c\right)\overline{Z} + \left(-\omega^2 d + i\omega e + g\right) = \overline{F} \tag{4.45a}$$

$$\left(-\omega^2 A + i\omega B + C\right)\overline{\theta} + \left(-\omega^2 D + i\omega E + G\right) = \overline{M} \tag{4.45b}$$

or,

$$\overline{Z} = \frac{(-a\omega^2 + ib\omega + c)\overline{M} - (-A\omega^2 + iB\omega + C)\overline{F}}{(-d\omega^2 + ie\omega + g)(-D\omega^2 + iE\omega + G) - (-a\omega^2 + ib\omega + c)(-A\omega^2 + iB\omega + C)} \tag{4.46a}$$

$$\theta = \frac{(-D\omega^2 + iE\omega + G)\overline{F} - (-a\omega^2 + ib\omega + c)\overline{M}}{(-d\omega^2 + ie\omega + g)(-D\omega^2 + iE\omega + G) - (-a\omega^2 + ib\omega + c)(-A\omega^2 + iB\omega + C)} \tag{4.46b}$$

If we choose to ignore coupling, the above expressions reduce to

$$\overline{Z} = -\frac{\overline{F}}{-a\omega^2 + ib\omega + c} \tag{4.47a}$$

$$\theta = -\frac{\overline{M}}{-A\omega^2 + iB\omega + C} \tag{4.47b}$$

If solutions of the equations of motion are obtained for a series of elements, or strips, along the length of the ship, a distribution of the *vertical dynamical loads* acting on the hull can be obtained for any instant in time.

7. Dynamic shear force and bending moment distributions

In order to determine the wave-induced (dynamic) shear force and bending moment distributions, it is necessary to first obtain a solution of the equations of motion. The bending moment can then be determined at any instant by integrating the longitudinal force acting on the hull girder. This is done in two stages. The first integration gives the dynamic shear force and the second the dynamic bending moment. The total load acting on the hull girder includes [25]:

1. The loads due to the static weight and still-water buoyancy distributions. These are assumed to be zero, since only deviations from the static condition are considered.
2. Loads due to changes in displacement distribution resulting from the wave pattern and pitching and heaving motions. These loads are equal to

$$-\rho g B^* (z + \xi\theta - \eta)$$

3. Loads due to the effect of the pressure gradient in the wave (Smith effect)

$$-\rho g B^* \eta \left(1 - e^{2\pi y/\lambda}\right)$$

4. Loads due to the relative water velocity of wave and ship motion:

$$\boxtimes\Bigg[\underset{1}{-N(\xi) + V_s\rho\frac{d(Sk_2k_4)}{d\xi}}\Bigg]\left[\boxtimes\dot{z} + \xi\,\dot{\theta} - \underset{2}{V_s}\theta - \dot{\eta}\,e^{2\pi y/\lambda}\right]$$

   Term 1 is the free-surface damping effect whereas term 2 is taken from potential flow development.

5. Loads due to (a) the inertia forces of the ship mass δm of a given length segment $\delta\xi$, (b) the virtual mass of entrained water, and (c) the inertial effect of water in waves:

$$\overbrace{-\delta m\left(\ddot{z}+\xi\,\ddot{\theta}\right)}^{a}\ -\rho(Sk_2k_4)\left[\overbrace{\ddot{z}+\xi\,\ddot{\theta}\ -2V_s\,\dot{\theta}}^{b}\ -\overbrace{\ddot{\eta}\,e^{2\pi y/\lambda}}^{c}\right]$$

The negative signs arise because inertial loads act in a direction opposite to that of the body accelerations (d'Alembert's principle). Allowing for the above terms leads to the following *dynamic load equation*:

$$\begin{aligned}\frac{df}{dx} &= -\delta m\left(\ddot{z}+\xi\,\ddot{\theta}\right)-\rho(Sk_2k_4)\left(\ddot{z}+\xi\,\ddot{\theta}\ -2V_s\,\dot{\theta}-\ddot{\eta}\,e^{2\pi y/\lambda}\right)\\ &\quad-\left(N(\xi)-V_s\rho\frac{d(Sk_2k_4)}{d\xi}\right)\left(\dot{z}+\xi\,\dot{\theta}\ -V_s\theta-\dot{\eta}\,e^{2\pi y/\lambda}\right)\\ &\quad-\rho gB^{*}\left(z+\xi\theta-\eta\right)\end{aligned} \tag{4.48}$$

where f is the total load due to motions of the waves and ship. This consists of the hydrodynamic force F due to waves and also the forces due to ship motions. This can be integrated to obtain the longitudinal distribution of dynamic shear forces and bending moments as in the case of the quasi-static wave loading, discussed in Sections 4.3 and 4.4. These quantities are sinusoidally varying quantities at a frequency equal to that of wave encounter.

Comparisons of dynamic bending moments with those calculated using quasi-static theory showed that the former are much smaller, even allowing for the Smith correction. This was confirmed using theoretical calculations, tests and full-scale measurements. The same conclusion was reached using much more rudimentary theory, as mentioned at the beginning of this section. This difference arises because strip theory takes into account the interaction between the ship and waves, which is neglected in quasi-static analysis. The more important conclusions are summarised below:

1. The midship bending moment is virtually a second-order effect that depends on small local variations in the longitudinal distribution of loads.
2. The weight distribution is very important and the bending moment is sensitive to changes in weight distribution, virtual mass and damping forces.
3. The body–wave interference effect is important and is the reason for the reduced bending moments predicted by linear strip theory.

The assumptions upon which strip theory is based give us an idea of its limitations as well. The theory described in this section relates only to ships moving head-on in regular seas, so that the only resultant motions are coupled pitching and heaving. The effect of angle of incidence of waves and the resultant coupling with other types of motions are not included. In addition, accurate results cannot be expected in cases in which:

- The wave amplitude exceeds a certain limit.
- Excessive motions of the ship itself arise, such as in particularly heavy seas.
- In way of the fore and aft ends of the ship where ship motions cease to be small, the hull form is no longer wall-sided and phenomena such as slamming arise.

Table 4.4 *Methods used for the prediction of hull girder response in waves*[20]

Degree of Nonlinearity	*Method*
1	Linear
2	Froude–Krylov nonlinear
3	Body nonlinear
4	Body exact
5	Fully nonlinear (smooth waves)
6	Fully nonlinear

When the angle of incidence of the wave system is non-zero, all six degrees of freedom have to be introduced, and coupling between them has to be accounted for. This requires the solution of the complete system of equations of motion. This is a subject that is beyond the scope of the present text, and the interested reader is advised to consult the relevant literature.[21]

Clearly, because of the number of calculations involved, strip theory lends itself to computer-based solutions, and since its inception a number of approaches and software packages have been developed.

Computer programs are now available that incorporate nonlinear hydrodynamics theory in order to allow for the effects of large motions, bow flare and slamming. These effects are particularly important in the case of fast ships such as containerships as well as ships with blunt bows such as oil tankers. Nonlinear methods have been classified according to the degree of nonlinearity introduced (Table 4.4) [44].

Solutions based on nonlinear theories have shown that, when considering maximum lifetime loads, the rule bending moment can be substantially underestimated, as demonstrated in Table 4.5 that has been abstracted from [18]. Other authors have also found that direct calculations indicate that higher loads arise than the design loads used by classification society rules.

The question arises as to whether, even though these extreme sea states do in fact exist, ships actually need encounter them. Masters can receive support from weather stations and are able to take avoiding action in the case of extreme seas[22] (Figure 4.27). As a result, ships do not need to be subjected to such extreme loads. The situation is different in the case of fixed offshore platforms and FPSOs.

In order to assess the results of ongoing research, a benchmark study was conducted by the members of ISSC Committee I.2 Loads (2000) in which hull girder response was modelled in a variety of ways and incorporated different kinds of nonlinear behaviour [17].

Significant differences were noted, as can be seen from Figure 4.28 that gives results for one particular case. On this basis it appears that more work is required before results from nonlinear theory can be used on a routine basis in design work or in the development of strength formulations.

[20] A detailed appraisal of these methods is included in [44].

[21] A comprehensive reference is [4]. Further references are [10, 29, 32, 50]

[22] It is in the interest of charterers that ships follow the fastest and safest route to their destination so that route-planning is implemented on a routine basis.

Table 4.5 *Midship design wave bending moment from different methods*

Method	*Wave bending moment ($\times 10^5$ kNm)*
Rules-hogging	1.5
Rules-sagging	1.9
Linear long term distribution, 24 years	4.1
Nonlinear long term distribution, 24 years-sagging	6.3
Nonlinear, 10 years design storm, 3 h duration-sagging	5.1
Nonlinear, 100 years design storm, 3 h duration-sagging	5.3
Nonlinear, coefficient of contribution method-sagging	6.1

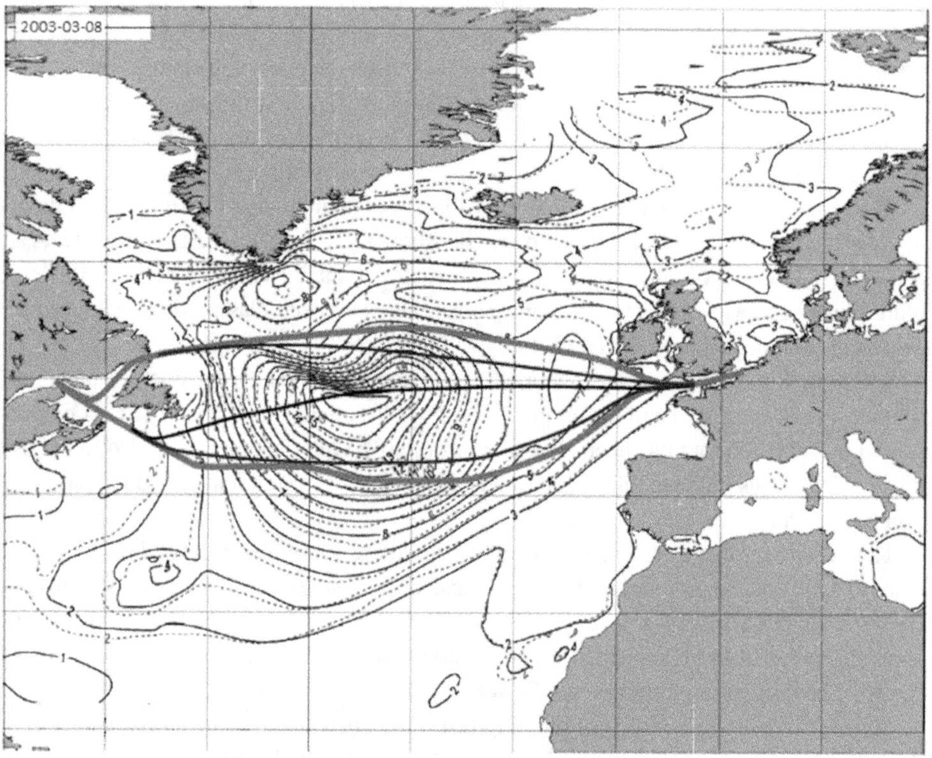

Figure 4.27 Avoidance of heavy weather in the North Atlantic on the basis of forecasts.

4.9 IACS Primary Load Estimation and Longitudinal Strength Requirements

Let us now consider the basis upon which primary loading of ship hulls and corresponding strength requirements are calculated in classification society rules. Before doing so, it is necessary to comment on the basic purpose of a standard such as that implemented by IACS. The aims of IACS, as stipulated in the Background Notes to the Common Structural

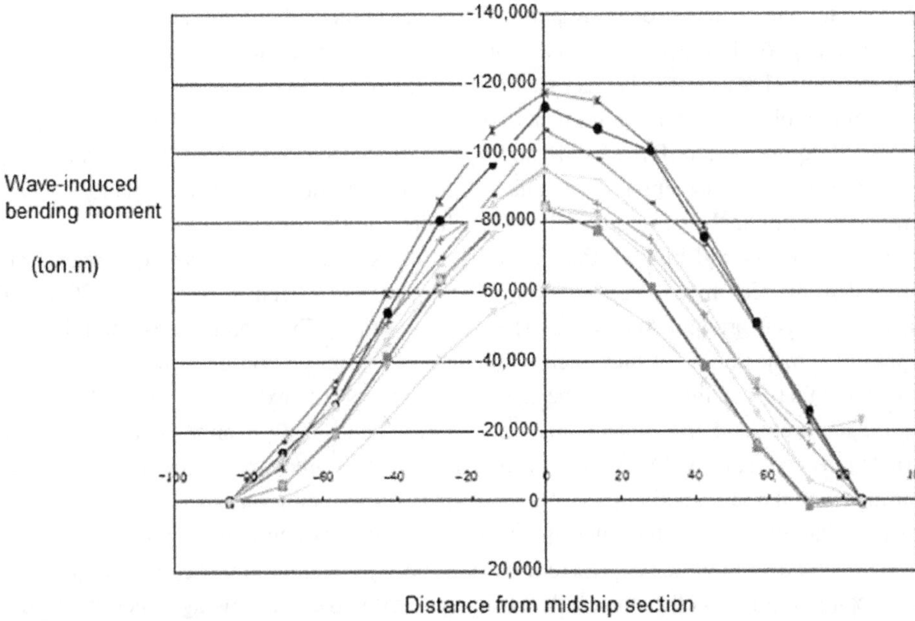

Figure 4.28 Longitudinal distribution of wave bending moment using nonlinear formulations (Basic hull, forward speed = 10 knots) [18].

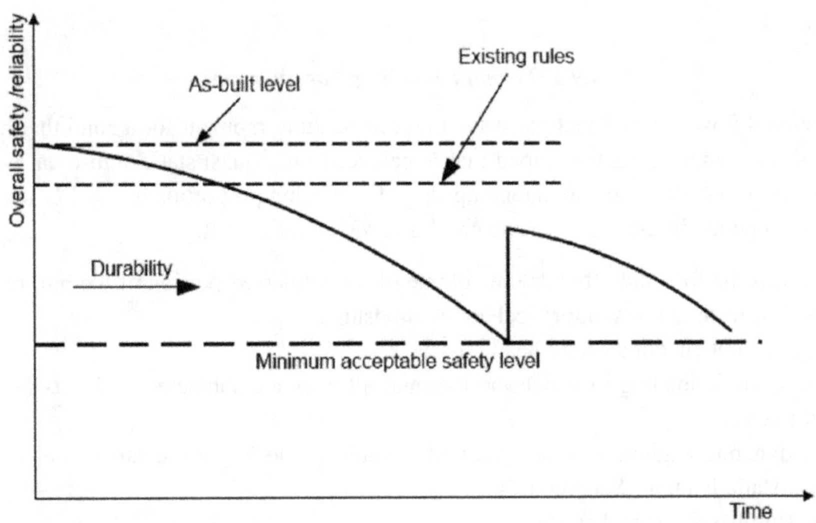

Figure 4.29 Structural safety and durability concepts [21]. Reprinted courtesy of IACS, London.

Rules for oil tankers [21], are to address the issues of *safety* on the one hand and *performance and durability* on the other (Figure 4.29).

Safety is achieved by ensuring that the ship has adequate strength and watertight integrity throughout its working life. Performance and durability are ensured by stipulating appropriate fatigue and strength requirements. Clearly then, structural design that is based on regulatory requirements does not aim at an 'optimum' structure, as understood by major stakeholders. This question is discussed further in Chapter 14.

Current global strength requirements for ocean-going ships are described in documents Unified Requirement (UR) S7 [24] and S11 [23] that cater to all ships whose length exceeds 90 metres, except for oil tankers and bulk carriers. More information on the Unified Longitudinal Strength Standard is included in [19, 30].

For oil tankers and bulk carriers whose length exceeds 150 metres, requirements are covered in the Common Structural Rules (CSR) that have been in force since 2006 and have since been brought together as Harmonised Rules. The Common Structural Rules represent a major effort at introducing a degree of rationalisation as well as an acknowledgement that ships that traverse the same sea lanes ought to exhibit similar qualities with regard to safety and the protection of the environment. They cannot therefore be constructed in accordance with different requirements. At the same time, transparency has been introduced to the rule-making process so that the maritime community is in a better position to appreciate the basis of the rules and also the direction of ongoing efforts.

We shall now turn to the expressions used to specify the minimum strength requirements for ocean-going ships other than bulk carriers and oil tankers, following which the results of a comparison between IACS UR S11 requirements and linear strip theory will be given. In the case of the CSR, the methodology followed its validation and use is described in several chapters of this text.

4.9.1 Primary Loading Formulation

In Section 4.5 we showed that the wave-induced bending moment for a solid that has the form of a rectangular parallelepiped that is balanced on a quasi-static sinusoidal wave is proportional to h_eBL^2, the corresponding shear force being proportional to h_eBL. Real ship hulls that operate in the sea environment, however, differ in that:

1. Towards the fore and aft ends the volume of the immersed portion of the hull reduces.
2. The hull form is not symmetrical about amidships.
3. Ships do not encounter only head-on seas.
4. The dynamic loading to which the fore and aft ends are subjected to differs in nature and intensity.
5. The dynamic loading to which the hull is subjected to is not the same as elementary quasi-static loading (Section 4.7).
6. The ship has a forward speed.

As a result, the loading will differ at the two ends and will vary along the length of the hull girder. The sagging bending moment will therefore not be equal to the hogging bending moment. In addition to nonlinearities due to hull form, events such as slamming and whipping as well as the conditions that prevail in the fluid itself, in particular in heavy seas (irregular and breaking waves, spraying), cause further divergence from linear response.

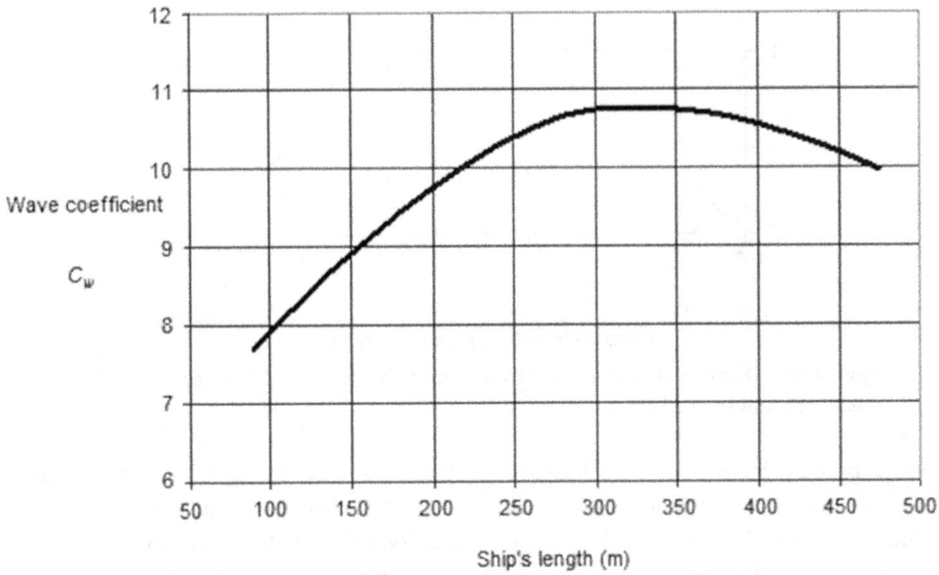

Figure 4.30 Wave coefficient C_w.

The Unified Requirement of IACS (S11) [23] includes the following two formulas for the wave-induced bending moment for ships that operate worldwide. The notional probability of failure is equal to 10^{-8} for a 25-year period of operation, based on wave statistics for the North Atlantic area.

For hogging (positive moment),

$$M_W = +190MC_wL^2BC_B\times10^{-3} \qquad \text{(kNm)} \tag{4.49a}$$

For sagging (negative moment),

$$M_W = -110MC_wL^2B(C_B+0.7)\times10^{-3} \qquad \text{(kNm)} \tag{4.49b}$$

where L, B are in metres, $C_B \geq 0.60$, C_w is a *wave coefficient* and M is a *distribution coefficient* whose value for each abscissa along the ship's length is indicated in Figure 4.31. The wave coefficient C_w is defined in terms of the ship's length L as follows (Figure 4.30):

$$C_w = 10.75 - \left(\frac{300-L}{100}\right)^{1.5} \qquad 90 \leq L \leq 300 \tag{4.50a}$$

$$C_w = 10.75 \qquad 300 < L < 350 \tag{4.50b}$$

$$C_w = 10.75 - \left(\frac{L-350}{150}\right)^{1.5} \qquad 350 \leq L \tag{4.50c}$$

These expressions are the result of a consensus between classification societies with regard to primary loading and are a step towards a rational basis for the estimation of hull

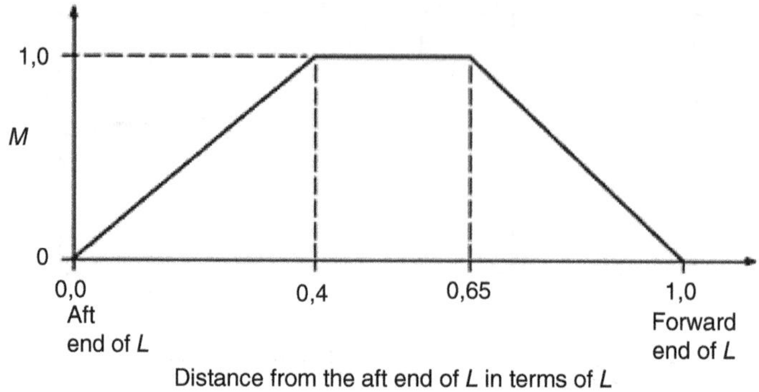

Figure 4.31 Distribution coefficient M for primary wave-induced bending moment [23]. Reprinted courtesy of IACS, London.

girder primary loads. They are based on expressions used by individual classification societies that have proven to be satisfactory, in the sense that ships that have an acceptable failure rate have been designed. In these expressions the block coefficient C_B has been introduced to allow for hull form nonlinearities whereas the wave coefficient C_w is introduced to allow for the dependency of wave loading on critical wave height, which is in turn a function of ship length L. Other nonlinearities such as those due to heavy seas effects are not directly catered for in these expressions. Ongoing and future research aims at (1) understanding further the effects of these nonlinearities, (2) defining sea data that are to be universally used for hull girder loading calculations, (3) developing and evaluating results from nonlinear formulations, and (4) studying the effects of form and mass distribution. A number of important questions arise, such as whether a single expression is suitable for all ship types and sizes. Until such issues are resolved, these formulas will continue to be used.

The sign conventions that is used by IACS for hull girder loading are shown in Figure 4.32. The corresponding expressions used to calculate the vertical shear force along the hull girder in conjunction with Equations (4.49a) and (4.49b) are as given below [22].

(a) In hogging:

$$Q_{wv-pos} = 0.52 f_{q-pos} f_p C_w L B C_B \qquad \text{(kN)} \qquad (4.51a)$$

(b) In sagging:

$$Q_{wv-pos} = -0.52 f_{q-neg} f_p C_w L B C_B \qquad \text{(kN)} \qquad (4.51b)$$

The values of the distribution coefficients $f_{q\text{-}pos}$ and $f_{q\text{-}neg}$ are given in Figure 4.33.

4.9.2 Primary Strength Requirements

Following the definition of hull girder primary loading, the engineering theory of beam bending can be used to specify strength and stiffness requirements. This is an elementary

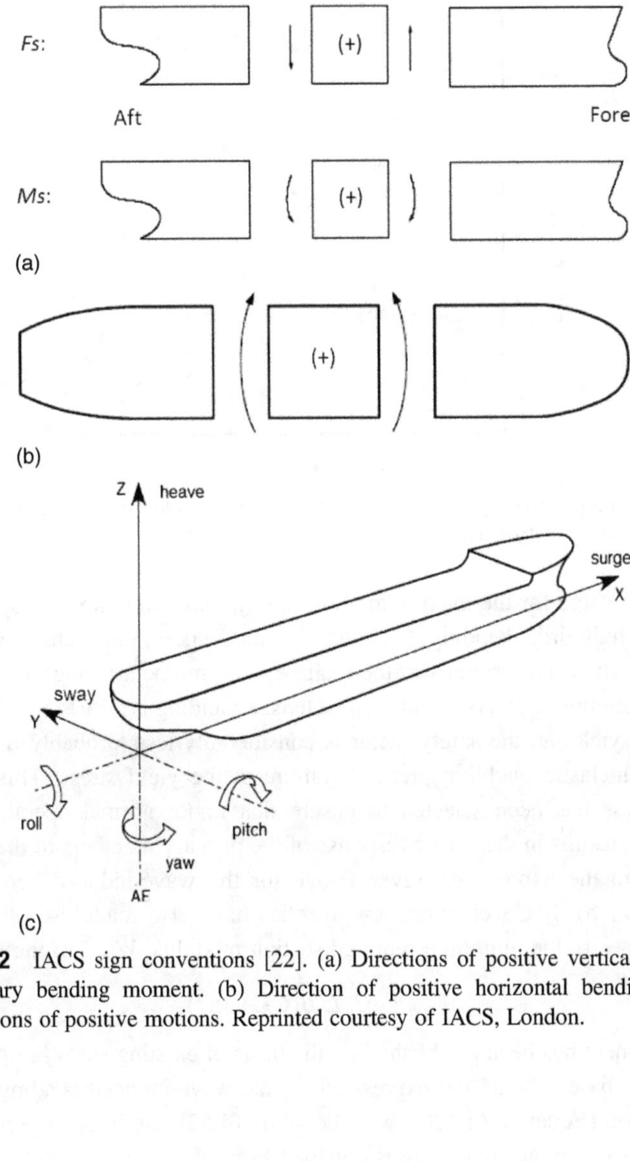

Figure 4.32 IACS sign conventions [22]. (a) Directions of positive vertical shear force and primary bending moment. (b) Direction of positive horizontal bending moment. (c) Directions of positive motions. Reprinted courtesy of IACS, London.

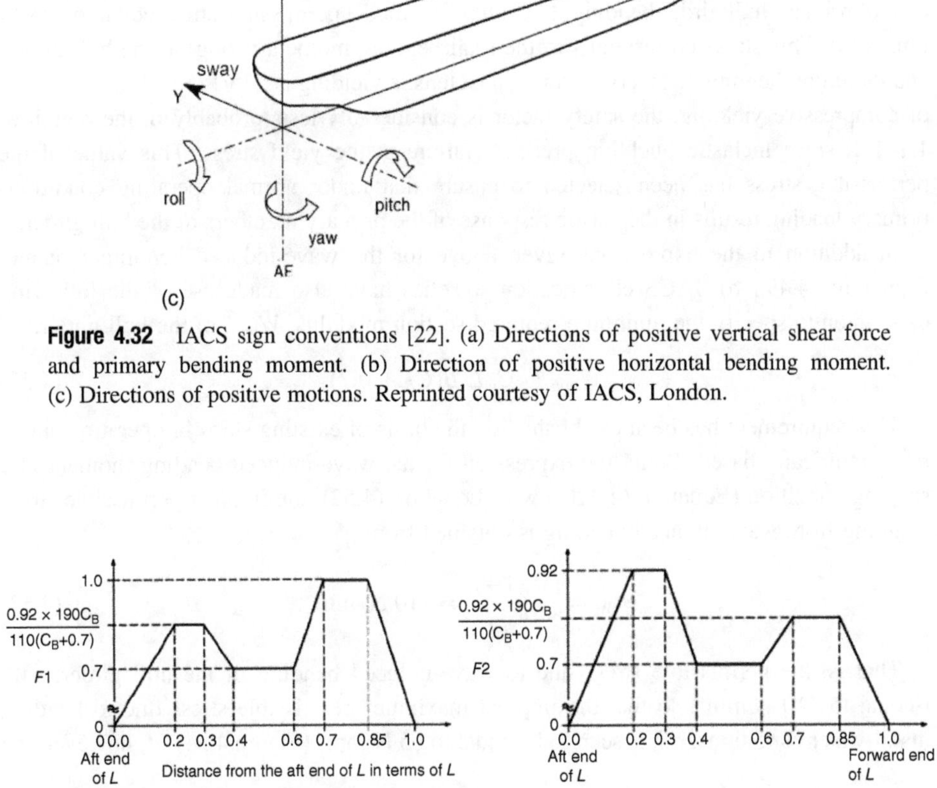

Figure 4.33 IACS distribution coefficients $f_{q\text{-}pos}, f_{q\text{-}neg}$ for wave-induced vertical shearing force [22]. Reprinted courtesy of IACS, London.

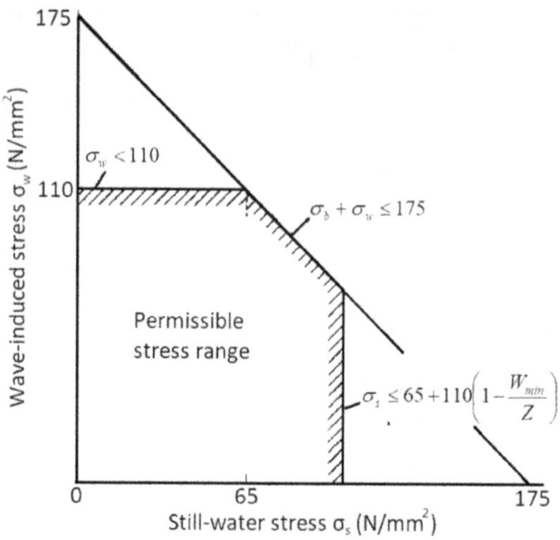

Figure 4.34 Permissible stresses for still-water and wave-induced bending amidship [31]. Reprinted courtesy of Elsevier.

step, provided a value for the maximum allowable (*permissible*) stress is specified. In the case of primary hull girder bending, IACS has specified a permissible stress equal to 175 N/mm^2 [31]. This stress corresponds to the total bending moment acting on the hull girder, and therefore the implied safety factor against tensile yielding is 245/175 = 1.4. In the case of compressive yielding, the safety factor is considerably less, probably in the region of 1.1–1.2, since inelastic buckling precedes failure at the yield stress. This value of the permissible stress has been selected to ensure that under normal operating conditions primary loading results in the elastic response of the primary members of the hull girder.[23]

In addition to the expressions given above for the wave-induced bending moment, Equations (4.49a–b), IACS classification societies have also made use of the following expression to specify the minimum required section modulus, W_{min}, of the hull girder:

$$W_{min} = C_w L^2 B(C_B + 0.7) \tag{4.52}$$

This requirement has been established on the basis of existing ships in operation and is thus empirically based. Using the expression for the wave-induced bending moment in a sagging condition (Equation (4.42b)) with Equation (4.52), the implied permissible stress resulting from wave-induced bending is obtained from:

$$\sigma_W = \frac{|M_W(-)|}{W_{min}} = 110\ \mathrm{N/mm^2} \tag{4.53}$$

That is, the permissible stress due to wave-induced bending of the hull girder, σW, isequal to 110 N/mm^2, so that the implied maximum permissible stress due to bending instill-water conditions, σS, is seen to be equal to 65 N/mm^2 (Figure 4.34). If, however, the

[23] As mentioned elsewhere (Table 4.5), occasions do arise in which the total bending moment specified by IACS rules is exceeded.

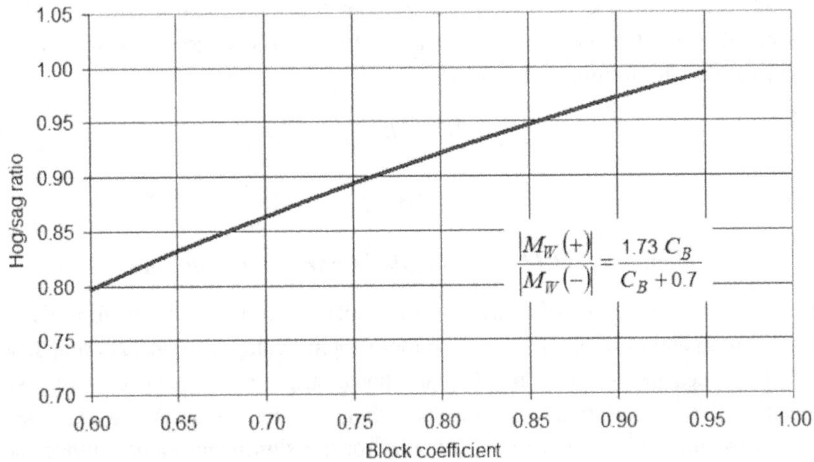

Figure 4.35 Hog/sag ratio versus block coefficient.

actual section modulus, Z, exceeds the minimum required, W_{min}, then the maximum permissible stress, σ_{SW}, is allowed to increase accordingly. The permissible stresses (in N/mm^2) resulting from primary loading of the hull girder are summarised below:

(a) Total primary bending stress, σ_T:

$$\sigma_T \leq 175 \qquad \text{and} \qquad \sigma_T = \sigma_S + \sigma_W \tag{4.54a}$$

(b) Normal stress due to wave-induced bending, σ_W:

$$\sigma_W \leq 110\ (\text{N/mm}^2) \tag{4.54b}$$

(c) Normal stress due to still-water bending, σ_S:

$$\sigma_S \leq 65 + 110\left(1 - \frac{W_{min}}{Z}\right) \tag{4.54c}$$

The relative contribution of the wave-induced bending moment when compared to that of the still-water bending moment increases with ship size. We see then that since fatigue life is dependent on the stress range amplitude, the fatigue life of larger ships will be much less than that of smaller ones that have been designed to the same total permissible stress. This deficiency is resolved in part by setting a relatively low permissible wave-induced stress σ_W of 110 N/mm^2.[24]

The implied hog/sag ratio from Equations (4.49a) and (4.49b) is:

$$\frac{|M_W(+)|}{|M_W(-)|} = \frac{1.73C_B}{C_B + 0.7} \tag{4.55}$$

which, as expected, approaches unity as the block coefficient increases (Figure 4.35).

[24] Nonetheless, experience bears out the fact that smaller ships have a longer fatigue life. Another factor that contributes to this is that in the case of smaller ships, it is local strength requirements that are critical so that an additional safety margin is thereby introduced.

We now need to consider stiffness requirements. Since the bending stiffness of the hull girder is equal to EI, the moment of inertia, I, of the cross-section of a steel ship can be specified to ensure minimum stiffness. Thus,

$$I_{min} = 3C_w L^3 B(C_B + 0.7) \tag{4.56}$$

4.9.3 Comparison of IACS Strength Requirements with Strip Theory

It is necessary to gain an idea of how the results obtained from the use of formulas such as Equations (4.49a) and (4.49b) compare with more sophisticated methods. A comparison was carried out between the S11 requirements and linear strip theory predictions for a series of containerships [28]. A selection of the results obtained is included in Figures 4.36 and 4.37.

Figure 4.36 summarises the results obtained for the significant values (twice the RMS values) of the wave-induced bending moments for a series of 14 ships operating at the indicated speeds in three different sea spectra.

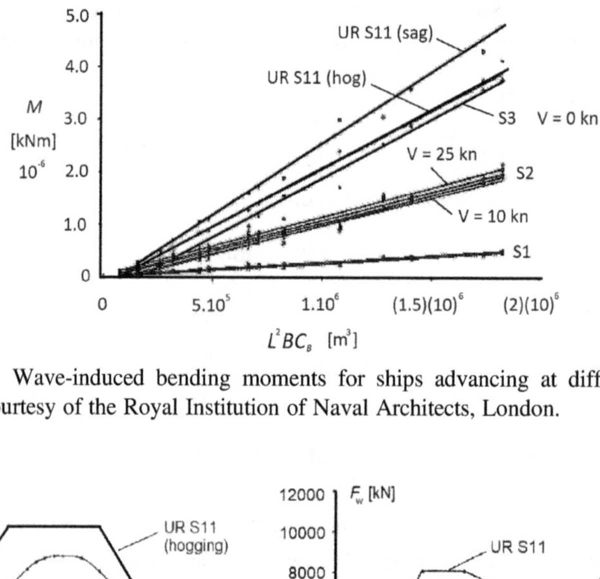

Figure 4.36 Wave-induced bending moments for ships advancing at different speeds. Reprinted courtesy of the Royal Institution of Naval Architects, London.

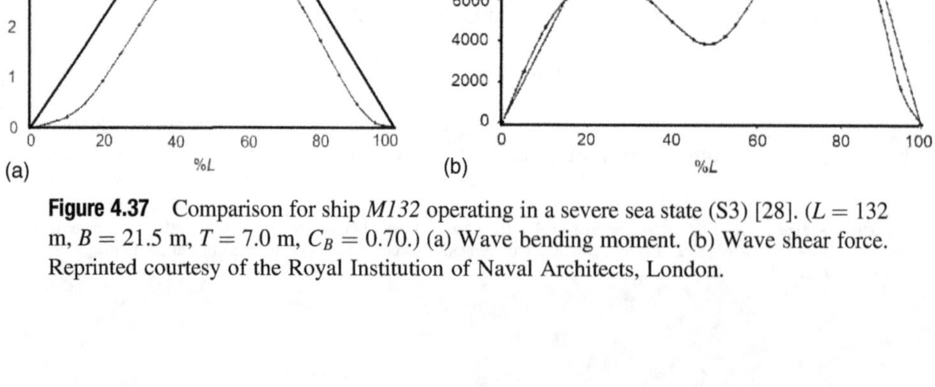

Figure 4.37 Comparison for ship *M132* operating in a severe sea state (S3) [28]. (L = 132 m, B = 21.5 m, T = 7.0 m, C_B = 0.70.) (a) Wave bending moment. (b) Wave shear force. Reprinted courtesy of the Royal Institution of Naval Architects, London.

The results are plotted as a function of L^2BC_B. The strip theory program used was based on the Savitsky–Tuck–Faltinsen formulation and the sea spectra considered were S1 ($H_s = 2$ m, $T_z = 9$ sec), S2 ($H_s = 7.5$ m, $T_z = 10.5$ sec) and S3 ($H_s = 16$ m, $T_z = 13$ sec). It is seen that in the case of S3, which is a severe sea state, the calculated wave-induced bending moment approaches and may exceed the values specified by UR S11.

Figure 4.37 includes data for a single ship operating in sea state S3. This is an extreme sea state having a probability of occurrence of 1 in 100,000 in the North Atlantic that can be used as a substitute for long-term prediction. It is seen that in this instance both the wave-induced bending moment and the vertical shear force are comparable to the S11 distributions with the IACS formulation providing results that are on the safe side with regard to bending strength.

References

1. Abrahamsen E. (1958) The Strength of Dry Cargo Ships. *Europ. Shbldg.* Vol. 7 No. 3 p. 62 & No. 4 p. 35.
2. Abrahamsen E. (1967) *Recent Developments in the Practical Philosophy of Ship Design.* Trans SNAME, Spring Mtg. Montreal.
3. American Bureau of Shipping (2018) *Rules for Building and Classing Steel Vessels.* Part 5C - Specific Vessel Types. Chap. 5: Vessels Intended to Carry Containers (130-450 meters in length) 619–950.
4. Beck R. F. Cummins W. E. Dalzell J. F. Mandel P. Webster W. C. (1989) Motions in Waves. In Lewis E. V. (Ed) *Principles of Naval Architecture (Second Revision) Vol. 3 Motions in Waves and Controllability.* Jersey City, NJ: SNAME 1–190.
5. Biles J. H. (1908) *The Design and Construction of Ships. Vol. 1 – Calculations and Strength.* London: C. Griffin & Co Ltd.
6. Clayton B. R. Bishop R. E. D. (1982) *Mechanics of Marine Vehicles.* London: E. & F.N. Spon.
7. Conn J. F. C. Miller N.S. (1961) The Effect of Various Factors on Wave Bending Moments. *Transactions NECIES* Vol. 77 107–190.
8. Det Norske Veritas (2005) *Hull Structure Course.*
9. Euler L. (1759) *Examen des efforts qu' ont à soutenir toutes les parties d' un vaisseau dans le roulis et dans le tangage ou recherches sur la diminution de ces mouvemens (Study of the efforts needed to sustain all parts of a vessel in rolling and pitching, or investigations on the reduction of these motions).* Paris: Acad. Roy. Sc. (published 1771).
10. Evans J. H. (Ed) (1975) *Ship Structural Design Concepts.* Centreville, MD: Cornell Maritime Press.
11. Evans J. H. (Ed) (1983) *Ship Structural Design Concepts – Second Cycle.* Centreville, MD: Cornell Maritime Press.
12. Faltinsen O. (1993) *Sea Loads on Ships and Offshore Structures* Cambridge: Cambridge University Press.
13. "File:Fjordn Surface Wave Boat.Jpg." n.d. Wikimedia.Org. Accessed January 6, 2021b. https://commons.wikimedia.org/wiki/File:Fjordn_surface_wave_boat.jpg.
14. Fukuda J.-I. (1967) Theoretical Determination of Design Wave Bending Moments. *Japan Shipb. & Mar. Eng.* Vol. 2 No. 3 13–22.

15. Gerstner F. J. (1802) *Theorie der Wellen (Theory of Waves)*. Prague: Abhand. Kön. Böhmischen Gesel. Wiss.
16. Grim O. (1953) Berechnung der durch Schwingungen eines Schiffskörper erzeugten hydrodynamischen Kräfte (Calculation of the hydrodynamic forces generated by oscillations of a ship's hull). *Jahr. Schiffbau. Gesell.* Vol 47 277–299.
17. Guedes Soares C. et al. (2000) Loads. Rept. of Technical Committee I.2 in: Ohtsubo H. Sumi Y. (Eds) *Proc. 14th Int. Ship & Offsh. Struct. Cong*. Vol. 1. Amsterdam: Elsevier 63–132.
18. Guedes Soares C. Fonseca N. Pascoal R. (2004) Long term prediction of non-linear vertical bending moments on a fast monohull. *Applied Ocean Research* Vol. 26 288–297.
19. Guedes Soares C. (1992) Discussion on Nitta A. et al.: Basis of IACS Longitudinal Strength Standard *Mar. Struct.* Vol. 5 559–561.
20. International Association of Classification Societies (1997) *Bulk Carriers. Guidance and Information on Bulk Cargo Loading and Discharging to Reduce the Likelihood of Over-stressing the Hull Structure.*
21. International Association of Classification Societies (2006) *Common Structural Rules for Oil Tankers. Background Document. Section 2 – Rule Principles.*
22. International Association of Classification Societies (2020) *Common Structural Rules for Bulk Carriers and Double Hull Oil Tankers.*
23. International Association of Classification Societies (2020) *Longitudinal Strength Standard. Unified Requirement S11 (Rev. 10) in: Requirements concerning Strength of Ships.*
24. International Association of Classification Societies (2010) *Minimum Longitudinal Strength Standards Unified Requirement S7 (Rev. 4) in: Requirements concerning Strength of Ships.*
25. Jacobs W. R. (1958) The Analytical Calculations of Ship Bending Moments in Regular Waves. *J. Ship Res.* Vol. 2 20–29.
26. Korvin-Kroukovsky B. V. (1955) Investigation of Ship Motions in Regular Waves. *Trans SNAME* Vol. 63 386–418.
27. Korvin-Kroukovsky B. V. Jacobs W. R. (1957) Pitching and Heaving Motions of a Ship in Regular Waves. *Trans SNAME* Vol. 65 590–615.
28. Motok M. D. Jovović J. (2007) Wave-induced Shear Force and Bending Moment for Service of Ships – Comparison & some Interpolation Procedures. In Developments in Classification and International Regulations. RINA Conf. 24–25 Jan; London; 111–118.
29. Newman J. N. (1977) *Marine Hydrodynamics*. Cambridge, MA: MIT Press.
30. Nitta A. Arai H. Magaino A. (1992) Basis of IACS Unified Longitudinal Strength Standard. *Mar. Struct.* Vol. 5 1–21.
31. Nitta A. et al. (1994) On C. Guedes Soares Discussion of Paper by A. Nitta et al. 'Basis of IACS Longitudinal Strength Standard'. Vol. 5 No. 1, 1992, 1–21. *Mar. Struct.* Vol. 7 567–572.
32. Ogilvie T.F. (1964) Recent Progress Towards the Understanding and Prediction of Ship Motions *Proc. 5th Symp. Naval Hydr.* Washington, DC: ONR.
33. Papanikolaou A. (1988) *Ship Design Vol. 1: Preliminary Design Methodology* (Μελέτη Πλοίου Τόμος Α: Μεθοδολογία Προμελέτης) Athens (in Greek).
34. Papazoglou V. Bardis L. (1989) *Ship Statics and Dynamics*. (Στατική και Δυναμική Πλοίου ΙΙ) Lecture notes Div. Mar. Struct. Sch. Naval Arch. & Mar. Eng. NTUA Athens.
35. Paulling J. R. (1988) Strength of Ships. In Lewis E. V. (Ed) *Principles of Naval Architecture Second Revision Vol. 1 Stability and Strength*. Jersey City, NJ: SNAME 205–299.

36. Pedersen P. T. (1985) Torsional Response of Containerships. *J. Ship Res.* Vol. 29 194–205.
37. Rawson K. J. (1976) *Structural Design Criteria*. Lecture Notes RCNC Office Dept. Mech. Eng. Univ. Coll. London Univ. of London.
38. Rawson K. J. Tupper E.C. (2001) *Basic Ship Theory Vol. 1 Hydrostatics and Strength*. 5th ed. Oxford: Butterworth Heinemann.
39. Reed E. J. (1871) "On the unequal distribution of weight and support in ships, and its effects in still-water, in waves, and in exceptional positions on shore". *Phil. Trans. R. Soc. Lond.* 161, 413–465.
40. Robb A. M. (1952) *Theory of Naval Architecture*. Glasgow: Charles Griffin & Co Ltd.
41. Salvesen N. Tuck E. O. Faltinsen O. (1970) Ship Motions and Sea Loads. *Trans SNAME* Vol. 78 250–287.
42. Smith W. E. (1883) Hogging and Sagging Strains in a Seaway as Influenced by Wave Structure. *Trans INA*. Vol 24 136–153.
43. Stokes G. G. (1849) On the Theory of Oscillatory Waves *Trans. Camb. Philos. Soc.* Vol. 8 441–455.
44. Temarel P. et al. (2009) *Loads*. Committee I.2 Rept. in Jang C. D. Hong S. Y. (Eds) *Proc. 17th Int. Ship & Offsh. Struct. Cong.* Vol. 1. Seoul: Seoul University Press 127–210.
45. Thomas G. (2011) Ship Global Response in Still-water. Lecture notes Australian Maritime College.
46. Ursell F. (1954) Water Waves Generated by Oscillating Bodies. *Quart. J. Mech. & Appl. Math.* Vol. 7 427–437.
47. Vedeler G. (1924) On the Torsion of Ships. *Trans. INA* 164.
48. Wah T. (Ed) (1960) *A Guide for the Analysis of Ship Structures*. Washington, DC: US Department of Commerce.
49. Weinblum G. St. Denis M. (1950) On the Motion of Ships at Sea. *Trans SNAME* Vol. 58 184–248.
50. Wehausen J. V. (1971) The Motion of Floating Bodies *Ann. Rev. Fl. Mech.* Vol. 3 267–268.

5

Hull Structure, Mechanical Equipment and Cargo-Related Loads

> The fewer moving parts, the better. Exactly. No truer words were ever spoken in the context of engineering.
>
> *Christian Cantrell*

In the previous chapter the primary loading for the typical operating conditions of a ship at sea or in port was considered in general terms. In the present chapter we shall outline several aspects of routine operations, as well as specific instances of hull girder loading that result in bending moment and shear force distributions that may be different from those that are normally expected. Some of these occur on a day-to-day basis, others infrequently, whereas others may never be encountered (Table 5.1). Nonetheless, it is important to appreciate the nature of the distributions of the longitudinal shear force and bending moment in these circumstances.

In this chapter we shall discuss, apart from major accidents and sloshing of liquid cargo in tanks, all the loads sources included in Table 5.1, in the following order:

1. Hull structure–related loads. We shall focus on the loads that arise on a number of specific occasions and have an effect on the entire hull girder as well as at a local level. In addition to these we shall consider several other loads that are related exclusively to the hull structure itself. Loads related to the environment have been described in Chapter 3, apart from hydrostatic and hydrodynamic pressure, which was discussed in Chapter 4.
2. Loads arising as a result of the operation of mechanical equipment
3. Loads related to the presence of cargo

In the last section of the chapter, load transfer within the hull girder is discussed and the most important load transfer mechanisms described.

5.1 Hull Structure–Related Loads

The following loads arise as a direct consequence of the layout and operation of the hull girder:

- Weight of structure (lightship)
- Inertial loads

Table 5.1 *Frequency of action of typical hull girder load sources*

Load Source	*Time Scale*
Still-water	Continuous (varies for each loading condition)
Sea environment	Continuous (varies for each voyage)
Dry cargo, wet cargo	Discontinuous (varies for each voyage)
Dry-docking	Occasional
Launching	Once-in-a-lifetime
Grounding	On average less than once-in-a-lifetime
Conversion	On average less than once-in-a-lifetime
Major accident	On average less than once-in-a-lifetime

- Fabrication process
- Dry-docking
- Launching
- Grounding and collision

5.1.1 Hull Weight

Hull weight is defined as the lightship minus the weight of the anchors, chains, windlasses, steering gear and main propulsion machinery. The lightship of ocean-going ships varies from approximately 1,000 to 12,000 metric tonnes, depending on ship size and type. For example, the lightship of a ship of a 30,000 dwt tonne bulk carrier lies in the range 5,000–6,000 tonnes. Larger ships such as VLCCs, very large ore carriers (VLOCs) and large passenger ships have considerably more lightship whereas smaller ships such as small coasters and tugs have a lightship that is less than 1,000 tonnes. The steel weight is the most important component of the lightship, which also includes the weight of fixed equipment (outfitting and machinery) but not that of fuel, consumables, cargo, crew and passengers. The hull weight is used when calculating the primary hull girder load distribution (Chapter 4).

In addition to being associated with hull girder longitudinal bending, the weight of the steel structure gives rise to in-plane and bending action of individual structural members at a local level. However, the resulting stresses are small and are ignored in strength calculations.

For structural design purposes, of importance is the *longitudinal weight distribution* for which it is assumed that weights are distributed symmetrically about the centreline plane. There are a number of different types of methods used to determine the longitudinal weight distribution of a ship.

If the weights of individual items and the positions of their longitudinal centres of gravity along the length are known, then the entire length can be subdivided into a number of stations (usually 20) and the weight of each assigned. The result, which is the *weight curve*, consists of a series of segments that extend along the full length of the ship.

Table 5.2 *Coffin diagram coefficients* a *and* b

Ship type	a/W_a	b/W_a
Oil tankers	0.75	1.125
Full cargo ships	0.55	1.225
Fine cargo ships	0.45	1.275
Large passenger ships	0.30	1.350

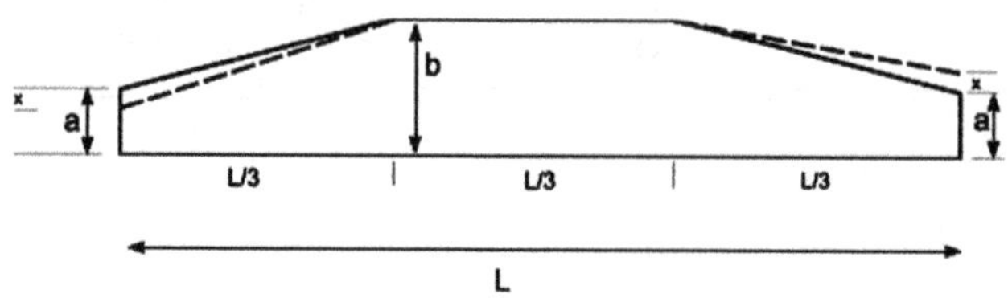

Figure 5.1 Coffin diagram approximation of longitudinal weight distribution.

If the weight distribution is not known and needs to be estimated, a number of different types of approximate methods can be used. These can be of use in the concept design phase and in early feasibility analysis. The most well-known approximate methods involve using a trapezoidal distribution or "coffin" diagram, whose exact geometry depends on the type of ship in question. Coffin diagrams are suitable for ships that have a parallel middle body, as do most cargo ships. An example is the distribution proposed by Prohaska, shown in Figure 5.1 [19].

The total area under the trapezoid equals the hull weight and thus the average weight per metre length, $W_a = W/L$. This is used in conjunction with the values of the coefficients a and b (Figure 5.1) to prepare a distribution for a particular type of ship. The values of a and b for various ship types are given in Table 5.2 [19].

Since the average weight remains constant and is equal to W_a, the values of a and b are of necessity related. Thus,

$$\frac{b}{W_a} = 1.5 - \frac{a}{2W_a} \tag{5.1}$$

In such approximations the area under the curve represents the total hull weight, and the longitudinal position of the centroid of the area coincides with that of the ship. It thus becomes necessary to be in a position to adjust the distribution accordingly. In the case of the Prohaska distribution, which is symmetrical about the centreline, the position of the longitudinal centre of gravity (LCG) of the ship may be corrected by adjusting the heights at the fore and aft ends of the coffin diagram by equal and opposite amounts (X), as shown in Figure 5.2.

In the case of ships that have no parallel middle body, such as sailing ships, a distribution of the form shown in Figure 5.3 can be used. In the case of the parabolic approximation, a geometrical technique termed "swinging the parabola" can be used to obtain a corrected position of the centroid [19]. This is illustrated in Figure 5.3.

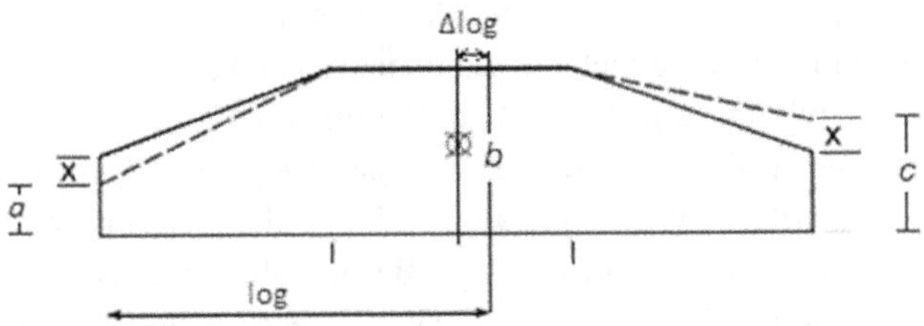

Figure 5.2 Adjustment of position of LCG.

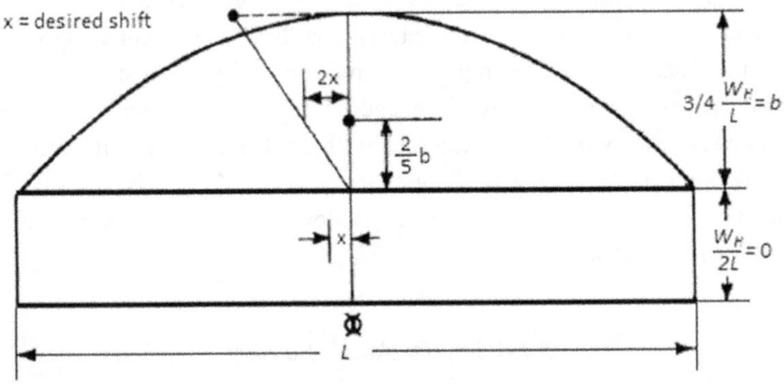

Figure 5.3 Parabolic approximation of weight distribution of ship with no parallel middle body [19].

In addition to the types of distributions mentioned above, there exist other methods that involve grouping of weight items such as the "bucket" method.

However, because the weights may correspond to masses that are distributed along the length of the ship, inaccuracies arise with regard to the distribution of individual weights. In closing, we may mention *direct distributions* that are based on distributing individual weights directly. Each individual weight has the form of a trapezoid, and thus it is necessary to know the weight, the distance over which the mass extends and the longitudinal position of its centre of gravity. The weight of the entire ship can then be obtained by summing individual weights.

5.1.2 Inertial Loads

Apart from gravity, individual members of the hull structure are subjected to inertial forces as a result of ship motions. The magnitude of these inertial forces will depend on the mass of the member in question and its position with respect to the longitudinal centre of gravity of the ship, about which linear and rotational accelerations take place. Clearly, inertial loads will be largest in the case of structural members that are distant from the LCG.

5.1.3 Fabrication Process

When structural members are welded to each other, distortions arise, and these are accompanied by *residual stresses* that have characteristic patterns. These *initial imperfections* are induced because firstly, welding takes place at an elevated temperature to be followed by cooling until the ambient temperature is reached; and secondly, structural connections are generally asymmetrical about the principal axis of bending (in particular stiffened plates). When a stiffener is welded to a flat plate, continuous fillet welds are used on either side of the stiffener web. Following welding, the weld material contracts as the temperature reduces to that of the surroundings. Because this contraction reduces the length of filaments in both the longitudinal and transverse directions, tensile stresses are induced in the weld material in these directions. The transverse contractions result in transverse bending of the attached plating, since on the other side of the plating there is no equilibrating stress field in the absence of a stiffener. This is the "hungry horse" type of distorted cross-section that characterises a thin plate to which several parallel stiffeners are attached. Longitudinal contraction also takes place and gives rise to a longitudinal bow of the stiffened plate. The stress fields that accompany these distortions result in strength reductions, and therefore the effects of the fabrication process have to be allowed for in both design and fabrication. In the latter case this is done by specifying acceptable tolerance limits for the induced imperfections.

5.1.4 Dry-Docking of a Ship

Ships are dry-docked at regular intervals throughout their working life, and the loading condition experienced is quite different from that which arises when the ship is afloat. There exist several different kinds of docks, but the two that are most commonly used are *floating docks* and *graving docks*. Of these the most challenging case is the *floating dock problem* that requires the analysis of an elastic beam (ship) with a longitudinally varying weight distribution supported at points (blocks) aligned on the deck of the floating dock, which has a finite bending stiffness; this in turn is supported by the buoyancy force distribution. A graving dock is much stiffer than a floating dock so that it can be treated as being rigid.

The vessel is supported along its length by a large number of blocks that are located in way of the flat of bottom. A near-continuous line of blocks is positioned in way of the centreline below the keel plate and extends from the stern post to the forward end of the bulbous bow, if one exists. Additional lines of blocks are positioned on either side of the centreline (Figure 5.4a). The vessel's docking plan indicates the positions at which appendages such as sea chests, pipe discharges etc. are located in order to ensure that no damage will occur to these items. No support is provided to the aft structure, and thus the stern structure overhangs (Figure 5.4b). Measurements have shown that the loads on individual blocks reduce after several hours as the hull girder settles into a permanent position. From the point of view of the ship, longitudinal strength is not a problem when the ship is in drydock. The important thing is to have the ship adequately supported so that damage does not occur at a local level.

In the case of the floating dock, the problem needs to be considered in stages. As a first approximation, it may be assumed that the dock and hull girder are both rigid. In this case the reactions at the positions of support can be determined without difficulty. If the LCG of

(a)

(b)

Figure 5.4 Ship in a floating dock. (a) Arrangement of blocks. (b) Sternpost.

the docked ship coincides with the longitudinal centre of the keel blocks, the reaction forces can be assumed, to a first-order approximation, to be uniformly distributed [11], so that (Figure 5.5a):

$$\text{Load per unit length} = \frac{W_s}{L_{keel}} \tag{5.2}$$

where

W_s = weight of docked ship

L_{keel} = keel-bearing length (Figure 5.5a)

If the *LCG* and the centre of the keel blocks do not coincide, the distribution of the reactions is assumed to be *trapezoidal* (Figure 5.5b). This trapezoidal distribution is determined using elementary mechanics, where the series of centreline blocks is assumed to be continuous and to extend over the full length of the keel line. This is treated as a short compressed column that is also subjected to bending due to the unequal distribution of the ship's weight. At the end of the keel line closest to the LCG, the load per unit length, w, is given by:

$$w = \frac{W_s}{L_{keel}} + 6W_s\frac{e}{{L_{keel}}^2} \tag{5.3a}$$

At the end of the keel line furthest from the LCG,

$$w = \frac{W_s}{L_{keel}} - 6W_s\frac{e}{{L_{keel}}^2} \tag{5.3b}$$

where e is the distance of the LCG of the docked ship from the centre of the keel blocks. Equations (5.3a) and (5.3b) are termed the *trapezoidal load equation* and are used to determine block and drydock loads.

If the weight distribution of the docked ship is available, a more accurate distribution of the loads can be obtained by assuming that the weight is transmitted directly onto the blocks and thereby onto the dock floor.

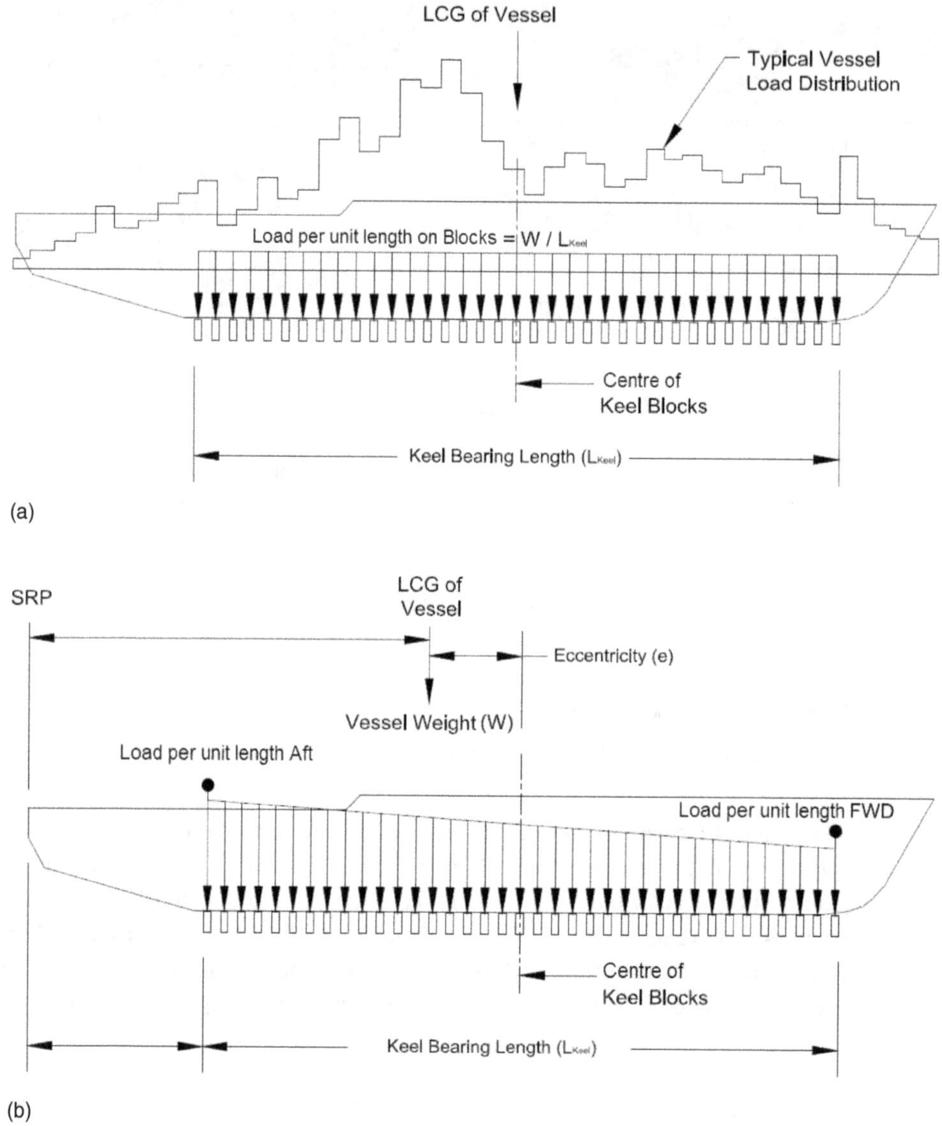

Figure 5.5 Distribution of load during docking [11]. (a) LCG and centre of keel blocks coincide. (b) LCG and centre of keel blocks do not coincide.

We note from Figures 5.4b and 5.5 that in way of the fore and aft ends, there are unsupported sections of the hull girder, especially in way of the aft end. The weight of the overhang is assumed to be distributed in a parabolic manner over a length that is twice that of the unsupported length (Figure 5.6) [3]. It can be shown that the additional weight per unit length, w, that has to be added at a distance x from the aft supported end is given by:

$$w = \left(\frac{3W_0}{4L_0^2}\right)\left[(2L_0 + 3L_G)\frac{x^2}{L_0^2} - 2(3L_0 + 4L_G)\frac{x}{L_0} + 4(L_0 + L_G)\right] \tag{5.4}$$

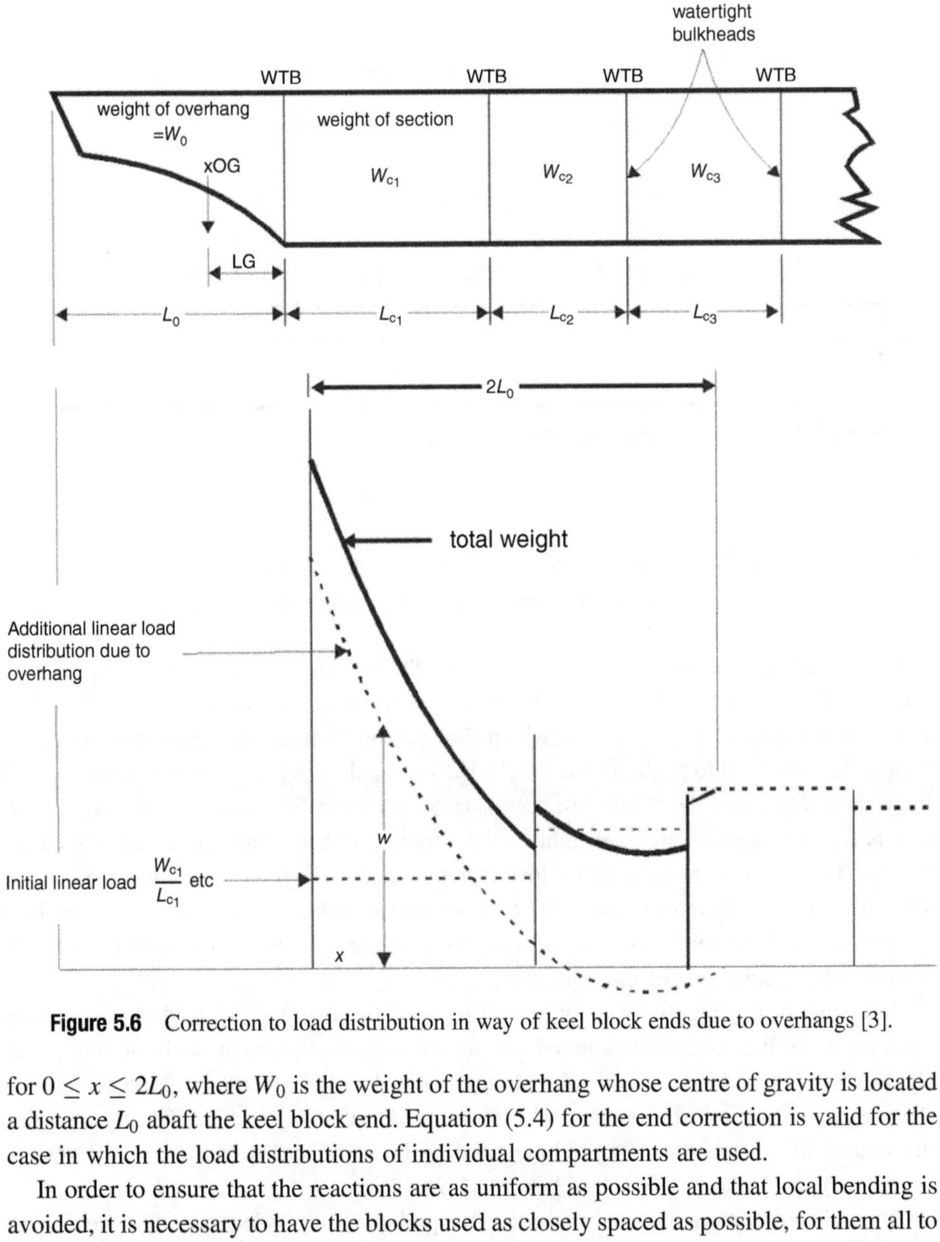

Figure 5.6 Correction to load distribution in way of keel block ends due to overhangs [3].

for $0 \le x \le 2L_0$, where W_0 is the weight of the overhang whose centre of gravity is located a distance L_0 abaft the keel block end. Equation (5.4) for the end correction is valid for the case in which the load distributions of individual compartments are used.

In order to ensure that the reactions are as uniform as possible and that local bending is avoided, it is necessary to have the blocks used as closely spaced as possible, for them all to have the same stiffness and for them to come into contact with the outer shell plating simultaneously. Side blocks are used to ensure that transverse bending is minimised and they also reduce the load carried by the keel blocks (Figure 5.4a). One other important role of the side blocks is to ensure the lateral stability of the hull girder. There arise circumstances in which tipping may occur, for example when lateral wind loads act, or when the dock is not absolutely level. In the case of graving docks, the possibility of earthquakes exists.

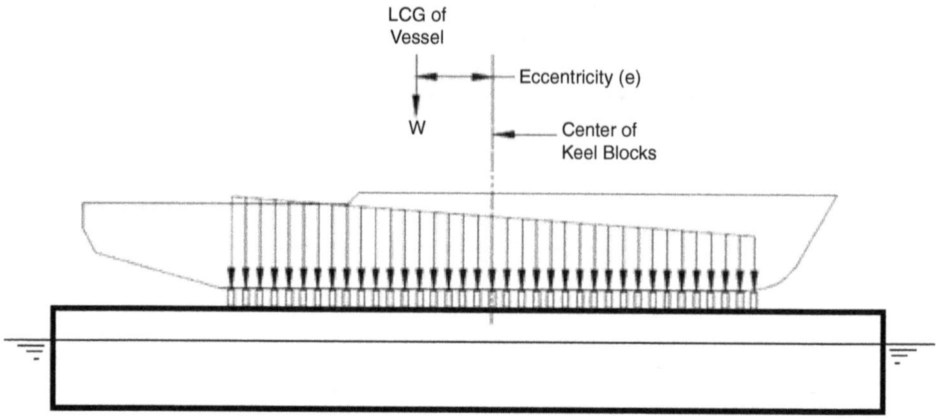

Figure 5.7 Ship–dock ensemble [11].

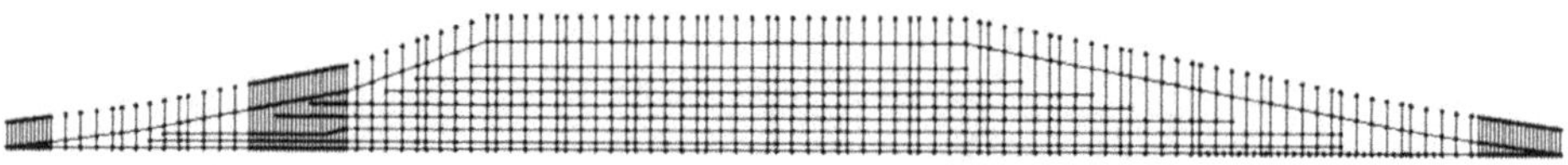

Figure 5.8 Grillage analysis of bottom plating to determine reaction loads [24].

For small and medium-sized vessels in which the spacing of transverse bulkheads does not exceed 20 metres, side blocks can be fitted following the discharging of the cargo space. In the case of large ships in which the spacing of transverse bulkheads is 40–50 metres, it is essential to position side blocks before dry-docking commences because of the larger beam. Because the blocks are not made of incompressible material, the side blocks that bear a larger share of the weight of the structure distort much more than the centre blocks. As a result, a greater reaction load is transmitted to the transverse bulkheads via the side structure. It is therefore reasonable to assume that the blocks along the centreline bear 50 per cent of the load whereas the side blocks bear the remaining amount, equally distributed on either side of the centreline.

Allowing for the longitudinal bending stiffness of the floating dock leads to the *floating dock problem*. It is generally assumed that the dock bends in tandem with the hull girder (Figure 5.7). The net load at any point along the keel length is then the difference between the buoyancy force and the weight of the dock added to that of the ship. The ship and dock are treated in unison when longitudinal bending is considered because it is not easy to separate the contribution of each to bending [19].

Efforts to obtain more accurate solutions to the problem have made use of 3D finite element analysis that, however, requires extensive time and the availability of data. In order to circumvent these difficulties, much simpler 2D analyses based on grillage theory have been used to calculate the loads on individual blocks [24]. Figure 5.8 shows the representation of the flat of the bottom of a container ship. Information is retrieved from structural drawings (docking plan), and the distributed loads are transferred to the positions of the blocks.

During docking the ship is invariably trimmed by a certain amount so that contact with the dock arises at one position, usually astern. Following touch-down, a point load termed the *knuckle reaction* acts until the ship comes into contact with the blocks along the length

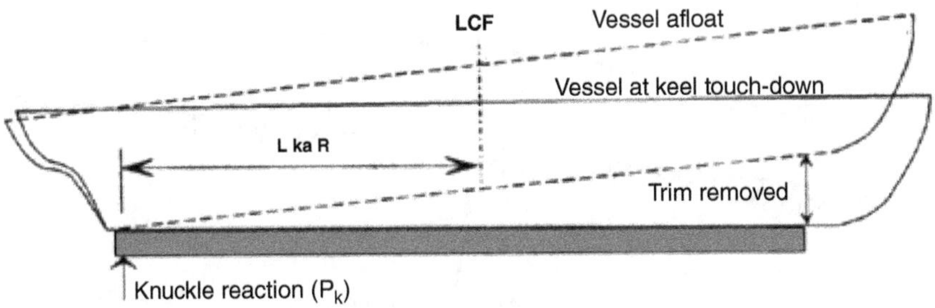

Figure 5.9 Knuckle reaction arising during docking [23].

of the dock. The knuckle reaction, P_k, is then equal to the trimming moment divided by the distance to the *LCF* (Figure 5.9).

The moment-to-change trim for the particular draught is found from the ship's hydrostatic data.

5.1.5 Launching

Launching can take place in a number of ways, these being stern-wise, side launching and lowering from dry-dock. In this section we shall focus on stern-wise launching, which is the important one from the point of view of longitudinal strength. During stern-wise launching the distribution of loads that act on the hull girder is different from, and may in certain circumstances exceed, those experienced under normal conditions. Furthermore, it has to be ensured that the launching procedure does not result in any damage to the ship. It is therefore necessary to calculate loads distribution and the resulting stresses in order to avoid damage. This may result from the inability of the ship to withstand these forces or else because of damage to the launching arrangements. Bending moment loads and stresses are calculated using a static analysis approach.

Usually two ground ways that are parallel to the centreline of the ship about one-third of the beam apart are used. The top of the ways is cambered, giving a slope at the top lower than the slope at the bottom end. This provides a margin against sinkage of the ways as the ship travels down, a slower start, faster finish and a deeper immersion. The lower ends of the ways extend to the low water mark. The height of the ground ways is arranged to provide access below the ship. The length of the ways reaches approximately 80 per cent of the length of the ship. These consist of long baulks of timber similar to ground ways, surmounted by a cradle that is shaped to the form of the ship and supporting it. At the ends the cradle is built-up. It consists of heavy wood struts, vertical or sloping and bearing on temporary shelves and angle lugs fitted to the outer shell plating. The end portions are the *poppets*. The slope of the ways is 1:24–1:16 (0.0417–0.0625), and the camber is approximately 1 in 300.

The Launching Procedure

The launching procedure is broken down into a number of stages during each of which a different set of forces acts on the ship. It is usually assumed that the slipway behaves as a

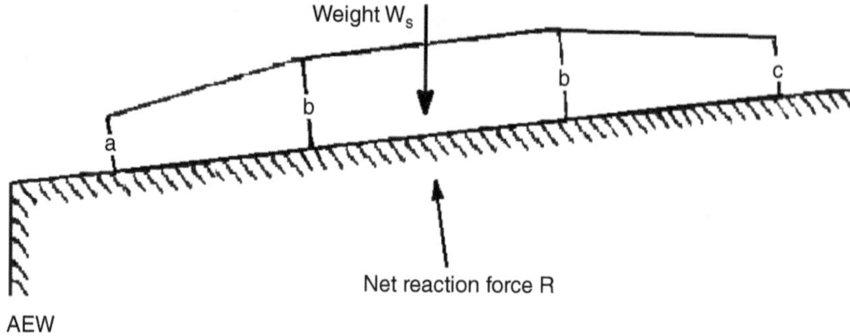

Figure 5.10 Forces acting while the ship is sliding down the slipway.

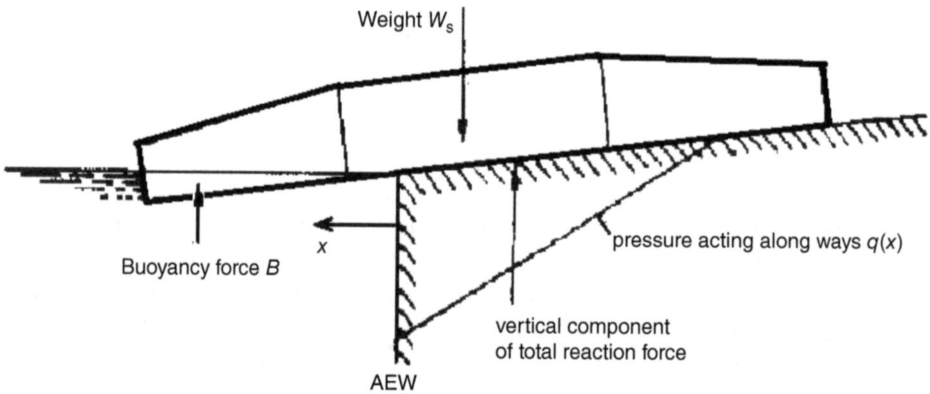

Figure 5.11 Forces acting on the ship following immersion and before rotation.

rigid structure. In a first approximate calculation of the shearing force and bending moment distributions, we can assume that a coffin diagram of the type shown in Figure 5.1 can be used to approximate the weight distribution.

During the first stage of launching, the ship, which is borne by a cradle, travels down the ways towards the water. Initially the hull is not in contact with the water. The forces acting on it are its weight that acts vertically downwards and the support reaction that acts perpendicular to the ways and is transmitted via the cradle to the hull (Figure 5.10). Equilibrium is achieved by including in the analysis the out-of-balance force that gives rise to the acceleration down the slipway.

After a certain speed of travel has been reached, the aft end comes into contact with the water and gradually becomes immersed. A buoyancy force whose magnitude increases with immersion then acts vertically upwards on the hull (Figure 5.11), resulting in a reduction in the net reaction force on the cradle which is perpendicular to the slipway and is assumed to increase linearly from zero to a maximum at the edge of the slipway.

Since the aft end of the ship overhangs the aft end of the ways (AEW), the situation is analogous to that of a cantilever with an elastic support. Initially the hull girder is in a hogging condition, but as the buoyancy force increases with immersion, the condition changes to sagging.

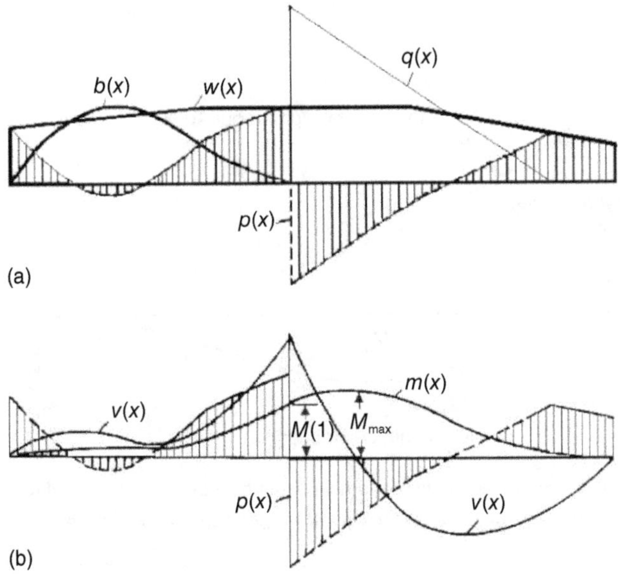

Figure 5.12 Initial stage of launching [1]. (a) Weight *w(x)*, buoyancy *b(x)*, distributed reaction force *q(x)* and net load *p(x)* distributions (per unit length). (b) Net load *p(x)*, shear force *v(x)* and bending moment *m(x)* distributions per unit length.

For the weight distribution shown in Figure 5.10, the moment of the weight of the ship about the AEW (Figure 5.11) is given by the following expressions:

$$\text{For } x \le L/3, \quad M(x) = \frac{x^2}{2}\left[a + (b-a)\frac{x}{L}\right] \tag{5.5a}$$

$$L/3 \le x \le 2L/3 \quad M(x) = \frac{L}{6}\left[\frac{3b}{L}x^2 - (b-a)(x - L/9)\right] \tag{5.5b}$$

$$\mathrm{x} \ge 2L/3 \quad M(x) = \frac{L}{6}\left[\frac{3b}{L}x^2 - (b-a)(x-L/9) - \frac{3}{L^2}(b-c)\left(x - \frac{2}{3}L\right)^3\right] \tag{5.5c}$$

where *a*, *b* and *c* are defined in Figure 5.10 and *x* is the length of the portion of the ship that extends beyond the AEW.

Figure 5.12a shows the distributions of weight, buoyancy, reaction force and load during the initial stage of launching, during which the moment of the buoyancy force about the AEW is less than the moment of the weight about the same position. Figure 5.12b shows the corresponding net load, shear force and bending moment distributions. As the ship translates down the slipway, the shear force and bending moment distributions change. The bending moment reaches a maximum value, $M_{max\,1}$, at some point along the length of the ship during the process of immersion. The change in the moments about the AEW as the ship translates until rotation occurs is shown in Figure 5.13.

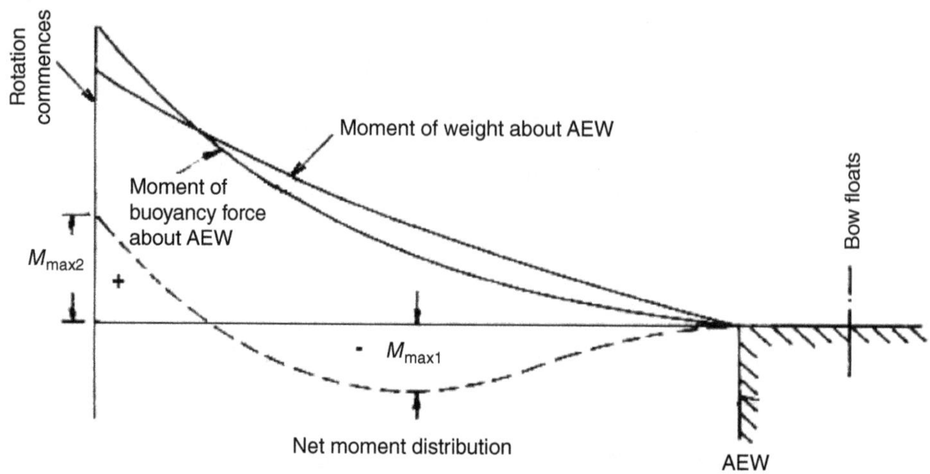

Figure 5.13 Change in magnitude of moments of buoyancy and weight and resultant moment about the AEW prior to rotation *(positive sign: hogging; negative sign: sagging)* [1].

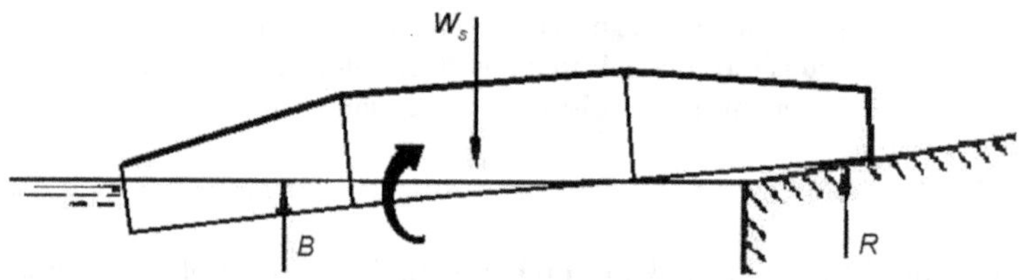

Figure 5.14 Resultant forces acting in the rotation phase.

It is seen that the resultant bending moment reaches a maximum value (M_{max1}) when the hull girder is in a hogging condition, following which a second maximum (M_{max2}) arises at the instant at which the stern begins to lift and the ship starts to rotate about the fore poppet (Figure 5.13). After this stage the resultant bending moment gradually reduces until it becomes equal to the still-water bending moment, when the ship is fully afloat and unsupported.

It is necessary to calculate the net bending moments M_{max1} and M_{max2}, as these are the critical ones during the entire launching procedure.

When the moment of the buoyancy force about the forward point of contact of the ship with the cradle exceeds the moment of the unsupported weight about the same position, the hull and cradle begin to rotate, and the hull is then supported by the buoyancy force and the reaction R at the fore end of the cradle (Figure 5.14).

Figure 5.15a shows the weight, buoyancy, and the resulting load distributions along the length whereas Figure 5.15b shows the load, shear force and bending moment distributions when the moment of the buoyancy moment about the fore poppet end is larger than the moment of the weight. The reaction force R acts at the point of rotation.

If during launching and when the longitudinal centre of gravity (LCG) is beyond the AEW the moment of the weight about the AEW exceeds the moment of the buoyancy

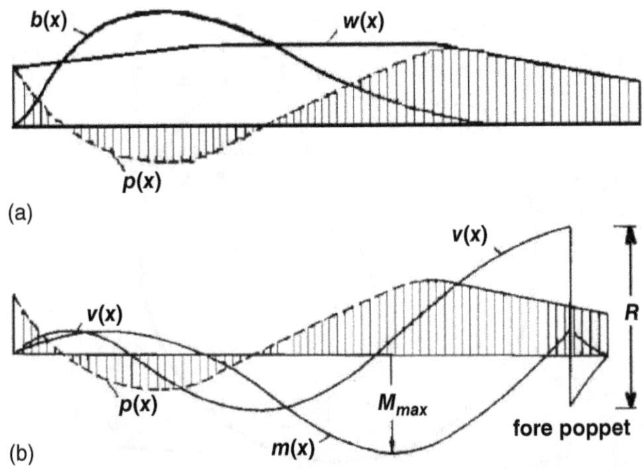

Figure 5.15 Bending moment, shear force and load distributions when rotation commences [1]. (a) Weight *w(x)*, buoyancy *b(x)*, reaction force *R*, and load distribution *p(x)* per unit length. (b) Distributed load *p(x)*, reaction *R*, shearing force *V(x)* and bending moment *M(x)* distributions.

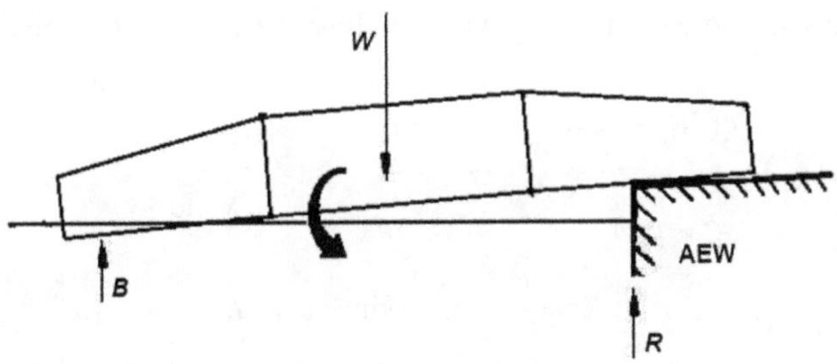

Figure 5.16 Resultant forces acting when tipping occurs.

about the same position, the aft end of ship will tip downwards (Figure 5.16). Support is then provided by the buoyancy force and a concentrated reaction *R* at the AEW whose magnitude is equal to *W–B*. This is not a desirable situation and is avoided by proper design of the ways in relation to the water level and the depth of water during launching.

Launching Curves

As seen previously, it is necessary to investigate a number of critical positions and loads that arise during the launching procedure. The results are presented in the form of *launching curves* that include displacement and moment curves for a number of positions along the ways. Draught calculations have to be carried out for each position. Corresponding displacement and longitudinal centre of buoyancy (LCB) values are then obtained from the Bonjean curves.

The launching curves are based on calculations performed for five positions along the travel. The initial travel is taken at the point just before the LCG passes over the AEW and the last point when the fore poppet is on the point of drop-off.

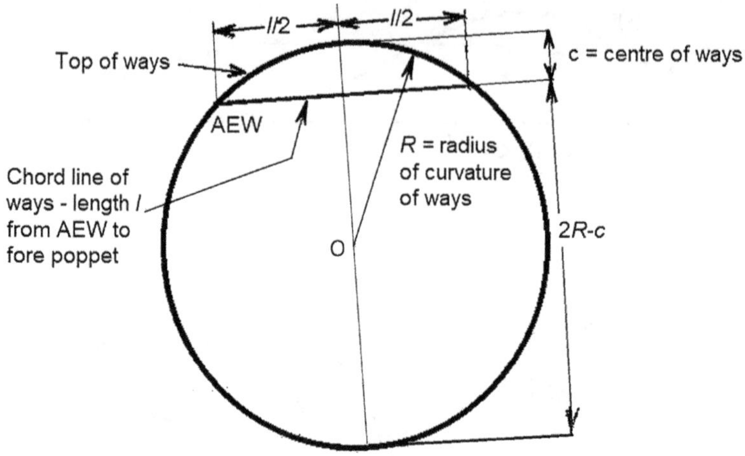

Figure 5.17 Geometry of launching ways.

In order to sketch the launching curves, it is necessary first to describe the geometry of the launching arrangements (Figure 5.17) as well as that of the position of the ship.

The ship translates in a circular arc whose radius of curvature R is given by:

$$c(2R-c)=\frac{l^2}{4} \qquad \text{or} \qquad 2cR=c^2+\frac{l^2}{4}$$

Hence,

$$R=\frac{l^2}{8c} \qquad \text{since } c \ll R \tag{5.6}$$

At travel x down the ways, we have for the height of the top of the ways above the chord line, y, from Figure 5.18:

$$y(2R-y)=(l-x)x$$

$$2Ry-y^2=(l-x)x$$

$$y=\frac{(l-x)x}{2R}+\frac{y^2}{2R}$$

where the second term may be ignored, since $y<<R$. Hence,

$$y\approx\frac{4cX(l-x)}{l^2}=\frac{4cx}{l}\left(1-\frac{x}{l}\right) \tag{5.7}$$

If the arc length is assumed to be equal to the chord length, the rotation θ of the ship for a travel x is given by:

$$\theta=\frac{x}{R}=\frac{8cx}{l^2} \tag{5.8}$$

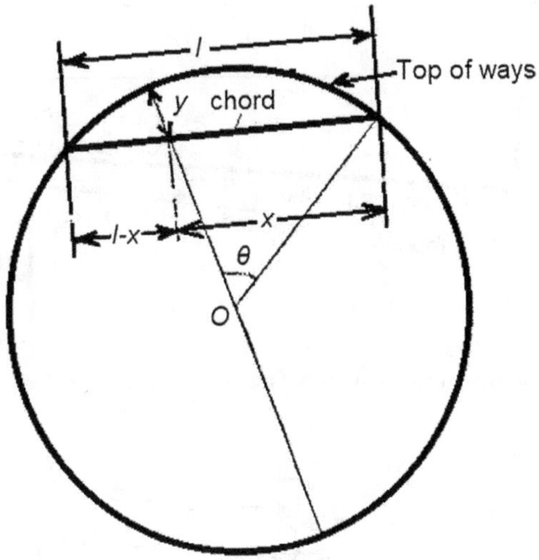

Figure 5.18 Height of top of ways, c, above chord line [16].

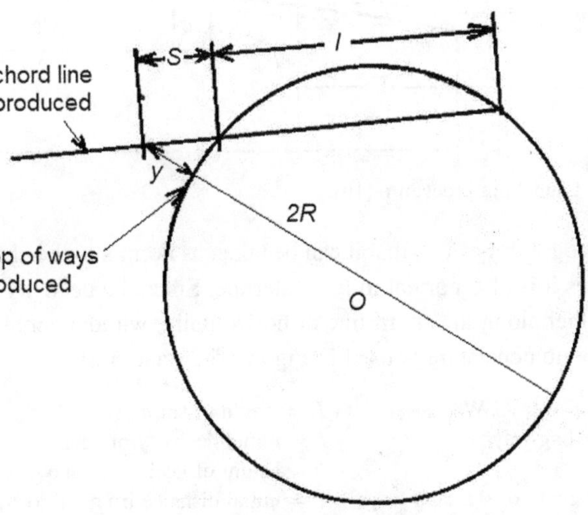

Figure 5.19 Drop below chord line produced of top of ways produced [16].

At a distance S beyond the AEW we have, from Figure 5.19,

$$y(2R+y) = S(l+S)$$
$$2Ry + y^2 = S(l+S)$$

and

$$\begin{aligned} y &= \frac{S(l+S)}{2R} - \frac{y^2}{2R} = \\ &= \frac{S(l+S)}{2R} = \frac{4cS}{l^2}(l+S) \end{aligned} \tag{5.9}$$

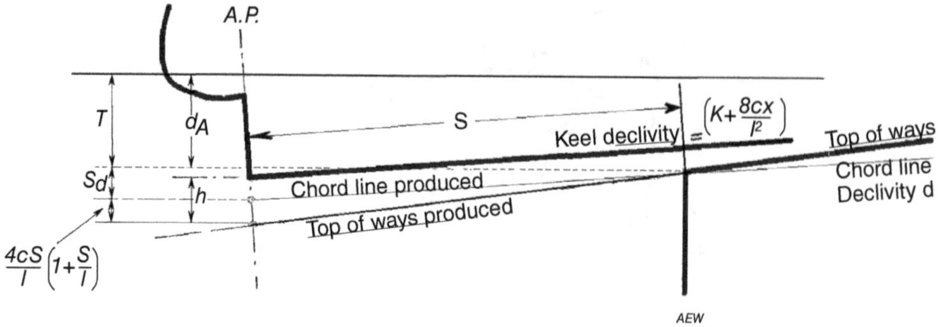

Figure 5.20 Draught after travel x with AP distant S from AEW [16].

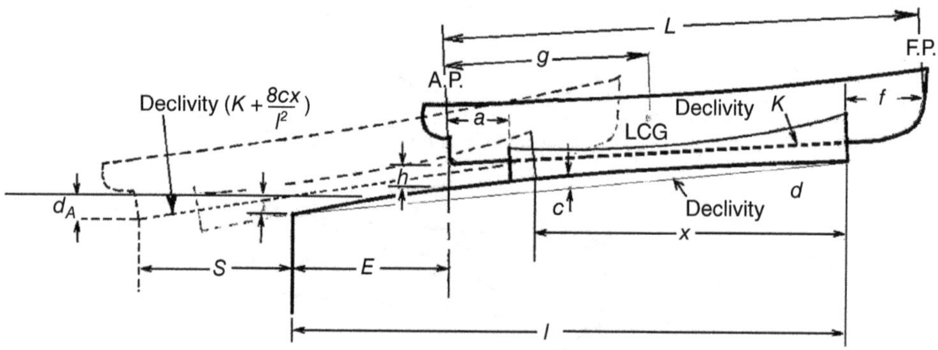

Figure 5.21 Launching procedure [16].

In the formulae for y, $y \ll R$ so that it can be taken as normal to the chord line, and for all practical purposes it is also normal to the waterline. Since the declivity is small, distances may be taken either along the chord line or horizontally, without appreciable error.

The following nomenclature is used in Figures 5.20 and 5.21.

T = depth of water over AEW
l = length of standing ways
c = camber
K = chord line declivity of standing ways
a = aft poppet forward of AP
x = total travel from initial position
g = distance of LCG forward of AP
L = length of ship
k = initial declivity of keel
h = height of keel at AP above standing ways
E = initial distance from AP to AEW
f = fore poppet aft of FP
S = travel of A.P. beyond AEW (=$x-E$)

$$
\begin{aligned}
d_A &= \text{draught at AP after travel } x = \\
&= T + (\text{Drop in chord line produced}) + (\text{Drop of ways produced}) - h = \\
&= T - h + Sd + \frac{4cS}{l}\left(1+\frac{S}{l}\right) \\
&= T - h + S\left(d + \frac{4c}{l}\left(1+\frac{S}{l}\right)\right) \\
&= [T-h] + S\left(\left[d+\frac{4c}{l}\right] + S\left[\frac{4c}{l^2}\right]\right)
\end{aligned}
\tag{5.10}
$$

Table 5.3 *Calculation of draughts at FP and AP* (d_F, d_A) for several travel distances x

No	Calculation	Travel distance
1	x	
2	E	
3	$S = x - E$ = Line 1 – Line 2	
4	S/l = Line 3/l	
5	$1 + S/l$ = 1.0 + Line 4	
6	$\frac{4c}{l}\left(1+\frac{S}{l}\right) = \frac{4c}{l}$ (Line 5)	
7	d	
8	$d + \frac{4c}{l}\left(1+\frac{S}{l}\right)$ = Line 7 + Line 6	
9	$S\left(d + \frac{4c}{l}\left(1+\frac{S}{l}\right)\right)$ = Line 3 × Line 8	
10	$T - h$	
11	d_A = Line 10 + Line 9	
12	$8cx/l^2$	
13	K	
14	$K + 8cx/l^2$ = Line 13 + Line 12	
15	$L(K + 8cx/l^2)$	
16	d_F = Line 11 – Line 15	

The values in square brackets are constant and S is the variable. The declivity initially is K, but after travelling in a circular arc, the declivity increases by $\theta = 8cx/l^2$.

Hence, the trim t in length $L = L(K + \theta) = L\left(K + \frac{8cx}{l^2}\right)$

The draught at the forepeak d_F, after travel x, is then

$$d_F = d_A - L\left(K + \frac{8cx}{l^2}\right) \tag{5.11}$$

From these draughts sectional areas for determining displacement and LCB can be read from the Bonjean curves. To plot the launching curves, calculations are made for about five values of the travel x; the maximum travel occurs when $x = l$ and the fore poppet is on the point of drop-off. Table 5.3 includes the calculation sequence for one value of the fore draught d_F; this can easily be programmed. The initial travel can be taken from just below the point at which the LCG passes over the AEW: that is, when $x = E - g$, where g is the distance of the LCG forward of the AP. The latter value of x gives zero point for the moment of the weight about the AEW.

A reliable estimate of the launching weight and position of the LCG is also required. The results are usually plotted as shown in Figure 5.22.

The intersection of the moment curves at A gives the position of stern lift, assuming static conditions. $(W_s - \Delta)$ = BC gives the corresponding load on the fore poppet, which must be designed to withstand it. It is assumed in the calculation that the ways are in complete contact, but this is not the case after the stern lifts. If the curve of buoyancy CD is

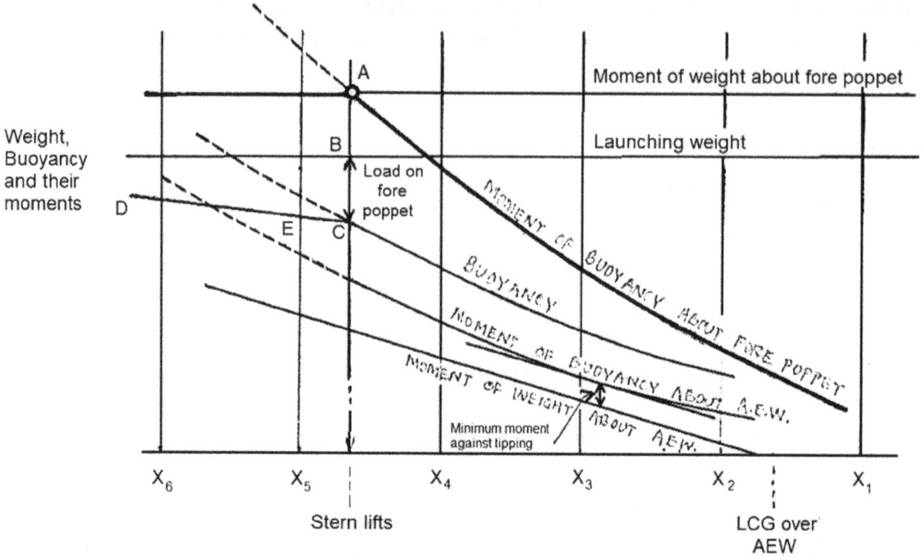

Figure 5.22 Launching curves [16].

required after the stern lifts, a further calculation is necessary. For example, for travel x, the moment of the buoyancy exceeds the moment of the weight about the fore poppet assuming the ways are in complete contact. The ship will therefore pivot about the fore poppet to attain equilibrium. To ensure that tipping does not occur, it is necessary to verify that the curve of the moment of buoyancy about the AEW lies above that of the moment of the weight about the AEW. The minimum distance between the two curves represents the minimum moment against tipping.

5.1.6 Grounding and Collision

Grounding and collisions constitute unwelcome events that involve ships coming into violent contact with either floating or fixed objects. If such contact arises in way of the bottom structure and the ship is immobilised, *grounding* is said to have taken place. If contact is made with any part of the side shell structure or indeed any other part of another vessel, the incident is classed as a *collision*. Ships are generally not designed to withstand impacts with solid bodies that may be moving or are fixed without sustaining a certain amount of damage.

Grounding may arise as a result of extreme weather conditions, inadequately charted waters or when navigating near shore for prolonged periods such as in port approaches and in rivers and lakes. In such cases ships may run aground because charts have not been updated or due to poor seamanship.

The probability of occurrence of grounding is higher than commonly expected, although a significant proportion of cases concern minor incidents in which no damage to the environment occurs and which, as such, do not attract media attention. Table 5.4 is

Table 5.4 *Frequency of occurrence of grounding incidents for the Greek flag fleet [25]* (ship-years per grounding)

Ship Type	*1995–2005*	*2001–2005*
Dry cargo	70	90
Tankers	144	297
Passenger and others	120	182
All ships	*110*	*148*

(a) (b) (c)

Figure 5.23 Various types of grounding damage. (a) Grounding with minor external damage to the hull structure. (b) Hull girder collapse. (c) Bottom damage of the M/T *Exxon Valdez.*

based on a statistical analysis of data relating to the Greek flag fleet for the period 1995–2005 [25].

The cases shown in Figure 5.23 indicate that the results of grounding vary a great deal. In considering its effect on the overall strength of the hull girder, it is necessary to distinguish between those cases in which any damage that may occur is localised (Figure 5.23a) and those in which the integrity of the hull girder is impaired (Figures 5.23b and 5.23c).

The extent of damage to the hull structure will depend on the energy imparted to it during impact. This is a function of the mass and speed of the ship. One other important factor is the nature of the seabed; if it is rocky, then tearing of the plating is to be expected, but if the seabed is sandy, this is less likely to occur. Damage may be confined to the region of the point of contact or may extend well beyond it.

In the cases of Figure 5.23a, the hull structure was subjected to hogging, although in these particular cases no serious damage was reported and the vessels were successfully refloated. In a series of experiments on grounding the sagging bending moment that resulted from the grounding of a ship against an upward slope was compared to the IACS design wave bending moment that is assigned a probability of occurrence of 10^{-8} during the ship's lifetime [12]. It was found that the shear force and longitudinal bending moment that result from grounding exceed the IACS design shear force and moment apart from cases in which the seabed is very steep. Therefore, should such an event occur, it may well lead to overall hull girder failure. It was also found that damage due to grounding is much more extensive in the case of larger ships, in agreement with collision and grounding impact theories. This can be attributed in part to the fact that smaller ships have comparably higher bottom strength [21].

For the cases in which damage is localised, this will affect the outer shell plating and the internal stiffening as well as any equipment located within the vicinity of the impact zone. If tearing of the plating does take place,[1] flooding may extend beyond the damaged plating and the possibility of environmental damage may also arise. Should the compartment in question be used as a fuel tank, then outflow of fuel will occur. Even worse, should outflow from the cargo tank of an oil tanker take place, the damage to the environment will be substantial.

Collisions may be classed as either of a minor nature (*contact damage*), in which case the integrity of the hull structure is not jeopardised; or major collisions as a result of which the vessel may founder and/or lives are lost. State-of-the-art technology is used in collision avoidance systems available to assist mariners. Collisions can be assessed using an energy approach in which the changes of momentum of the striking object and the struck object are found.

Although the number of incidences relative to the world fleet has been going down, a number of new potential collision scenarios have arisen in recent years (ships with offshore platforms, offshore wind generating farms, FPSOs, bridges).

Ship collisions may lead to all or some of the following unwelcome events:

1. Structural damage, be localised or at some distance from the point of contact (Figure 5.24).
2. Stability damage, leading to possible foundering of the ship.
3. Loss of life.
4. Damage to the environment resulting from the outflow of cargo or fuel.

Each of the above is a serious consequence in itself, and for this reason an extensive effort has been made by the maritime community to reduce the probability of any of these events occurring. Of direct interest are current and future legislation and rules of construction as well as proposals concerning structural design layouts that aim at avoiding the

[1] Considered to be a "major" incident as opposed to a "minor" incident in which plate tearing does not occur [13].

(a) (b)

Figure 5.24 Ship collisions. (a) Collision between two cargo ships. (b) Bow damage resulting from collision.

consequences of ship collisions. Double hull structures have been considered to be the answer to oil pollution and as such have been imposed by the 1990 US Oil Pollution Act. They are, however, not the only answer, nor necessarily the best one in all circumstances. Collision studies have shown that it is preferable to allow the bow of a striking ship to absorb the energy transferred during a collision incident by deforming substantially (*buffer bow*) rather than to allow it to cause significant damage to the side structure of the struck ship, since side structures cannot be strengthened enough in order to provide sufficient protection. In an effort to provide increased resistance against side collisions, a variety of honeycomb and sandwich-type structures have been proposed.

If we wish to assess the behaviour of a ship involved in a grounding or collision incident, we have to perform an *accidental limit state* analysis which involves the testing of the ship against a number of accident scenarios. The purpose of this is to assess the ship's performance and to decide whether it is of a satisfactory design (Figure 5.25). The procedure involves [18]:

1. Specifying grounding and/or collision scenarios of relevance to the particular ship.
2. Establishing acceptance criteria in relation to loss of human life, environmental pollution and financial loss.
3. Performing an analysis of structural response.
4. Comparing ship performance against acceptance criteria.

Collision events are of a large variety and are complicated hydro-mechanical processes. In collision mechanics the analysis is split into two parts: *internal mechanics* and *external mechanics*. The internal mechanics deal with the response of the hull girder, which includes the damage sustained in way of the point of contact. In addition to this, energy is dissipated in global hull girder vibrations, which for large ships is a small proportion of the total energy absorbed by the struck ship. For this reason it is possible to ignore the elastic energy and to consider only the contribution of energy converted to plastic work. The external mechanics deal with the ship's motion and the associated hydrodynamic forces that act on the outer hull surface as a result of the impact.

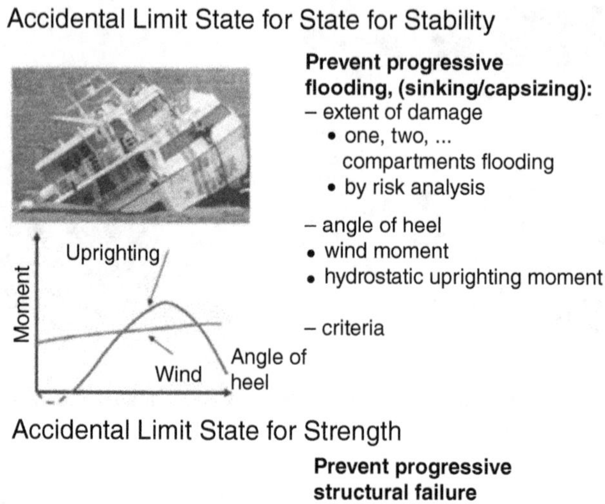

Figure 5.25 Accidental limit state analysis [18].

The internal mechanics analysis can be performed using simple formulae, simplified analytical approaches or numerical approaches (linear or nonlinear finite element method (FEM)). Collisions involve plate rupture that is modelled using FEM. The external mechanics is modelled using rigid body motion theory. In order to model the contact force, a nonlinear FEM analysis is performed, which is then combined with the ship motion analysis.

5.1.7 Example Conversions and Major Repairs

Conversions and major repairs may result in important changes to external as well as internal (cargo) loads that act on the hull structure. They are generally carried out on older vessels that are in good condition and that fulfil the requirements set by relevant authorities, apart from owners'. A large number of conversions from single hull tankers to:

- Double hull tankers and chemical tankers.
- Double hull bulk carriers and ore carriers.
- Floating production storage and offloading units (FPSOs) and floating storage and offloading units (FSOs).
- Other specialised types of vessels such as heavy lift and cattle carriers.

took place a few years ago. In addition to these, one other type of conversion that has been carried out in several cases is the lengthening of various types of ships with an emphasis on cruiseships.

From the point of view of longitudinal strength, the conversions of VLCCs to FPSOs and of VLCCs to VLOCs are the most important. In the case of the former, the ship no longer transports oil but is stationed at a particular location and operates as an offshore installation. As a result, it is subjected to external loading conditions on a permanent basis

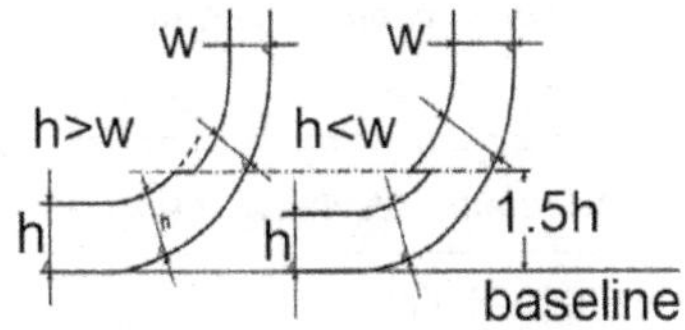

Figure 5.26 Definition of variables for spacing of inner hull.

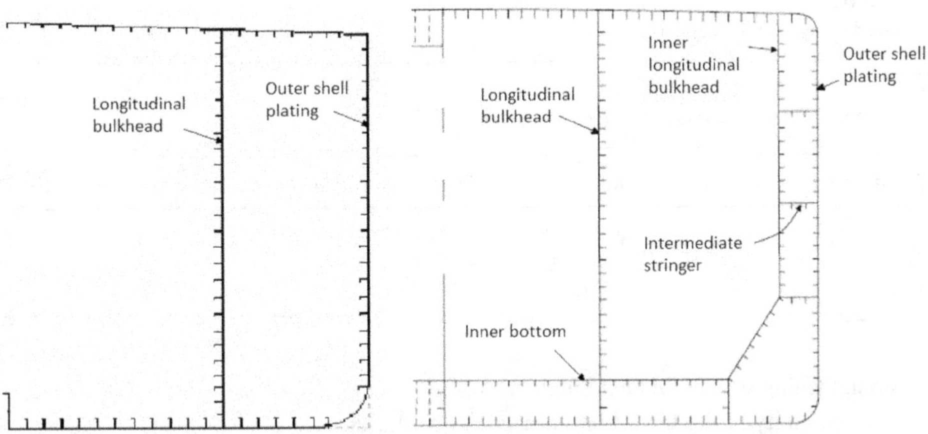

Figure 5.27 Conversion from single to double hull oil tanker (100,000 dwt tonnes) [8].

in the selected location. FPSOs have been employed in oil extraction in the North Sea where the weather conditions can be exceedingly harsh.

Conversion of Single Hull to Double Hull Oil Tanker

The conversion of single to double hull tankers or FPSOs has been dictated by market requirements, and many such conversions have been carried out. Oil tankers require the installation of a double bottom and/or sides in accordance with the existing arrangement and rule requirements. The primary strength of the modified hull is enhanced when compared to the initial one due to the addition of longitudinally effective material.

Rule requirements specify the spacing between the inner and outer hull as follows:

$w = 0.5 + \text{DWT}/20{,}000$ or 2 m whichever is less (h>1m)
$h = B/15$ or 2 m whichever is less (h>1m)

where w and h are defined in Figure 5.26.

Figure 5.27 shows the most important modifications carried out on a 100,000 dwt tonne tanker. The most important consequences of such conversions are summarised in Table 5.5 for selected cases [8].

In addition to the installation of the inner bottom and the inner longitudinal bulkhead, other important items that need to be considered are:

- Strengthening of existing bottom floors.
- Additional intermediate floors may need to be installed.

Table 5.5 *Steelweight and changes in cargo capacity resulting from the conversion of a single hull tanker*

Vessel Type	*Additional Steelweight (mt)*	*Change in Cargo Capacity (m^3)*
Single hull VLCC	7,000–8,000	Increase by 8,000–10,000
Double-side only Aframax	1,000	Decrease by 10,000
Double-side only Handysize (30k dwt)	450	Decrease by 6,000–7,000
Standard single hull Suezmax	4,300	Increase by 8,000

Table 5.6 *Still-water loads for converted tanker FPSOs*

Quantity	*Units*	*Rule*	*Finite Element Analysis*	*Percentage Difference*
M_{SW} (hogging)	kNm	5,123,200	5,072,000	−1.0
M_{SW} (sagging)	kNm	4,740,000	4,050,000	−17.0
V_{SW}	kN	80,550	142,250	+77.0

- Strengthening of longitudinal bulkhead(s).
- The continuity of the inner bottom needs to be preserved.
- Changes to deck stiffening may be required with respect to section modulus requirements, given that the bottom structure is strengthened substantially.
- New hatches and coamings.

Conversion of Single Hull Tanker to FPSO

Floating production storage and offloading units, on the other hand, operate in a different mode from trading vessels, and due to the continuous loading and discharging of oil, the draught of the vessel changes continuously. As a result, the still-water bending moment also changes, in contrast to ships that trade, for which it is not the critical loading parameter.

In the case of FPSOs the still-water loading should thus be treated as a stochastic process. Strength assessment of FPSOs has been carried out using both rules and direct methods in which a diffraction analysis is used to determine hydrodynamic loading, and this has been combined with finite element analysis. Results from the two approaches are compared in Tables 5.6 and 5.7 [10]. The variation in still-water bending moment in way of the midship section is shown in Figure 5.28 [28].

Conversion of VLCC to VLOC

Very large ore carriers resulted from the conversion of single hull VLCCs that were due for withdrawal from operation into bulk ore carriers. Market forces led owners to consider such conversions as well as new ship construction in this class [27]. The result of the conversion is an ore carrier with a deadweight capacity of approximately 250,000 tonnes, which is significantly higher than that of capesize bulk carriers. VLOCs that are currently purpose-

Table 5.7 *Permissible still-water bending moment (rule and direct calculation)*

	Still-water Bending Moment (kNm)(10⁶)				
Vessel Dimensions	*(sagging)*		*(hogging)*		*Sag/hog ratio*
(L × B × D) (m)	*Rule*	*Direct*	*Rule*	*Direct*	*Rule/Direct*
334 × 43 × 27	4.74	4.05	5.12	5.07	0.92/0.8
266 × 44 × 23	0.18	–	0.17	–	1.06
233 × 42 × 21	3.139	–	0.77	–	4.08

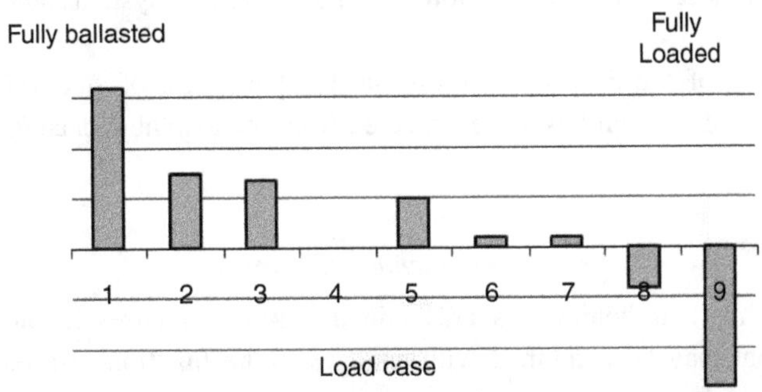

Figure 5.28 Variation in still-water bending moment i.w.o. midship section [28].

built for trading on the Brazil–China route have an even larger capacity of 400,000 tonnes deadweight.

In the case of VLOCs that originate from converted VLCCs, only the centre tanks are used for the transportation of cargo with the side tanks being used for ballast or as void spaces. A homogeneous loading arrangement is the only one possible because of the large capacity of each hold. For loading and discharging purposes, hatch openings have to be made on the main strength deck, which causes a loss of longitudinal strength as well as of torsional stiffness of the hull girder. The resulting deficiencies in strength are restored by introducing *saddle tanks* that are located in the underdeck position to port and starboard of the hatch opening. The saddle tanks extend in the transverse direction as far as the side longitudinal bulkheads (over approximately 10% of the ship's beam in each position) and in the longitudinal direction over the full length of the hold. Their cross-section has the form of a trapezium, and they are constructed as closed cellular structures. They can thus fulfil a number of roles. Their torsional rigidity provides support to the main deck which carries the weight of the hatch cover and the structural members that support it. More importantly, their torsional stiffness resists the twisting of the deck in way of the hatch coaming under torsional loading. Finally the sides of the saddle tanks that extend in the longitudinal direction contribute to the longitudinal strength of the hull girder.

The existing bottom structure has to be suitably modified in order to provide the strength necessary to support the increased weight of cargo. The frames are cut back, longitudinal side girders are added, and the centreline girder is suitably modified. Intermediate floors are also introduced to provide further support to the inner bottom that is retrofitted, and in this way a rigid structure capable of supporting the cargo load is created. To avoid giving rise to local stress concentrations in way of the connection of transverse stiffening elements with the longitudinal bulkheads, these are extended into the wing tanks.

Apart from strength considerations, other questions such as the entrapment of cargo on the horizontal girders of the transverse bulkheads have to be resolved. To avoid this, shedder plates are fitted between the edge of each frame and the adjoining bulkhead. Additional strengthening of the bulkhead is then required, which is done by increasing the thickness of plate elements of the lower stool. Other necessary modifications relate to questions of access, the installation of suitable alarm systems and new statutory requirements.

The sum total of these changes results in a net increase of steelweight of approximately 4,500 tonnes, which has to be deducted from the available deadweight capacity of the ship.

Lengthening of Ships

The issue of ship lengthening has arisen in a number of cases in the past, and is a modification that may be combined with changes to the function of a particular ship, in which case it becomes a complicated engineering exercise. In this section we will outline how the basic longitudinal strength requirements may be determined, following the approach described in [5]. It has to be added that lengthening will bring about changes to the hull form and proportions so that hydrodynamic performance, propulsion, cargo capacity and outfitting are affected, in addition to strength requirements. 'Jumboisation', a term commonly used, has been carried out for several ship types, notably passenger ships and containerships.

At the level of execution, the lengthening of a cruiseship is a complex operation because of the large number of decks and systems (Figure 5.29). The contribution of each deck to the longitudinal strength depends on a number of factors, in particular the length over which it extends in the longitudinal direction and the distance of the longitudinal external bulkheads from the side shell plating.

Figure 5.29 Lengthening of the cruiseship *Enchantment of the Seas*.

Table 5.8 *Lengthening of cruiseships [29]*

	Before Lengthening			After Lengthening		
Ship	GT	LOA (m)	No of berths	GT	LOA (m)	No of berths
Enchantment of the Seas	74,136	279.5	1,995	81,278	301.8	2,252
Thomson Dream	42,092	203.9	1,132	53,872	243.2	1,506
Superstar Aquarius	39,217	189.9	1,246	51,309	229.8	1,750
Norwegian Dream	39,217	189.9	1,246	50,764	229.8	1,750
Balmoral	34,242	187.8	990	43,537	217.9	1,340
Louis Majesty	32,396	173.4	1,056	40,876	207.3	1,460
Black Watch	21,847	177.7	536	28,613	205.4	804
Boudicca	21,891	177.7	536	28,388	205.4	839
Albatros	21,800	177.7	536	28,078	205.4	884
Braemar	19,089	163.7	727	24,344	195.1	968
Aquamarine	18,400	169.8	724	23,149	194.2	1,050
Festival	18,416	169.8	724	23,005	194.2	1,050

Table 5.8 gives an idea of the extent of lengthening that has been carried out on a number of cruise-ships, shown in order of size.

From the point of view of longitudinal strength, the increase in ship's length is the most important factor because it results in an increase in bending moment. Therefore, the scantlings of the added section will have to be able to contribute effectively to the longitudinal strength of the lengthened hull. The IACS rules stipulate that the section modulus of a transverse section amidships is to have a minimum value given by Equation (4.52).

An increase in ship's length will result in changes in block coefficient and draught as well. As a result, it is necessary to calculate the new values of these parameters. Ideally an iterative procedure involving at least structural and stability requirements is required, but in the approach described here it is assumed that C_B and T can be determined explicitly. From the geometry of the added section, the new values of C_B and T are found using the lines plan and hydrostatics curves.

From Equation (4.52) it is seen that the required section modulus, W_{min}, is a function of L, B and C_B. Assuming that the breadth B remains unchanged, the change in the required section modulus per unit length, dW_{min}/dL, is given by:

$$\frac{dW_{min}}{dL} = \frac{\partial W_{min}}{\partial L} + \frac{\partial W_{min}}{\partial C_B}\frac{dC_B}{dL} \tag{5.12}$$

For the interval L_o to L_n,

$$\delta W_{min} = \int_{L_o}^{L_n} \frac{\partial W_{min}}{\partial L} dL + \int_{L_o}^{L_n} \frac{\partial W_{min}}{\partial C_B}\frac{dC_B}{dL} dL \tag{5.13}$$

The value of dCB/dL is a function of ship type and its form factors; a rough estimate can be obtained for the particular ship type being considered. If it is assumed constant, from Equation (4.52) the required change (increase) in section modulus is given by

$$\delta W_{min} = \left[W_{min} + \frac{dC_B}{dL} \frac{B}{100} \left(322.5L^3 + 0.02\lambda^{4.5} - 15.42\lambda^{3.5} + 3240\lambda^{2.5} \right) \right] \tag{5.14}$$

where $\lambda = 300 - L$. This expression is based on the use of the formula for the wave coefficient C_w for ships with $L < 300$ m. It is seen from Equation (5.14) that the increase in section modulus depends on the value of dCB/dL. The value of dCB/dL is larger when the increase in displacement for the required increase in length is large, and in such cases the lengthening process would necessitate extensive changes in strength requirements. In the case of volume-critical ships such as passenger/vehicle ferries and container ships, the value of dCB/dL is smaller, and as a result, the additional longitudinal strength requirements are not as extensive.

The actual section modulus, Z, of the as-built ship is greater than W_{min}, and there is a margin of safety against ultimate bending collapse of the hull girder. It may then be possible to proceed with the increase in length maintaining the original scantlings of the midship section. If this path were to be permissible and was followed in practice, the safety margin, W_{margin}, would be correspondingly reduced. The safety margin is given by

$$W_{margin} = Z - W_{min} \tag{5.15}$$

If, however, the same value of the safety margin, W_{margin}, were to be maintained for the modified ship, we would have

$$\frac{\delta Z}{Z} = \frac{\delta W_{min}}{W_{min}}$$

The new value of the required section modulus, W_n, would then be given by

$$W_n = Z\left(1 + \frac{\delta Z}{Z}\right) \tag{5.16}$$

Assuming that the added length is located in the midship region of the ship, the midship region can then be redesigned in order to satisfy the new requirements. However, the strength requirements extend to 0.4 L about the midship section, and it is probable that the length of the added section does not extend over such a proportion of the ship's length. In this case the existing structure will need to be modified for its section modulus to satisfy the new requirement.

In the next step it is necessary to estimate how much additional material is required, a measure of this being its cross-sectional area. From the point of view of longitudinal strength, the greatest contribution is provided by material that is located furthest from the neutral axis of the cross-section, this being the strength deck and other upper decks of the ship. It is desirable that the same configuration be maintained over the added portion, and this concerns the stiffening system, the spacing of frames and their cross-section as well as the types of longitudinal stiffeners and their spacing. In this manner the new section can be introduced seamlessly into the existing structure. In addition to longitudinal strength

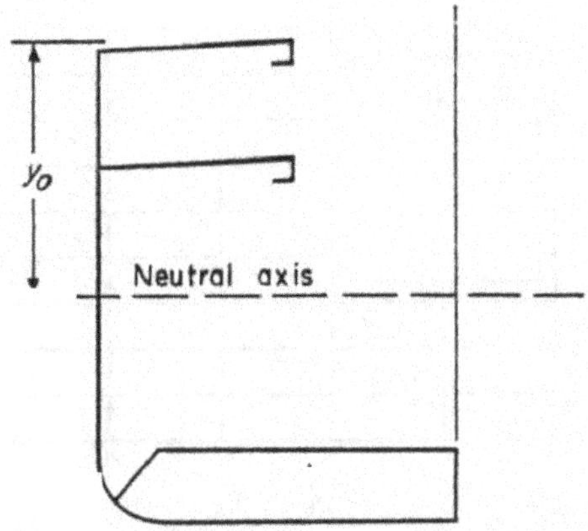

Figure 5.30 Idealised midship section of cargo ship. Reprinted courtesy of Alexandria Engineering Journal, Alexandria.

requirements, local buckling strength needs to be checked. It is expected that in the case of a volume-critical ship, lengthening would result in a small increase in draught and hence a corresponding small increase in the design loads used in local strength calculations so that the local strength requirements would continue to be satisfied.

If it thus proves possible to satisfy the strength requirements in general by modifying the strength deck alone, an expression for the required added area will be obtained.

We consider the idealised midship section shown in Figure 5.30. For this section, the shift in neutral axis, δy_o, resulting from the addition of material to the strength deck is given by

$$\delta y_o = \frac{a y_o}{A_o + a}$$

where a is the added area and A_o is the original area of the cross-section. The new position of the neutral axis, y_n, is then obtained from

$$y_n = y_o - \delta y_o = y_o\left(1 - \frac{a}{A_o + a}\right) = y_o\left(\frac{A_o}{A_o + a}\right) \tag{5.17}$$

The moment of inertia of the modified cross-section, I_n, is

$$I_n = I_o + a{y_o}^2 - \frac{(a y_o)^2}{(A_o + a)} \tag{5.18}$$

The section modulus of the modified cross-section, W_n, is then

$$W_n = \frac{I_n}{y_n} = Z\left(1 + \frac{a}{A_o}\right) + a y_o \tag{5.19}$$

The proportional change in section modulus, $(W_n - Z/Z)$, is given by

$$\frac{W_n - Z}{Z} = \frac{a}{A_o}\left(1 + \frac{A_o y_o}{W_o}\right)$$

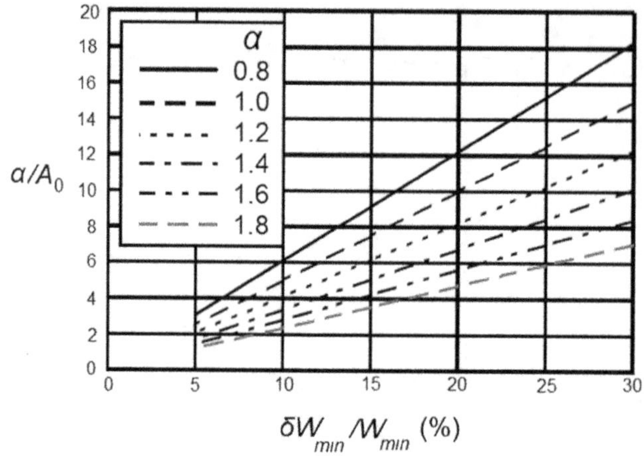

Figure 5.31 Effect of hull section index α on required material, a. Reprinted courtesy of Alexandria Engineering Journal, Alexandria.

also,

$$\frac{A_o y_o}{Z} = \frac{A_o y_o^2}{I_o} = \left(\frac{y_o}{r_o}\right)^2$$

where r_o is the radius of gyration of the original cross-section. Letting y_o/r_o be equal to α, the following expression for the required proportional change in cross-sectional area, a/A_o is obtained:

$$\frac{a}{A_o} = \frac{W_n - Z}{Z(1 + \alpha^2)} \tag{5.20}$$

where α stands for the *hull section index*. This is a measure of the properties of the hull section and the distribution of the longitudinally effective material. Its value depends on the configuration of the midship section and the main dimensions of the ship; in particular, it depends on the extent of longitudinally effective material in the bottom structure. From calculations for representative ship types it has been found that $0.8 < \alpha < 1.8$, depending on ship type. In the case of cargo ships that are fitted with longitudinally stiffened double bottoms such as double hull tankers, bulk carriers and container ships, $\alpha \approx 1.4$. Equation (5.20) is used to estimate the area to be added to the strength deck within $0.4L_n$ from the midship section, and is plotted in Figure 5.31.

It is seen that when α is small, a larger additional area of cross-section is required. Structural optimisation techniques can be brought to bear on this problem. The important issues that need to be considered are the extent to which an actual increase in area is required given that higher-strength steel may be used instead, as well as the distribution of material between the strength deck and the attached longitudinally effective members.

5.2 Mechanical Equipment Loads

These loads result from the operation of *mechanical equipment* that does not form an integral part of the hull structure.

- Weight.
- Inertial loads arising from ship motions and equipment operation (reciprocating or rotational motion).
- Loads borne by particular items (for example mooring-line loads, anchor and chain loads, thrust loads acting on thrust stool.
- Vibration arising from equipment operation.
- Thermal gradients.

These loads are generally of a local nature and are due to the presence and operation of mechanical equipment. Their understanding is necessary in order to design foundations and supports, especially for larger items such as main propulsion machinery.

Equipment weights vary considerably, with the most important one being that of the main propulsion machinery. Other large items are related machinery equipment within the engine room such as the propeller shaft and bearings and auxiliary machinery items such as boilers. In each of these cases seating arrangements are purpose-designed in order to carry the weight as well as the other aforementioned loads. Table 5.9 includes the most important items that form part of the outfitting and impose additional loads to the hull structure.

To gain an idea of the magnitude of forces involved, we quote the following expression that is used to provide a weight estimate of the main propulsion equipment [33]:

$$W = 12\left(\frac{\text{MCR}}{\text{RPM}}\right)^{0.84} \tag{5.21}$$

where $W =$ weight, MCR $=$ maximum continuous power rating in kilowatts and RPM $=$ engine revolutions per minute. For slow-speed marine diesel engines the weight is approximately 0.037 tonnes/kW or 22–28 kW/tonne. For a ship fitted with a 7,500 kW engine the

Table 5.9 *Major equipment*

Engine room	Main propulsion machinery (diesel engine, main turbine)
	Propeller, propeller shaft, shaft bearings
	Boilers (main boiler, auxiliary boiler)
	Other equipment (generators, compressors, heat exchangers, purifiers, pumps etc.)
Structural items	Stern-frame
	Platform decks
	Masts
	Casings
	Deck machinery and equipment (cargo and supplies handling equipment, mooring equipment)
Outfit items	Appendages (rudder, rudder stock, shaft brackets)
	Hatch covers

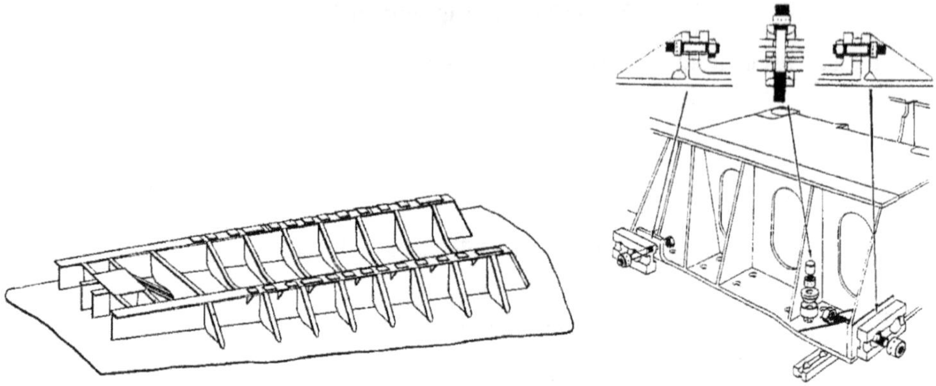

Figure 5.32 Main diesel engine foundation and details [20].

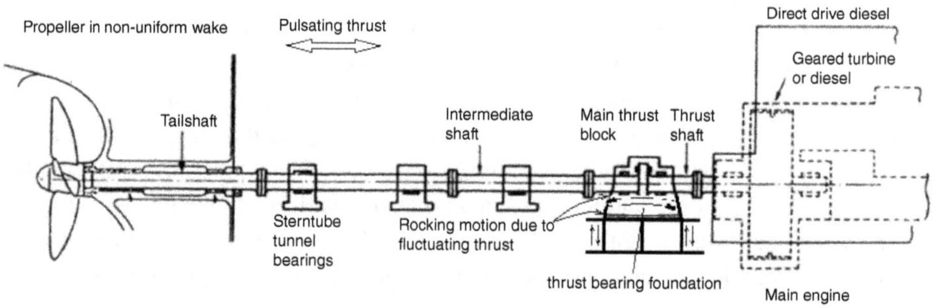

Figure 5.33 Loads acting on thrust stool [15]. Reprinted courtesy of the Society of Naval Architects and Marine Engineers.

main engine weight will be around 280 tonnes and will be distributed over the seating, an example of which is shown in Figure 5.32 [20]. In the case of diesel engines the seat is embedded in the double bottom structure and forms an integral part of it. Scantlings are substantially increased, and the engine bedplate is attached to the seat using holding down bolts. Furthermore, in order to distribute the loads properly, the structural members in both longitudinal and transverse directions have to be continuous.

In the case of deck equipment such as anchor windlasses, bollards, etc., the loads on the deck and supporting structure have to be carefully calculated given the particular role these items fulfil. The same applies to cargo-handling equipment.

Figure 5.33 shows how the thrust created by a propeller that is operating in a non-uniform wake leads to a rocking motion, which is in turn damped by the supporting structure. The fluctuating vertical load in the double bottom of the hull girder that causes alternate bending of the outer shell may, if attention is not paid to detailed structural design, lead to the premature appearance of fatigue cracks.

Inertial loads arising from ship motions vary in the same manner as in the case of hull structural items. However, in addition to these, inertial loads due to equipment operation must be considered.

Equipment Vibration

Vibration is the most challenging of all load types that result from the operation of machinery equipment found on board ships (Figure 5.34). It arises because many different types of equipment include components that perform rotational or reciprocating motion that imposes cyclic loading on supports and structural members. Vibration can affect the hull structure in several ways (fatigue, equipment malfunction) as well as cause passenger discomfort. As a result, equipment, their supports and the structure have to be designed so as to limit vibration to acceptable levels. The most important sources of vibration excitation are the propeller and the main propulsion machinery. Since both of these are in continuous operation when a ship is at sea, the excitation that they generate in the hull structure, at both a global and a local level, is classified as *steady state*.

Propeller-Induced Vibration

The propeller accounts for a large proportion of the vibration-related problems that have arisen in the past. The problem stems from the fact that a propeller is fitted with a finite number of blades each of which provides thrust, resulting in a fluctuating pressure field within the surrounding fluid and thereby in way of the stern hull structure (Figures 5.35a and 5.35b). In Figure 5.35b the trace of the pressure pulse acting on the hull just above the propeller is shown. The longitudinal distribution of these forces per unit length along the

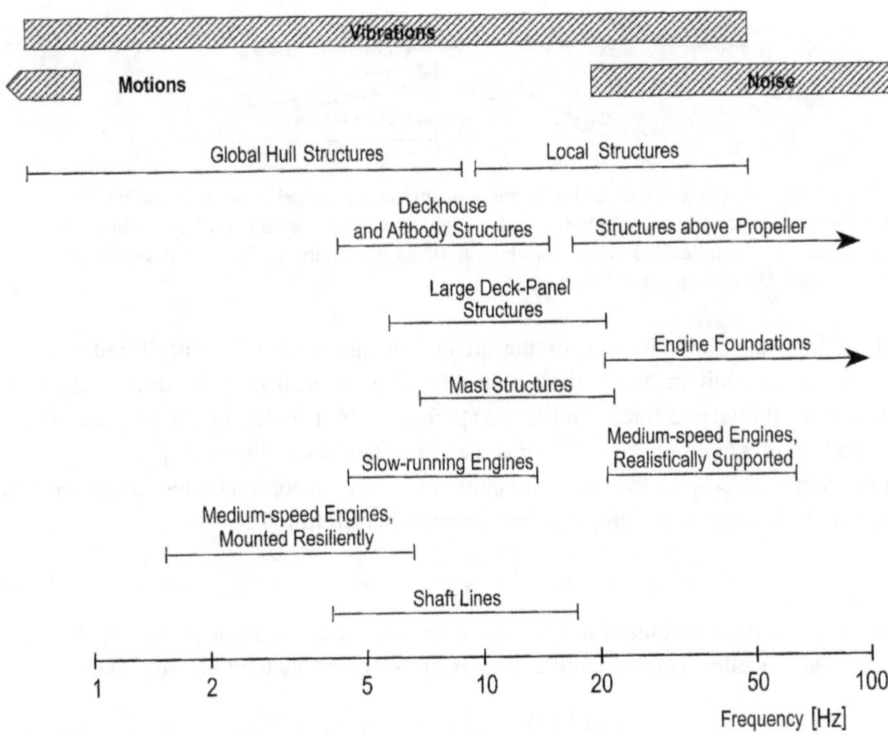

Figure 5.34 Natural frequencies of vibration of individual items on board ships.

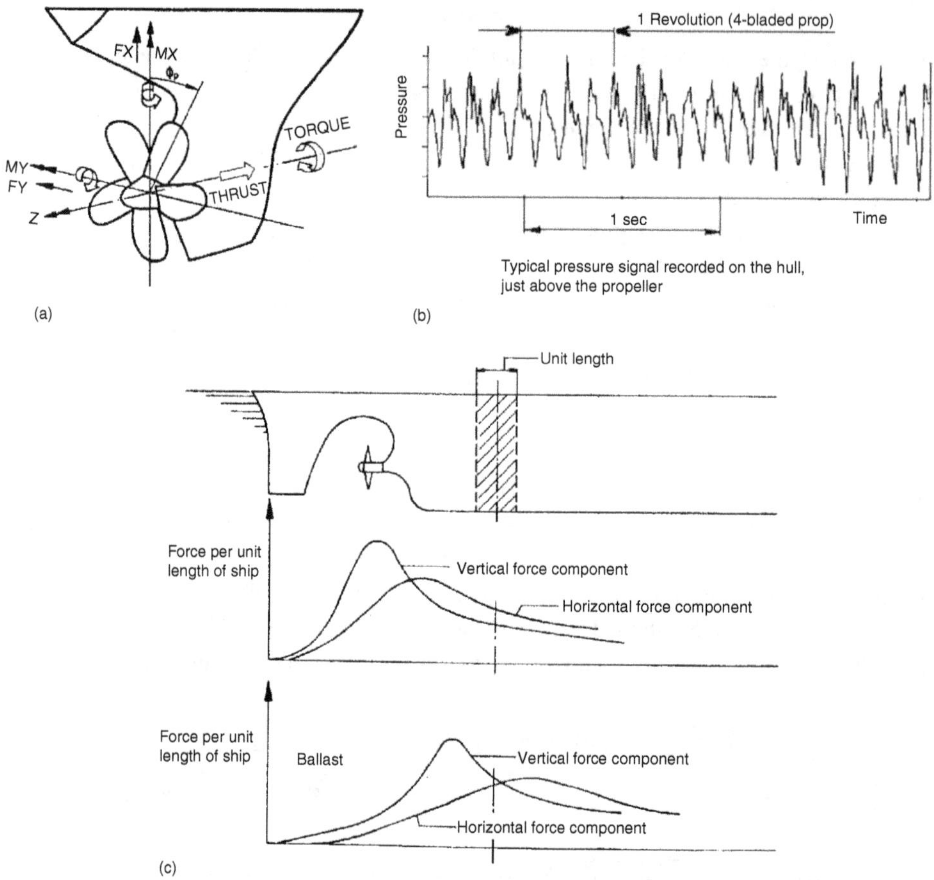

Figure 5.35 Dynamic forces and moments transmitted by propeller to shaft and hull [31]. (a) Force components. (b) Pressure pulse signal recorded on the hull just above the propeller. (c) Longitudinal force distribution (loaded condition). Reprinted courtesy of Det Norske Veritas, Høvik.

length of the hull girder in way of the aft end at any instant in time is indicated in Figure 5.35c. In addition, the hydrodynamic loading on each propeller blade results in fluctuating axial dynamic forces and moments that are transmitted by the propeller shaft; these result in rocking motion of the thrust bearing foundation (Figure 5.33).

For design purposes, the total pressure pulse that acts on a portion of the submerged hull surface in the vicinity of the propeller can be evaluated from [31]:

$$\Delta p_z = \sqrt{\Delta p_c{}^2 + \Delta p_0{}^2} \tag{5.22}$$

where Δp_c is the pressure pulse for a cavitating propeller and Δp_o is the pressure pulse for a conventional propeller. These quantities are given (in $\mathrm{N/m^2}$) by the following expressions:

$$\Delta p_c = \frac{(\mathrm{RPM}.D)^2}{160} \frac{V_s(w_{T_{\max}} - w_e)}{\sqrt{h_a + 10.4}} \left[\frac{1}{d/R}\right]^{\kappa_c} \tag{5.23a}$$

$$\Delta p_0 = \frac{(\text{RPM}.D)^2}{70} \frac{1}{Z^{1.5}} \left[\frac{1}{d/R}\right]^{\kappa_0} \tag{5.23b}$$

where

$\text{RPM} =$ propeller revolutions per minute
$D =$ propeller diameter (m)
$R =$ propeller radius (m)
$V_s =$ ship forward velocity (m/s)
$Z =$ number of blades
$h_a =$ depth to propeller shaft axis (m)
$r =$ radial distance from propeller shaft axis along blade (m)
$d =$ distance from $r/R = 0.9$ to the position on the submerged hull when a blade is in the top position
$\kappa_c = 1.7 - 0.7(d/R)$
$= 1.0 \quad (d/R > 1)$
$\kappa_o = 1.8 + 0.4(d/R) \quad (d/R < 2)$
$w_{Tmax} =$ maximum wake peak ($0.8 > w_{Tmax} > 0.5$ for single screw ships)
$w_e =$ effective wake

Criteria for acceptance of pressure fluctuations are based on (a) the magnitude of local vibrations of the afterbody and the ensuing probability of appearance of cracks and (b) criteria for global vibration levels within the ship's accommodation spaces.

Vibration produced by the propeller and/or other equipment located in the engine room can propagate to the superstructure. In the case of passenger ships, classification societies have introduced class notations that denote an acceptable level of vibration, noise and indoor climate for passenger comfort.

Listed below are measures that may be implemented to reduce propeller-induced vibration:

- Ensure that the natural frequency of vibration of hull structure components differs from the natural frequency of propeller-induced pressure pulses and axial excitation forces in order to avoid resonance.
- Modify the hull form in way of the propeller to improve wake. For existing ships this can be done using fins, vortex generators and duct systems.
- Modify the number of propeller blades (in general, an odd number of blades induce smaller thrust fluctuations).
- Use a propeller of a different geometry (highly skewed).
- Reduce the propeller RPM or diameter.

The selection of the appropriate measure will be based on practical considerations in each case.

Machinery-Induced Vibration

The operation of internal combustion engines involves the reciprocating motion of the piston assembly and the rotational motion of the crankshaft (Figure 5.36a). The vertical thrust (P) generated by the gas explosion within the cylinder chamber produces an axial force on the connecting rod (S) and a horizontal reaction force (G) on the cylinder wall in

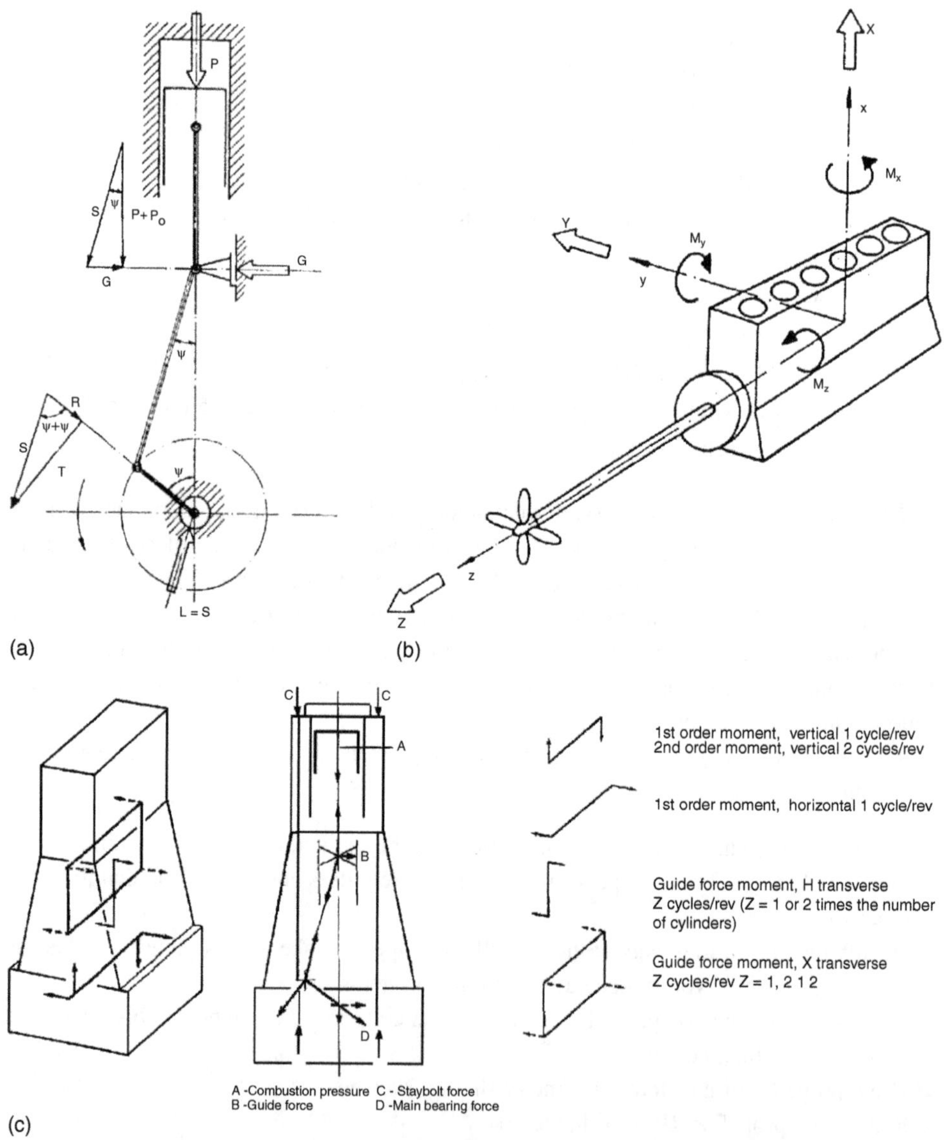

Figure 5.36 Main diesel engine forces and moments [31]. (a) Piston-crankshaft assembly forces. (b) Reaction forces and moments. (c) External forces and moments. Reprinted courtesy of Det Norske Veritas, Høvik.

way of the crosshead (Figure 5.36b). The magnitudes of the two force components change as the angle of the connecting rod alternates between $\pm\psi_{max}$, where ψ_{max} is the maximum angle to the vertical (Figure 5.36a).

The force that acts on the connecting rod (S) is transmitted to the crankpin and can be resolved into two components with respect to the crankshaft, a radial one (R) and a tangential one (T). The radial component leads to bending of the crankshaft whereas the tangential one may result in *torsional shaft vibration*. These forces act on internal members of the engine, and since the engine has a finite rigidity, deformations of its structure as well as its seating will arise. Equilibrium is achieved by the presence of reaction forces and moments. Vibration will tend to shake the engine itself and its seating, which forms part of the hull structure in way of the engine room double bottom. In multi-cylinder engines the free moments can be limited by appropriate engine design (number of cylinders, firing order, use of counterweights and moment compensators). These loads are cyclic in nature, and their frequency depends on engine RPM, so that first-order moments occur at a frequency of 1 cycle/rev, second-order moments at 2 cycles/rev, etc. (Figure 5.36c). In order to evaluate the net effect of all components of the excitation force, a *harmonic vibration analysis* has to be performed. The excitation source is then represented as a sum of excitations operating at different frequencies that are multiples of the engine's rotation frequency. This can be expressed as

$$F = F_1 x \cos(\alpha + \phi_1) + \cdots + F_n x \cos(\alpha + \phi_n) + \cdots \tag{5.24}$$

where α is the crank angle and φ_n are the phase angles. The first term is called the *first-order force* because it acts once per each engine revolution. Harmonic analysis then involves identifying the vibration characteristics of the exciting force. The natural frequencies of vibration of the structure also must be calculated. If it is found that if the excitation frequencies coincide with the natural frequencies of vibration of the structure, resonance will occur with a significant amplification of deflections and stresses. The high-frequency fluctuation of stresses will result in fatigue that eventually will lead to the appearance of cracks. To prevent this from happening, either the magnitude of the relevant exciting force or moment has to be reduced to an acceptable level or the natural frequencies of vibration of the structure have to be modified. The former is done by using counterweights or moment compensators whereas the latter may be possible by rearranging the structure and by introducing stiffeners at appropriate positions. If the structure is strengthened in way of peak amplitude positions, the vibration mode changes and the natural frequencies increase.

Thermal gradients arise as a result of machinery operation and produce changes in local stresses.

5.3 Cargo-Related Loads

These include:

- Cargo weight.
- Associated inertial forces.
- Thermal gradients.

5.3.1 Cargo and Ballast Sea-Water

The question of distribution of the various weights has to be considered with respect to fixed weights as well as with respect to those weights that change during the lifetime of the ship. Strictly speaking, fixed weights include the lightship and all machinery and outfitting items (Figure 4.9). In the case of cargo, the magnitude and position of the weights will vary. Certain deadweight items (constants) will always be present on board, regardless of whether or not the ship is in a loaded condition.

Cargo ships have their machinery located near the aft end, and thus the cargo spaces extend over a substantial proportion of the length, specifically from the forward engine room bulkhead to the forward collision bulkhead. Due to reduced buoyancy at the ends and a fairly uniform distribution of the lightship given the substantial proportion of the length over which the cargo spaces extend, such vessels will tend to be in a hogging condition, especially when unloaded.

Upon delivery to her owners, each ship is accompanied by a loading manual that sets out in detail the standard and recommended loading conditions. When preparing the loading manual, consideration is given to both global and local strength.

In the case of cargo ships, another important source of load changes is the loading, discharging and exchanging of ballast seawater. Ships are ballasted with seawater to improve stability characteristics and motion response in waves when not carrying cargo as well as to provide sufficient immersion for the propeller. The improvement in motion response involves increasing the roll period and reducing the probability of bow slamming in waves. Clearly, to achieve these aims, fairly large quantities of ballast water are required (up to 30–40% of the deadweight capacity and more, depending on vessel size). On the other hand, *permanent ballast* is fitted in order to improve buoyancy and stability (heel and trim) characteristics. This is done in the case of ships whose displacement changes only marginally (passenger ships, pleasure craft).

Ballasting operations are carried out on a routine basis, that is, during loading and discharging operations and also whilst at sea. The quantities involved are not insignificant, and if not planned properly, ballasting and de-ballasting can bring about significant changes in stress levels, generally at a local level. Ships are provided with water ballast management plans that set out in detail the sequence to be followed in each case. An important requirement of water ballast management planning is the minimisation of hull girder loading, at a global as well as a local level.

The ballast water carried by cargo ships is stored in dedicated spaces whose layout differs for each type of ship. Bulk carriers carry ballast water in the fore and aft peak tanks, which facilitates trim corrections, in way of the double bottom, and also in the hopper spaces. When in a ballasted condition, hydrostatic pressure is exerted on all tank surfaces, although when calculating the net pressure, it is necessary to allow for any compensating pressure acting on the other side of the bulkhead such as in the case of the double bottom and the lower hopper spaces.

Oil tankers sail in a ballasted condition for half their working life, since they do not transport cargo following completion of each loaded voyage. On board double hull tankers the ballast tanks are used to form a barrier between the cargo tanks and the outer shell structure. They are thus U- or L-shaped, depending on whether or not they extend around the girth of the hull girder. As in bulk carriers, the fore and aft peak tanks carry ballast water. When fully loaded, the cargo tanks are nearly full, since the existence of segregated ballast tanks increases the buoyancy when compared to the previous generation of ships. As a result, the hydrostatic pressure acting on the inner bottom is directly related to tank size.

Chemical tankers, on the other hand, are much smaller in size, and so are the corresponding hydrostatic pressure loads. In general they have similar ballast tank arrangements to crude oil tankers, although the type of products transported determines to a large extent the tank layout, which, for a number of other reasons, is much more complicated.

As an example, we consider the case of a handysize bulk carrier shown in Figure 5.37 for which two loading cases are shown (homogeneous and alternate loading condition. Bending moment (BM) and shear force (SF) curves are included in Figure 5.38. This figure includes the BM (dashed line) and SF (full line) envelope curves and corresponding curves for two of the loading conditions that are included in the loading manual. In the case of the homogeneous loading condition (Figure 5.37a), the maximum shear force arises in way of the forward bulkhead of the engine room. This is caused by the imbalance of weight and buoyancy for the section of the ship that extends from the aft extremity to this position. In this case the out-of-balance load is to a large extent borne by the bulkhead in question. Forward of this position, the near-uniform weight distribution that consists predominantly of lightship and cargo is counteracted by the constant buoyancy force so that the net load on the hull girder is minimal in the still-water condition.

In the case of the alternate hold loading condition (Figure 5.37b), the shear force and bending moment are substantially higher than in the previous case and, for this particular ship, approach the permissible values. We can see from the shear force curve that in way of the transverse bulkheads the change in load manifests itself in the form of a discontinuity in

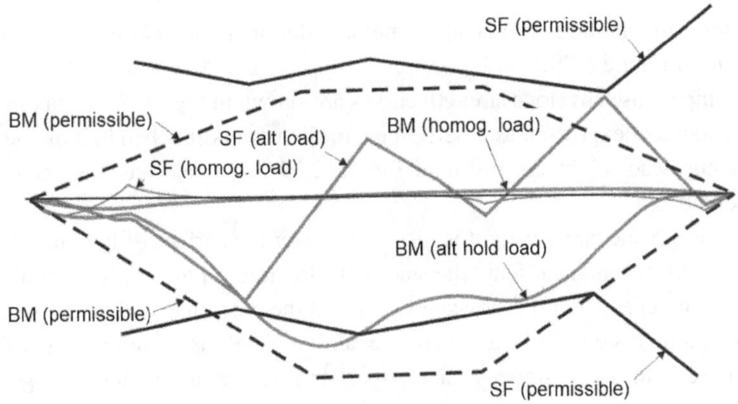

Figure 5.37 Bending moment and shear force distributions (handysize bulk carrier). (a) Homogeneous loading condition. (b) Alternate hold loading condition.

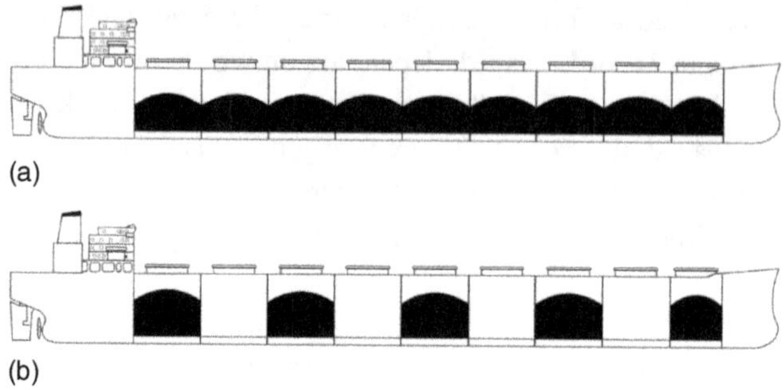

Figure 5.38 Example loading conditions for bulk carriers.

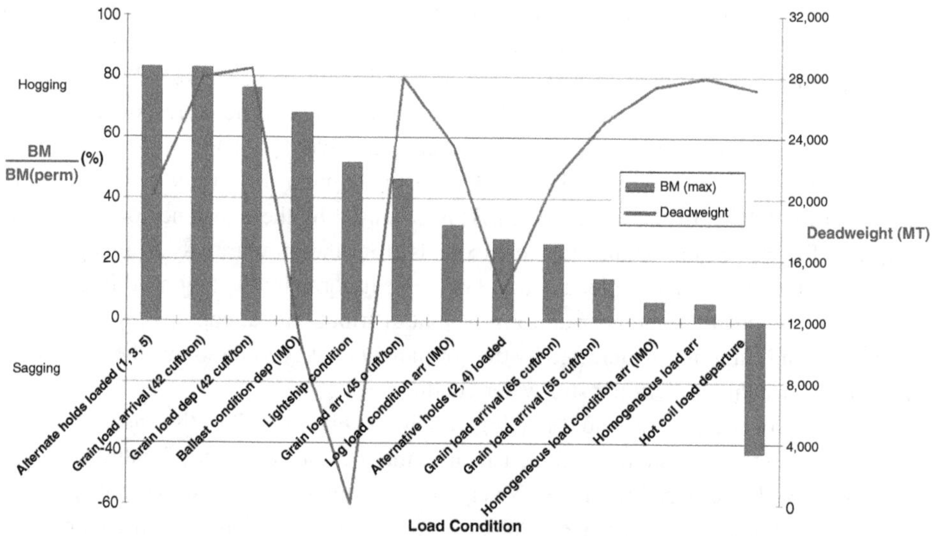

Figure 5.39 Maximum bending moment and corresponding deadweight (handysize bulk carrier).

the shear force distribution. The bending moment distribution is correspondingly affected, as is evident in Figure 5.39.[2]

For the ship whose envelope strength curves are shown in Figure 5.38, maximum values of bending moment expressed as a percentage of the permissible bending moment and the corresponding deadweight are given in Figure 5.39 for a selected number of standard loading conditions.

In bulk carriers the cargo weight is distributed over the surface of the inner bottom, the lower regions of the bulkheads and the side shell structure. In the case of granular cargoes, the distribution depends on the stowage factor and the recommended angle of repose. For cargoes with a high stowage factor such as coal, the weight is more uniformly distributed than for those with a low stowage factor such as iron ore and minerals in general. The

[2] Results of a study on the strength of bulk carriers in the alternate hold loading condition are included in Chapter 13.

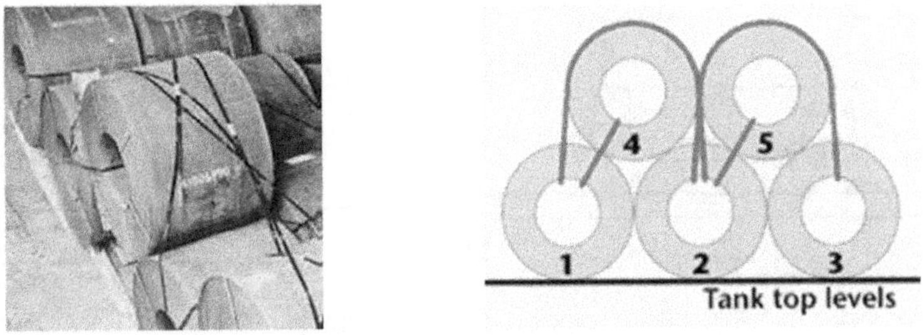

Figure 5.40 Arrangement of steel coils in bulk carrier hold [14].

height to which low-density cargoes reach approaches the upper levels with the hold, thereby exerting a near-uniform pressure on the inner bottom, the transverse bulkheads as well as the side shell structure. When fully loaded, the ship floats at its maximum draught. The net pressure acting on the side shell structure is then the difference between the internal and external pressures.

A similar situation arises in the case of transverse bulkheads when consecutive holds are loaded (homogeneous loading condition) (Figure 5.38a).

The cargoes that impose the largest loads on the inner bottom structure are not granular cargoes but steel products such as coils and slabs. Steel coils can weigh up to 15 tonnes each and are arranged as shown in Figure 5.40. The maximum pressure, expressed in metric tonnes/m^2, which the inner bottom can support is a measure of the cargo-carrying capacity of the ship.

Since the weight is transferred to the inner bottom structure over a small area, in order to avoid any damage, the coils are placed on *dunnage* that consists of a number of layers of wooden boards. In this way the weight is distributed much more uniformly over the area of the inner bottom.

In ore carriers the cargo holds are, by comparison, long and narrow, with the height of cargo extending up to the upper region of the hold, since adequate buoyancy is provided by the empty ballast spaces on either side of the hold and within the double bottom. As a result, pressure is exerted on the inner bottom, the inclined longitudinal hold bulkheads and the transverse bulkheads.

In the case of container ships, the weight of individual containers is transmitted via strengthened pads at the corners of each container. Within each cargo hold the exact position of each container is specified, and thus it is necessary to ensure sufficient strength at the positions of support within the inner bottom.

One other type of load is vehicles that are transported by ro-ro ships (ferries, pure car carriers and MPPs). In this case the load is applied over the area of the tyre imprint, and the most severe condition is the case in which the centre of the area of the imprint coincides with that of the largest plate panel (Figure 5.41).

An important feature of chemical tankers is the use of stainless steel for the cargo compartments because of its corrosion-resistant properties. Stainless steel is also used in *gas carriers* (LNG, LPG), which have even more complicated cargo carrying compartments because of the need to transport their cargo under pressure and/or low temperature.

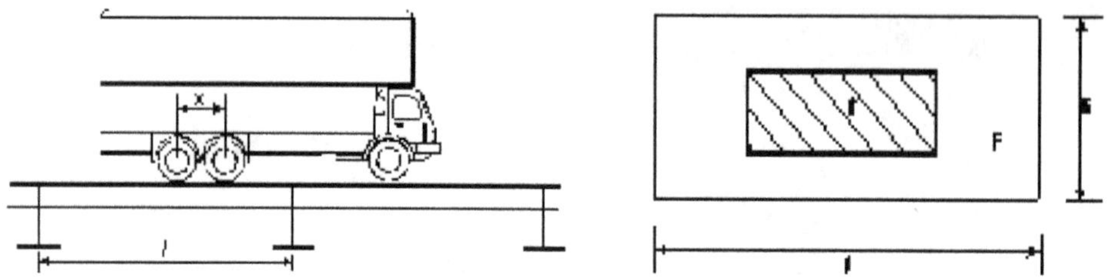

(a) (b)

Figure 5.41 Vehicle loading on car deck. (a) Truck on longitudinally stiffened car deck. (b) Tyre imprint on plate panel.

A variety of arrangements are now in existence, each of which is suitable for the carriage of a particular type of cargo. Pressurised cargo tanks are used and are of cylindrical or spherical shape. They have to be able to withstand vapour pressures in excess of 1.7 MPa. A variety of steel or non-steel alloys can be used. In the case of pressurised and refrigerated cargoes, the design pressures are 0.5–0.7 MPa, and the carriage temperatures are between –33°C and –104°C, depending on the type of cargo. Such tanks are built to conventional pressure vessel codes.

5.3.2 Inertial Loads and Cargo Shifting

When motion of a ship is restricted to pitching and heaving, the inertial loads that act on the hull girder as a result of the presence of cargo will not normally cause problems, provided they are allowed for at the design stage. If, however, the cargo is subjected to inertial accelerations in the lateral direction, as would be the case in severe sea states, *cargo shifting* may occur. In the case of granular cargoes this will result in asymmetrical loading of the hull structure, leading to the formation of longitudinal torsional moments and possibly damage. *Liquefaction* is another process that can cause cargo shifting (Figure 5.42).

A number of ships have been lost in which cargo shifting has been a major contributory cause. Ships are lost through a variety of sequences of events. Extreme angles of heel result in loss of transverse stability, and at the same time abnormally high loads are imposed on the hull structure. Since structural members are not normally designed to withstand such loads, failure may result. Water ingress may take place from openings on the main deck, leading to capsizing of the vessel. It is also possible that under a large angle of heel, if the side shell structure is already weakened as in aged ships, cracks may appear, leading to water ingress in cargo holds.

Damage may also result from deck cargo shifting, such as timber. Stanchions and rails used to retain timber may suffer damage, leading to a loss of cargo overboard from one side, causing a listing of the ship. In the case of containerships, containers impacting the internal structure of cargo holds during loading may cause damage leading possibly to rupture of the plating and ingress of water from a ballast tank.

Some of the most serious incidents that have occurred as a result of cargo shifting have taken place on board ro-ro ships. In at least one case, uncontrolled movement of large

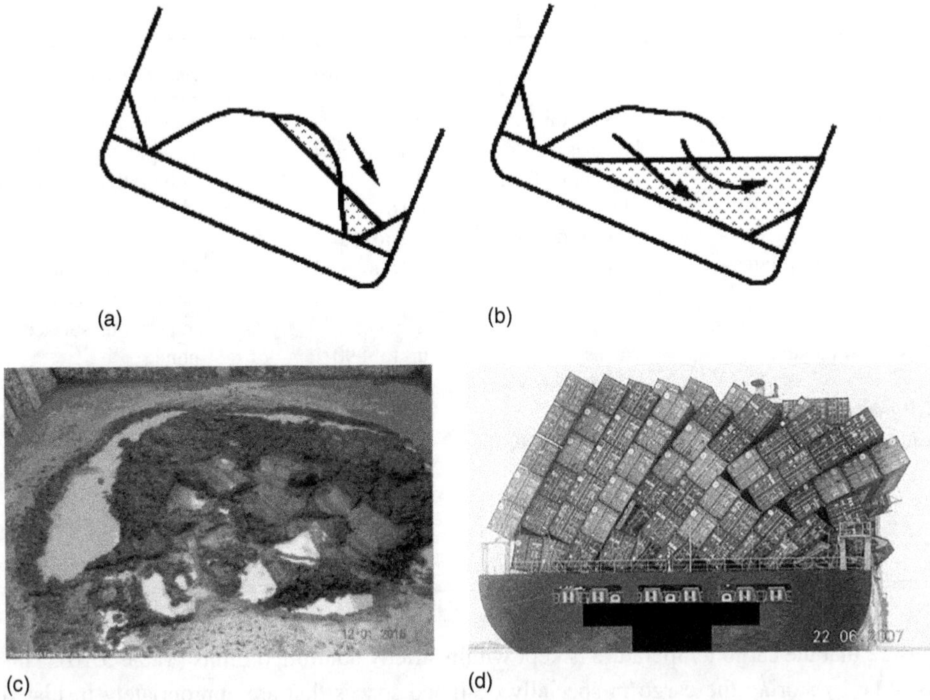

Figure 5.42 Accidents involving cargo shifting and movement. (a) Sliding failure [30]. (b) Liquefaction [30]. (c) Cargo (bauxite) liquefaction [17]. (d) Cargo shifting due to heavy weather.

vehicles that impacted the side structure in heavy seas caused serious structural damage that resulted in the loss of the ship.[3] In the case of passenger ferries, accidents have in most cases occurred after the ingress of water and as a result of the existence of the large free surfaces that are a functional necessity in the cargo spaces. In several well-documented cases this has resulted in rapid capsizing.

In the case of wet cargoes and ballast water, inertial forces result in *sloshing*, that is, the motion of liquids that results in them impacting the tank bulkheads. Since this occurs continuously and even if immediate damage does not occur, over an extended period of time fatigue of structural members may result. Sloshing loads can be minimised by ensuring that when filling individual tanks, depths in the range 25–75 per cent are avoided.

5.3.3 Thermal Gradients

From the point of view of the hull structure it is *temperature differences* that are important because these result in thermal gradients and consequent thermal stress fields. There exist several different kinds of cargo that necessitate the use of arrangements and equipment to

[3] Passenger ferry *Heraklion*, December 1966, Aegean Sea.

Table 5.10 *Cargoes transported at controlled temperatures*

Ship Type	*Cargo*	*Temperature (°C)*	*Insulation Type*
Refrigerated cargo ships	citrus fruit	+4 to +5	rockwool or fibreglass slabs or polyurethane foam
	bananas	+13 to +14	
	perishable fruits	0	
	meat	−18	
	dairy products	−2 to +5	
Factory ships, trawlers	fish	−18 to −30	fibrous with vapour barrier or plastic with impervious surface
Chemical tankers	waxes	up to +90	none
	other cargoes	cooled	none
Oil tankers	crude oil	+55	none
Liquefied gas carriers	methane (LNG)	−163	polyurethane foam
	propane (LPG)	−43	
	butane (LPG)	−1	
	ethylene	−104	
	ammonia	−34	

ensure that the cargo temperature is kept within strictly controlled limits (Table 5.10). This is done by storing the cargo in specially designed spaces that are appropriately insulated or heated.

A variety of foods such as meat, fish, fruits and dairy products are transported by sea and are refrigerated so that they do not perish during the voyage and prior to transhipment. For these products, insulation is provided in order to protect them rather than the ship's structure.

Apart from liquefied gases, discussed later, no effort is made to thermally protect the surrounding structure from the cargo, such as in the case of oil tankers in which *heating coils* are installed over the surface of the inner bottom of cargo and slop tanks. The cargoes are heated so that they can achieve the required degree of viscosity necessary during discharging. The cargo is heated by pressurised steam that is passed through the coils, with the heating achieved by convection (Figure 5.43). It has to be added that the persistence of temperature gradients and differences in moisture content within the ullage spaces and the air above the strength deck may result in accelerated corrosion of the underdeck structure.

In the case of chemical tankers, cargoes are heated using heating coils within tanks as well as by deck-mounted heat exchangers. A limited number of chemical cargoes need to be cooled in order to limit evaporation and the development of high pressures in the tanks. Because of the corrosive nature of chemical cargoes, stainless steel (grade 316 and, more recently, *duplex*[4]) is used for cargo storage whilst mild steel is used for other parts of the

[4] *Duplex* stainless steel is an austenitic-ferritic alloy with higher corrosion resistance and strength properties than other marine-grade stainless steels. It is becoming more extensively used, especially for cargo tank plating.

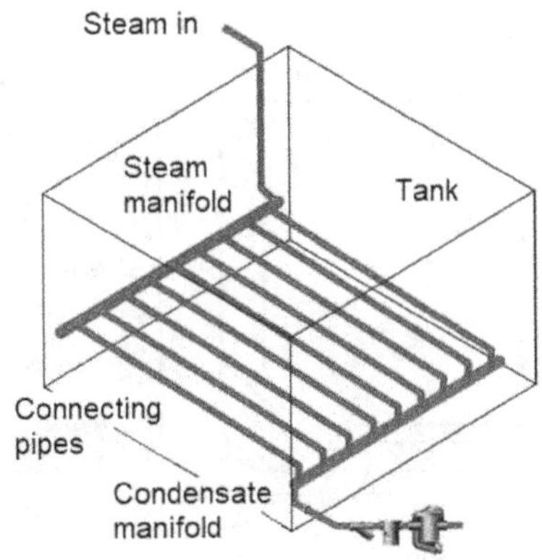

Figure 5.43 Steam heating-coil arrangement in a tank used for carriage of crude oil [26].

hull structure. As a result, at elevated temperatures thermal stresses may be significant, because the thermal expansion of stainless steel is significantly greater than that of mild steel, causing differential expansion. A further problem that arises is that elevated temperatures bring about deterioration in the mechanical properties of both stainless steel and mild steel alloys (reduction in yield strength).

Accelerated corrosion has been found to occur in the underdeck region of crude oil tankers as a result of the thermal gradients and the high moisture content in the ullage spaces.

Arguably, the transportation of liquefied gases poses the most severe challenges with regard to thermal insulation and protection of the surrounding structure from the cargo. Gas carriers comprise of two groups of ships: *liquid petroleum gas* (LPG) and *liquid natural gas* (LNG) carriers. As their names indicate, they are used for the transportation of petroleum or natural gas in liquid form. At ambient pressure, the boiling point of these gases ranges from –5°C for butadiene to –161.6°C for methane, which accounts for 75–95 per cent of natural gas. Therefore, these gases need to be stored under pressure and low temperature or a combination of both. The primary technical challenge is then for the tanks used to transport liquefied gases to be able to (1) withstand the internal pressure, (2) ensure a sufficiently low temperature and (3) provide effective insulation to prevent boil-off. The structure of the ship also has to be protected from the low-temperature environment. Additionally, there arises one other technical problem in relation to the operation of these tanks, namely that of cargo sloshing. Two types of tanks are currently used: *self-supporting* and *membrane* (Figure 5.44).

Self-supporting (or independent) tanks do not form part of the hull girder and as such do not contribute to its strength. There are three types of self-supporting tanks, designated as A, B and C. Type A are of prismatic form, and in these the maximum permitted vapour pressure is 0.07 MPa, so that the cargo has to be fully refrigerated at ambient pressure. Type B tanks can be either prismatic or else are pressure vessels with a revolution configuration. Their design vapour pressure has to be less than 0.07 MPa. These tanks are subjected to a rigorous strength analysis, including fatigue and crack propagation studies, and are used in LNG carriers. Type C tanks are spherical with a design vapour pressure greater than 0.2

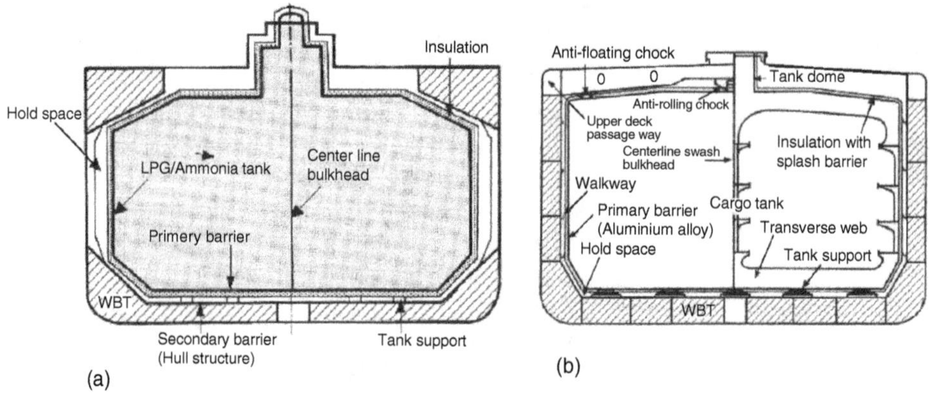

Figure 5.44 Self-supporting tanks used in gas carriers [6]. (a) Type A (LPG/ammonia carrier). (b) Type B (LNG carrier). Reprinted courtesy of the Society of Naval Architects and Marine Engineers.

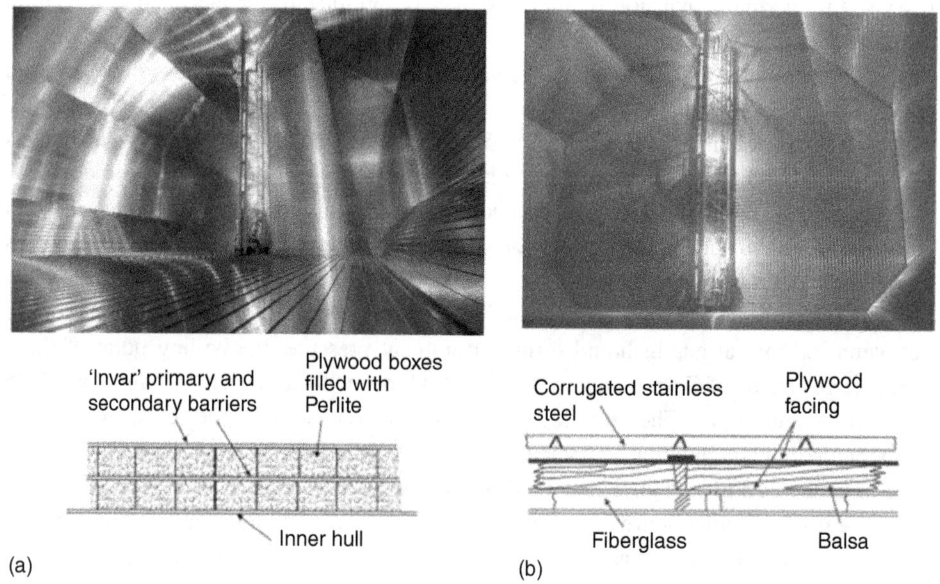

Figure 5.45 Membrane tank wall insulation systems [7]. (a) Invar. (b) Chrome/nickel stainless steel alloy.

MPa. These tanks are used in pressurised and combined pressurised and refrigerated ships. The tanks are built to pressure vessel codes, and an exact stress analysis can be performed. For this reason no secondary barrier is required.

Membrane tanks form part of the hull structure and are connected to it through a thin membrane. They consist of a thin layer of metal that acts as a primary barrier, insulation, a secondary membrane barrier, and further insulation in a sandwich construction (Figure 5.45). The inner layer is either a high-nickel-content steel (Invar) or else a stainless steel alloy that contains 18 per cent chrome and 8 per cent nickel. Invar is used because it

Table 5.11 *ABS requirements for structures contiguous to cargo containment system [22]*

Minimum Temperature	*Steel Thickness*	*Steel Grade*[5]
0°C to –10°C	All	As per ABS rules
+10°C to –25°C	t ≤ 12.5 mm	B
	12.5 mm < t ≤25.5 mm	D
	t > 25.5 mm	E
–25°C and below	t ≤12.5 mm	D
	t > 12.5 mm	E

has a very low coefficient of thermal expansion, and thus when expansion or contraction occurs, the resulting stresses are small. The Invar membrane is only 0.5–0.7 mm thick, which makes for a very light structure. Insulation consists of plywood boxes filled with perlite. In membrane tanks the design vapour pressure should not exceed 0.025 MPa, and this type of tank is used in LNG carriers.

The hull structure of LNG carriers, on the other hand, is constructed of steel, as in all ocean-going ships, and in order to prevent brittle fracture from occurring should cargo leakage take place, a suitable grade has to be chosen (Table 5.11). Under normal operating conditions the temperature of the steel structure within the vicinity of the cargo tanks is at a temperature that lies between the ambient temperature of the environment and that of the outer surface of the tank insulation. It is a classification society requirement that the temperature of the structural members surrounding the tanks be calculated assuming that the primary insulating barrier of the cargo is not functioning. In doing so, due account has to be taken of the temperature gradient between the cargo and the external environment (Figure 5.46) [22].

5.4 Loads and Their Relative Importance

Having described the loads that act on ship structures, we are in a better position to identify some of their properties that will be useful in evaluating the effect they have on the structure. Table 5.12 summarises the most important of these. We see that the picture that emerges is that of a complex system of loads that vary in extent of application, frequency of application, direction of action and duration. In order to appreciate the relative imortance of each of these it is necessary to consider their properties, which are summarised in Table 5.12; in addition to these we need to consider the peak value the load may attain in each case. For some of these the magnitude can be easily determined (e.g. hydrostatic pressure) whereas for others, such as hydrodynamic forces, it has proven to be a more difficult task. It is even more difficult to assess the resultant loading of a structural member when more than one load types act.

When first considering the loads that act on ship structures, we are most interested in those that act on the entire hull girder, namely those included in column 1 of Table 5.12. These include weight, external hydrostatic pressure, hydrodynamic loads, launching loads and

[5] The properties of these grades of steel are listed in Table 1.7.

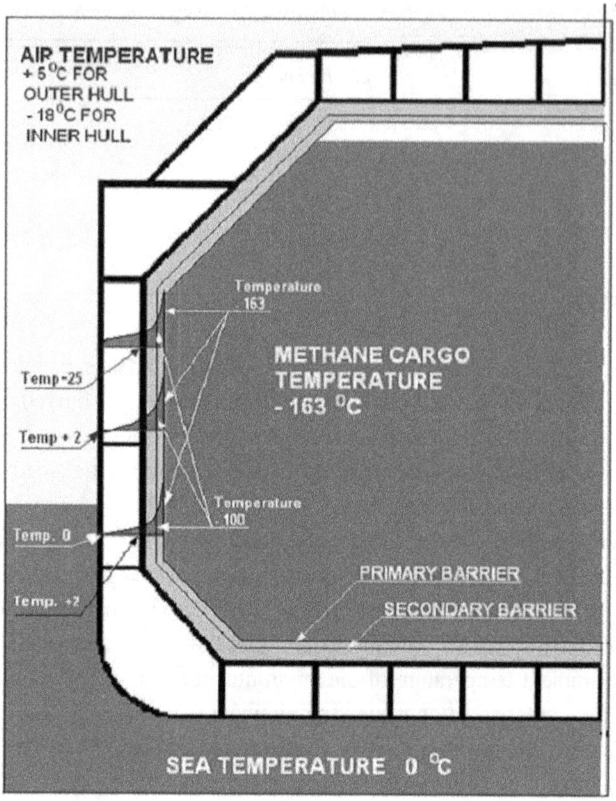

Figure 5.46 Temperatures within the vicinity of a LNG tank [22].

dry-docking loads. Of these the latter two arise on rare occasions; we therefore need to concentrate on the remaining loads, namely weight, hydrostatic pressure and hydrodynamic loads. The weight includes the steel hull, mechanical equipment, outfitting as well as consumables. It is possible to determine in an accurate manner the weight and the distribution of hydrostatic pressure using available information once the hull geometry has been determined.

The nature of the sea surface does not, however, permit us to do the same with hydrodynamic loads, since the elevation varies continuously in space and time. For this reason a different approach is needed that will allow us to describe the interaction of the ship with the liquid environment in order to specify critical load values that will enable a satisfactory design of the structure. A brief outline of the formulation for design loads that has been developed and incorporated in the IACS Common Structural Rules is included in Chapter 10. In the next section we shall consider the question of load action.

5.5 Load Action in Ship Structures

An understanding of the manner in which loads act on ship structures is necessary because it clarifies how they are borne and transferred between adjoining members. In order to perform this task, Newton's Third law is applied.

Table 5.12 *Properties of loads*

Loads	Extent of Application			Duration			Direction of Action		Frequency		
	Global	*Port*	*Local*	*Static*	*Slowly-varying*	*Dynamic*	*Normal*	*Other*	*Continuous*	*Occasional*	*Rare*
1. Environment – related loads											
Environmental thermal gradients	■			■				■	■		
Hydrodynamic loads	■	■			■		■	■	■	■	
Wave transient impact loads		■				■	■	■		■	
Wind loads		■			■			■		■	
2. Hull structure–related loads											
Still-water buoyancy force	■			■			■	■	■		
Weight	■	■	■								
Imperfections (residual stresses, etc.)			■	■				■	■		
Loads arising from port manoeuvring		■		■				■		■	
Inertial loads		■			■		■			■	
Launching	■				■		■				■
Dry-docking	■			■			■				■
3. Mechanical equipment loads											
Weight		■		■			■		■		
Inertial loads arising from motions		■			■		■			■	
Inertial loads arising from equipment operation		■			■		■			■	

Table 5.12 (*cont.*)

Loads	Extent of Application			Duration			Direction of Action		Frequency		
	Global	*Port*	*Local*	*Static*	*Slowly-varying*	*Dynamic*	*Normal*	*Other*	*Continuous*	*Occasional*	*Rare*
Loads borne by particular items		■		■			■	■	■	■	
Vibration arising from equipment operation		■				■	■	■		■	
Propeller fluctuating loads		■				■	■	■	■		
Cargo-handling gear operations		■		■			■	■		■	
Thermal gradients		■		■			■	■	■		
4. Cargo-related loads											
Weight of cargo		■		■			■		■		
Inertial loads		■			■		■		■		
Impact loads on inner bottom of holds			■		■		■			■	
Weight of ballast water and other liquids		■		■			■		■		
Weight of consumables		■		■			■		■		
Thermal gradients		■		■			■		■		
5. Other loads											
Grounding damage		■				■	■				■
Collision damage		■				■		■			■

5.5.1 Principal Loads Acting on Ship Structures

The loads that act on ship structures have components in the direction of each of the three orthogonal axes and about them, giving rise to three forces and three moments (six-degrees-of-freedom system shown in Figure 5.47). What characterises ship structures is that:

- They float on water and are at equilibrium under vertically acting weight and buoyancy forces.
- Their predominant dimension is length.
- They are unrestrained at their fore and aft extremities.
- They are constructed of stiffened plating that is very thin in comparison to their principal dimensions.

For these reasons, the most important forces and moments from the point of view of overall strength are (Figure 5.47):

- M_y – vertical bending moments acting about the y-axis
- V_z – vertical shearing forces acting in the direction of the z-axis
- M_t – torsional moments acting about the x-axis

Other loads that act and that are of lesser importance are:

- M_z – horizontal bending moments acting about the z-axis
- V_y – horizontal shearing forces acting in the direction of the y-axis

The latter are coupled with the corresponding vertical bending moments and vertical shear forces and result in increased values of these. Ship structural response is thus the outcome of the loads that act in these directions. Before attempting to consider the response to combined loading it is necessary to consider each type of load in isolation.

Our first task is then to describe the nature and magnitude of loads and load effects in the structure on a systematic basis. It is important to recognise that loads affect the structure at different levels, as noted in Chapter 3. It is also important to realise that in most cases

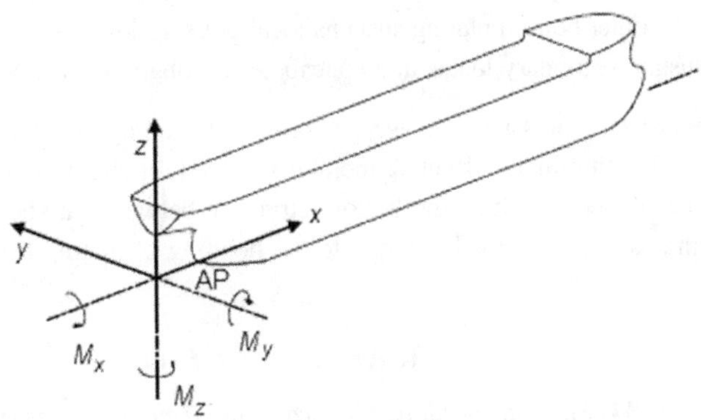

Figure 5.47 Co-ordinate axis system for ship structures [12]. Reprinted courtesy of IACS, London.

individual members are subjected to complex (multiaxial) systems of loading. This is considered in the companion volume (*Local Strength of Ship Structures*) in the case of the secondary and tertiary structures.

In the next two subsections we shall outline how the primary vertical bending and shear force loads are transmitted throughout the structure at these different levels. The response to torsional moments is discussed in Chapters 6 and 7.

5.5.2 Load Systems That Act on Ship Structures

There are three major load systems that coexist in conventional mono-hull ship structures. The hull structure can then be seen to consist of three corresponding types of members:

1. *Primary (longitudinal bending) loads.* The primary loads result from the difference between the longitudinal weight and buoyancy distributions, and they give rise to in-plane loading of the *primary structure* as well as vertical deflections of the entire hull girder. The primary structure consists of all longitudinally effective material, the main items being the strength deck, the side shell and bottom structures and the longitudinal bulkheads (if any).
2. *Secondary loads.* These act normal to the surfaces of plated structures, regardless of whether these are primary or secondary, and they result in out-of-plane bending. We may subdivide the secondary structures into *intermediate* and *local.* Examples of secondary intermediate structures are the transverse bulkheads, the superstructure located at the aft end of cargo ships and other items that do not extend for most of the length of the hull girder. The secondary local structures are in general the stiffened plating that is encountered throughout the hull girder and is loaded in the normal direction. An example of a secondary load is the weight of cargo. This acts normal to the inner bottom plating (primary and secondary structure) as well as the transverse bulkheads (secondary structure).
3. *Tertiary loads* act normal to plating and result in bending of the plating between its supports. The plating constitutes the *tertiary structure*. It also forms part of the secondary structure, since all plating is stiffened, and it may also form part of the primary structure. For example, the outer bottom plating sustains local pressure loads, works in tandem with stiffeners to sustain secondary loads, and contributes to longitudinal strength.

We see that hydrostatic pressure acts as a primary load, leading to the in-plane compression and tension of longitudinally effective members of the hull girder; as a secondary load, leading to bending of items such as the bottom structure between bulkheads and the side shell structure; and as a tertiary load, leading to the bending of plating between supports.

References

1. Antoniou A. (1984) *Strength of Ships* (Αντοχή του Πλοίου) Athens (in Greek).
2. Buermann T. M. (1969) Structural Components. In d' Arcangelo A. M. (Ed) *Ship Design and Construction* Jersey City, NJ: SNAME 191–250.

3. Chalmers D. W. (1993) *Design of Ships' Structures*. London: HMSO Publications.
4. Det Norske Veritas (2005) *Hull Structure Course*.
5. El-Kilani H. S. Ramadan R. (2003) Structural modifications for the longitudinal strength of lengthened ships. *Alex. Eng. J.* Vol. 42 No. 4 389–396.
6. Emi H. Fujitani T. Abe A. (2004) Liquefied Gas Carriers. In Lamb T. (Ed) *Ship Design and Construction* Vol. II. Jersey City, NJ: SNAME 32.1–32.41.
7. Eyres D. J. (1994) *Ship Construction* 4th ed. Oxford: Butterworth-Heinemann.
8. Firefly Digital Inc W. F. Cc and 800-397- "Home : Aries Marine Corporation" *Ariesmarine.com*. [Online]. Available: www.ariesmarine.com/site.php (accessed 19 December 2020).
9. Guedes Soares C. Fonseca N. Pascoal R. Guenther F. Clauss C. Schmittner E. Janou H. (2004) Analysis of wave-induced loads on a FPSO due to abnormal waves. *Proc OMAE Spec. Conf. Integ. Float. Prod. Stor. Offl. (FPSO) Sys.* Houston, TX.
10. Hamdan F. H. Burgan B. A. White R. (2003) Margins of safety in FPSO hull strength ASME 2002 *21st Int. Conf. Offsh. Mech. & Arc. Eng. (OMAE2002)* 23–28 June 2002 Oslo 1–9.
11. Heger R. (2008) Calculating block load by the moment-area method. *Drydock Conference* Norfolk, VA.
12. International Association of Classification Societies (2020) *Common Structural Rules for Bulk Carriers and Oil Tankers*.
13. Jones N. (1979) A literature review on the collision and grounding protection of ships. Ship Str. Comm. Rept SSC-283 Washington, DC.
14. Jørgensen H. (2008) *Carriage of Steel Cargoes (Skuld)*. Edinburgh: Witherbys Seamanship International.
15. Lamb T. (Ed) (2003) *Ship Design and Construction* Vol. 2 Jersey City, NJ: SNAME.
16. Livingstone M. (1974) *Naval Architecture III*. Lecture Notes, Dept. Naval Arch. & Ocean Eng. Univ. Glasgow.
17. "Loading of bauxite and other cargoes that may liquefy" *Gard.no*. [Online]. Available: www.gard.no/web/updates/content/20894756/loading-of-bauxite-and-other-cargoes-that-may-liquefy (accessed 19 December 2020).
18. Moan T. (2004) Marine Structures for the Future. Inaugural Keppel Lecture Nat. Univ. of Singapore.
19. Munro-Smith R. (1967) *Applied Naval Architecture*. London: Longmans Greene & Co.
20. NTUA (2010) *An Introduction to Vibration Aspects of Two-stroke Diesel Engines in Ships*. Athens: Lab Mar. Eng. Sch. Naval Arch. & Mar. Eng., Nat. Technical Univ.
21. Pedersen P. T. (1994) Ship Grounding and Hull Girder Strength. *Mar. Struct.* (1st Charles Smith Memorial Lecture) Vol. 7 1–29.
22. Piciocchi R. (2006) *Gas Carriers Training Course*. Houston, TX: American Bureau of Shipping.
23. Rawson K.J. Tupper E.C. (2001) *Basic Ship Theory Vol. 1 Hydrostatics and Strength*. 5th edition. Oxford: Butterworth Heinemann.
24. Ryu C.-H. Kim S.-C. Lee J.-H. (2013) Parametric Study for Assessment of Reaction Forces on Ship Docking Supports. *Jnl Korean Society of Marine Environment and Safety*. Vol. 19 290–301.
25. Samuelides M. S. Ventikos N. P. Gemelos I. C. (2009) Survey on Grounding Incidents: Statistical Analysis and Risk Assessment. *Ships and Offshore Structures* Vol. 4 No. 1 55–68.
26. Sapra R. (2016) Cargo oil heating practices. *Standard Safety Tankers*. London: The Standard Club 16–18.

27. *Steamshipmutual.com.* [Online]. Available: www.steamshipmutual.com/pdf.htm?id=262617&pdf=true (accessed 19 December 2020].
28. Sun H.-H. Bai Y. (2003) Time-variant reliability assessment of FPSO hull girders *Mar. Struct.* Vol. 16 219–253.
29. *Thecruisepeople.ca*. [Online]. Available: http://thecruisepeople.ca/cruiseblog/?p=574 (accessed 19 December 2020].
30. *U-tokyo.ac.jp*. [Online]. Available: http://underwater.iis.u-tokyo.ac.jp/research/bulk/bulk-chp1-e.html (accessed 19 December 2020).
31. Veritec A. S. (1985) *Vibration Control in Ships* Høvik: Veritec.
32. Wang G. et al. (2006) Collision and Grounding. Committee V.1 Rept. In Frieze P. A. Shenoi R. A. (Eds) *Proc. 16th Int. Ship & Offsh. Struct. Cong.* Vol. 2 Univ. of Southampton 1–56.
33. Watson D. G. M. (1998) *Practical Ship Design*. Amsterdam: Elsevier Ocean Engineering Book Series.

6

Linear Response to Primary Loading

The noblest pleasure is the joy of understanding.

Leonardo da Vinci

6.1 Introduction: Definition of Primary Structure

In the previous chapter we considered the primary loads that act on hull girders under a variety of conditions; from the point of view of structural analysis and design, the most important ones arise when a ship is underway at sea. A range of distributions of bending moments, shear forces and torsional moments need to be considered, and each of these has particular characteristics. In this chapter and the two that follow we will focus on the response of the hull girder to these loads.

In the present chapter the response of the hull girder to these loads that are treated as being independent of one another is discussed, and the resulting geometrical distortions and stress distributions are obtained. The combined action of these loads is also considered, given that the superposition of displacements, strains and stresses is permissible when linear theory can be used. This is the case for bending of beams (normal and shear) stresses, because it is permissible to use superposition. In the case of torsion it is permissible only when coupling of the load effects does not arise. This arises when the axes of bending (horizontal and vertical neutral axes) and the centre of twist are close to each other, as in the case of closed sections. In the case of open sections the centre of twist, which coincides with the centre of shear, is located below the baseline, and as a result, strong coupling arises. This problem therefore requires special treatment and is discussed in the next chapter.

In addition to developing suitable methods of analysis for existing ship structures, the contents of this chapter can be used to establish design requirements that are expressed in terms of the required strength and stiffness.

In the next chapter more complex types of response are dealt with. These concern coupled loads (bending and torsion) as well as cases in which material nonlinearity arises. The latter include the calculation of the residual strength of the hull girder following an accident as well as the important question of ultimate strength of the hull girder, which involves elasto-plastic buckling.

In Chapter 9 we will present the probabilistic description of hull girder response based on theoretical results obtained in recent years. The material included in this chapter is a necessary prerequisite to the reliability-based treatment of the hull girder, discussed in Chapters 11 and 12. In this manner we will have at our disposal as complete a description of hull girder response as is possible at this point in time; specific design issues will be discussed at the end of individual sections as required, in addition to Chapters 10 and 14 that deal exclusively with this topic.

Before embarking on the description of hull girder response, it is helpful to write down the general equation of motion of a body that is subjected to a time-varying, general external load $\boldsymbol{F}(t)$:

$$\mathbf{m}\ddot{\mathbf{x}} + \mathbf{c}\dot{\mathbf{x}} + \mathbf{kx} = \mathbf{F}(t) \tag{6.1}$$

In Equation (6.1) $\boldsymbol{x}$ is a generalised variable that includes six degrees of freedom. On the left-hand side of this equation are the response terms, which consist of acceleration, velocity and deflection. In the case of a ship at sea, the nature of the response depends on the spatial and temporal distribution of the excitation $\boldsymbol{F}(t)$ and of the properties of the ship itself ($\boldsymbol{m}$, $\boldsymbol{c}$, $\boldsymbol{k}$).

Since our aim is to consider the stress and deflection distributions that arise in the hull girder as a result of primary loading, we propose to initiate our study with the case of a ship in a still-water (*harbour*, as termed in IACS rules) condition. In this condition all features of the different types of response are observable, and we do not need to include dynamic loading effects.[1] In general terms, the response of a structure to a specific loading condition depends on its material properties (stiffness and range of elastic behaviour), its spatial distribution, its dimensions and its actual condition at the time of loading.[2] Since we are dealing with response to primary loading, we shall be concerned with the beam-type behaviour of the hull girder. The distinguishing features of hull girder primary loading are the form of the vertical bending moment, shear force and torque distributions. The ship behaves as a free-free beam, as shown in Figure 4.8.

Before considering the effect these loads have on a real ship, it is useful to replace the original section with a simplified structural model (Figure 6.1a). In the case of an oil tanker that has two longitudinal bulkheads, a corresponding box girder model is shown in Figure 6.1c. The model includes all members of the structure that carry primary loads and is therefore termed the *primary structure*. This includes all material that is aligned in the longitudinal direction and extends over a significant proportion of the length of the ship. Various complications arise because of discontinuities, etc., but for the time being we need not be concerned with these. In replacing the exact section with the box girder model (Figure 6.1c) we introduce two further simplifications, which consist firstly of merging the stiffeners attached to longitudinal elements with the plating (Figure 6.1b) and secondly of replacing the inner longitudinal bulkheads and girders with a single bulkhead (Figure 6.1c). The first of these can be done by modifying the

[1] There have been several instances of hull girder collapse in such a condition. We can therefore gain valuable insight by studying response to still water loading before considering more complex situations such as wave-induced bending, bending about other axes, torsion, etc.

[2] Effects of ageing (corrosion, local buckling, yielding)

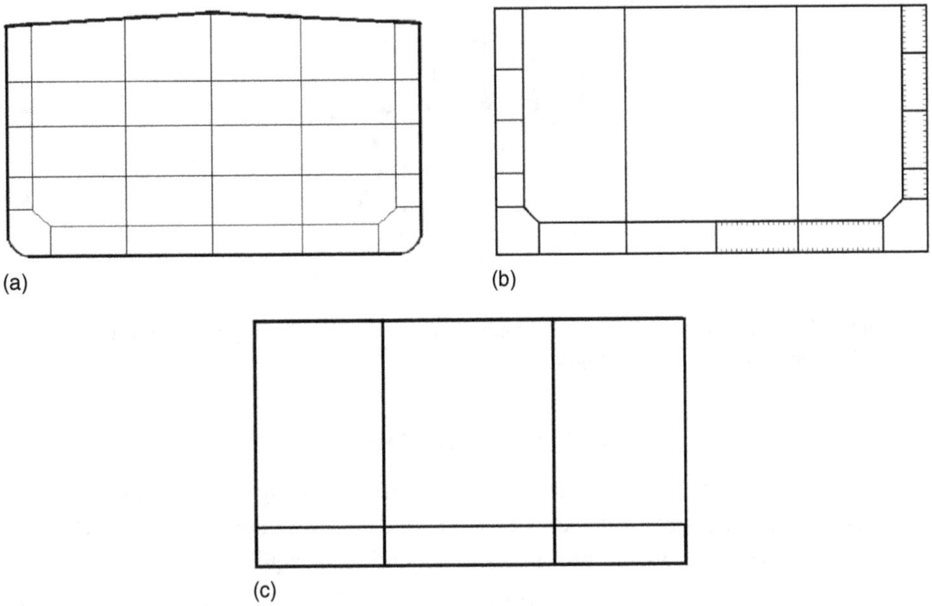

Figure 6.1 Transverse section and box girder models for primary loading. (a) Oil tanker with two longitudinal bulkheads. (b) Primary structure members. (c) Simplified box girder model of primary structure.

area of the plating to take into account that of the stiffeners. An equivalent plate thickness is thus obtained for each span of plating from the following formula:

$$t_{eq} = \frac{A_p + A_s}{s} \tag{6.2}$$

where t_{eq} is the equivalent plate thickness, A_p, A_s are the original plate and stiffener cross-sectional areas and s is the span of the plating between supports. The second simplification is based on merging the material that is aligned both vertically and horizontally in such a way that the bending stiffness of the cross-section does not change. By making use of the box girder model we simplify the problem without obscuring its essential features, which for the present purposes are its global bending characteristics (stiffness and strength). In a later section we will consider structural models relevant to torsional loading.

6.2 Vertical and Horizontal Bending of the Hull Girder

In this section we shall consider the axial and shear stress fields that arise as a result of vertical bending of the hull girder. Initially these will be treated separately. The combined action of axial and shear stresses will then be described, leading to the shear lag effect. Consideration will then be given to horizontal bending. This involves the simultaneous action of vertical and horizontal bending stresses.

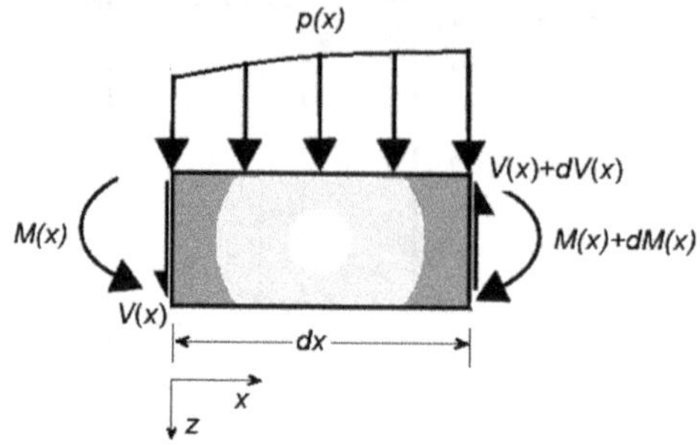

Figure 6.2 Element of hull girder of length dx and loads acting on it (positive directions for M, V as per IACS sign convention).

6.2.1 Axial Normal Stresses

In order to determine the internal reactions to the vertical bending moment and shear force, we consider equilibrium of an element of the hull girder of length dx (Figure 6.2). Since we are treating this as a two-dimensional problem, the following equilibrium conditions apply:

$$\sum_i M_y(x) = 0 \tag{6.3a}$$

$$\sum_i V_z(x) = 0 \tag{6.3b}$$

where $M_y(x)$ is the bending moment about the y-axis and $V_z(x)$ is the shear force that acts in the z-direction. For rotational equilibrium about any axis parallel to the y-axis, the external (applied) bending moment $M_{ext}(x)$ is balanced by an internal couple of equal magnitude and opposite sign, $M_{int}(x)$:

$$M_{ext}(x) + M_{int}(x) = 0 \tag{6.4}$$

The internal couple or stress resultant is the sum of the moments of all normal stresses that are induced in the elements of the cross-section that contribute to longitudinal strength (Figure 6.1b). The following relationship holds:

$$M_{int}(x) = \int_{A(x)} \sigma_x(x, z) z dA \tag{6.5}$$

The stress resultant $M_{int}(x)$ varies along the length of the hull girder as the external moment. It is evident that the magnitude and distribution of the longitudinal stresses over any transverse section can be changed, provided Equation (6.5) is satisfied. This can be achieved by:

- Selecting appropriate material(s). If, for example, steel weight has to be reduced, high-strength materials that will carry higher loads leading to higher stresses can be used.
- Modifying the spatial arrangement and/or the scantlings of the cross-section.

Either way the moment of inertia of the cross-section $I(x)$ will be affected.

Having said that, let us briefly turn our attention to hull girder deflections. In accordance with beam theory, the vertical bending deflections $w_b(x)$ depend on the vertical bending stiffness, $EI(x)$, which is the product of the moment of inertia of the cross-section $I(x)$ and the modulus of elasticity of the material, E. In general the modulus of elasticity does not change in the longitudinal direction, although it may do so in the vertical direction if different materials are used. For constant E,

$$M(x) = -EI(x)\frac{d^2 w_b(x)}{dx^2} \tag{6.6}$$

Therefore, when we rearrange the material of the cross-section, we affect the bending stiffness and consequently the deflections wb. From Equation (6.6) we have that $d^2 w_b/dx^2$ is inversely proportional to the bending stiffness, $EI(x)$. We shall consider hull girder deflections in Section 6.2.4.

6.2.2 Shear Stresses

For vertical equilibrium, the vertical shear force $V_{ext}(x)$ at each section along the length is balanced by an internal shear force (stress resultant) of equal magnitude and opposite sign such that

$$V_{ext}(x) + V_{int}(x) = 0 \tag{6.7}$$

The internal shear stress resultant, $V_{int}(x)$, is the integral of the shear stresses τ_{xy} that act on the surface of each transverse section. Since this is a thin-walled structure, we can assume that the shear stress is constant through the thickness of each member.[3] We can therefore write down the following expression:

$$V_{int}(x) = \int_{A(x)} \tau_{xy}(x,z)dA \tag{6.8}$$

We consider now a rectangular element $ABCD$ that forms part of a longitudinal vertically aligned web, as shown in Figure 6.3. The longitudinal web is part of a rectangular hollow section that is at equilibrium under the action of a vertical bending moment and a vertical shear force.

The normal stresses $\sigma_x(x,z)$ vary according to the Euler-Bernoulli hypothesis and increase linearly from the neutral axis towards the outermost fibres.

In addition to the shear stresses distributed along each vertical section, a *complementary shear stress field*, τ_{xz}, arises on horizontal planes in the longitudinal direction. The justification for these stresses can be based on grounds of both geometry and equilibrium. With regard to geometry, and more specifically *material compatibility*, we see that if we

[3] For thin-walled members, these assumptions enable solutions to be obtained that are sufficiently accurate for practical purposes. A description of this solution is included in [4]. The exact solution to this problem was given in [27] of Chapter 2.

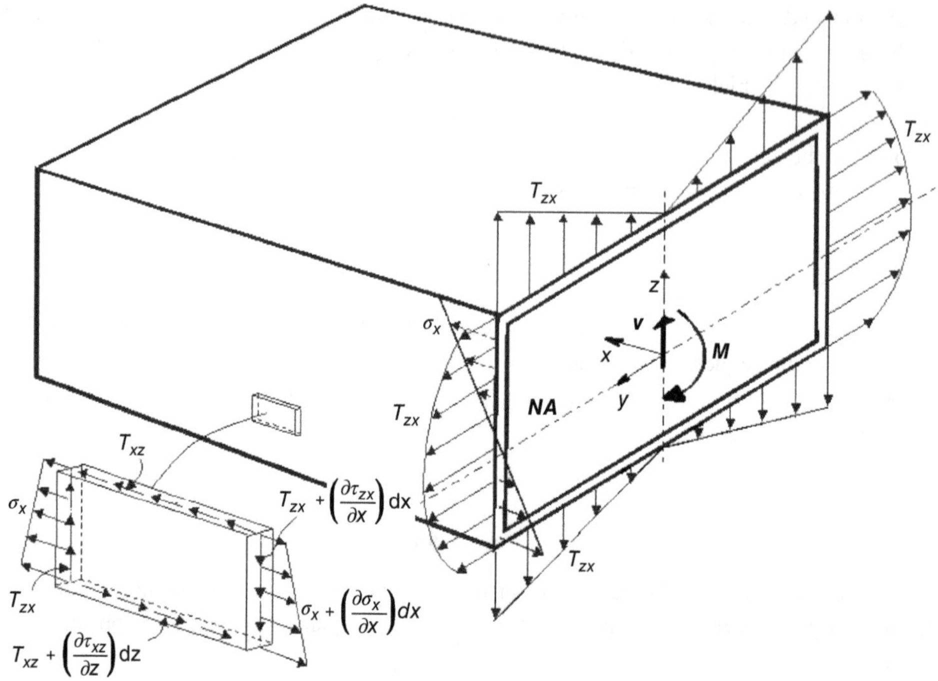

Figure 6.3 Normal and shear stresses in web of a rectangular hollow section subjected to longitudinal bending.

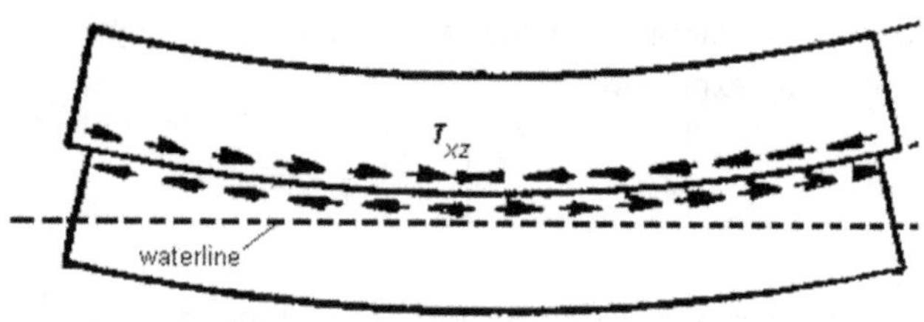

Figure 6.4 Hull girder in sagging condition in still water. (Horizontal shear stresses arise as a consequence of material continuity along planes that are parallel to the neutral axis of each cross-section.)

were to replace a beam of a particular depth with a series of beams whose total depth equals the depth of the first beam, in the absence of the longitudinal shear stresses, sliding would occur along each interface when the beam is subjected to vertical bending (Figure 6.4).[4]

With regard to equilibrium, the complementary shear stressses τ_{xz} and $\tau_{xz} + (\partial\tau_{xz}/\partial z)dx$ are necessary to ensure rotational equilibrium about an axis normal to the element surface.

[4] The deflection in the case of the multiple n-beam arrangement would be much greater than that of the original beam because $d^3 > n(d/n)^3$ where d is the beam depth and $n > 1$.

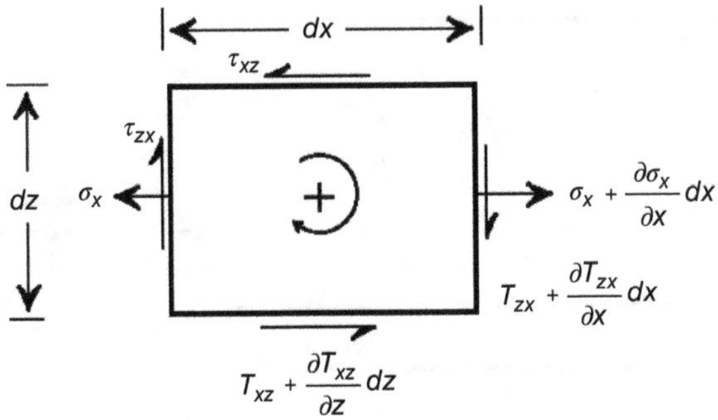

Figure 6.5 Farces acting on plate element $dxdz$ in the longitudinal direction.

In order to determine the direction and magnitude of the complementary shear stresses, we take moments of all stresses about the centroid of the plate element (Figure 6.5).

The moments of the normal stresses σ_x, $\sigma_x + (\partial\sigma_x/\partial x)dx$ that arise on the transverse edges of the element about the centroid are zero as $dz \to 0$.

$$\left(\tau_{zx} + \frac{\partial \tau_{zx}}{\partial x}\right) dz.t\left(\frac{dx}{2}\right) + (\tau_{zx} dzt)\left(-\frac{dx}{2}\right) + \left(\tau_{xz} + \frac{\partial \tau_{xz}}{\partial z}\right) dxt\left(\frac{dz}{2}\right)$$

$$+ (\tau_{xz} dxt)\left(-\frac{dz}{2}\right) = 0$$

Hence,

$$\frac{\partial \tau_{zx}}{\partial x} + \frac{\partial \tau_{xz}}{\partial z} = 0 \tag{6.9}$$

As $dx, dz \to 0, \tau_{zx} = \tau_{xz}$.

The direction of the complementary shear stresses is such as to ensure rotational equilibrium of the element (Figure 6.5).

6.2.3 Vertical Bending Shear Stress Distributions in Closed Sections

We will now turn our attention to the shear stress fields that arise in sections of the types encountered in ship structures. These are either *closed* or *mixed* (open and closed) sections. Examples of the former are oil tankers, passenger ships and gas carriers, whereas examples of the latter are MPP ships and bulk carriers.

The general theory of shear stresses in multi-cell closed sections with application to ship structures has been presented in [2]. We consider first equilibrium of the thin-walled open section shown in Figure 6.6. This is subjected to external loading that consists of a bending moment and a vertical shearing force that gives rise to normal stresses σ_x and shearing stresses τ_s. The normal stresses are found using beam bending theory, and we need to determine the shear stress τ_s. For a thin-walled section both normal and shear stresses are constant through the thickness, which is here assumed to be constant in the x direction but

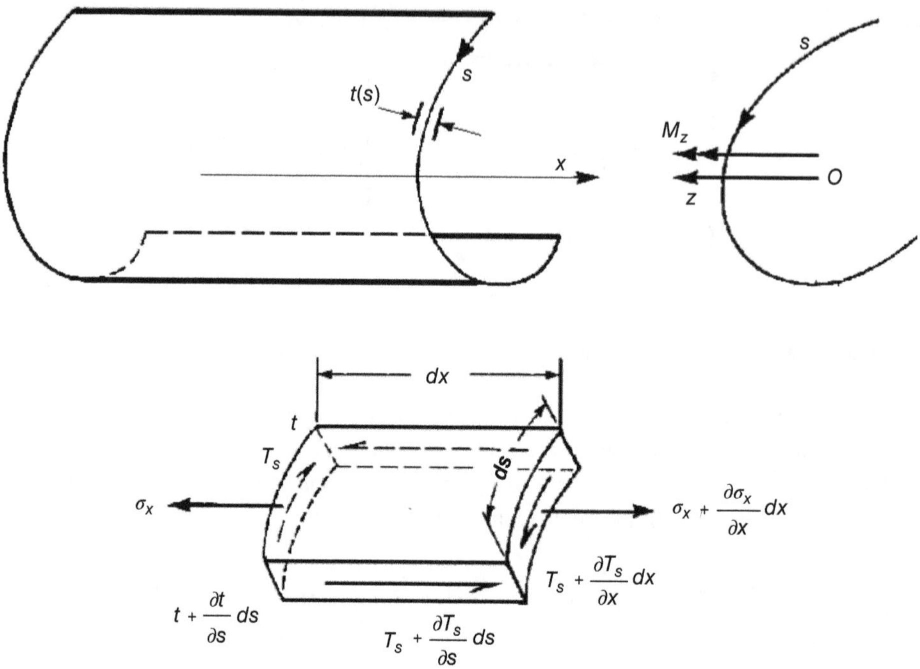

Figure 6.6 Shear stresses in an open section [2].

varies along the girth of the section. The shear stress is tangential to the thickness and for equilibrium along the section, with $\partial t/\partial x = 0$, requires

$$\left(\tau_s + \frac{\partial \tau_s}{\partial x} dx\right)\left(t + t + \frac{\partial t}{\partial s} ds\right)\frac{ds}{2} - \tau_s\left(t + t + \frac{\partial t}{\partial s} ds\right)\frac{ds}{2} = 0$$

or

$$\frac{\partial q_s}{\partial x} = 0 \tag{6.10}$$

where q_s, the *shear flow*, is equal to $\tau_s t$. Equation (6.10) shows that q_s is constant in the x direction. For longitudinal equilibrium of the element,

$$\left(\sigma_x + \frac{\partial \sigma_x}{\partial x} dx\right)\left(t + \frac{1}{2}\frac{\partial t}{\partial s} ds\right) ds - \sigma_x\left(t + \frac{1}{2}\frac{\partial t}{\partial s} ds\right) ds +$$

$$\left(\tau_s + \frac{\partial \tau_s}{\partial s} ds\right)\left(t + \frac{\partial t}{\partial s} ds\right) dx - \tau_s t dx = 0$$

Substituting $\frac{\partial t}{\partial x} = 0$ gives

$$\frac{\partial (t\sigma_x)}{\partial x} + \frac{\partial (t\tau_s)}{\partial s} = 0 \tag{6.11}$$

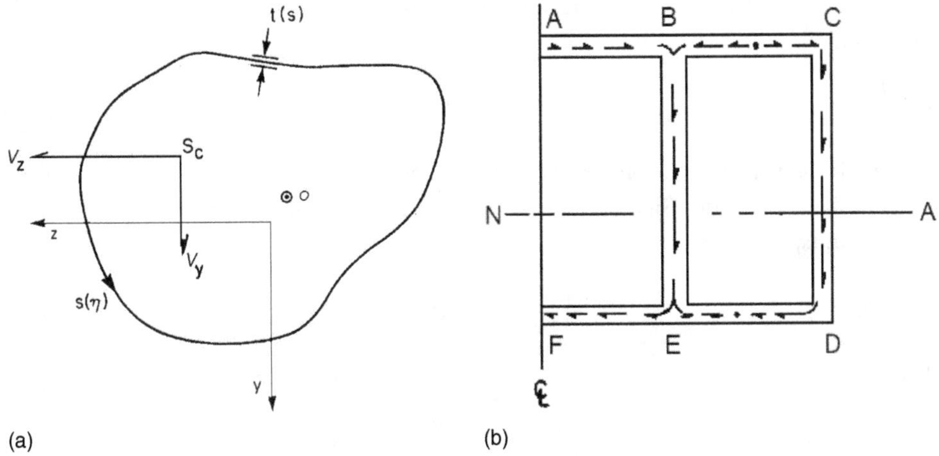

Figure 6.7 Single- and multi-cell closed sections [2]. (a) Single-cell section. (b) Multi-cell section.

This can be rewritten as

$$q_s = -\int_0^s \frac{\partial(t\sigma_x)}{\partial x} ds \tag{6.12}$$

since $\tau_s = 0$ on the edge, where $s = 0$. Equations (6.10) and (6.11) are the equations of equilibrium in the transverse and longitudinal directions, respectively.

We will now consider closed sections. The analysis of mixed-type sections is a simple extension of closed section theory that allows for the presence of open branches. The analysis that follows is based on the fundamental assumption that pure bending with no twist takes place. For this to be the case, the vertical shearing force has to pass through the centre of shear of the section. This is a realistic assumption for ship sections provided the ship does not encounter oblique waves, since geometrical symmetry about the centreline plane exists. Thus, when the hull girder is subjected to bending alone, the shear force will act in the vertical direction in way of the centreline plane.

However, Equation (6.11) does not generally hold for closed sections (Figure 6.7a), because the position at which $s = 0$ is not known. To overcome this difficulty, a cut is introduced at an arbitrary position, thereby specifying the position where $s = 0$. We can then calculate the shear stress as for an open section. In effect we are transforming the statically indeterminate problem into a determinate one. The cut we introduce permits the two sides to deflect relative to each other. However, since this does not take place in the real structure, it is necessary to impose a *corrective shear flow* q_o to restore continuity in the displacement field. The corrective flow is introduced in the part of the section that is indeterminate and in which the cut has been introduced. The actual shear flow, q_s, is then the sum of the flow in the determinate structure, q_{sc}, and the corrective flow, q_0:

$$q_s = q_{sc} + q_0 \tag{6.13}$$

We shall now determine the corrective shear flow, q_0. In the single-cell section shown in Figure 6.7, the shear strain is

$$\gamma_{sx} = \frac{\partial \eta}{\partial x} + \frac{\partial u}{\partial s} \tag{6.14}$$

where η is the tangential displacement in the s direction and u is the longitudinal displacement in the x direction. Since the shear force acts through the centre of shear, no twist arises, that is,

$$\frac{\partial \eta}{\partial x} = 0 \tag{6.15}$$

Thus,

$$\gamma_{sx} = \frac{\partial u}{\partial s} \tag{6.16}$$

Since $\tau_{sx} = G\gamma_{sx}$

$$\frac{q_s}{t} = G\frac{\partial u}{\partial s} \tag{6.17}$$

From Equations (6.13) and (6.17) we have

$$\frac{q_{sc} + q_0}{t} = G\frac{\partial u}{\partial s} \tag{6.18}$$

Integrating around the perimeter of a closed section,

$$\oint \frac{q_{sc}}{t} ds + q_0 \oint \frac{ds}{t} = G \oint \frac{\partial u}{\partial s} ds \tag{6.19}$$

since q_0 is constant. However, for compatibility,

$$\oint \frac{\partial u}{\partial s} ds = 0 \tag{6.20}$$

so that the corrective shear flow is obtained from Equation (6.19) as follows

$$q_0 = \frac{-\oint \frac{q_{sc}}{t} ds}{\oint \frac{ds}{t}} \tag{6.21}$$

The same approach is followed in the case of a multi-cell section (Figure 6.7b).

Let us now consider a section that consists of a number of closed cells, as shown in Figure 6.8. This in-line configuration is encountered in the double bottom of most ocean-going ships as well as in the side shell structure of containerships.

The shear force V is applied through the centre of shear S_c so that no twisting arises. Cuts are introduced in every closed section, and the section becomes statically determinate. For the jth cell, Equation (6.18) becomes

$$q_{sj} = q_{scj} + q_{0j} \tag{6.22}$$

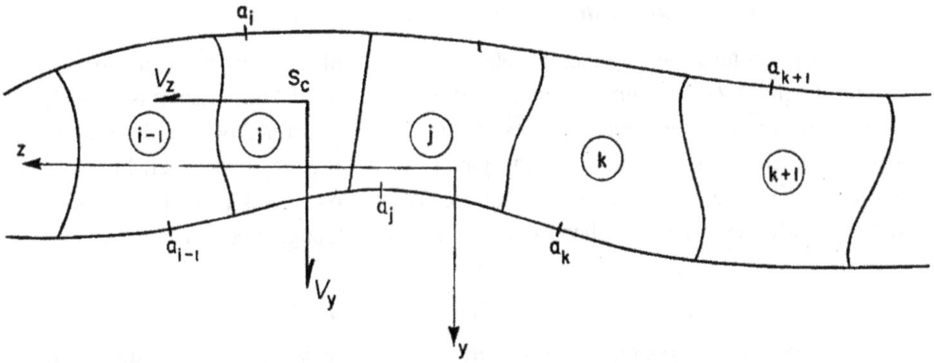

Figure 6.8 Multi-cell section [2].

where q_{oj} is the corrective shear flow in cell *j* with all other cells remaining open. The final total flow in the section when all the cells are closed consists of the shear flows in the determinate structure to which are added the corresponding corrective shear flows. For cells *i*, *j* and *k* Equations (6.14), (6.15) and (6.22) give

$$\frac{q_{sj}}{t} = G\frac{\partial u}{\partial s} \tag{6.23}$$

Performing the line integration around the perimeter of cell *j* on both sides of Equation (6.23), we obtain

$$\oint_{sj} \frac{q_{sj}}{Gt} ds = 0 \tag{6.24}$$

since

$$\oint_{sj} \frac{\partial u}{\partial s} ds = 0$$

for compatibility. Substituting Equation (6.22) into Equation (6.24) gives

$$\frac{1}{G}\oint_{sj} \frac{q_{scj}ds}{t} + \frac{1}{G}\oint_{sj} \frac{q_{oj}ds}{t} - \frac{1}{G}\int_{sji} \frac{q_{oi}ds}{t} - \frac{1}{G}\int_{sjk} \frac{q_{ok}ds}{t} = 0 \tag{6.25}$$

In this equation the first integral represents the shear flow in cell *j* when cut whereas the second integral represents the corrective shear flow q_{oj} in cell *j* with all other cells open. These two integrals correspond to those included in Equations (6.19) and (6.21) for an isolated cell section. The third and fourth integrals in Equation (6.25) are the corrective flows in adjacent cells *i* and *k*, which affect the shear flow in the cell along their common boundaries. In the arrangement shown in Figure 6.8, therefore, the shear flow in cell *j* is unaffected by that in other cells.

6.2.4 Hull Girder Deflections Due to Vertical Bending

It is necessary to control the longitudinal deflection of the hull girder in order to avoid problems of both a structural and a commercial nature. If excessive deflections arise, it is possible that local buckling may occur in critical locations of the primary structure. Systems that are attached to structural members such as piping and propulsion shafting may be subjected to excessive distortion of their foundations and suffer damage. The cargo-carrying capacity of a cargo ship may also be adversely affected when excessive hogging or sagging occurs.

Vertical Bending Moment–Induced Deflections

The most important component of these deflections is associated with longitudinal normal stresses. Integrating the equation for the primary bending moment $M(x)$ along the length of the hull girder gives the slope $dw_b(x)/dx$ that results from bending:

$$\frac{dw_b(x)}{dx} = \int_{AP}^{x} \frac{M(x)}{EI(x)} dx + C_3 \tag{6.26a}$$

A further integration gives the bending deflection $w_b(x)$:

$$w_b(x) = \int_{AP}^{x} \int_{AP}^{x} \frac{M(x)}{EI(x)} dxdx + C_3 x + C_4 \tag{6.26b}$$

Since the slope is not known at any position along the length of the ship, it is not possible to evaluate C_3. The bending deflection, $w_b(x)$, can be set equal to zero at the neutral axis position in way of the fore and aft perpendiculars. In this way deflections along the length are evaluated with respect to the axis joining these two positions. It should be added that this axis will not necessarily remain horizontal as the vessel may trim. Constants C_3 and C_4 can then be evaluated. Since $w_b(AP) = 0$, from Equation (6.26b),

$$C_4 = 0 \tag{6.27a}$$

Since $w_b(FP) = 0$,

$$\begin{aligned} C_3 &= -\frac{1}{L} \int_{AP}^{FP} \frac{dw_b(x)}{dx} dx = \\ &= -\frac{1}{L} \int_{AP}^{FP} \int_{AP}^{FP} \frac{M(x)}{EI(x)} dxdx \end{aligned} \tag{6.27b}$$

Substituting in Equations (6.26a) and (6.26b), the following expressions are obtained:

$$\frac{dw_b(x)}{dx} = \int_{AP}^{x} \frac{M(x)}{EI(x)} dx - \frac{1}{L} \int_{AP}^{FP} \int_{AP}^{FP} \frac{M(x)}{EI(x)} dxdx \tag{6.28a}$$

$$w_b(x) = \int_{AP}^{x}\int_{AP}^{x} \frac{M(x)}{EI(x)}dxdx - \frac{x}{L}\int_{AP}^{FP}\int_{AP}^{FP} \frac{M(x)}{EI(x)}dxdx \tag{6.28b}$$

The question of accuracy in predicting actual deflections using Equation (6.28b) arises. It has been found that using beam theory bending deflections are obtained with satisfactory accuracy. In real hull girders the cross-section changes along the length, and thus the height of the neutral axis also changes, with the distance from the keel increasing towards the vessel's extremities as the cross-section ceases to be wall-sided. If changes in the cross-section are taken into account, the bending deflection can be predicted with a fair degree of accuracy.

Vertical Shear Force–Induced Deflections

In addition to deflections due to the action of primary bending stresses, deflections also arise because of the action of vertical shearing forces. To determine these deflections we refer to Equation (6.63). The fourth term in Equation (6.63), which is a restoring linearly varying stress distribution equivalent to an equilibrating bending moment, gives rise to an additional vertical deflection. The magnitude of this *shear deflection* can be determined by direct integration from

$$\frac{d^2 w_s}{dx^2} = -\frac{M_s}{EI} \tag{6.29}$$

The deflection due to shear, w_s, is then given by

$$w_s(x) = \int_{AP}^{x}\int_{AP}^{x} \frac{M_s(x)}{EI(x)}dxdx \tag{6.30}$$

where $M_s(x)$ is given by Equation (6.61b). Substituting in Equation (6.30) and integrating,

$$\begin{aligned} w_s(x) &= \frac{1}{EI}\int_{AP}^{x}\int_{AP}^{x}\left(\oint Ey\frac{de_s}{dx}ydA\right)dxdx = \\ &= \frac{1}{I^2 G}\int_{AP}^{x}\int_{AP}^{x}\left(\oint\left(\int_{A}\frac{\sum A\bar{y}}{t}d\xi\right)wydA\right)dxdx \end{aligned} \tag{6.31}$$

The total vertical deflection of the hull girder, $w(x)$, at any section can then be obtained by adding the shear deflection $w_S(x)$ to that due to bending, $w_B(x)$. Thus, the total deflection $w(x)$ is given by

$$w(x) = w_B(x) + w_S(x) \tag{6.32}$$

Shear deflections have been measured on ships and found to be generally less that 20 per cent of the bending deflections (Figure 6.9). The application of the Euler–Bernoulli assumption that plane sections remain plane to the calculation of shear deflections has

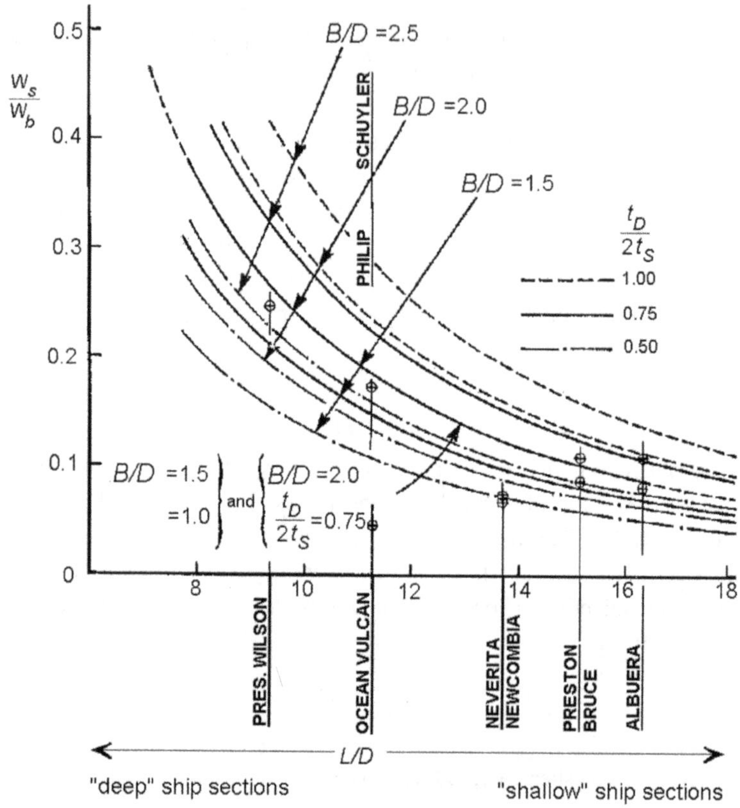

Figure 6.9 Relative magnitudes of shear, thermal and bending deflections (w_S, w_T, w_B) [2].

been found to be accurate than in the case of bending deflections. This is understandable, since vertical shearing forces vary along the length of the hull girder giving rise to distortions of transverse sections, so that plane sections do not remain plane. One way of estimating shear deflections is to allow for variations in section geometry and to sum the effect of individual sections using an expression such as [2]:

$$w_S(x) = \sum_{AE}^{x} \frac{V(x)l}{A_w G} \tag{6.33}$$

where A_w is the web area of individual sections (that includes all major vertical structural components). In this expression a *nominal shear stress* is used $(V(x)/A_w)$. Nonetheless, even the use of such an equation does not necessarily provide accurate answers.

Bending and shear deflections for a number of ship hull forms whose data are included in Table 6.1 are compared in Figure 6.9 for a range of values of *L/D*, *B/D* and $t_D/2t_S$ (t_D = deck structure thickness, t_S = side structure thickness) and the following conclusions reached:

1. For the ship geometries considered, $0.40 > w_S/w_B > 0.10$.
2. As *L/D* increases (shallower ships), w_S/w_B decreases.

Table 6.1 *Principal dimensions of ships considered in assessment of global bending and shear deflections [2]*

Ship	*L* (m)	*B* (m)	*D* (m)	*B/D*	w_S/w_B	w_s/w_T
Ocean Vulcan	126.8	17.3	11.4	1.52	0.05	0.05
Clan Alpine	126.8	17.3	11.4	1.52	0.13	0.12
Neverita	140.2	18.0	10.4	1.74	0.07	0.07
Newcombia	140.2	18.0	10.4	1.74	0.08	0.07
Preston	94.5	9.4	6.3	1.50	0.09	0.08
Bruce	94.5	9.4	6.3	1.50	0.11	0.10
Philip Schuyler	126.8	17.3	11.4	1.52	0.18	0.15
President Wilson	174.7	23.0	18.7	1.23	0.25	0.20
Albuera	108.2	12.3	6.7	1.83	0.08-0.11	0.08–0.10

3. As *B/D* increases, (beamier ships) w_S/w_B increases.
4. As $(t_D/2t_S)$ increases, w_S/w_B increases.

When launched, ships tend to sag, especially when construction starts at amidships and proceeds toward the fore and aft ends. This may necessitate corrective calculations and possibly the repositioning of draught marks.

Distortions of the hull girder due to *thermal fields* generally affect the structure in all directions and are not symmetrical about a longitudinal axis.

Deflections in the longitudinal direction due to thermal fields have been considered to depend on the temperature difference between the deck plating and the seawater, which implies that the bottom plating is at ambient temperature. We therefore expect the hull girder to hog. Thermal deflections increase with the length to depth ratio *L/D* as well as with the temperature difference.

The following semi-empirical formula due to Burtner & Tingey has been proposed for deflections due to thermal stress fields [2]:

$$w_T = (9)(10)^{-6}\frac{L^2}{D}\varphi_D \tag{6.34}$$

where w_T is the maximum deflection due to temperature effects in inches, L and D are given in feet, and φ_D is the temperature difference in degrees Fahrenheit.

As mentioned previously, deflections of the hull girder have to be controlled for a variety of reasons. The dependence of the bending deflection on the bending stiffness *EI* is evident from Equation (6.28b), and thus the only means at the disposal of the designer is to check the moment of inertia of the cross-section, given the properties of a particular material and length of ship. In the past, classification societies imposed limits to the *L/D* ratio.

6.2.5 Validation of Hull Girder Longitudinal Bending Theory

The beam theory described in previous sections needs to be tested in order to assess its validity and the reasons for possible discrepancies between actual measurements and

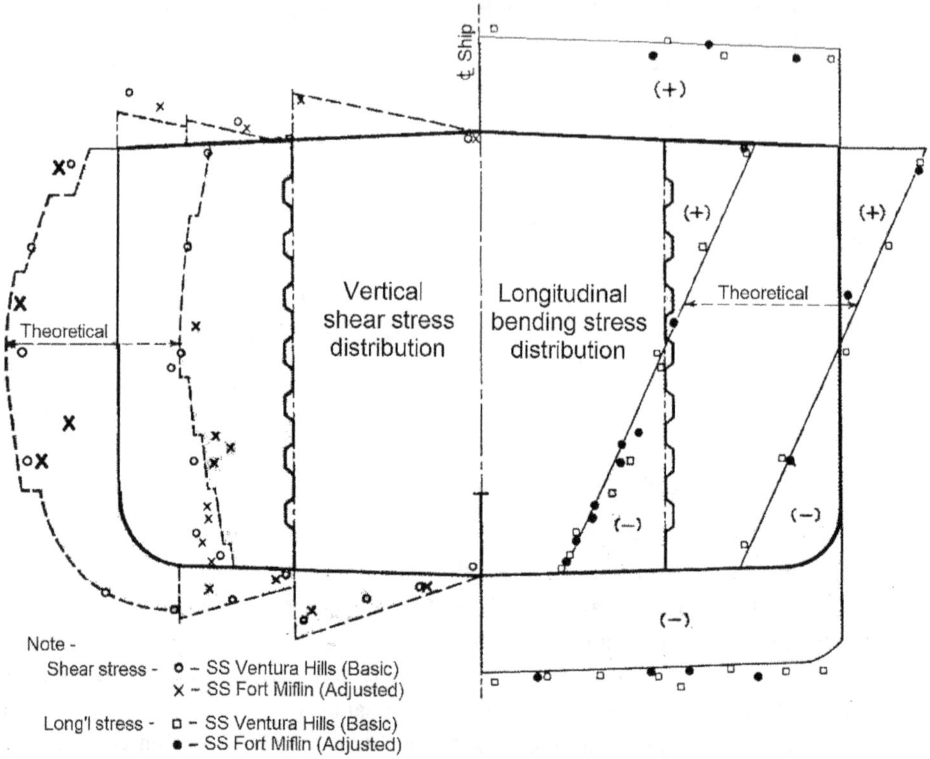

Figure 6.10 Hull girder primary bending and shear stress measurements and linear elastic beam theory predictions (SS *Ventura Hills*, SS *Fort Miffin*) [12]. Reprinted courtesy of the Society of Naval Architects and Marine Engineers.

predicted values. Test data also bring into focus other factors that may have a greater or lesser influence on structural behaviour.

During the period 1930–1950, extensive full-scale measurements and tests were performed on ships in the United States and the United Kingdom [12]. Tests were performed on 19 ships, including 3 to destruction, that is, collapse of the hull girder. The ships involved include three dry cargo ships, eight tankers, four Great Lakes ore carriers, one passenger ship and three destroyers. The purpose of the tests was to check the overall response of the hull girder to known bending moments and to study the strength of components of the hull girder under specified static loads. The structural components considered were:

- Bottom structure
- Hatch corners
- Corrugated and flat bulkheads
- Openings in plating

Figure 6.10 shows comparisons of stress measurements with predictions of linear elastic bending theory. Results for both longitudinal stresses and shear stresses are shown, and it is

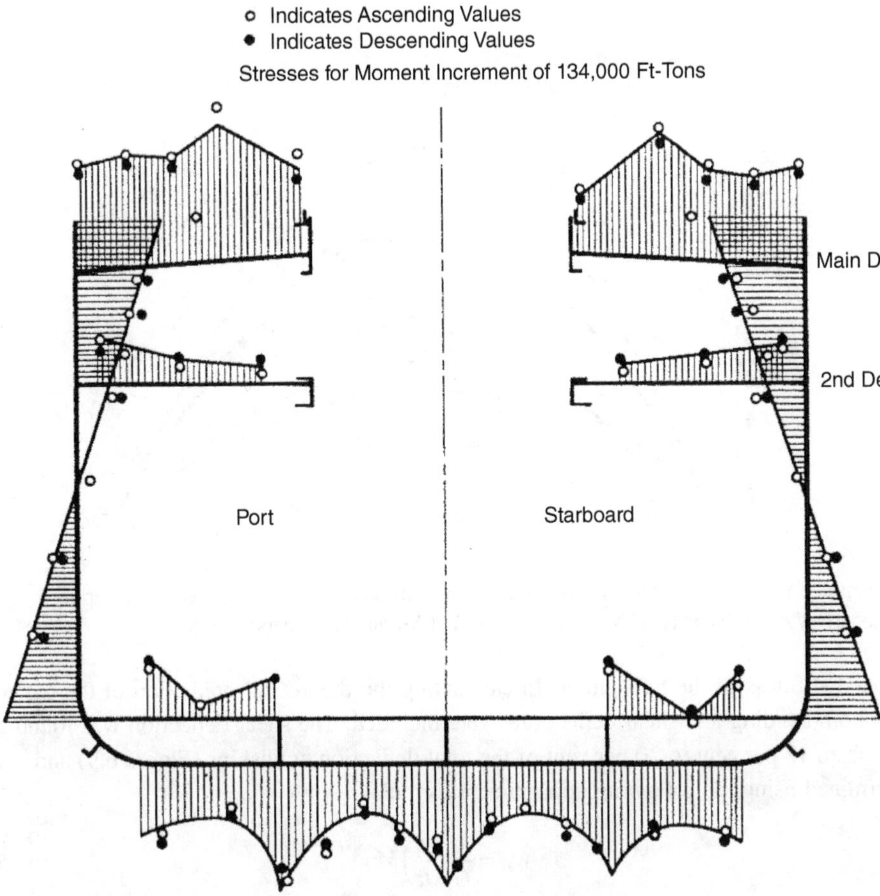

Figure 6.11 Longitudinal bending stresses in post-buckling range (SS *Philip Schuyler*) (compressive buckling of outer bottom plating) [12]. Reprinted courtesy of the Society of Naval Architects and Marine Engineers.

seen that for the section considered, linear theory provides an accurate prediction of longitudinal bending stresses. A question that needed to be investigated was whether a longitudinal bulkhead contributes to longitudinal shear strength as predicted by theory. The test results showed that if buckling is avoided, longitudinal bulkheads perform as expected. It is concluded that longitudinal bulkheads may contribute to both longitudinal bending strength and shear strength, provided they are designed to withstand local buckling. The effect of buckling is clearly evident in the stress distributions measured on board *SS Philip Schuyler* (Figure 6.11). It is seen that the outer bottom stresses decrease at the mid-bay positions.

Figure 6.12 shows longitudinal distributions of hull girder deflections for five different bending moments. It is seen that the deflections are very accurately predicted for the lower three bending moments whereas for the two higher values discrepancies of increasing magnitude arise. This is due to nonlinear behaviour arising from buckling that leads to

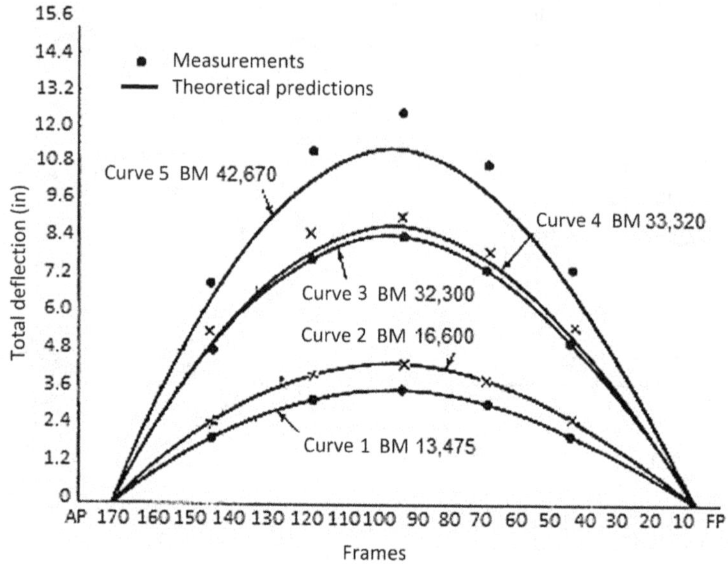

Figure 6.12 Calculated and measured hull girder deflections (USS *Preston*) [12]. Reprinted courtesy of the Society of Naval Architects and Marine Engineers.

reduced stiffness of the hull girder. In calculating the theoretical prediction of the deflection, both bending and shear deflections were included. The shear deflection was found to vary from 12 per cent to 20 per cent of the total deflection amidships (Figure 6.9) and was determined using the following approximate formula:

$$w_s = \frac{1}{G}\left(\frac{m}{It}\right)M \tag{6.35}$$

where the quantity in brackets is determined amidships. It was confirmed that, in the linear elastic range, the total deflection is the sum of the bending deflection and the shear deflection.

As regards global torsional strength, tests that were performed on three ships with closed sections indicated that the resulting shear stresses are small and do not affect longitudinal bending strength. Open-deck ships such as containerships and bulk carriers experience much higher shear stresses due to torsion.

6.2.6 Initial Design Considerations for Longitudinal Bending Strength

In this section we shall discuss the ideas behind the formulas that are implemented in classification society rules to ensure that the hull girder has sufficient vertical bending strength. From elementary beam theory,

$$\frac{\sigma_L}{y} = \frac{M_{tot}}{I_{NA}}\left(= \frac{E}{R}\right) \tag{6.36}$$

Table 6.2 *Permissible hull girder bending stress [5]*

Operation	*Design Load*[5]	*Permissible Hull Girder Bending Stress* σ_{perm} *(N/mm²)*				
		≤ 0.1	$0.1 \leq x/L$ $x/L \leq 0.3$	$0.3 \leq x/L$ $x/L \leq 0.7$	$0.7 \leq x/L$ $x/L \leq 0.9$	>0.9
Seagoing	*S+D*	140/*k*	*LI*[6]	190/*k*[7]	*LI*	140/*k*
Harbour/sheltered	*S*	105/*k*	*LI*	143/*k*	*LI*	105/*k*
Flooded condition at sea for bulk carriers (*L*>150 m)	*A:S+D*	140/*k*	*LI*	190/*k*	*LI*	140/*k*

where for a hull girder, M_{tot} is the total longitudinal bending moment acting at a particular section $(= M_{SW} + M_{WV})$, I_{NA} is the second moment of area of that section about its neutral axis, y is the distance from the neutral axis to a particular member, and σ_L is the longitudinal bending stress induced in that member. The bending moment can be calculated from rule formulas or else using approved software. Hence, if a permissible maximum stress, σ_{perm}, is defined, which in the case of a hull girder is specified by classification society rules to be less that the material yield stress, $(\sigma_{perm} < \sigma_0)$, we have

$$\frac{I_{NA}}{y} = \frac{M_{tot}}{\sigma_{perm}} = Z \tag{6.37}$$

where Z is the *section modulus* of any transverse section of the hull girder. For a given applied bending moment M, therefore, the *required section modulus,* W_{min} for a member of the relevant section that is located a distance y from the neutral axis will be

$$W_{min}\left(=\frac{I}{y}\right) \geq \frac{M_{tot}}{\sigma_{perm}} \tag{6.38}$$

Therefore, for the deck and bottom structure of a particular section, the design requirements are

$$W_{deck} = \frac{I}{y_{deck}} \geq W_{min} \qquad W_{bot} = \frac{I}{y_{bot}} \geq W_{min} \tag{6.39}$$

where y_{deck}, y_{bot} are the distances of the centroids of the deck and bottom structures from the neutral axis. W_{deck} and W_{bot} are thus the required section moduli for the deck and bottom structures in order that $\sigma_{deck}, \sigma_{bot} < \sigma_{perm}$ for a given value of total bending moment, M_{tot}.

Rule Requirements for Vertical Bending Strength

With regard to *permissible stresses*, the IACS Common Structural Rules for bulk carriers and oil tankers (2019) include Table 6.2. In Table 6.2 k is a material factor used to allow for

[5] '*S*': static load case '*D*': dynamic load case. [6] *LI*: linear interpolation.
[7] In the case of all other ship types, the IACS UR S11 stipulates a value of 175/*k* (Chapter 4).

Table 6.3 *Material factor,* k *[5]*

Minimum Yield Stress (N/mm²)	*k*
235	1.00
315	0.78
355	0.72
390	0.68

Table 6.4 *Normal stress* σ_L *[5]*

Operation	*Normal Stress,* σ_L $(\times 10^{-3})$ *(N/mm²)*		
	At any point located below Z_{VD}	At bottom⁽¹⁾	At deck⁽¹⁾
Seagoing	$\frac{M_{SW}+f_\beta M_{WV}}{Z_{A-n50}}$	$\frac{M_{SW}+f_\beta M_{WV}}{Z_{B-n50}}$	$\frac{M_{SW}+f_\beta M_{WV}}{Z_{D-n50}}$
Harbour/sheltered water	$\frac{M_{SW-P}}{Z_{A-n50}}$	$\frac{M_{SW-P}}{Z_{B-n50}}$	$\frac{M_{SW-P}}{Z_{D-n50}}$
Flooded condition at sea for bulk carriers (L > 150 m)	$\frac{M_{SW-f}+M_{WV}}{Z_{A-n50}}10^{-3}$	$\frac{M_{SW-f}+M_{WV}}{Z_{B-n50}}10^{-3}$	$\frac{M_{SW-f}+M_{WV}}{Z_{D-n50}}10^{-3}$

(1) The values of σ_L in the bottom and deck structure correspond to the application of the formula given for any point, calculated at the equivalent deck line and at the baseline

the additional strength as well as the effect of using high tensile steel in ship structures (reduced stiffness and fatigue strength). Numerical values for *k* for different values of yield stress are given in Table 6.3.

The *normal stresses*, σ_L, that arise in the hull girder as a result of the action of vertical bending moments can be calculated using the still-water and wave-induced bending moments and actual section moduli of the relevant sections using the expressions included in Table 6.4.

$$\sigma_L \leq \sigma_{perm} \tag{6.40}$$

The following IACS definitions are relevant to Tables 6.2 and 6.4.

M_{SW} Permissible hogging and sagging vertical still-water bending moment in intact seagoing condition, in kNm, at the hull transverse section considered

M_{SW-p} Permissible hogging and sagging vertical still-water bending moment for harbour/ sheltered water operation, in kNm, at the hull transverse section considered

M_{SW-f} Permissible hogging and sagging vertical still-water bending moment in flooded condition at sea, in kNm, at the hull transverse section considered

M_{WV} Vertical wave bending moment in sea-going condition, in kNm, in intact or flooded conditions at the hull transverse section considered
V_D Vertical distance to the equivalent deck line, in m
Z_n Z coordinate, in m, of horizontal neutral axis of the hull transverse section
Z_{VD} Z coordinate, in m, taken equal to $V_D + Z_n$
Z_{A-n50} Net section modulus, in m^3, at any point of the hull transverse section
Z_{B-n50} Net section modulus, in m^3, at bottom
Z_{D-n50} Net section modulus, in m^3, at deck
f_β Heading correction factor, to be taken as
= 1.05 for seagoing conditions
= 1.0 for ballast water exchange at sea, harbour/sheltered water and accidental flooded design load scenarios

The permissible stress requirement is relaxed in way of the fore and aft ends of a ship because of the reduction in longitudinal bending moment there. For any hull girder section, it is necessary to determine the height of the neutral axis, the second moment of area, and thereby the required section moduli of the critical regions of the section for longitudinal (primary) loading, which are the deck and bottom structure. In doing so it is necessary to include all members that contribute to the longitudinal strength of the hull girder. These are all longitudinally aligned members that extend over a significant proportion of the ship's length, and are in most cases located on or below the strength deck. It is important to select a critical section when determining strength properties, and this will be the one with the smallest cross-sectional area in the midships region of the hull girder. Allowance has to be made for deck openings and the reduction in stress in way of their fore and aft boundaries. The distances b_1 and b_2 are to be deducted when considering the transverse section indicated in Figure 6.13.

The IACS Common Structural Rules also stipulate that *net scantlings* are used in strength calculations (Figure 6.14). Net scantlings refer to thicknesses that are required for strength purposes alone; the *corrosion allowance*, t_c, is added to the *required net*

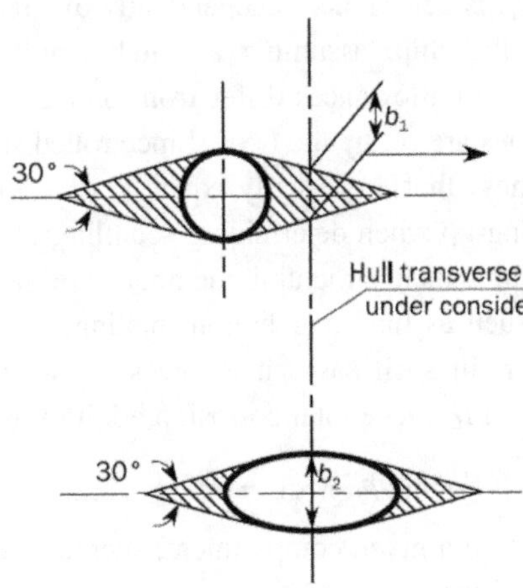

Figure 6.13 Allowance for deck openings when determining longitudinally effective structure [5]. Reprinted courtesy of IACS, London.

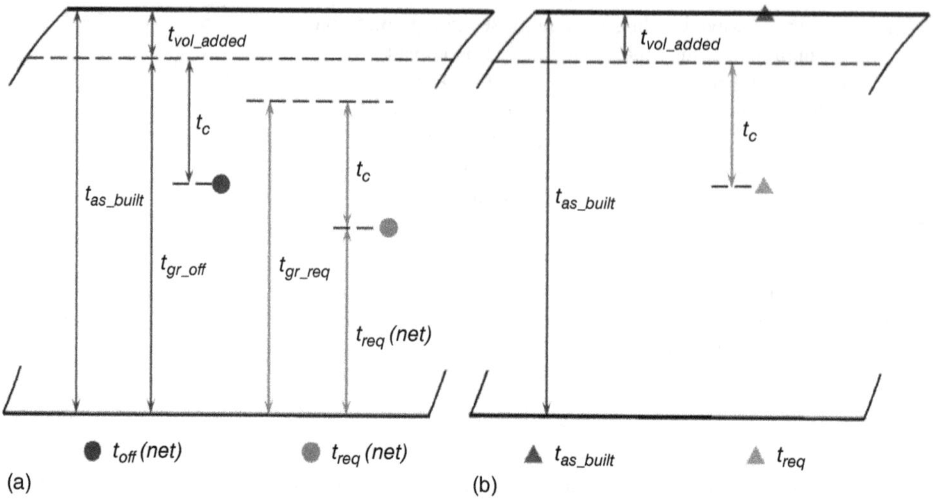

Figure 6.14 The net scantling approach used by IACS. (a) Design. (b) In service. Reprinted courtesy of IACS, London.

thickness, $t_{net\_required}$, of a particular member, rounded to the closest half millimetre, to obtain the *gross required thickness*, $t_{gross\_required}$. Since the actual thickness proposed by the shipyard, in line with classification society requirements, may differ, it becomes necessary to define the actual gross thickness provided at the newbuilding stage, termed the *gross offered thickness*, $t_{gross\_offered}$. By subtracting the corrosion allowance from the gross offered thickness, the *net offered thickness*, $t_{net\_offered}$, is obtained.

There may also be an additional thickness, $t_{vol\_addition}$ (*voluntary addition*), this being the owners' optional extra margin against corrosion. The *as-built thickness*, t_{as_built}, is the actual thickness, in mm, and includes $t_{vol\_addition}$.

The corrosion allowance, t_c, is determined independently of strength requirements. It is based on a 25-year life of the ship, assuming a suitable hull structure maintenance programme is in place. Corrosion allowances differ from one region of the hull structure to another, and minimum values are set by the CSR. Since rolled steel plating is manufactured in increments of thickness that are usually equal to or greater than 0.5 mm, it is necessary to allow for rounding-up when determining scantlings.

There are many regions in which structural members are subjected to a corrosive environment on both sides, such as the outer bottom plating of a double bottom that is used to store ballast seawater. In such cases it is necessary to allow for the action of corrosion on both sides, t_{C1} and t_{C2}. The total corrosion addition is given by

$$t_C = Roundup_{0.5}(t_{C1} + t_{C2}) + t_{reserve} \tag{6.41}$$

For an internal member within a given compartment, such as the floor within a double bottom, the total corrosion addition is given by

$$t_C = Roundup_{0.5}(2t_{C1}) + t_{reserve} \tag{6.42}$$

Table 6.5 *Corrosion addition on one side of structural members*

Compartment Type	*Structural Member*		*Corrosion Addition* t_{C1} *or* t_{C2}*(mm)*	
			BC-A or BC-B Ships with L ≥ 150 m	Other
Ballast water tank	Face plate of primary members	Within 3 m below the top of the tank	2.0	
		Elsewhere	1.5	
	Other members	Within 3 m below the top of the tank	1.7	
		Elsewhere	1.2	
	Transverse bulkhead	Upper part	2.4	1.0
		Lower stool sloping and top plate	5.2	2.6
		Other parts	3.0	1.5
Dry cargo hold	Other members	Upper part	1.8	1.0
		Webs and flanges of the upper end brackets of side frames of single side bulk carriers		
		Webs and flanges of the lower end brackets of side frames of single side bulk carriers	2.2	1.2
		Other parts	2.0	1.2
	Sloped plate of hopper tank, inner bottom plating	Continuous wooden ceiling	2.0	1.2
		No continuous wooden ceiling	3.7	2.4

Table 6.5 gives the corrosion allowance for ballast tank and dry cargo hold spaces in bulk carriers, according to IACS CSR.

Section Modulus Calculation for a Hull Girder Section

The calculation of the section modulus of a section is carried out in standard tabular form, as shown in Table 6.6.

All items that contribute to longitudinal strength are listed in column 2 and their scantlings, areas and distances from a chosen baseline are included in columns 3–5. Subsequent calculations relate to the first and second moments of the areas about the baseline of the section, which is usually the keel-line. The last column includes the second moment of area of each item about its own neutral axis. The contents of columns 4 and 5–7 are summed and the following data obtained.

Height of neutral axis of the cross-section above the baseline from

$$Z_N = \sum_{i=1}^{n} a_i z_i / \sum_{i=1}^{n} a_i \tag{6.43}$$

Table 6.6 *Section modulus calculation*

No	Item	Scantlings (mm)	Area $a_i (mm^2)$	Height above Baseline z (mm)	1st Moment $a_i z_i$ (mm^3)	2nd Moment $a_i z_i^2$ (mm^4)	Local i_{na} (mm^4)
1							
2							
.							
.							
n							
			$\sum_{i=1}^{n} a_i$		$\sum_{i=1}^{n} a_i z_i$	$\sum_{i=1}^{n} a_i z_i^2$	$\sum_{i=1}^{n} i_i$

Second moment of area of the section about the baseline:

$$I_{BL} = \sum_{i=1}^{n} i_i + \sum_{i=1}^{n} a_i z_i^2 \tag{6.44}$$

Second moment of area of the section about the neutral axis, using the *parallel axis theorem*:

$$I_{NA} = I_{BL} - AZ_N^2 \tag{6.45}$$

Having obtained the second moment of area, I_{NA}, the section moduli for the deck and bottom structure at this location along the length of the hull girder can be calculated from

$$Z_{deck} = \frac{I_{NA}}{y_{deck}} \qquad Z_{bot} = \frac{I_{NA}}{y_{bot}} \tag{6.46a–b}$$

For the given bending moment M,

$$\sigma_{deck} = \frac{M}{Z_{deck}} \qquad \sigma_{bot} = \frac{M}{Z_{bot}} \tag{6.47a–b}$$

These stresses can then be compared to the permissible ones (σ_{perm}) included in Table 6.2, for given operating conditions and material yield stress. If required, scantlings can be modified. It has to be borne in mind that in addition to global strength requirements, it is necessary to satisfy local strength requirements. This involves performing buckling and fatigue strength calculations for specific structural members.

To ensure sufficient *bending stiffness*, the second moment of area of the hull girder amidships has to exceed a minimum value. The IACS CSR (2020) formula is

$$I_{yR} = 2.7 C_w L^3 B (C_B + 0.7) 10^{-8} \qquad (\text{m}^4) \tag{6.48}$$

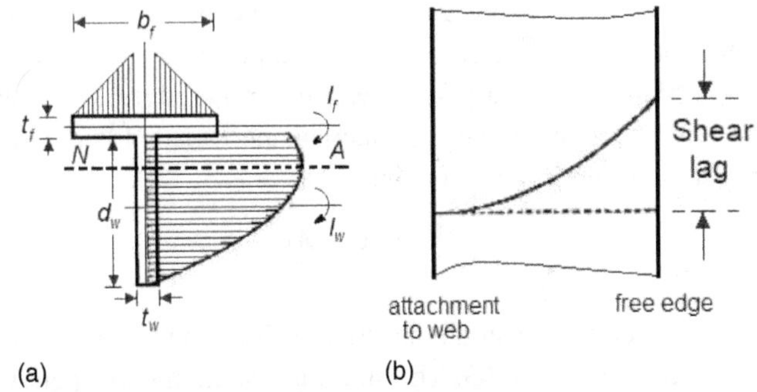

Figure 6.15 Shear stresses in T-bar section [9]. (a) Shear stresses (and strains). (b) Elongations along breadth of flange (plan view).

6.2.7 Axial Normal Stresses Due to Combined Vertical Bending and Shear

The first part of this chapter has dealt with the axial normal and shear stresses that arise when longitudinal (primary) hull girder vertical bending takes place. That treatment was based on the linear elastic bending theory of beams, according to which the stresses induced by the vertical bending moment and the transverse shear force can be treated independently of one another, and that no interaction between the two arises. However in real structures interaction between the two stress fields does arise, and in certain cases it is important to allow for it. The most important consequence of interaction is the development of additional axial normal stresses due to the vertical shear force distribution. These have to be added to the axial normal stresses that are calculated using linear beam theory. As will be seen, the distribution of normal stresses is no longer constant at any distance from the neutral axis of bending so that the Euler-Bernoulli theory does not hold. As a consequence, plane sections do not remain plane as *flexural warping* of hull girder transverse sections arises.

Axial Normal Stresses Induced by Vertical Shear Force

Before considering the case of a box girder section, which is similar to a wall-sided ship, it is instructive to consider a simple T-bar section subjected to linear elastic bending. The bending load is assumed to act normal to the flange surface, so that the neutral axis of bending is as shown in Figure 6.15a.

This section consists of two members that are welded to each other along the line of attachment. Their bending stiffness differs substantially, since for the web $I_w = td_w{}^3/12$ and for the flange $I_f = b_f t_f{}^3/12$. Hence, $I_w \gg I_f$. As a result, when a bending load is applied, the web, which is the stiffer member, prevents the flange from deflecting. Since the two members are welded to each other, compatibility of longitudinal strains is required along the line of attachment $(\varepsilon_w = \varepsilon_f)$[8]. It is thus seen that a shear stress develops along the

[8] If the two members were not welded to each other, they would translate in the axial direction with respect to one another, and no axial shear strains would develop. In this case the interaction between the two would be limited to a normal load distribution along the line of attachment.

line of attachment in order to ensure geometrical compatibility when a load is applied. Along the free edges of the flange no forces act so that the shear stresses are zero there. The shear stresses in the flange thus vary from zero along the edges to a finite value at the junction with the web. From mechanics of materials the variation is found to be linear, as shown in Figure 6.15a. The relevant equation is

$$\tau_{xz} = \frac{V(x)\sum A\bar{y}}{It} \tag{6.49}$$

In the case of a linear elastic material, the strain distribution has an identical form, since the constitutive relation is $\gamma_{xz} = \tau_{xz}/G$. The strain distribution corresponds to a particular distortion pattern which can be deduced from the definition of strain, so that [11]

$$\frac{\partial e}{\partial z} = \gamma_{xz} \tag{6.50}$$

where e is the *extension* of a longitudinal fibre of the flange.

The value of e is then obtained by integrating Equation (6.50), in this case across the breadth of the flange, so that

$$e(z) = \oint_{z} \gamma_{xz} dz \tag{6.51}$$

In Equation (6.51) the circular integration is performed from the neutral axis of the section along the girth of the member to the fibre in question, as is done when calculating shear strain. Since the shear strain varies linearly across the breadth of the flange (Figure 6.15a), the corresponding elongation will vary quadratically (Figure 6.15b). It is seen that the flange distorts in its own plane in a nonlinear manner and this distortion is termed *shear lag*.

In the web, fibre extension takes place as well, but in this case the degree of nonlinearity is much reduced and the end effect is small in comparison to the axial deflections due to bending. The distribution of the total extension is quasi-linear, but compatibility of axial deflections along the line of attachment is required, as in the case of strains $(\varepsilon_w = \varepsilon_f)$.

Exactly the same behaviour as that encountered in the T-bar section arises in the rectangular box girder section shown in Figure 6.16. In this case there are two lines of attachment for each flange and each web so that shear lag arises with respect to both of these.

We shall proceed to obtain the distribution of the axial extensions associated with the shear stress field that arises in a rectangular solid section of depth t and breadth b (Figure 6.17). This will be carried out for two cases. In the first case the beam is loaded under a constant vertical shear force ($V =$ const) (Figure 6.17a), and in the second the shear force varies over the length of the beam being considered (Figure 6.17b).

An example of the first case is a beam that is supported at two or more pin-joints with concentrated load(s) acting at one or more positions along its length. In this case the vertical shear force remains constant between the points at which the loads act, and there a discontinuity arises in the shear force distribution. Figure 6.17a shows an element of length

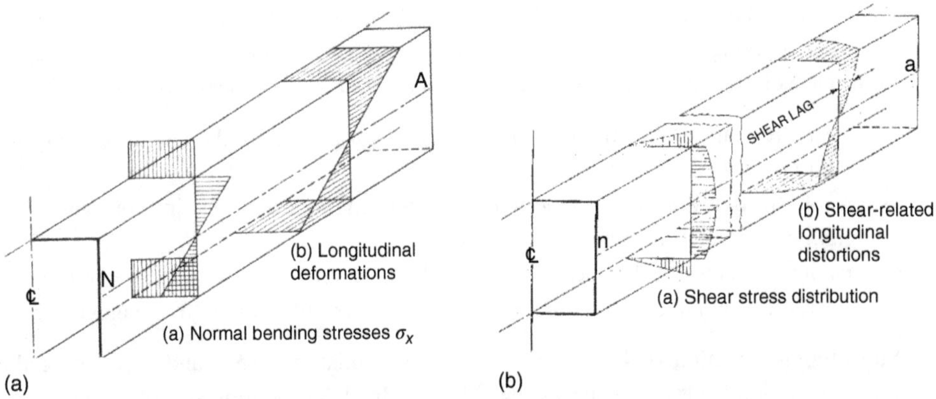

Figure 6.16 Stresses and deformations resulting from bending of a rectangular section [2]. (a) Normal stresses and axial deformations. (b) Shear stresses and axial distortions.

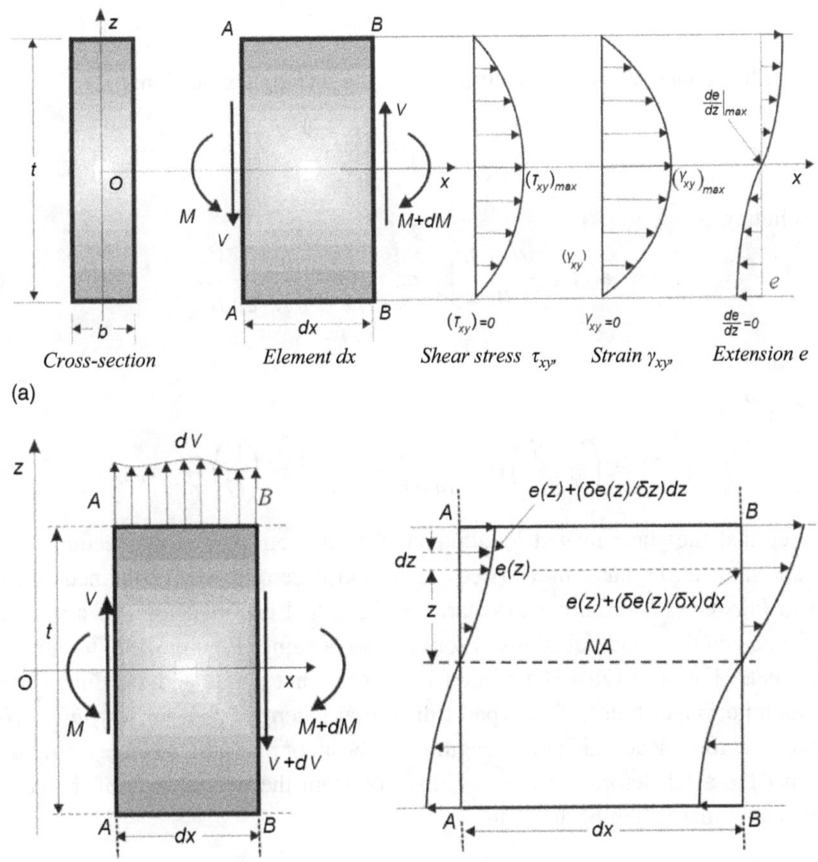

Figure 6.17 Axial shear deformations of a beam due to vertical shear force $V(x)$ [1]. (a) Constant vertical shear force ($V =$ const). (b) Varying vertical shear force $V(x)$.

dx of a longitudinal section over which $V = \text{const}$. This figure also includes the shear stress obtained by applying Equation (6.49) as well as the strain distribution.

In the case of a solid rectangular section subjected to a constant shear force ($V = \text{const}$):

- The shear stress and shear strain distributions are constant along the length of the beam for which the vertical shear force V is constant.
- As a result, the distribution of the longitudinal deformations $e(z)$ remains constant along the length of the beam for which $V = \text{const}$.
- The maximum values of the shear stresses and strains arise in way of the neutral axis.
- The shear stress and shear strain are zero in way of the outer fibres (free edges).

Substituting Equation (6.49) and $\gamma_{xz} = \tau_{xz}/G$ in Equation (6.51) and performing the circular integration leads to the shear-related longitudinal deformation $e(z)$ of a fibre at any distance z from the neutral axis, along the perimeter of the section (Figure 6.17a),

$$e(z) = \oint_z \left(\frac{V(x) \sum_i A\bar{y}}{GIt} \right) dz \tag{6.52}$$

At a girth distance z from the neutral axis of the rectangular section,

$$\sum_i A\bar{y} = (bz)(z/2)$$

Substituting,

$$e(z) = \frac{V(x)}{GIt} \frac{b}{2} \oint_z z^2 dz = \left(\frac{V(x)b}{2GIt} \right) \left[\frac{z^3}{3} \right]_0^z \tag{6.53}$$

For $z = 0$, $e(0) = 0$

For $z = t/2$,

$$e(t/2) = \left(\frac{V(x)b}{6GIt} \right) t^3 = \frac{Vb}{6G(bt^3/12)t} t^3 = \left(\frac{V}{G} \right) \frac{2}{t} = e_{max} \tag{6.54}$$

We see that the shear-related longitudinal deformation $e(z)$ of the section reaches a maximum value at the outer fibres (free edges) and is zero in way of the neutral axis, as shown in Figure 6.17a. In order to determine the actual distorted form of a section it is necessary to add the longitudinal distortions due to bending (Figure 6.15a) to those due to vertical shear (Figure 6.17a). The section is therefore not plane, and the Euler–Bernoulli assumption no longer holds. This type of distortion is termed *flexural warping*.[9] When a constant shear force V acts along the length of the beam or part of it, we note from Equation (6.53) that the axial deformation at any distance from the neutral axis of the section is constant along that length of the beam.

[9] Flexural warping arises as a result of longitudinal flexure. If a beam is subjected to a torque, this gives rise to *longitudinal warping*.

If the vertical shear force $V(x)$ varies along part of the length of the member, the shear deformation $e(z)$ of the section will also vary in the longitudinal direction in accordance with Equation (6.52), so that at a section BB located a distance dx from section AA, at a distance z from the neutral axis, the longitudinal shear deformation will be $e(z) + (\delta e(z)/\delta x)dx)$ (Figure 6.17b).

Since the shear-induced deformation changes in the axial direction, a corresponding *shear-induced axial strain*, ε_s, arises. For a fibre located a distance z from the neutral axis this is given by

$$\varepsilon_s(z) = \frac{e(z) + \left(\frac{\delta e(z)}{\delta x}\right)dx - e(z)}{dx} = \frac{de(z)}{dx} \tag{6.55}$$

The corresponding *shear-induced axial stress,* $\sigma_s(z)$, is then [11]

$$\sigma_s(z) = E\varepsilon_s(z) = E\left(\frac{de(z)}{dx}\right) \tag{6.56}$$

This additional axial stress has to be added to the stress due to the bending moment, $\sigma_b(y)$. When the shear force $V(x)$ does not vary substantially along the length of the beam, the longitudinal stress $\sigma_s(z)$ can be disregarded as being negligible. It should be noted that $\sigma_s(z)$ varies along the section with z, and not with y, the perpendicular distance to the neutral axis, as the bending stress $\sigma_b(y)$ does. (The distinction becomes clearer if we consider a section other than a rectangular bar, for which z and y are identical.) As mentioned previously, the total longitudinal deformation (extension) of any longitudinal fibre consists of a bending stress-related component and a shear stress-related component. Since linear theory is used, superposition is permissible and the two can be added.

Bending of Ship-type Rectangular Sections and Shear Lag

In the previous section we saw that when a beam that has a rectangular solid section is subjected to a vertical shear force $V(x)$ that varies along its length, a longitudinal shear stress related component, σ_s, given by Equation (6.56), arises in addition to the axial bending stress, σ_b. We will now apply this result to the case of a rectangular hollow section, which resembles the cross-section of a hull girder (Figure 6.18a).

Substituting in Equation (6.56) from Equation (6.51) and then from Equation (6.49), we obtain [11]:

$$\begin{aligned}\sigma_s(z) &= E\frac{d}{dx}\left(\oint_z \gamma_{xy}dz\right) = \\ &= E\frac{d}{dx}\left(\oint\left(\frac{V(x)\sum A\bar{y}}{GIt}\right)dz\right)\end{aligned} \tag{6.57}$$

$\sigma_s(z)$ is the axial stress that arises as a result of the action of the transverse shear force, and varies along the perimeter of the section.

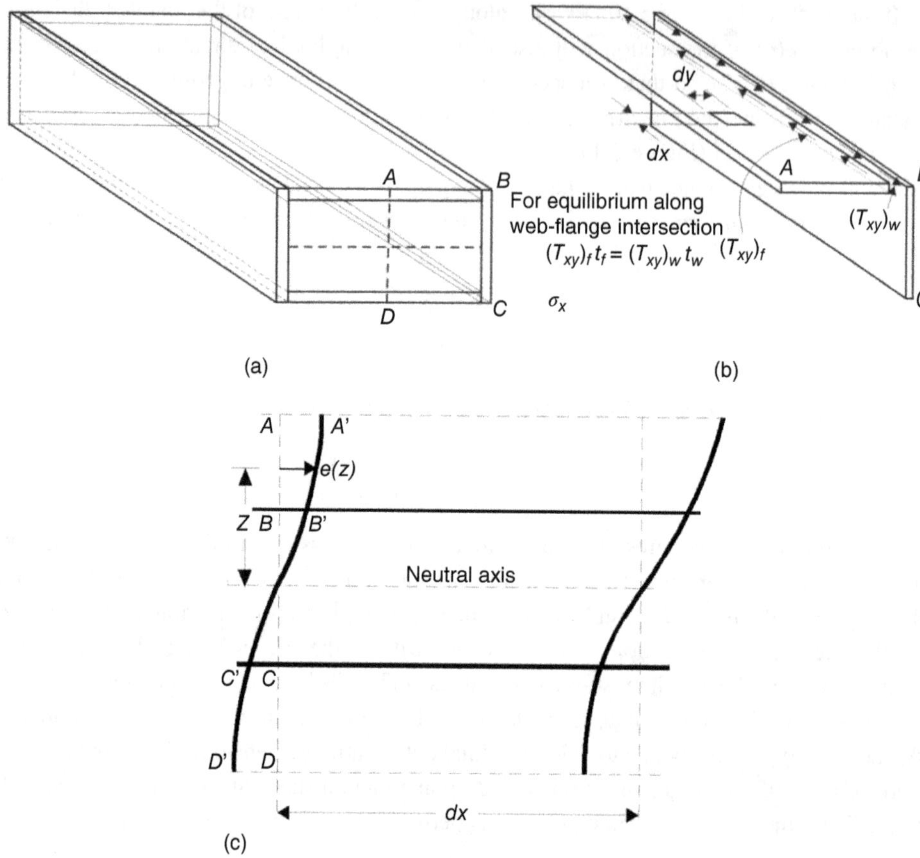

Figure 6.18 Axial distortion of a box girder section due to shear forces [1]. (a) Rectangular hollow section. (b) Equilibrium along a web–flange intersection. (c) Unfolded view of half-section A–B–C–D.

Figure 6.18b illustrates the longitudinal equilibrium condition along the web-flange intersection whereas Figure 6.18c is an unfolded view of the box girder section shown in Figure 6.18a that shows the distribution of transverse shear force-related longitudinal distortions in way of two sections located a distance dx apart. In this view B and C are the upper and lower web–flange intersections, respectively, whereas A and D are the intersections of the centreline plane with the upper and lower flanges (Figure 6.18a). We note the continuity in the deflection field along z as well as the different magnitudes in the two sections.

Equation (6.57) is valid for any prismatic section. Since $A\bar{y}$, I and t remain constant [11],

$$\sigma_s(z) = \frac{E}{GIt}\left(\frac{dV(x)}{dx}\right)\oint\left(\sum A\bar{y}\right)dz \tag{6.58}$$

We note that $\sigma_s(z)$ is proportional to $dV(x)/dx$, the rate of change of $V(x)$. If, therefore, $V(x)$ remains constant, $\sigma_s(z) = 0$ as seen in the case of the rectangular beam previously.

If a concentrated force acts, however, at that point $\sigma_s(z)$ will be large. In theory it will be infinite, since a singularity arises. In practice its magnitude will depend on the finite area over which it is distributed.

In ship structures the longitudinal vertical shear force varies gradually along the length of the hull girder, as seen in Figure 4.8, so that $\sigma_s(z)$ cannot be expected to be large. However, when the hull girder is supported at one position, such as when a ship is grounded, or else a point load acts in a particular location, $\sigma_s(z)$ needs to be taken into account.

The additional stress field $\sigma_s(z)$ due to the effect of shear stresses on axial stresses disturbs the longitudinal and rotational equilibrium of any section. This is clear if the resultant axial force and bending moment resulting from $\sigma_s(z)$ are calculated. Thus, the longitudinal force dF_s that acts on an element of area dA that is located a distance z from the neutral axis is

$$dF_s = \int \sigma_s(z) dA = \frac{2(1+\nu)}{It} \frac{dV}{dx} \int \left[\oint \left(\sum A\bar{y} \right) dz \right] dA \tag{6.59a}$$

The corresponding bending moment of the element about the neutral axis, dM_s, is given by

$$dM_s = \int \sigma_s y dA = \frac{2(1+\nu)}{It} \frac{dV}{dx} y \int \left(\oint \left(\sum A\bar{y} \right) dz \right) dA \tag{6.59b}$$

Integrating these two quantities over the whole area of the cross-section, the total force F_s and bending moment M_s are obtained:

$$F_s = \oint_A E \frac{de_s}{dx} dA \tag{6.60}$$

$$M_s = \oint_A \left(E \frac{de_s}{dx} \right) y dA \tag{6.61}$$

The conditions of equilibrium that have to be satisfied at the ith section are

$$\sum_i F_x = 0 \qquad \sum_i M_y = 0 \tag{6.62a–b}$$

To restore equilibrium, the additional force F_s and moment M_s that result from the stresses $\sigma_s(z)$ need a corresponding force and a corresponding moment to be imposed. Thus we require:

1. A uniformly distributed stress over the area of the section acting in a direction opposite to that of F_s of magnitude $-F_s/A$ and
2. A linearly varying bending stress of magnitude $M_s(y/I)$ acting in a direction opposite to that of σ_s at each section.

The resulting stress distribution is then:

$$\sigma = \frac{My}{I} \pm \left[E \frac{de_s}{dx} - \frac{1}{A} \oint_A E \frac{de_s}{dx} dA - \frac{y}{I} \oint_A E \frac{de_s}{dx} y dA \right] \tag{6.63}$$

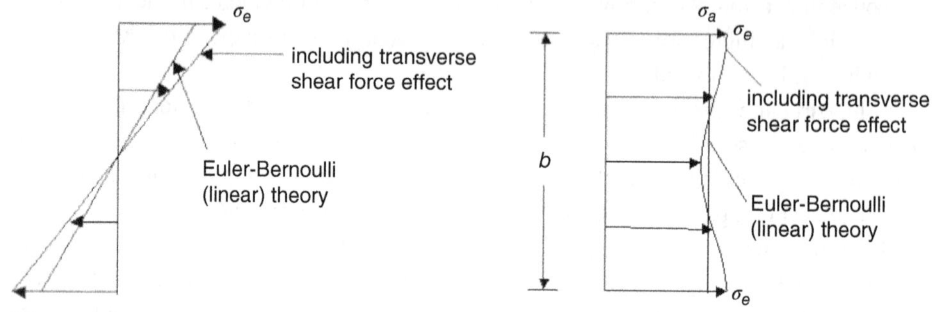

(a) (b)

Figure 6.19 Normal stresses in box girder including transverse shear force effect [1]. (a) Web. (b) Flange.

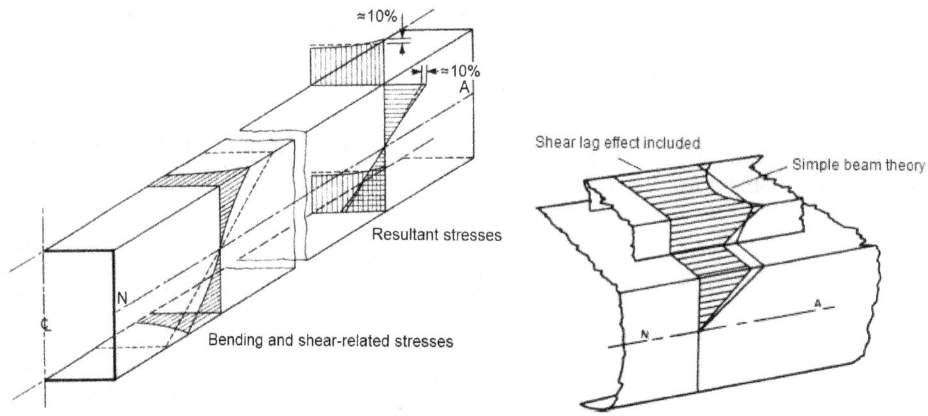

(a) (b)

Figure 6.20 Effect of shear lag on axial normal stress distribution [2]. (a) Rectangular hollow section. (b) Deckhouse arrangement.

where the first term is the Euler–Bernoulli bending stress and the first term in the brackets is the transverse shear force–related component. Equation (6.63) gives the final distribution of axial normal stresses at a transverse section, including the effect of a longitudinally varying shear force. The sign of the quantity in brackets depends on the signs of the bending moment and the load at the particular section. The general form of the stress distributions in a rectangular hollow section (Figure 6.18a) is shown in Figures 6.19a and 6.19b.

We note that in the web, the normal stresses are much higher than those predicted by Euler–Bernoulli theory, and they increase linearly up to a maximum value at the web–flange junction. In the flange, the normal stress is no longer linear. In way of the junction with the web it is much larger than the stress predicted by linear theory, whereas along the centreline plane of the section it is smaller. The normal strains ε_x follow a similar pattern. The final distribution of stresses in a box-girder section is shown in Figure 6.20.

The shear lag effect is important:

(a) at the intersection of members whose bending stiffness differs substantially, so that the bending load is transmitted by shear to a flange from a stiff web; and
(b) where significant changes in the transverse shear force distribution arise.

In hull girders, condition (a) arises at the intersection of horizontal decks with vertically aligned structure and is therefore expected to be accentuated in multi-deck ships (Figure 6.20b). It has been seen that the transverse shear-related normal stress $\sigma_s(z)$ depends on $dV(x)/dx$, the gradient of the vertical curve. The positions at which this occurs are readily identifiable in the vertical shearing force distribution along the ship's length (e.g., Figure 4.8). It is also dependent on the line integral of the first moment of area about the neutral axis, and is therefore proportional to the ratio of the flange breadth to the web depth.

The resulting increase in normal bending stress at an intersection can reach 10 per cent, which is non-negligible (Figure 6.20a). It is most accentuated in way of the midship section, where the rate of change of the transverse shear force is greatest. In practical ship design the increases in thickness of the sheer strake and of the adjoining deck strake over those of neighbouring strakes compensate for this effect. In ship sections that have several longitudinal bulkheads the shear lag effect diminishes as shear is transferred to decks at multiple points of support. In mathematical terms this brings about a reduction in the value of the line integral of the first moment of area, since z does not extend as far as it would in the absence of bulkheads.

Full-scale measurements have shown that shear lag effects in hull girders, where the loading is uniformly distributed, are of secondary importance [12]. This is evident from Figure 6.21 that includes longitudinal stress measurements across the breadth of the main deck of an oil tanker. It is seen that the linear elastic beam theory predicts stress values accurately and that there is no noticeable increase in stress in way of the deck/side shell junction in relation to the centreline stress. Similar conclusions were reached when tests were performed on another ship (SS *Ocean Vulcan*) specifically in order to assess the effect of shear lag. It may be concluded that shear lag is only of importance when point loads act on the hull girder in way of the vertical webs (side shell, longitudinal bulkhead).

Design of Structures Using the Effective Breadth Concept

In ship structures arguably the most important case of shear lag arises at a local level in laterally loaded stiffened plating, especially when the load is concentrated (Figure 6.22).

From Figures 6.20 and 6.22 it is observed that the effect of shear lag is an increase in normal stress in way of a line of attachment of a vertically aligned member with a horizontally aligned one and a decrease towards the centreline plane of a box girder section. From equilibrium considerations, the total force in each transverse section of the structure has to be the same as before. Since for design purposes we are interested in the maximum stress that arises in the section, which is the edge stress, σ_e, we can define the properties of a notional plate that has the same thickness as the original plate and an *effective breadth* over which the stress is constant

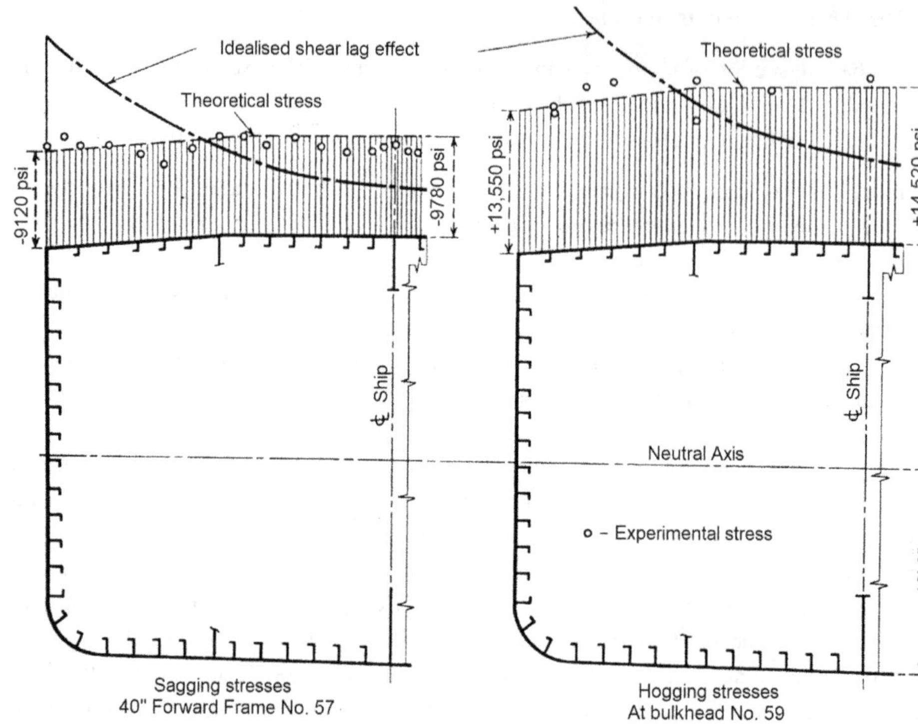

Figure 6.21 Longitudinal stress measurements in upper deck and predictions using linear elastic bending theory and nonlinear theory, including shear lag effects (SS *Shiloh*) [12]. Reprinted courtesy of the Society of Naval Architects and Marine Engineers.

and equal to the edge stress (Figure 6.23). The effective breadth is then determined such that the total load in the notional plate is equal to that in the original one.

In Figure 6.23 σ_a is the average stress in the actual plate (deck) and σ_e is the maximum stress, which arises at the plate (deck) edges where the plate is attached to a web (side shell structure). The plate is replaced by the notional plate that consists of two strips of width $b_e/2$ each. Since the longitudinal force ought to be the same in both cases,

$$\int_{-b/2}^{+b/2} \sigma(y)dy = \sigma_e b_e \tag{6.64}$$

where

$\sigma(y)$ = normal bending stress (nonlinear distribution)
σ_e = edge (maximum) stress in flange (deck plating)
b = breadth of flange (deck plating)
b_e = effective breadth of flange (deck plating)

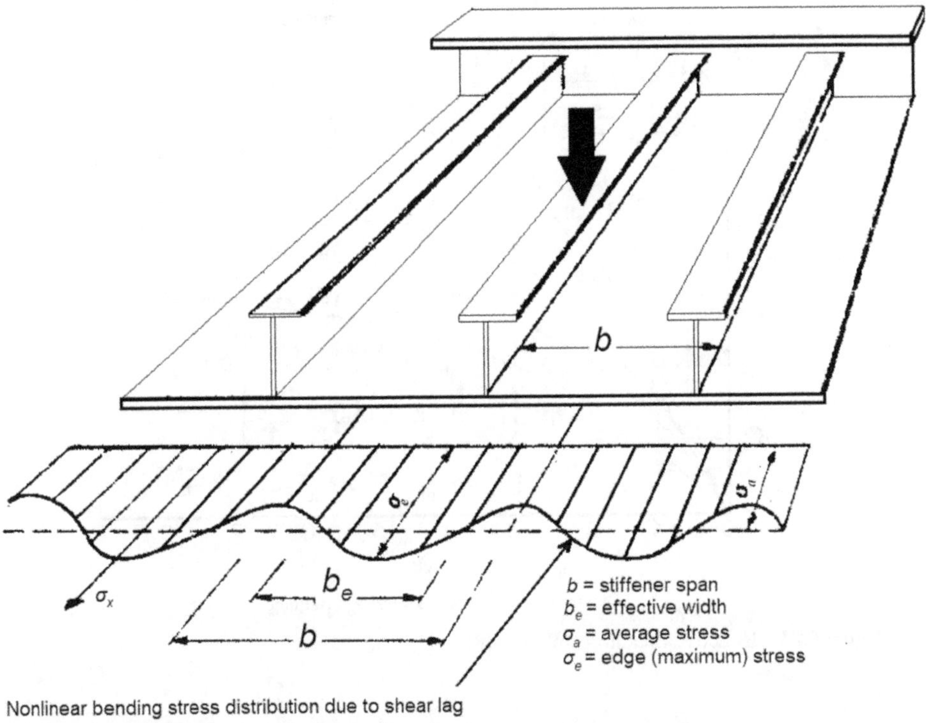

Figure 6.22 Shear lag effect in plate bending stress distribution in laterally loaded stiffened plating.

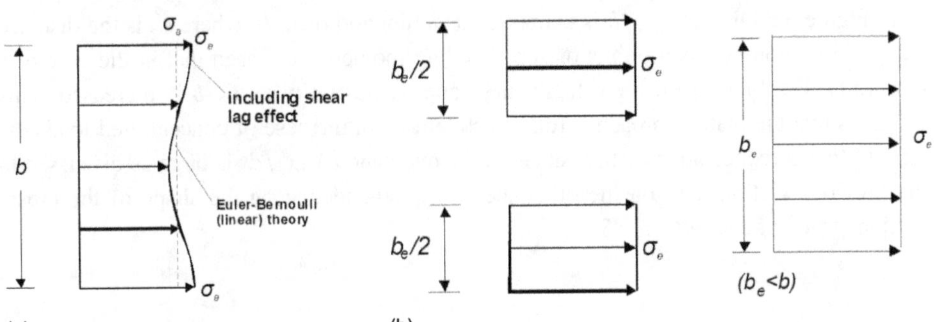

Figure 6.23 Definition of effective breadth of plating [1]. (a) Original distribution. (b) Notional distribution.

Hence,

$$\frac{b_e}{b} = \frac{P/A}{\sigma_e} = \frac{\sigma_a}{\sigma_e} \tag{6.65}$$

The above relationship provides a definition of the effective breadth that can be used to replace the actual structure with one in which the stress is uniformly distributed. In practice,

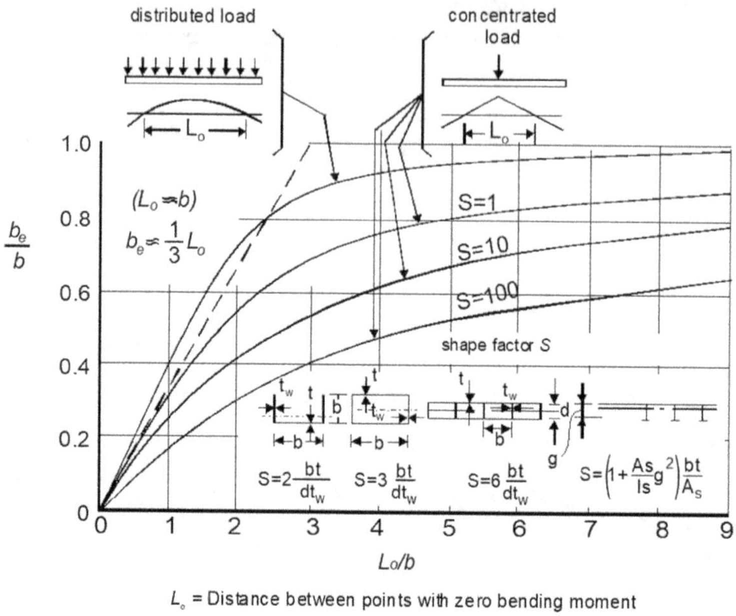

Figure 6.24 Effective breadth of plating [8].

the value of b_e is found using charts such as the one included in Figure 6.24, which gives values of b_e for two different types of loading (uniformly distributed and concentrated point load) and a variety of structural configurations.

In Figure 6.24 the ratio b_e/b is expressed as a function of L_o/b where L_o is the distance along the section between points of zero bending moment. It is seen that in the case of a uniformly distributed load the effectiveness approaches unity as L_o/b is increased. This indicates that the plate approaches full effectiveness. In the case of concentrated loads the ratio b_e/b can reduce appreciably, since on the one hand $dV(x)/dx$ is large as discussed in the previous section, and on the other there is a dependence on the shape of the cross-section (value of *shape factor S*).

6.2.8 Horizontal Bending of the Hull Girder

When discussing the bending response of ships, the terms 'vertical' and 'horizontal' refer to axes that are, respectively, perpendicular and parallel to the neutral axis of vertical bending and can thus be termed 'structural axes'. If the ship is floating in a flat calm sea, the axes are also perpendicular and parallel to the sea surface. When a ship is in a seaway, its motion involves rolling, and as a result the vertical axis is no longer perpendicular to the surface of the sea, nor is the horizontal axis parallel to it. For present purposes we consider the terms 'vertical' and 'horizontal' with reference to the structural axes so that the term *horizontal bending* refers to bending about the centreline plane of the ship.

Pure Horizontal Bending

A floating body undergoes horizontal bending as a result of different water pressure distributions acting on its sides. In order for pure horizontal bending to occur – that is, without vertical bending or torsion occurring simultaneously – the body has to remain upright. For this it is necessary that its structure is symmetrical about the neutral axis of horizontal bending, and the horizontal loading is uniformly distributed with respect to the centreline plane. The first of these conditions does hold for most ships whereas the second does not. The end result is that horizontal bending does not occur on its own but is coupled with vertical bending and/or torsion. Before considering these more realistic cases, we need to consider pure horizontal bending of a ship.

Figure 6.25 shows a ship in an upright position (no roll) encountering a regular wave system at an angle β. In this case there is no assumed relationship between any and all of the following variables:

a) the wavelength λ and the wave height H
b) the ship's length L, breadth B and draught T and
c) the angle β and the frequency of wave encounter ω.

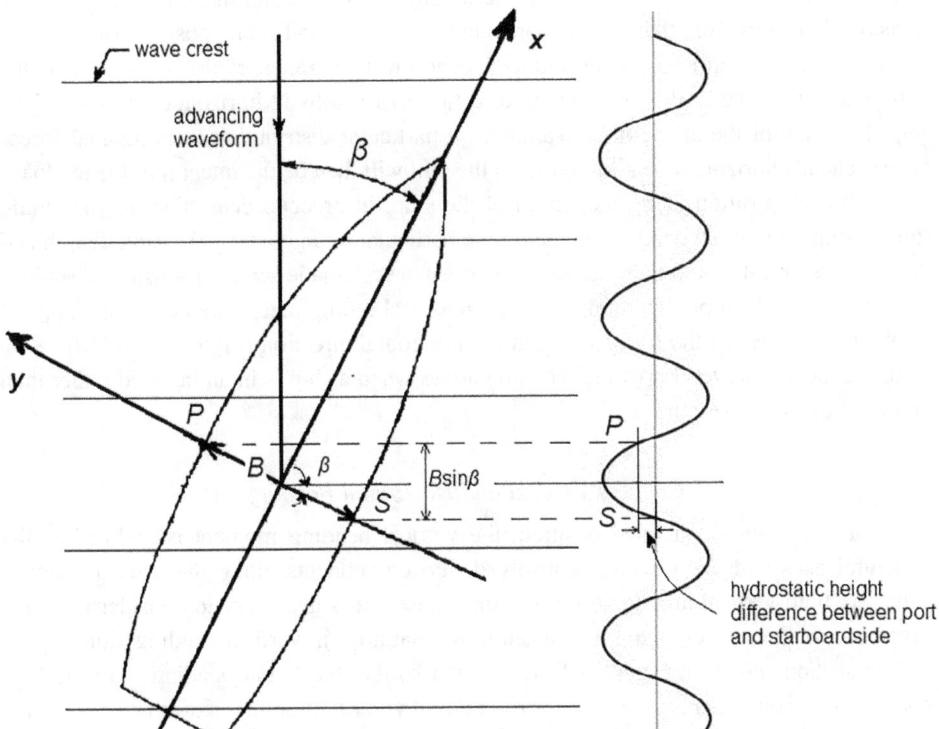

Figure 6.25 Horizontal bending force acting on a ship in an upright position encountering a regular wave system at an angle β.

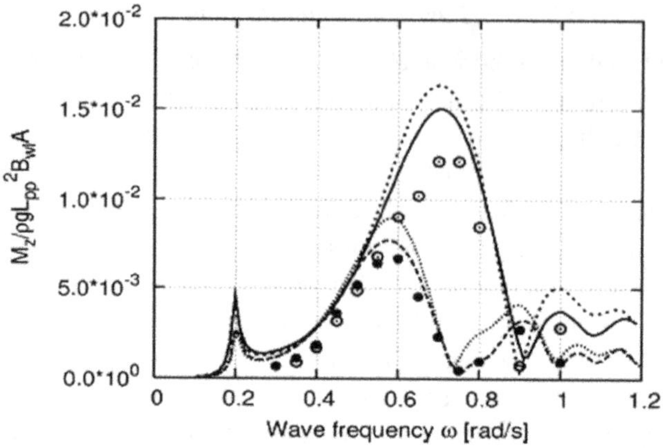

Figure 6.26 Linear horizontal bending moment for wave directions 60° and 30° amidships ___/- - -: Rankine source method (60°/30°). - - - /. . . . . . : zero-speed wave Green function method (60°/30°). ⊙/●: model tests (60°/30°) [10].

It is seen that at any position along the length, the hydrostatic height experienced on portside (*P*) differs from that on starboardside (*S*). The net load at any position varies along the length of the ship and is expected to depend on the parameters listed above. It is the variation in net load in the longitudinal direction that results in horizontal bending of the ship. For each of the above-listed variables, a particular distribution of horizontal forces arises. The net horizontal bending force on the hull will then be the integral of the net loads acting at each position along the length of the ship. In order to determine the maximum (three-point) horizontal bending moment for a particular hull geometry (breadth, length and draught of ship), it would be necessary to perform a systematic series of sensitivity studies.

A number of load prediction methods such as seakeeping theory are now being applied to obtain estimates of the loads acting in the horizontal direction (Figure 6.26) [10].

In practice, however, horizontal bending arises when a ship is in an inclined rather than an upright position (Figure 6.27).

Coupled Vertical and Horizontal Bending

When a ship is in an inclined position, the vertical bending moment is inclined to the structural axes and, as a result, is resolved into components along the axes. In such a situation, both vertical and horizontal bending arises. It is necessary to consider this load case because the resulting stresses exceed those that arise in vertical bending alone.

The section shown in Figure 6.27 is inclined so that the initial bending moment M is resolved into components about the horizontal and vertical structural axes that are inclined at an angle θ to the upright position. These are given by

$$M_y = M\cos\theta \qquad \text{(horizontal axis)} \tag{6.66a}$$

$$M_z = M\sin\theta \qquad \text{(vertical axis)} \tag{6.66b}$$

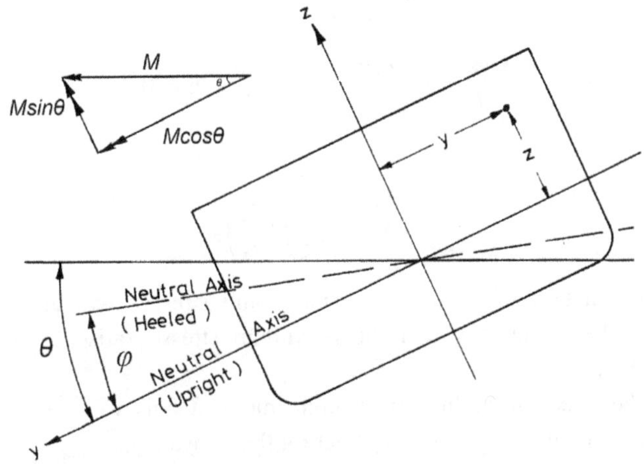

Figure 6.27 Inclined bending of hull girder section.

At a point of the section whose coordinates are (y, z), the total stress σ is the result of bending about the horizontal (y) as well as vertical (z) axes. The stress that results from bending about the horizontal axis in the inclined position, σ_y, is given by

$$\sigma_y = \frac{M_y}{I_y/z} = \frac{M\cos\theta}{I_y}z \tag{6.67a}$$

Also,

$$\sigma_z = \frac{M_z}{I_z/y} = \frac{M\sin\theta}{I_z}y \tag{6.67b}$$

The total stress is then

$$\sigma = \frac{M\cos\theta}{I_y}z + \frac{M\sin\theta}{I_{CL}}z \tag{6.68}$$

To determine the position of the neutral axis in the inclined position, we set $\sigma = 0$. Then,

$$\frac{\cos\theta}{I_y}y = -\frac{\sin\theta}{I_z}z$$

or

$$\frac{z}{y} = -\frac{I_y}{I_z}\tan\theta = \tan\phi \tag{6.69}$$

The neutral axis is thus inclined to the original *NA* at an angle φ. The angle θ at which maximum stresses arise as a result of combined bending is found using

$$\frac{d\sigma}{d\theta} = 0 \tag{6.70}$$

where σ is given by Equation (6.68). Thus,

$$\frac{d\sigma}{d\theta} = -\frac{M \sin\theta}{I_y} z + \frac{M \cos\theta}{I_z} y = 0$$

or

$$\tan\theta = \left(\frac{y}{z}\right)\left(\frac{I_y}{I_z}\right) \qquad (6.71)$$

Since σ is a function of y and z, the maximum stresses will arise at the corners of the section. The angle at which the maximum stresses arise as the ship rolls is approximately 35°.

As regards the effect of heeling on bending moment, this will not affect the weight distribution and will have only a minor effect on the buoyancy.

In the preceding analysis it was assumed that the vertical and horizontal bending moments are in phase so that the maximum values occur simultaneously. In practice, however, they are not, and the phase difference that arises is substantial, so that this has to be allowed for when calculating the resulting stress.[10] If the ship advances in a regular sinusoidal wave system at an arbitrary angle of incidence, then

$$M_y = M_{ymax} \cos(\omega t + \varepsilon_y) \qquad (6.72a)$$

$$M_z = M_{zmax} \cos(\omega t + \varepsilon_z) \qquad (6.72b)$$

where M_y and M_z are the horizontal and vertical bending moments, M_{ymax} and M_{zmax} are the corresponding maximum values, ε_y and ε_z are the phase angles and ω is the frequency of encounter. The stresses are then found from

$$\sigma_y = \frac{M_y}{I_y/z} = \frac{M_{ymax} \cos(\omega t + \varepsilon_y)}{I_y/z} = \frac{M_{ymax} \cos(\omega t + \varepsilon_y)}{Z_y} \qquad (6.73a)$$

$$\sigma_z = \frac{M_z}{I_z/y} = \frac{M_{zmax} \cos(\omega t + \varepsilon_z)}{I_z/y} = \frac{M_{zmax} \cos(\omega t + \varepsilon_z)}{Z_z} \qquad (6.73b)$$

where Z_y and Z_z are the actual section moduli with respect to the horizontal and vertical axes. Thus,

$$\sigma_y = \sigma_{ymax} \cos(\omega t + \varepsilon_y) \qquad (6.74a)$$

and

$$\sigma_z = \sigma_{zmax} \cos(\omega t + \varepsilon_z) \qquad (6.74b)$$

From Figure 6.28 the resultant stress in critical positions of a transverse section can be found. The maximum stress is given by

[10] This question is discussed further in Chapter 9.

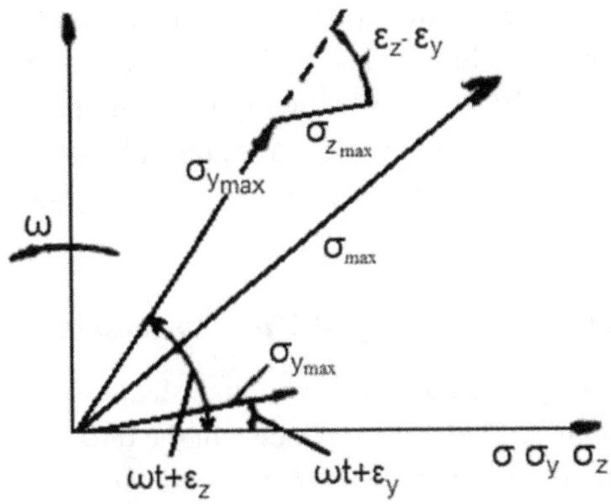

Figure 6.28 Addition of vertical and horizontal stresses allowing for phase angle difference [4].

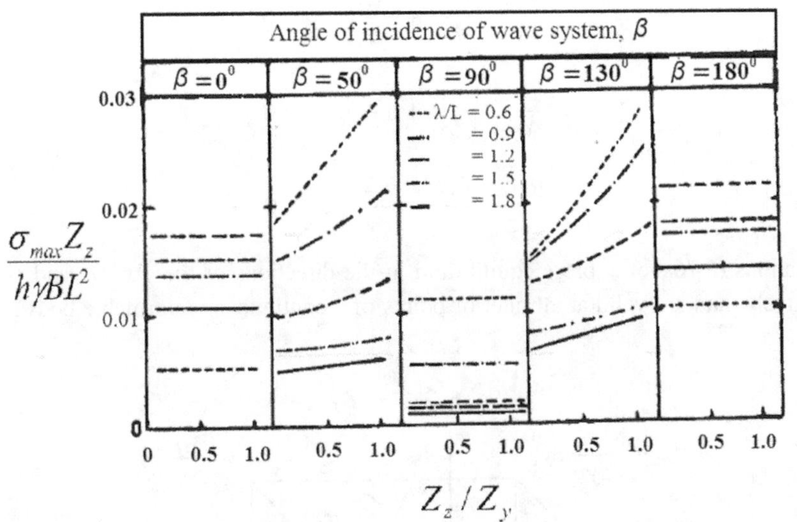

Figure 6.29 Maximum combined stress and the effect of several parameters [4].

$$\sigma_{max} = \sqrt{\sigma_{ymax}{}^2 + \sigma_{zmax}{}^2 + 2\sigma_{ymax}\sigma_{zmax}\cos\left(\varepsilon_z - \varepsilon_y\right)} \tag{6.75}$$

Figure 6.29 shows the effect of the ratio of the moduli, Z_z/Z_y, the angle of wave incidence β and that of the wavelength λ on the maximum stress, σ_{max}.

The presence of horizontal bending stresses can result in an increase in the total bending stress by up to 30 per cent.

6.3 Linear Response of Thin-Walled Structures Subjected to Torsional Loading

In this section an introduction to the torsional behaviour of structural members will be given. The first thing to note is that, when considering problems involving the torsional loading of structures, it is necessary to refer to the complete set of the equations of elasticity that describe equilibrium, deformations and the constitutive relations of a three-dimensional body. This is because the simplifying assumptions relevant to the plane stress state used in thin-walled theory are not applicable when a body is subjected to torsional loading. The relevant equations are given below without derivations.

6.3.1 Equations of Elasticity of Three-Dimensional Solid Bodies

An element of a elastic body extending in three-dimensional space subjected to an arbitrary external force system with *X*, *Y* and *Z* components in the directions of coordinate axes *Ox*, *Oy* and *Oz*, respectively, is shown in Figure 6.30.

The *equations of equilibrium* are obtained by considering summations of forces in the directions of the three coordinate axes. The equations are given below:

$$\frac{\partial \sigma_x}{\partial x} + \frac{\partial \tau_{xy}}{\partial y} + \frac{\partial \tau_{xz}}{\partial z} + X = 0 \tag{6.76a}$$

$$\frac{\partial \sigma_y}{\partial y} + \frac{\partial \tau_{yx}}{\partial x} + \frac{\partial \tau_{yz}}{\partial z} + Y = 0 \tag{6.76b}$$

$$\frac{\partial \sigma_z}{\partial z} + \frac{\partial \tau_{zx}}{\partial x} + \frac{\partial \tau_{zy}}{\partial y} + Z = 0 \tag{6.76c}$$

Equations (6.76a–c) express equilibrium in the directions of the *Ox*, *Oy* and *Oz* axes, respectively, and are valid at all interior points of a continuous, deformable body.

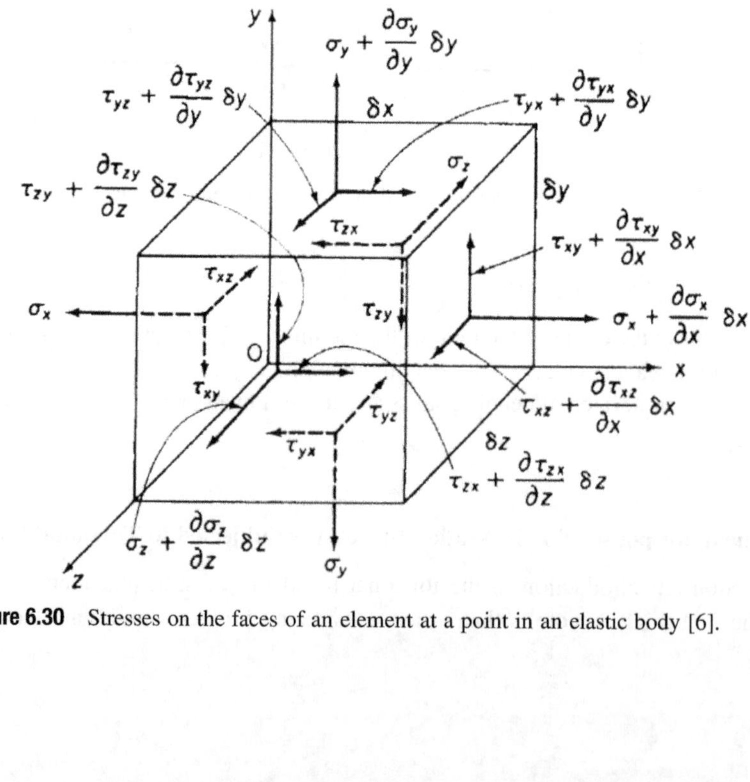

Figure 6.30 Stresses on the faces of an element at a point in an elastic body [6].

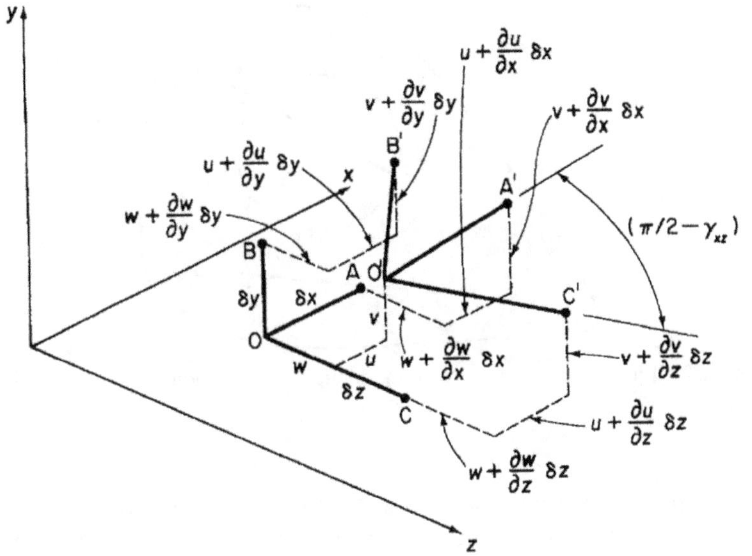

Figure 6.31 Displacements of components of any arbitrary deformation [6].

With regard to deformations of a three-dimensional body, the displacements of the components along the three coordinate axes, OA, OB and OC of any deformation, are shown in Figure 6.31.

From the definitions of strain and shear strain, bearing in mind we are dealing with infinitely small strains, the following relations result:

$$\left.\begin{aligned}\varepsilon_x &= \frac{\partial u}{\partial x}\\ \varepsilon_y &= \frac{\partial v}{\partial y}\\ \varepsilon_z &= \frac{\partial w}{\partial z}\end{aligned}\right\}\text{ normal strains}\qquad \left.\begin{aligned}\gamma_{xz} &= \frac{\partial w}{\partial x}+\frac{\partial u}{\partial z}\\ \gamma_{xy} &= \frac{\partial v}{\partial x}+\frac{\partial u}{\partial y}\\ \gamma_{yz} &= \frac{\partial w}{\partial y}+\frac{\partial v}{\partial z}\end{aligned}\right\}\text{ shear strains}\qquad (6.77\text{a–f})$$

These are termed the *strain–displacement* or *kinematic* relations of a linear elastic body. In order to ensure continuity, it is assumed that each element of the body is connected to neighbouring elements without any gaps or overlaps (Figure 6.32). Material *continuity* (*compatibility*) is ensured if certain compatibility relations hold. These can be expressed in mathematical form, and in infinitesimal strain theory they are derived from Equations (6.77a–f).

For continuity, the components (u, v, w) of an arbitrary deformation have to be continuous, single-valued functions, that is,

$$u = f_1(x, y, z) \qquad (6.78\text{a})$$

$$v = f_2(x, y, z) \qquad (6.78\text{b})$$

$$w = f_3(x, y, z) \qquad (6.78\text{c})$$

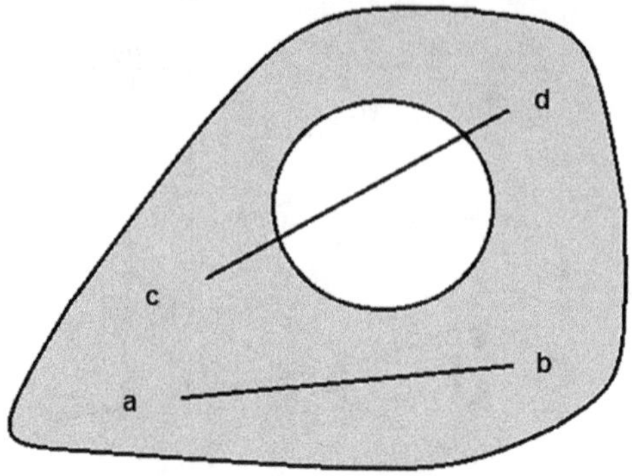

Figure 6.32 Line elements a–b, c–d in contiguous and non-contiguous regions of solid body.

If voids were present or were created as a result of external loading, the displacements would need to be expressed by different functions of x, y, z. These three single-valued functions ensure *material compatibility* or *continuity*.

The six equations of compatibility obtained by combining the kinematic relations, Equations (6.77a–f), are

$$\left.\begin{aligned}
\frac{\partial^2 \gamma_{xy}}{\partial x \partial y} &= \frac{\partial^2 \varepsilon_y}{\partial x^2} + \frac{\partial^2 \varepsilon_x}{\partial y^2} \\
\frac{\partial^2 \gamma_{yz}}{\partial y \partial z} &= \frac{\partial^2 \varepsilon_y}{\partial z^2} + \frac{\partial^2 \varepsilon_z}{\partial y^2} \\
\frac{\partial^2 \gamma_{xz}}{\partial x \partial z} &= \frac{\partial^2 \varepsilon_z}{\partial x^2} + \frac{\partial^2 \varepsilon_x}{\partial z^2} \\
2\frac{\partial^2 \varepsilon_x}{\partial y \partial z} &= \frac{\partial}{\partial x}\left(-\frac{\partial \gamma_{yz}}{\partial x} + \frac{\partial \gamma_{xz}}{\partial y} + \frac{\partial \gamma_{xy}}{\partial z}\right) \\
2\frac{\partial^2 \varepsilon_y}{\partial x \partial z} &= \frac{\partial}{\partial y}\left(\frac{\partial \gamma_{yz}}{\partial x} - \frac{\partial \gamma_{xz}}{\partial y} + \frac{\partial \gamma_{xy}}{\partial z}\right) \\
2\frac{\partial^2 \varepsilon_z}{\partial x \partial y} &= \frac{\partial}{\partial z}\left(\frac{\partial \gamma_{yz}}{\partial x} + \frac{\partial \gamma_{xz}}{\partial y} - \frac{\partial \gamma_{xy}}{\partial z}\right)
\end{aligned}\right\} \qquad (6.79\text{a–f})$$

Finally, it is necessary to relate strains to stresses. The required relations, termed *constitutive relations*, are based on experimental results. In the case of a linear elastic isotropic material, the application of a uniformly distributed normal load gives rise to:

a) A normal strain ε_x that is uniformly distributed over the loaded surface
b) A corresponding uniform direct stress σ_x given by

$$\sigma_x = E\varepsilon_x \qquad (6.80)$$

E is a material constant termed the *modulus of elasticity* of the material (*Young's modulus*).

c) Lateral strains $\varepsilon_y, \varepsilon_z$. These are related to the normal strain ε_x by the following relations

$$\varepsilon_y = -\nu\varepsilon_x \qquad \varepsilon_z = -\nu\varepsilon_x \tag{6.81a–b}$$

where ν is a material constant termed *Poisson's ratio.*

As the body is compressed or extended in the direction of the applied load, it distorts in the lateral directions in order to compensate for the compression or extension accordingly.

d) No shear distortion of the material is observed. Hence, $\gamma_{xy} = \gamma_{xz} = \gamma_{yz} = 0$.

For a linear elastic body subjected to external loads that give rise to stresses $\sigma_x, \sigma_y, \sigma_z$, the following constitutive relations hold:

$$\left.\begin{aligned} \varepsilon_x &= \frac{1}{E}\left[\sigma_x - \nu(\sigma_y + \sigma_z)\right] \\ \varepsilon_y &= \frac{1}{E}\left[\sigma_y - \nu(\sigma_x + \sigma_z)\right] \\ \varepsilon_z &= \frac{1}{E}\left[\sigma_z - \nu(\sigma_x + \sigma_y)\right] \end{aligned}\right\} \tag{6.82a–c}$$

6.3.2 Torque Applied to Prismatic Members

Before discussing the response of particular classes of structural members, it is necessary to clarify certain terms and definitions that arise in torsion theory. Figure 6.33 shows a uniform structural member subjected to an external torque, and Figure 6.34 shows typical structural members and their structural parameters (centroid and centre of shear). It is seen that in the case of closed sections the two coincide, whereas in open sections their distance may differ substantially, depending on the type of section. In the case of the channel section the distance between the two is greatest.

We need to define two particular geometric quantities concerning the application of the torque, these being the *centre of shear* and the *centre of twist*. The centre of shear refers to the stress distribution whereas the centre of twist refers to the distortion of the member, and both of these lie in the plane of any cross-section of a structural member. As seen in Section 2.3, if a shear force is applied through the centre of shear of a section, this will result in bending only and no twisting.

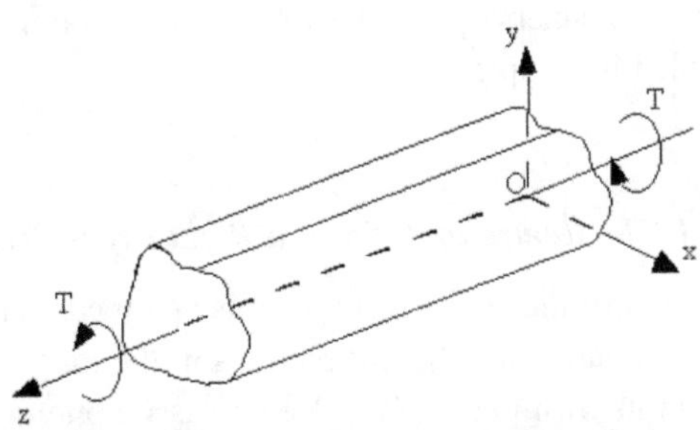

Figure 6.33 A prismatic member subjected to an external torque at its ends.

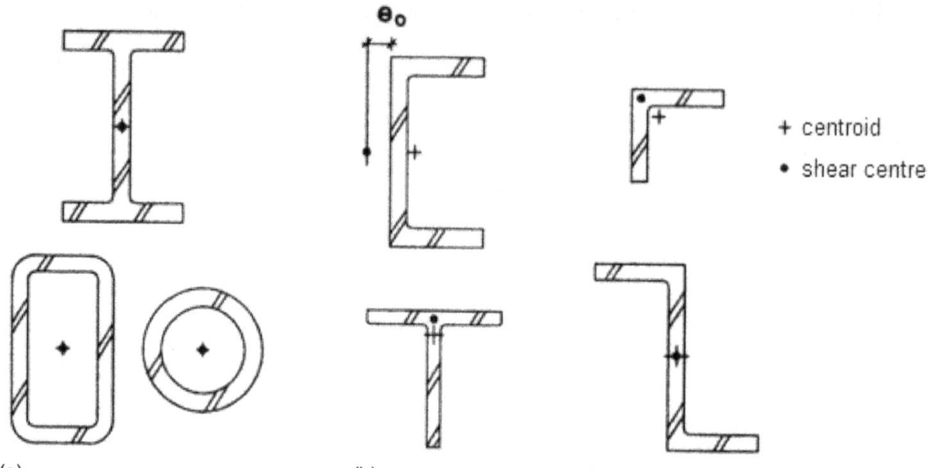

Figure 6.34 Centroid and centre of shear of various types of sections. (a) Doubly symmetric. (b) Singly symmetric. (c) Unsymmetric.

By invoking *Maxwell's reciprocal theorem*, we see that if a torque acts at the centre of shear of a section, it will not produce any translation but only twist of the section. The centre of shear then acts as the centre of twist. If a member undergoes *non-uniform torsion*, then the centre of twist does not coincide with the centre of shear. In this case the position of the centre of twist may be obtained using the method described below [6].

Ship structures have a large variety of structural configurations, but they have certain important characteristics in common. They consist of thin-walled members that are welded to each other. To a first approximation, they can be treated as being rigid, meaning that the cross-sections do not distort when a torque is applied. For present purposes it is important to classify ship-type thin-walled sections as being either *open* or *closed.* These two structural types differ in important ways in their response to torsional loading, and in the following sections the background to the behaviour of these types of sections will be given. At the hull girder level, which is our present concern, their response can be treated as being either that of 'closed' sections that in reality consist of multiple closed sections (as in the case of oil tankers, passenger ships, etc.) or of 'mixed' sections, being assemblies of multiple closed sections (containerships, bulk carriers, etc.). Examples of these are included in several figures included in Chapter 1.

6.3.3 The Centre of Twist of a Rigid Cross-Section

In determining the position of the centre of twist, it is assumed that the cross-section does not distort as it rotates in relation to adjacent sections in the longitudinal direction. It thus rotates as a rigid body in its own plane (Saint Venant assumption). The translation of any point is defined by the angle of rotation θ and translations v, w along the y and z axes,

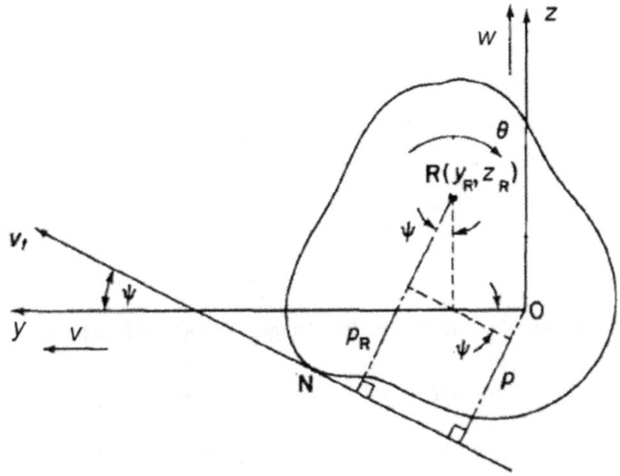

Figure 6.35 Rotation of cross-section about centre of twist $R(y_R, z_R)$ through angle θ [6]. (a) Rotation through θ. (b) Translation through (v, w).

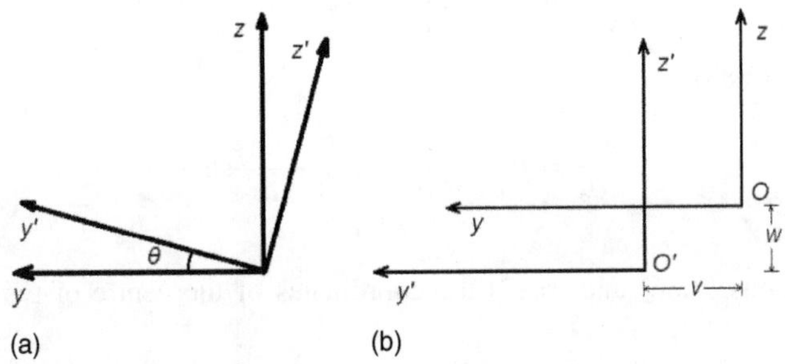

Figure 6.36 Rotation and translation of the coordinate axis system.

respectively. The cross-section of a body subjected to a torque T about a point $R(y_R, z_R)$ is shown in Figure 6.35.

An arbitrary origin O of the coordinate axis system is chosen, and the coordinates of the centre of twist of the section R, (y_R, z_R), need to be determined. The section rotates about R through an angle θ. As a result, the coordinate axes rotate about R through this angle and translates through (v, w), these being displacements in the y and z directions, respectively (Figure 6.36).

The total displacement of a point lying on the surface of the body – say, N – is then

$$v_t = p\theta + v\cos\psi + w\sin\psi \tag{6.83}$$

The rotation of point N about the centre of twist R is given by

$$v_t = p_R\theta \tag{6.84}$$

where p_R is the distance of the tangent at the surface at N from the centre of twist R.

From Figure 6.37, $RA = z_R\cos\psi$ $OB = y_R\sin\psi$

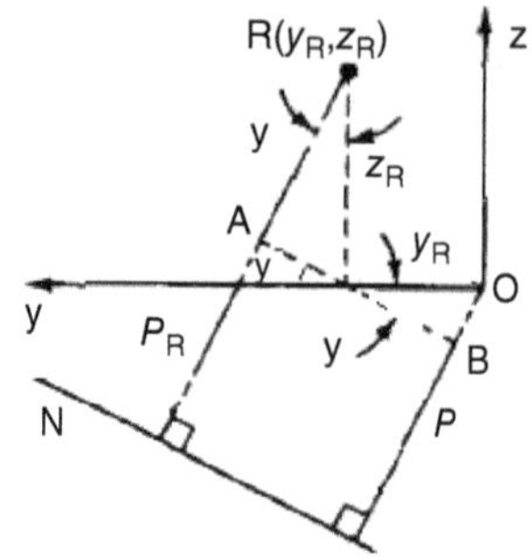

Figure 6.37 Distance of centre of twist from tangent to body surface passing through N [6].

Hence, $$p_R = p - OB + RA = p - y_R \sin\psi + z_R \cos\psi$$

Thus $$v_t = p_R\theta = p\theta + z_R\theta\cos\psi - y_R\theta\sin\psi \tag{6.85}$$

Differentiating with respect to x,

$$\frac{\partial v_t}{\partial x} = p\frac{\partial\theta}{\partial x} + z_R\frac{\partial\theta}{\partial x}\cos\psi - y_R\frac{\partial\theta}{\partial x}\sin\psi \tag{6.86}$$

From Equation (6.83)

$$\frac{\partial v_t}{\partial x} = p\frac{d\theta}{dx} + \frac{dv}{dx}\cos\psi + \frac{dw}{dx}\sin\psi \tag{6.87}$$

From Equations (6.86) and (6.87) the coordinates of the centre of twist, (y_R, z_R), are obtained:

$$y_R = -\frac{dw/dx}{d\theta/dx} \qquad z_R = \frac{dv/dx}{d\theta/dx} \tag{6.88a–b}$$

6.3.4 Torsion of a Cylindrical Section

Before considering ship-type sections, it is necessary to consider the problem of the torsion of a cylindrical section, which is the only one that can be expressed in closed form. The basic assumption on which the solution is based is that adjacent transverse sections rotate with respect to one another along the length of the member, this being the only form of distortion (Saint Venant assumption). That is, transverse sections do not distort in their own plane.

Consider two sections located a distance dx from each other. The angle of rotation of one with respect to the other, φ, is termed the *angle of twist* of the section. In this case, the *angle of twist per unit length*, θ, or *rate of twist*, is given by

$$\theta = \frac{d\phi}{dx} \tag{6.89}$$

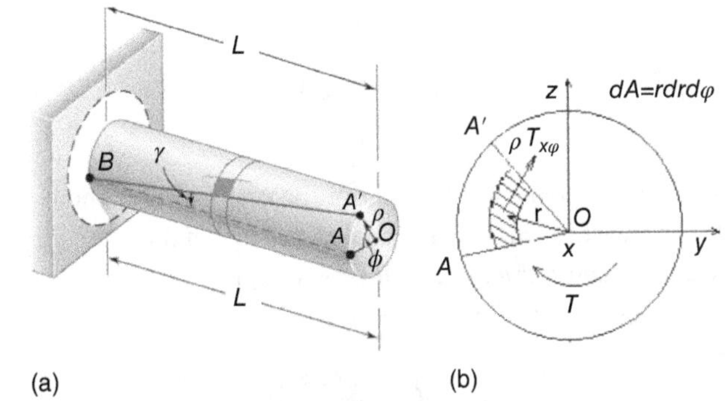

Figure 6.38 Shear stress $\tau_{x\varphi}$ on an element of area dA $(= rd\varphi dr)$ a distance r from the centre of twist (the angle of twist φ and the shearing strain γ share the same length AA'). (a) Circular section subjected to torque. (b) Transverse section at distance x from origin.

For two sections located a distance dx from each other consider two adjacent points lying at a distance ρ from the centre of twist (Figure 6.38a). Following application of a torque (twisting moment), the points rotate in the tangential direction with respect to each other through a distance $\rho d\varphi$. The resulting shear strain γ is then (Figure 6.38b):

$$\gamma_{x\theta} = \rho\frac{d\phi}{dx} = \rho\theta \tag{6.90}$$

For a linear elastic isotropic material undergoing shear loading, the constitutive relation is

$$\tau = \gamma G \tag{6.91}$$

where G is the *shear modulus* of the material, given by

$$G = \frac{E}{2(1+\nu)} \tag{6.92}$$

The internal moment due to the shear force resultant corresponding to an area dA, located a distance r from the centre of twist, is then $(\tau_{x\theta}dA)\rho$, where $\tau_{x\theta}\,dA$ is the elemental internal shear force resultant. For equilibrium, the internal resisting moment acting over area dA is equal and opposite to the applied elemental torque dT. Thus,

$$dT = (\tau_{x\theta}dA)\rho = (\gamma GdA)\rho = \rho^2\left(\frac{d\phi}{dx}\right)GdA$$

For the whole area of the cross-section,

$$T = \int_A \rho^2\frac{d\phi}{dx}GdA$$

Since $d\varphi/dz$ and G are constants,

$$T = \frac{d\phi}{dx}G\int_A \rho^2 dA = GI_p\frac{d\phi}{dx} \tag{6.93}$$

where

$$I_p = \int_A \rho^2 dA \tag{6.94}$$

The quantity I_p is the *polar moment of inertia* of the cross-section about the perpendicular axis at the centre of twist. In a manner analogous to bending theory, we may define *torsional rigidity* as GI_p and *torsional stiffness* as GI_p/L.

The shear stress is then found from

$$\tau = \gamma G = \rho \left(\frac{d\phi}{dx} \right) G$$

From Equation (6.93),

$$\frac{d\phi}{dx} = \frac{T}{GI_p} \qquad \text{and} \qquad T = GI_p \frac{d\phi}{dx} \tag{6.95}$$

Hence,

$$\tau = \frac{T\rho}{I_p} \tag{6.96}$$

The torque-induced shear stress at a point in a circular cross-section is thus seen to be proportional to the radial distance of that point from the centre of twist. This formula is analogous to $\sigma = My/I$ used in bending theory, where y is the distance of a fibre from the neutral axis of bending.

6.3.5 Uniform Torsion of Sections of Arbitrary Form

Let us now turn to the case of a non-circular section that is subjected to a torque at both ends that are free to distort in the axial direction (Figure 6.39a). What distinguishes the behaviour of this member from the cylindrical shaft considered previously is that the cross-sections no longer remain plane but deform (*warp*) in the out-of plane direction. With two exceptions,[11] the only case in which warping does not arise is when the cross-section is circular. When this is the case, the angle subtended between two sections some distance apart by any pair of points lying on a perimeter is constant, provided the arc length is the same.

In order to understand why warping arises in non-circular cross-sections, we turn to Figure 6.39a. This figure shows two points A and B located on different parts of the perimeter of the cross-section, whose distances ρ_A, ρ_B from the centre of twist differ. When the member twists through an angle β, these two points translate by different amounts ($AA' \neq BB'$). As a result, differences in circumferential shear strain arise along the perimeter of the cross-

[11] These are the square hollow section with a constant thickness along all sides [3] and the *Neuner* rectangular section for which $at_b = bt_a$ where a, b are the dimensions of the sides and t_a, t_b are the corresponding thicknesses [6].

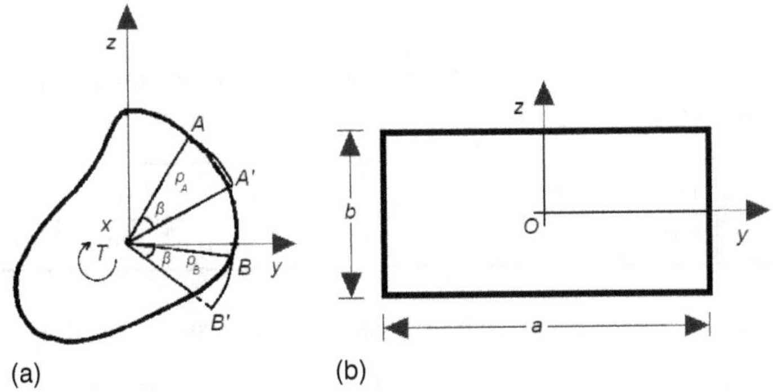

Figure 6.39 Rotation of solid cross-sections. (a) Arbitrary form. (b) Rectangular section.

section. For this reason, the complementary shear strains that act in the longitudinal direction will also differ along the perimeter, and indeed over the whole area of the cross-section, because of material continuity requirements. This is concomitant with longitudinal (out-of-plane) distortions of each cross-section and is termed *longitudinal* or *torsional warping*. We are now in a position to justify the use of the term *uniform torsion*. It is used to signify that cross-sections along the length of the member all deform in an identical manner.

In order to calculate the polar moment of inertia I_p about an axis perpendicular to the plane of other cross-sections, the *perpendicular axis theorem* is invoked:

$$I_p = \int_A \rho^2 dA = \int_A (y^2 + z^2) dA = \int_A y^2 dA + \int_A z^2 dA = I_y + I_z \tag{6.97}$$

where I_y and I_z are the area moments of inertia with respect to the in-plane perpendicular axes, Oy, Oz (Figure 6.39b). This result is based on the application of Pythagoras' theorem. For a rectangular solid section,

$$I_y = \int_A y^2 dA = \int_{-b/2}^{+b/2} \int_{-a/2}^{+a/2} y^2 dy dz = b \int_{-a/2}^{+a/2} y^2 dy = \frac{ba^3}{12} \tag{6.98a}$$

$$I_z = \int_A z^2 dA = \int_{-b/2}^{+b/2} \int_{-a/2}^{+a/2} z^2 dy dz = a \int_{-b/2}^{+b/2} z^2 dz = \frac{ab^3}{12} \tag{6.98b}$$

Using Equations (6.98a) and (6.98b),

$$I_p = I_y + I_z = \frac{ba^3}{12} + \frac{ab^3}{12} = \frac{ba}{12}(b^2 + a^2) \tag{6.99}$$

Equation (6.96) was derived for circular sections in which no warping arises, and as a result it is not valid for non-circular sections, for which approximate expressions are used. For a rectangular solid section the following expression is used:

$$I_p = \beta a b^3 \tag{6.100}$$

where β depends on the aspect ratio a/b and a and b are the lengths of the long and short sides of the cross-section, respectively.

a/b	1.0	1.5	2.0	2.5	3.0	4.0	5.0	10.0	∞
β	0.141	0.196	0.229	0.249	0.263	0.281	0.291	0.312	0.333

It is seen that for a thin section ($a/b>10$) consisting of a single member

$$I_p = \frac{1}{3} a b^3 \tag{6.101}$$

6.3.6 Uniform Torsion of Closed Thin-Walled Sections

In this section the response of a closed hollow section to a torque will be considered. In the case of rectangular or other prismatic sections, simple exact solutions cannot be obtained as in the case of the circular shaft discussed previously. It is instead necessary to consider the equations of equilibrium, of geometrical compatibility and the constitutive relations. In this section a simplified analysis will be presented, based only on satisfying the equilibrium conditions. The following assumptions are made:

1. Axial constraints are not present. The member is thus free to distort in the axial direction and thus uniform torsion arises.
2. Shear stresses normal to the beam surface may be neglected since they are zero at every point of each surface and the wall is thin. The term 'thin' denotes that the length of each member that forms part of the perimeter is large compared to the thickness of that member.
3. Shear stresses acting on planes normal to the surface of the member are constant through the thickness of the member.
4. Direct stresses normal to the section surfaces, both interior and exterior, are zero as are the stresses through the thickness of the member.
5. The direct stresses σ_x and σ_s are not necessarily zero. However, σ_s can be seen to be equal to zero from Figure 6.40. Regarding σ_x, in the case of uniform torsion we have that the axial stress resultant is zero. We may therefore conclude that the stress in the axial direction is of minor importance, and we can assume that it is zero along the perimeter of the member.
6. The beam is prismatic. That is, the thickness may vary along the girth of a section but remains constant in the axial direction of the member.
7. When calculating section constants, squares and higher powers of the thickness t are neglected.

We proceed by considering an element $\delta s \delta x$ of the member. As a consequence of the assumptions made above, the element is at equilibrium under the system of internal forces

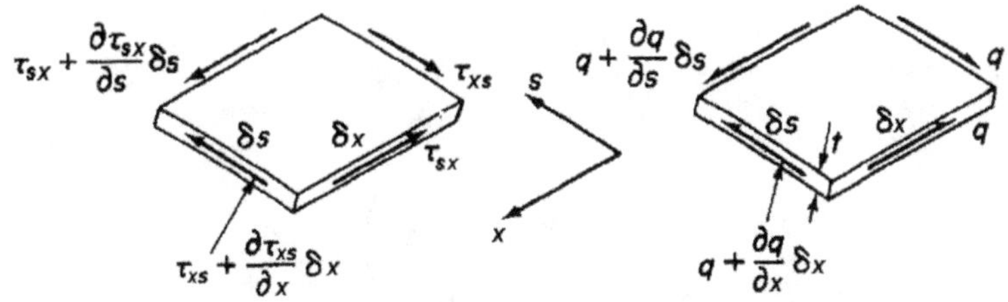

Figure 6.40 Shear stresses and shear flows on an element of a thin-walled section in torsion [6].

shown in Figure 6.40. At a convenient point of the transverse section an origin is chosen for a parameter s that traces a path around the girth of the member.

The internal stress resultant that ensures equilibrium at any section is the integral of the shear stresses τ_{xs}. From considerations of rotational equilibrium of the element, complementary shear stresses τ_{sx} arise in the longitudinal direction of the member. From the equations of equilibrium (6.76a) and (6.76c), $\tau_{xs} = \tau_{sx}$. The analysis is simplified if it is carried out in terms of the shear flow q. Along the girth of the section,

$$\left(\tau_{sx} + \frac{\partial \tau_{sx}}{\partial s}\delta s\right) t\delta x - \tau_{sx} t\delta x = 0$$

or,

$$\left(q + \frac{\partial q}{\partial s}\delta s\right)\delta x - q\delta x = 0$$

and

$$\frac{\partial q}{\partial s} = 0 \tag{6.102}$$

Hence the shear flow remains constant around the cross-section. This is the case if the torque is resisted by shear stresses only. The equation of equilibrium in the x direction, Equation (6.76a), gives

$$\frac{\partial q}{\partial x} = 0 \tag{6.103}$$

Hence the shear flow remains constant in the x direction as well.

In order to determine the magnitude of the shear flow it is necessary to apply the equilibrium condition between the (external) applied load T_{ext} and the (internal) stress resultant T_{int}. In this case the internal stress resultant is the integral of the moments of the shear stresses all along the perimeter of the wall of the section, about the centre of twist. Therefore,

$$T_{ext} + T_{int} = 0 \tag{6.104}$$

where

$$T_{ext} = T$$

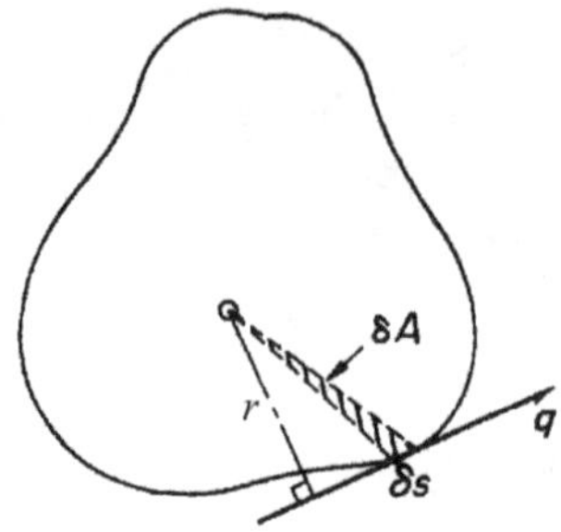

Figure 6.41 Swept area δA corresponding to an element δs of the wall section [6].

and

$$T_{int} = \oint rqds \tag{6.105}$$

r is the perpendicular distance from the tangent to the wall to the centre of twist (Figure 6.41). The circular integral denotes integration around the wall of the section (s being the distance measured from a selected origin). Therefore,

$$T + \oint rqds = 0 \tag{6.106}$$

This equation can be written as

$$T = \oint rqds \tag{6.107}$$

provided account of the direction of the shear flow q is taken.

In Figure 6.41 an element of the section wall of length ds located at a distance r, over which a shear flow of magnitude q is acting, is considered. The swept area δA is found from

$$\delta A = \frac{1}{2}(r\delta s)$$

thus,

$$\oint dA = \frac{1}{2}\oint rds$$

or,

$$\oint rds = 2A \tag{6.108}$$

Since q is constant,

$$T = 2Aq \tag{6.109}$$

This expression, termed the *Bredt-Batho* formula, relates torque to the resulting shear flow for a member subjected to uniform torsion. The shear flow is thus determined from

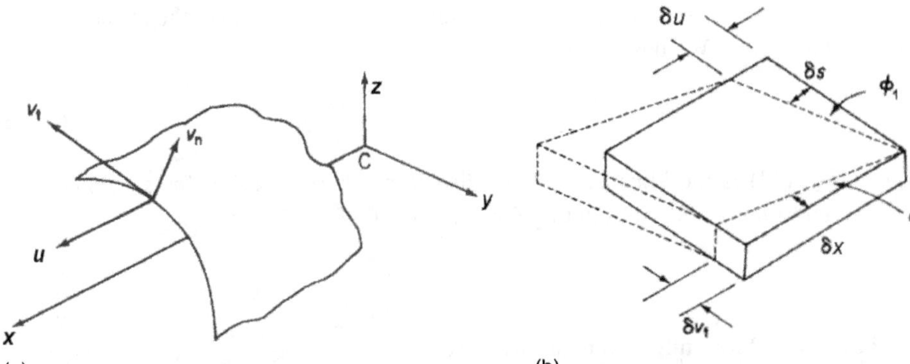

Figure 6.42 Definition of displacement components and shear strain of a member subjected to axially unrestrained (pure) torsional loading [6]. (a) Axial, tangential and normal. (b) Shear strain γ in terms of axial and displacement components (w, v_t, v_n) tangential displacement components.

$$q = \frac{T}{2A} \tag{6.110}$$

The shear stress at any point along the girth, τ_{sz} is then

$$\tau_{sz} = \frac{T}{2At} \tag{6.111}$$

We shall now obtain expressions for the kinematic (strain–displacement) relationships for the member considered.

The constitutive relation Equation (6.91) can be used to relate the shear strain $\gamma \ (= \gamma_{sz} = \gamma_{zs})$ in the member to the shear stress $\tau_{sx} \ (= \tau_{xs})$. Alternatively, both axial and shear strains (ε_z, γ) can be obtained in terms of the three components of the displacement of a point in the section wall (Figure 6.42a). These components are:

- v_t, which is a tangential displacement in the yz plane, taken to be positive in the direction of increasing s.
- v_n, which is a normal displacement in the yz plane, positive outwards.
- u, which is a displacement in the axial ($x-$) direction. This is termed the *warping displacement*.

The direct strain in the axial direction ε_x is obtained directly from Equation (6.77a):

$$\varepsilon_x = \frac{\partial u}{\partial x} \tag{6.112}$$

For axially unrestrained torsion, $\varepsilon_x = \sigma_x = 0$. This means that longitudinal fibres do not change in length, as they would if restrained at the ends. However, they translate relative to one another, resulting in longitudinal warping of each cross-section. In uniform torsion all transverse sections along the length of the member undergo the same form and degree of distortion.

The shear strain γ_{sx} is obtained by considering the shear distortion of the element of area $\delta s \delta x$ (Figure 6.42b). We have

$$\gamma_{sx} = \phi_1 + \phi_2 = \frac{\partial u}{\partial s} + \frac{\partial v_t}{\partial x} \tag{6.113}$$

Equation (6.113) is the kinematic (strain–displacement) relationship for a member subjected to uniform torsion. Substituting for v_t in Equation (6.113), we get

$$\gamma_{sx} = \frac{\partial u}{\partial s} + r\frac{\partial \phi}{\partial x} \tag{6.114}$$

Since $\partial w/\partial x = 0$ for uniform torsion, we have

$$\oint dw = \frac{q}{G}\oint \frac{ds}{t} - \frac{d\phi}{dx}\oint rds \tag{6.115}$$

where all circular integrals are taken around the girth of the section. The first integral must be zero, since there will be no relative displacement at the origin of integration. The third integral is equal to 2*A*. Thus,

$$\phi = \frac{q}{2AG}\oint \frac{ds}{t} = \frac{T}{4A^2G}\oint \frac{ds}{t} \tag{6.116}$$

This is known as *Bredt's second equation*, and it relates the angle of twist φ to the applied torque *T* for any particular type of cross-section. The quantity $4A/\int ds/t$ is constant and is characteristic for any particular section. It is termed the *torsion constant*, *J*, of the section:

$$J = \frac{4A}{\oint \frac{ds}{t}} \tag{6.117}$$

If *t* is constant,

$$J = \frac{4A^2t}{S} \tag{6.118}$$

where *S* is the girth (peripheral) length.

To assess the effect of shape on torsional constant and hence on torsional rigidity, consider two sections whose cross-sectional area is the same (*A*). The first section is square and for the second the aspect ratio is (4a)/(a/4) = 16.

The shear stress τ_{sz} is calculated from Equation (6.111) and is thus seen to be the same in both cases. However, for the square box, the angle of twist is

$$\theta = \frac{T}{4a^4G}\frac{4a}{t} = \frac{T}{a^3Gt} \tag{6.119a}$$

whereas for the rectangular box,

$$\theta = \frac{T}{4a^4G}\frac{8a + a/2}{t} = \frac{17}{8}\frac{T}{a^3Gt} \tag{6.119b}$$

which is more than twice that of the square box. Thus long narrow boxes are torsionally much more flexible and are thus inefficient in resisting torsional deformation. If the structure is a combination of one closed cell and a number of open sections, the torsional constant J is given by

$$J = \frac{4A^2}{\oint \frac{ds}{t}} + \sum_{i=1}^{n} \frac{1}{3} bt^3 \tag{6.120}$$

The first term relates to the closed section whereas the second refers to the open sections. The rate of twist $\partial\varphi/\partial x$ is obtained from $\partial\varphi/\partial x = T/GJ$ and is the same for all cells and open members comprising the section. The shear stresses in the open sections are calculated from

$$\tau_i = \pm Gt\frac{\partial\phi}{\partial x} \tag{6.121}$$

In the case of a single cell the shear stress is found from Equation (6.111). In this case the value of T is that borne by that cell alone, T_c.

6.3.7 Warping of an Open Thin-Walled Section

We shall now consider the effect that the application of a torque has on the geometry of a uniform open-channel section that is free to distort in the axial direction at its ends. This may be treated as a first approximation to the response of the cargo region of an open deck ship. A more accurate approximation would need to take into account the presence of transverse bulkheads and ancillary structure (cross-deck beams, etc.) as well as closed deck structures, although it has been found that the most important contribution of transverse bulkheads is the retention of form. Transverse bulkheads do not produce any significant reduction in twisting angle, in contrast to the closed deck structures that are located at the fore and aft ends and the engine room spaces.

Before considering the channel section it is necessary to describe the method used to determine the distribution of longitudinal warping deflections in open sections subjected to torsional loading.

A general expression for the longitudinal warping of an open section, u, will be obtained, starting from the basic kinematic assumptions made concerning the nature of the distortions that arise when a torque is applied. These are indicated in Figure 6.42, and the corresponding expression for the shear strain γ_{sx} is Equation (6.114).

If αx is the infinitesimal angle of rotation about the centre of twist, O, (Figure 6.43) the tangential displacement η is given by

$$\eta \cong r\alpha x \tag{6.122}$$

From Equations (6.114), (6.91) and (6.122), the following expression is obtained:

$$\frac{\partial u}{\partial s} = \frac{\tau_{sx}}{G} - r\alpha \tag{6.123}$$

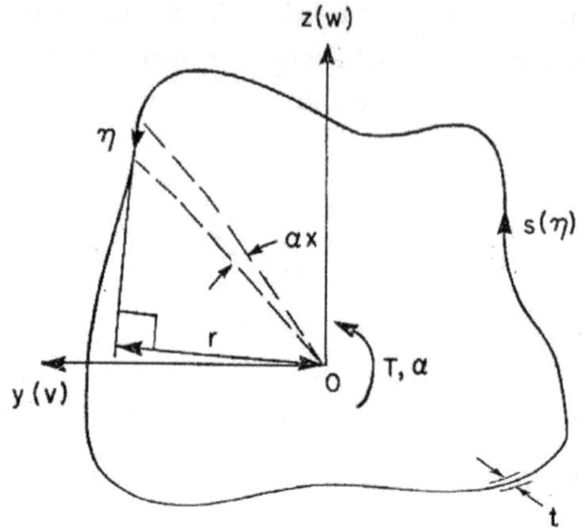

Figure 6.43 Definition of angle of twist α, tangential displacement η and normal to tangent r [2].

Multiplying both sides by ds and integrating gives

$$\int du = \int \frac{\tau_{sx}}{G} ds - \alpha \int r ds$$

or

$$u = \frac{1}{G} \int \tau_{sx} ds - 2\alpha\omega(s) + u_O \tag{6.124}$$

where $\omega(s)$ is a *sectorial area* defined by the following expression:

$$\omega(s) = \frac{1}{2} \int_0^s r ds \tag{6.125}$$

and u_O is the warping displacement at $s = 0$. If therefore the origin is chosen at the point at which $u = 0$, $u_O = 0$.

The sectorial area $\omega(s)$ is a quantity that corresponds to each point of the section at a distance s along the periphery from the origin ($s = 0$) (Figure 6.44). It is the area swept out by a generator that joins a chosen point (*pole*), such as the centre of twist, which in this case is the centre of shear (SC), to each point of the periphery. The sectorial area is bounded by the generator, the tangent to the point in question and the axis joining the pole to the origin ($s = 0$). The quantity r is the distance from the pole (SC) along the normal to the tangent to the tangent (point A) (Figure 6.44).

The sign of the warping displacement u for a given direction of torque depends on the sign of $\omega(s)$, which in turn depends on the sign of r. This depends on that of the torque, which is assumed to be positive in the anticlockwise direction. Therefore r, and, hence, $\omega(s)$ are positive if the movement of the foot of r (point A) along the tangent in the assumed direction of s leads to an anticlockwise rotation of r about the centre of twist.

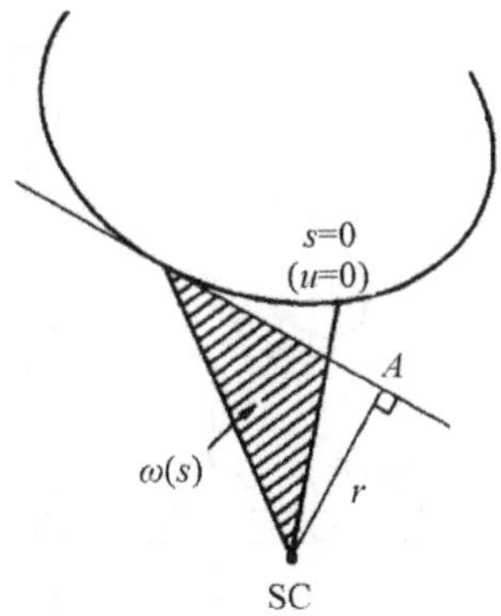

Figure 6.44 Sectorial area $\omega(s)$ in an open section.

We introduce u_r as the *relative longitudinal warping displacement* with respect to the plane that is perpendicular to the x-axis and contains the origin of the s coordinate, such that

$$u_r = u - u_0 \tag{6.126}$$

giving

$$u_r = \frac{1}{G}\int \tau_{sx} ds - 2\alpha\omega(s) \tag{6.127}$$

The first term on the right-hand side of Equation (6.127) is a shear term and the second is a twisting term,[12] so that the warping displacement is seen to be the result of these two load effects. In the case of open sections the shear strain γ_{sx} is small and the twisting term dominates the response. In contrast, in the case of closed sections this is not the case, and the resulting warping displacement is significantly smaller. In the case of open sections we can disregard this term so that

$$u_r = -2\alpha\omega(s) \tag{6.128}$$

is an accurate approximation of the warping displacement.

For an open section the positive direction of s may be chosen arbitrarily, since for a given torque the sign of u depends only on the sign of $\omega(s)$. Also, the pole used to trace the generator which in this case is the centre of shear can be any point on the plane. At a different location r and $\omega(s)$ would differ (say, from r to r_I and from $\omega(s)$ to $\omega_I(s)$). Thus, new values of the (relative) warping displacement would be obtained, $u_1 - u_0{}'$, where

$$u_1 - u_0 = \frac{1}{G}\int \tau_{sx} ds - 2\alpha\omega(s)$$

The actual warping displacement must be the same in each case so that

$$u_1 = u + (u_0{}' - u_0) + 2\alpha(\omega(s) - \omega_1(s))$$

Since $u_O{}', u_O$ are constants, $u_O{}' - u_O$ is simply a rigid body motion displacement of the point $s = 0$ in the x direction. Similarly the term $2\alpha(\omega(s) - \omega(s_I))$ is a longitudinal

[12] This is clear from Equation (6.110) from which Equation (6.122) has been derived.

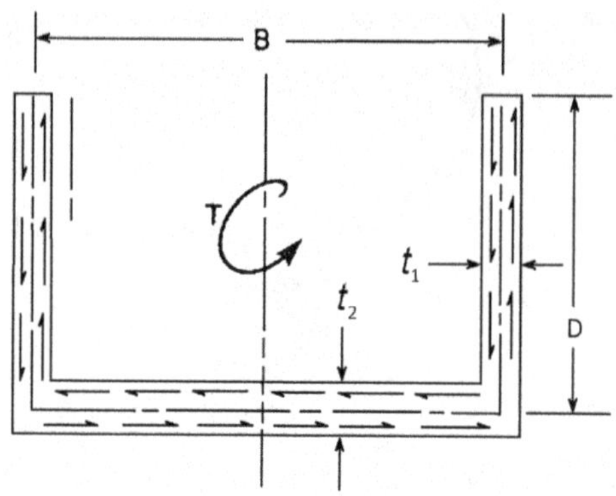

Figure 6.45 Channel section acting as a structural model of an open-deck ship.

displacement due to a rigid body displacement of the cross-sectional plane. Thus u_l and u differ only by rigid body displacements and not as a result of changes of stress, so that the choice of axis is arbitrary and we can work with a relative warping displacement.

6.3.8 Warping of a Channel Section

Figure 6.45 shows a structural model that can be adopted for the section of an open-deck ship undergoing torsion. In this model the structure is represented by solid members of constant thickness along the bottom (web) and sides (flanges). In a containership the side and bottom structures have a cellular construction, the purpose of this being to ensure stability at a local level. As discussed in Section 7.2, the cellular construction does not enhance the torsional rigidity of the hull girder cross-section to any significant extent. It is therefore possible, in principle, to model the structure with sections of lengths D, B and thicknesses t_l, t_2, respectively, such that they have a torsional rigidity equal to that of the real structure.

When a torque T is applied to a channel section, significant longitudinal warping of the section arises, since the centre of shear and the neutral axis of bending do not coincide (Figure 6.34). From strength of materials the following expression for the distance of the centre of shear from the baseline is obtained:

$$e = \frac{3D^2 t_1}{6Dt_1 + Bt_2} \tag{6.129}$$

The torsional constant of the section is given by

$$J = \frac{1}{3} B{t_2}^2 + \frac{2}{3} D{t_1}^3 \tag{6.130}$$

Hence,

$$\alpha = \frac{T}{GJ} = \left(\frac{3}{G\left(2D{t_1}^3 + B{t_2}^3\right)} \right) T \tag{6.131}$$

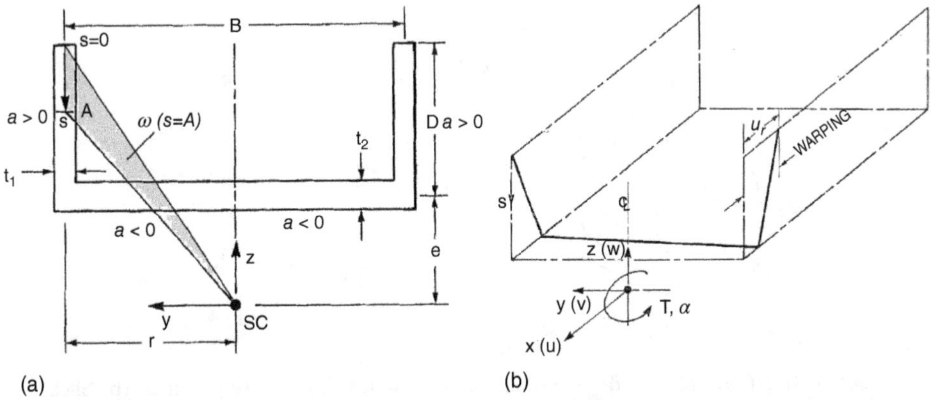

Figure 6.46 Longitudinal warping of channel section subjected to torque T [2]. (a) Sectorial area $\omega(s = A)$ and sign of α. (b) Distribution of longitudinal warping displacements.

Substituting Equations (6.131) and (6.125) in Equation (6.128), the following expression for the relative warping displacement of the channel section is obtained:

$$u_r = -\left(\frac{3}{G(2Dt_1{}^3 + Bt_2{}^3)}\right)T\int_0^s r\,ds \qquad (6.132)$$

In this problem the origin of s is specified to be at the upper deck on portside (Figure 6.46a). Applying the sign convention described earlier, the sign of α in each member of the section is as indicated in Figure 6.46a.

The sectorial area $\omega(s = A)$ of point A lying in the range $(0 \le s \le D)$ is indicated by the shaded area in Figure 6.46a. For all such points $r = B/2$, and thus their relative warping displacement is given by

$$u_r = -\left(\frac{3}{G(2Dt_1{}^3 + Bt_2{}^3)}\right)T\left(s\frac{B}{2}\right) \qquad (6.133a)$$

Correspondingly,

$$u_r = -\left(\frac{3}{G(2Dt_1{}^3 + Bt_2{}^3)}\right)T\left(\frac{BD}{2} - (s - D)e\right) \qquad (6.133b)$$

when $D \le S \le B + D$.

$$u_r = -\left(\frac{3}{G(2Dt_1{}^3 + Bt_2{}^3)}\right)T\left(\frac{BD}{2} - Be + (s - B - D)\frac{B}{2}\right) \qquad (6.133c)$$

when $B + D \le s \le B + 2D$.

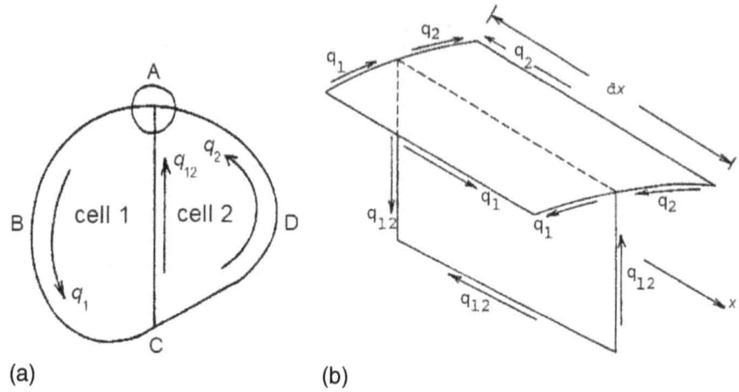

Figure 6.47 Two-cell closed section and shear flow [8]. (a) Two-cell section. (b) Shear flow along junction *A*.

We note that the relative warping displacement varies linearly along all members of the section and depends on *s*. The distribution of u_r is shown in Figure 6.46b and is seen to be anti-symmetrical about the centreline plane, at which point $u_r = 0$.

6.3.9 Uniform Torsion of a Section Consisting of Two Cells

In this section we will consider an arrangement encountered in a large number of ship-type cross-sections, albeit in simplified structural models. The arrangement shown in Figure 6.47a involves a cross-section that consists of two hollow closed *cells* (1 and 2) that have a common boundary (*AC*).

The analysis of this problem differs from that of the single hollow closed section, because this is a statically indeterminate problem which necessitates the use of at least one additional equation in order to determine the stress distribution in the section. The purpose of the analysis is to determine the shear flow in each wall of the section and thence the rate of the angle of twist of the entire section, $\partial\theta/\partial x$. It is assumed that the section is fixed at one end and that it is constrained to maintain its form all along the length of the member. As a result, all cells rotate through the same angle, and the rate of twist remains constant for all cells along the length of the member (uniform torsion).

The shear flow in this two-cell section consists of flows q_1, q_2 and q_{12} along the common boundary. It is therefore necessary to obtain three equations to solve this problem. From longitudinal equilibrium considerations it is seen that the shear flow *q* remains constant along paths *ABC*, *AC* and *CDA*. For equilibrium along junction *A* (Figure 6.47b),

$$q_2 + q_{12} = q_1 \tag{6.134}$$

That is, the total incoming shear flow is equal to the total outgoing shear flow along this junction. It is thus possible to determine the actual shear flow in each wall of the cross-section by superposition of the shear flows of the two individual cells treated separately (Figure 6.48).

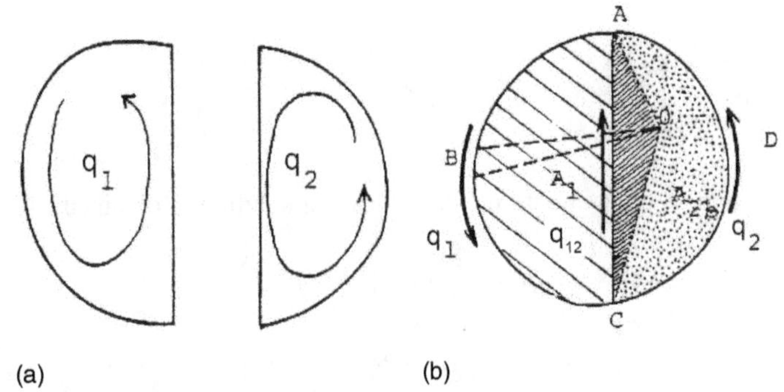

Figure 6.48 Individual cells and resulting shear flows [8]. (a) Shear flows in each cell. (b) Superposition of shear flows.

The total torque about any point O lying within the cross-section is then, from Equation (6.107),

$$T = q_1 \oint_{ABC} rdS + q_2 \oint_{CDA} rdS - q_{12} \oint_{CA} rdS$$

or else, using the parameters of Figure 6.48b and Equation (6.134),

$$\begin{aligned} T &= q_1 2(A_1 + 2A_{2a}) + q_2 2A_{2b} - (q_1 - q_2)2A_{2a} \\ &= q_1(2A_1 + 2A_{2a} - 2A_{2a}) + q_2(2A_{2b} + 2A_{2a}) \\ &= 2A_1 q_1 + 2A_2 q_2 \end{aligned} \tag{6.135}$$

At this stage we have obtained two of the three equations that are needed (Equations (6.134) and (6.135)). The third equation will be based on the assumption[13] that the angles of twist per unit length of each cell are equal to each other and to that of the whole section. That is,

$$\frac{\partial \phi_1}{\partial x} = \frac{\partial \phi_2}{\partial x} = \frac{\partial \phi}{\partial x} \tag{6.136}$$

where the angles of twist per unit length are given by Equations (6.116) and (6.110).

The third equation is then

$$\frac{1}{2GA_1}\left[q_1 \oint_{ABC} \frac{ds}{t} + q_{12} \oint_{CA} \frac{ds}{t}\right] = \frac{1}{2GA_2}\left[q_2 \oint_{CDA} \frac{ds}{t} - q_{12} \oint_{CA} \frac{ds}{t}\right] \tag{6.137}$$

Hence,

$$\frac{1}{A_1}(q_1\eta_1 + q_{12}\eta_{12}) = \frac{1}{A_2}(q_2\eta_2 - q_{12}\eta_{12}) \tag{6.138}$$

[13] This is known as the Saint Venant assumption concerning torsion of thin-walled members.

where

$$\eta_1 = \oint_{ABC} \frac{ds}{t} \qquad \eta_2 = \oint_{CDA} \frac{ds}{t} \qquad \eta_{12} = \oint_{CA} \frac{ds}{t} \tag{6.139a–c}$$

Solving the system of three equations, the following solution is obtained:

$$\left.\begin{aligned} q_1 &= \frac{T}{2H}(\eta_2 A_1 + \eta_{12} A) \\ q_2 &= \frac{T}{2H}(\eta_1 A_2 + \eta_{12} A) \\ q_{12} &= \frac{T}{2H}(\eta_2 A_1 - \eta_1 A_2) \end{aligned}\right\} \tag{6.140a–c}$$

where

$$\left.\begin{aligned} H &= \eta_1 {A_2}^2 + \eta_{12} A^2 + \eta_2 {A_1}^2 \\ A &= A_1 + A_2 \end{aligned}\right\} \tag{6.141a–b}$$

From the equations for the rate of twist of each cell,

$$\frac{\partial \phi_i}{\partial x} = \frac{q_i}{2GA_i}\eta_i \tag{6.142}$$

we have

$$\left.\begin{aligned} 2GA_1 \frac{\partial \phi_1}{\partial x} &= q_1 \eta_1 + q_{12}\eta_{12} \\ 2GA_2 \frac{\partial \phi_2}{\partial x} &= q_2 \eta_2 - q_{12}\eta_{12} \end{aligned}\right\} \tag{6.143a–b}$$

Adding these expressions and noting Equation (6.136), the angle of twist per unit length of the cross-section is obtained:

$$\frac{\partial \phi}{\partial x} = \frac{1}{2GA}[q_1 \eta_1 + q_2 \eta_2] \tag{6.144}$$

6.3.10 Uniform Torsion of an N-Cell Section

The procedure described in the previous section can be extended to the case in which there are more than two cells, making use of the following principles:

1. Along any junction, the total incoming shear flow is equal to the total outgoing flow. That is,

$$\sum_{i=1}^{n} q_i = 0 \tag{6.145}$$

where n is the number of branches at a particular junction.

2. The equation of equilibrium between external and internal forces is

$$T = 2\sum_{i=1}^{N} q_i A_i \tag{6.146}$$

where N is the total number of branches comprising a section.

3. Saint Venant's assumption concerning compatibility of angles of twist is valid. This results in the following N-1 equations:

$$\left.\begin{aligned} \frac{\partial \phi_1}{\partial x} &= \frac{\partial \phi_2}{\partial x} \\ \frac{\partial \phi_2}{\partial x} &= \frac{\partial \phi_3}{\partial x} \\ &\cdots\cdots \\ \frac{\partial \phi_{N-1}}{\partial x} &= \frac{\partial \phi_N}{\partial x} \end{aligned}\right\} \tag{6.147}$$

On this basis, the following procedure may be implemented in the case of a multi-cell section:

Determination of angle of twist of multi-cell section

1. Number the cells from 1 to N.
2. For the ith cell assume a shear flow q_i acting in the counter-clockwise direction.
3. At common cell junctions the shear flow is set equal to the algebraic sum of that of adjacent cells (principle of superposition).
4. For the N unknown shear flows write down N Equations (6.146) and (6.147).
5. Solve the system of N equations for the unknown shear flows $q_1, q_2, \ldots, q_N$. If the resulting sign of any shear flow q_i is negative, this indicates that the direction of flow is opposite to that originally assumed.
6. Calculate the angle of twist per unit length $\partial\varphi/\partial x$ as a function of the torque T and the shear stresses $\tau_i = q_i/t_i$

Example

The section shown in Figure 6.49 is a simplified model of a ship cross-section. Calculate the rate of angle of twist and the shear stresses in each branch of the section.

Solution

The enclosed areas in each cell are

$$A_1 = 4S^2$$

$$A_2 = 8S^2$$

$$A_3 = 4S^2$$

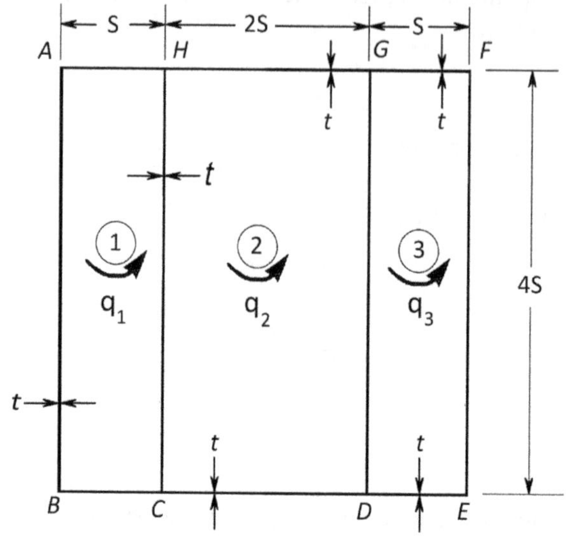

Figure 6.49 Simplified model of ship cross-section.

The internal-external equation of equilibrium is, using Equation (6.146),

$$T = 2\left(4S^2 q_1 + 8S^2 q_2 + 4S^2 q_3\right) \tag{a}$$

The rates of the angles of twist of each cell are

$$\frac{\partial \phi_1}{\partial x} = \frac{1}{2GA_1}\left[q_1 \oint_{HABC} \frac{ds}{t} + (q_1 - q_2)\oint_{CH} \frac{ds}{t}\right]$$

$$\frac{\partial \phi_2}{\partial x} = \frac{1}{2GA_2}\left[q_2\left(\oint_{GH} \frac{ds}{t} + \oint_{CD} \frac{ds}{t}\right) - (q_1 - q_2)\oint_{HC} \frac{ds}{t} + (q_2 - q_3)\oint_{DG} \frac{ds}{t}\right]$$

$$\frac{\partial \phi_3}{\partial x} = \frac{1}{2GA_3}\left[q_3 \oint_{DEFG} \frac{ds}{t} + (q_2 - q_3)\oint_{GD} \frac{ds}{t}\right]$$

These can be written in the following form:

$$2GA_1 t\frac{\partial \phi_1}{\partial x} = 6Sq_1 + 4S(q_1 - q_2) \tag{b}$$

$$2GA_2 t\frac{\partial \phi_2}{\partial x} = 4Sq_2 - 4S(q_1 - q_2) + 4S(q_2 - q_3) \tag{c}$$

$$2GA_3 t\frac{\partial \phi_3}{\partial x} = 6Sq_3 - 4S(q_2 - q_3) \tag{d}$$

The equations of compatibility for the angles of twist are

$$\frac{\partial \phi_1}{\partial x} = \frac{\partial \phi_2}{\partial x} \qquad \frac{\partial \phi_2}{\partial x} = \frac{\partial \phi_3}{\partial x}$$

Thus,

$$3q_1 - \frac{5}{2}q_2 + \frac{1}{2}q_3 = 0 \tag{e}$$

$$-\frac{1}{2}q_1 + \frac{5}{2}q_2 - 3q_3 = 0 \tag{f}$$

Solving the system of equations comprising of (a), (e) and (f) leads to the following solution:

$$q_1 = q_3 = 0.0260\frac{T}{S^2} \qquad q_2 = 0.0365\frac{T}{S^2} \tag{g}$$

Since

$$\frac{\partial\phi_1}{\partial x} = \frac{\partial\phi_2}{\partial x} = \frac{\partial\phi_3}{\partial x} = \frac{\partial\phi}{\partial x}$$

from any of equations (b)–(d) we obtain

$$\frac{\partial\phi}{\partial x} = 0.0143\frac{T}{GtS^3}$$

and thus

$$T = 69.818GtS^3\frac{\partial\phi}{\partial x}$$

The torsion constant is $J = 69.818tS^2$

The shear stresses in the section can now be readily calculated:

$$\tau_1 = \tau_3 = 0.0260\frac{T}{tS^2}$$

$$\tau_2 = 0.0365\frac{T}{tS^2}$$

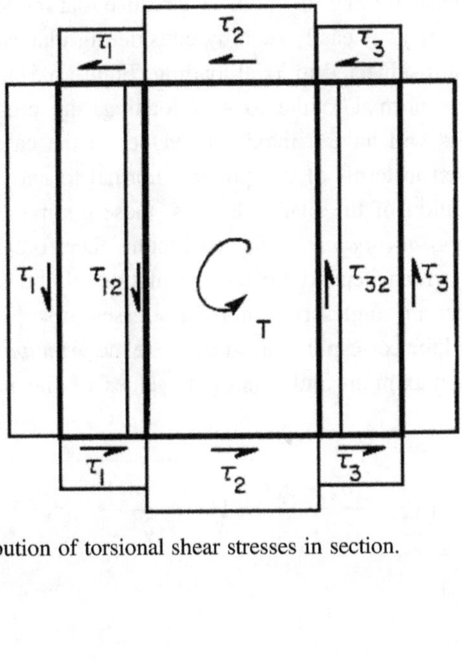

Figure 6.50 Distribution of torsional shear stresses in section.

$$\tau_{12} = \frac{q_{12}}{t} = \frac{q_1 - q_2}{t} = -0.0105\frac{T}{tS^2}$$

$$\tau_{32} = \frac{q_3 - q_2}{t} = \frac{q_1 - q_2}{t} = \tau_{12} = -0.0105\frac{T}{tS^2}$$

It is seen from Figure 6.50 that the largest shear stresses due to torsion arise in the upper and lower branches of the centre cell. In contrast to shear stresses due to bending, the shear stresses due to torsion remain constant between junctions.

6.4 Determination of Critical Regions of the Hull Girder for Longitudinal Strength

Elementary strength of materials theory shows that in thin-walled structures, stress fields consist of normal as well as shear stresses, so that the stress-at-a-point results from the combined action of these. In order therefore to be able to assess whether yielding has occurred at any particular position within the structure it is necessary to calculate the stress-at-a-point at that position. This stress can then be compared with the yield stress of the material in order to check whether the structure has failed by yielding or not.

6.4.1 Principal Stresses

The elementary definition of *stress* is ratio of force acting normally per unit area of a body. This is useful when considering problems such as pin-jointed trusses that consist of bars that are subjected to axial loading. In such cases the stress within each truss is simply the load divided by its cross-sectional area. However, if we analyse any point of an *extended structure* subjected to external loading and consider the state of stress with respect to a number of planes that pass through this point, we realise that the stress components differ for each of these planes. This is easily seen by considering planes that are inclined to the axis of loading of a beam subjected to axial loading (Figure 6.51).

For the plane that is normal to the axis of loading, the cross-sectional area is the minimum and the stress will have a maximum value. In the case of an inclined plane, the load can be analysed in terms of components normal to and tangential to the plane. As the angle of inclination of the plane changes, these components vary in magnitude and direction. The *stress-at-a-point* within a structure therefore has to be described in terms of components, which depend on the orientation of the surface with respect to the applied load. At certain angles only normal stresses arise. These angles are called *principal directions*, and the corresponding stresses are the *principal stresses*. The principal stresses σ_1, σ_2 are the maximum and minimum values of the normal stresses and are given by

$$\sigma_{1,2} = \frac{(\sigma_x + \sigma_y)}{2} \pm \sqrt{\frac{(\sigma_x - \sigma_y)^2}{2} + \tau_{xy}{}^2} \tag{6.148a}$$

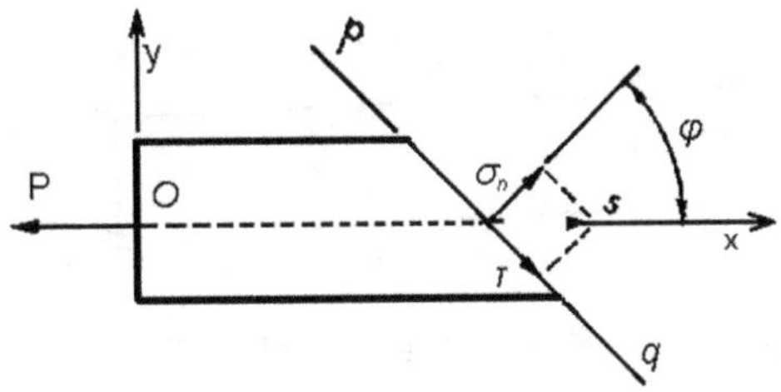

Figure 6.51 Force components acting normally and tangentially to ail inclined plane [3].

The maximum shear stress is given by

$$\tau_{max} = \frac{\sigma_1 - \sigma_2}{2} \tag{6.148b}$$

The principal angle is given by

$$\tan 2\theta_p = -\frac{2\tau_{xy}}{\sigma_x - \sigma_y} \tag{6.148c}$$

6.4.2 Plane Stress Condition: Yield Criterion

In the case of a hull girder of usual geometry, the normal stress σ_y can be ignored, since the weight and buoyancy effectively cancel each other out over a large proportion of the length. Equations (6.148) then become

$$\sigma_{1,2} = \frac{\sigma_x}{2} \pm \frac{1}{2}\sqrt{\sigma_x{}^2 + 4\tau_{xy}{}^2} \tag{6.149a–b}$$

$$\tau_{max} = \frac{\sigma_1 - \sigma_2}{2} \tag{6.149c}$$

$$\tan 2\phi = -\frac{2\tau_{xy}}{\sigma_x} \tag{6.149d}$$

Tensile test results show that steel fails when the shear stress reaches a critical value at 45° to the axis of loading. For such a material the most accurate theory of failure is that of von Mises. It is important to establish, for individual ship geometries, the critical location at which material failure will first occur under longitudinal bending, especially when calculating the ultimate longitudinal strength of the hull girder, since yielding will not necessarily arise in way of the midship section where the bending stresses are at their maximum values.

6.4.3 Principal Stress Fields in a Box Girder

Hull girder sections are approximated by box girders or I-beams (Figure 6.52). For such cross-sections, the shear stress distribution becomes more complex. The principal stresses will have lower values because of the increased moment of inertia of the section.

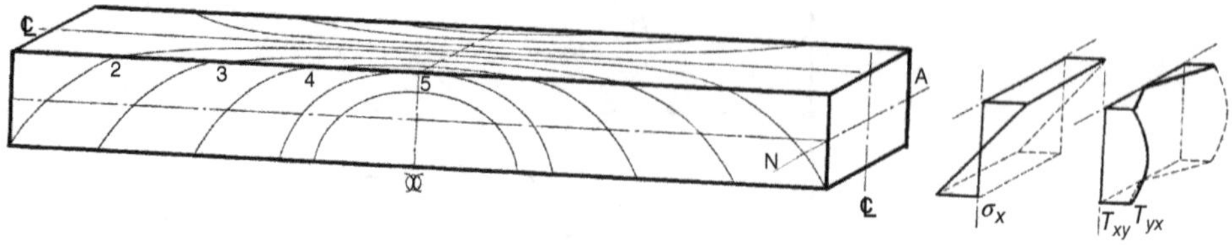

Figure 6.52 Principal stress field in a box girder in bending [2].

A study of principal stresses in box girders reveals that:

1. If the decks and sides have the same thickness, the normal stresses σ_x and the shear stresses τ_{xy} are equal in both members. As a result, the deck edge angles and the magnitudes of the principal stresses are equal in the deck and side along the intersection.
2. Orthogonal shear stresses τ_{xy} and τ_{yx} are at their maximum along the neutral axis. Thus the principal trajectories cross the neutral axis at 45°. In way of the deck they are parallel to the centreline.
3. Principal stresses are uniform and high in way of the deck plating amidships.

The maximum shearing stress, which is a measure of resistance to yielding, may arise in way of the sections with maximum vertical shearing forces (near the quarter-length positions) or in way of maximum bending moments (near amidships). This latter observation holds not only for box girders but also for ship-type sections. An investigation into the effect of ship geometry for ship-type sections showed that [2]:

1. Critical locations may arise at (a) the deck amidships, (b) the deck at quarter-length positions or (c) the neutral axis at quarter-length positions.
2. The actual critical location is dependent on ship geometry. Thus, for ships with low *L/D* ratio ('deep' ships), the critical location is the quarter length position in way of the neutral axis, provided that the deck area is greater than the side shell area.
3. Changes to the prismatic form do not produce any noticeable changes.
4. For 'shallow' ships the deck at the quarter-length positions are more critical than the neutral axis position at quarter-length until area ratios become large.
5. Very small *L/D* values make the critical stress locations fall other than at the deck edge amidships. The same occurs with large flange (deck) areas or low web (side shell) areas.

6.4.4 Design Considerations

The simplest way to avoid structural failure is by ensuring that the stresses will not exceed a certain value at all points of a transverse section. This limit is usually a fraction of the yield strength of the material, which in the case of ships is normally some type of steel alloy. In principle, failure can also arise from instability (local buckling), but since plate slenderness in ship structures is such that elasto-plastic failure predominates, it is inappropriate to consider elastic instability. Thus, if it can be ensured that the stresses are limited to the elastic range, and below the inelastic buckling level, failure can be avoided. This can easily be done by imposing limits to the section modulus along the length of the hull girder. This is the traditional approach and follows naturally from beam theory. Thus,

$$Z_{des} = \frac{I}{y} \geq \frac{M_{des}}{\sigma_{des}} \tag{6.150}$$

where Z_{des} is the section modulus of the complete section at any point along the length of the hull girder. Classification society rules have been based on Equation (6.37) given that a permissible (design) stress has been specified and that the bending moment calculation includes an appropriate factor of safety.

References

1. Caridis P. A. (2017) *Ship structures – local strength problems*) (in Greek) (Η Μεταλλική Κατασκευή του Πλοίου. Θέματα Τοπικής Αντοχής) 2nd ed. Athens.
2. Evans J. H. (1983) *Ship Structural Design Concepts. Second Cycle*. Centreville MD: Cornell Maritime Press.
3. Gere J. M. Timoshenko S. P. (1986) *Mechanics of Materials* 2nd ed. Delhi: CBS Pub. & Distributors.
4. Getz J. R. (1971) *Longitudinal Strength of the Hull Girder*. Translated by F.C. Michelsen Rept. No. 119 Dept Nav. Arch. Mar. Eng. Univ. of Michigan.
5. International Association of Classification Societies (2020) *Common Structural Rules for Bulk Carriers and Oil Tankers*.
6. Megson T. H. G. (1999) *Aircraft structures for engineering students*. 3rd ed. Oxford: Butterworth Heinemann.
7. Moe J. (1971) *Analysis of Ship Structures. Part 2: Bending Torsion and Buckling*. Transl. Michelsen F.C. Rept. No. 070 Dept Naval Arch. & Mar. Eng. Univ. of Michigan.
8. Papazoglou V. Bardis L. (1990) *Statics and Dynamics of the Ship* I (Στατική και Δυναμική Πλοίου I) (in Greek). Athens: Dept Naval Arch. & Mar. Eng. Nat. Tech. Univ.
9. Robb A. M. (1952) *Theory of Naval Architecture*. Glasgow: Charles Griffin and Co.
10. Shigunov V. von Graefe A. Moctar O. (2014) Calculation of horizontal sectional loads and torsional moment. *J. Offsh. Mech. Arc. Eng.* Vol. 137 Issue 2 021603-1.
11. Taylor J. L (1924) The theory of longitudinal bending of ships. *Trans NECIES* Vol. XLI 123–144.
12. Vasta J. (1958) Lessons Learnt from Full-Scale Ship Structural Tests. *Trans SNAME* Vol. 64, 165–243.

7

Nonlinear Response to Primary Loading

Everything must be made as simple as possible. But not simpler.

Albert Einstein

7.1 Introduction

This is the second chapter in which hull girder response to primary loading is described. In the previous chapter longitudinal bending, shear and torsion were treated as individual load effects. In practice, however, in many cases they are *coupled*, and in order to develop an understanding that is closer to real conditions it is necessary to treat them as such. Coupling arises if the reference point, or axis, differs for each of the load types. An example we shall consider in this chapter is the torsional loading of an open-deck ship. Torsional loading of a hull girder is the result of the changing value and sign of the net horizontal force that acts at each section along the length of the hull girder (Chapter 4). As a result, torsion is accompanied by horizontal bending, which takes place about the corresponding vertical neutral axis. It is seen that each load system has an effect on the other, and coupling of the two load systems occurs. The response is *geometrically nonlinear*, which is reflected in the mathematical modelling of the problem; a system of coupled partial differential equations needs to be solved.

In addition to geometrical nonlinearity, *material nonlinearity* can also arise, but in circumstances that are unrelated to torsional loading. These concern events that a ship is not expected to encounter throughout its operating life but are nevertheless part of the reality of ship operations. They concern, firstly, serious *grounding* and *collision accidents* in which part of the bottom or side shell structure is no longer able to support external loads; and secondly, *hull girder collapse*, which involves progressive elasto-plastic buckling and yielding. In the former case it is the *residual strength* of the hull girder that is calculated.

The practical need to perform such calculations is, in the case of residual strength, for classification societies and other bodies to be able to provide assistance to ship operators should an accident occur. Having an accurate estimate of the ultimate strength of the hull girder permits a designer to be able to assess the true safety margin against overall collapse of a ship. Additionally, techniques such as the ones described in the later sections of this chapter can contribute to safer and more robust ships, if accounted for at the design stage.

7.2 Torsion of the Hull Girder

The sources of torsional loading of ships were discussed in Chapter 4, and in the previous chapter it was mentioned that the most important issue with regard to the torsional strength of a hollow, thin-walled structure is whether it is open or closed. Both types arise in ship structures, as is evident from Chapter 1. As a result, torsional strength varies considerably and in certain cases is of critical importance. This is particularly so in the case of dry cargo ships such as large containerships whose holds have very little deck structure, in contrast to liquid cargo carriers whose strength deck extends over the full length and breadth of the cargo region.

7.2.1 Structural Arrangements and Hull Girder Torsional Strength

From the elementary theory of torsion Equation (6.90), we see that the twisting angle θ is proportional to the applied torque and inversely proportional to the torsional rigidity GJ. When a ship is at sea, the applied torque depends on the angle of incidence of the waves, the speed and other variables that are unrelated to the structural arrangement. The torsional rigidity, on the other hand, depends exclusively on the structural arrangement and for this reason can vary considerably. It is instructive to compare the torsional constant of a closed ship section with that of an open section with a double bottom and that of an open section with wing tanks, using the simple structural models shown in Figure 7.1.

Using the closed section as a benchmark and applying Equation (6.94), the relative stiffness of each section is 1:0.00067:0.00185 (Table 7.1). The torsional stiffness thus reduces considerably (to 1/1,000th of the benchmark value) in the case of the open section. The addition of closed sections in way of the side shell structure does little to mitigate the problem. Some other solution is required.

We shall now take a closer look at the structure of the entire hull girder. Figure 7.2 includes a longitudinal view of a containership. The sections in way of the cargo holds are

Table 7.1 *Torsional constant of ship-type cross-sections*

Ship Type	*Torsional Constant* ($\times 10^9$ mm^4)
Closed section	135,000
Open section with double bottom	900
Open section with db and wing tanks	2,500

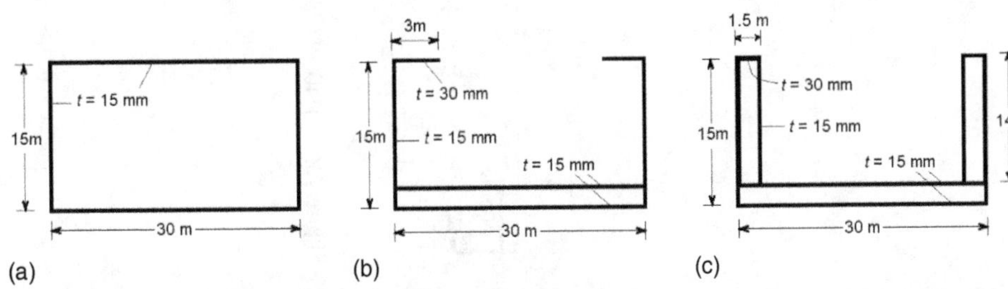

Figure 7.1 Structural models of ship-type cross-sections. (a) Closed section. (b) Open section with double bottom. (c) Open section.

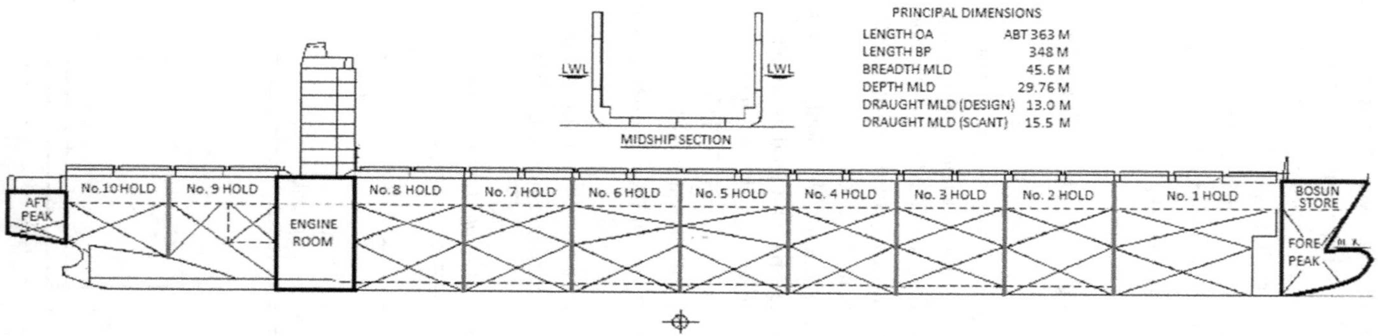

Figure 7.2 11,400 TEU containership profile (*black lines* denote closed deck structures; *vertical grey lines* denote transverse bulkheads).

open and thus torsionally weak. The transverse bulkheads that are located at the ends of each cargo hold, apart from fulfilling other operational and safety roles, enhance the retention of form of the cargo spaces – that is, they prevent the double bottom and sides from distorting. The cargo spaces extend over a large proportion of the length of the ship, rendering it a torsionally weak structure. In this ship type, there are three regions that are closed-deck structures: the fore and aft ends of the vessel and the machinery spaces. These regions, in conjunction with the transverse bulkheads, contribute to the torsional rigidity of the entire hull girder. Further strengthening of the hull girder in a torsional sense, although of limited effect as illustrated earlier, is achieved by detailed local structural design that involves the use of a cellular layout (torsion boxes) , as well as suitable connections of the longitudinal structure with the stiffer regions. The aim is to satisfy structural design requirements, which are twofold: strength (limitation of stresses) and stiffness (limitation of deformations). This question is discussed further at the end of this section.

At positions along the length of the ship where a torsionally flexible member is attached to a torsionally stiffer member significant stresses arise in the stiffer member whereas significant distortions arise in the more flexible member. It is thus necessary to select suitable structural arrangements that will satisfy stiffness as well as strength requirements for the entire hull girder. Indeed, such arrangements need to be able to withstand repetitive loading that can lead to fatigue failure. Lastly, they have to be able to be fabricated efficiently and effectively by shipyards.

Up to this point in our discussion, we have considered torsion as being the only type of loading that acts on the hull girder. The hull girder is of course subjected to longitudinal bending as well, so that longitudinal bending and shear stresses arise in addition to those related to torsion. The question then arises as to how the cumulative effect of these loads ought to be treated. From elementary mechanics of solids we can stipulate that in the linear elastic range – that is, before the onset of nonlinear geometrical or material behaviour – and provided coupling does not arise, the principle of superposition can be applied. An example of the manner in which these loads are combined is described in Appendix C that includes relevant extracts of the Nippon Kaiji Kyokai Rules.

7.2.2 Effect of Geometry on Position of Centre of Shear and on Torsional Stiffness

The positions of the centres of shear and twist of a containership subjected to torsional moments along its length are shown in Figure 7.3 [18]. It is seen that in way of the open-deck cargo spaces, the centre of shear lies below the baseline (negative values) whereas in way of the enclosed fore and aft ends it is above (positive values). In the open sections, where the centre of shear is below the baseline, the moments of the horizontal forces due to the hydrostatic pressure acting on the side shell structure on the port and starboard sides about the centre of shear are large, whereas in the closed sections the moments are very much smaller. As a result, the torque acting in the open section is significantly larger than that acting in the closed section.

It is noted that a discrepancy arises between the positions of the centre of shear and the centre of twist, for a significant proportion of the ship's length. This is because non-uniform

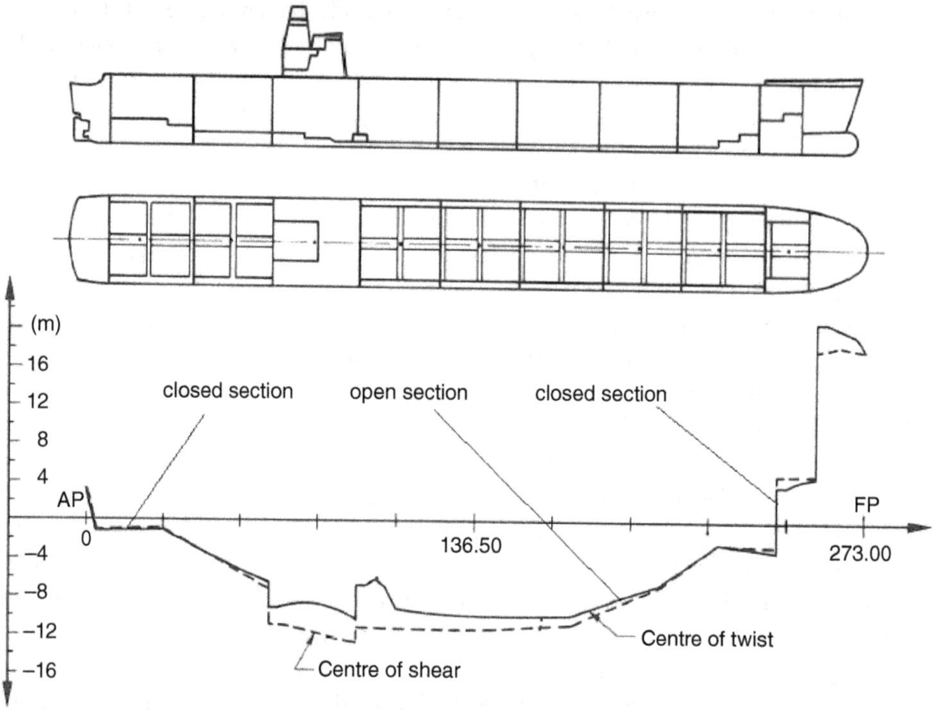

Figure 7.3 Longitudinal positions of the centres of shear and twist of a containership (adapted from [18]). Reprinted courtesy of the Society of Naval Architects and Marine Engineers.

torsion takes place and the influence of shear stresses on the torsional stresses is not negligible there [20]. The cargo region of the ship is subdivided by the transverse bulkheads into compartments over which shear effects on torsion become important, because the length of each compartment is comparable to the breadth and depth of the hull girder.

The other important observation is that, in way of the engine room, discontinuities (increases) arise in the distributions of both the centre of shear and the centre of twist. This is because the engine room is an enclosed space, and as a result the forward and aft engine room bulkheads act as rigid supports. Significant longitudinal warping stresses arise there.

7.2.3 Free, Restrained and Constrained Warping

In the previous section, it was mentioned that when a member is subjected to uniform torsion, since there are no axial restraints at its ends, the member warps. If axial constraints arise, as in the case of the hollow rectangular section that is fully fixed at one end and to which a torque is applied at the other, significant longitudinal normal stresses arise at the built-in end (Figure 7.4). These are larger than the shear stresses that result from uniform torsion.

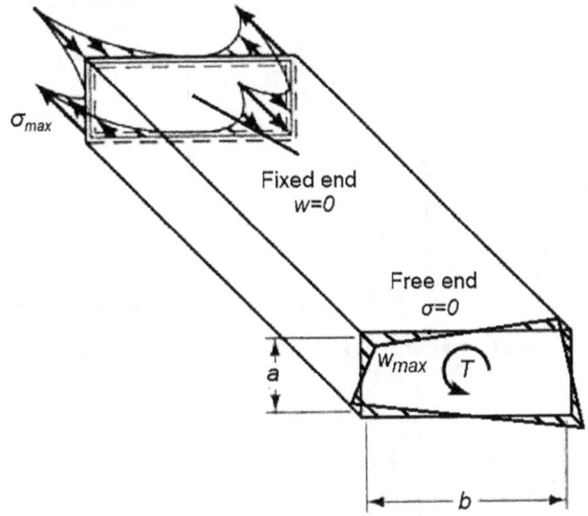

Figure 7.4 Normal stresses in a rectangular hollow section with restrained warping at one end and free warping at the other (adapted from [1]).

The following diagrams include the distributions of normal and shear stresses that arise in a hollow rectangular section to which a concentrated torque is applied at the ends (Figure 7.5a). The section is attached in such a way that warping is fully restrained and the distributions are evaluated on the basis of expressions obtained using linear torsion and bending theory [15].

Figures 7.5b and 7.5c show the shear stress distributions due to torsion and bending. It is seen that for the section considered, uniform torsion arises and the magnitude of the two shear distributions are similar.

From Figure 7.6a it is seen that the longitudinal warping stresses increase in way of the rigid supports located at the ends, whereas in way of the mid-length position where the cross-section is free to rotate, they are negligible. Contrariwise, the warping shear stresses are zero at the fixed ends and are non-negligible in way of the mid-length position (Figure 7.6b). Figure 7.7 includes distributions of normal and warping stresses in the section and also the resultant distribution (Figure 7.7c).

In ship structures, uniform torsion does not take place because of the topology of the structure and because warping is prevented (constrained), as a result of the structural arrangements. From the point of view of torsional strength, the weakest regions of the hull girder of open-deck dry cargo ships are the side shell and bottom structure in way of the deck openings (hatches). The open cross-section is prevented from distorting primarily by the presence of transverse bulkheads and closed deck structures such as the fore and aft ends and the machinery spaces and secondarily by detailed structural design in way of the attachments to these structural components. However, as is clear from Figure 7.8, warping does arise, as the relative stiffness of the adjoining structure is finite.

The torsionally weak members are welded to, and supported by, the adjoining transverse structure. Because the transverse bulkhead and deck are elastic structures, they also distort,

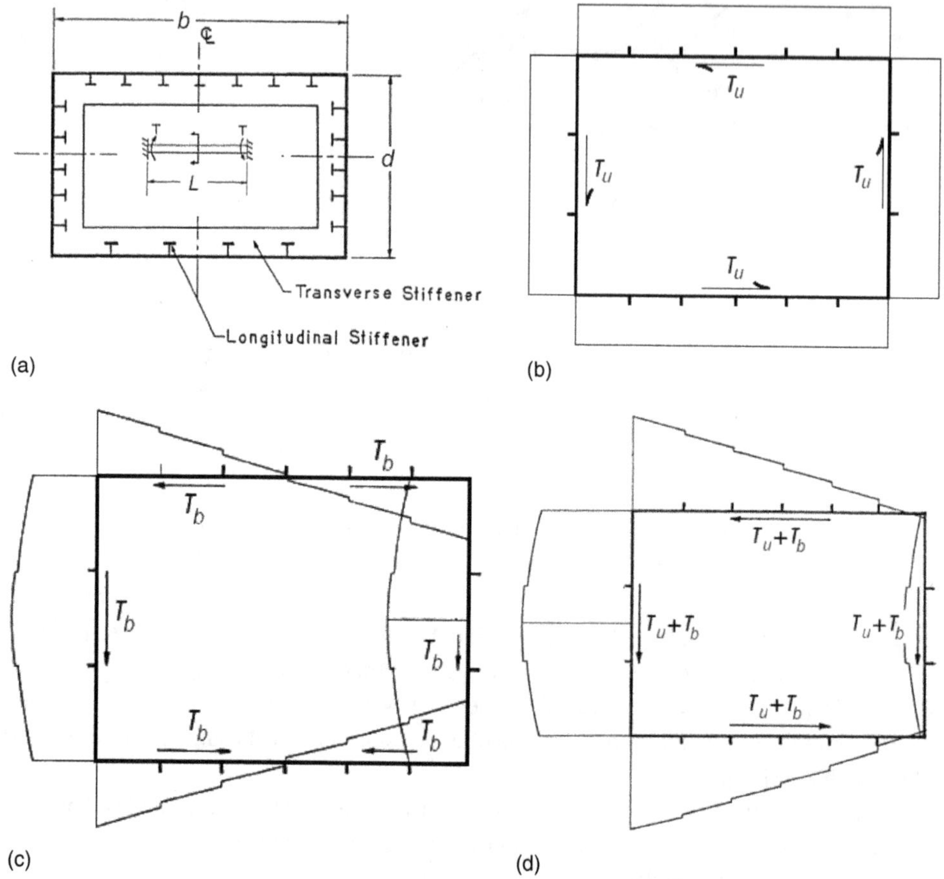

Figure 7.5 Shear stresses in a transverse section of a hollow girder subjected to a torque [15]. (a) Hull girder test model. (b) Shear stress due to uniform torsion (τ_u). (c) Shear stress due to bending. (d) Resultant shear stress.

although to a much lesser extent than the attached members, thereby providing support to them. As a result, longitudinal warping stresses are induced in the side shell and bottom structure, and *constrained warping* arises. The magnitude of the warping stresses increases as the stiffness of the supporting structure increases, and thus in constrained warping both warping distortions and stresses are present.

Summarising, we see that the application of a torque to an open-deck hull girder results in:

1. Rotation of each cross-section along the length of the hull girder with respect to adjacent sections, as the entire hull girder twists along its length.
2. Constrained longitudinal warping of members supported by torsionally stiff structural units at their extremities in the longitudinal direction.
3. Shear stresses associated with uniform torsion. These shear stresses are tangential to every transverse section, and by the principle of complementary shear stresses, shear stresses arise along longitudinal planes as well so that warping is induced.

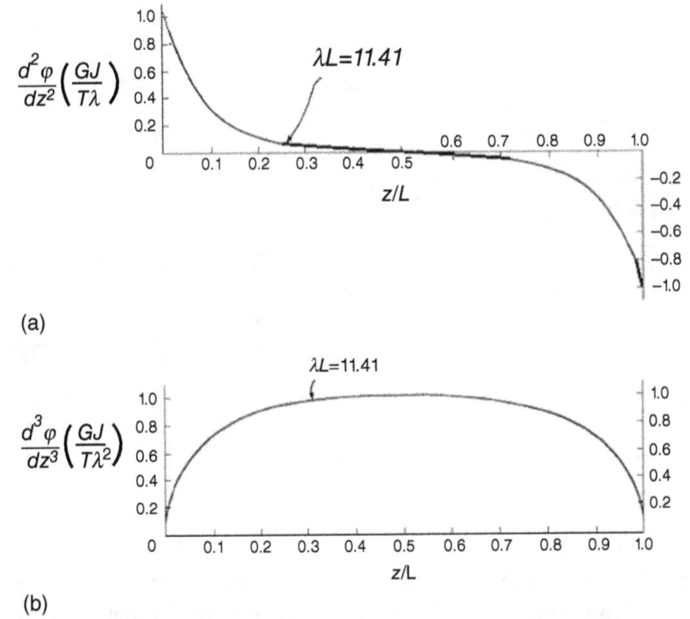

Figure 7.6 Longitudinal distributions of warping stresses ($\lambda = GJ/E\Gamma$) [15]. (a) Warping normal stresses. (b) Warping shear stresses.

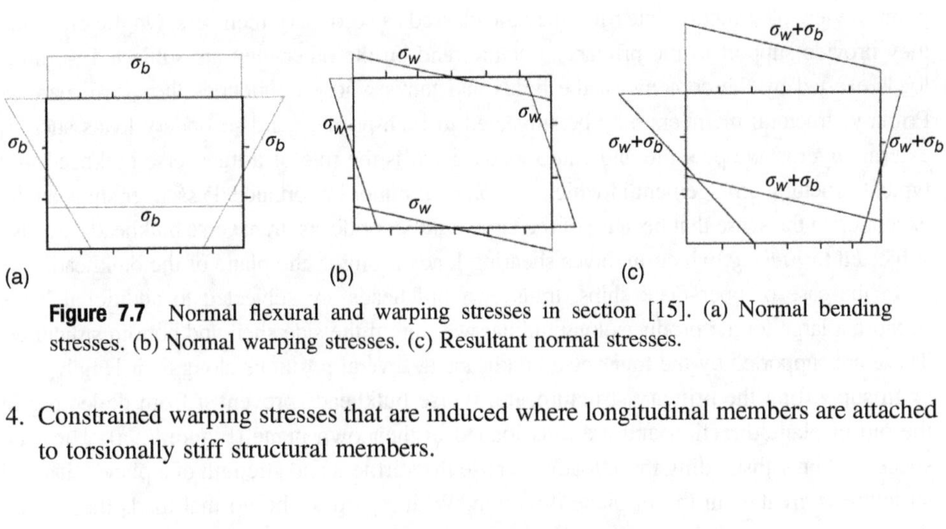

Figure 7.7 Normal flexural and warping stresses in section [15]. (a) Normal bending stresses. (b) Normal warping stresses. (c) Resultant normal stresses.

4. Constrained warping stresses that are induced where longitudinal members are attached to torsionally stiff structural members.

7.2.4 Effect of Discontinuities

It is necessary to consider the role of the parts of the structure of any open-deck ship that inhibit the twisting of the hull girder, especially when the structure is torsionally weak. These elements are the *transverse bulkheads* and the *closed deck structures* at the fore and aft ends of the ship. Closed-deck structures may also exist along the length of the hull girder, such as the machinery spaces in containerships.

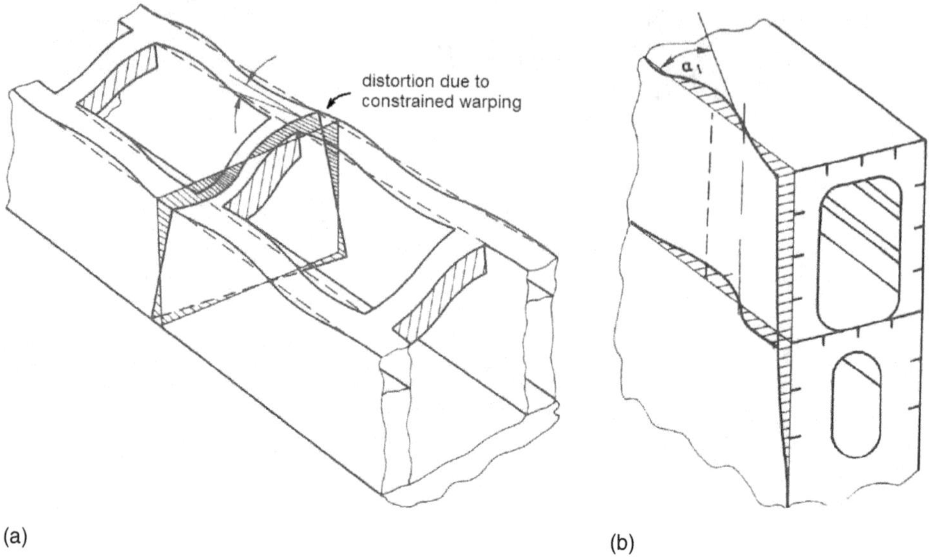

Figure 7.8 Warping in open-deck ship and interaction effects [18]. (a) Warping at junction with transverse structure. (b) Distortion of transverse deck. Reprinted courtesy of the Society of Naval Architects and Marine Engineers.

Transverse bulkheads fulfil important roles in the structure of any type of ship. From the point of view of structural integrity, they are classed as secondary members. On the one hand they provide support to the primary structure, and on the other they are subjected to direct loads exerted by the contents of the holds and tanks whose boundaries they form part of. Primary structural members may be subjected to both primary and secondary loads such as external hydrostatic pressure and cargo loads. Such is the role of a transverse bulkhead in a typical structural arrangement in which torsion is of minor importance. Passenger ships are an exception in the sense that because of the large number of decks, transverse bulkheads are also subjected to *racking*, which involves shearing loads acting in the plane of the bulkhead.

In the case of open-deck ships, transverse bulkheads are subjected to additional loads because a large torque produces longitudinal warping of the side shell and bottom structures. These are supported by the transverse bulkheads at several positions along their length.

In supporting the primary structure, transverse bulkheads prevent it from deflecting in the out-of-plane direction and are thus loaded in their own plane (Figure 7.9a). They are successful in withstanding these loads because the stiffness and strength of a plane stiffened structure is greatest in the in-plane direction. With regard to the normal loads that act on them as a result of the presence of cargo or ballast water, transverse bulkheads are supported along their boundaries by the lower and upper stools as well as the side shell structure. Furthermore, they are designed to withstand these loads with the assistance of stiffening systems such as vertical stiffeners in conjunction with horizontal stringers or corrugations (Figure 7.9a).

In the case of torsionally weak structures, transverse bulkheads are additionally subjected to longitudinal warping loads by the side shell and bottom structure (Figure 7.9b).

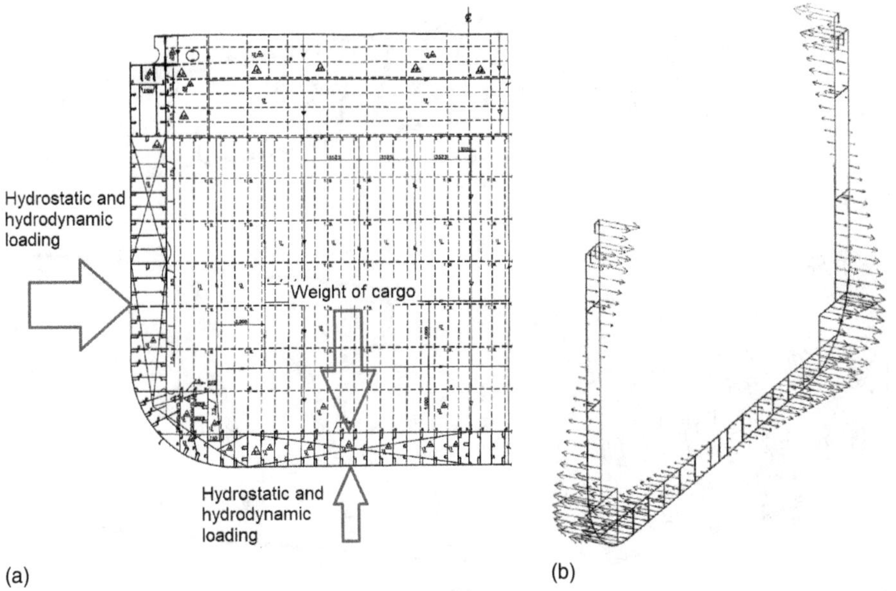

Figure 7.9 Loads on a watertight transverse bulkhead fitted to a 7,800 TEU containership. (a) Primary and secondary loads (excluding torsion). (b) Warping function [20].

Warping loads are normal to the plane of the bulkhead, and therefore the out-of-plane stiffness of the bulkhead is of critical importance. However the out-of-plane stiffness of plane stiffened structures is not large, so that when large warping loads act, transverse bulkheads cannot withstand the longitudinal twisting of the side shell and bottom structure. In this role they are much more successful, since their in-plane stiffness in significant.

The above statements are supported by the findings of FE studies, selected results of which are included in Figures 7.10 and 7.11. Figure 7.10a shows a hull girder that includes open-deck compartments, closed-deck structures at the fore and aft ends and the enclosed engine room structure. A visual examination of these figures reveals the following:

1. The transverse bulkheads distort in the longitudinal direction due to their low resistance to longitudinal warping loads (Figures 7.10a, 7.10c).
2. The transverse bulkheads retain their form, thereby limiting the distortion of the side shell and bottom structure (Figure 7.10b). Figure 7.11 presents results of a comparison performed using FE analysis of an open section with one that is fitted with transverse bulkheads. It is seen that the twist angle of the hull girder is reduced, and the figure given for the reduction was approximately 10 per cent [19].
3. The closed-deck structures retain their rigidity and are successful in withstanding the longitudinal warping loads. They also provide support against twisting. As a result, high warping stresses are expected to arise there.
4. The enclosed engine room structure distorts much less than the transverse bulkheads, and thus provides effective support against longitudinal warping distortions, as do the fore and aft ends. Similarly significant warping stresses are expected there.

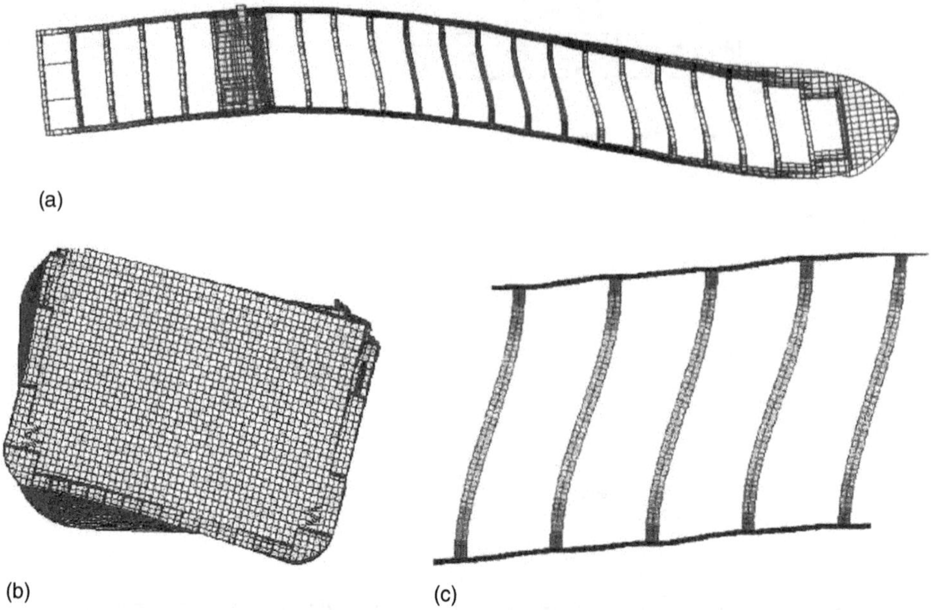

Figure 7.10 Results of FE analysis of open deck ship subjected to torsional loading [13]. (a) Distortion of open deck hull girder subjected to torsional loading (plan view). (b) Twisting of transverse section. (c) Longitudinal warping.

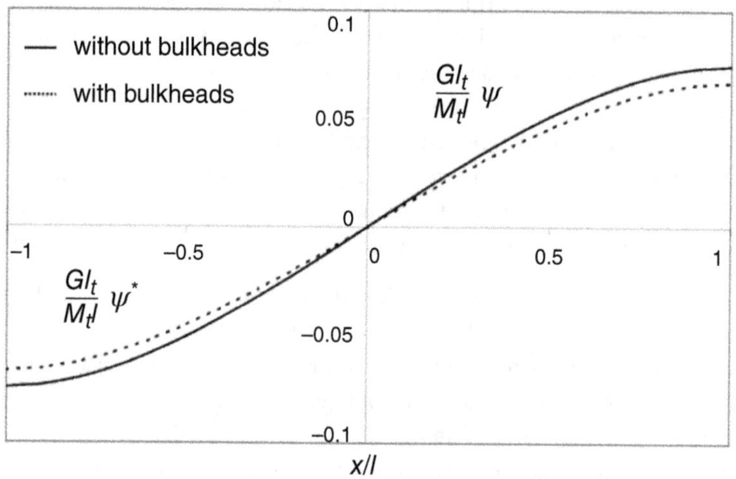

Figure 7.11 Effect of presence of transverse bulkhead on twisting angle, ψ [19].

7.2.5 Design Issues

It is clear that in the case of open-deck ships the application of a torque gives rise to some challenging questions with regard to hull girder response. These need to be addressed to prevent failure of any type during the lifetime of the ship. The two fundamental issues that

Figure 7.12 Longitudinal torsion boxes in a containership section.

arise relate to strength and stiffness of the hull girder, at both a global and a local level. Adequate strength is necessary to ensure that the induced stresses are such that yielding and buckling do not arise. Adequate stiffness is required to ensure that deflections are below specified limits. Strength is related to the ultimate limit state whereas stiffness is related to the serviceability limit state.

In the case of open-deck ships, the following issues arise:

1. The side shell structure and the bottom structure twist along the full length of the hull girder and are restrained only by closed-deck structures. Global distortions of the hull girder may give rise to damage in piping systems that run the length of the ship, as well as to hatch coamings in way of deck openings. In the latter case, damage may occur to hatch covers as well as loss of watertightness of the cargo holds. For such reasons limitations are imposed on the distortion of the hull girder.
2. It is also seen that longitudinal warping stresses arise where torsionally weak structural members are attached to stiff structural members. In order to determine the stress state at any of these positions it is necessary to add the warping stresses to the longitudinal bending stresses and to allow for the effect of shear stresses due to vertical bending and torsion. These stresses are periodic, depending on the matching of the wave loading with the hull girder response. The end-effect is possibly yielding or buckling as well as reduced fatigue life. The most susceptible region of the hull girder to such failures is the upper region of the side shell structure, where the total stresses are greatest. In order to provide a solution to this issue, *torsion boxes* are used in both the longitudinal and transverse directions (Figure 7.12). A torsion box is a cellular structure designed to resist high torsional loads. It was shown in the rudimentary calculations carried out to compare the torsional stiffness of different types of ship sections that even if closed cellular structures are used, the torsional stiffness of an open cross-section increases only marginally. Therefore, closed cellular structures are not effective in reducing the twisting of the entire hull girder. They are, however,

effective in preventing local damage from occurring due to yielding, buckling or fatigue.

3. If 'synchronous'[1] torsional loading arises, torsional vibrations may occur that will result in reduced fatigue life of specific structural members.

It becomes clear that the twin demands of adequate stiffness and strength compete with one another, since the higher the stiffness, the higher the warping stresses and, hence, the total stress, and vice versa. The designer has to satisfy requirements for both, and the following outline design procedure is proposed:

1. Define permissible flexibility requirements for the particular ship in accordance with the serviceability limit state.
2. Ensure as much flexibility of the structure as is permissible in way of regions with significant changes in torsional stiffness (fore and aft ends, machinery space bulkheads, transverse bulkheads). This may require novel arrangements of the structure at a local level.
3. Consider measures to reduce longitudinal warping stresses.

The issues that arise concern design at a local structural level, but this is a matter that cannot be dealt with in the present text.

7.2.6 Coupled Horizontal Bending and Torsion

The quantification of the joint action of horizontal bending and torsion requires either a linear or a nonlinear analysis. A linear analysis may be used if it is the outcome of two separate load effects, in which case each of which has no influence on the other. The task thus consists of determining the outcome of the joint action of the two, this being termed a *combined loading condition*, and in this case superposition of stresses is permitted. In the latter case each load effect has an influence on the other, so that we speak of a *coupled loading condition*. In this case it is necessary to solve the relevant differential equations as a system, and these include cross-coupling terms. Coupling arises in one case and not the other because in the latter case the reference axes of the two load systems do not coincide. As a result, the distance between the two reference axes acts as a lever, and a couple involving each load system about the reference axis of the other load system acts. To clarify the difference between the two cases we shall refer to an elementary problem from strength of materials theory [21] and then discuss coupled horizontal bending and torsion in ships.

Combined Bending and Torsion

Consider a hollow cylinder that is subjected to bending about the y-axis and torsional loading about the longitudinal (x-) axis (Figure 7.13). We note that the centre of gravity and

[1] *Synchronous* torsional loading involves the matching of the wave encounter frequency with the natural frequency of torsional vibration.

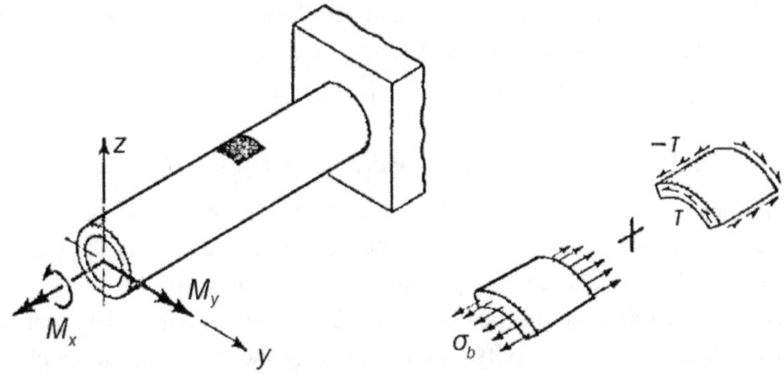

Figure 7.13 Hollow cylinder under combined bending and torsion [21].

the centre of shear coincide, and therefore there is no interaction between the two load effects, so that linear theory can be used. It is therefore permissible to use superposition in calculating elastic stresses, strains and deflections.

The stress distributions for the two load effects are given by

$$\sigma_x = \pm \frac{M_y R}{I_y} \qquad \text{(bending)} \tag{7.1a}$$

$$\tau = \frac{M_x R}{I_p} = \frac{M_x R}{2I_y} \qquad \text{(torsion)} \tag{7.1b}$$

From plane stress theory, the maximum shear stress is given by

$$\tau_{max} = \sqrt{\left(\frac{\sigma_x - \sigma_y}{2}\right)^2 + \tau_{xy}^2} \tag{7.2a}$$

The maximum and minimum normal (principal) stresses are given by

$$\sigma_{max,\, min} = \frac{\sigma_x + \sigma_y}{2} \pm \tau_{max} \tag{7.2b}$$

Substituting Equations (7.1a) and (7.1b) in Equations (7.2a) and (7.2b) gives

$$\tau_{max} = \frac{R\sqrt{M_x^2 + 2M_y^2}}{2I_y} \tag{7.3a}$$

$$\sigma_{max} = \frac{R\left(2M_y + \sqrt{M_x^2 + 2M_y^2}\right)}{2I_y} \tag{7.3b}$$

These equations can be used to design a hollow circular cylindrical section subjected to bending and torsion. For a non-circular section,

$$\sigma_x = \frac{M_y z}{I_y} - \frac{M_z y}{I_z} \tag{7.4}$$

where y and z are the principal axes. If the centre of shear and the axis of bending coincide in the non-circular section, combined action arises. Otherwise coupled bending and torsion arise.

Coupled Horizontal Bending and Torsion in Ship Hull Girders

In ship-type sections, symmetry exists about the centreline plane but not about the horizontal axis of bending, and as a result, the centre of gravity and the centre of shear do not coincide. This leads to coupling of horizontal bending with torsion.

In the case of open-deck ships, the centre of shear is located below the baseline, and the distance between the centre of gravity and the centre of shear is therefore much larger than in closed-deck ships. As a result, strong coupling of these two load effects occurs. In addition to this, a number of other complications need to be overcome in order to obtain results that can be of use in analysis and design. These concern the following.

1. The Definition of Wave Loading

It is possible to consider simplified load models, and classification societies propose standard distributions along the length of a ship (Appendix C). In reality, however, the wave loading is nonlinearly distributed along the ship's length, and in order to model it, models that are more accurate than linear theory are needed. To this end, *3D Rankine source theory*, which takes into account the interaction of the steady nonlinear flow due to the forward speed of the ship with the flow produced by incoming waves and ship motions, has been used. A comparison of results using linear strip theory and 3D Rankine source theory is included in Figure 7.14. These relate to towing tank tests for models whose particulars are included in Table 7.2 whereas the wave conditions in the tank tests are summarised in Table 7.3 [10].

It is seen that the response functions for torsional moments vary nonlinearly with λ/L, the ratio of the wavelength of the incident wave to that of the ship's length. For an encounter angle of 120° (stern quartering seas), the largest torsional moments arise at SS 2.5. The differences in response for different wave heights are more clearly observable at SS 7.5, where the largest incident wave height (15 m) gives rise to much larger torsional moments. Linear strip theory does not predict the observed moments as well as the 3D Rankine source theory does, especially at SS 2.5.

2. The Geometry of the Hull Girder

In real ship structures, the calculation of coupled horizontal bending and torsion is complicated for the following reasons:

a) The change of hull girder form along its length (Figure 7.15).
b) The presence of transverse bulkheads (Figure 7.2).
c) The coexistence of open-deck and closed-deck arrangements in open-deck ships (Figure 7.2).

As a result of (a) and (b), the hull girder is subjected to *non-uniform torsion*. The torsional response is influenced by both the stiffness properties of the hull girder

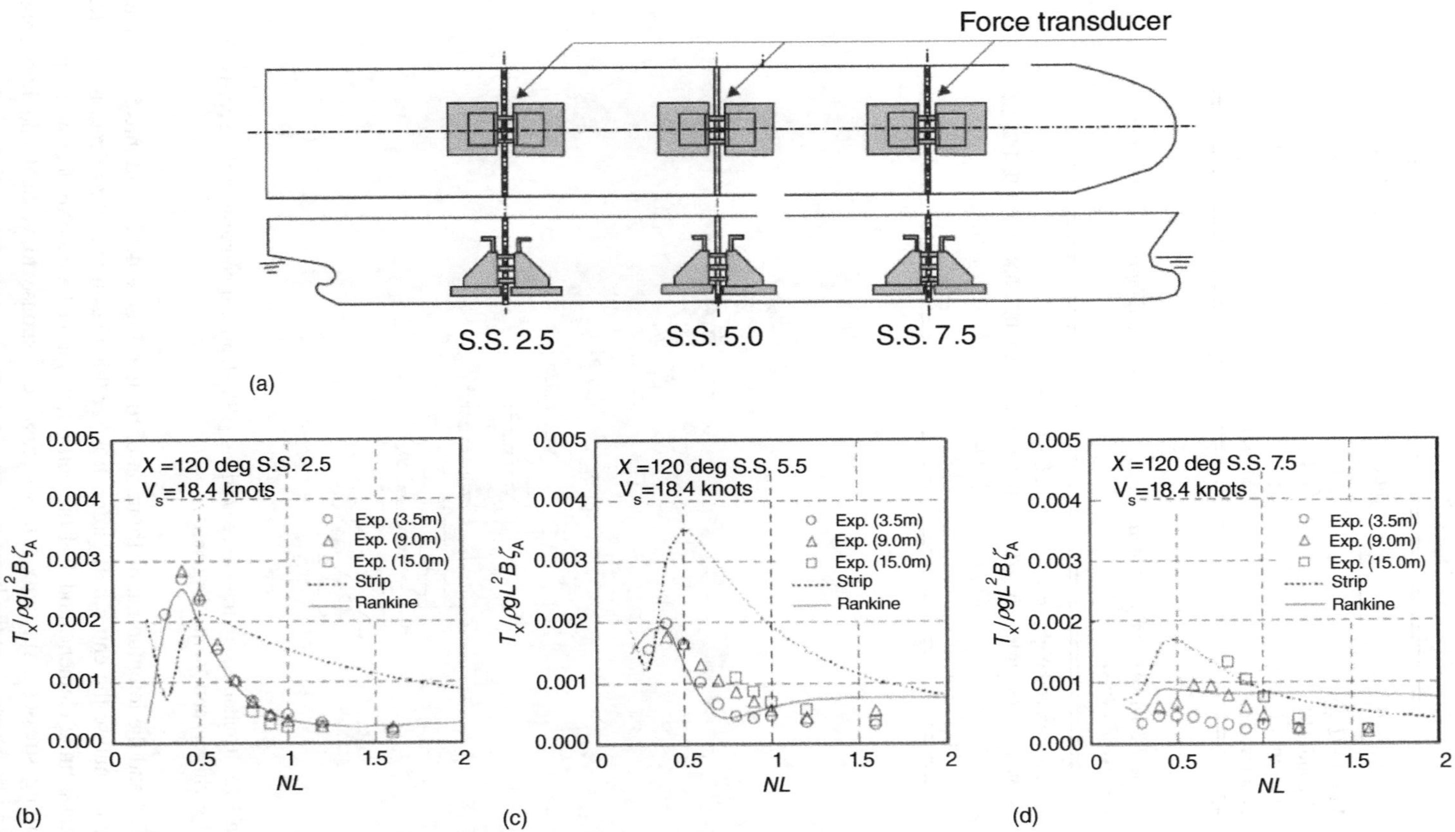

Figure 7.14 Comparisons of response functions of torsional moments T_X obtained by numerical methods with those obtained by tank tests for an encounter angle $\chi = 120°$ [10]. (a) Positions of force transducers on model; (b) at SS 2.5; (c) at SS 5.0; (d) at SS 7.5. Reprinted courtesy of Elsevier.

Table 7.2 *Principal particulars of containership model*

Description	*Ship*	*Model*
Length (L_{pp}) (m)	284.0	5.000
Breadth (B) (m)	43.0	0.754
Draught (d) (m)	14.0	0.247
Displacement (Δ)	109,480 (tonnes)	584.1 (kgf)

Table 7.3 *Wave conditions of tank tests (results for bold figures are included in Figures 7.14b–d)*

Incident wave height (m)	**3.5, 9.0, 15.0**
Wavelength/Ship length (λ/L)	**0.3, 0.4, 0.5, 0.6, 0.7, 0.8, 0.9, 1.0, 1.2, 1.6**
Encountering angle χ (deg)	0, 30, 60, 90, **120**, 150, 180
Ship speed (knots)	**18.4**, 24.5

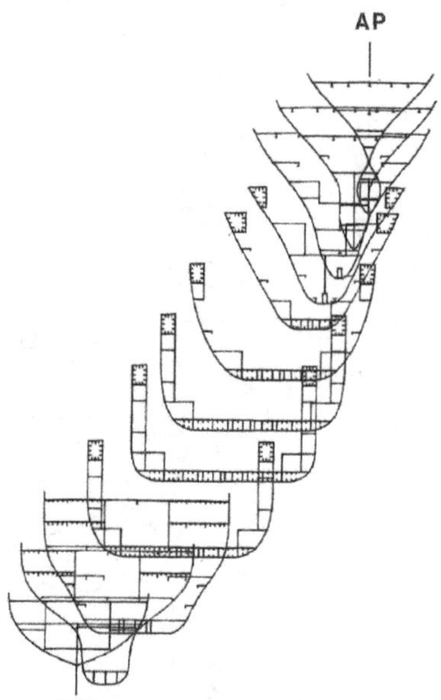

Figure 7.15 Containership transverse sections [18]. Reprinted courtesy of the Society of Naval Architects and Marine Engineers.

(Figure 7.3) and the longitudinal distribution of torque (Figure 4.17), and thus it becomes necessary to analyse the entire hull girder. It is for this reason that classification societies require that a finite element model of the entire hull girder be used to determine longitudinal warping stresses (Appendix C), whereas the response to vertical and horizontal bending can be assessed on the basis of a single, critical, transverse section.

An advanced procedure that caters for the problems mentioned above is described in two publications [17, 18]. This procedure is an energy-based approach in which Hamilton's variational principle is used to describe the static and dynamic behaviour of a hull geometry that varies gradually in the longitudinal direction. The presence of transverse bulkheads and closed deck structures is allowed for through the use of discontinuities. Such a procedure has to be linked to appropriate loading distributions for practical use. The effect that the closed engine room structure has on hull girder stiffness is considered in [19].

Classification societies generally require the use of direct strength procedures for the entire hull girder in order to assess torsional strength (stress and distortions). In the absence of results from direct strength calculations, ABS requires a pair of curves to be obtained in way of the critical position of the wave-induced torsional moment, such as, for example, in way of the forward bulkhead of the engine room. These curves are used to obtain the peak torque and the peak value of the torque-induced shear. Apart from the engine room region, it is necessary to consider at least two other positions: amidships and in way of the forward quarter length position.

Nippon Kaiji Kyokai, on the other hand, requires the use of a procedure outlined in Figure C.1. This procedure requires the calculation of stresses resulting from vertical bending, horizontal bending and torsion in still-water condition and in waves. The expressions used to calculate the stress in each case as well as for the addition (superposition) of stresses are included in Appendix C.

7.3 Reserve Strength of the Hull Girder Following Damage

In this section the question of reserve strength following major damage, in which a part of the hull girder is incapacitated and can no longer contribute to the load-carrying capability, will be considered. In such circumstances it is important to assess whether the residual strength is sufficient for survivability of the vessel and to plan emergency procedures based on the conclusion. In the case of grounding these may involve refloating. In addition, it will be necessary to plan temporary repairs and the redistribution of cargo and ballast water should the vessel be considered able to sail to a yard for repairs.

7.3.1 Redundancy of Continuous Structures

When dealing with engineering structures, the term *redundancy* is used in various ways, and it is helpful to make reference to them in order to clarify how it will be used in the present context.

In strength of materials, redundancy is closely linked to *indeterminacy*. An indeterminate structure is one for which the number of unknowns exceeds the number of equations obtained by applying conditions of external equilibrium (applied loads). To overcome this difficulty, internal equilibrium has to be considered as well.

In structural design, on the other hand, redundancy refers to the fact that one or more members of an assembly are not essential for survivability, so that the structure will not collapse if they are removed. This will be the case when there exist alternate load paths that are utilised should the need arise. In this case it is *structural redundancy* that is being considered. The presence of redundant members may be necessary for different reasons, such as to achieve a certain safety margin against collapse, or else if the structure is loaded during its lifetime in a manner different to that expected. It is also possible that they fulfil a role that is not related to structural issues. An example from the field of ship structures is the orthogonally stiffened grillage, which not only carries in-plane loads but also provides protection to enclosed spaces. They are thus not redundant in a functional (operational) sense, although they may be redundant in part in a structural sense. With regard to structural behaviour, a member such as a beam is designed to carry a load in bending, but it also has compressive and torsional strength. The ability to carry compressive and torsional loading is thus not normally required, so that the beam is redundant with respect to these types of load actions.

In order to obtain a quantitative measure of redundancy, a *redundancy index, RI,* has been proposed [14]:

$$RI = \frac{P_u - \overline{P_u}}{\overline{P_u}} \tag{7.5}$$

where

P_u = ultimate load of the *parent structure* for a specific failure mode
$\overline{P_u}$ = ultimate load of the as-designed structure for the same failure mode

The parent structure is that in which all members that are not absolutely necessary for stability have been removed.

7.3.2 Redundancy at a Local Level and a Global Level

In the case of stiffened plating, if the design load does not exceed the plate buckling load, the parent structure consists of the flat plate alone. In order to examine the effect adding members to a parent structure has on the redundancy index, a rectangular flat plate that was loaded in compression had stiffening added to it in stages (Figure 7.16) [4].

The structure was studied in the elastic range, and the critical load was calculated by performing an eigenvalue analysis using a series of finite element models. In each case the *RI* was calculated, and the results are included in Table 7.4.

In order to gain an idea of how redundancy varies with structural topology, a series of sections have been included in Table 7.5, in which redundancy with respect to vertical shear and torsion is considered. The first of these sections is a simple I-beam, which has good bending and shearing capacity but low torsional strength. The open section that follows could relate to a very simple ship section in way of a cargo hatchway; this has satisfactory bending and shear strength but low torsional resistance. The hull girder generally consists of a series of open and/or closed cells, and as the number of cells increases, so does the corresponding degree of redundancy.

It is seen that the redundancy index increases with the number of longitudinal stiffeners used. The effect of the transverse stiffener is particularly noticeable.

Table 7.4 *Redundancy indices (RI) for different geometries of stiffened plating*

Structural Configuration	Volume of Plate (cm^3)	Total Volume of Stiffeners (cm^3)	Total Volume (cm^3)	Cross-sectional Area (cm^2)	Critical Load (kg)	RI
a	2,754	0	2,754.00	27	26,756	0
b	2,754	220.32	2,974.32	29.16	36,318	0.36
c	2,754	440.64	3,194.64	31.32	42,729	0.60
d	2,754	660.96	3,414.96	33.48	35,509	0.33
e	2,754	336.96	3,090.96	29.16	98,508.4	2.68
f	2,754	557.28	3,311.28	31.32	115,536	3.32
g	2,754	777.60	3,513.60	33.48	117,699	3.40

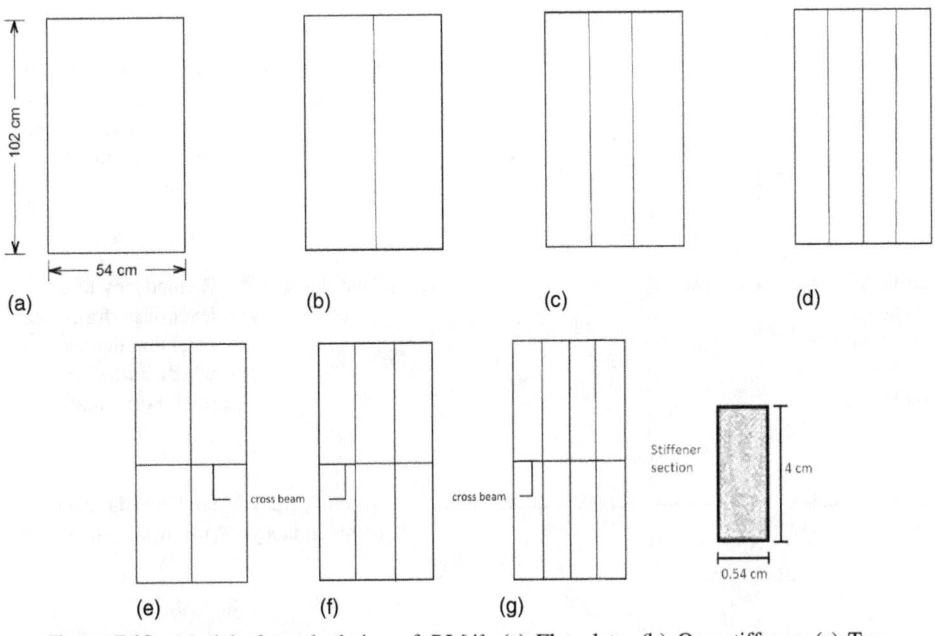

Figure 7.16 Models for calculation of *RI* [4]. (a) Flat plate. (b) One stiffener. (c) Two stiffeners. (d) Three stiffeners. (e) Stiffener and crossbeam. (f) Two stiffeners and crossbeam. (g) Three stiffeners and crossbeam. Reprinted courtesy of the Ship Structure Committee, Washington, DC.

It is clear that an exact determination of redundancy in ship structures is a complex task requiring significant effort as well as expertise in the use of numerical tools.

7.3.3 Reserve Strength and Residual Strength

If the applied load does not exceed the design load and $RI > 0$, the structure will have excess capacity, that is, *reserve strength*. The reserve strength may thus be defined as the

Table 7.5 *Structural and hull girder cross-sections and degrees of redundancy in shear and torsion [4]*

Cross-section		Vertical Shear	Torsion
Typical section beam		No real redundancy except for losing flange	No real redundancy except for losing flange
Simple, thin-walled open section		No real redundancy except for losing part of a flange or web	No real redundancy
Hollow rectangular box beam		One degree of redundancy	Some redundancy, but the percentage loss of capability is so large that it should be ignored
Hollow rectangular box with one vertical bulkhead		Two degrees of redundancy	Redundancy to one level of performance reduction depends on which surface fails (incl. horizontal)
Two vertical bulkheads		Three degrees of redundancy	Higher redundancy (follows from above)
Inclined tank bulkheads and inner side shell		Three degrees of redundancy	Multicellular, high level of redundancy
Multicellular structure	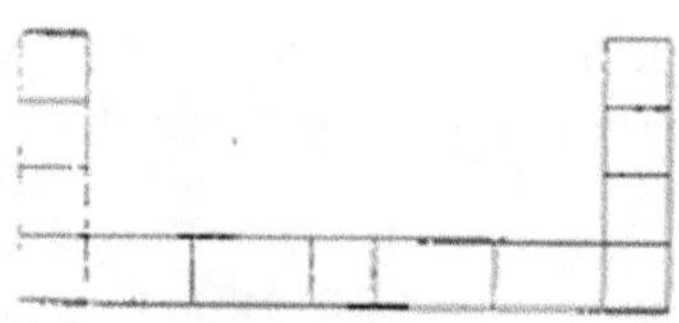	Multiple degrees of redundancy	Multiple degrees of redundancy

difference between the ultimate load-carrying capacity of the component or system and the actual maximum applied load. The magnitude of the reserve strength depends on the safety margin(s) implicit in the design and is therefore code-dependent.

A *reserve strength index* may be defined as the ratio of the actual ultimate strength, P_{ult}, to the design load, P_{des}, of the structure:

$$RSI = \frac{P_{ult}}{P_{des}} \tag{7.6}$$

If $RSI = 1$, the structure has no reserve strength; if $RSI > 1$, the structure does have reserve strength; if $RSI < 1$ the structure will fail if the design load is applied. In the case of a beam in bending, if the safety factor against initial yielding is γ_0 (say, equal to 1.70), the reserve strength index is found from

$$RSI = \frac{P_{ult}}{P_{des}} = \frac{P_{ult}}{P_0/\gamma_0} = \frac{P_{ult}\gamma_0}{P_0} \tag{7.7}$$

If, on the other hand, the structure does not collapse when one or more members fail, then it has *residual strength*. Since residual strength depends on the availability of other load paths or other modes of carrying load (e.g. compression to bending), it is a measure of the capability of the structure to continue to carry load after damage. It has to be determined with respect to a particular damage type in a specified location of a particular ship, under given circumstances. A *residual strength index* (*RIF*) was proposed by several authors [11, 14] as follows:

$$RIF = \frac{P_r}{P_u} \tag{7.8}$$

where

P_r = residual strength of the damaged structure

P_u = ultimate strength of the intact structure

In the absence of residual strength, $RIF = 0$, whereas if the damage does not in any way affect the ultimate strength, $RIF = 1$. If it is required to assess the effect of damage on the ultimate vertical bending strength of a hull girder, P_r will be the ultimate bending moment in the damaged condition ($= M_{dam}$), whereas P_u will be the ultimate bending moment of the intact ship ($= M_{int}$) [6]. Thus,

$$RIF = \frac{M_{dam}}{M_{\text{int}}} \tag{7.9}$$

Damage may occur to primary, secondary and tertiary members of the hull girder and to combinations of these. In this section we shall consider damage to members that form part of the primary structure of the hull girder, which affects overall integrity. Such cases affect not only structural behaviour but also other aspects such as stability and possibly propulsion. In this section we shall limit our discussion to structural issues. Damage to the primary structure arises as a result of:

(a) Contact with an exterior body
(b) Explosions and fire

(c) Hydrodynamic loading
(d) Loading and discharging operations
(e) Degradation with time (thinning due to corrosion, fatigue cracks, minor buckling).

Apart from (a) and (b), the remaining causes of damage generally result in local buckling and/or fractures, and for this reason we shall not discuss them in any detail[2]. In case (a), damage to the bottom structure may arise as a result of *grounding* and to the side shell and bow structure due to *collisions* with moving or fixed objects. They may be classified as *symmetrical* or *unsymmetrical* with respect to the ship's centreline axis.

Figure 7.17 illustrates the case of a US warship that was damaged during the Second World War.

An example of symmetrical damage is grounding over part of the flat-of-bottom. From the point of view of hull girder strength two cases may be identified, depending on whether the ship remains in contact with the seabed or not. If following grounding the vessel is no longer in contact with the seabed, the consequences will be:

1. A reduction in the area of the hull girder cross-section. For the same external bending moment the remaining area will be more highly stressed.
2. A reduction in moment of inertia. This implies a loss of overall bending stiffness and hence larger deflections of the hull girder.
3. A vertical shift of the neutral axis of the cross-section. This will be away from the bottom structure and will accentuate hogging of the hull girder. The members that lie below the neutral axis will thus be more highly stressed, both because of the increased lever to the neutral axis and because of the reduced area in compression. The effect on members that lie above the neutral axis will depend on the distribution of material within the cross-section and the extent of damage.
4. A change in draught, depending on the extent of flooding and thus also the buoyancy curve.

The gravity of the situation will depend on the position of damage in the longitudinal sense. The effect on hull girder longitudinal bending stresses will be most acute amidships, whereas at the quarter-length positions, where the peak values of the vertical shearing force distribution arise, the effect on shear stresses will be greatest.

If rupture does not take place, the most likely failure mode of the compressed flat panels that form part of bottom and deck structures is *interframe buckling*. This failure mode generally precedes lateral-torsional buckling and overall grillage buckling, which involves bending of transverse members. The compressive strength of stiffened panels may be substantially reduced by deformations and residual stresses resulting from elasto-plastic bending of stiffeners.

[2] This does not make them any less important. On the contrary, the manner in which local damage may or may not affect global response is a subject that requires special consideration. The effects of local behaviour need to be integrated at the global level, and the method used has to be able to predict the correct load paths, as the damage will have brought about changes those followed in the intact condition [7]. Also, the direction and magnitude of the external loading will change as damage propagates. Hydrostatic stability may also be affected to a greater or lesser extent.

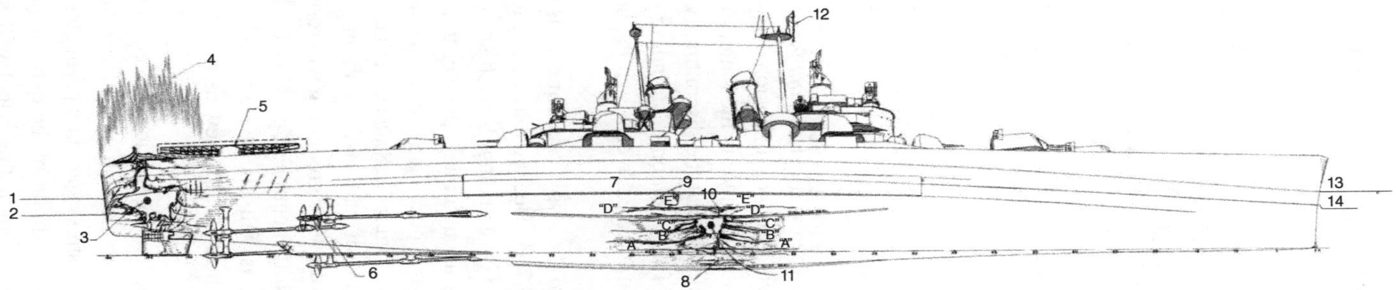

1: WL after 2nd torpedo hit with 60° list
2. WL before damage
3: Approximate position of hit on 16 October
4: Gasoline fire
5: Starboard catapult was removed after midships hit and before stern hit
6: No. 1 shaft moved aft about 5.5 feet
7: Armor belt
8: Port side of bottom buckled and cracked
9: Cracks in “E” strake
10: Crack
11: Approximate position of hit on 14 October
12: SK radar damaged
13: WL after 2nd torpedo with 60° list
14: WL before damage

Figure 7.17 USS *Houston*[a] following torpedo damage in 1944 shown listing to port at 30° [12].

[a]The analysis of this case represents an early attempt at determining the residual strength of a hull girder following extensive damage and subsequent repairs. It was estimated that the area of the midship section was reduced by 24 per cent as a result of torpedo explosion. The section moduli to the keel and main deck were reduced to 75 per cent and 53 per cent of the intact condition, respectively. Following repairs, the section moduli were restored to 91 per cent and 58 per cent of their respective intact values.

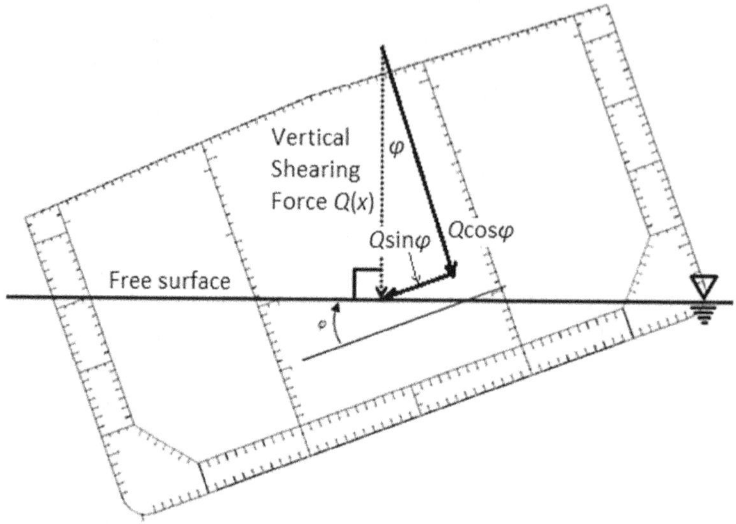

Figure 7.18 Loading of heeled section.

If the vessel is *stranded*, in addition to the above, a vertical *contact force* will act in way of the damaged region. If the hull girder is treated as a two-dimensional object, the contact force will act as a point support and will bring about changes to the longitudinal shear force and bending moment distributions. It has to be added that the discontinuity in the longitudinal shear force distribution that arises in way of the position of stranding will give rise to pronounced local shear lag effects, depending on the area over which the bottom structure is in contact with the sea bed.

If the damage is unsymmetrical with respect to the ship's centreline, as is most probable, the vessel will heel, as shown in Figures 7.18 and 7.19. Heeling may be due to penetration of the outer shell leading to flooding or else because of stranding. In such situations the vessel will continue to be subjected to vertical bending about an axis parallel to the free surface. This will result in the vertical shearing force and bending moment having components normal to and in the plane of the centreline plane (Figure 7.18), as described in Section 6.2.

The change in structural geometry due to loss of part of the bottom structure that is asymmetric with respect to the centreline plane following grounding will result in a rotation of the neutral axis, in addition to the vertical shift mentioned previously. In general, the angle of heel and the rotation of the neutral axis will not be equal. This will result in bending about the horizontal and the vertical plane about the new neutral axis as shown in Figure 7.18.

The shear force component perpendicular to the NA in the inclined position is, from Figure 7.20, $V\cos\varphi\cos\theta + V\sin\varphi\sin\theta$. The shear force component perpendicular to this is $V\cos\varphi\sin\theta + V\sin\varphi\cos\theta$. In order to determine the stress state in a particular component of the cross-section it is necessary to perform vector addition of the two components.

If the stress in any stiffened panel increases enough, this panel will suffer inelastic buckling. This will cause a gradual shift and rotation of the hitherto elastic neutral axis

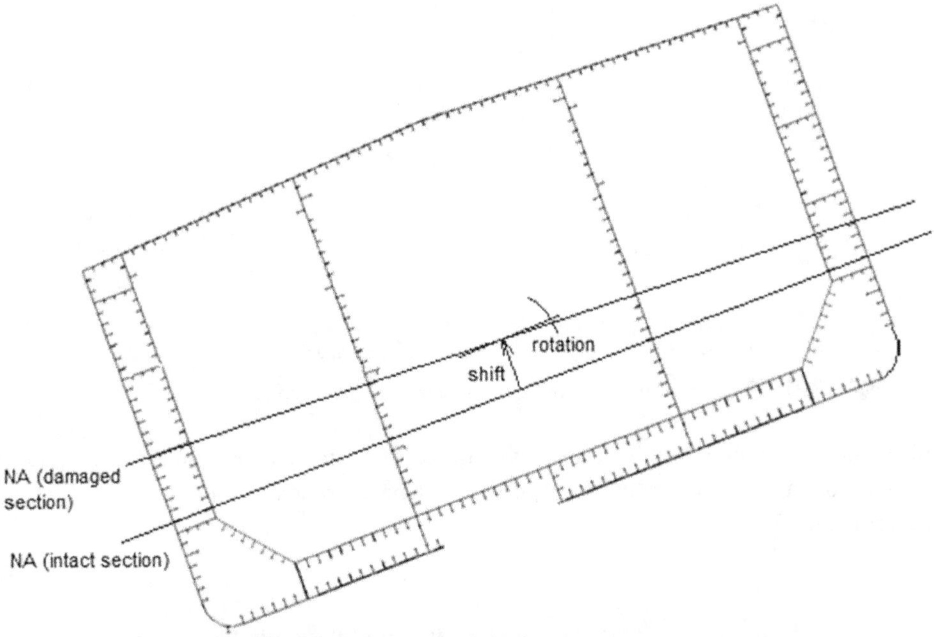

Figure 7.19 Position of neutral axis (NA) of asymmetrically damaged section.

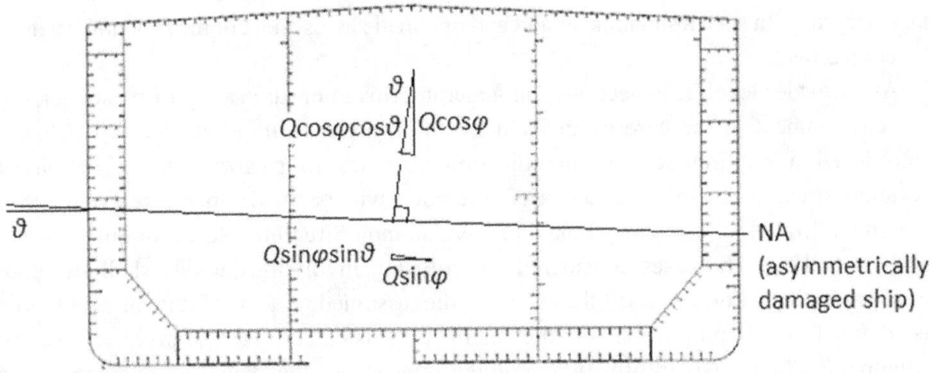

Figure 7.20 Shear force components acting about neutral axis of asymmetrically damaged, heeled vessel (elastic stresses).

(ENA) in order to satisfy conditions of longitudinal equilibrium. If the fibres furthest from the ENA lie on the strength deck, failure will occur there, and thus the shift of the neutral axis will be toward the bottom structure (Figure 7.21).

It is thus seen that when calculating the residual strength of the hull girder, it is necessary to allow for both translation and rotation of the neutral axis. In calculating the ultimate strength using the incremental-iterative approach, which forms the basis of the IACS approach, it was assumed that inelastic buckling is symmetric with respect to the centreline axis, in which case the neutral axis does not rotate. From the preceding

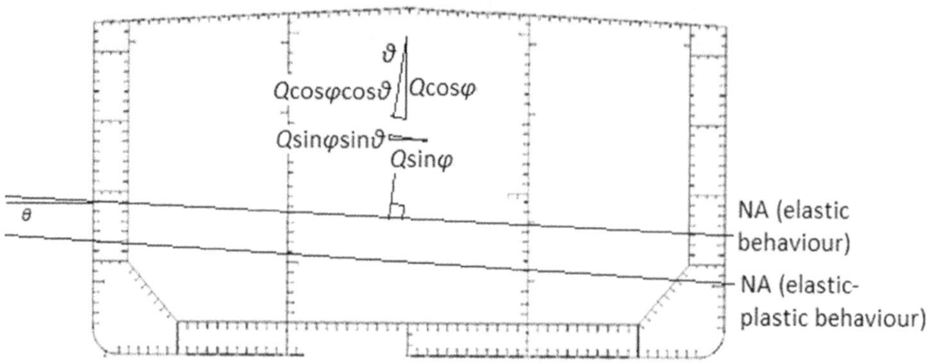

Figure 7.21 Position of neutral axis of asymmetrically damaged section.

discussion it is seen that it is necessary to allow for neutral axis rotation, particularly in the case of an asymmetrically damaged vessel. One approach to solving this problem is described in [3].

7.3.4 The DNV-GL Approach to Residual Strength Assessment

As mentioned previously, an immediate, practical need for the determination of residual strength arises when classification societies are requested to provide guidance in the case of an emergency. In such a situation a rapid and pragmatic assessment of the situation needs to be carried out.[3]

At a broader level, it is necessary in general terms to ensure safety of personnel and the environment in the case of an incident. In an effort to improve safety, IACS has introduced a requirement for classification societies to perform an estimation of residual strength. In this section brief reference will be made to the relevant DNV document [5], which is closely linked to the Common Structural Rules for bulk carriers and oil tankers. The cases considered are collision involving the side shell structure and grounding. For each of these cases the assumed extent of the affected area is defined as a proportion of the main dimensions of the transverse section (Figures 7.22a–b). No information is given concerning the extent of damage in the longitudinal direction so that presumably a two-dimensional analysis is carried out. The strength of the hull girder is examined in a damaged condition and compared with an acceptability criterion.

The underlying safety criterion of the CSR is that the probability of failure of the hull girder in a damaged condition ought not exceed the probability of failure of the intact ship. This means that following damage, the resulting buckling, yielding and permanent deflections are acceptable provided the hull girder does not collapse and break in half, with plate tearing and flooding of compartments.

[3] Examples of the application of the finite element method to residual strength assessment are discussed in Chapter 13.

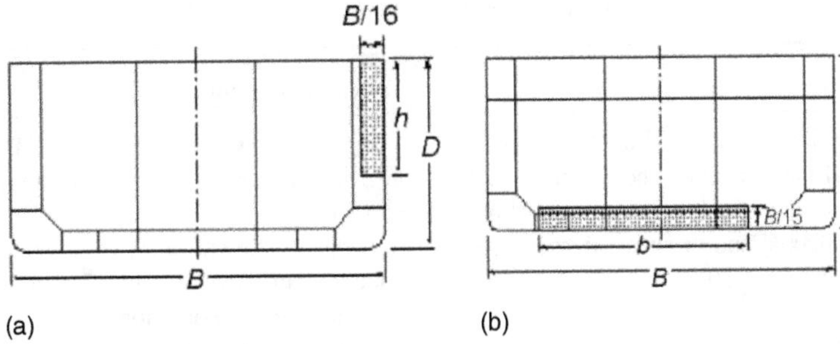

		Extent of damage	
		Single side/bottom	*Double side/bottom*
Side collision	Height (h/D)	0.75	0.60
	Length (l/L)	0.10	0.10
Grounding	Breadth (b/B)	0.75	0.55
	Length (l/L)	0.50	0.30
L = ship's length, l = length of damage			

Figure 7.22 Damaged regions of a hull girder transverse section for residual strength assessment purposes in accordance with DNV CSA-2 criteria [5]. (a) Damaged region in side shell collision. (b) Damage due to grounding. Reprinted courtesy of Det Norske Veritas, Høvik.

Section 2 of the Common Structural Rules (Rule Principles) stipulates that either the working stress design method or the partial safety factor method (PSF) may be used to assess hull girder integrity in any condition. The general format of the PSF method is

$$\gamma_{stat-1} W_{stat} + \gamma_{dyn-1} W_{dyn} \leq \frac{R}{\gamma_R} \tag{7.10}$$

where

γ_{stat-1} = partial safety factor that accounts for uncertainties related to static loads
W_{stat} = simultaneously occurring static loads
γ_{dyn-1} = partial safety factor that accounts for uncertainties related to dynamic loads
W_{dyn} = simultaneously occurring dynamic loads
R = characteristic structural capacity
γ_R = partial safety factor that accounts for uncertainties related to structural capacity

The acceptance criteria are such that consistent and acceptable safety levels for all combinations of hull girder integrity are achieved. It is on this basis that the partial safety factors are determined.

In the case of hull girder ultimate strength, the DNV document provides the following general format:

$$\gamma_S M_S + \gamma_W M_W \leq \frac{M_U}{\gamma_M} \tag{7.11}$$

where M_S, M_W are the still-water and wave-induced bending moments, M_U is the ultimate bending moment in vertical bending, and γ are partial safety factors reflecting uncertainties and ensuring the required target safety margin. The still-water and wave bending moments

Table 7.6 *Partial safety factors for intact strength and damaged strength conditions [5]*

Intact Strength	Damaged Strength
γ_{W1} = p.s.f. for environmental loads (= 1.1) γ_M = material factor in intact condition (=1.15) γ_M = material factor for hogging checks and designs with biaxial and or shear stresses in the bottom structure (= 1.30)	γ_S = p.s.f. allowing for increase in M_S in the presence of damage that leads to hold flooding (= 1.1) γ_{W2} = p.s.f. allowing for hydrodynamic load reduction corresponding to a 3-month exposure to world climatic conditions (=0.67) γ_M = 1.0 in general γ_M = material factor in damaged condition for hogging checks and designs with biaxial and or shear stresses in the bottom structure (=1.10)

Table 7.7 *Residual strength assessment of hull girders based on CSR*

Source		Percentage Reduction in				
Authors	Ship Type	Section Area	M_H	M_S	Z_{bot}	Z_{deck}
Hussein et al. [8]	tanker (SH)	13	29	7		
	tanker (DH)	9.4	14.1	4		
	tanker (DH)	10	2.7	2.2	4.4	4.0
Wang et al. [24]	bulk carrier	10	2.7	1.5	4.2	0.9
	tanker SH	10	4.5	3.9	6.7	1.7

used are the peak values in each case. Applying this format to the intact and damaged case, the following conditions of acceptability are obtained:

$$\text{Intact:} \qquad M_S + \gamma_{WI} M_W \leq \frac{M_{UI}}{\gamma_M} \tag{7.12a}$$

$$\text{Damaged:} \qquad \gamma_S M_S + \gamma_{W2} M_W \leq \frac{M_{UD}}{\gamma_M} \tag{7.12b}$$

where M_{UI}, M_{UD} are the ultimate bending moments in the intact and damaged conditions, respectively, and the partial factors assume the values included in Table 7.6.

In order therefore to determine the residual strength and to check Equation (7.12b), it is necessary to calculate M_{UI} and M_{UD}. In the case of M_{UI} the recommended approach is Smith's incremental-iterative method, whereas when calculating M_{UD} it is necessary to allow also for the rotation of the neutral axis as discussed in the previous section. It may be possible to use nonlinear finite element analysis although this is a time-consuming and expensive task. In this case further details of the damage geometry will be required.

A limited number of studies have been performed to estimate the residual strength of damaged ships. Table 7.7 includes results from two such recent series of studies.

7.3.5 Residual Strength of Oil Tankers Using the IACS Rules

A closer look will be taken at a study carried out by Hussein and Guedes Soares [8] in order to gain a better acquaintance with a number of aspects of the behaviour of damaged hull girders.

The residual strength of a series of double hull tankers was evaluated following CSR requirements, assuming a 50 per cent reduction in corrosion margin. Particulars are included in Table 7.8. As suggested in the rules, the progressive (iterative-incremental) collapse method was used and the damage was also as stipulated in the rules. The effect of position and extent of damage on residual strength was studied in the cases of side shell impact and grounding.

Figure 7.23 depicts the effect that the extent of damage has on the deck and bottom section moduli. In general, the section moduli are largely insensitive to the extent of damage. However, the safety margins differ according to location of damage and also size of ship. In the case of side shell damage the lowest margin occurs in the case of the smaller ship (Figure 7.23a). The bottom section modulus exceeds minimum requirements for all ship sizes whereas the deck section modulus reduces below the minimum acceptable for the larger vessels (Figure 7.23b). Similar conclusions can be drawn in the case of grounding damage, although safety may be compromised only in the case of extensive damage (Figure 7.23c–d).

In Figure 7.24 the effect of damage location on ultimate bending moment is shown, and it is clear that the effect is greatest when damage occurs at the fibres furthest from the neutral axis. Also, as expected, the greater the extent of damage, the larger is the reduction in ultimate bending strength. The corresponding residual strength index with those of the other tanker types are plotted in Figure 7.25. It is seen that the RIF (defined by Equation 7.9) decreases to approximately 0.95 and is largely insensitive to the damage size to depth ratio.

7.4 Ultimate Strength of the Hull Girder in Vertical Bending

In this section we shall address the question of ultimate strength of the hull girder in longitudinal vertical bending. Collapse under longitudinal bending represents an

Table 7.8 *Tanker principal dimensions and hull girder intact strength*

Quantity		Tanker 1	Tanker 2	Tanker 3
L_{BP}	m	320	264	233
B	m	58	45.1	42
D	m	31	23.8	20.0
C_b	–	0.82	0.83	0.84
M_U	MNm	19,996	11,062	7,823
Z_{deck}	m^3	70.49	39.31	24.25
Z_{bot}	m^3	122.48	48.6	34.54

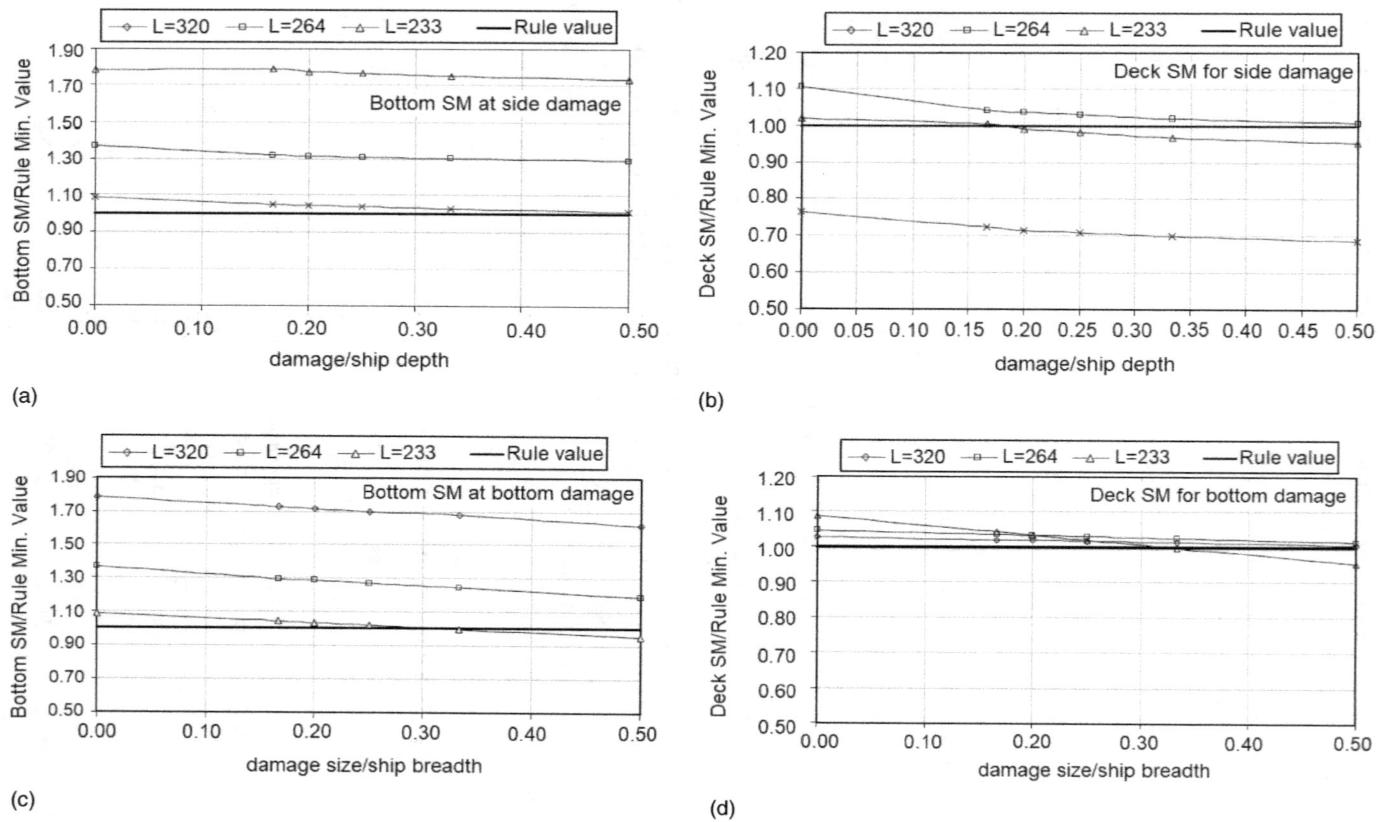

Figure 7.23 Effect of damage on deck and bottom section moduli.

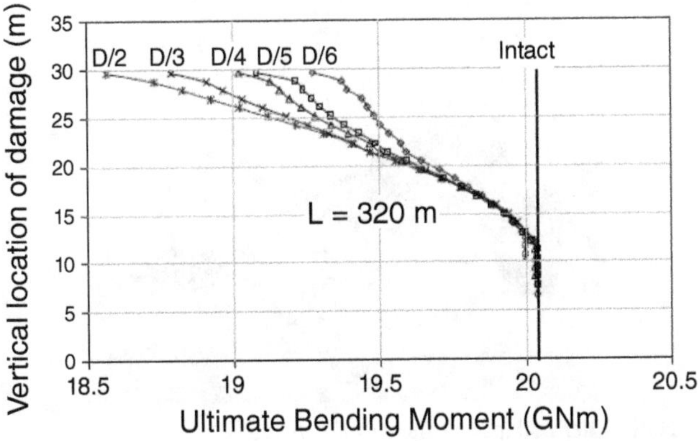

Figure 7.24 Effect of location and extent of damage on residual bending moment (Tanker 1).

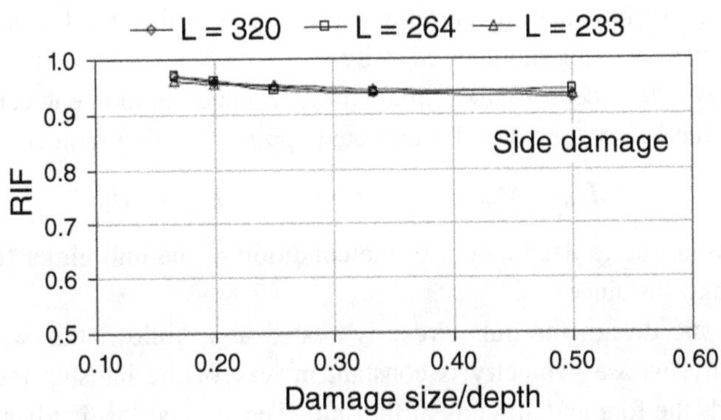

Figure 7.25 Residual strength indices (RIF) for three tankers (worst-case scenario of side shell damage).

extreme type of failure, but it has occurred on several occasions and for a number of reasons (Figure 7.26).[4]

7.4.1 Introduction

The most important case of loading of the hull girder is that of longitudinal vertical bending, although in principle, failure due to horizontal or combined vertical and horizontal bending may also arise. The latter cases will be considered in a later section.

Collapse of the hull girder due to vertical bending will occur if the applied bending moment, M_{ext}, exceeds the actual ultimate bending moment, M_u:

$$M_{ext} \geq M_u \tag{7.13}$$

[4] Two more cases are illustrated in Figure 1.9.

Figure 7.26 Hull girder collapse of the *Trade Daring*.

This may occur due to unintended overloading at sea, or in port during loading/discharging operations such that M_{ext} exceeds the normally expected applied bending moment by a significant extent (even allowing for the partial safety factor γ_R between the design and ultimate bending moment capacity).

Collapse may also occur if the actual ultimate bending moment capacity M_{uc} has reduced to a value below the normally expected applied bending moment, M_{ext}, that is,

$$M_{ext} \geq M_{uc} \qquad \text{where} \qquad M_{uc} < M_u \tag{7.14}$$

This may occur due to degradation of the condition of the hull girder (corrosion, local buckling, damage sustained).

In linear elastic design the hull girder is treated as a hollow, thin-walled section of some complexity, whose geometry is constant in way of the midship region but which changes towards the fore and aft ends of the ship. The applied loading has both static and dynamic components, although the assumed load condition is quasi-static. The design value of the hull girder section modulus, Z_d, is then determined using classification society rules, which make use of synthetic formulas that are based on elastic design principles and simplifying assumptions regarding nonlinear effects. The rule section modulus is thus a means of ensuring that the stresses at any point of the hull girder remain elastic throughout the operating life of the ship. However, two hull girder sections that are designed to the same value of maximum permissible elastic stress will not necessarily have the same ultimate strength because their structural arrangements and scantlings differ. As a result, the ultimate strength of individual members will also differ at a local level. The partial safety factor that reflects the true safety margin against collapse of the hull girder will therefore not be the same for ships that are designed elastically.

Since vertical bending of the hull girder results in in-plane loading of individual hull girder components, the failure modes will be relevant to tension and compression. Failure in tension may occur by yielding or possibly by brittle fracture (Figure 7.27) whereas failure in compression will involve inelastic buckling (Figure 7.28). Brittle fracture need not be considered, as it has long since been eliminated as a possible failure mode from the

(a) (b)

Figure 7.27 Observed hull girder failure modes (side shell structure). (a) Brittle fracture (M/T *Ponaganset*, T2-SE-A2 tanker, 1947). (b) Elasto-plastic buckling and tearing (VLCC *Energy Concentration*, 1981).

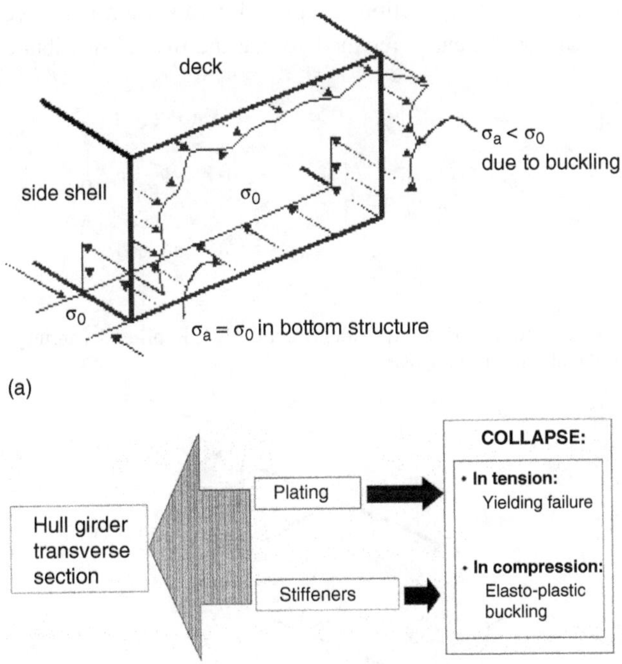

Figure 7.28 Ultimate failure of the hull girder. (a) Failure mechanism of a ship section in sagging [2]. (b) Failure modes of hull girder. Figure 7.28a reprinted courtesy of the Royal Institution of Naval Architects, London.

standpoint of material behaviour. The failure modes that do need to be considered are yielding in tension and buckling in compression (Figure 7.28b). Since in ship structures the slenderness of individual members is such that compressive failure occurs after the initiation of yielding, buckling does not take place in the elastic range but is *inelastic.* Figure 7.27b shows the side shell structure of the VLCC *Energy Concentration* that failed in compression during a discharging operation in port. In this case the strength deck remained intact with failure occurring in the bottom and side shell structure.

Therefore, in order to determine the ultimate strength of a hull girder we need to allow for yielding in tension and inelastic buckling in compression. These failure modes are summarised in Figure 7.28b.

7.4.2 Upper Bound Estimate of the Ultimate Strength

A first approximation to the ultimate strength that represents an upper bound can be obtained by ignoring local buckling failure. This facilitates the calculation considerably, and in this case the stress distribution is as shown in Figure 7.30. In this section we shall describe Caldwell's approach to this problem [2].

We consider the hull girder section shown in Figure 7.29 that comprises the strength deck, vertical sides and bottom structure. The plating is assumed to be longitudinally stiffened throughout. The actual section is replaced by a section that is equivalent to it in that the area of each component is retained so that the overall distribution of material is

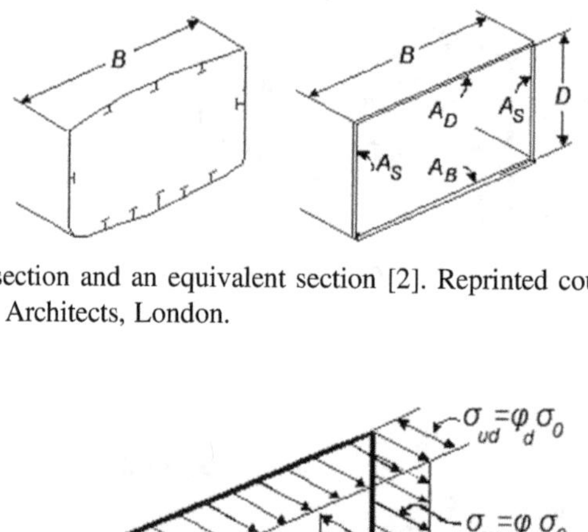

Figure 7.29 Real section and an equivalent section [2]. Reprinted courtesy of the Royal Institution of Naval Architects, London.

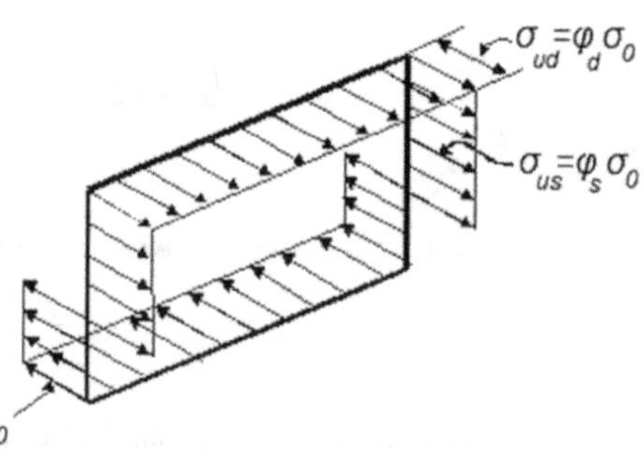

Figure 7.30 Section in fully plastic state. Reprinted courtesy of the Royal Institution of Naval Architects, London.

correct. The plate thickness is increased in order to allow for the area of the stiffener cross-sections. The total area of the cross-section is

$$A = A_d + A_b + 2A_s \tag{7.15}$$

where

A_d = area of strength deck[5]
A_b = area of bottom structure
A_s = area of side shell structure

The material is assumed to be ductile, exhibiting a sharp yield point and a distinct yield stress σ_o in both tension and compression (elastic perfectly plastic response). In ship structures the neutral axis of the section generally lies closer to the bottom structure so that failure initiates in the strength deck. Following failure of the strength deck, as the applied loading is increased, other parts of the section will also fail. After deck failure there will be a shift in the neutral axis towards the bottom structure to ensure equilibrium in the longitudinal direction. As failure spreads throughout the cross-section, the neutral axis translates so that in the fully plastic state, the plastic neutral axis divides the section into two parts that have equal areas (Figure 7.30).

For the plastic neutral axis (PNA), we have

$$A_d + \frac{2c_d}{D}A_s = \frac{2(D - c_d)}{D}A_s + A_b \tag{7.16}$$

$$\frac{c_d}{D} = \frac{2A_s + A_b - A_d}{4A_s} \tag{7.17}$$

The fully plastic bending moment, M_p, is then

$$\begin{aligned} M_p = {} & \sigma_o A_d c_d + \sigma_o A_b (D - c_d) + \\ & 2A_s \sigma_o \left\{ \frac{c_d}{D}\frac{c_d}{2} + \frac{D - c_d}{D}\frac{D - c_d}{2} \right\} \end{aligned} \tag{7.18}$$

or

$$M_p = \sigma_o Z_p \tag{7.18}$$

where Z_p, the plastic section modulus is given by

$$Z_p = A_d c_d + A_b (D - c_d) + 2A_s \left\{ \frac{c_d c_d}{D2} + \frac{D - c_d}{D}\frac{D - c_d}{2} \right\} \tag{7.19}$$

Writing

$$A_d/A = \alpha_d, A_s/A = \alpha_s, A_b/A = \alpha_b, c_d/D = \gamma$$

[5] *d*: deck; *b*: bottom; *s*: side.

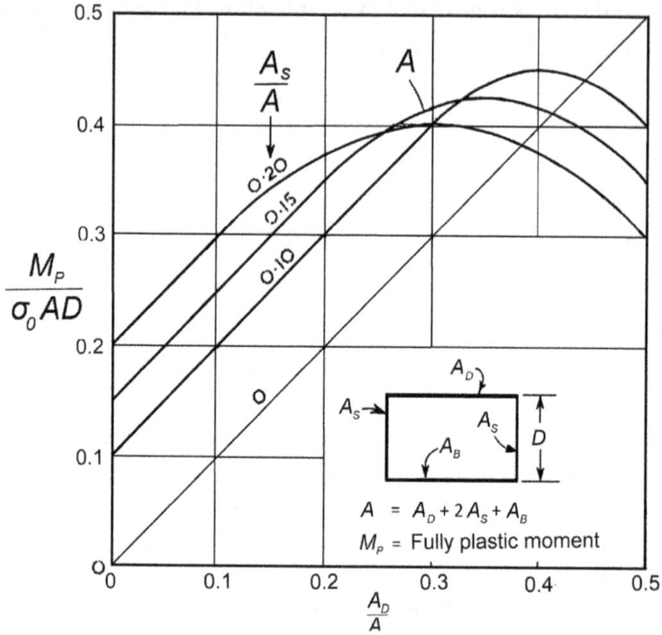

Figure 7.31 Variation of plastic collapse with geometric parameters [2]. Reprinted courtesy of the Royal Institution of Naval Architects, London.

we obtain

$$M_p = \sigma_o AD \left\{ \alpha_d \gamma + \alpha_b (1 - \gamma) + 2\alpha_s \left(\frac{1}{2} - \gamma + \gamma^2 \right) \right\} \tag{7.20}$$

where

$$\gamma = \frac{2\alpha_s + \alpha_b - \alpha_d}{4\alpha_s} \tag{7.21}$$

From Figure 7.31 the following conclusions are reached:

1. 'Structural efficiency', defined as the ratio of strength to weight of the section, increases when α_s reduces.
2. For any given value of α_s, efficiency is maximised when the area of the cross-section of the deck is equal to that of the bottom structure ($\alpha_d = \alpha_b$). Therefore,

$$\begin{aligned} \alpha_D &= \alpha_B = \frac{1}{2}(1 - 2\alpha_S) = \frac{1}{2} - \alpha_S \\ \gamma &= \frac{1}{2} \\ \frac{M_p}{\sigma_o AD} &= \left(\frac{1}{2} - \alpha_s \right) + 2\alpha_s \left(\frac{1}{2} - \frac{1}{2} + \frac{1}{4} \right) = \\ &= \frac{1}{2}(1 - \alpha_s) \end{aligned} \tag{7.22}$$

From Equation (7.22) it is seen that the plastic collapse moment increases when the area of the side shell structure is reduced. The ideal optimum section therefore consists of two flanges of equal areas and no webs.

7.4.3 Progressive Collapse Analysis

In the approach described in the previous section it was assumed that overall collapse occurs when each and every component of the hull girder section reaches its squash load. This is an oversimplification, and in order to obtain a more accurate solution it is necessary to develop an understanding of the nature of collapse of each part of the hull girder as well as to estimate relevant parameters. We therefore need to consider the possible collapse modes that arise at a local and at an intermediate structural level, since these will affect the nature of collapse of the entire cross-section.

Test results have shown that in strength decks of normal geometrical proportions, elasto-plastic buckling of stiffened plates takes place between transverse frames (*interframe* buckling), whereas in the case of lightweight decks overall buckling of the entire grillage occurs. Since the overwhelming majority of structural arrangements are of the former type, we shall focus our attention on interframe elasto-plastic buckling of stiffened plates.

It is useful at this point to compare an actual strength curve of a stiffened plate with that representing the elastic perfectly plastic response in which buckling does not occur. Figure 7.32 includes curves for several different types of response. The case involving no buckling is represented by the dotted line that extends upwards, whereas the true response that includes elasto-plastic buckling is represented by the continuous line.

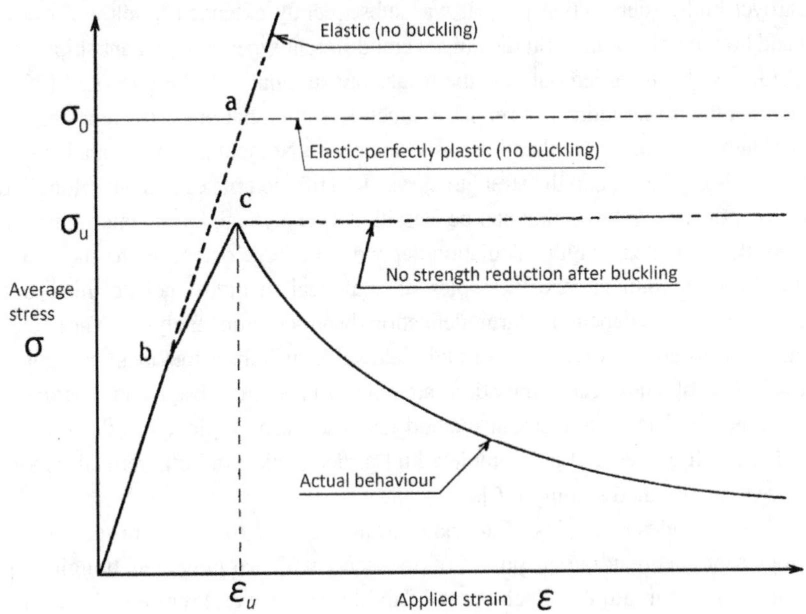

Figure 7.32 Comparison of idealised (no-buckling) response with elasto-plastic buckling strength curve of stiffened plate section [25].

The following observations are made:

- The ultimate strength of the stiffened plate section is less than the yield strength of the material ($\sigma_u < \sigma_0$).
- Following failure, the load-carrying capability of the section diminishes.

One other fact that has to be borne in mind is that individual stiffened plate units will fail when the applied compressive strain reaches a critical value (ε_u). This strain is not applied at the same instant in time for all stiffened plate elements, but in a sequential manner, as the curvature of the section increases. Hence, for a given curvature, the load carried by individual stiffened plate elements differs, depending on their distance from the instantaneous neutral axis of the cross-section as well as on the form of the individual strength curve. Failure of the hull girder section thus takes place in a sequential manner so that the entire process is one of *progressive collapse*. Overall failure of the hull girder is the end-result of failure of all elements that comprise the section.

The important calculation that has to be performed is the determination of the strength curve of each type of stiffened plate that forms part of a particular hull cross-section. Once this is done, it is necessary to simulate the response of the entire cross-section in order to calculate the ultimate strength. This is done by treating the section as an assembly of stiffened plate and other elements, each having a particular load-deflection curve and each being allocated a particular position (distance from neutral axis) and orientation in the cross-section.

This approach forms the basis of the method introduced by IACS in the Harmonised Common Structural Rules for bulk carriers and oil tankers. Details of the IACS procedure are included in Appendix D. The method was first applied to the calculation of the ultimate strength of a destroyer hull girder section [23]. It was subsequently extended to allow for combined vertical and horizontal bending. Studies of several different types of merchant ships, shown in Figure 7.33, have been carried out, and the results are summarised in Figure 7.34 [25]. It was confirmed that the actual ultimate strength is much less than that predicted ignoring buckling and post-collapse unloading, so that these aspects need to be included in any analysis.

The methods used to obtain the strength curves for stiffened plates and other longitudinally effective components of the section can be classified as analytical, hybrid and numerical. The accuracy of the ultimate strength calculation depends to a large extent on the method used to obtain the local strength curves. Examples of analytical methods include those based on analytical solutions (Ostapenko), large-deflection beam-column analysis (Yao) and rigid-plastic mechanism analysis (Rutherford and Caldwell). In hybrid methods, computer-based iterative solutions of analytical formulations are obtained and are subsequently incorporated in Smith's procedure. The finite element method has also been employed in the modelling of both local strength curves and the complete hull girder section in both vertical bending and several cases of combined loading [16].

Figure 7.35 includes examples of ultimate strength curves for an entire section. The two curves represent the bending response of the section in sagging and hogging, plotted against the applied curvature χ. Each curve is divided into several regions. When curvature is first introduced to the section, the response is linear elastic. At some point yielding commences in one or more parts of the section. As yielding spreads, a loss of stiffness of

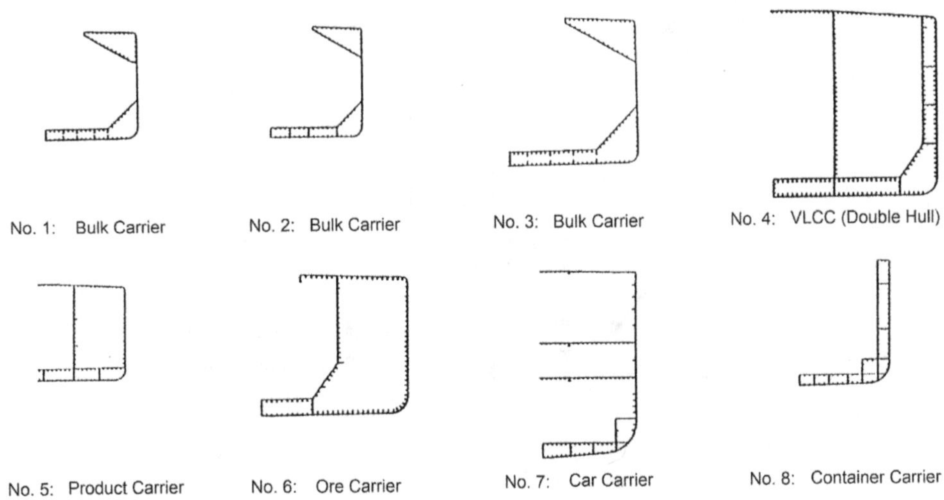

Figure 7.33 Cross-sections considered in reserve strength factor calculation [25].

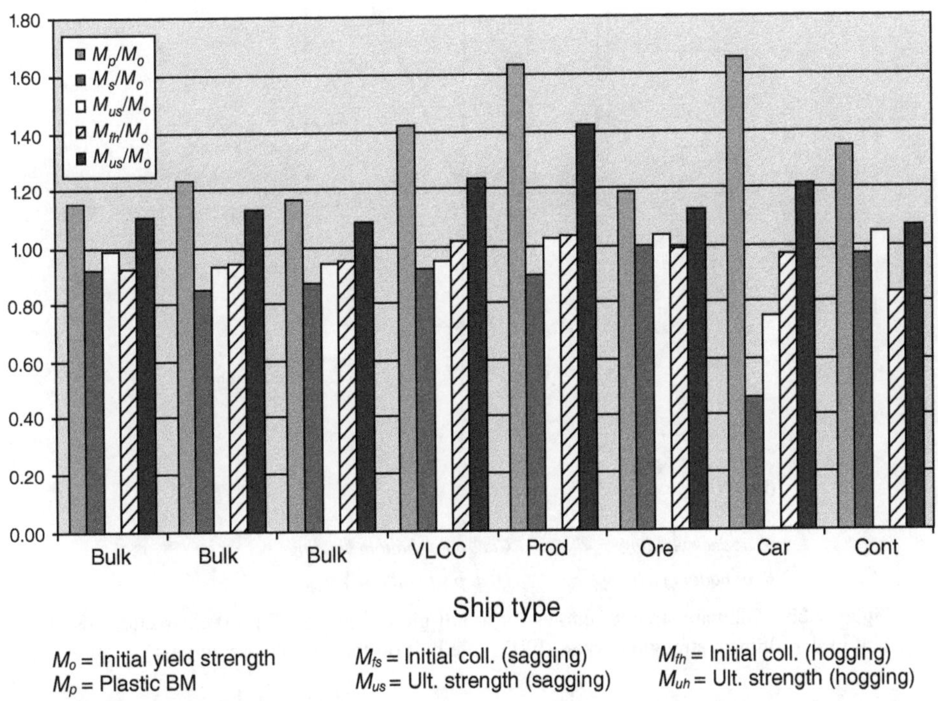

Figure 7.34 Reserve strength factors for ultimate strength of various ship types [25].

the cross-section results, as an increasing proportion of the material is no longer capable of sustaining further load. The stiffness eventually reaches zero, so that the section cannot carry any additional load; this is in effect the point of collapse of the hull girder. If the curvature is increased beyond this point, unloading occurs, that is, load formerly carried by the section is shed to adjoining intact members.

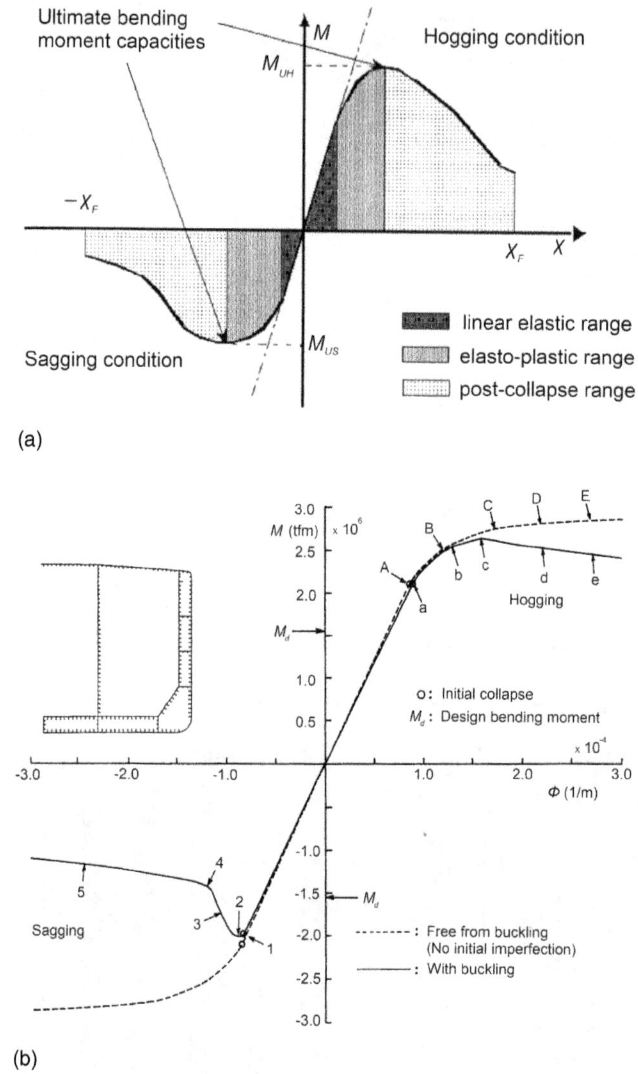

A, a: deck yielding *C, c: inner bottom yielding*

B, b: bottom yielding *D, d: post-collapse range*

Figure 7.35 Ultimate strength curves for a hull girder section [27]. (a) Characteristics of hull girder ultimate strength curves. (b) Double hull oil tanker collapse curves.

It is evident from Figure 7.35a and from the results included in Figure 7.35b that hogging and sagging responses differ. This is because of the asymmetry of the distribution of material within the section. Figure 7.35b shows strength curves for an oil tanker section. It is seen that, as expected, collapse occurs at a higher bending moment when in a hogging condition. The figure also includes curves for the no-buckling condition, and it is seen that in the sagging condition sharp unloading occurs following collapse. This is because of the instantaneous loss of stiffness of all members comprising the strength deck. In the hogging

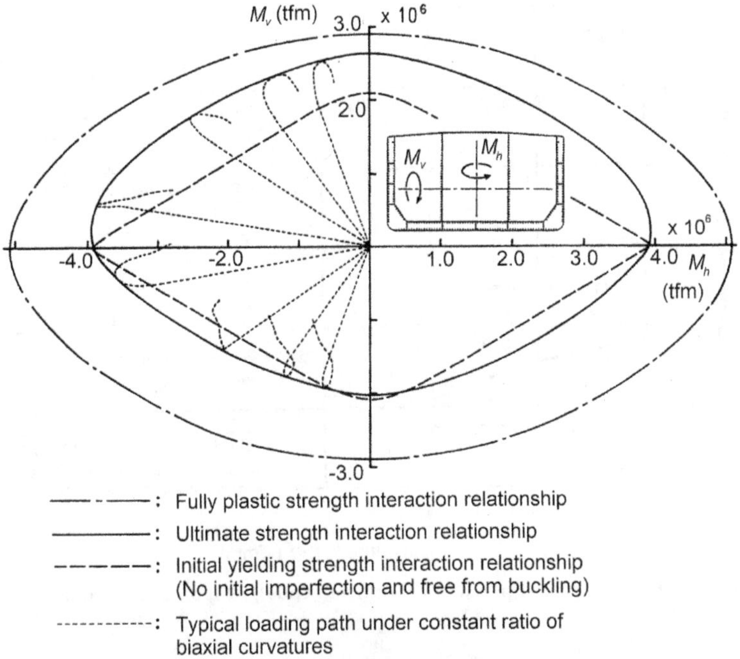

Figure 7.36 Interaction relationship for ultimate hull girder strength under combined vertical and horizontal bending (double hull tanker) [27].

condition the structure exhibits a greater degree of redundancy, as the bottom structure consists of the inner and outer bottom shell structure.

When a ship sails in rough seas and is subjected to significant rolling, both vertical and horizontal longitudinal bending moments act. As seen in Section 6.2.9, under these conditions the longitudinal bending stresses increase, with the maximum value occurring in way of the intersection of the strength deck with the side shell plating. It is therefore necessary to assess the effect this has on ultimate strength. In order to do so it is necessary to consider a range of horizontal to vertical applied curvature ratios. The results of a study on the response of a double hull tanker are summarised in Figure 7.36 [27].

In Figure 7.36 the dotted lines represent horizontal/vertical load paths for different curvature ratios. The solid line is the envelope of these paths and is the interaction curve for ultimate hull girder strength under combined loading. The dashed line represents the interaction at initial yielding, assuming the absence of initial imperfections and ignoring buckling. The region between the two curves therefore represents the reserve of strength from initial yielding up to collapse. The outer chained line represents fully plastic collapse. From the diagram we note that the collapse bending moment in pure horizontal bending exceeds that in vertical bending. This is because $B > D$ for this and the overwhelming majority of ship sections.

A series of studies on the effect of coupled loading on ultimate strength using nonlinear finite element software have been performed and are described in [16].

The effect of lateral pressure on ultimate strength has been studied in the case of a 313,000 dwt metric tonne double hull tanker with two longitudinal bulkheads [16]. Results

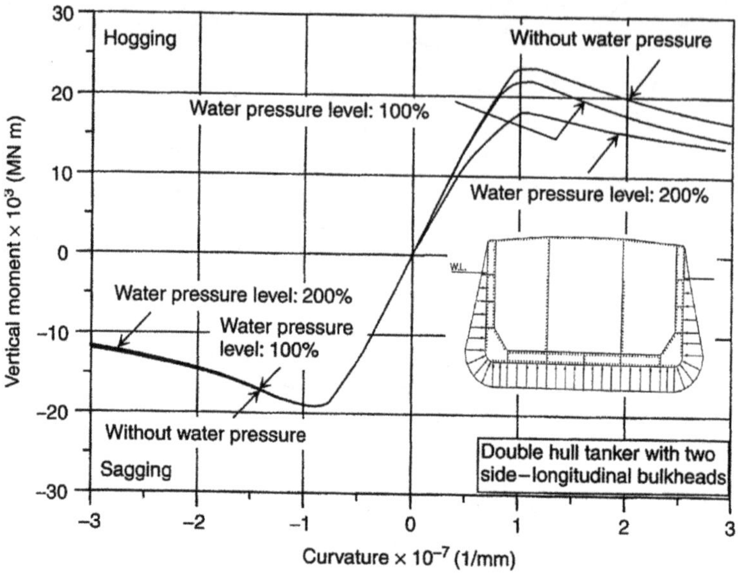

Figure 7.37 Effect of lateral pressure on vertical bending response (double hull tanker).

were obtained for zero, 100 per cent and 200 per cent water pressure (Figure 7.37). The plating was assumed to have average imperfections in the direction of application of the pressure load. The effect that lateral pressure has on stiffened plating that is loaded in end-compression depends on the magnitude of the pressure in comparison to the axial load. If the pressure is significant, it produces a clamped condition along the edges, whereas if it is small, the boundary condition is closer to a simple support. It was found that under hogging, the pressure effect can be important, whereas in sagging the effect is small. This is because the highest pressure is applied to the outer bottom plating that is loaded in compression when the ship is in a hogging condition.

In several cases in which ships have broken in half, this has been due to a reduction in its bending capacity due to corrosion below the applied bending moment when the accident took place. It is therefore important to assess the effect corrosion has on ultimate strength in vertical bending. In order to do so the effect is studied at a local level and then the adjusted local strength curves are used in the global strength calculation. A study carried out [9] on two VLCCs and two bulk carriers showed that a loss of thickness in the bottom plating largely affects hogging strength whereas a loss of thickness in the deck plating affects sagging strength. In another study it was found that a 30 per cent overall reduction in thickness leads to a strength reduction of 34 per cent and 38 per cent in sagging and hogging, respectively [27].

In order to gain a clear idea of the relative importance of each of the factors that affect the ultimate strength of the hull girder, a sensitivity analysis was performed by the ISSC Special Task Force on Ultimate Hull Girder Strength [26]. In this study the sensitivity, S_i, was defined as

$$S_i = \frac{\partial M_u}{\partial x_i} \tag{7.23}$$

Table 7.9 *Design variables for ultimate strength sensitivity analysis and their statistical properties*

Design Variable		Mean Value	COV
Plate thickness	t_p	nominal values	0.6
Stiffener thickness	t_s	nominal values	0.6
Mild steel yield stress	σ_{MS}	285 MPa	9.0
HT32 yield stress	σ_{HT32}	364 MPa	7.0
HT36 yield stress	σ_{HT36}	402 MPa	4.3
HT40 yield stress	σ_{HT40}	440 MPa	3.0
Compressive residual stress	σ_{rc}	calculated value	30.0
Initial deflection in panels	w_{p0}	calculated value	50.0
Initial deflection in stiffeners (flexural mode)	w_{s0}	$0.0005L$	20.0
Initial deflection in stiffeners (tripping mode)	v_{s0}	$0.0005L$	20.0

The sensitivity can be normalised by multiplying by μ_{xi}/μ_{Mu} where μ_{xi} and μ_{Mu} are the mean values of the *i*th variable and the ultimate bending moment respectively. Thus,

$$\overline{S_i} = \frac{\mu_{x_i}}{\mu_{M_u}}\frac{\partial M_u}{\partial x_i} = \frac{\partial\left(M_u/\mu_{M_u}\right)}{\partial\left(x_i/\mu_{x_i}\right)} \tag{7.24}$$

The design variables considered and their respective statistical properties are listed in Table 7.9. In Table 7.9 $L =$ frame spacing.

The mean values of the ultimate strength in each case are evaluated using the mean values of each design variable:

$$\overline{S_i} = \frac{\mu_{x_i}}{\mu_{M_u}}\frac{\partial M_u}{\partial x_i} = \frac{\left(M_u{}^{+5\%} - M_u{}^{-5\%}\right)/\mu_{M_u}}{\left(1.05\mu_{x_i} - 0.95\mu_{x_i}\right)/\mu_{x_i}} = \frac{M_u{}^{+5\%} - M_u{}^{-5\%}}{0.1\mu_{M_u}} \tag{7.25}$$

where $M_u{}^{-5\%}$ and $M_u{}^{+5\%}$ are the ultimate hull girder strength by changing the design value x_i by $\pm 5\%$. The variance of the ultimate strength is found from

$$\begin{aligned}\sigma_{M_u}^2 = \sum\left(\frac{\partial M_u}{\partial x_i}\sigma_{x_i}\right)^2 &= \left(\frac{\partial M_u}{\partial t_p}\sigma_{t_p}\right)^2 + \left(\frac{\partial M_u}{\partial t_s}\sigma_{t_s}\right)^2 + \left(\frac{\partial M_u}{\partial \sigma_{MS}}\sigma_{\sigma_{MS}}\right)^2 + \\ &+ \left(\frac{\partial M_u}{\partial \sigma_{HT32}}\sigma_{\sigma_{HT32}}\right)^2 + \left(\frac{\partial M_u}{\partial \sigma_{HT36}}\sigma_{\sigma_{HT36}}\right)^2 + \left(\frac{\partial M_u}{\partial \sigma_{HT40}}\sigma_{\sigma_{HT40}}\right)^2 + \\ &+ \left(\frac{\partial M_u}{\partial \sigma_{rc}}\sigma_{\sigma_{rc}}\right)^2 + \left(\frac{\partial M_u}{\partial w_{op}}\sigma_{w_{op}}\right)^2 + \left(\frac{\partial M_u}{\partial w_{os}}\sigma_{w_{os}}\right)^2 + \left(\frac{\partial M_u}{\partial v_{os}}\sigma_{v_{os}}\right)^2\end{aligned} \tag{7.26}$$

Each term on the right-hand side of Equation (7.22) is calculated from

$$\frac{\partial M_u}{\partial x_i}\sigma_{x_i} = \frac{\partial\left(M_u/\mu_{M_u}\right)}{\partial\left(x_i/\mu_{x_i}\right)}\frac{\mu_{M_u}}{\mu_{x_i}}\sigma_{x_i} = \frac{\partial M_u/\mu_{M_u}}{\partial x_i/\mu_{x_i}}\frac{\mu_{M_u}}{\mu_{x_i}}\mu_{x_i}COV_{x_i} = \frac{\partial M_u/\mu_{M_u}}{\partial x_i/\mu_{x_i}}\mu_{M_u}COV_{x_i} \tag{7.27}$$

Table 7.10 includes the normalised values of the sensitivities of selected variables that have a bearing on the value of ultimate strength. The most influential values can be determined for each ship type and conclusions with regard to all ship types. In the case of the double hull VLCC, of all parameters, the yield stress is found to be the most

Table 7.10 *Normalised sensitivities for hull girder ultimate strength*

Variable	Double Hull VLCC		Containership		Bulk Carrier		Energy Concentration	
	Sag	Hog	Sag	Hog	Sag	Hog	Sag	Hog
t_p	0.06640	0.04157	0.8972	1.1366	1.1297	1.0096	1.0880	0.7943
t_s	0.2638	0.3438	0.2387	0.1847	0.3159	0.2486	0.3678	0.3615
σ_{MS}	–	–	–	–	–	–	0.00646	0.01545
σ_{HT32}	0.7836	0.8450	0.3038	0.6036	0.01578	0.3300	0.7248	0.9134
σ_{HT36}	0.01687	0.03834	0.4296	0.2412	0.1023	0.2205	–	–
σ_{HT40}	–	–	–	–	0.5848	0.2784	–	–
σ_{rc}	−0.04922	−0.04738	−0.02149	0.003143	−0.03705	−0.02123	−0.03157	−0.01097
w_{p0}	−0.01272	−0.01548	0.01882	−0.02356	−0.0144	−0.00798	−0.02816	−0.00075
w_{s0}	−0.02125	−0.01604	−0.02209	−0.00666	−0.06975	−0.02156	−0.00980	−0.00283
v_{s0}	0.000259	0.00141	0.00085	−0.000002	0.000162	−0.00144	0.000079	−0.00237

influential one. In the case of other ship types it is the plate and stiffener thickness that are most important. Ultimate strength is not found to be affected by welding-induced imperfections. It is also seen that sensitivities differ when the ship is in a hogging and a sagging condition.

A study concerned with the elastic design of ocean-going ships based on IACS requirements showed that design bending moments in hogging are smaller than those in the sagging condition. On the other hand, ultimate bending moments in hogging are larger than those in sagging condition. The safety margin in hogging is thus much larger than in the sagging condition.

References

1. Boresi A. P. Schmidt R. P. (2003) *Advanced Mechanics of Materials* 6th ed. Chichester: John Wiley & Sons Inc.
2. Caldwell J. B. (1965) Ultimate Longitudinal Strength *Trans RINA* Vol. 107 411–418.
3. Choung J. Nam J.-N. Ha T.-B. (2012) Assessment of Residual Ultimate Strength of an Asymmetrically Damaged Tanker Considering Rotational and Translational Shifts of Neutral Axis Plane. *Mar. Struct.* Vol. 25 71–84.
4. Das P. K. Garside J. F. (1991) Structural Redundancy for Continuous and Discrete Systems. Ship Struct. Comm. Rept. SSC-354 Washington, DC.
5. Det Norske Veritas (2009) CSA – Direct Analysis of Ship Structures. Classification Note No. 34.1.
6. Fang C. Das P. K. (2004) Hull girder ultimate strength of damaged ships. *Proc. 9th Conf. Prac. Des. Ships & Float. Struct. (PRADS2004)* Lübeck-Travenmünde 309–316.
7. Ghose D. J. Nappi N. S. Wiernicki C. J. (1994) Residual Strength of Damaged Ship Structures. Ship Struct. Comm. Rept. SSC-381 Washington, DC.
8. Hussein A. W. Soares G. (2009) Reliability and residual strength of double hull tankers designed according to the new IACS common structural rules. *Ocean Engineering* Vol. 36 1446–1459.
9. Jensen J. J. et al. (1994). Ductile Collapse. Rept. of Committee III.1. In Jeffery J. E. Kendrick A. M. (Eds) *Proc. 12th Int. Ship & Offsh. Struct. Cong*. Vol. 1 229–387.
10. Iijima K. Shigemi T. Miyake R. Kumano A. (2004) A practical method for torsional strength assessment of container ship structures. *Mar. Struct.* Vol. 17 355–384.
11. Lloyd A. R. Clawson W. C. (1985) Reserve and residual strength of pile founded offshore platforms. *Proc. Int. Symposium: The Role of Design Inspection and Redundancy in Marine Structural Reliability*. Williamsburg, VA.
12. *Navy.mil*. [Online]. Available: www.history.navy.mil/library/online/wardamagereport no53.htm. (accessed 19 December 2020).
13. Nguyen N. D. (2012) *Torsional Hull Girder Response Of Containerships – Feasible with Cargo Hold Models?* Univ. Rostock.
14. de Oliveira J. G. Zimmer R. A. (1983) Redundancy considerations in the structural design of floating offshore platforms. *Proc. Int. Symposium: The Role of Design Inspection and Redundancy in Marine Structural Reliability*. Williamsburg, VA.
15. Ostapenko A. Chen Y. F. (1982) Effect of Torsion on Strength of Ship Hulls. Fritz Eng. Lab. Rept. No. 468.11. Lehigh Univ.
16. Paik J. K. Thayamballi A. K. (2003) *Ultimate Limit State Design of Steel-Plated Structures*. Chichester: John Wiley & Sons.

17. Pedersen P. T. (1983) A Beam Model for the Torsional-Bending Response of Ship Hulls. *Trans. RINA* Vol. 125 171–182.
18. Pedersen, P. T. (1985) Torsional Response of Containerships. *J. Ship Res.* Vol. 29 194–205.
19. Senjanović I. Vladimir N. Tomić M. (2011) Effective Stiffness of the Engine Room Structure in Large Container Ships. *Brodogradnja* Vol. 62 No. 1 15–27.
20. Senjanović I. Rudan S. Vladimir N. (2009) Influence of Shear on the Torsion of Thin-walled Girders. *Transactions FAMENA* XXXIII-2 35–50.
21. Shanley F. R. (1957) *Strength of Materials*. New York: McGraw Hill Book Co.
22. Smith C. S. (1977) Influence of Local Compressive Failure on Ultimate Longitudinal Strength of a Ship's Hull. *Proc. Int. Symposium on Practical Design in Shipbuilding (PRADS)*. Tokyo 17–21 October 73–79.
23. Smith C. S. Dow R. S. (1981) Residual Strength of Damaged Steel Ships and Offshore Structures *J. of Constructional Steel Research* Vol. 1 No. 4 2–15.
24. Wang G. Chen Y. Zhang H. Shin Y. (2000) Residual strength of damaged ship hull. In *Ship Structures for the New Millenium: Supporting Quality in Shipbuilding*. Arlington, VA.
25. Yao T. (1995) Investigation into longitudinal strength of ship hull; historical review and state of the art. *30th JTC Meeting* Singapore 1–40.
26. Yao T. et al. (2000) Ultimate hull girder strength. Special Task Committee VI.2 Rept. In Ohtsubo H. Sumi Y. (Eds) *Proc. 14th Int. Ship & Offsh. Struct. Cong*. Vol. 2. Amsterdam: Elsevier 321–391.
27. Yao T. Fujikubo M. Kondo K. Nahagama S. (1994) Progressive collapse behaviour of double hull tanker under longitudinal bending. *Proc 4th ISOPE* Vol. VI Osaka 570–577.

8
Hull Girder Vibration

No problem can withstand the assault of sustained thinking.

Voltaire

8.1 Introduction

Ship hull girders are large structures, their mass is not uniformly distributed in the longitudinal direction, and they float on the surface of the sea with a significant proportion of their mass immersed. They are subjected to a variety of loads that can result in a static and a dynamic response, at a local or a global level. In earlier chapters the low-frequency loads to which a ship is subjected were assumed to produce a static or, more strictly speaking, a quasi-static response. In general, the response consists, on the one hand, of distortions and stresses in the hull girder and, on the other, of motions that involve translations and rotations about the three coordinate axes. The latter are the subject matter of *ship dynamics,* in which the ship is treated as a rigid body.

Upon reflection, it is seen that from this viewpoint matters are interpreted from the perspective of the ship, and the effect the ship has on the body of fluid that surrounds it is disregarded. This consists of accelerations and motions of the fluid, which in turn result in changes to the hydrodynamic loading of the ship, and thereby to the initially predicted response of the ship itself. The response includes stress and displacement fields (structural response) as well as motions. The study of the body–fluid interaction is the subject matter of *hydroelasticity theory* that was formulated in the 1970s [5] and is now being applied much more widely in practical ship design.

In this chapter we shall be considering the global vibration response of the hull girder, which is, strictly speaking, a problem in ship dynamics. However, as we shall see, the effect of the surrounding water is important, and needs to be taken into account, as it affects the vibration behaviour of the ship. Vibration is important at a local level as well, and in certain cases interaction between global and local vibration arises. This topic as well as the measurement of vibration and that of design issues lie outside the scope of this chapter.

In order to understand global ship vibration, the subject has to be approached in a number of stages. It is necessary to start from a simple and well-understood model and proceed towards real ship conditions in an orderly manner. In this way it is possible to assess the importance of the assumptions made at each stage. The following stages will be considered:

1. Elementary vibration theory of a single-degree-of-freedom (SDOF) system (lumped mass)
2. Undamped free vibration of the Euler–Bernoulli free-free beam (continuous structure)
3. Energy method for linear vibration of uniform beam
4. Free vibration of Timoshenko beam with free ends
5. Hull girder vibration using empirical formulas based on elementary theory and the energy method (1–4)
6. Derivation of differential equation for the vertical vibration of the hull girder
7. Derivation of system of differential equations for coupled horizontal and torsional vibration of the hull girder
8. The use of the FEM to determine hull girder natural frequencies and vibration modes (Section 13.2.2 and Appendix E).

The dynamic response of complex structures such as that of a ship is most effectively studied using computer-based methods. These include the *direct integration* method and the *mode superposition* (*modal analysis*) method. Of these the latter is the one that is used in ship vibration problems, the choice being made on the grounds of numerical efficiency. The application of the finite element method to solve ship vibration problems is discussed briefly in Chapter 13. The modal analysis method, which is used to determine the natural frequencies of vibration of any continuous structure such as a ship, is presented in Appendix E.

8.2 Elementary Vibration Theory: The Undamped SDOF System

The direct experience of vibration that most people have is that of machinery that resonates, or else by listening to a stringed musical instrument or a tuning fork. It is possible that seamen and/or passengers on board a ship have experienced vibration, in which case they will have gained direct experience of the extent to which it interferes with activities. A vibration is in effect a small amplitude oscillation about a rest position of a body that is supported in a particular way. In order to introduce the most important characteristics and quantitative measures of the vibration of solid bodies, we turn to the case of a single-degree-of-freedom system consisting of a mass m and spring of stiffness k to which a load $F(t)$ is applied. The degree of freedom of the system is the displacement of the mass in the direction of action of the force.

In this case the system consists of a spring of length l attached to a rigid horizontal surface to whose lower end a mass m is attached, so that following attachment, the length of the spring becomes $l+L$, L being the extension due to the weight of the mass, mg. This is the *equilibrium* position of the system about which subsequent displacements are measured. The displacement from the rest position is denoted by w, and this assumes positive values in the downward direction and negative values in the upward direction.

From Newton's second law,

$$m\frac{d^2w}{dt^2} = \sum_i F_i(t) \tag{8.1}$$

where $F_i(t)$ are the force components in the vertical direction and w is the resulting vertical instantaneous displacement of the body, whose mass m remains constant. Rearranging, we obtain

$$0 = \sum_i F_i(t) - m\frac{d^2w}{dt^2} \tag{8.1'}$$

Equation (8.1′) is now an equation of *dynamic equilibrium* in which the total applied external force, represented by the first term, is balanced by an *inertial force* of equal magnitude acting in the opposite direction, represented by the second term. This is *d' Alembert's principle*, which enables equations of motion to be viewed as equations of dynamic equilibrium.

The forces acting on the mass are:

1. The gravitational force, mg
2. The spring force. For a spring that obeys Hooke's law, at any instant the force will be $-k(L+w)$, where k is the spring constant (force per unit extension) and $(L+w)$ is the extension at any point in time during motion. The minus sign is used to ensure that the force acts in the correct direction, depending on the sign of $(L+w)$. It acts on the body to restore static equilibrium.
3. External force (*forcing function*) $F(t)$ that may be of any form (continuous, periodic, impact, rectangular pulse).

Substituting in Equation (8.1), the equation of motion becomes

$$m\frac{d^2w}{dt^2} = mg - k(L+w) + F(t)$$

Rewriting,

$$m\frac{d^2w}{dt^2} + kw = mg - kL + F(t) \tag{8.2}$$

For static equilibrium,

$$mg - kL = 0$$

Thus, Equation (8.2) becomes

$$m\frac{d^2w}{dt^2} + kw = F(t) \tag{8.3}$$

If the force $F(t)$ is removed, the equation of motion becomes

$$m\frac{d^2w}{dt^2} + kw = 0 \tag{8.4}$$

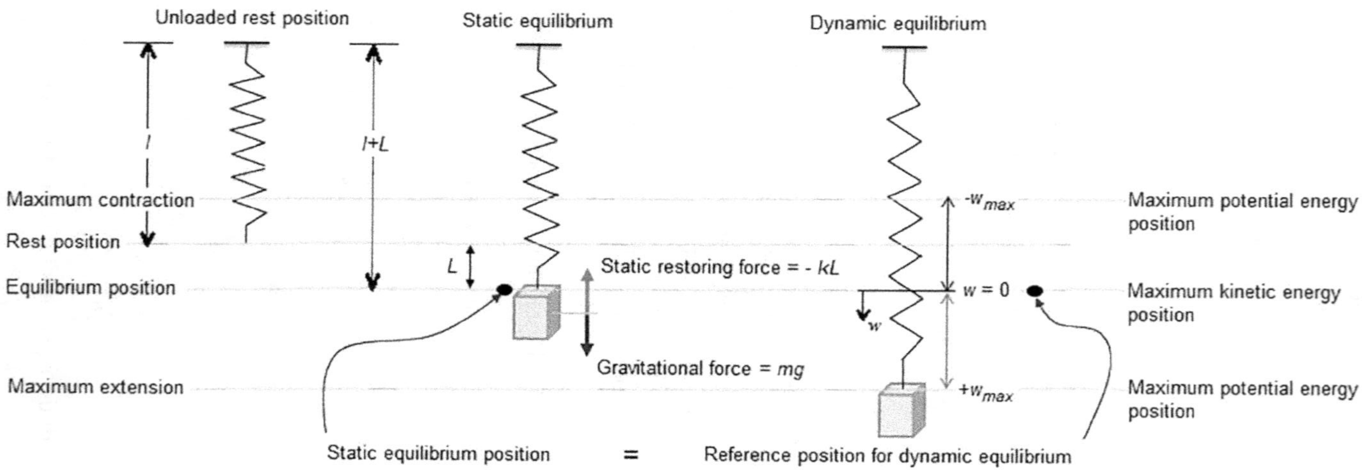

Figure 8.1 Free vibration of mass-spring system (adapted from [12]).

The mass-spring system now undergoes *free vibration*. Equation (8.4) is a second-order ordinary differential equation whose solution has the form $w(t) = Ge^{st}$. Substituting this in Equation (8.4), we obtain

$$\left(ms^2 + k\right)Ge^{st} = 0 \tag{8.5}$$

Dividing through by mGe^{st} and letting

$$\omega^2 = \frac{k}{m} \tag{8.6}$$

Equation (8.5) becomes

$$s^2 + \omega^2 = 0$$

Hence, $s \pm \omega$. The response is then

$$w(t) = G_1 e^{i\omega t} + G_2 e^{-i\omega t} \tag{8.7}$$

where the two terms result from the two values of s and the constants G_1, G_2, represent the amplitudes of motion. If we use Euler's equations,

$$e^{\pm i\omega t} = \cos \omega t \pm i \sin \omega t$$

Equation (8.6) can be written as

$$w(t) = A \sin \omega t + B \cos \omega t \tag{8.8}$$

We see that the motion varies in time in accordance with the trigonometric quantities sinωt, cosωt, where ω is the *circular frequency*. The values of the constants A, B are defined by the initial conditions, that is, the displacement w and the velocity dw/dt at $t = 0$. Letting $t = 0$ in Equation (8.8) and in the corresponding expression for $dw/dt|_{t=0}$ we obtain $A = dw/dt|_{t=0}/\omega$, $B = w|_{t=0}$. Substituting these in Equation (8.8) we obtain

$$w(t) = \frac{1}{\omega}\frac{dw}{dt}\bigg|_{t=0} \sin \omega t + w|_{t=0} \cos \omega t \tag{8.9}$$

This expression gives the displacement of the body from the equilibrium position at any time t and represents *simple harmonic motion*. If $w|_{t=0} = 0$, the response is shown in Figure 8.2.

It is seen that the body executes a repetitive oscillation whose *period T* is the time it takes to undergo one complete cycle. The *natural frequency f* is then given by $f = 1/T$. Since $T = 2\pi/\omega$, substituting from Equation (8.6), the natural frequency of vibration of the mass-spring system shown in Figure 8.2 is

$$f = \frac{\omega}{2\pi} = \frac{1}{2\pi}\sqrt{\frac{k}{m}} \tag{8.10}$$

We see that the natural frequency depends on m and k alone. These are properties of the mass-spring system that define vibration response. This is an important formula that has formed the basis of several early empirical formulas used to predict the natural frequency of vibration of ships. In these allowance has been made for ship conditions, and such formulas

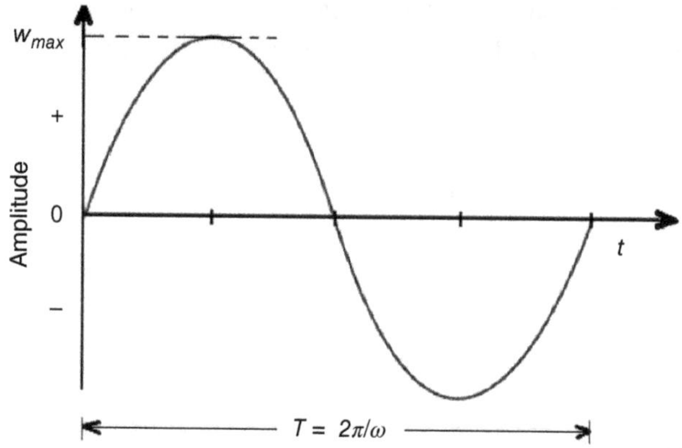

Figure 8.2 Simple harmonic motion.

can still be used at the early stages of design. In the case of ships, vibration arises about the horizontal neutral axis (vertical vibration), the vertical neutral axis (horizontal vibration) and the longitudinal axis (torsional vibration). In each case the appropriate stiffness has to be used. Also, the mass has to be suitably adjusted for the effect of added mass of water, which varies with mode (vertical, horizontal). Both of these issues are addressed in later sections of this chapter.[1]

Alternatively, vibration response can be described by considering the energy content of the body in question at each stage throughout each oscillation (Rayleigh). If a body performs oscillations as described in Figure 8.2, the total energy content at any point in time will consist of kinetic and potential energy components and, in a conservative system, will remain constant throughout each cycle. The kinetic energy is a function of velocity whereas the potential energy is a function of position. When the peak amplitude is reached, the velocity reduces to zero, and thus the kinetic energy is also equal to zero. At this position the potential energy content reaches its maximum value (Figure 8.1). At the position of static equilibrium, the kinetic energy is a maximum and the potential energy is zero.

If the excitation force does not cease and is periodic, as in the case of propeller-excited hull vibration, *forced vibration* occurs. This will cause oscillations of the body, in this case the hull girder, at a frequency that is unrelated to its natural frequency.

If to the mass-spring system is added a device that produces *damping*, specified to be proportional to the instantaneous velocity of the body, this will result in a loss of energy content. This is because energy is dissipated to overcome damping, which may be present for a number of reasons. *Structural damping* (internal friction) arises in all elastic bodies, whereas in the case of bodies that are immersed in water, as ships are, *hydrodynamic damping* also arises. This is because when a ship vibrates, the surrounding water is also set in motion. As a result, a net energy transfer from the body to the fluid takes place. If the excitation force is removed, the vibration of the body–fluid system will eventually die out.

[1] For detailed appraisals of empirical formulas used to predict ship vibration, see Section 8.4.4, appendix A.3.2 of [48] and the ABS document *Guidance Notes on Ship Vibration*.

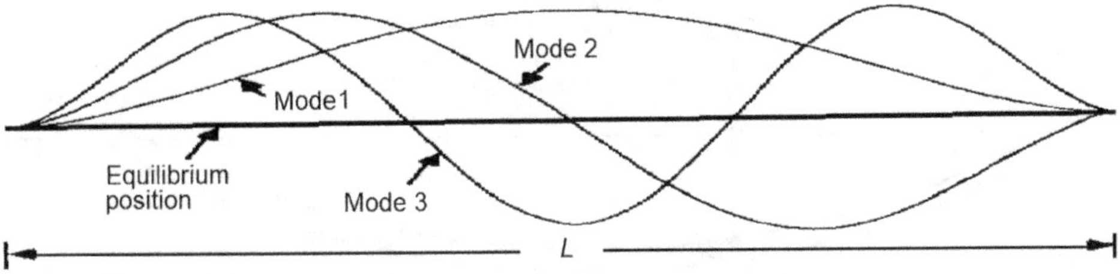

Figure 8.3 First three modes of a slender beam clamped at both ends.

Instead of the rigid body whose response was treated up to now as a 'lumped mass', we will consider a *continuous elastic body* such as a beam. The mass of such an elastic body is distributed in space in a uniform or possibly a non-uniform manner. The number of degrees of freedom then corresponds to all possible translations and rotations of each and every particle of which the mass consists. Such a body will have an infinite number of degrees of freedom to which correspond an equal number of natural frequencies, for each of which there arises a different deflected shape. Each particular shape is termed a *vibration mode* (Figure 8.3). Of the infinity of natural frequencies that a body possesses, a small number are of practical interest, the lowest of which is termed the *fundamental natural frequency.* The magnitudes of the natural frequencies (in cycles/sec or Hertz) depend on structural form, material properties and the nature of the supports.

When an elastic body vibrates, it may do so in one or more modes, each of which may have any amplitude. The magnitudes of the amplitudes of vibration in each mode and the phase relations between modes depend on the nature of the excitation force that acts prior to free vibration. The kinematic relationships relate deflections to strains. Thus, it is seen that the strains will increase when the amplitude increases. Furthermore, the stresses in the body that are related to strains through constitutive relations in the linear elastic range will also increase when the amplitude of vibration increases.

The possible deformations are flexural (bending), extensional and torsional. Of these, in general, the flexural deformations are of most interest because in this case the natural frequencies are lower than for the extensional and torsional modes. As a result, the resulting amplitudes and stresses are usually greater.

In the next section we will consider the transverse vibration of the Euler–Bernoulli beam that is free at both ends. This is the boundary condition that corresponds to a ship floating on the surface of the sea.

8.2.1 Free Vibration of the Euler–Bernoulli Free-Free Beam

In this highly idealised case, the properties of the hull girder (stiffness and weight distribution), its response as well as the environment have been simplified. This case will act as a starting point and will enable us to approach the real problem in stages that will follow. We term the study of beam behaviour in air as the *dry mode*, to distinguish it from the ship that operates in the sea environment (*wet mode*). The latter approximates real conditions much more accurately. At this stage the structural model that is used is, as in the

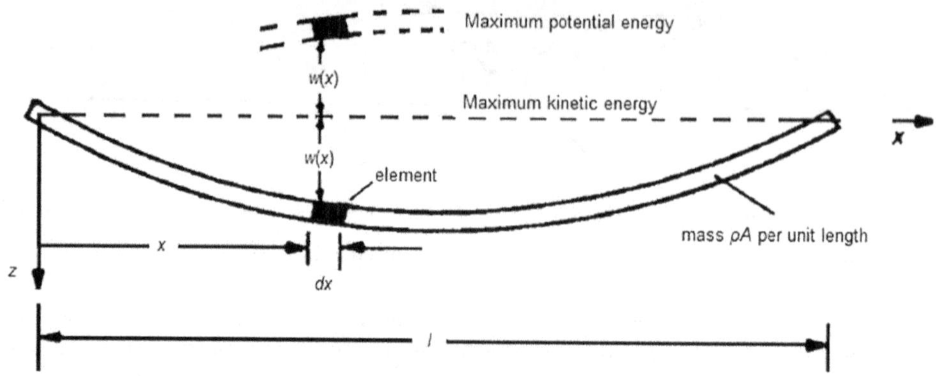

Figure 8.4 Uniform beam undergoing transverse vibration (adapted from [4]). Reprinted courtesy of Elsevier.

case of quasi-static loading, that of a prismatic, free-free beam whose weight and stiffness properties are constant along its length. This is essentially a simplified one-dimensional version of the problem. In the general case, the equation of motion (dynamic equilibrium) is

$$m\frac{d^2w}{dt^2} + c\frac{dw}{dt} + kw = p(t) \tag{8.11}$$

where w = lateral deflection, m = mass, c = damping coefficient, k = stiffness and $p(t)$ = forcing function. Since here we assume $c = 0$ and $p(t) = 0$, the equation of motion is

$$m\frac{d^2w}{dt^2} + kw = 0 \tag{8.12}$$

This is identical to Equation (8.4). In this case it is assumed that the mass m is uniformly distributed and possesses elastic properties and is not a lumped, rigid mass as previously assumed. We shall now obtain the equation of motion of the beam in terms of the beam properties, following which we will proceed to obtain the solution of the equation for undamped free vibration with free-free boundaries.

Consider a thin uniform beam of length l, cross-sectional area A, flexural rigidity EI and material density ρ undergoing transverse vibration (Figure 8.4).

The translation in the z-direction of a point at a distance x from the origin at time t, $w(x,t)$, is assumed to vary sinusoidally with time so that

$$w(x,t) = w(x)\sin\omega t \tag{8.13}$$

Differentiating,

$$\frac{\partial w}{\partial t} = \dot{w} = \omega w(x)\cos\omega t \tag{8.14}$$

and

$$\frac{\partial^2 w}{\partial t^2} = w'' = -\omega^2 w(x)\sin\omega t \tag{8.15}$$

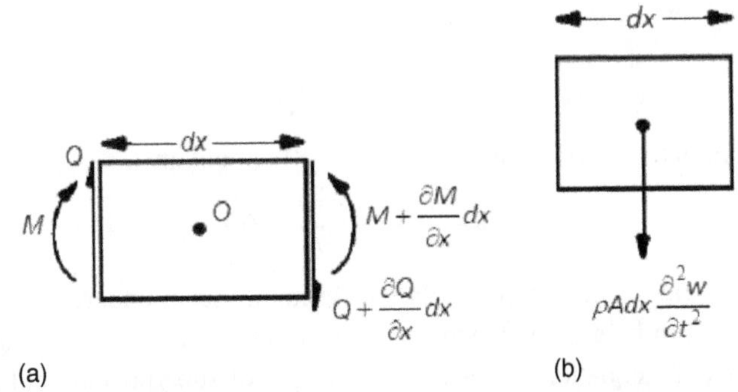

Figure 8.5 Forces acting on a beam element of length dx undergoing free vibration [4]. (a) Stress resultants acting on beam element. (b) Inertial force. Reprinted courtesy of Elsevier.

Equations (8.14) and (8.15) are, respectively, the velocity and acceleration of a point at a distance x from the origin at time t. Consider a free body diagram of an element of length dx (Figure 8.5).

The mass of this element, $dm = \rho Adx$. The element is in motion and is thus subjected to an inertial force $\rho Adx\left(\partial^2 w(x,t)/\partial t^2\right)$. Additionally at the edges stress resultants act, consisting of bending moments M, $M + dM$, and shear forces V, $V + dV$. In order to establish the equation of dynamic equilibrium we consider rotational and vertical dynamic equilibrium of the element. Consider first rotational equilibrium of the element about point O:

$$M + V\frac{dx}{2} + V\frac{dx}{2} + \left(\frac{\partial V}{\partial x}dx\right)\frac{dx}{2} - \left(M + \frac{\partial M}{\partial x}dx\right) = 0$$

Hence,

$$V = \frac{\partial M}{\partial x} \tag{8.16}$$

For dynamic equilibrium in the vertical direction (d'Alembert's principle),

$$\begin{aligned}\left(V + \frac{\partial V}{\partial x}dx\right) - V - \rho Adx\frac{d^2 w}{dt^2} = 0 \\ \frac{\partial V}{\partial x} = \rho A\frac{d^2 w}{dt^2}\end{aligned} \tag{8.17}$$

where $\partial V/\partial x$ is the net force acting on the beam element in the vertical direction and $-\rho A\partial^2 w/\partial t^2$ is the inertial force necessary for dynamic equilibrium. Substituting Equation (8.16) in Equation (8.17) we obtain

$$\frac{\partial^2 M}{\partial x^2} = \rho A\frac{d^2 w}{dt^2} \tag{8.18}$$

For a prismatic beam, EI is constant. Hence,

$$M = -EI\frac{\partial^2 w}{\partial x^2}$$

Differentiating this expression twice we obtain

$$\frac{\partial^2 M}{\partial x^2} = -EI\frac{\partial^4 w}{\partial x^4} \tag{8.19}$$

Substituting Equation (8.19) into Equation (8.18) we obtain

$$EI\frac{\partial^4 w}{\partial x^4} + \rho A\frac{\partial^2 w}{\partial t^2} = 0 \tag{8.20}$$

This is the equation of motion of an Euler–Bernoulli beam undergoing undamped free vibration. It is seen to be a homogeneous linear partial differential equation with constant coefficients. Since the equation is linear, the principle of superposition holds. This means that the motion of the beam will take place as though each component oscillation with a certain frequency existed alone, not influenced by the occurrence of oscillations taking place at other frequencies. This constitutes a significant simplification which allows us to be guided by physical intuition without having to rely exclusively on mathematical methods. The application of physical intuition in this case is given in the next paragraph. This equation can be solved by using the method of *separation of variables* which enables a partial differential equation to be replaced by a number of ordinary differential equations, whose solution is obtained much more easily. Since the equation involves a second-order derivative with respect to time and a fourth-order derivative with respect to displacement, two initial conditions with respect to time and four edge boundary conditions are needed to find a unique solution for $w = w(x,t)$.

The lateral displacement of the beam assumes the form of a *stationary* (*standing*) *wave*, which thus does not change (the number of half-sine waves along its length remains constant) and whose amplitude at any point fluctuates in time. A stationary wave results from the superposition of two identical waves travelling in opposite directions. That is,

$$\begin{aligned} w(x,t) &= w_+(x,t) + w_-(x,t) = \\ &= W_0 \sin(\omega t - kx) + W_0 \sin(\omega t + kx) = \\ &= 2W_0 \sin(kx)\cos(\omega t) = \\ &= W(x)W(t) \end{aligned} \tag{8.21}$$

The solution of Equation (8.20) can thus be written as $w(x,t) = W(x)W(t)$, so that the spatial and temporal components are separated. Substituting this in the equation of motion and rearranging gives two ordinary differential equations each of which involves one independent variable (x or t). Because of this, the differential equation can be satisfied for arbitrary (x, t) only if each equation is equal to the same constant. Thus,

$$\left(\frac{EI}{\rho A}\right)\frac{1}{W(x)}\frac{d^4W(x)}{dx^4} = -\frac{1}{W(t)}\frac{d^2W(t)}{dt^2} = a = \omega^2 \tag{8.22}$$

where a is a positive constant.[2] These two equations can be termed the *mode equation* and the *phase equation*, each of which involves one variable (x, t respectively):

[2] Before assigning a sign to the constant, it is necessary to investigate this in order to ascertain that a non-trivial solution of the equation is obtained.

$$\frac{d^4W(x)}{dx^4} - \omega^2\left(\frac{\rho A}{EI}\right)W(x) = 0 \tag{8.23}$$

$$\frac{d^2W(t)}{dt^2} + \omega^2 W(t) = 0 \tag{8.24}$$

We can now solve each equation separately. Equation (8.24) becomes

$$\frac{d^2W(t)}{dt^2} = -\omega^2 W(t) \tag{8.25}$$

This is the equation of motion of an undamped SDOF system whose solution has the following form (Equation (8.8)):

$$W(t) = A\sin\omega t + B\cos\omega t \tag{8.26}$$

The constants A and B depend on the initial conditions, so that

$$B = T|_{t=0} \tag{8.27a}$$

and

$$A = \frac{1}{\omega}\frac{dT}{dt}\bigg|_{t=0} \tag{8.27b}$$

Substituting in Equation (8.26) gives

$$W(t) = \frac{1}{\omega}\frac{dT}{dt}\bigg|_{t=0}\sin\omega t + T|_{t=0}\cos\omega t \tag{8.28}$$

The solution of the mode equation, Equation (8.23), has the following form:

$$W(x) = Ce^{sx} \tag{8.29}$$

Substituting in Equation (8.23), the following relationship is obtained:

$$(s^4 - k^4)Ce^{sx} = 0$$

Hence, $s = \pm k$ and $s = \pm ik$. Substituting these values in Equation (8.29) gives

$$W(x) = C_1e^{ikx} + C_2e^{-ikx} + C_3e^{kx} + C_4e^{-kx}$$

Converting the exponential functions to trigonometric functions, the solution of the mode equation is obtained in the following form:

$$W(x) = C_1\sin kx + C_2\cos kx + C_3\sinh kx + C_4\cosh kx \tag{8.30}$$

The values of $C_1 - C_4$ are specified by the boundary conditions. For a free-free beam the boundary conditions are

$$\text{At } x = 0, \qquad M = -EI\frac{\partial^2 w}{\partial x^2} = 0 \qquad V = \frac{\partial}{\partial x}\left(\frac{\partial^2 w}{\partial x^2}\right) = 0 \tag{8.31a–b}$$

$$\text{At } x = L, \qquad M = -EI\frac{\partial^2 w}{\partial x^2} = 0 \qquad V = \frac{\partial}{\partial x}\left(\frac{\partial^2 w}{\partial x^2}\right) = 0 \tag{8.31c–d}$$

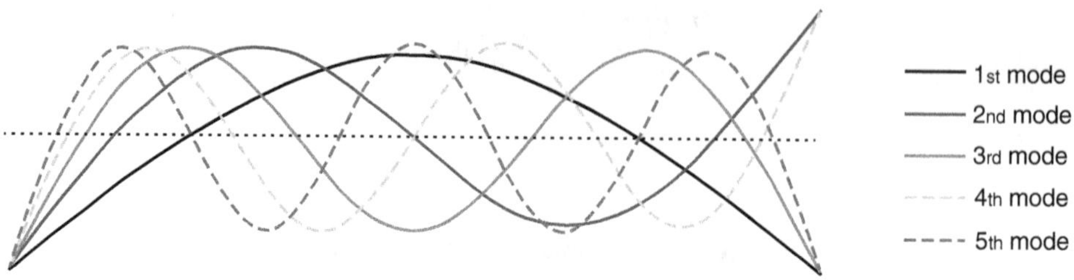

Figure 8.6 First five modes of a free-free beam [17].

The following relations for $C_1 - C_4$ are obtained from Equations (8.31a) to (8.31d), respectively:

$$-C_2 + C_4 = 0 \tag{8.32a}$$

$$-C_1 + C_3 = 0 \tag{8.32b}$$

$$-C_1 \sin(kL) - C_2 \cos(kL) + C_3 \sinh(kL) + C_4 \cosh(kL) = 0 \tag{8.32c}$$

$$-C_1 \cos(kL) + C_2 \sin(kL) + C_3 \cosh(kL) + C_4 \sinh(kL) = 0 \tag{8.32d}$$

Substituting Equations (8.32a) and (8.32b) into Equations (8.32c) and (8.32d), the following system of equations is obtained in C_1 and C_2:

$$\begin{bmatrix} \sinh(kL) - \sin(kL) & \cosh(kL) - \cos(kL) \\ \cosh(kL) - \cos(kL) & \sinh(kL) + \sin(kL) \end{bmatrix} \begin{bmatrix} C_1 \\ C_2 \end{bmatrix} = \begin{bmatrix} 0 \\ 0 \end{bmatrix}$$

For a non-trivial solution ($C_1 \neq 0, C_2 \neq 0$) the determinant of the matrix has to be equal to zero. Multiplying out we obtain

$$\cosh(kL)\cos(kL) = 1 \tag{8.33}$$

This is a transcendental equation that can be solved using the Newton–Raphson method to give the eigenvalues that correspond to the modes of vibration. The multiplicity of solutions is indicated by introducing an integer subscript n (=1, 2, ...) to the wave number, k:

$$\cosh(k_nL)\cos(k_nL) = 1 \tag{8.33'}$$

The lowest five values are included below and are plotted in Figure 8.6.

Mode order n	0[3]	1	2	3	4	5
k_nL	0	4.730	7.853	10.995	14.137	17.278

Substituting these in Equation (8.33′) gives the mode shapes that correspond to the natural frequencies of vibration ω_n.

$$W_n(x) = [\sinh(k_nx) + \sin(k_nx)] + \frac{\sin(k_nL) - \sinh(k_nL)}{\cosh(k_nL) - \cos(k_nL)}[\cosh(k_nx) + \cos(k_nx)] \tag{8.34}$$

[3] Rigid body mode.

Table 8.1 *Percentage deviates from experimental values [47]*

Beam model	1st natural frequency	2nd natural frequency
Euler–Bernoulli	+14% to +26%	+78% to +133%
Timoshenko	–1% to +2%	–1% to +6%

Equation (8.34), which gives the natural frequencies of vibration of an Euler–Bernoulli beam, can be used with acceptable accuracy for the lower frequencies. In contrast to static behaviour in which the fundamental mode dominates response, in vibration response the higher modes can contribute to a significant extent. However, in the case of the higher frequencies, discrepancies arise when using the Euler–Bernoulli assumptions for beam behaviour. In order to understand why this is so, we turn to Figure 8.6, which includes data for the first five modes. We observe that as the frequency increases, the distance between the nodes decreases. This distance may be termed the *effective length in dynamic response* of the beam, to adapt a term used in column buckling theory. The ratio of this length to the radius of gyration of the cross-section ($r^2 = I/A$) reduces as the frequency increases. As a consequence, and as in the case of static response, it becomes necessary to include the effect of transverse shear since we are dealing with a "shorter" and effectively deeper beam. This is disregarded in Euler–Bernoulli theory.

Table 8.1 includes data from a comparison of different models of vibration response with experimental values [6]. It is seen that the Euler–Bernoulli model is inaccurate by approximately 100 per cent for the second natural frequency so that it becomes clear that it cannot be used.

For high-frequency vibration response a more refined theory is therefore needed. In fact, in addition to shear effects, it is necessary to allow also for *rotatory inertia* effects, since each element of the beam is accelerating in both the linear as well as the rotational sense. These two effects are allowed for in the *Timoshenko beam*, which is the topic of the next section.

8.2.2 Undamped Free Vibration of the Timoshenko Beam

Before proceeding with the analysis of the Timoshenko beam, it is useful to be able to understand in qualitative terms in what way structural response will be affected. By allowing for rotatory inertia and transverse shear we are in effect introducing additional degrees of freedom. When this is done, this will involve additional types of displacements and/or rotations that lead to increased flexibility as well as reductions in induced stresses in the original degrees of freedom. Conversely, if one or more degrees of freedom were to be removed, a structure would become "stiffer" and the deflections in the remaining degrees of freedom would reduce. Since the applied loads are borne by a smaller number of degrees of freedom, the induced stresses in these will increase.

Vibration theory has its origins in the 18th century with the work of D. Bernoulli. During the 19th century, Rayleigh introduced the effect of rotatory inertia [42]. However,

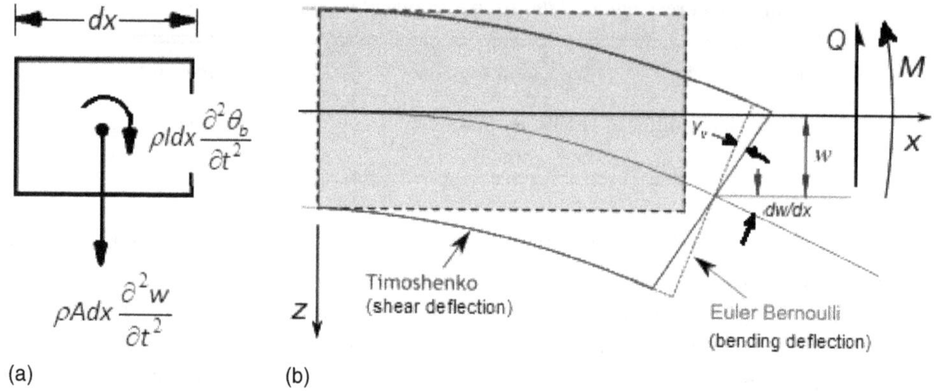

Figure 8.7 Nonlinear dynamic effects in beam vertical response. (a) Rotational and linear inertial forces. (b) Effect of transverse shear on deformation (Timoshenko theory).

this did not succeed in predicting a higher-frequency response with sufficient accuracy, and Timoshenko later included the effect of transverse shear [44, 45].

In this treatment we shall consider these two effects sequentially. Inclusion of the rotatory inertial force will be carried out first. It is assumed that one principal axis lies in the plane of vibration so that bending takes place in a direction of maximum or minimum flexural rigidity. From Figure 8.7 it is seen that the element of length dx translates and also rotates. The angle of rotation θ_b is equal to the slope of the deflection curve, dw/dx. The corresponding angular velocity and angular acceleration are then

$$\frac{\partial}{\partial t}\left(\frac{dw}{dx}\right) = \frac{\partial \theta_b}{\partial t} \qquad \text{and} \qquad \frac{\partial^2}{\partial t^2}\left(\frac{dw}{dx}\right) = \frac{\partial^2 \theta_b}{\partial t^2}$$

The rotational force of the element of length dx is

$$\rho I \frac{\partial^2 \theta_b}{\partial t^2} dx$$

where I is the moment of inertia about the axis through its centre of gravity and perpendicular to the xz plane. For equilibrium, therefore, the equation

$$\frac{dM}{dx} = V \qquad \text{becomes} \qquad \frac{dM}{dx} = V + \rho I \frac{\partial^2 \theta_b}{\partial t^2} \tag{8.35}$$

Equation (8.20) becomes

$$EI\frac{\partial^4 w}{\partial x^4} + \rho A \frac{\partial^2 w}{\partial t^2} + \rho I \frac{\partial^2 \theta_b}{\partial t^2} = 0 \tag{8.36}$$

This is the equation of motion (dynamic equilibrium) of a beam in which linear and rotatory inertia are allowed for.

We shall consider next the inclusion of *transverse shear* effects. In static loading conditions these are of importance in deep beams, that is, those with a low span-to-depth

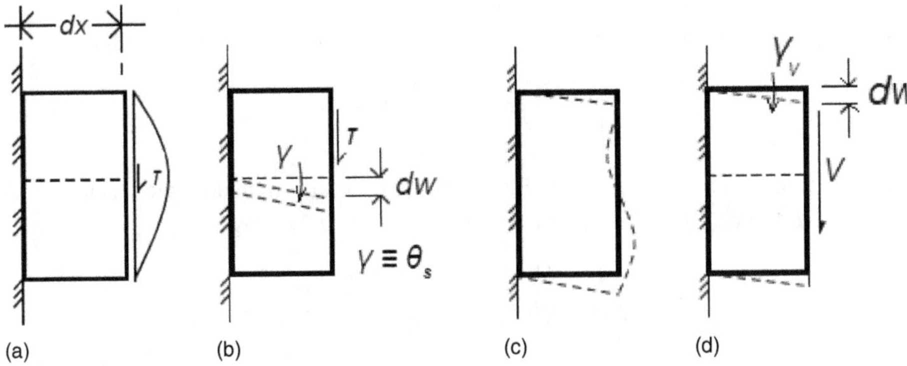

Figure 8.8 Transverse shear stresses, strains and deformations [32]. (a) Transverse shear stresses. (b) Fibre rotation at neutral axis deformation. (c) Actual. (d) Assumed deformation.

ratio. In this case a significant proportion of a transverse load is borne by transverse stresses, in contrast to long beams in which the transverse load induces predominantly bending stresses. In dynamic loading conditions, as seen in the previous section, at high frequencies we are dealing with a "deep" beam, since the distance between the vibration nodes reduces considerably (Figure 8.6). The inclusion of transverse shear effects is thus seen to be necessary in order to predict high-frequency response accurately.

The transverse shear stress distribution in a rectangular section is shown in Figure 8.8a and corresponding transverse strains and defections at the neutral axis in Figure 8.8b. The actual deformed profile is shown in Figure 8.8c. It is seen that the deformed section is no longer normal to the neutral axis (Figure 8.8b) and is also distorted (Figure 8.8c).

In Timoshenko beam theory the Euler–Bernoulli assumption that cross-sections are normal to the neutral axis is relaxed so that the cross-sections are no longer normal to the neutral axis. However, the section does not distort (warp) as shown in Figure 8.8c but is assumed to remain plane (Figure 8.8d). Thus the Timoshenko beam is an approximation to the exact behaviour, which can be predicted using elasticity theory. The vertical deflection of the beam is such that the work done in the assumed, average deformation (Figure 8.8d) is equal to the work done by all longitudinal fibres in the warped section (Figure 8.8c), as predicted by elasticity theory. From Figures 8.8b and 8.8d,

$$\int_A dw\tau dA = dw_v V \tag{8.37}$$

Letting $\theta_s \equiv \gamma$ and substituting the kinematic relationships $dw = \gamma dx, dw_v = \gamma_v\ dx$ on both sides of Equation (8.37) gives

$$\int_A (\theta_s dx)\tau dA = (\gamma_v dx)V$$

The shear stress–shear strain relationship is $\tau = \gamma G$. Substituting,

$$\int_A \left(\frac{\tau}{G}dx\right)\tau dA = \left(\frac{\tau_v}{G}dx\right)V$$

The shear stress τ_v acts over the portion of the cross-section that is subjected to shear stress, so that $\tau_v = V/A_v$, where A_v is an *effective shear area* to be defined. Thus,

$$\int_A \left(\frac{\tau}{G}dx\right)\tau dA = \left(\frac{V}{A_v}\right)\frac{1}{G}dxV \tag{8.38}$$

The shear stress as given by Euler–Bernoulli theory is substituted on the left hand side and if $A_v = \beta A$, where β is the *shear correction coefficient*, we have

$$\int_A \frac{1}{G}dx\left(\frac{V}{It}\right)^2 dA = \left(\frac{V}{\beta A}\right)\frac{1}{G}dxV$$

Solving for β gives

$$\beta = \frac{I^2}{A\int_A \left(\frac{V}{t}\right)^2 dA} \tag{8.39}$$

The shear correction coefficient can be evaluated for each type of cross-section.[4] The rotation due to transverse shear, θ_s, is obtained from

$$\theta_s = \frac{\tau}{G} = \frac{V}{\beta AG} \qquad \text{or} \qquad V = \beta AG\theta_s \tag{8.40}$$

Thus the total rotation of the neutral axis is no longer θ_b but

$$\theta = \theta_b + \theta_s \tag{8.41}$$

Differentiating V (Equation (8.40)) with respect to x and substituting in Equation (8.17) gives

$$\frac{\partial V}{\partial x} = \frac{\partial(\beta AG\theta_s)}{\partial x} \qquad \text{and} \qquad \beta AG\frac{\partial\theta_s}{\partial x} = \rho A\frac{d^2w}{dt^2} \tag{8.42}$$

The moment–curvature relationship, expressed in terms of the total angle of rotation θ, is

$$M = EI\frac{\partial\theta}{\partial x}$$

Substituting in Equation (8.35) the derivative of this expression and Equation (8.40) for the transverse shear force gives

$$EI\frac{\partial^2\theta}{\partial x^2} = \beta AG\theta_s + \rho I\frac{d^2\theta}{dt^2} \tag{8.43}$$

From Equation (8.41), $\theta_s = \theta - \theta_b$. Substituting for θ_s in the two dynamic equilibrium relationships, Equations (8.42) and (8.43), gives

$$\beta AG\frac{\partial}{\partial x}(\theta - \theta_b) + \rho A\frac{d^2w}{dt^2} = 0 \tag{8.44a}$$

[4] The shear correction coefficient as originally determined by Timoshenko was found to give erroneous values for the natural frequencies of vibrating beams. The Timoshenko approach makes use of what is termed the *first-order shear deformation theory*. More accurate shear deformation theories have been proposed, known as *higher-order shear deformation theories*.

Table 8.2 *Effect of rotatory inertia and transverse shear on beam vibration response*

Mode	1	2	3	4	5
Change in frequency (%)	–0.4	–1.6	–3.7	–6.6	–10.3

$$EI\frac{\partial^2\theta}{\partial x^2} = \beta AG(\theta - \theta_b) + \rho I\frac{\partial^2\theta}{\partial t^2} \tag{8.44b}$$

The first of these expressions is used to obtain $\partial\theta/\partial x$. Substituting this in Equation (8.42), an equation is obtained with w, the transverse displacement, as the only dependent variable. From Equation (8.40a),

$$\frac{\partial\theta}{\partial x} = \frac{\partial^2 w}{\partial x^2} - \frac{\rho A}{\beta AG}\frac{\partial^2 w}{\partial t^2}$$

Differentiating Equation (8.40) with regard to x and substituting for $\partial\theta/\partial x$ and its derivatives from above, the following equation is obtained:

$$\underbrace{\frac{EI}{\rho}\frac{\partial^4 w}{\partial x^4}}_{1} - \underbrace{\left[I + \frac{EI}{\beta G}\right]\frac{\partial^4 w}{\partial x^2 \partial t^2}}_{2} + \underbrace{\frac{\rho I}{\beta G}\frac{\partial^4 w}{\partial t^4}}_{3} + A\frac{\partial^2 w}{\partial t^2} = 0 \tag{8.45}$$

This is the equation of motion (dynamic equilibrium) of a uniform Timoshenko–Rayleigh beam undergoing undamped free vibration. Term 1 is the Euler–Bernoulli term, term 2 includes the effects of rotatory inertia and transverse shear, and term 3 accounts for the coupling of the latter two effects.[5]

A solution of this equation for a general loading case, in which the equation is non-homogeneous, was obtained by employing the Laplace transform method in conjunction with the method of *variation of parameters* that was used to obtain the particular integral [13]. For simply supported and free ends, the solution is discussed in Section 8.4.4.

The results of a comparison obtained excluding and including the effects of rotatory inertia and transverse shear for a long simply supported beam with a rectangular cross section are included in Table 8.2 [6].

It is seen that as the mode number increases, the discrepancy in predicted frequency increases. Transverse shear effects account for three-quarters of this. Figure 8.9 shows the comparative effect of transverse shear, shear lag and longitudinal inertia. Shear lag is included in this comparison in order to allow for the differences between a beam and a hull girder. The latter is a plated structure, and as seen in Chapter 6, the axial normal stress distributions are affected by the transverse shear stress distributions, leading to shear lag. It is clear that the differences increase (1) with frequency, (2) with inclusion of secondary effects (transverse shear, shear lag and longitudinal inertia) and (3) as the aspect ratio reduces. A reduction in the aspect ratio implies a reduction in node spacing.

[5] A comparison of frequencies calculated with and without the coupling term has shown that this can be neglected, as being of minor importance [15].

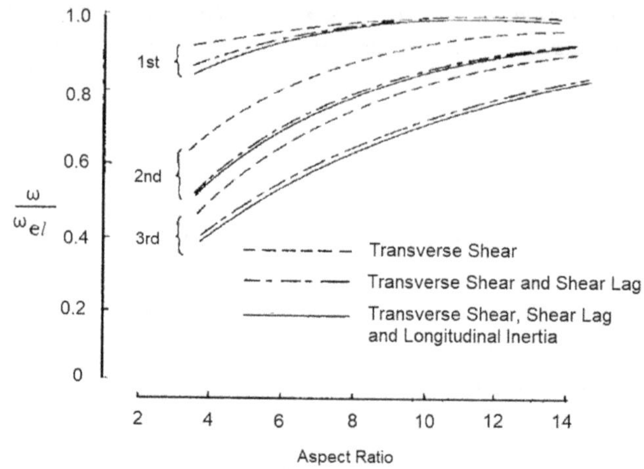

Figure 8.9 Influence of various secondary effects on the vibration frequencies in bending of a symmetric free-free thin-walled uniform rectangular beam with $B/D = 3.649$ [50] (originally from [18]). Reprinted courtesy of the Ship Structure Committee, Washington, DC.

8.2.3 Natural Frequency of Vibration of a Beam Using the Energy Method

We shall now consider an alterative approach to formulating and solving problems in structural dynamics. *Rayleigh's energy method* provides the basis for a number of the early formulas used in calculating hull girder vertical and horizontal vibration frequencies.

In an undamped, freely vibrating conservative system, the kinetic energy stored in the mass of a structure is continually and totally transformed into potential energy, and vice versa, as the oscillation executes one complete cycle. From the principle of conservation of energy, the total energy content of a conservative system, which consists of kinetic and potential energy, remains constant. When the displacement reaches its maximum value, the velocity reduces to zero and the potential energy then reaches its maximum value. When the structure returns to its equilibrium position, the displacement is zero and the kinetic energy reaches its maximum value. For a conservative system, there are no energy losses and all potential energy is converted into kinetic energy, and vice versa. Therefore, the maximum kinetic energy equals the maximum potential energy. If these quantities are equated, the fundamental natural frequency of vibration of a structure can be obtained. This in essence is what *Rayleigh's principle* states.

In the preceding statements it is implied that we know the exact displacement of any position in the structure at any point in time. The use of simple mathematical expressions, however, means that we are treating the structure as a single-degree-of-freedom system and not as a continuous structure. We need to bear this in mind as we consider any problem of this nature. Further comments are made later on in this section.

The kinetic energy T is given by

$$T = \frac{1}{2} m \dot{x}^2 \tag{8.46}$$

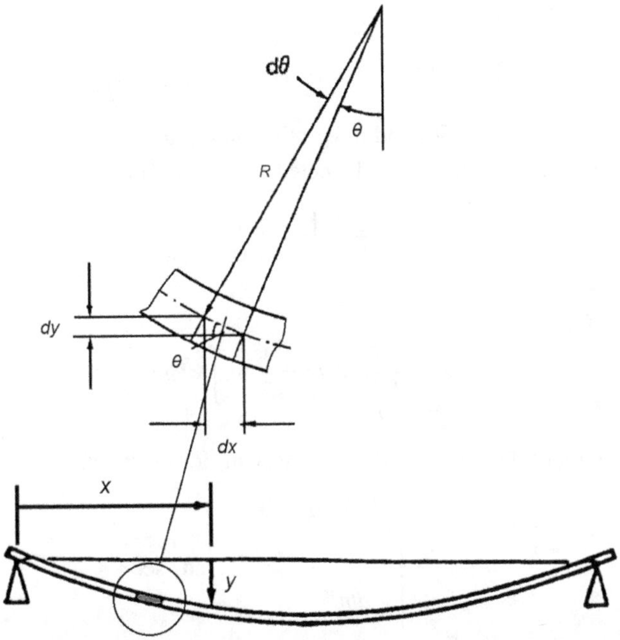

Figure 8.10 Transverse vibration of beam [4]. Reprinted courtesy of Elsevier.

where m is the mass and $\dot{x}$ is the velocity of the body at any instant in time. If I is the mass moment of inertia, the kinetic energy of a body executing simple harmonic motion is given by

$$T = \frac{1}{2} I \omega^2 \tag{8.47}$$

where $\dot{\omega}$ is the angular velocity of the body.

The potential energy U stored in an elastic structure is given by

$$U = \frac{1}{2} k x^2 \tag{8.48}$$

where x is the displacement of a point from the equilibrium position at any point in time during the vibration cycle and k is the stiffness.

Let us now consider the problem of a vibrating beam, shown in Figure 8.10.

For small deflections, from elementary beam theory we have

$$\frac{1}{R} = \frac{d\theta}{dx} = -\frac{d^2 y}{dx^2} \qquad \text{and} \qquad M = \frac{EI}{R} \tag{8.49}$$

where M is the bending moment, EI is the flexural (bending) stiffness of the beam and $1/R$ is its curvature.

The potential (bending strain) energy U stored in an element of length dx is then

$$dU = \frac{1}{2} M \left(\frac{1}{R} \right) dx \tag{8.50}$$

For a linear elastic beam,

$$M = -EI\frac{d^2y}{dx^2} \qquad \text{and} \qquad \frac{1}{R} = \frac{d^2y}{dx^2}$$

Substituting in Equation (8.50) and integrating along the length of the beam, we obtain the potential energy U in terms of the deflected form, $y = f(x)$,

$$U_{max} = \frac{1}{2}\int_L EI\left(\frac{d^2y}{dx^2}\right)^2 dx \tag{8.51}$$

The kinetic energy T is

$$T = \frac{1}{2}\int_L I\omega^2 dx = \frac{1}{2}\omega^2\int y^2 dm \tag{8.52}$$

Apply Rayleigh's principle, $U_{max} = T_{max}$. Solving for ω^2 we obtain

$$\omega^2 = \frac{\int_L EI\left(\frac{d^2y}{dx^2}\right)^2 dx}{\int y^2 dm} \qquad (\text{rad/ sec})^2 \tag{8.53}$$

The frequency, f (in Hz), is obtained from Equation (8.10). This expression gives the fundamental frequency of oscillation of a beam that is treated as a single-degree-of-freedom system. In order to obtain an actual value it is necessary to select a suitable expression for y such that the deflected form of the beam is compatible with the boundary conditions. Of necessity, only simple mathematical expressions can be used, and these will exclude other vibration modes. As a result, the beam will appear stiffer than it actually is, and the predicted frequency will be higher than the true one, albeit by only a few percentage points. An exact result is obtained only if all (an infinite number of) vibration modes are included in the deflection model. We note that the bending stiffness EI is included in the integral of the numerator and that the denominator involves integration with respect to the mass per unit length. In the case of a uniform beam these calculations are readily performed. For a non-uniform beam, however, it will be necessary to perform numerical integrations.

8.3 Ship Vibration

In this section we shall explore how the theory of vibration applied to elementary bodies can be applied to the ship hull girder considered in its entirety. In order to do so it is necessary to discuss the types of vibration encountered on board ships as well as the differences between ships and elementary bodies when subjected to vibration excitation.

8.3.1 Classification and Consequences of Ship Vibration

The hull structure is subjected to a variety of loads, some of which can result in vibratory response. These can be classified in several ways. One way is to consider the duration of

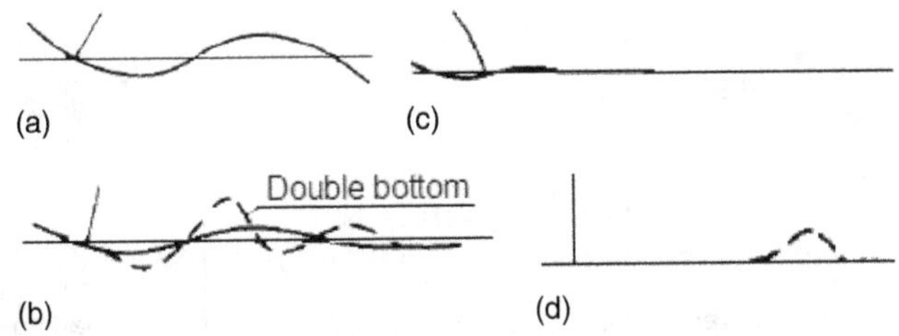

Figure 8.11 Types of ship vibration [40]. (a) Pure hull vibration. (b) Integrated hull vibration. (c) Superstructure – aft body vibration. (d) Local vibration. Reprinted courtesy of Elsevier.

the excitation, and on this basis we speak of *transient* and of *steady-state* excitation. The most important of these are the latter, since steady-state loads act throughout the time a ship is at sea. Transient excitation is the result of motion of a ship in a particular seaway and is due to wave impact, slamming, shock and other effects.

Alternatively, ship vibration can be classified on the basis of the extent of the structure involved. In a number of cases the whole hull girder is affected, in which case we speak of *global vibration*. The geometry of the hull girder makes it possible to apply the results of beam theory as discussed in previous sections, suitably modified to allow for the particularities of ship geometry and the environment within which ships operate. In other cases only one particular region of the hull girder is affected, in which case we speak of *local vibration*. Usually this has a higher frequency and lower amplitude than global vibration. It is difficult to predict local vibration in advance, but it is possible to control it following construction of the ship, in contrast to global vibration, which is difficult to control once a ship has been built. Local vibration cannot in general be assessed using beam theory because it affects structural members of widely differing geometries and also because at a local or regional level vibration can propagate in one or more directions. An example is the case of local vibration propagating throughout a superstructure and interacting with global hull girder vibration.

The distinction between local and global is therefore not always clear-cut (Figure 8.11). In quantitative terms, the observed frequency ranges and corresponding vibration types are [40]:

1. Pure hull vibration, frequency $f < 3$–6 Hz (depending on type of ship). The hull vibrates as a beam with narrow resonant peaks.
2. Integrated hull vibration, 3–6 Hz $< f <$ 6–12 Hz. The hull vibration modes interact with such subsystems as the double bottom in bulk carriers and longitudinal bulkheads in tankers, giving coupled resonant frequencies.
3. Superstructure–aft body vibration, 6–8 Hz $< f <$ 12–14 Hz.
4. Local vibration of panels, $f =$ 10–40 Hz.

Vibration of the hull girder has consequences at the following two levels:

- Reduced fatigue life of structural components
- Intolerance (physical strain) of humans on board

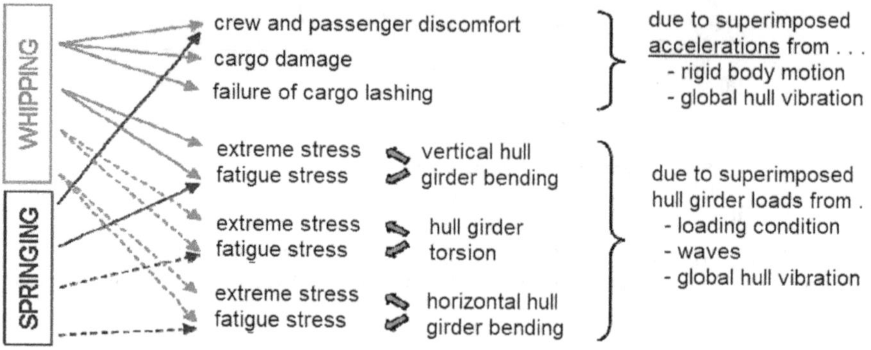

Figure 8.12 Consequences of wave-induced action on hull girder and humans [33]. Reprinted courtesy of the Ship Structure Committee, Washington, DC.

A study reported in the proceedings of ISSC 2018 was performed to assess fatigue damage accumulation on board an LNG carrier, based on full-scale measurements over a period of 12 months. During this period the vessel was in actual operation for less than half the time. The fatigue damage rate was discovered to be lower than that predicted by component stochastic fatigue analysis, and the fatigue life was estimated to be longer than the ship's design life of 40 years. However, the contribution of vibration to fatigue damage was found to be significant, and accounted for 30–50 per cent of the total damage accumulated. The highest fatigue damage occurred in head seas, in following seas, in rough seas and when the ship was in a fully loaded condition. It was found that speed reduction in rough seas contributed to a reduction in fatigue damage to below that occurring at constant service speed. Torsional and horizontal vibration were also observed. Vibration was found to contribute significantly to side shell fatigue, but overall side shell fatigue loading was estimated to be at an acceptable level.

The consequences of vibration resulting from springing and whipping are summarised in Figure 8.12.

Whipping, in addition to steady-state excitation, may also contribute to fatigue damage, depending on ship type. In the case of long bulk carriers this contribution has been estimated to account for up to 50 per cent of accumulated fatigue damage. Furthermore, the combination of whipping and the primary nonlinear wave bending moment arising in heavy seas was found in several investigations to exceed IACS rule values. It was also found that the vibratory response of the hull to waves may be important in the torsional and horizontal bending modes, in addition to the vertical bending mode.

Local vibration on the other hand may result in:

- Stresses in the hull structure, in the equipment and/or supports (fatigue)
- The need to ensure trouble-free operation of machinery or other equipment that might be jeopardized by the malfunction or degradation of individual components
- Physical strain on humans (habitability and performance)
- Deleterious effects on instruments, machinery, etc. within the vicinity of the source.

In the remainder of this chapter we shall consider global ship vibration response.

8.3.2 From the Uniform Beam to the Ship at Sea: The Factors That Characterise Ship Vibration Response

In previous sections of this chapter several different idealised structural models of ship vibration response that are based on beam theory were described. At this stage it is necessary to point out the differences these have with a ship at sea. These differences can be grouped as follows:

1. Structural layout
2. Ship design and operational aspects
3. The marine environment.

Structural Layout

From the point of view of structural layout, a ship differs from a beam in that:

1. Its weight is not uniformly distributed along its length and in the vertical direction.
2. The weight distribution varies during a ship's lifetime, in particular that of bulk cargo ships, since the quantity and the distribution of the cargo transported changes from one voyage to the next (Figure 8.13c).
3. As a consequence of (1), the second moment of area (Figure 8.13b), and hence the bending and shear stiffness, also varies along the length about both the horizontal and vertical axes (Figure 8.13a). The effect of shear stiffness on the frequency response of a vibrating beam was noted in Section 8.2.2. In the case of ships, the following correction formula has been proposed for higher order vibration modes [48]:

$$N = N_0 \frac{1}{\sqrt{1+\gamma_s}} \tag{8.54}$$

where

N = natural frequency of hull girder flexural vibration allowing for shear deflection

N_0 = natural frequency based on Euler–Bernoulli theory

γ_S = shear correction factor (Table 8.3).

4. The torsional stiffness varies along the ship's length.
5. The ship, being a plated structure, is not fully effective due to the shear lag effect (Chapter 6). The shear lag effect results in a reduced effectiveness of horizontal structures (decks and bottom structure) in contributing to vertical bending stiffness. This is important in the case of high-frequency modes in which the wavelength reduces, and thus the distance between zero displacements correspondingly reduces. In effect, the plate effectiveness is frequency dependent. Figure 8.14 may be used to determine the effective breadth for a rectangular hull section vibrating in the first five modes. It is one of a set of figures included in [48].
6. There are structural items along its length that do not contribute or may partially contribute to its elastic response (all partially effective or ineffective material, such as superstructures and deckhouses).

Table 8.3 *Shear correction factor* $\sqrt{1+\gamma_s}$

Type of vibration	Dry cargo ship	Oil tanker[6]
2-Node vertical	1.11	1.055
2-Node horizontal	1.2–0.075(L/100)	1.07
3-Node vertical	1.31	1.05+0.1(L/100)
3-Node horizontal	1.57–0.2(L/100)	1.32

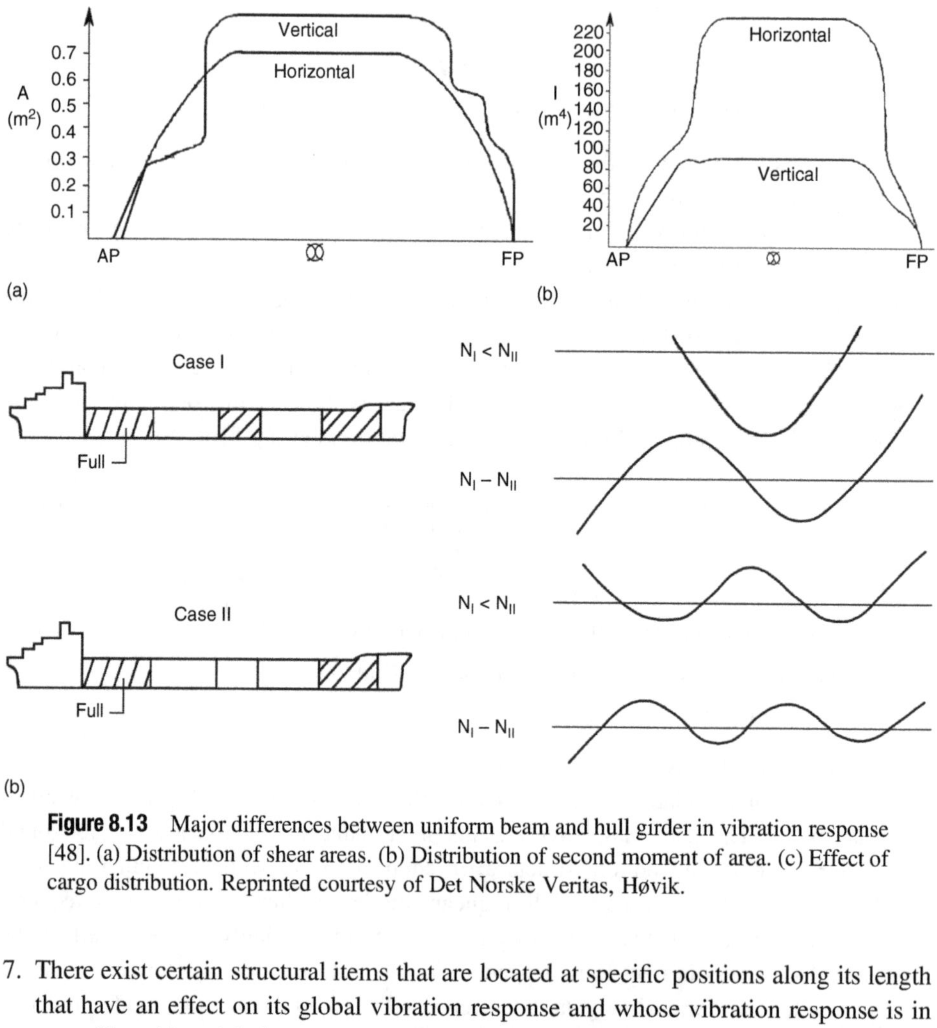

Figure 8.13 Major differences between uniform beam and hull girder in vibration response [48]. (a) Distribution of shear areas. (b) Distribution of second moment of area. (c) Effect of cargo distribution. Reprinted courtesy of Det Norske Veritas, Høvik.

7. There exist certain structural items that are located at specific positions along its length that have an effect on its global vibration response and whose vibration response is in turn affected by global response, such as aft–end superstructures.

[6] These figures apply to single hull tankers. For double hull tankers they need to be modified accordingly.

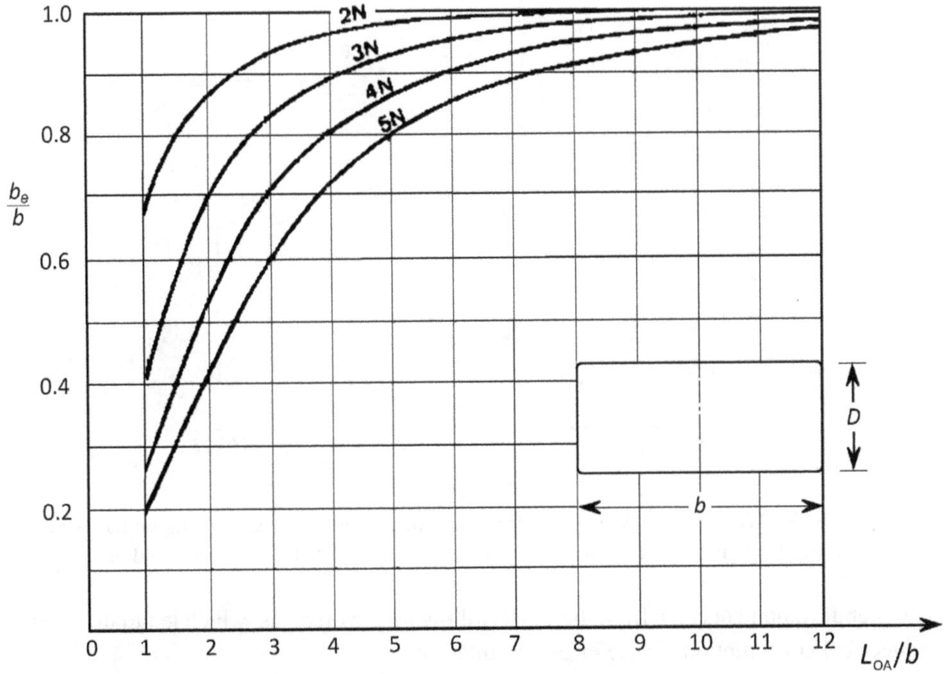

Figure 8.14 Effective breadth ratio for rectangular hull section in vertical vibration [48]. Reprinted courtesy of Det Norske Veritas, Høvik.

Ship Design and Operational Aspects

All of the above introduce complications when assessing ship vibration response. In addition to these factors, the situation has become exacerbated as a result of developments in ship design. First and foremost is the increase in size of oil tankers from a deadweight capacity of around 50,000 metric tonnes to more than 200,000 metric tonnes during the 1960s. Vibration was observed on the larger ships that had not been experienced on smaller ships. The empirical formulas used to calculate the natural frequency of vertical vibration,[7] based on elementary vibration theory, show that the frequency reduces with increases in displacement as well as ship's length. Comparing the natural frequency of vertical vibration with the form of wave energy spectra such as the ISSC spectrum shows that the energy content available to excite vertical vibration increases in the case of large ships [16].

Figure 8.15 shows three ISSC wave energy spectra and the natural frequencies of vertical two-noded vibration of three ships sailing at 16 knots in head seas with a significant wave height of 3 metres (10 ft). It is seen that for the smaller ship (50,000 tonnes dwt), the frequency lies at the tail end of the three spectra, whereas as displacement increases, the frequency reduces and coincides with sea states with a higher energy density.

[7] These are discussed in Section 8.4.4.

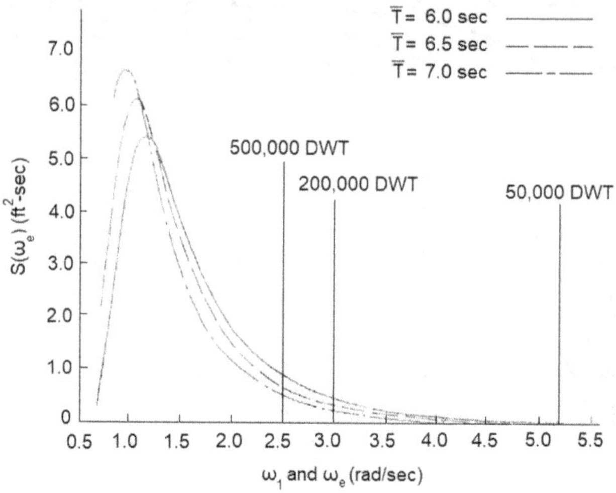

Figure 8.15 Wave energy density and frequency of two-noded vertical vibration (adapted from [16]). Reprinted courtesy of the Royal Institution of Naval Architects, London.

Other developments date from the 1970s following the oil crisis, which led to the need to reduce fuel consumption. These efforts resulted in:

- Modifications to the hull form in order to improve propulsive efficiency
- Reductions in main engine fuel consumption
- Improvements of heat efficiency by recovery of the heat content of flue gases.

The changes in hull form that were introduced led to a reduction in powering requirements, which also resulted in a reduction in propeller-induced vibration. In contrast, efforts to reduce fuel consumption resulted in a deterioration of the situation, because the adoption of long-stroke engines having a smaller number of cylinders gave rise to longitudinal vibration of the main engine block. Furthermore, larger-amplitude torque fluctuations have induced correspondingly larger thrust fluctuations. The relocation of the engine room to the stern region, which up to the 1960s was located amidships, has also resulted in more acute propeller-induced vibration. Lastly, with the accommodation superstructure also located in the stern region and thus directly above the propeller, propeller-excited deckhouse vibration has become a problem to be considered. The interaction of deckhouse response with that of the hull girder affects both structures.

There are particular types of ships for which it is necessary to apply strict vibration control measures for functional reasons, such as passenger ships and cruiseships. In these cases the vibration response has to be accurately estimated and appropriate vibration attenuation measures implemented.

Finally, adverse operating conditions frequently result in significant increases in vibration amplitudes. Some relevant factors:

1. Sea conditions. Under ideal conditions (flat calm sea, straight ahead) hull vibration will modulate from maximum to minimum by a factor of 2 to 1.
2. Hard turns to port or starboard
3. Shallow water

4. Light draught condition. An increase in hull vibration by 25 per cent may be experienced in ballast condition. To minimise hull vibration it is recommended that the ship be fully ballasted with aft peak tanks filled.

The Marine Environment

Perhaps the most important difference between a uniform beam vibrating in air and a ship is the fact that the ship is immersed in water. When a vibrating body is immersed in water, it sets in motion the particles of the surrounding fluid as well. The density of water is considerably higher than that of air, and thus to succeed in setting the fluid in motion, a non-negligible quantity of energy has to be transmitted from the solid body to the fluid particles. Otherwise put, the vibrating surface exerts a force on the surrounding water that is proportional to its own acceleration at any instant in time. An opposing force acts on the solid body whose magnitude depends on the inertia of the water and its acceleration. This produces a change in the vibration response of the body. The natural frequency of vibration of the body thus has to be recalculated when it is immersed in a fluid of appreciable density, such as water.

The problem can also be viewed by considering a body of increased mass. From elementary vibration theory (Equation (8.10)), the natural frequency of vibration of a body, f, depends on its mass m and stiffness k:

$$f \propto \sqrt{\frac{k}{m}} \tag{8.55}$$

where the stiffness will be that which is relevant to the particular type of motion (flexural or torsional). The effect of the surrounding water on the frequency response of the body can be accounted for by increasing the mass of the body accordingly. The increase in mass is termed the *added virtual mass* or *hydrodynamic added mass*.

In order to appreciate this, we shall address an example described in [46] that illustrates the concept of added mass in a simple and direct manner. Consider the case of an infinitely long cylinder of radius b, moving at right angles to its length with velocity U in an inviscid liquid that is at rest at infinity. Potential flow theory gives the following expression for the kinetic energy of the liquid per unit length of cylinder:

$$K.E. = \frac{1}{2}\rho\pi b^2 U^2 \tag{8.56}$$

where ρ is the mass density of the fluid. Letting $\rho\pi b^2 = M_f$ we see that M_f is the mass of fluid displaced per unit length of the cylinder, so that

$$K.E. = \frac{1}{2}M_f U^2 \tag{8.57}$$

The effect of the presence of the fluid can thus be allowed for by adding a mass M_f to the mass per unit length of the cylinder. In the case of rectilinear motion when an external force X per unit length acts on the cylinder, applying Newton's second law we have

$$\left(M + M_f\right)\frac{dU}{dt} = X \tag{8.58}$$

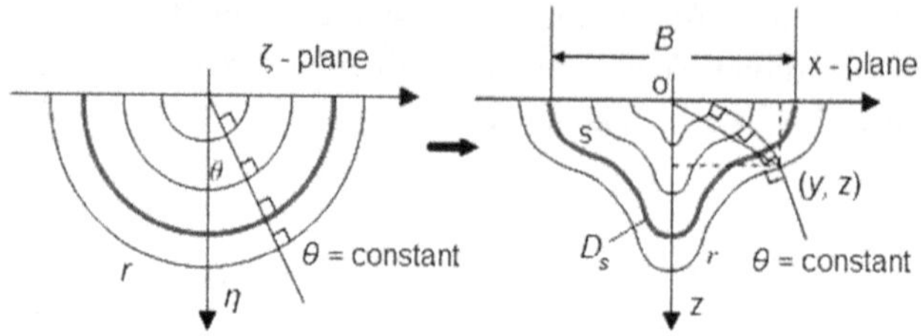

Figure 8.16 Mapping of half-circle onto ship hull section [38].

where M is the mass of the cylinder per unit length. Writing Equation (8.58) as follows,

$$M\frac{dU}{dt} = X - M_f\frac{dU}{dt} \tag{8.59}$$

we see that the presence of the fluid is equivalent to a force $M^1(dU/dt)$ per unit length opposing the motion. In steady motion (constant velocity), $dU/dt = 0$, and thus the force due to the presence of fluid is zero. This effect is therefore present only in accelerated motion, as in the case of vibration. If we write

$$M_f = k\,(\text{mass of fluid displaced by body}) \tag{8.60}$$

k is termed the *added virtual mass coefficient* and depends on the type of motion and the shape of the body. For a sphere of radius b,

$$K.E. = \frac{1}{2}\frac{2}{3}\rho\pi b^3U^2 = \frac{1}{2}M_fU^2 \tag{8.61}$$

Thus the inertia coefficient of the sphere is one-half that of the cylinder, due to the effects of the three-dimensional motion of the fluid.

The added virtual mass can be determined independently of the structural model using available analytical solutions. For body forms such as an ellipsoid of revolution, this is possible. In the case of ship hull forms, however, it is necessary to transform the underwater profile to a known shape (Figure 8.16). This can be achieved by *conformal mapping* [28].[8]

The procedure consists of two steps. First, the added mass of equivalent two-dimensional cross-sections, $m(x)$, along the ship hull is determined. These are given by

$$m(x) = \left(\frac{\pi}{8}\right)\rho B^2(x)C(x)J \tag{8.62}$$

where ρ is the density of water, $B(x)$ is the beam of the section in question, $C(x)$ is the two-dimensional added mass coefficient of the section and J is a *longitudinal inertia coefficient*, defined in this case as the ratio of the actual kinetic energy of the surrounding fluid to that of the fluid if motion is restricted to transverse planes (Figure 8.17a). It is thus introduced to allow for three-dimensional flow effects.

[8] Other methods, such as the Frank close-fit method, have also been applied to this problem.

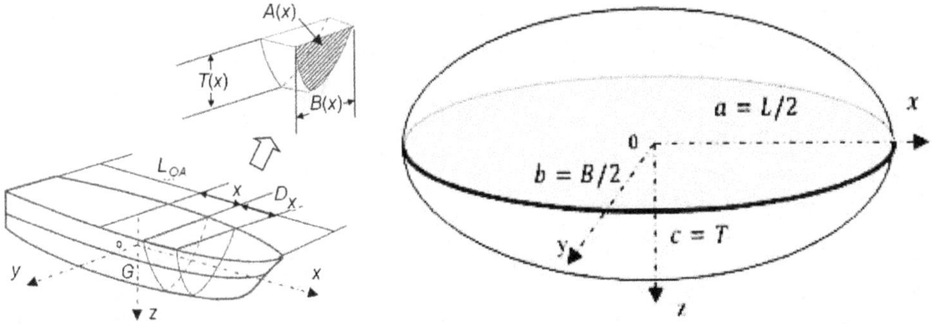

(a) (b)

Figure 8.17 Representation of the hull girder in the Lewis transformation method [38]. (a) Subdivision of the hull girder into 2D sections. (b) Use of ellipsoid of revolution for *J* coefficient.

The transformation from the unit circle to the ship profile is performing using

$$Z = z + \frac{a(x)}{z} + \frac{b(x)}{z^3} \tag{8.63}$$

where $a(x)$ and $b(x)$ are coefficients determined such that the section area coefficient and the beam/draught ratio are preserved in the transformation. By specifying a and b, the ideal fluid solution for the unit circle provides the solution for the two-dimensional added mass of the ship section at any position x along the ship's length. Figure 8.18a can be used to determine $C(x)$ for any section area coefficient and beam to draught (B/T) ratio along the ship's length.

Lewis assumed that the value of J for a ship is equal to that of a spheroid, which is an ellipsoid of revolution, shown in Figure 8.17b, for which the exact value is available. On this basis Figure 8.18b can be used to find the longitudinal inertia (3D added mass) coefficient. The value of J depends to a certain extent on the form of the wetted surface, but mostly on the global vibration mode. As the beam/length ratio decreases, J approaches unity. Substituting in Equation (8.62), the added mass for each section along the length of the ship is found. This is then added to the structural mass at each section in order to estimate the vibration response of the ship.

The discussion up to this point concerns vertical vibration; ships, however, vibrate in the horizontal and torsional modes as well, and it is necessary to calculate the corresponding values of the added virtual mass. For horizontal and torsional vibration the earliest work done is reported in [26] and [25], respectively.

Lewis's conformal mapping method has been developed further to model real ship sections more accurately and efficiently. This procedure for determining the added mass gives satisfactory results for conventional hull forms that have a pronounced strength deck and well-defined global vibration modes. For passenger and Ro-Ro ships, however, it cannot be applied directly because the global vibration modes of the hull are not well defined and the J factor cannot be determined reliably. To produce solutions for such

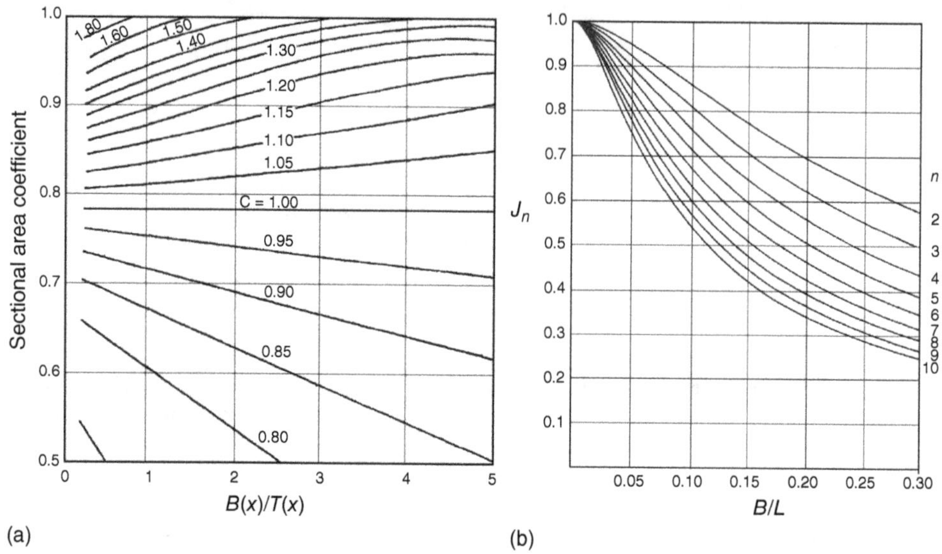

Figure 8.18 2D and 3D added mass coefficients for Lewis forms [49]. (a) 2D added virtual mass coefficient $C(x)$. (b) Longitudinal inertia coefficient, J (n = number of nodes). Reprinted courtesy of the Society of Naval Architects and Marine Engineeers.

problems, numerical methods used to solve problems of fluid–structure interaction, such as the boundary element method (BEM) or the Reynolds-Averaged Navier–Stokes (RANS) method, are used. Often a FE model of the entire ship is combined with a BEM model of the fluid.

8.4 Vibration of the Hull Girder Resulting from Steady-State Excitation

The sources of excitation of the hull girder can be described as *steady-state*, *random* and *transient*. The steady-state excitation that results from the operation of the main engine and the propeller is discussed in Chapter 5, whereas transient excitations are discussed in Chapter 3. Random excitations are discussed in Chapter 9 that deals with the probabilistic modelling of hull girder response.

8.4.1 General

Table 8.4 summarises the sources of excitation and the consequent nature of the vibration response. Of most importance to hull girder response is whether its natural frequencies are close to or coincide with those of the excitation source. An overall picture of the natural frequencies of sources of steady-state excitation and those of the hull structure is given in Figure 5.34. It is seen that a degree of overlap of the natural frequencies of the hull girder arises with those of slow-running engines as well as those of shaft lines. It is therefore important to ensure that the performance of the equipment does not result in resonance

Table 8.4 *Cause-effect relationships in hull girder vibratory behaviour*

Nature of excitation	*Type of excitation*	*Response*	*Global/local vibratory response*
Steady state (periodic)	Propeller action (pressure pulses on stern structure, longitudinal vibration of shaft, radial pulses on bearings)	Vertical and longitudinal vibration	Both
	Diesel engine reciprocating motion (vertical, lateral forces)	Vertical vibration	Both
	Forces on thrust bearing due to longitudinal vibration of main engine	Longitudinal vibration	Both
	Wave impact on entire bottom structure	Springing (vertical, horizontal and torsional modes)	Global
Random (non-periodic)	Wave impact on ship side structure	Horizontal and torsional vibration (fatigue)	Both
	Cargo (container) inertialloads due to vessel heeling	Torsional vibration	Global
Transient (step)	Bow flare impact	Whipping (fatigue)	Both
	Bottom slamming	Whipping (and local structural damage)	Both

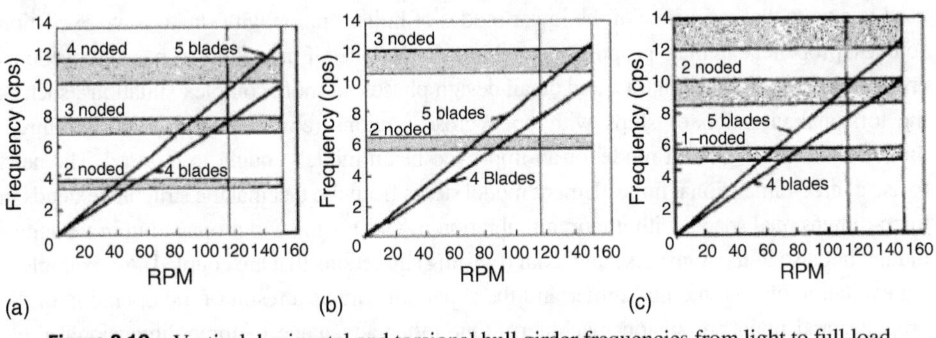

Figure 8.19 Vertical, horizontal and torsional hull girder frequencies from light to full load compared with four-bladed or five-bladed propeller frequencies[9] [33]. (a) Vertical vibration. (b) Horizontal vibration. (c) Torsional vibration. Reprinted courtesy of the Ship Structure Committee, Washington, DC.

effects at either a local or a global level. This can be done, once hull girder frequencies have been calculated, by plotting hull frequencies versus shaft RPM in a *hull resonance* diagram. An example of such a diagram that was prepared for an icebreaker is included in Figure 8.19 [33]. It is seen that under normal operating conditions (115 RPM), resonance

[9] In each shaded area the top line corresponds to the light load and the bottom line to the full load condition.

may arise with vertical vibration in the case of a four-bladed propeller. So if a five-bladed propeller is selected, the other vibration modes need to be checked.

From Table 8.4 it is evident that it is necessary to develop an understanding of the following types of global vibration response:

- Vertical vibration
- Horizontal vibration
- Torsional vibration
- Coupled torsional-horizontal vibration
- Springing
- Whipping.

Longitudinal vibration of the hull girder has rarely been observed and for this reason will not be considered. The most important type of vibration is associated with vertical and horizontal flexure.

The methods that can be used to calculate natural frequencies include:

1. Mathematical formulas, in certain cases modified to allow for complex effects.
2. Beam models of varying degrees of complexity. Beam models can be used when only bending is of concern as a limited number of modes is modelled.
3. Finite element models.

In this section we shall present the derivation of mathematical expressions that may be used to obtain natural frequencies of vibration. The use of the finite element method is discussed in Chapter 13. At the preliminary design stage, available formulas and simple beam models can be used to ensure that resonance of the lower modes of hull vibration with steady-state excitation sources (propeller and main propulsion machinery) is avoided. Finite element-based procedures are best reserved for the contract and detail design phases. In more complex situations, such as the torsional vibration of ships with open cross sections (containerships), either a three-dimensional finite element model or an improved beam model[10] ought to be used. The need to use a three-dimensional finite element model stems from the fact that the structure extends in three-dimensional space, with important vibration modes arising and propagating in the longitudinal but also in the transverse and axial (twisting) directions that are coupled. An example is the excitation of both the hull girder and the superstructure as a result of the operation of the propeller and main propulsion machinery. One other advantage of three-dimensional finite element models is that stresses and deformations are determined at any position [37].

The fundamental elements of a vibrating system include its basic mass-elastic properties as well as damping and excitation forces. In order to control or limit the vibratory response, it is in principle necessary to modify the mass-elastic properties of the system either (a) by increasing the damping, (b) by reducing the amplitude of the excitation forces or (c) by changing the excitation frequencies. Increasing the damping may be useful in the solution of local structural vibration problems and in certain machinery and equipment problems but is not a practical solution for the reduction of hull girder vibration. Instead, what is done is

[10] A model that includes torsion-related shear.

to first calculate the natural frequencies of vibration of the hull girder and then design the main propulsion engines and propeller so that resonance is avoided. A detailed study of the main propulsion system, comprising of the main engines, propeller shaft and propeller, as well as the hull form in way of the aft section needs to be carried out [49]. This is the only practical way, because once the layout of an entire hull girder has been defined, it is not possible to modify its vibration characteristics.

Concern with vibration extends beyond the global level to both regional (major substructure) and local levels, as well as the coupling between different levels, such as between the hull girder and a superstructure. This is an important topic that will not be covered in this text.

8.4.2 Vertical Vibration of the Hull Girder

The vertical vibration of a beam-type structure occurs about its transverse neutral axis of bending, all along its length. In the formulas we will be comparing in this section this effect is included. Finally, it is possible that coupling of the higher modes of vertical vibration of the hull girder with those of the superstructure occurs.

The response of the hull girder may be resonant, leading to significant amplification of the vibration. This occurs when the hull girder natural frequency coincides with or is sufficiently close to the excitation frequency through the first five or six modes of vibration when driven by the propeller shaft, or if propeller frequencies are present. It is likely to be non-resonant (forced) above the fifth or sixth mode. In this case the hull girder vibrates approximately in proportion to the generated forces.

The principal exciting frequencies are the propeller blade frequency and the propeller blade frequency harmonics. Hydrodynamic forces may also stimulate the resonant frequency of hulls, rudders or struts excited by hydrodynamic flow over the appendage. Figure 8.20b shows the response spectrum measured at the extreme aft end of a UK Type 42 frigate. It is seen that the constituent components are multiples of the propeller shaft rotational speed (240 rpm) [51].

As a general rule, the fundamental (first) vertical mode of vibration may be as low as 1 Hz while the higher modes will follow the fundamental frequency by the ratios 2, 3, 4, etc. (Figure 8.20). The longer the hull and the more flexible its midship section is, the lower will be its fundamental natural frequency.

Details for oil tankers and bulk carriers having a displacement in the range 2,000–200,000 metric tonnes have been plotted in Figure 8.21 [49]. We see that:

1. For each vibration mode, the natural frequency reduces with an increase in displacement (length).
2. Higher modes exhibit higher-vibration frequencies (as expected).
3. As the mode number increases, the reduction in frequency that accompanies displacement also increases.
4. The standard deviation of the data increases as the mode increases. The prediction accuracy therefore correspondingly reduces.

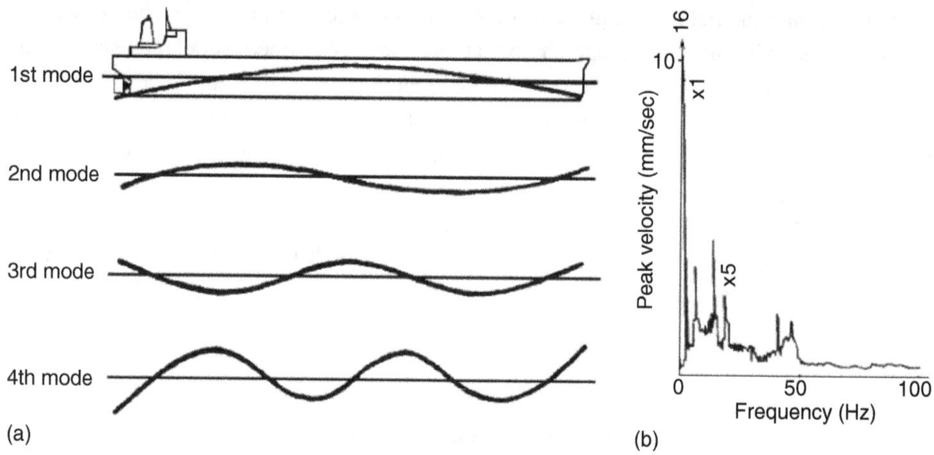

Figure 8.20 Vertical vibration of the hull girder. (a) Modes [33]. (b) Measured spectrum [51]. Figure 8.20a reprinted courtesy of the Ship Structure Committee, Washington, DC.

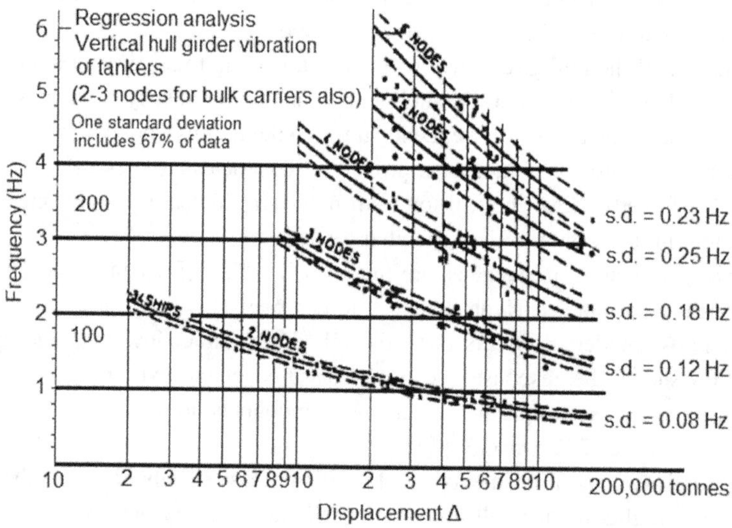

Figure 8.21 Vertical hull girder vibration frequencies for first through fourth modes of oil tankers and bulk carriers [49]. Reprinted courtesy of the Society of Naval Architects and Marine Engineeers.

5. The fundamental frequency (first mode) lies in the range 0.7–2.0 Hz (from larger to smaller ships).
6. These conclusions hold for both types of ships.

8.4.3 Differential Equation of Vertical Hull Girder Vibration

In this section a differential equation for the vertical vibration of a beam subjected to the loads a hull girder is subjected to will be obtained, following closely the derivation

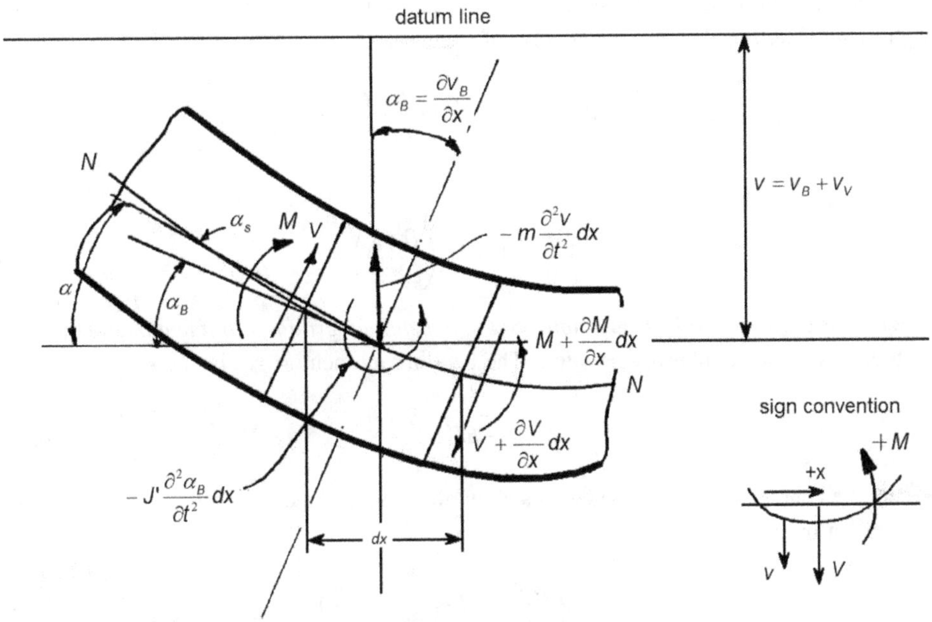

Figure 8.22 Beam section of ship-type element [34].

included in [34]. In this the effects of hydrodynamic inertia, rotatory inertia and shear deformations are included.

As mentioned in Section 4.3.2, a major difference between the response of a beam and that of a hull girder is due to shear lag. When performing a study of a hull girder, it is necessary to allow for this by determining the dimensions of the equivalent, fully effective beam.

Figure 8.22 shows an element of a beam of length *dx* and mass *m* per unit length that is subjected to the loads that act on a hull girder in vertical bending. These are the bending moment *M*, the vertical shear force *V* and the vertical inertia force, and these result in a rigid body dynamic response of the section. In this treatment damping is ignored. In this figure (v_B, α_B) and (v_V, α_S) are the displacement and angle of rotation related to bending and shear respectively.

Dynamic equilibrium in the vertical direction is achieved for the beam element considered when

$$\frac{\partial V}{\partial x}dx = mdx\left(\frac{\partial^2 v}{\partial t^2}\right)$$

or

$$\frac{\partial V}{\partial x} = m\left(\frac{\partial^2 v}{\partial t^2}\right) \tag{8.64}$$

In Equation (8.64) $\partial V/\partial x$ is the net load acting on the beam element in the vertical direction, which produces a vertical acceleration that is equal to $\partial^2 v/\partial t^2$. For dynamic equilibrium in a rotational sense moments are considered about the element face on the right-hand side. For the sign convention of Figure 8.22,

$$\frac{\partial M}{\partial x}dx = Vdx - J'dx\left(\frac{\partial^2}{\partial t^2}\left(\frac{\partial v_B}{\partial x}\right)\right)$$

or

$$\frac{\partial M}{\partial x} = V - J'\left(\frac{\partial^3 v_B}{\partial x \partial t^2}\right) \tag{8.65}$$

where J' is the *ship and hydrodynamic mass moment of inertia per unit length* and the term in brackets is the angular acceleration. The bending moment M is given by

$$M = -EI(x)\frac{\partial^2 v_B}{\partial x^2} \tag{8.66}$$

where the bending stiffness $EI(x)$ varies along the length of the hull girder. The shear force V is given by

$$V = k_2\frac{\partial v_V}{\partial x} = k_2\left(\frac{\partial v}{\partial x} - \frac{\partial v_B}{\partial x}\right) \tag{8.67}$$

where k_2 is the *shear stiffness*. This is obtained from

$$k_2(x) = \frac{V(x)}{y_V(x)} = K(x)A(x)G \tag{8.68}$$

where K depends on the geometry of the cross-section. (For a rectangular section, $K = 0.84$).

Variables are separated by expanding deformations and loads using infinite series of eigenfunctions (mode shapes) (Appendix E):

$$v = \sum_{j=1}^{\infty} v_j e^{i\omega_j t} \tag{8.69a}$$

$$v_B = \sum_{j=1}^{\infty} v_{Bj} e^{i\omega_j t} \tag{8.69b}$$

$$v_V = \sum_{j=1}^{\infty} v_{Vj} e^{i\omega_j t} \tag{8.69c}$$

$$V = \sum_{j=1}^{\infty} V_j e^{i\omega_j t} \tag{8.69d}$$

$$M = \sum_{j=1}^{\infty} M_j e^{i\omega_j t} \tag{8.69e}$$

where

$$\omega_j = \text{natural frequency of degree } j$$
$$v, v_B, v_V = \text{amplitudes of the vertical deflections}$$
$$V_j, M_j = \text{loads of the jth mode}$$

These expressions are substituted in Equations (8.64)–(8.67) to give, for any frequency ω,

$$\frac{dV}{dx} = -\omega^2 m v \tag{8.70a}$$

$$\frac{dM}{dx} = V + \omega^2 J' \frac{dv_B}{dx} \tag{8.70b}$$

$$M = -EI(x)\frac{d^2 v_B}{dx^2} \tag{8.70c}$$

$$V = k_2\left(\frac{dv}{dx} - \frac{dv_B}{dx}\right) \tag{8.70d}$$

The variables are now uncoupled. The equation of dynamic equilibrium needs to be expressed in terms of a single dependent variable, the total deflection v. It is therefore necessary to eliminate the deflection due to bending, v_B, so that M and V can then be expressed in terms of v alone and its derivatives. For the bending moment M, Equation (8.70d) is solved for dv_B/dx, differentiated with respect to x and substituted in Equation (8.70c):

$$\begin{aligned} \frac{dv_B}{dx} &= \frac{dv}{dx} - \frac{V}{k_2} \\ \frac{d^2 v_B}{dx^2} &= \frac{d^2 v}{dx^2} - V\frac{d}{dx}\left(\frac{1}{k_2}\right) - \frac{1}{k_2}\left(\frac{dV}{dx}\right) \end{aligned} \tag{8.71}$$

Hence, Equation (8.70c) becomes

$$M = -EI(x)\left[\frac{d^2 v}{dx^2} - V\frac{d}{dx}\left(\frac{1}{k_2}\right) - \frac{1}{k_2}\left(\frac{dV}{dx}\right)\right] \tag{8.72}$$

The bending moment M is now expressed in terms of d^2v/dx^2. An analogous procedure is followed for the shear force V. Expression in Equation (8.70d) is differentiated with respect to x and set equal to Equation (8.69e):

$$\frac{dM}{dx} = -\left(EI(x)\frac{d^3 v_B}{dx^3} + \frac{dEI(x)}{dx}\frac{d^2 v_B}{dx^2}\right) = V + \omega^2 J' \frac{dv_B}{dx} \tag{8.73}$$

The third derivative of v_B is obtained by differentiating Equation (8.71):

$$\frac{d^3 v_B}{dx^3} = \frac{d^3 v}{dx^3} - V\frac{d^2}{dx^2}\left(\frac{1}{k_2}\right) - \frac{dV}{dx}\frac{d}{dx}\left(\frac{1}{k_2}\right) - \frac{d^2 V}{dx^2}\frac{1}{k_2} - \frac{dV}{dx}\frac{d}{dx}\left(\frac{1}{k_2}\right)$$

Using Equation (8.70a),

$$\frac{d^2 V}{dx^2} = -\omega^2 m\frac{dv}{dx} - \omega^2\frac{dm}{dx}v \tag{8.73'}$$

The following expression is finally obtained for the third derivative of v_B:

$$\begin{aligned}\frac{d^3 v_B}{dx^3} = \frac{d^3 v}{dx^3} - V\frac{d^2}{dx^2}\left(\frac{1}{k_2}\right) + \omega^2 mv\frac{d}{dx}\left(\frac{1}{k_2}\right) + \frac{\omega^2 m}{k_2}\frac{dv}{dx} + \\ + \omega^2\frac{dm}{dx}v\left(\frac{1}{k_2}\right) + \omega^2 mv\frac{d}{dx}\left(\frac{1}{k_2}\right)\end{aligned} \tag{8.74}$$

Substituting Equation (8.74) into Equation (8.73) and using Equation (8.71), a differential equation is obtained for V in terms of v and its derivatives:

$$-EI(x)\left\{\frac{d^3 v}{dx^3} - V\frac{d^2}{dx^2}\left(\frac{1}{k_2}\right) + \omega^2 mv\frac{d}{dx}\left(\frac{1}{k_2}\right) + \frac{\omega^2 m}{k_2}\frac{dv}{dx} + \omega^2\frac{dm}{dx}v\left(\frac{1}{k_2}\right) + \right.$$

$$\left. + \omega^2 mv\frac{d}{dx}\left(\frac{1}{k_2}\right)\right\} - \frac{dEI(x)}{dx}\left\{\frac{d^2 v}{dx^2} - V\frac{d}{dx}\left(\frac{1}{k_2}\right)\right\} = V + \omega^2 J'\left(\frac{dv}{dx} - \frac{V}{k_2}\right)$$

Rearranging terms, the following expression is obtained:

$$V = \frac{\frac{d}{dx}\left[EI\left(\frac{d^2 v}{dx^2} + \frac{\omega^2 m}{k_2}v\right)\right] + \omega^2\frac{dm}{dx}v\left(\frac{1}{k_2}\right) + \omega^2 mEI(x)v\frac{d}{dx}\left(\frac{1}{k_2}\right) + \omega^2 J'\frac{dv}{dx}}{-1 + \frac{\omega^2 J'}{k_2} + \frac{d}{dx}\left(EI(x)\frac{d}{dx}\left(\frac{1}{k_2}\right)\right)} \tag{8.75}$$

Equations (8.72) and (8.75) link the mode shape v to the load V. The equation of motion (dynamic equilibrium) can then be obtained by eliminating both M and V. This is done by equating Equation (8.70a) to the derivative of Equation (8.70d) and substituting for V in the resulting expression from Equation (8.75) to give

$$\frac{d}{dx}\left[\frac{\frac{d}{dx}\left[EI(x)\left(\frac{d^2 v}{dx^2} + \frac{\omega^2 m}{k_2}v\right)\right] + \omega^2\frac{dm}{dx}v\left(\frac{1}{k_2}\right) + \omega^2 mEI(x)v\frac{d}{dx}\left(\frac{1}{k_2}\right) + \omega^2 J'\frac{dv}{dx}}{-1 + \frac{\omega^2 J'}{k_2} + \frac{d}{dx}\left(EI(x)\frac{d}{dx}\left(\frac{1}{k_2}\right)\right)}\right] + \omega^2 mv = 0 \tag{8.76}$$

This is a fourth-order linear homogeneous differential equation with variable coefficients. The dependent variable v is expressed in terms of the structural properties and the applied load. The four integration constants are determined by applying the boundary conditions for a free-free beam:

$$V(0) = V(L) = M(0) = M(L) = 0 \tag{8.77}$$

An estimate of the mode shape allows values for the natural frequency to be obtained. Several methods to solve Equation (8.76) have been described in the literature, and these as well as other approaches to determine natural frequencies and mode shapes are compared in [34]. A list follows:

1. The discrete disk method (Csupor)
2. The stepped beam method
3. The method of successive approximations (Stodola)

4. The energy theorem and Rayleigh's principle
5. Empirical formulas (Schlick, Todd, Burrill, IRCN).

We shall not extend our discussion further in this direction, as our aim, which has been to indicate the most important parameters that affect the mode and frequency of vertical vibration of the hull girder, has been achieved. Empirical formulas are discussed in the next section.

8.4.4 Prediction of Natural Frequency of Vertical Vibration Using Approximate Formulas

In this section we shall compare the results obtained from the use of the Timoshenko–Rayleigh formula, derived in Section 8.2.2, as modified in accordance with ship vibration data, with formulas currently in use in rules and handbooks for predicting the vertical vibration frequency of ship hull girders. The Timoshenko–Rayleigh formula (Equation (8.36)) was used in conjunction with accurately calculated values for the beam shear coefficient [52]. The shear coefficient is given by the following formula [8]:

$$\beta = \frac{10(1+\nu)(1+3m)^2}{(12+72m+150m^2+90m^3)+\nu(11+66m+135m^2+90m^3)+10n^2[(3+\nu)m+3m^2]} \tag{8.78}$$

where

$$m = \frac{Bt_1}{Ht} \qquad n = \frac{B}{H}$$

B = height of beam
H = width of beam
t_1 = thickness of upper and lower flanges
t = web thickness
ν = Poisson's ratio

For simply supported boundary conditions, the solution of Equation (8.45) has the following form:

$$w = A_j \sin\frac{j\pi x}{l}\sin(\omega_j t + \varphi_j) \tag{8.79}$$

Substituting Equation (8.79) into Equation (8.45) and ignoring the cross-coupling term, we obtain

$$EI\left(\frac{j\pi}{l}\right)^4 - \rho A\omega_j^2 - \rho I\left(1+\frac{E}{kG}\right)\left(\frac{j\pi}{l}\right)^2\omega_j^2 = 0 \tag{8.80}$$

Table 8.5 *Coefficient C_n* for hull girder natural frequency of vertical vibration

Ship type	First-order coefficient	Second-order coefficient	Third-order coefficient
Bulk carrier	620.6	1,212.2	1,735.6
Oil tanker	639.5	1,373.8	2,082.7
Containership	560.6	1,111.7	1,591.8

For simply supported ends, the Euler–Bernoulli beam natural frequency is

$$\omega = \frac{j^2\pi^2}{L^2}\sqrt{\frac{EI}{\rho A}} \tag{8.81}$$

For free ends, the Euler–Bernoulli beam natural frequency is

$$\omega = \frac{(j+0.5)^2\pi^2}{L^2}\sqrt{\frac{EI}{\rho A}} \tag{8.82}$$

On the basis of Equations (8.80)–(8.82), an approximate formula for a beam with free ends considering the effects of shear deformation and moment of inertia can be obtained:

$$\omega = \frac{(j+0.5)^2\pi^2}{L^2}\sqrt{\frac{EI}{\rho A}}\sqrt{\frac{1}{1+\frac{(j+0.5)^2\pi^2 I}{L^2 A}\left(1+\frac{E}{kG}\right)}} \tag{8.83}$$

This equation needs to be modified in order to (1) allow for the effect of varying moment of inertia along the length of the beam, (2) allow for transverse shear deformation and (3) introduce empirical constants to allow for the vibration characteristics of actual ships. The following expression is subsequently obtained for the natural frequency of a hull girder:

$$f_n = C_n\sqrt{\frac{BD^2}{\Delta_V L^2}} \tag{8.84}$$

where B and D are the breadth and depth of the ship and Δ_L is the virtual displacement, which includes the added mass. This is given by the Todd formula:

$$\Delta_V = (1+\tau)\Delta \qquad \text{where} \qquad \tau = 0.2 + \frac{B}{3d} \ (d = \text{draught})$$

The coefficient C_n is obtained from Table 8.5.

A comparison of results obtained using Equation (8.84) with formulas proposed by different authorities for a series of representative ship types is included in Figure 8.23. The particulars of the ships selected in this study are included in Table 8.6.

Equation (8.84) is based on Timoshenko's equation of uniform beam vibration that allows for rotatory inertia and shear. This has been combined with statistical data for a number of ships and is thus a semi-empirical formula. The results indicate that such an approach can provide useful predictions for the natural frequency of the lower modes of vertical vibration.

Table 8.6 *Principal particulars of ship considered in vertical vibration frequency study*

Ship type	Length (m)	Breadth (m)	Depth (m)	Moment of inertia (m^4)	Displacement (mt)
Bulk carrier	183.0	27.4	14.8	115.46	19,900
Bulk carrier	144.8	20.4	12.2	50.23	19,720
Tanker	255.8	37.1	18.4	403.97	102,810
Tanker	218.0	32.9	15.6	212.03	66,289
Tanker	320.0	58.0	31.0	1,454.62	342,507
Containership	319.0	42.8	24.6	549.60	136,701

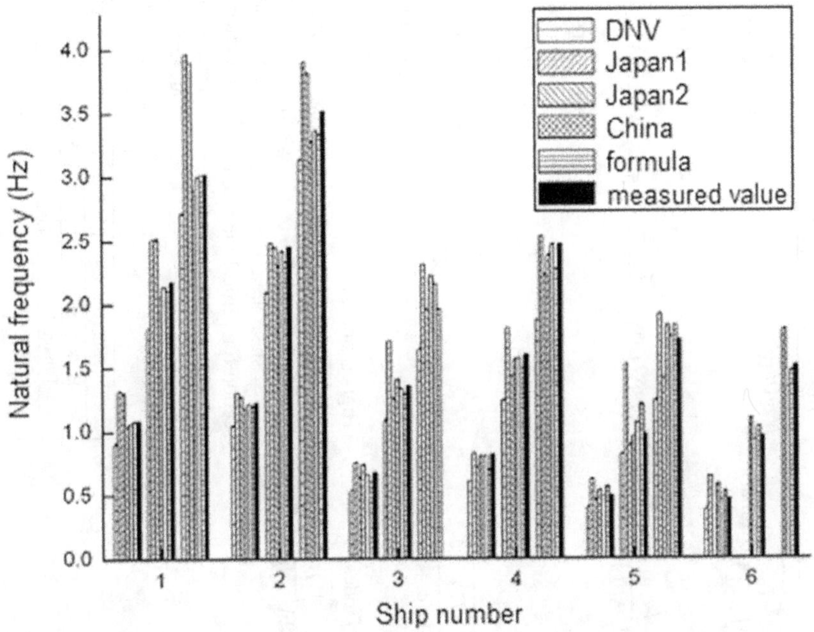

Figure 8.23 Comparison of predictions of the natural frequency of vertical vibration using selected formulas (Table 8.7) [52].

The expression used by DNV [10][11] was obtained by Kumai and was based on Rayleigh's method.[12] For an elastic hull girder of variable cross-section floating in water, the maximum potential energy, U_{max}, is equal to the maximum kinetic energy, T_{max}. Therefore, [25]

$$\int_L EIy''^2 dx = \omega^2 \int_L (m + Jm_e) y^2 dx \tag{8.85}$$

[11] A comparison of selected empirical formulas (Schlick, IRCN, Kumai) with results obtained using the FEM for a selected number of ships led to the conclusion that the Kumai and the IRCN formulas give the best results [48].

[12] Section 8.2.3.

Table 8.7 *Selected formulas for natural frequencies (Hz) of vertical vibration of ship hull girders*

Authority	*1st vibration mode*	*Higher vibration modes*	*Coefficients*
DNV [10]	$(1.61)(10^6)\sqrt{\frac{I_V}{\Delta_V L^3}}$ (Kumai)	$N_{2V}(n-1)^{\mu_v}$	$\Delta_V = \left(1.2 + \frac{B}{3d}\right)\Delta$ $\mu_V = 1.02$ for tankers $\mu_V = 1.00$ for bulk carriers
Japan shipbuilding handbook[1]	$(27.1)(10^5)\sqrt{\frac{I_V}{\Delta_V L^3}} + 14.5$	$(38.8)(10^5)\sqrt{\frac{I_V}{\Delta_V L^3}} + 58.5$	$\Delta_V = \left[1 + 0.3\left(\frac{B}{T}\right) - 0.033\left(\frac{B}{T}\right)^2\right]\Delta$ for cargo ships $\Delta_V = \left[1 + 0.3\left(\frac{B}{T}\right) - 0.033\left(\frac{B}{T}\right)^2\right]\Delta$ for tankers
Japan (incl. effect of shear deformation and moment of inertia)[1]	$c_n\sqrt{\frac{I_V}{\Delta_V L^3[1+(\alpha+\beta)n^2\pi^2]}}$		$c_2 = (4)(10)^6$, $c_3 = (9.2)(10)^6$, $\alpha = \frac{EI_V}{GA_S L^2}$ A_s = effective shear area, $\beta = \frac{r_0^2}{L^2}$ where r_0 = turning radius
China ship design handbook[2]	(1st and 2nd modes) $K_{ib}\left(A_{iv}K_{iv}E_{iv}\sqrt{\frac{I_V}{\Delta_L L^3}} + B_{iv}\right)$	(Higher modes) $A_{iv}\sqrt{\frac{A_{OV}}{\Delta_L L}} + B_{iv}$ A_{OV} = shear area of midship section for vertical vibration	i = no of nodes A_{iv}, B_{iv}= regression coefficients K_{ib}= reduction coefficient of bending stiffness for vertical vibration K_{iv} = influence coefficient caused by change of moment of inertia along the length for natural frequency of vertical vibration E_{iv}= superstructure influence coefficient
Timoshenko-type formula [52]	$C_n\sqrt{\frac{BD^2}{\Delta_L L^2}}$		

Ship type	*1st order coefficient*	*2nd order coefficient*	*3rd order coefficient*
Bulk carrier	620.6	1,212.2	1,735.6
Tanker	639.5	1,373.8	2,082.7
Containership	560.6	1,111.7	1,591.8

[1]Okada Y. (1983) *Shipbuilding design handbook*, Tokyo: The Kansai Society of Naval Architects
[2]Zhao D.Y. (1993) *Calculation method of natural frequency for ship's overall vibration*, Beijing: China Biaozhun Publishing House.

where

$\omega =$ natural circular frequency of flexural vibration

$EI =$ flexural stiffness

$m =$ mass per unit length

$m_e =$ two-dimensional added mass of water per unit length

$x =$ coordinate axis along length

$y =$ assumed deformation mode

$J =$ three-dimensional local reduction factor of m_e

Solving for ω^2, the following expression is obtained:

$$\omega^2 = \frac{EI_0 \int_L \left(\frac{I}{I_0}\right) y''^2 dx}{\overline{m} \int_L \frac{m}{\overline{m}} y^2 dx \left(1 + \frac{\int_L J m_e y^2 dx}{\int_L m y^2 dx}\right)} \tag{8.86}$$

where I/I_0 is the distribution function of the moment of inertia along the length, $\overline{m}$ is the mean mass per unit length, and m$/\overline{m}$ is the mass distribution function. We see that Equation (8.85) is similar to Equation (8.53), with two important differences. The first is that a non-uniform beam is assumed, so that the stiffness *EI* is included in the integral on the left-hand side of the equation. Secondly, the added mass of water has been included on the right-hand side of the equation.

The following expression is obtained from Equation (8.86) by introducing c_1, a variable section coefficient based on Prohaska's work:

$$\omega = c_n \sqrt{\frac{gEI_0}{\Delta_V L^3}} \tag{8.87}$$

where

$$c_n = \frac{L^2}{n^2 \pi^2} \sqrt{\frac{\int_L \frac{I}{I_0} y''^2 dx}{\int_L \frac{m}{\overline{m}} y^2 dx}}$$

For higher vibration modes, Kumai proposed the following empirical formula [25]:

$$N_{Vn} = N_{V2}(n-1)^{\mu} \tag{8.88}$$

where

$N_{Vn} =$ frequency of vertical vibration with n nodes

$N_{V2} =$ frequency of vertical vibration with two nodes

$n =$ number of nodes

$\mu = 1.02$ for tankers

$= 1.0$ for bulk carriers

$= 0.845$ for cargo ships

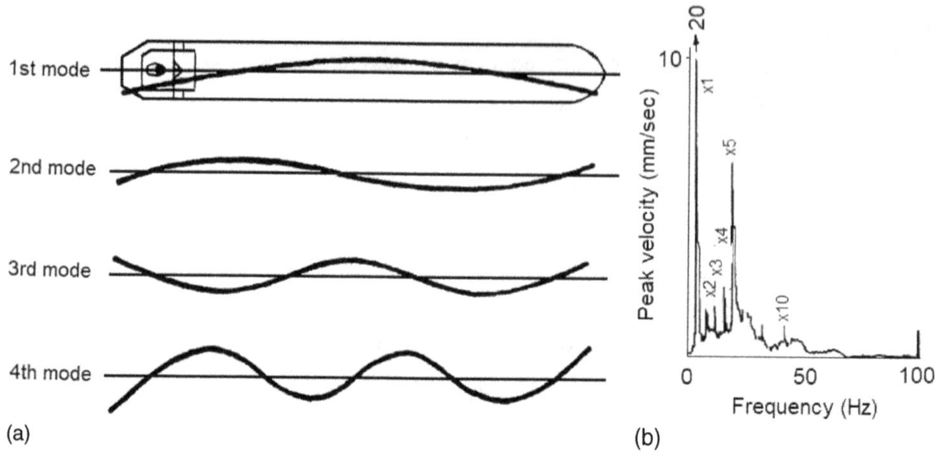

Figure 8.24 Horizontal vibration. (a) The first four modes [33]. (b) Measured spectrum [51]. Figure 8.24a reprinted courtesy of the Ship Structure Committee, Washington, DC.

8.4.5 Horizontal Vibration

Horizontal vibration follows a similar pattern to that of vertical vibration, as shown in Figure 8.24a. In general the fundamental (two-noded) frequency has been found to be approximately 50 per cent higher than the fundamental frequency of vertical vibration. The narrower the beam, and consequently the lower the stiffness about the vertical axis lying in the centreline plane, the lower the horizontal vibration frequency. In the case of twin-screw ships, significant excitation of horizontal modes may occur due to resonance of propeller out-of-balance forces. Figure 8.24b shows the vibration spectrum measured at the extreme aft end of a UK Type 42 class frigate [51]. As such, the vibration is propeller-excited, and individual components of the spectrum are multiples of the propeller shaft rotational speed (240 rpm) as indicated.

In principle, it is possible to derive the frequency of horizontal vibration, N_H, from the value obtained for vertical vibration (N_V). Assuming that the distributions of stiffness and moment of inertia are similar in the vertical and horizontal modes (Figure 8.13a), the frequency of horizontal vibration is given by

$$N_H \approx N_V \sqrt{\frac{I_{XH}}{I_{XV}} \frac{\Delta_V{}'}{\Delta_H{}'}} \tag{8.89}$$

where

I_{XV}, I_{XH} = vertical and horizontal moments of inertia
$\Delta_V{}', \Delta_H{}'$ = displacement including added mass effect in vertical and horizontal directions

For two-noded vibration it has been found that $1.5 > N_H/N_V > 1.3$. For higher modes, on the basis of calculations for individual ships, the following rule is recommended: 1:2:3:4 [46].

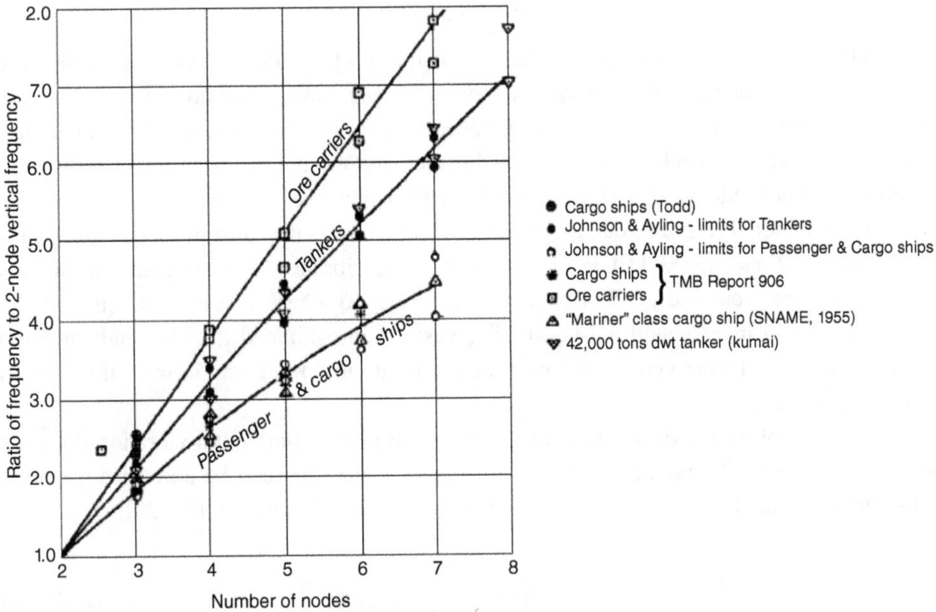

Figure 8.25 Horizontal vibration frequencies [46].

The frequency of horizontal vibration can also be determined using Figure 8.25 or

$$N_{H2} = 2.614\sqrt{\frac{I_H}{\Delta_{1H}\sqrt{1+rL^3}}} \qquad \text{(Hz)} \qquad \text{(Schlick)} \qquad (8.90)$$

where

$$\Delta_{1H} = \left(1.2 + \frac{T}{3B}\right) = 1.32\Delta \qquad \text{and} \qquad \sqrt{1+r} = 1.07$$

The following formula can also be used, obtained by IRCN[13]:

$$N_{H2} = 76K'\sqrt{\frac{I_{OH}}{\Delta_i L^3}} + 29 \qquad \text{(cpm)} \qquad (8.91)$$

where

I_{OH} = transverse moment of inertia amidships $(\mathrm{mm^2 m^2})$
Δ_i = displacement including added virtual mass = $(1 + d/B)\Delta$ (tons)
K' = 1 for ships with short superstructures. For other ships, K' depends on the deckhouse arrangement.

[13] IRCN: Institut Recherche Construction Navale, France.

8.4.6 Torsional Vibration

The calculation of torsional vibration has become important in recent years because large containerships that have open deck sections are being built. Vibration per se does not necessarily cause damage but results in reduced fatigue life. In a number of recent studies the extent of vibration-related damage has been quantified and has been found to be non-negligible. Thus, Storhaug and Kahl confirmed an increase of about 5 per cent in the maximum torsional response due to torsional vibration measured on a transverse deck strip on two (8,400 and 8,600 TEU) containerships. The vibration damage contribution was about 15–16 per cent and 52–55 per cent, respectively, on these vessels. In another study, Ki et al. (2015) investigated a 14,000 TEU vessel and estimated the torsional vibration damage as being 13 per cent on the hull girder and up to 25 per cent in way of the hatch corners [37].

In the case of closed-deck ships, in which coupling with horizontal vibration does not arise, the fundamental natural frequency of torsional vibration can be found from Horn's empirical formula that is based on the elementary theory of vibration [21, 30]:

$$N = 60k\sqrt{\frac{g}{\Delta}\frac{k_3}{\left(B^2 + D^2\right)L}} \tag{8.92}$$

where

B = moulded beam
D = depth
L = length
Δ/g = mass
k_3 = torsional stiffness
k = empirical coefficient

An approximate (low) value for the torsional stiffness k_3 is given by Equations (6.114) or (6.116). The value obtained does not take into account all internal longitudinally aligned members such as decks, bulkheads as well as the bottom structure. A more accurate value is obtained by implementing the method described in Section 6.3.10.

For the empirical coefficient k the values of 1.58, 3.0 and 4.57 were calculated for the first three modes, respectively.

8.4.7 Differential Equations of Coupled Horizontal and Torsional Vibration

As described in Chapter 7, torsional loads are coupled with horizontal bending loads in the case of open-deck ships where the centre of shear, which acts as the centre of twist in uniform torsion, is located below the keel-line. Analytical solutions to this problem have been obtained and presented in [24] and [39], and a finite difference-based solution is described in [30]. This section presents the differential equations for coupled torsion and horizontal bending derived in [34].

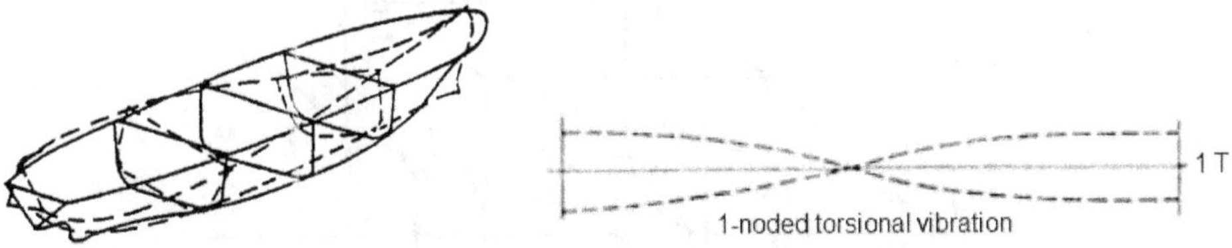

Figure 8.26 Global torsional vibration of the hull girder [29, 33]. Figure 8.26a reprinted courtesy of the Ship Structure Committee, Washington, DC. Figure 8.26b reprinted courtesy of the Society of Naval Architects and Marine Engineeers.

The approach followed is analogous to that in vertical vibration. It is assumed that the form of the hull girder varies slowly along the length of the ship, so that each transverse section has different geometrical properties (centre of shear, centre of mass).

Figure 8.27 shows the forces acting on an element of length dx of an open-deck ship that is vibrating in the coupled torsional-horizontal mode. The two transverse sections at the ends of this element are located at x and $x + dx$ (Figure 8.27a). The centres of mass (gravity) of each transverse section are denoted C_1 at x and C_2 at $x + dx$. The distance between the two resulting from the curvature of the neutral axis is $d\xi$. The datum point K coincides with the ship's keel-line. The centres of shear of the two sections are S_1 and S_2. They do not coincide because the centre of shear line is assumed curved, so that the vertical distance between the centres of shear is $d\eta$ (Figure 8.27b).

Any combination of shear force and torque acting on an element may be resolved into a shear force V acting through the centroid C_1 and a free torque M_T. At C_2 a resultant shear force $V + dV$ acts. The inertial force dV acting on the element of length dx acts through the centre of gravity of the element $C_1 - C_2$ and the rotational inertia produces a free couple. Since the centre of shear does not coincide with the centre of gravity, particularly in the case of open-deck ships, any horizontal shear force V will be associated with a torsional couple. The extent of coupling depends on the distance between the centre of shear and the centre of gravity. It is seen from Figure 8.27b that $\rho(x) = \eta(x) + \xi(x)$. For rotational equilibrium,

$$dM_T = J'' \frac{\partial^2 \varphi}{\partial t^2} dx + \rho dV + V d\eta \tag{8.93}$$

where

J'' = torsional moment of inertia, including hydrodynamic effects
φ = torsional displacement
dV = resultant inertial force acting on element of length dx
$= mdx(d^2 v/dx^2)$
ρdV = torsional couple
$Vd\eta$ = free moment due to curvature of centre of shear line

The other two equilibrium conditions are as in the case of vertical vibration. As previously, variables are separated and one normal mode is considered:

$$\phi(x, t) = \phi(x) e^{i\omega t} \tag{8.94}$$

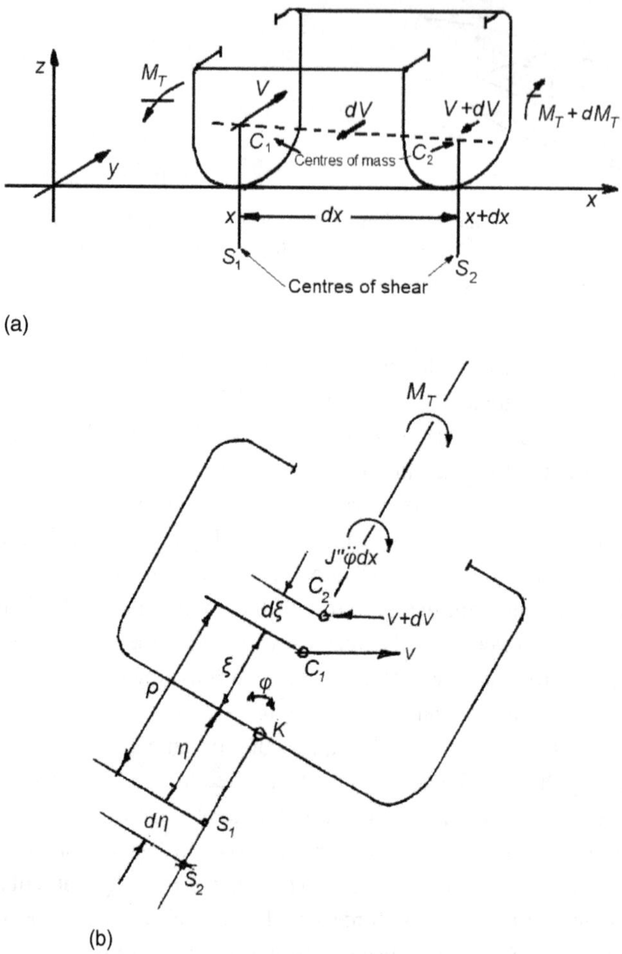

Figure 8.27 Forces acting on hull girder element of length dx [34]. (a) Longitudinal view. (b) Transverse section.

Equation (8.93) becomes

$$\frac{dM_T}{dx} = -\omega^2 J''\phi + \rho\frac{dV}{dx} + V\frac{d\eta}{dx} \tag{8.95}$$

As in the case of vertical vibration (Equations (8.70a) and (8.70b)), we have

$$\frac{dV}{dx} = -\omega^2 mv \tag{8.96a}$$

$$\frac{dM}{dx} = V + \omega^2 J'\frac{dv_B}{dx} \tag{8.96b}$$

The displacement v is now given by

$$v = v_B + v_V + v_T \tag{8.97}$$

where v_T is the displacement due to torsion and

$$dv_T = \phi d\xi + \rho d\phi$$

or

$$\frac{dv_T}{dx} = \phi \frac{d\xi}{dx} + \rho \frac{d\phi}{dx} \tag{8.97'}$$

The load-deformation relationships are expressed by Equations (8.64), (8.65) and

$$M_T = k_3 \frac{d\phi}{dx} \tag{8.98}$$

Equations (8.73), (8.73′) and (8.95), which are the dynamic equilibrium conditions in the vertical, horizontal and torsional direction, respectively, and the load-deformation relationships Equations (8.64), (8.65) and (8.98) constitute a system of six simultaneous differential equations. This system can be solved, given Equations (8.97) and (8.97′) for the free vibration problem provided boundary conditions are specified. For free ends,

$$M = V = M_T = 0 \qquad \text{at } x = 0, \ \ x = L \tag{8.99}$$

As in the case of vertical vibration, all deformation variables apart from v and φ are eliminated to obtain expressions for M, V and M_T in terms of these variables. These expressions are as follows:

$$M = -EI(x)\left[\frac{d^2v}{dx^2} - v\frac{d}{dx}\left(\frac{1}{k_2}\right) + \frac{\omega^2 m}{k_2}v\right] - \frac{d}{dx}\left(\phi\frac{d\xi}{dx} + \rho\frac{d\phi}{dx}\right) \tag{8.100a}$$

$$V = \frac{1}{\frac{d}{dx}\left(EI(x)\frac{d}{dx}\left(\frac{1}{k_2}\right)\right) + \omega^2\frac{J'}{k_2} - 1}\left\{\frac{d}{dx}\left(EI(x)\left(\frac{d^2v}{dx^2} + \frac{\omega^2 m}{k_2}v\right)\right) + \right.$$
$$\left. + \omega^2 mEI(x)\frac{d}{dx}\left(\frac{1}{k_2}\right)v + \omega^2 J'\frac{dv}{dx} - \frac{d}{dx}\left[EI(x)\frac{d}{dx}\left(\phi\frac{d\xi}{dx} + \rho\frac{d\phi}{dx}\right)\right] - \omega^2 J'\left(\phi\frac{d\xi}{dx} + \rho\frac{d\phi}{dx}\right)\right\} \tag{8.100b}$$

$$M_T = k_3\frac{d\phi}{dx} \tag{8.100c}$$

The two coupled equations of dynamic equilibrium are given below:

$$0 = \omega^2 mv + \frac{d}{dx}\left\{\frac{1}{\frac{d}{dx}\left(EI(x)\frac{d}{dx}\left(\frac{1}{k_2}\right)\right) + \frac{\omega^2 J'}{k_2} - 1}\left\{\frac{d}{dx}\left[EI(x)\left(\frac{d^2v}{dx^2} + \omega^2\frac{mv}{k_2}\right)\right] + \right.\right.$$
$$\left.\left. + \omega^2 mEI(x)\frac{d}{dx}\left(\frac{1}{k_2}\right)v + \omega^2 J'\frac{dv}{dx} - \frac{d}{dx}\left[EI(x)\frac{d}{dx}\left(\phi\frac{d\xi}{dx} + \rho\frac{d\phi}{dx}\right)\right] - \omega^2 J'\left(\phi\frac{d\xi}{dx} + \rho\frac{d\phi}{dx}\right)\right\}\right\} \tag{8.101a}$$

$$0 = \frac{d}{dx}\left\{\frac{1}{d\eta/dx}\left(\frac{d}{dx}\left(k_3\frac{d\phi}{dx}\right) + \omega^2 J''\phi + \omega^2 m\rho v\right)\right\} + \omega^2 mv \tag{8.101b}$$

If $\rho = 0$ and $\xi = 0$, the equations are uncoupled. The equations can be solved using the methods proposed in the case of vertical vibration.

8.4.8 Hydroelastic Resonance (Springing)

Springing vibration may arise in the vertical bending, horizontal bending and torsional modes. Figure 8.28 shows an overall vertical bending stress pattern and its two components, these being the wave-induced stress component and the springing component. Springing stresses appear as a narrow-banded noise component that damps out after a significant number of cycles.

Depending on the type of ship, springing may contribute to fatigue damage to quite a significant extent. In the case of long bulk carriers this contribution has been estimated to account for up to 50 per cent of the accumulated fatigue damage.

8.5 Vibration of the Hull Girder Resulting from Transient Loading

In the discussion of slamming in Chapter 3 only local structural behaviour was considered. Slamming impacts at the fore and aft ends of a ship, that depend on hull form, draught, sea conditions, ship speed and heading, can generate a transient vibration of the entire hull girder.

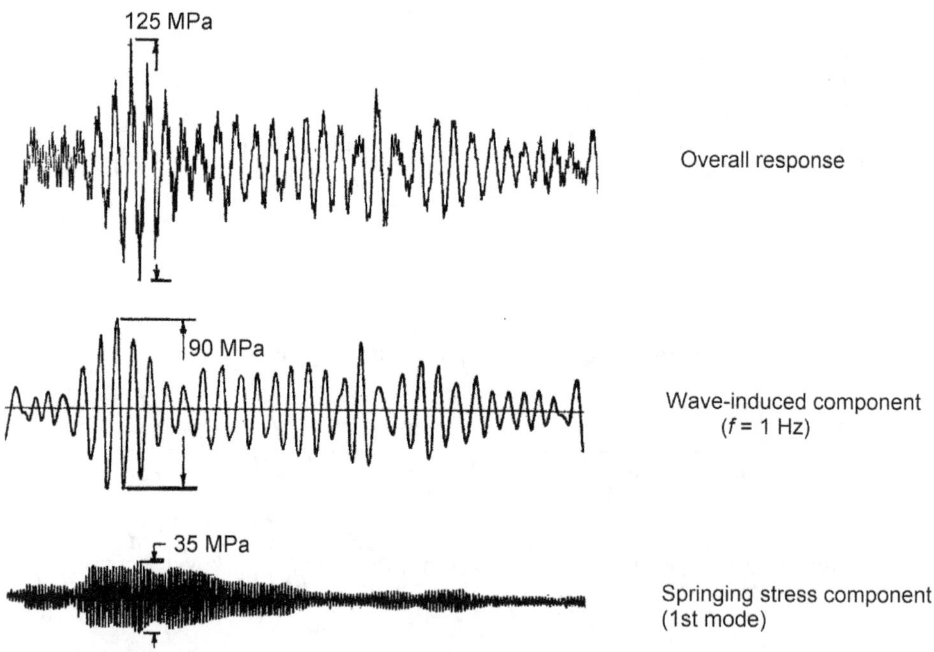

Figure 8.28 Stress profile and components during springing [27]. Reprinted courtesy of the Ship Structure Committee, Washington, DC.

8.5.1 Slam-induced Whipping

It is possible for vibration excitation of the hull girder to take place as a result of localised impacts in way of the bottom forward, in which case the phenomenon is called *whipping*. In this case the two- and three-node vertical vibration modes are the most important.

Whipping is a high-frequency stress fluctuation that can arise as a result of even mild slamming. The magnitude of the whipping stresses is of the order of the wave-induced, low-frequency bending stresses (Figure 8.29a–b). The vibration persists for a large number of cycles, the rate of decay being small. Complete oscillations of the hull take place while the bow is immersed as well as possibly during periods when the forefoot is clear of the water.

The additional loading imposed on the hull girder has to be taken into account when assessing the ultimate limit state as well as the fatigue limit state, because this high-frequency flexural vibration contributes to stress increases above the values corresponding to quasi-static wave bending (light-gray Figure 8.30) as well as to fatigue damage. There exist ship types for which a substantial proportion of all fatigue damage is attributable to the consequences of slamming.

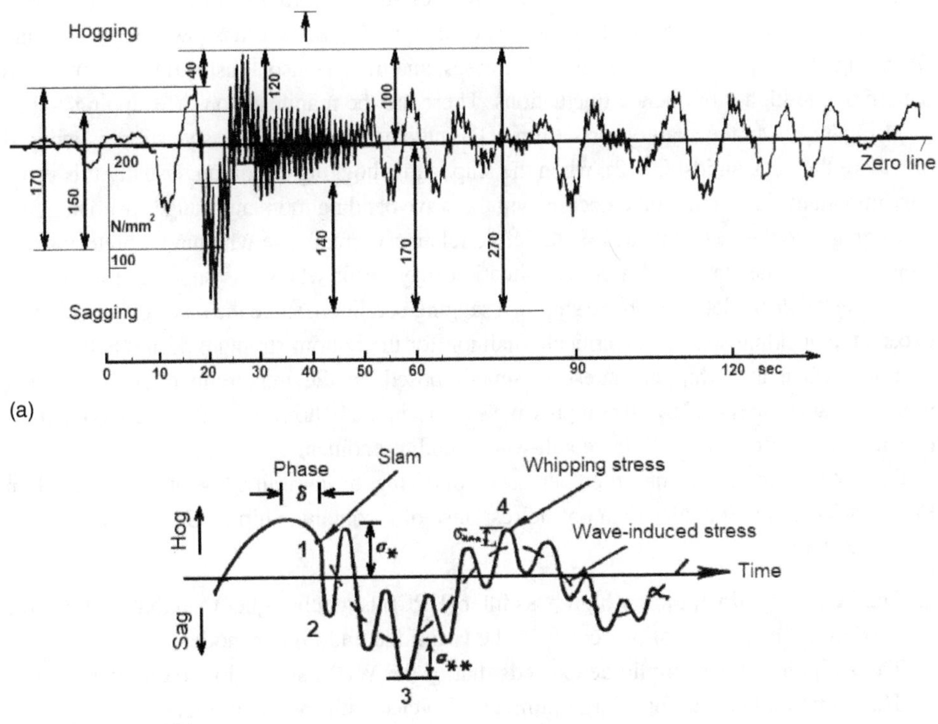

Figure 8.29 Slam-induced whipping. (a) Measured stress-time history following a slam [2]. (b) Features of stress-time history following slamming [35]. Figure 8.29b reprinted courtesy of the Society of Naval Architects and Marine Engineeers.

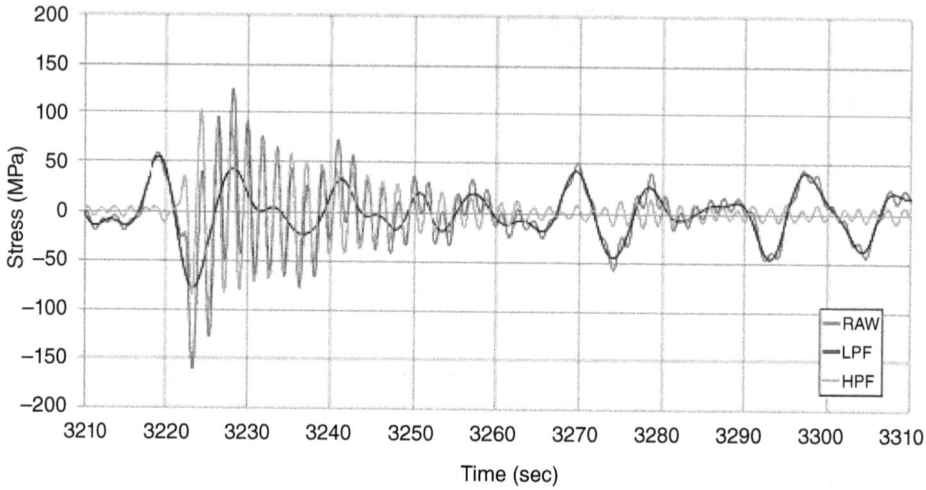

Figure 8.30 Containership bending stress history [7]. WBM frequency = 0.1 Hz; whipping frequency = 0.56 Hz.

Figure 8.29a includes a measured stress-time history of an actual slam event, whereas Figure 8.29b shows the main features of a slam stress-time record, obtained at one position of the hull structure. The trace of stresses included in Figure 8.29a shows that following slamming, the amplitude of the stresses increases, and there is also a distinctly different stress pattern due to higher-frequency fluctuations. These are the manifestation of whipping.

In Figure 8.29b, the slam occurs at time (1). Initially a large compressive stress is induced in way of the deck amidships (2), when the ship is in a hogging condition. Usually this is not very important because it would occur when the wave bending moment changes from hogging to sagging, and the resultant stress is therefore relatively small. The whipping vibratory stress continues after the slam, and a critical situation may occur when it is superimposed on the stresses in the deck plating with the ship in a sagging condition, since the total stress (σ_{**}) may exceed the buckling stress. The critical condition for the bottom structure occurs half a cycle later (4), when the whipping stress is superimposed on the maximum hogging bending moment that induces the largest compressive stresses in the bottom structure. These conditions become more precarious with large still-water bending moments.

One other record of measured stresses following a slamming event is included in Figure 8.30, which includes data on the response of a containership (*MOL Comfort* class). It is seen that:

1. The frequency of whipping (high pass filter (HPF)) is much higher than that of the wave bending moment. Vibrations occur in the two-node and higher modes.
2. The whipping stress amplitude exceeds that of the WBM stress (low pass filter (LPF)).
3. The vibration persists for a large number of cycles with low damping.

It is clear from Figure 8.30 that whipping vibration continues for a period corresponding to several cycles in wave bending moment, during which the bow is immersed and also when the forefoot is possibly clear of the water.

(a) (b)

Figure 8.31 External and internal damage visible resulting from slamming [11].[14] (a) Plating set in and punctured. (b) Plating set in. Longitudinals tripped. Web frames buckled between access openings. Reprinted courtesy of Det Norske Veritas, Høvik.

In order to obtain a complete picture of the events that follow a slam, an understanding of the spatial distribution of the stress fluctuations is required, in addition to time records, such as Figure 8.30. To this end the response of the hull girder is subdivided into the following three stages [9]:

1. Local response, which has already been discussed
2. Transitional response
3. Overall response of the hull girder

Following occurrence of a slam, the region of the hull girder that is loaded directly suffers the consequences of impact. This includes the outer shell plating, the internal stiffening members and the adjoining members to which these are attached. This initial stage is followed by a transitional stage during which momentum is transferred to the nearby outer hull plating and interior members.

Loads are thereby transmitted to the neighbouring structure. At this point the hull girder is not aware of the slamming event. During the last stage, if sufficient momentum has been transferred, vibration propagates along the entire length of the hull girder and may be vertical, horizontal, torsional or any combination of these, depending on the exact location of impact.

The effects of slamming are evident at both a local and a global level. At a local level, damage can be sustained in the form of deformed plating, distorted stiffening members, buckled longitudinals and frames and fatigue cracks in transverse frames. Damage is generally local and affects the outer shell plating and attached internal stiffeners (Figure 8.31). The stiffening system (secondary structure) may suffer damage not only as a result of the direct impact but also because of the vibration associated with whipping. Damage has also been noted far from the point of impact, in way of deckhouse connections

[14] Figure 8.31a shows damage to the bottom plating of a small ship in ballast condition whereas Figure 8.31b shows damage to a 220,000 dwt tonne Capesize bulk carrier.

where rupture has been known to occur. In the case of Figure 8.31b the most important consequence of the tripping of longitudinal stiffeners is that they cannot provide any further support to the plating at a local level.

The physical parameters that affect whipping response are similar to those mentioned in the case of slamming. In addition to these we ought to mention:

- Hull girder stiffness and mass distribution. The dynamic response of the hull girder due to the slamming loads depends on the dynamic properties of the hull. These in turn depend on the mass and stiffness distributions.
- Damping is normally of less importance for extreme whipping response, since the peak response occurs during the first vibration cycle.

We shall not discuss overall hull girder response to slam-induced whipping further, since this is a purely dynamical problem. In early work the ship was treated as a free-free beam loaded at one end and various assumptions were made concerning loading, damping and stiffness properties of the hull girder.

Many of the same parameters are important for both springing and whipping. In the case of slam-induced whipping, the parameters that influence the slamming loads are important.

- Bow (and stern) geometry. Wider shapes with pronounced flare give higher slamming loads. Wide and flat hull bottoms may also give large slamming forces.
- Draught. Low draught at the bow (or stern) results in more frequent and normally more severe slamming.
- Ship speed. Higher speed gives more frequent and normally more severe slamming.
- Ship motions. Large heave and pitch motions will result in more slamming.
- Wave condition. Waves that give large heave and pitch motions will give more slamming. Steep waves generally give more severe bow slamming loads.
- Hull girder stiffness and mass distribution. The dynamic hull girder response due to the slamming loads depends on the dynamic properties of the hull. These properties are given by the stiffness and mass distribution.
- Damping is normally of less importance for extreme whipping response, since the peak response occurs during the first vibration cycle.

Whipping and springing response are monitored using model and full scale tests as well as theoretical methods. Figure 8.32 includes samples of traces of the vertical and horizontal hull girder bending moments measured on board the containership *Victoriaborg* [1].

Measurements consisting of global strains amidships were made over a period of one year in order to determine the effect of whipping on fatigue. Some of the findings were as follows:

- The short-term extreme whipping response and loads are linear with the short-term extreme low frequency wave-induced response and loads.
- The whipping amplitude fluctuated with peak values reaching 70 per cent of the wave-induced amplitude.

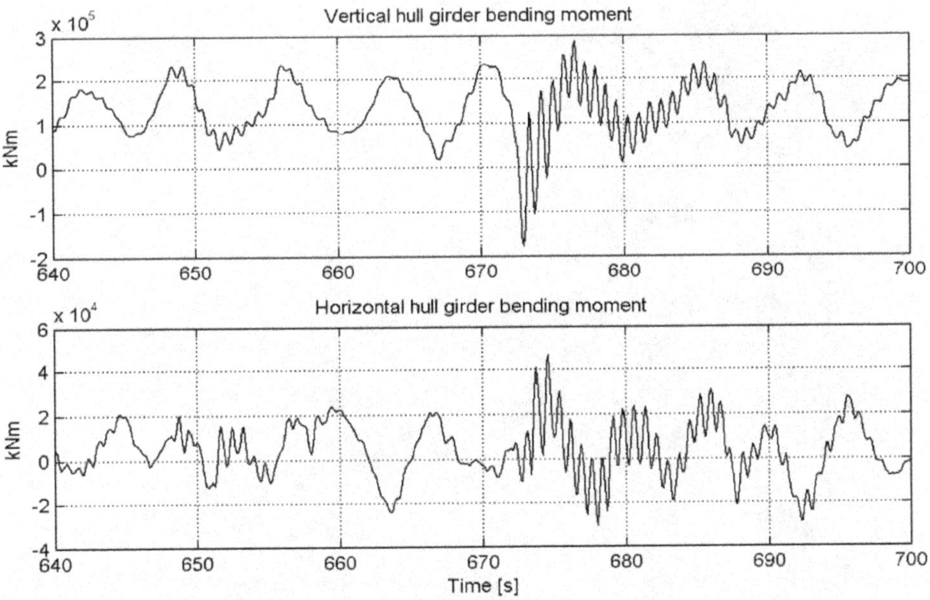

Figure 8.32 Bending moments amidships including extreme whipping response [1].

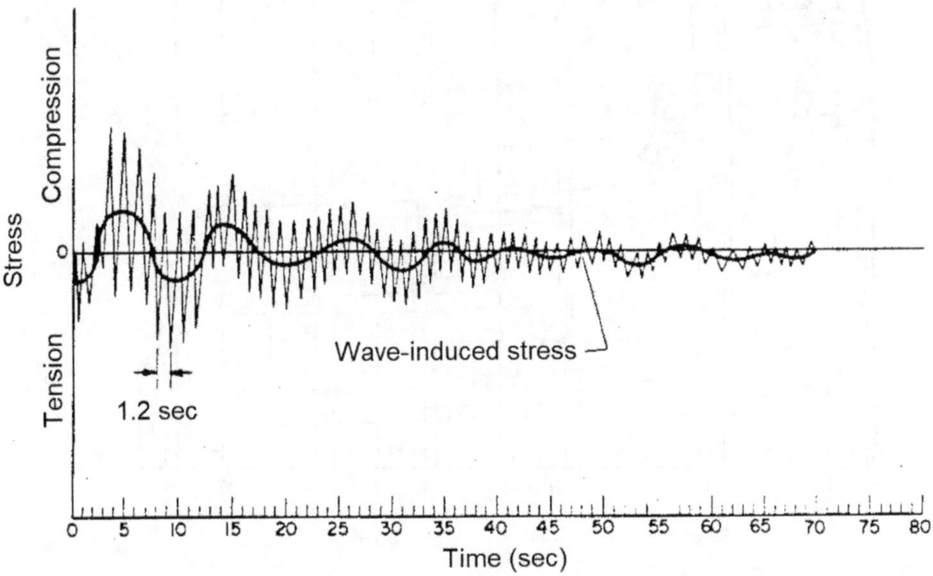

Figure 8.33 Whipping stress superimposed on wave bending moment following bow flare impact (USS *Essex*) [19].

(a)

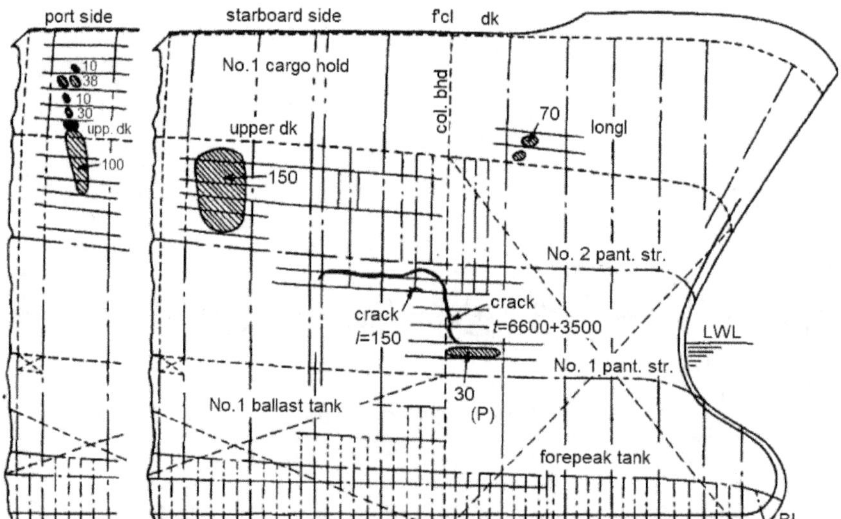

(b)

Figure 8.34 Damage sustained as a result of bow flare impact. (a) USS *Ticonderoga*. Buckled hangar deck. (b) Damage in way of fore-end [23]. Figure 8.34b reprinted courtesy of the Royal Society, London.

- The peak value of the combined wave bending (low frequency) and whipping response reaches approximately 150 per cent of the low frequency response.
- Whipping occurs most often in the speed range of 8–12 knots.
- Whipping accounts for an increase of fatigue damage of up to 30 per cent.

8.5.2 Whipping Induced by Bow Flare Impact

This type of response was identified during the 1950s when measurements were made on board ships of the US and Dutch navies in order to understand what caused buckling of the hangar deck and the double bottom structure amidships of several aircraft carriers. The location of damage was at a significant distance from the point of impact, which was the bow region.[15]

A plot of whipping stresses obtained on board the USS *Essex* is reproduced in Figure 8.33. These measurements indicate the presence of a whipping stress whose frequency of 0.83 Hz is approximately equal to the natural frequency of vertical vibration of the hull girder (0.87 Hz). Prior to these measurements it was thought that whipping stresses are related exclusively to bottom slamming. The records obtained demonstrated that bow emergence is not required for whipping. The magnitude of the whipping stress was found to be significantly higher than the wave-induced stress so that the validity of the quasi-static wave bending moment (Chapter 4) was brought into question. It was concluded that since the induced vibration is in the fundamental mode of the hull girder, this can be treated as a single-degree-of-freedom system. Therefore, if the duration of impact is greater than half the natural period of vibration, the maximum compressive stress in the deck will be equal to the sum of the wave-induced stress and the total peak-to-peak variation in whipping stress. The total effect may be found by adding the two maximum values, since the phase angle between them is negligible. The effect of bow flare immersion is thus a nonlinear addition to the sagging wave bending moment amidships [19].

References

1. Aalberts P. J. Nieuwenhuis M. W. (2006) Full scale wave and whipping induced loads. *Hydroelasticity in Marine Technology* Vol. 11 65–78.
2. Aertssen G. (1979) An estimation of whipping vibration stresses on slamming data of four ships. *Int. Shipbuilding Progress* Vol. 26 No. 294 23–31.
3. American Bureau of Shipping (2015) *Guidance Notes on Ship Vibration*.
4. Beards C. F. (1996) *Structural Vibration: Analysis and Damping*. Amsterdam: Elsevier Science and Technology.
5. Bishop R. E. D. Price W. G. (1979) *Hydroelasticity of Ships*. Cambridge: Cambridge University Press.
6. Clough R. W. Penzien J. (1975) *Dynamics of Structures*. New York: McGraw Hill.

[15] Damage due to bow flare whipping was first recognised as such on the *Essex* class aircraft carriers of the US Navy during the 1950s. Buckling of the hangar deck and the double bottom structure arose amidships (Figure 8.34a).

7. Committee on Large Container Ship Safety (2013) *Interim Report of Committee on Large Container Ship Safety* Japan.
8. Cowper G. R. (1966) The shear coefficient in Timoshenko's beam theory. *J. of Applied Mechanics* Vol. 33 335–340.
9. Daidola J. C. Mishkevitch V. (1995) Hydrodynamic Impact Loading of Displacement Ship Hulls: An assessment of the state of the art. Ship Struct. Comm. Rept. SSC-385 Washington, DC.
10. Det Norske Veritas (1983) *Prevention of Harmful Vibration in Ships.*
11. Det Norske Veritas (2005) *Hull Structure Course.*
12. "Differential equation – modeling – spring and mass" *Sharetechnote.com.* Available: www.sharetechnote.com/html/DE_Modeling_Example_SpringMass.html (accessed 19 December 2020).
13. Doshi C. (1979) On the Analysis of the Timoshenko Beam Theory with and without Internal Damping. MS Thesis, Dept Mech. Eng., Rochester Institute of Technology.
14. *Edu.au.* Available: www.varg.unsw.edu.au/Assets/link%20pdfs/Beam_vibration.pdf. (accessed 19 December 2020).
15. Egle D. M. (1969) An Approximate Theory for Transverse Shear Deformation and Rotatory Inertia in Vibrating Beams Effects. NASA Contr. Rep. CR-1317, Washington, DC.
16. Goodman R. A. (1970) Wave-Excited Main Hull Vibration in Large Tankers and Bulk Carriers. In *Proc. RINA Mtg* London: Royal Institute of Naval Architects. 167–184.
17. Han S. M. Benaroya H. Wei T. (1999) Dynamics of transversely vibrating beams using four engineering theories. *J. Sound Vibr.* Vol. 225 No. 5 935–988.
18. Hedgepeth J. M. (1956) Recent Research on the Determination of Natural Modes and Frequencies of Aircraft Wing Structures. AGARD Mtg (NACA 41873). Washington, DC: Nat. Adv. Comm. Aeron.
19. Heller Jr. S. R. Kammerer J. T. (1967) Buckling of long slender ships due to wave-induced whipping. In Herrmann G. (Ed) *Dynamic Stability of Structures.* Proc. Int. Conference Evanston Ill. 1965. Pergamon Press 129–156.
20. Hermundstad O. A. (2007) *Springing and whipping of ships.* Lecture notes in hydro-elasticity Marintek Trondheim.
21. Horn F. (1925) Horizontal und Torsionsschwingungen von Frachtschiffen (Horizontal and Torsional Vibration of Cargo Ships) *Werft Reederie und Hafen.* Berlin: Springer Verlag.
22. Kaminski M. L., et al. (2006) Dynamic Response. Rept. of Committee II.2. In Frieze P. A. Shenoi R. (Eds) *Proc. 16th Int. Ship & Offsh. Struct. Cong.* Vol. 1 Southampton 263–358.
23. Kapsenberg G. K. (2011) Slamming of ships: where are we now? *Phil Trans. R. Soc. Lond. A* Vol. 369 2892–2919.
24. Kumai T. (1956) *Coupled Torsional-Horizontal Bending Vibrations of Ships.* Fukuoka: Research Institute of Applied Mechanics, Kyushu University.
25. Kumai T. (1958) Added mass moment of inertia induced by torsional vibration of ships. *European Shipbuilding* 93–100.
26. Landweber L. de Macagno M. C. (1957) Added mass of two-dimensional forms oscillating in a free surface. *J. Ship Res.* 20–30.
27. Lewis E. V. Hoffman D. Maclean W. M. van Hooff R. Zubaly R. B. (1973) Load Criteria for Ship Structural Design. Ship Struct. Comm. Rept. SSC-240 Washington, DC.
28. Lewis F. (1929) The inertia of water surrounding a vibrating ship. *Trans SNAME* 1–20.

29. Lewis F. M. (1967) Hull Vibration of Ships. In Comstock J. P. (Ed) *Principles of Naval Architecture* (Revised edition). New York: SNAME 718–751.
30. McGoldrick R. T. Gleyzal A. N. Hess R. L. Hess G. K. (1953) *Recent Developments in the Theory of Ship Vibration*. DTMB Rept. No. 739 D. Taylor Mod. Bas. Washington, DC.
31. Mumm H., et al. (2009) Dynamic Response. Committee II.2 Rept. In Jang C. D. Hong S. Y. (Eds) *Proc. 17th Int. Ship & Offsh. Struct. Cong*. Vol. 1. Seoul: Seoul National University 289–374.
32. N.d. Ubc.Ca. Accessed 7 January 2021. http://civil-terje.sites.olt.ubc.ca/files/2020/01/Timoshenko-Beams.pdf.
33. Noonan E. F. (1989) Ship Vibration Design Guide. Ship Struct. Comm. Rept. SSC-350 Washington, DC.
34. Nowacki H. (1970) Ship Vibrations. Lecture Notes Rept. No. 045. Dept. Naval Arch. Mar. Eng. Univ. Ann Arbor, MI.
35. Ochi M. K. Motter L. E. (1973) Prediction of Slamming Characteristics and Hull Responses for Ship Design. *Transactions SNAME Annual Meeting* 144–176.
36. Okumoto Y. Takeda Y. Mano M. Okada T. (2009) *Design of Ship Hull Structures*. Berlin: Springer.
37. Ringsberg J., et al. (2018) Quasi-static Response Rept. of Committee II.1. In Kaminski M. Rigo P. (Eds) *Proc. 20th Int. Ship & Offsh. Struct. Cong*. IOP Press 171–253.
38. Sen D. T. Vinh T. C. (2016) Determination of added mass and inertia moment of marine ships moving in 6 degrees of freedom. *Int. J. of Transportation Engineering and Technology* Vol. 2 No. 1 8–14.
39. Senjanović I. Tomašević S. Grubišić R. (2007) Coupled horizontal and torsional vibrations of containerships. *Brodogradnja* Vol. 58 No. 4 365–379.
40. Skaar K. T. Carlsen C. A. (1980) Modelling aspects of finite element analysis of ship vibration. *Computers and Structures* Vol. 12 409–419.
41. St. Denis M. Pierson W. J. (1955) On the Motion of Ships in Confused Seas. *SNAME Annual Meeting* 281–357.
42. Strutt J. W. (Baron Rayleigh) (1877) *The Theory of Sound Vol. 1* London: Macmillan.
43. Timoshenko S. (1937) *Vibration Problems in Engineering*. 2nd ed. New York: D Van Nostrand Co.
44. Timoshenko S. P. (1921) On the correction for shear of the differential equation for transverse vibration of prismatic bars. *London Phil Mag.* Vol. 41 744–746.
45. Timoshenko S. P. (1921) On the transverse vibration of bars of uniform cross-section. *London Phil. Mag.* Vol. 43 125–131.
46. Todd F. H. (1962) *Ship Hull Vibration*. London: Edward Arnold (Pub.) Ltd.
47. Traill-Nash R. W. Collar A. R. (1953) The effects of shear flexibility and rotatory inertia on the bending vibrations of beams. *Quart J. Mech. & App. Math.* Vol. VI 186–222.
48. Veritec (A/S) (1985) *Vibration Control in Ships*. Høvik.
49. Vorus W. S. (1988) Vibration. In Lewis E. V. (Ed) *Principles of Naval Architecture Vol. II Resistance Propulsion and Vibration Second Revision*. New Jersey: SNAME 255–316.
50. Wah T. (Ed) (1960) *A Guide for the Analysis of Ship Structures*. Washington, DC: US Dept of Commerce.
51. Ward G. (1986) Vibration of Ship Structures with Particular Reference to Warships. In Smith C. S. Clarke J. D. (Eds) *Advances in Marine Structures* Int. Conf. Adm. Res. Est. Dunfermline. London: Elsevier Applied Science Pub. Ltd 63–88.
52. Yin Y. Cui H. Zhao D. Hong M. (2014) Predicting method of natural frequency for ship's overall vertical vibration. *Brodogradnja* Vol. 65 No. 3 49–58.

9

Probabilistic Modelling of Primary Loading and Hull Girder Response

If your experiment needs statistics, you ought to have done a better experiment

Ernest Rutherford

9.1 The Surface of the Sea

Waves are disturbances that result in a continuous change in the elevation of the surface of the sea that does not follow a predictable pattern or direction at any instant in time. As the surface becomes more disturbed, the loading on the hull girder increases. It is thus necessary to know the expected maximum load to which a ship may be subjected to throughout its lifetime, in order to design a safe structure that will not fail due to overloading. Furthermore, ships are subjected to sea loads throughout their active life so that the continuous, repetitive loading will eventually result in fatigue of elements of the hull girder. For both of these reasons we need to develop an understanding of the nature of wave loads.

Before attempting to quantify the sea loads that act on surface ships, it is necessary to have an understanding of the behaviour of the sea itself. We shall therefore provide a brief description of the features of the ocean environment that are relevant to our purpose. In this we will mention the *linear wave theory* that was developed during the 19th century by Airy. There are available many descriptions in a wide range of publications, both books and journal articles.

It is appropriate to first comment on the mechanism by which waves form. Of the power spectrum shown in Figure 9.1, only a part is relevant to the types of loads that act on structures that are floating freely on the surface of the sea. The frequency range that is of relevance to ship wave loading is such that it is *wind-generated* waves that are of importance. The surface of the sea is in direct contact with the atmosphere, and when winds develop, they give rise to waves through mechanisms that include the action of friction.

The growth of waves includes two distinct phases: that of the formation of *initial wavelets* and, subsequently, of *short choppy waves* [47].

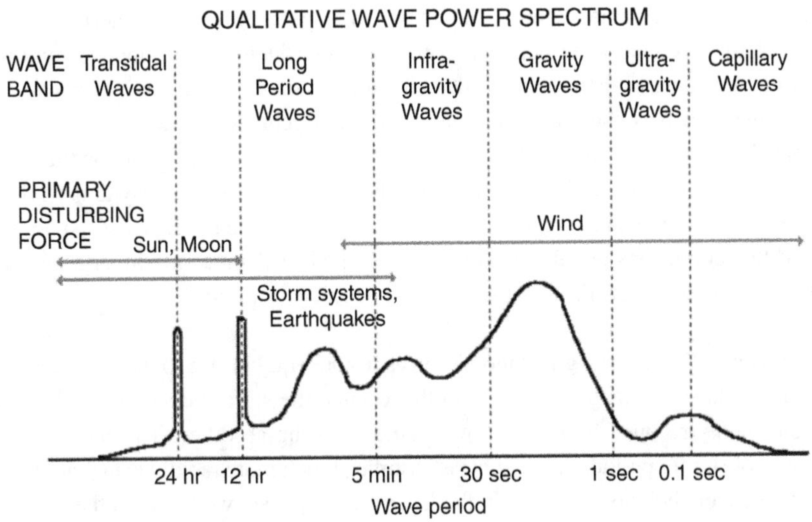

Figure 9.1 Qualitative wave power spectrum [18].

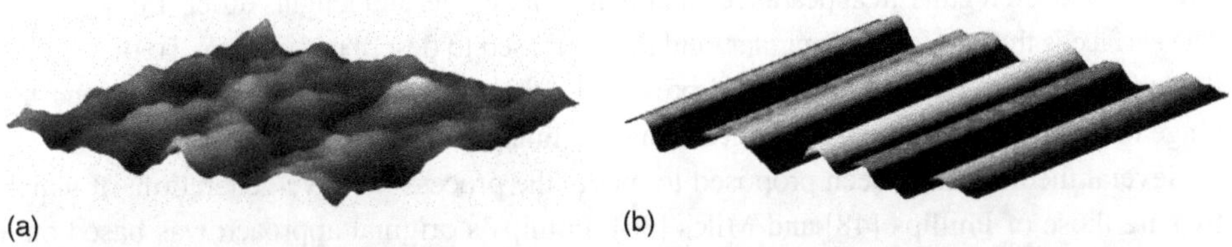

Figure 9.2 Types of seas. (a) Short-crested sea (irregular). (b) Long-crested sea.

Initial wavelets are tiny ripples that form approximately regular arcs of long radius. The height of these ripples increases rapidly until breaking processes become evident. Wavelets then develop into waves that have distinct periods and wavelengths of increasing magnitude as further energy is imparted by the wind. The surface then assumes an irregular appearance. The development of a sea state beyond the stage of wavelet formation depends on the wave-breaking mechanisms of waves of different geometrical characteristics that involve energy dissipation and transfer from waves that have a particular geometry to those of another. Waves continue to grow if energy transferred from the wind exceeds that lost through the various energy transfer mechanisms, which include turbulence and wave-breaking, in addition to the energy transfer from one particular wave type to other ones. This is termed a *short-crested* sea (Figure 9.2a). The surface of the sea absorbs wind energy by forces acting in both the normal and the tangential direction. Winds act over a particular region of the surface that may change in area and also may translate, so that over the expanse of an ocean one region may be severely disturbed whereas the remainder may be calm.

If the duration of wind action is long enough and the region over which it acts is large enough, the irregular sea reaches a *fully developed state*. The energy transmitted by the wind to the sea surface is then such that the waves reach their maximum height and speed (*celerity*) for that particular wind speed. Also, their geometrical characteristics are independent of the distance over which the wind blows (*fetch length*) and its duration [4]. In these circumstances it is possible to describe the sea state by treating it as consisting of a large number of waves each of which has a particular geometry and for that particular sea state contributes to a certain extent. A measure of the extent of contribution is the frequency of occurrence of each particular wave profile. The particular sea state may thus be described by a *power spectrum*, this being a measure of the energy of the particular sea state. It has to be added that conditions pertaining to fully developed seas arise only in large expanses of the sea such as the world's oceans. In more restricted waterways these conditions do not arise.

When wind action ceases, gravitational forces restore equilibrium so that the appearance of a flat calm surface returns. However, until this occurs, the surface continues to be disturbed whilst the energy retained by all wave geometries is being absorbed. The energy associated with waves of shorter periods and heights is absorbed much more rapidly than that of waves of longer periods and heights. As a result, the latter can propagate well beyond the region of the disturbance, which may be a storm. The appearance of these waves is quite different from that of the irregular sea that arises in storm conditions, as they appear to propagate from one direction and are regular in appearance, although their heights and lengths differ (Figure 9.2b). The surface is thus no longer irregular, and the term used to describe it is *swell.* To distinguish it from the short-crested sea described previously, this is called a *long-crested* sea, and the range of wave periods (frequencies) it contains is limited.

Several theories have been proposed to model the process of wave generation, of which two are those of Phillips [48] and Miles [36]. Phillips's original approach was based on a *resonance mechanism* according to which turbulent wind pressures moving over the surface at a specific velocity excite wave components that have the same free velocity. For waves of frequency f travelling in a direction θ_c relative to the wind direction, the resonance condition is

$$\theta_c = \cos^{-1}\left(\frac{g}{2\pi f}\right) \tag{9.1}$$

Miles, on the other hand, suggested that the presence of waves causes turbulent pressures in the wind, which encourages further wave growth. This results in an exponential growth of wave energy, which is limited by wave-breaking. Phillips argued on dimensional grounds[1] that the wave energy grows until it is limited by a spectrum that is proportional to f^{-5}. At this stage equilibrium is reached between wave growth and wave-breaking. Data analysis led to the proposal of an equilibrium spectrum of the form

$$S(f) = ag^2f^{-5} \qquad \text{where } a \approx (7.5)(10)^{-6} \tag{9.2}$$

Miles later combined the two mechanisms, with behaviour being described by the Phillips model in the early stages of wave generation and subsequently behaviour following the Miles model [31].

[1] For an application of dimensional analysis to ocean wave theory, see [4].

A nonlinear mechanism was proposed by Longuet-Higgins [26] in which it was suggested that the breaking of short sea waves on the forward slopes of long waves imparts momentum and hence energy to those waves. The energy of the longer waves increases at a rate proportional to the square of time. To this day there remain unanswered questions with regard to the nature of wind–wave interaction.[2]

9.2 Short-Term Representation of the Sea Environment

The short-term representation of the marine environment is the necessary first step to understanding how a ship is affected when travelling in a seaway. Once a satisfactory description of the sea environment is obtained, it is possible to consider the interaction that arises when a ship is resting on the surface as well as when it is moving forward.

9.2.1 Deterministic Modelling

Measurements of the elevation and period of surface waves have led to the conclusion that it is not possible to depict the sea surface in an exact manner on the basis of past measurements. *Linear wave theory* shows that the velocity of propagation of a wave, its *celerity*, C, is proportional to the square root of its wavelength λ ($C^2 = g\lambda/2\pi$). Therefore, in a seaway that consists of waves of different lengths, the longer waves will be overtaking the shorter ones, producing a continually varying surface. This continually varying, irregular sea surface can be analysed by considering it to be the sum of a number of regular sinusoidal waves that have different lengths and heights. Figure 9.3 shows two different models of an irregular wave pattern that is synthesised using a finite number of sinusoidal waves in two and three dimensions, respectively. These approaches are *deterministic* models of the sea surface.

Alternative approaches would be to describe sea conditions in a particular geographical location and time using statistical properties, or else as a *stochastic process*, since it has been found that they obey such laws [31, 35, 49]. In a stochastic process waves in deep water are categorised as a Gaussian random process for which the probability distribution function of wave elevation obeys the normal probability law.

Short-Term Analysis and Statistics Descriptors of Ocean Wave Records

The study of the sea surface is based on data collected from sources such as measurements by instruments and observations by humans. The data invariably takes the form of wave records whose duration is finite, so that they are termed *short-term* records. An example of a record over a 150-second period is included in Figure 9.4.

In order to analyse and interpret in a meaningful manner the information contained in a wave record it is necessary to introduce a number of parameters that may be used to describe the data. The most important wave parameters used to define the properties of

[2] A review is included in [39].

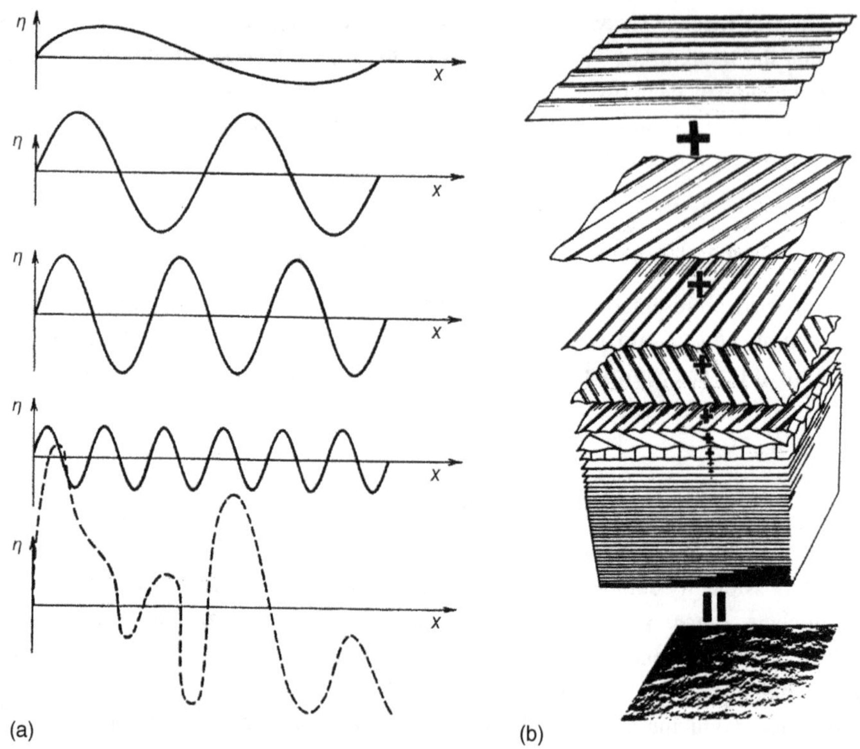

Figure 9.3 Representation of an irregular seaway using regular sinusoidal wave. (a) Two-dimensional model [2]. (b) Three-dimensional model [4].

individual waves are indicated in the enlarged portion of the wave record of Figure 9.4. These are:

(a) *Wave elevation* $\eta(t)$, being the distance from the datum line to a wave trace at time t
(b) *Wave amplitude a,* being the distance from the datum line to a peak
(c) *Wave height H,* being the distance from a trough to the following peak[3]
(d) *Zero up-crossing period,* T_z, being the time elapsed between two subsequent up-crossings of the datum line
(e) *Up-crossing period for a specific height A* for a particular wave, $T_z(A)$.

The datum line that is shown in a wave record such as Figure 9.4 corresponds to zero elevation ($\eta = 0$), or a flat, calm sea. A table is prepared that includes the wave number by specifying each up-crossing so that from a pair of adjacent up-crossings the height and period of a particular wave are determined. The encircled numbers in Figure 9.4 are the wave numbers included in the first column of Table 9.1. Once this is determined, the wave height and period corresponding to each wave are added in the second and third columns, respectively.

[3] The wave height H may also be defined as the distance between a wave peak and the trough that follows.

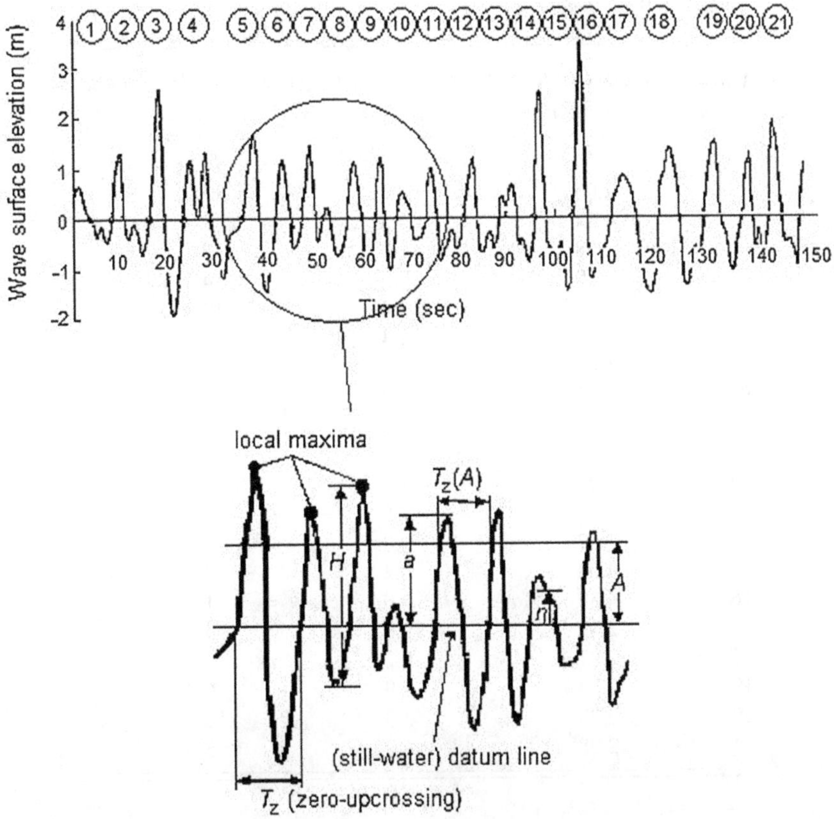

Figure 9.4 Short-term wave record from a random sea.

Using this information, the mean zero up-crossing period, $\overline{T_z}$, and the mean zero up-crossing frequency, $\overline{f_z}$ $\left(= 1/\overline{T_z}\right)$, are readily obtained. For a level A, (Figure 9.4), the mean up-crossing frequency is given by

$$\overline{f_z}(A) = \frac{1}{\overline{T_z}(A)} \tag{9.3}$$

The wave height data included in Table 9.1 can also be used to determine the mean wave height, but this is inconsequential, as it does not convey useful information, since the height of most waves will exceed it by far. Of more importance are two additional parameters, the *root mean square wave height,* H_{rms}, and the *significant wave height*, H_s, defined by

$$H_{rms} = \sqrt{\frac{1}{J}\sum_{j=1}^{J} H_j^2} \tag{9.4}$$

Table 9.1 *Wave heights and periods for the wave record of Figure 9.4*

Wave number	*Wave height H (m)*	*Wave period T (sec)*	*Order number (m)*
1	0.54	4.2	21
2	2.05	8.0	12
3	4.52	6.9	2
4	2.58	11.9	8
5	3.20	7.3	4
6	1.87	5.4	17
7	1.90	4.4	16
8	1.00	5.2	20
9	2.05	6.3	13
10	2.37	4.3	10
11	1.03	6.1	19
12	1.95	8.0	15
13	1.97	7.6	14
14	1.62	7.0	18
15	4.08	8.2	3
16	4.89	8.0	1
17	2.43	9.0	9
18	2.83	9.2	7
19	2.94	7.9	6
20	2.23	5.3	11
21	2.98	6.9	5

$$H_s = \frac{1}{J/3} \sum_{m=1}^{J/3} H_m \tag{9.5}$$

respectively. In the case of the root mean square wave height it is noted that the larger wave heights are given more weight in the averaging process. The significant wave height is the most widely used statistical parameter for short-term wave records and is defined as the mean (or average) of the largest one-third of the waves in a wave record (H_s or $H_{1/3}$).

In order to calculate H_s it is necessary to select the waves with the largest one-third wave heights and use Equation (9.5). This is readily done using the information included in the last column of the table, in which each individual wave is ranked by wave height. This column includes the *rank order*, *m*, of each wave by wave height. The wave with the largest wave height is ranked $m = 1$ and so on. It is possible to perform the same for the wave periods, in which case the significant wave period, T_s (or $T_{1/3}$), is obtained. These two parameters can be used to specify the properties of a wave that can be used for design purposes.

From a wave record such as that of Figure 9.4 additional information can be obtained. If wave amplitudes a are added to the above list, the *mean amplitude* $\overline{a}$ is obtained from

$$\overline{a} = \frac{1}{N} \sum_{n=1}^{N} a_i \qquad \text{(m)} \tag{9.6}$$

where N is the number of observations. The *variance of the amplitude relative to the mean*, m_0, is

$$m_0 = \frac{1}{N}\sum_{n=1}^{N}(a_n - \overline{a})^2 \qquad (\mathrm{m}^2) \tag{9.7}$$

The standard deviation or root mean square value of the amplitude relative to the mean is then

$$\sigma_0 = \sqrt{m_0} \qquad (\mathrm{m}) \tag{9.8}$$

It has to be added that in order for such data to produce statistically significant results at least 100 pairs of peaks and troughs are required so that the duration of the record would extend to 20–30 minutes. Very long records run the risk of including data from different weather conditions due to wind speed and direction changes, and thus they ought to be avoided.

Description of a Sea State

Up to this point we have discussed the means at our disposal to describe and compare individual wave records. It is necessary, however, to develop tools that can be used to describe and characterise the condition of the sea at a broader level, in order to provide information that is of practical use. The condition of the sea surface is affected by atmospheric conditions (wind) over large areas, and it is necessary to have accurate descriptions for these areas. Storms that have particular features develop, with calm conditions prevailing elsewhere. The condition of a large body of water with respect to waves and swell, at a certain location and time, is termed a *sea state*.

Our task is therefore to describe the properties of a sea state, and in order to do so we need to overcome the problem of the irregularity of individual waves. We can circumvent this difficulty by making use of a characteristic property of a sea state: its *energy content*.

For an individual wave, the energy content is the sum of the kinetic and potential energy of the particles of which the wave consists, resulting from their circular motion (Figure 9.5). By summing the energy content of all waves that make up a particular sea state we determine the energy content of the sea state itself. This is a measure of its severity and, as we shall see, is closely related to the sea loads that act on a ship that sails under these particular conditions.

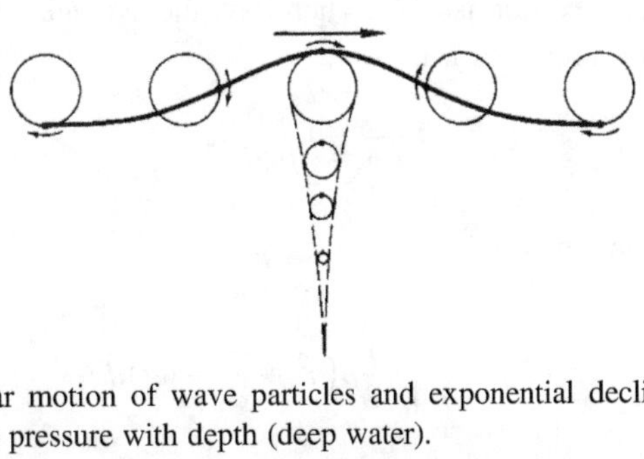

Figure 9.5 Circular motion of wave particles and exponential decline of particle orbit radius and dynamic pressure with depth (deep water).

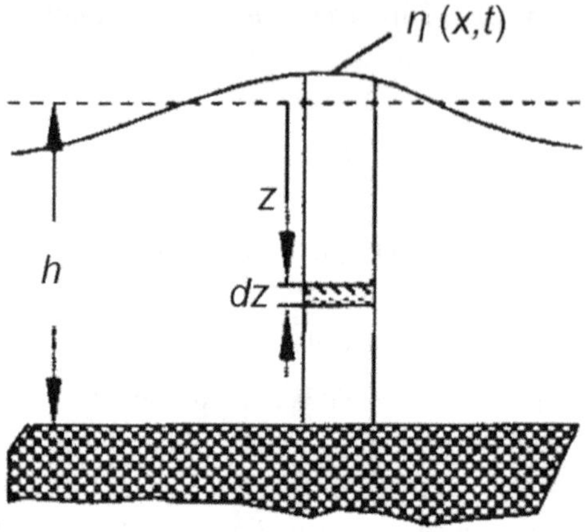

Figure 9.6 Elemental column of water of area dA and depth $(h+\eta)$ [20].

Consider a column of water with cross-sectional area dA as shown in Figure 9.6 [20]. The still-water condition is indicated by the dotted line, and the corresponding depth is h. The elevation (depression) above the still-water condition is $\eta(x, t)$, and the distance of any layer of thickness dz from the still-water position is z. The potential energy of this column of water of area dA, dE_p, is given by

$$dE_p = \int_{z=-h}^{\eta} \rho g z dV = dA \int_{z=-h}^{\eta} \rho g z dz = \\ = dA\rho g \frac{\eta^2 - h^2}{2} \tag{9.9}$$

Subtracting the energy content of the column in the still-water condition ($\eta = 0$) from the total we obtain the potential energy per unit area resulting from the wave disturbance:

$$\frac{dE_p - dE_p(\eta = 0)}{dA} = \frac{1}{2}\rho g \eta^2 \tag{9.10}$$

where η is the instantaneous elevation. For a sinusoidal wave with amplitude a it can be shown that the average elevation is $a^2/2$. Therefore, the average potential energy per unit area of a plane wave of amplitude a is

$$\overline{E_P} = \frac{1}{4}\rho g a^2 \tag{9.11}$$

The kinetic energy is obtained from

$$dE_k = \int_{z=-h}^{\eta} \frac{1}{2}\rho\left(u^2 + v^2 + w^2\right) dV \tag{9.12}$$

where u, v and w are the component velocities in the orthogonal directions. In the case of a plane wave of the form

$$\eta(x,t) = a\sin(\omega t - \mathbf{kx}) \tag{9.13}$$

where a is the amplitude, $\boldsymbol{k}$ is the wave number vector and $\boldsymbol{x}$ is a position vector, we obtain

$$u^2 + v^2 + w^2 = (\omega a)^2 e^{2kz}$$

Therefore,

$$\overline{E_K} = \rho\omega^2 a^2 \int_{z=-h}^{\eta} \frac{1}{2} e^{2kz} dz \approx \rho\omega^2 a^2 \int_{-\infty}^{0} \frac{1}{2} e^{2kz} dz = \frac{1}{2}\rho\omega^2 a^2 \frac{1}{2k} = \frac{1}{4}\rho g a^2$$

The total average energy per unit area is then

$$E = \overline{E_P} + \overline{E_K} = \frac{1}{2}\rho g a^2 \tag{9.14}$$

where ρ = density of sea water, g = gravitational constant and a^2 = mean square value of the amplitude. We note that the total average energy is independent of time and frequency and thus remains constant.

We shall consider now a sea state that consists of a large number of waves of different geometrical properties, termed an *irregular seaway* (Figure 9.3b). Since the energy of a single wave is $\rho g a^2/2$ as per Equation (9.14), the total energy per unit area of a region of the sea containing n waves of heights $a_1, a_2, \ldots, a_{\mathrm{n}}$ at different frequencies $\omega_1, \omega_2, \omega_{\mathrm{n}}$ is given by

$$E = \frac{1}{2}\rho g \sum_{i=1}^{n} a_i^2 \ \mathrm{E} = \frac{\rho g}{2} \sum_{\mathrm{i=1}}^{\mathrm{i=n}} \mathrm{h_{w_i}} \tag{9.15}$$

where a_{i}^2 is the mean square value of all amplitudes at a particular frequency ω_i. The frequencies are given by

$$\omega_n = \frac{2\pi n}{T_H} \qquad n = 1, 2, \ldots, \infty \qquad (\text{rad/sec}) \tag{9.16}$$

where T_H is the time history. If a_{i}^2 is plotted against frequency ω_i for a range of frequencies, a curve that is characteristic of that sea state is obtained (Figure 9.7a). From Equation (9.14) it is evident that the area under this curve is a measure of the energy content of the sea state.

If the wave amplitude $a(t)$ at a particular location is the resultant of a number of waves of different amplitudes a_i, frequencies ω_i and phase angles φ_i, a Fourier series representation can be used:

$$a(t) = \bar{a} + \sum_{n=1}^{\infty} a_i \cos(\omega_n t + \phi_n) \qquad (\mathrm{m}) \tag{9.17}$$

For the frequency range $\omega_a - \omega_b$ shown in Figure 9.7a, the shaded area is proportional to the total energy of all wave components (Figure 9.7b) within this particular range of

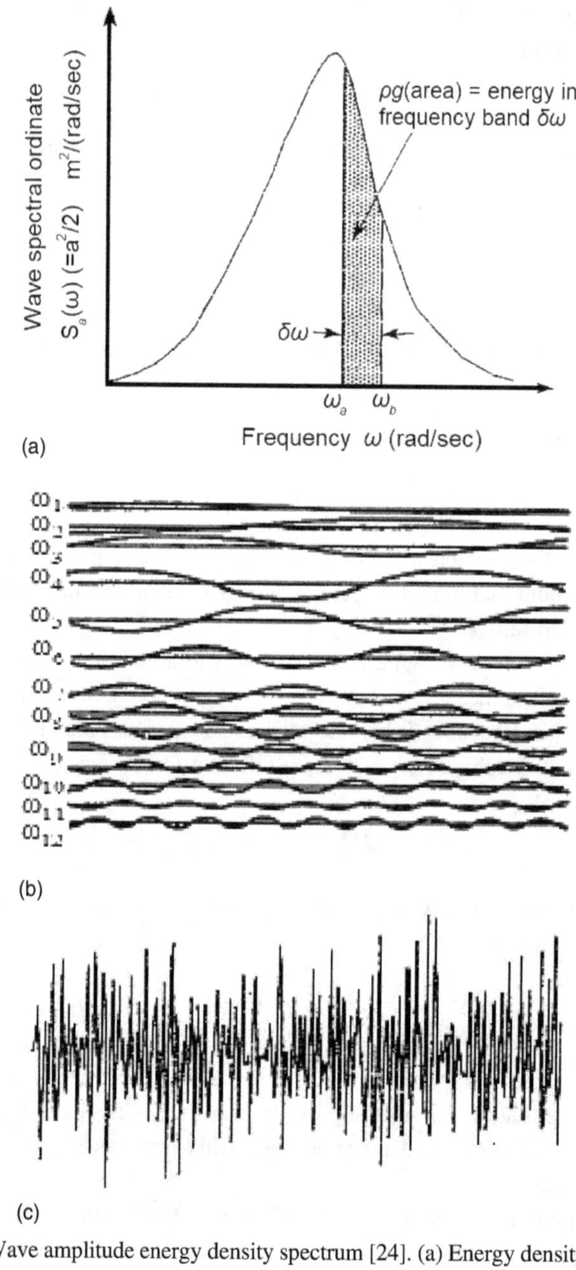

Figure 9.7 Wave amplitude energy density spectrum [24]. (a) Energy density spectrum. (b) Component waves represented by incremental frequency bands. (c) Sum of component waves of time history of wave elevation.

frequencies. For this frequency range there will be only one frequency component ω_n, and the wave amplitude spectral ordinate $S_a(\omega_n)$ corresponding to frequency ω_n is

$$\rho g S_a(\omega_n)\delta\omega = \frac{1}{2}\rho g {a_n}^2$$

so that for frequency ω_n the *spectral ordinate* is

$$S_a(\omega_n) = \frac{1}{2}\frac{{a_n}^2}{\delta\omega} \qquad (\mathrm{m}^2/(\mathrm{rad/\,sec})) \tag{9.18}$$

Rearranging, we obtain the amplitude of each sine wave component for frequency ω_n:

$$a_n = \sqrt{2S_a(\omega_n)\delta\omega} \qquad (\mathrm{m}) \tag{9.19}$$

If the time history T_H has a zero mean and the number of observations is large, the definition of variance given by Equation (9.7) can be written as follows:

$$m_0 = \frac{1}{T_H}\int_0^{T_H} (a(t))^2 dt \qquad (\mathrm{m}^2) \tag{9.20}$$

A time history represented by Equation (9.17) therefore has a variance

$$m_0 = \frac{1}{T_H}\int_0^{T_H} \left(\sum_{n=1}^{\infty} a_n \cos(\omega_n t + \phi_n)\right)^2 dt \ (\mathrm{m}^2) \tag{9.21}$$

Since the frequencies are chosen in accordance with Equation (9.16), this becomes

$$m_0 = \frac{1}{2}\sum_{n=1}^{\infty} {a_n}^2 \qquad (\mathrm{m}^2) \tag{9.22}$$

and from Equation (9.19),

$$m_0 = \sum_{n=1}^{\infty} S_a(\omega)\delta\omega = \int_0^{\infty} S_a(\omega)d\omega \qquad (\mathrm{m}^2) \tag{9.23}$$

The variance of the amplitudes of the waves, m_0, of the irregular wave time history is therefore equal to the area under the wave energy spectrum.

Differentiating Equation (9.17) with respect to time, the vertical velocity of the wave surface at time t is obtained:

$$a'(t) = \sum_{n=1}^{\infty} -\omega_n a_n \sin(\omega_n t + \phi_n) \tag{9.24}$$

The 'energy' of the wave velocity system is

$$E = \frac{1}{2}\rho g \sum_{n=1}^{\infty} \frac{1}{2}(\omega_n){a_n}^2 \tag{9.25}$$

Therefore, the *velocity spectrum*, S_v, will be

$$S_v(\omega) = \omega^2 S(\omega) \tag{9.26}$$

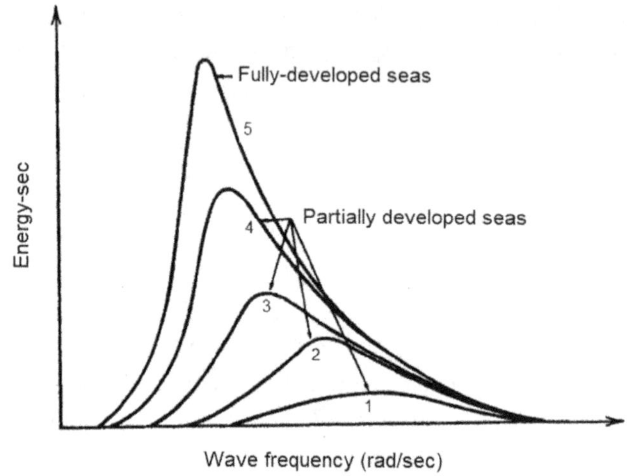

Figure 9.8 Effect of formation of fully developed sea on wave energy spectrum shape.

The area under the vertical velocity spectrum is given by

$$S_v = \int_0^\infty S_v(\omega)d\omega = \int_0^\infty \omega^2 S(\omega)d\omega = m_2 \tag{9.27}$$

where m_2 is the second moment of the spectrum $S(\omega)$ about the origin. Similarly, for the vertical acceleration, $a''(t)$,

$$S_{a''}(\omega) = \omega^4 S(\omega) \text{ and } S_{a''} = m_4 \tag{9.28}$$

The shape of the wave spectrum depends primarily on wind speed. It also depends on wind duration, distance (*fetch*) and the location of other storms, which may generate swell.[4]

As a result of the effect atmospheric conditions have on the sea surface, the condition of the sea changes, and for a given wind speed, after a specific time period, the amplitude of surface waves ceases to increase and only the wave period increases. When this condition arises, we have what is termed a *fully developed sea*. A sea is fully developed at a given wind speed, when all possible wave components in the spectrum between $f = 0$ and $f = \infty$ are present with their maximum amount of spectral energy [46].

For this to occur a certain amount of time and fetch are required. With the passage of time, longer waves are formed, and these have shorter frequencies. The effect is a corresponding shift in the energy spectrum (Figure 9.8). Table 9.2 includes data for each sea state, that is, the resulting significant wave height (feet) and period (secs) with the minimum time required (in hours) for a fully developed sea to form. For example, for a

[4] As mentioned previously, a *swell* is a series of waves that are not generated by local wind conditions but by distant weather systems, where wind has blown for sufficient time over a fetch of water. At large distances from the wave generation area waves that have long periods appear, as these survive energy dissipation and interactions with other waves. Waves whose periods are shorter than ≈10 sec interact with wind and waves during their travel. As a result, their energy is dissipated relatively quickly so that they do not manage to cover long distances.

Table 9.2 *Conditions for the formation of fully developed seas (wind speed in knots)*

Wind speed	*Fetch length (nautical miles)* 50	75	100	150	200	300	400	500	600	700	800	900
20	8@4.5 13	8@5 14	8@5.5 17	8@6 22	8@6.5 28	9@7.5 38	9@7.5 46	9@8.5 56	-	-	-	-
25	9@4.5 9	11@5.5 12	12@6 16	12@6.5 20	12@7.3 26	13@8 34	14@8.5 50	14@9.2 50	14@10 56	-	-	-
30	10@4.5 8.5	14@6 12	15@6.3 14	16@7.2 19	18@8 24	18@8.8 32	18@9.4 38	19@10 46	20@10.5 52	20@11 58	-	-
35	15@5.6 8	17@6.2 10	20@7 13	22@7.8 18	23@8.4 22	24@9.5 30	25@10 36	25@10.5 42	25@11 48	25@11.7 54	26@12.3 60	-
40	18@6 7	21@6.5 10	23@7.2 12	25@8 16	28@9 20	30@10 28	31@10.5 34	32@11.4 40	33@12 46	34@12.5 51	34@13 56	35@13.5 61
45	20@6.5 7	25@7 10	27@7.7 12	31@8.5 16	33@9.5 19	36@10.5 26	38@11.4 32	40@12 38	40@12.5 43	41@13 48	42@13.5 54	42@14 58
50	23@6.9 6.5	27@7.5 9	30@8 11	35@9 15	38@9.5 18	43@11 25	45@11.8 30	46@12.5 36	49@13 41	50@13.8 46	50@14.3 51	50@14.8 56
55	27@7.2 6.5	29@7.5 8.5	33@8.2 10	40@9.4 14.5	43@10 18	48@11.3 24	53@12.3 29	55@13.2 35	55@13.8 39	58@14.3 44	60@14.9 49	60@15.4 53
60	25@7.5 6	30@8 8	37@8.5 10	45@9.5 14	50@10.5 17	57@12.5 22	60@13 28	62@13.5 33	65@14 38	67@15 42	67@15.5 47	70@16 50

Source: www.stormsurf.com/page2/papers/seatable.html.

wind blowing at 35 knots over a fetch of 150 nautical miles, a fully developed sea in which $H_s = 22$ ft (6.7 m) and $T_p = 7.8$ sec will form if these conditions prevail for 18 hours. In fact, fully developed seas can arise for wind speeds of up to 32 knots, since for higher speeds the required duration exceeds that which is practically observed.

9.2.2 Wave Height Spectra

Statistical theory – that is, the central limit theorem – tells us that the distribution of wave amplitudes follows the normal (*Gaussian*) distribution [35]. This gives the probability that the amplitude at a given location at a particular point in time has a certain value. The normal distribution therefore refers to the complete wave profile.

Figure 9.9 shows two records of wind-generated waves. It is seen that the profiles differ, since in the former case[5] there is a distinctly defined cycle about the mean value, involving a maximum and a minimum value, and the frequency is more or less constant. For this reason this is termed a *narrow-band process*. This process can be described as a sine wave of varying amplitude, in which a peak is followed by a zero down-crossing and then by a trough and a zero up-crossing. Peaks below the datum line and troughs above it are rare, so that the average period of the peaks is almost equal to the average up-crossing period [25].

In the latter case it is seen that intermediate peaks and troughs arise during each cycle, so that the frequency varies much more so than in the former case. This is termed a *wide-band process*.

The ratio of the average period of the peaks to the average zero up-crossing period can be regarded as a measure of how narrow-band a wave record and the corresponding spectrum are. In [25] the *bandwidth parameter* ε, was introduced:

$$\varepsilon = \sqrt{1 - \frac{\overline{T_p}^2}{\overline{T_z}^2}} \tag{9.29}$$

The *mean period of the peaks* and the *mean zero up-crossing period* are given by [35]:

$$\overline{T_p} = 2\pi\sqrt{\frac{m_2}{m_4}} \qquad \overline{T_z} = 2\pi\sqrt{\frac{m_0}{m_2}} \tag{9.30}$$

The bandwidth parameter can therefore be written as

$$\varepsilon = \sqrt{1 - \frac{m_2{}^2}{m_0 m_4}} \qquad (1 \geq \varepsilon \geq 0) \tag{9.31}$$

When $\varepsilon \to 0$, the wave spectrum is Rayleigh-distributed, whereas when $\varepsilon \to 1$, the wave spectrum is normally distributed (Figure 9.10).

[5] In the case of wind-generated waves, *moderate seas* are those that do not exceed sea state 4 on the Douglas scale (wave height <2.5 m). (The scale ranges from 0, corresponding to a calm sea, to a value of 9, corresponding to seas for which wave heights exceed 14 metres.)

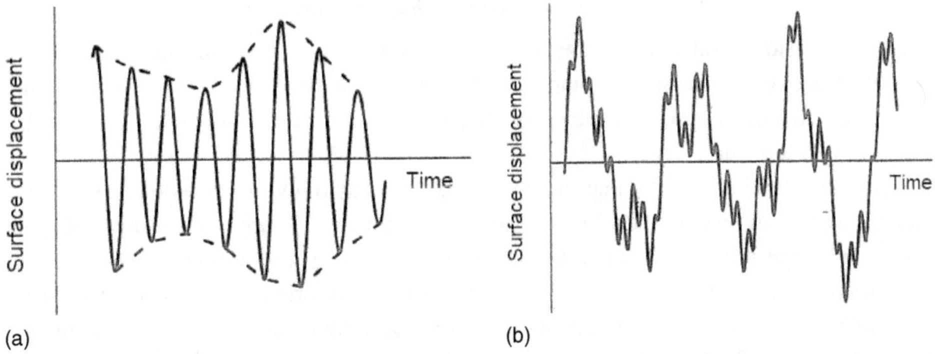

Figure 9.9 Surface elevation in wave records [33]. (a) Moderate seas (narrow-band). (b) Rough seas (wide-band).

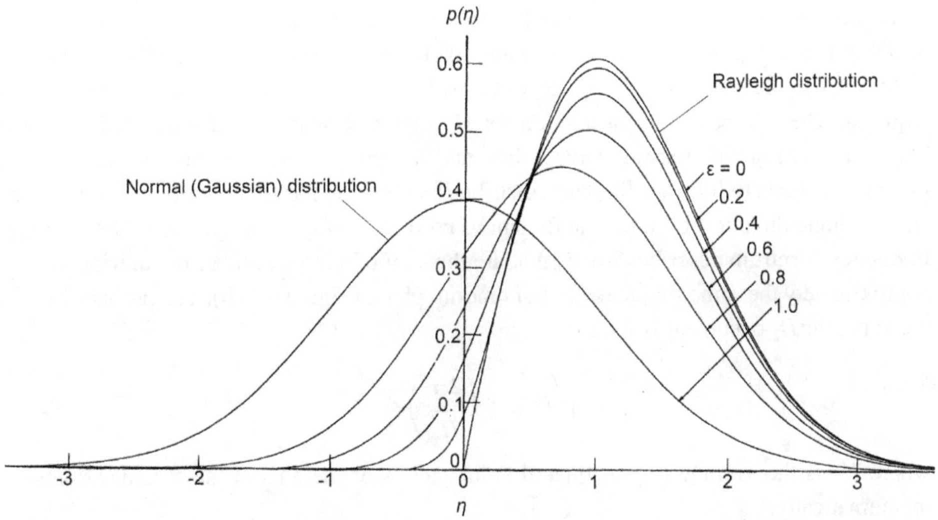

Figure 9.10 Probability distribution of wave height maxima η $\left(= a/\sqrt{m_0}\right)$ for different values of the bandwidth parameter ε [6].

It has been shown [6] that the significant wave height can be related to the area under the wave energy spectrum by

$$\overline{H_s} = 4.00\sqrt{m_0}\sqrt{1 - \frac{\varepsilon^2}{2}} \qquad \text{(m)} \tag{9.32}$$

so that for a narrow-band sea, ($\varepsilon = 0$),

$$\overline{H_s} = 4.00\sqrt{m_0} \qquad \text{(m)} \tag{9.33}$$

This relationship between the significant wave height and the area under the wave energy spectrum is valid in the case of moderate seas.

Wave Height Spectra for Moderate Sea States

We shall consider moderate sea states first. Narrow-band processes are stationary ergodic random processes whose mean square spectral density $S(\omega)$ has significant values only in a range of frequencies whose width is small compared with the magnitude of the central frequency of the band, ω_0.

We note from Figure 9.9a that the upper and lower amplitudes (elevations and depressions) differ from one another. In order to account for this variation, instead of using the upper and lower amplitudes, *wave height* (peak-to-trough) values (H) are used.

It has been proven theoretically [25] and demonstrated by Bretschneider that in the case of moderate sea states, which are narrow-band processes, the distribution of wave heights is described by the *Rayleigh distribution.*[6] In order to establish the type of distribution that represents a wave record, the heights of all waves in the record are measured and the percentage of occurrence calculated. This is done by dividing the number of wave heights lying within a particular range (for example, $1\text{m} < H < 2\text{m}$) by the total number of wave heights in the record. The percentages for each range are then plotted against wave height, resulting in a histogram; this can then be approximated by a suitable probability distribution.

In order to measure wave height in wave profiles in a consistent manner, a definition is required: The International Association for Hydraulic Research (1989) has defined wave height as 'zero-down-crossing wave height and the trough occurs before the wave crest'. For wave heights to follow a Rayleigh distribution the wind has to blow over a calm surface from a single direction for a time sufficient to produce a fully developed sea. The resulting frequency spectrum describes waves that are the sum of a large number of contributions of approximately the same frequency but of random phase (Figure 9.11b). The probability of a wave height H_i occurring is then

$$p(H_i) = \left(\frac{2H_i}{\overline{H^2}}\right) e^{-\frac{H_i^2}{\overline{H^2}}} \tag{9.34}$$

where $\overline{H^2}$ is the average of the squared values the wave heights in the record. Expressed mathematically,

$$\overline{H^2} = \frac{1}{N}\sum_{i=1}^{N} H_i^2 \tag{9.35}$$

where N is the total number of wave heights in the record.

Equation (9.34) gives the percentage of times that a wave of a particular height (H_i) (± 0.5 m) will occur in that record. It can be used to describe a record instead of having to refer to individual values. From previous discussion it is clear that if the area under a spectrum curve is known, it can be related to the Rayleigh distribution formula (Equation (9.34)), and from this a variety of probabilities of occurrence of different wave heights can be determined.

[6] A *Rayleigh distribution* can be interpreted as the distribution of the length of a vector in a plane Cartesian coordinate system, the coordinates of which are independent and normally distributed, with parameters 0 and σ^2 [42].

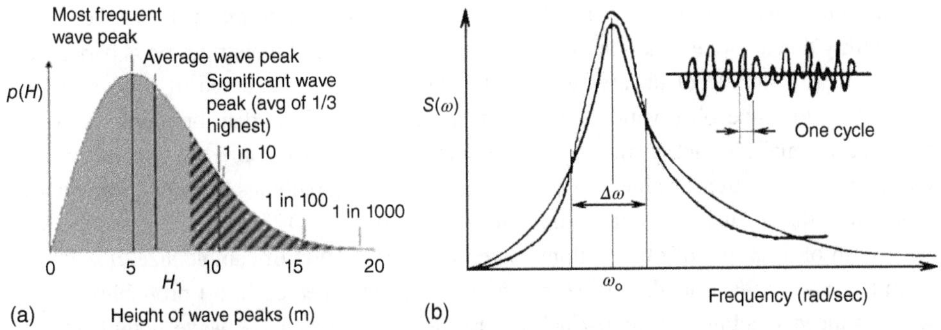

Figure 9.11 Probability density function of short-term distribution of wave peaks. (a) Rayleigh distribution and exceedance probability. (b) Spectrum of a narrow-band process [29]. Reprinted courtesy of the Ship Structure Committee, Washington, DC.

The most frequent wave peak is identical to the maximum value of the distribution. This is found by setting

$$\frac{d}{dH}p(H) = 0$$

Differentiating Equation (9.34),

$$e^{-\frac{H^2}{\overline{H^2}}}\left(\frac{2}{\overline{H^2}} - \frac{4H^2}{\overline{H^2}^2}\right) = 0$$

The most frequent wave peak, H^f, is then given by $\mathrm{f_p(a)} = \frac{\mathrm{a}}{\sigma_x^2}\exp\left[-\frac{1}{2}\left(\frac{\mathrm{a}}{\sigma_x}\right)^2\right]$

$$\frac{H^f}{\sqrt{\overline{H^2}}} = \frac{1}{\sqrt{2}} \qquad \text{and} \qquad H^f = 0.7671\sqrt{\overline{H^2}} \tag{9.36}$$

The probability of the wave height exceeding a particular value, say H_i, $P(H > H_i)$, is the area of the shaded region under the distribution curve in Figure 9.11a. This is given by

$$P(H > H_i) = 1 - \int_0^{H_i} \frac{2H_i}{\overline{H^2}} e^{-\frac{H_i^2}{\overline{H^2}}} dH = e^{-\frac{H_i^2}{\overline{H^2}}} \tag{9.37}$$

That is, for a record of N wave heights, there will be $N\exp\left(-H_i^2/\overline{H^2}\right)H_i$ waves whose height exceeds H_i.

To find the mean height of the qN highest waves, H^q, where $1 > q > 0$, the proportion q of the wave heights H that exceed a certain value H_1 is equal to the value of $P(H)$ for $H_1 < H < \infty$. That is,

$$q = P[\infty > H > H_1] = e^{-H_1^2/\overline{H^2}} \tag{9.38}$$

It has been discovered, however, that the values located at the extreme ends of a distribution such as the Rayleigh distribution do not obey the probability laws that hold for the bulk of the values of the distribution. It therefore becomes necessary to consider distributions that describe the probability laws that govern *extreme values*. Clearly the probability that any one observation lies within the extremes of a distribution is particularly small, but if a sample of observations of increasing size is considered, it is seen that the probability that one observation lies within the extreme of the distribution is no longer particularly small; in fact, as the sample size increases, the probability that one member of the sample is an extreme value also increases. Figure 9.12a includes the probability density function of the normal distribution function whereas Figure 9.12b shows how the expected maximum of a sample of observations increases as a function of sample size, n.

In the case of wave loading of ships, the ratio of the expected (most probable) extreme wave-induced bending moment, $E[M_{WVe}]$, to the rms value of the wave height, H_{rms}, is plotted on log-normal paper against the number of encounters, N (Figure 9.13) [27]. It is

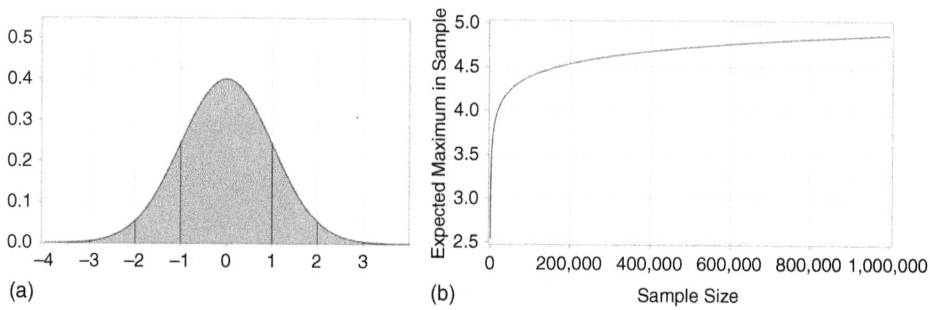

Figure 9.12 Expected value of the maximum of a sample of observations. (a) Normal distribution. (b) Weibull distribution for wave heights.

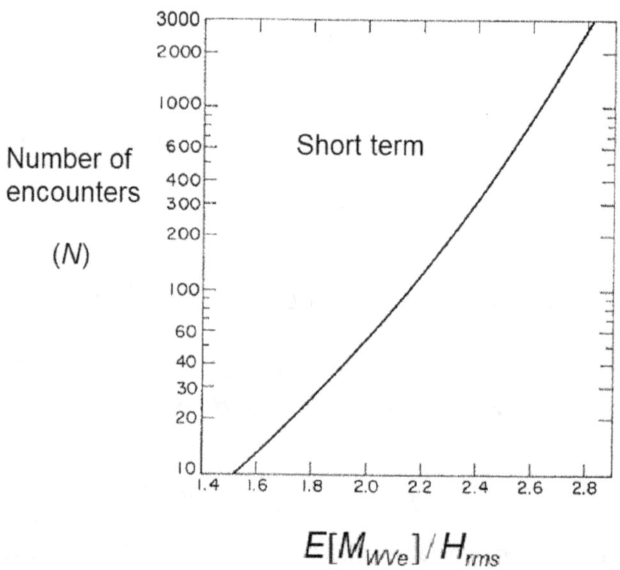

Figure 9.13 Expected extreme wave beding moment versus sample size [27]. Reprinted courtesy of the Society of Naval Architects and Marine Engineers.

seen that for $N = 60$, $E[M_{WVe}] = 2.03\sqrt{H_{rms}}$ whereas for $N = 200$ this value rises to $2.30\sqrt{H_{rms}}$.

In ship structural design extreme values are of particular interest because it important to be able to calculate with accuracy the probability of exceedance of a particular extreme value during a given storm as well as during the operating life of a ship. The issue is even more important in the case of a regulatory body that monitors the behaviour of many, possibly thousands of ships.

Wave Height Spectra for Rough Sea States

We shall now examine the nature of the wave energy spectrum of rough seas, in which case it is wide-banded (Figure 9.9b). As in the case of moderate seas, the duration of the loading does not exceed more than a few hours, and since conditions are assumed constant, the process is stationary ergodic. In this case, however, the Rayleigh distribution is not suitable and a more generalised distribution is required. Such a distribution ought to be identical to the Rayleigh distribution and the normal distribution upon selection of appropriate values of a critical parameter. In the case of the generalised distribution, the parameter used is the bandwidth parameter; the density function of the peaks is [50]:

$$\begin{aligned} f_p(H,\varepsilon) = {} & \frac{\varepsilon}{\sqrt{2\pi m_0}} \exp\left(-\frac{(H/2)^2}{2\varepsilon^2 m_0}\right) \\ & + \sqrt{1-\varepsilon^2}\,\frac{(H/2)}{m_0} \exp\left(-\frac{(H/2)^2}{2m_0}\right) \phi\left[\frac{\sqrt{1-\varepsilon^2}}{\varepsilon}\frac{(H/2)}{\sqrt{m_0}}\right] \end{aligned} \tag{9.39}$$

This density function is plotted in Figure 9.10 for several values of the bandwidth parameter. Integrating Equation (9.39) leads to the cumulative distribution function:

$$F_p(H,\varepsilon) = \phi\left[\frac{H/2}{\varepsilon\sqrt{m_0}}\right] - \sqrt{1-\varepsilon^2}\exp\left(-\frac{(H/2)^2}{2m_0}\right)\phi\left[\frac{\sqrt{1-\varepsilon^2}}{\varepsilon}\frac{(H/2)}{\sqrt{m_0}}\right] \tag{9.40}$$

where

$$\phi(u) = \int_{-\infty}^{u} \frac{1}{\sqrt{2\pi}} \exp\left(-\frac{z^2}{2}\right) dz \tag{9.41}$$

As $\varepsilon \rightarrow 1$ (wide-band process), Equations (9.39) and (9.40) approach the normal (Gaussian) distribution so that the distribution of the peaks reduces to that of the surface itself (Figure 9.10).

Wave Height Spectra for Ship Structural Design

To assist designers, a number of wind-generated wave spectra have been proposed, based on Phillips's and Miles's theories mentioned previously. A well-established one, proposed for fully developed seas, is that of Pierson and Moskowitz, which assumes the following form:

$$S(\omega) = (8.10)(10)^{-3} g^2 \omega^{-5} \exp\left[-0.74\left(\frac{\omega U}{g}\right)^{-4}\right] \quad (9.42)$$

The Pierson–Moskowitz spectrum was derived by making use of the *similarity theory* of Kitaigorodskii [19]. According to this, when plotted in a certain dimensionless way, the power spectrums for all fully developed seas should be of the same shape [46]. This is based on the hypothesis that functional relationships exist among the non-dimensional parameters that describe a physical system. The approach is empirical, and once the fundamental variables are established, measurements are made to determine the exact form of the expression.

In the case of wind-generated wave spectra, the number of variables that are expected to govern the development of wave spectra is large. In order to make effective use of the dimensional analysis approach, Kitaigorodskii simplified the model so that the following four variables were considered [33]: wind speed U_∞, fetch length F, wind duration t and the gravitational constant g. Therefore,

$$S(\omega) = F(\omega, g, U_\infty, t) \quad (9.43)$$

where U_∞ is the mean wind velocity at the upper limit of the boundary layer. In the case of a fully developed unidirectional sea the non-dimensional spectrum depends only on the non-dimensional velocity:

$$\frac{S(\omega) g^3}{U^5} = \varphi\left(\frac{\omega U}{g}\right) \quad (9.44)$$

Pierson and Moskowitz determined the value of φ so that Equation (9.44) is transformed into Equation (9.42). The wave spectrum that has been selected by IACS as a basis for the design sea states used in the Common Structural Rules is the *ISSC spectrum,* shown in Figure 9.14. It is based on the Pierson–Moskowitz (P–M) spectrum, which is applicable to fully developed seas. However, because the need for fully developed seas is too restrictive, Bretschneider adapted the P–M spectrum so that it can be used in the case of seas that are not fully developed. This is the spectrum illustrated in Figure 9.14; it is also known as the *Bretschneider spectrum* and as the *modified Pierson–Moskowitz spectrum.*

The ISSC spectrum is intended to be used with data included in wave scatter diagrams (observed significant wave heights and significant periods, the latter being the average periods of significant waves). It is thus a *two-parameter spectrum,* and is given by the following expression:

$$S(\omega) = AB\omega^{-5} e^{-B\omega^{-4}} \quad (9.45)$$

where

$$A = \left(\frac{H_{1/3}}{2}\right)^2 \quad (9.46a)$$

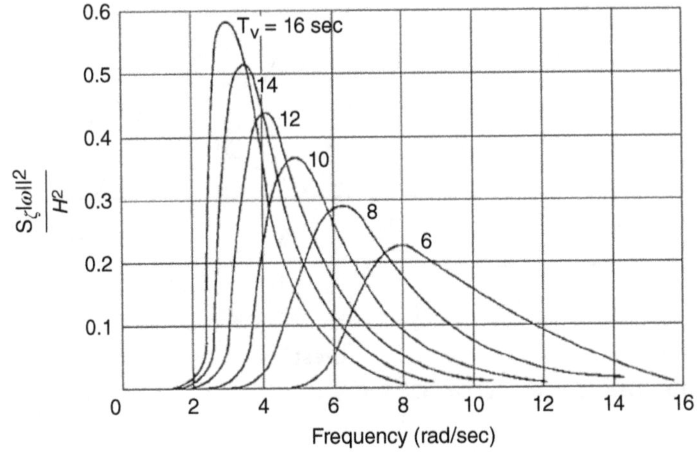

Figure 9.14 ISSC wave spectra for different periods (16 sec $\geq T_V \geq$ 6 sec).

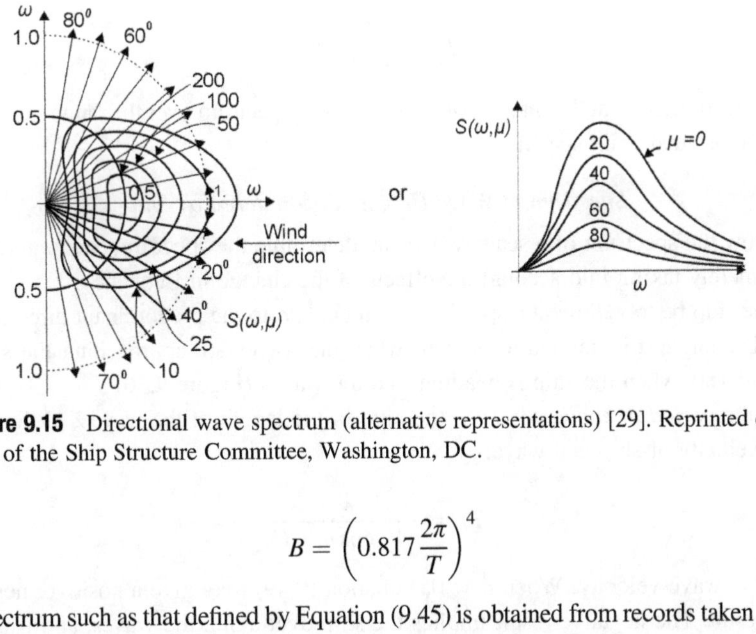

Figure 9.15 Directional wave spectrum (alternative representations) [29]. Reprinted courtesy of the Ship Structure Committee, Washington, DC.

$$B = \left(0.817 \frac{2\pi}{T}\right)^4 \tag{9.46b}$$

A spectrum such as that defined by Equation (9.45) is obtained from records taken at a fixed point (*point spectrum*) with no indication of the direction of the wave components. Thus no indication of the energy contribution of wave components in each direction is apparent. In order to provide this additional information, measurements in a number of directions have to be made. The representation of such measurements takes the form of graphs such as Figure 9.15.

A *directional wave spectrum* $S(\omega,\mu)$ is then specified by the frequency ω and μ, the angle between the direction of the wave component and that of the prevailing wind. Such a spectrum can be approximated by two independent functions:

$$S(\omega, \mu) = S(\omega) f(\mu) \tag{9.47}$$

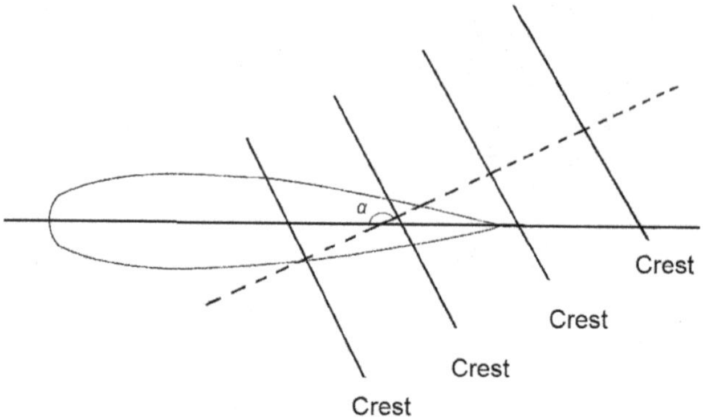

Figure 9.16 Definition of angle α [32].

where $f(\mu)$ is termed the *spreading function*. This can be assumed to have the following form:

$$f(\mu) = \left(\frac{2}{\pi}\right)\cos^2\mu \tag{9.48}$$

It is easy to verify that the integral of the directional spectrum over the range $-\pi/2$ *to* $+\pi/2$ returns the original point spectrum.

Spectrum of Wave Heights Encountered by Ship

If we know the spectrum of a seaway, we can determine the spectrum encountered by the ship by merely taking into account the effects of the change in frequency.

Let the ship be travelling at a speed V_s at an angle α to the predominant direction of the waves. The angle α is defined to be zero when the waves are approaching the ship from astern and 180° when the ship is heading into the waves (Figure 9.16).

Then the *period of encounter* T_e is the ratio of the length of the wave λ to the resolved relative velocity of ship and wave:

$$T_e = \frac{\lambda}{-V_s\cos\alpha + C} \tag{9.49}$$

where $C =$ wave velocity. When $\alpha = 0$, Equation (9.49) may give a positive, negative or infinite result. The negative result implies the ship overtaking the waves, and it is infinite when the speed of the ship is equal to that of the wave. This difficulty over sign makes the practical application of the above theory very difficult when α is less than $\pm 45^\circ$, since in practice some waves will be overtaking the ship and the ship will be overtaking the shorter waves.

Accepting this restriction we can proceed and obtain the wave spectrum encountered by the ship. Dividing numerator and denominator by C,

$$T_e = \frac{\lambda/C}{-(V_s/C)\cos\alpha + 1} = \frac{T}{-(V_s/C)\cos\alpha + 1}$$

where T is the period of the wave. Since

$$\omega_e = \frac{2\pi}{T_e}$$

the encounter frequency ω_e is given by

$$\omega_e = \omega\left(1 - \frac{V_s}{C}\cos\alpha\right) \tag{9.50}$$

where ω is the wave frequency. From wave theory $C = g/\omega$ so that

$$\omega_e = \omega\left(1 - \frac{\omega V_s}{C}\cos\alpha\right)$$

Hence,

$$S(\omega_e) = S(\omega)\frac{d\omega}{d\omega_e} \tag{9.51}$$

$$d\omega_e = d\omega\left(1 - \frac{2\omega V_s\cos\alpha}{g}\right)$$

Substituting,

$$S(\omega_e) = S(\omega)\left(1 - \frac{2\omega V_s\cos\alpha}{g}\right) \tag{9.52}$$

The effect is a distortion of the spectrum, as indicated in Figure 9.17. It should be noted that the area under all curves represents the energy of the wave system, which thus remains constant.

9.2.3 Response of Ships to Wave Loading

Having described the short-term representation of the sea environment, we can consider the response of a ship when in a irregular seaway.

Linear Response

Figure 9.3 shows that an irregular seaway can be decomposed into a large number of sinusoidal waves that have different orientations and geometrical as well as dynamical properties. In order to perform an analysis of ship response to such a composite loading condition, St. Denis and Pierson introduced the assumption that the ship behaves *linearly* [51]. That is, the response, whether it is motions or stresses, can be described by the superposition of the responses to all wave components that make up the irregular sea. It has been confirmed experimentally that this assumption is valid for sea states up to moderate levels. Figure 9.18 includes histograms of stress measurements that are seen to be approximated by the Rayleigh distribution, which as seen previously, is also used to model wave heights.

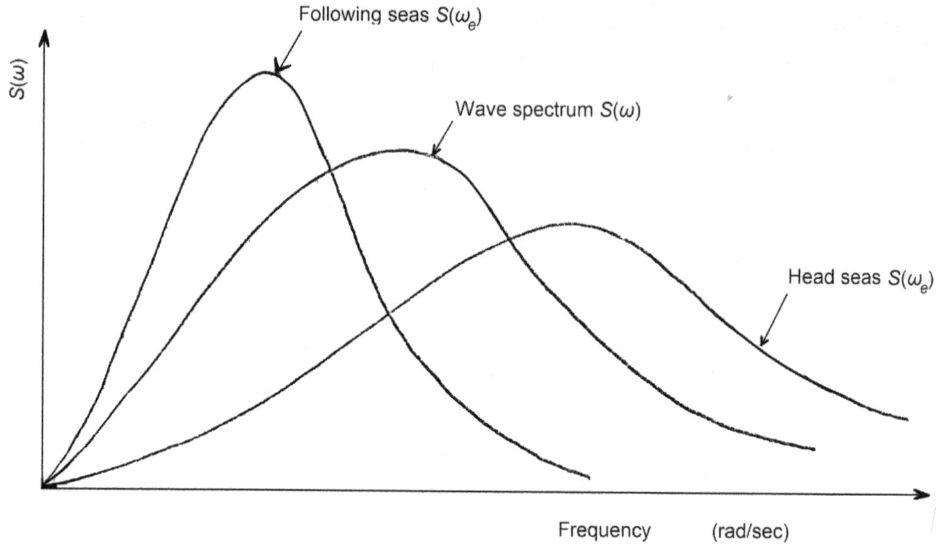

Figure 9.17 Encounter wave spectra in head seas, following seas and stationary vessel [38].

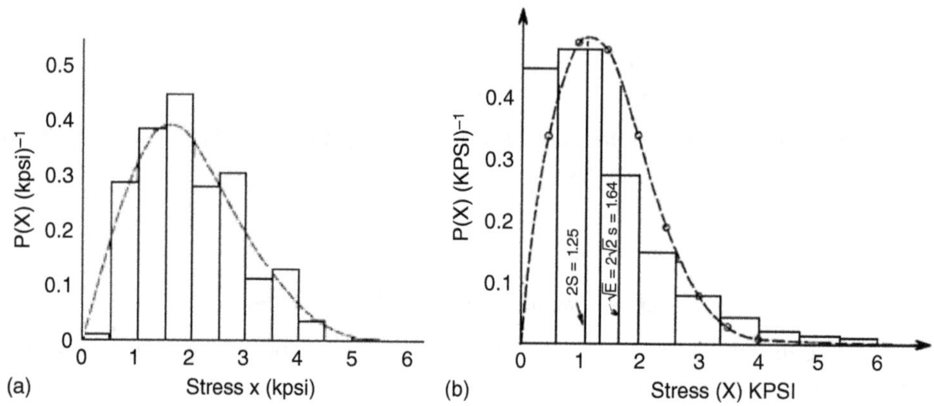

Figure 9.18 Stress measurements obtained at sea (SS *Wolverine State*) [12]. (a) Stress histogram for a typical record. (b) Histogram for 941 record intervals. Reprinted courtesy of the Ship Structure Committee, Washington, DC.

From Equations (9.23), (9.26) and (9.28) it is seen that the relationship of the wave spectra and each of these quantities (displacement, velocity and acceleration) has the following form:

$$A^2 \times \text{Input (Wave height) Spectrum} = \text{Output Spectrum} \tag{9.53}$$

where A^2 is the numerical value of the parameter relating the quantity in question to the wave height.

The linearity assumption allows us to relate a particular sea state to the applied loading as well as the motion of a ship. It is convenient to treat the environment–ship problem as an

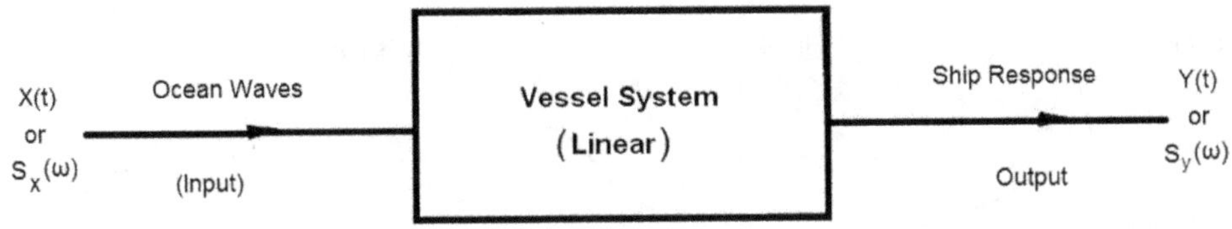

Figure 9.19 Environment–ship system [29]. Reprinted courtesy of the Ship Structure Committee, Washington, DC.

input/output system, for which the environment (ocean waves) represents the input, the ship itself acts as the filter, and the response, which may be motions and/or loads, is the output (Figure 9.19).

The fundamental relationship between the three system components is [7, 49]:

$$S_x(\omega)|H(\omega)|^2 = S_y(\omega) \tag{9.54}$$

where

$S_x(\omega)$ = amplitude of the input
$S_y(\omega)$ = amplitude of the output
$H(\omega)$ = transfer function for the required output

In the case of a ship at sea, the system is *time-invariant*. This means that the output is the same irrespective of the time at which the input is specified. In the case of linear, time-invariant systems, linear differential equations with constant coefficients are used to describe behaviour. Thus, a single-degree-of-freedom system is described by a single second-order differential equation.

The input and output can be expressed in the frequency domain, ω, or the time domain, t. For computational purposes a frequency-domain representation is preferable, although there arise situations in which a time-domain representation is required.

The modulus of $H(\omega)$, $|H(\omega)|$, is termed the *Response Amplitude Operator* (*RAO*). In Equation (9.54) the output is expressed in terms of circular frequency, ω. It is apparent, therefore, that in order to obtain loading data for a particular ship it is necessary first to have a suitable representation of the sea environment and then to determine the response amplitude operators for the particular output variables. The *RAO* is a measure of the output for unit input, and its dimensions will therefore be m^2sec.

Figure 9.20 shows the general form of the input, transfer function (RAO) and output spectra in the case of a ship at sea and their relationships.

It is now necessary to determine the value of the response amplitude operator, $|H(\omega)|$, that will enable the calculation of ship loads and motions. The simplest case to deal with is that of a SDOF system [49]. This corresponds to the excitation produced by a wave consisting of a single sinusoid:

$$\ddot{q}(t) + 2C\omega_1\dot{q}(t) + \omega_1{}^2q(t) = Q(t) \tag{9.55}$$

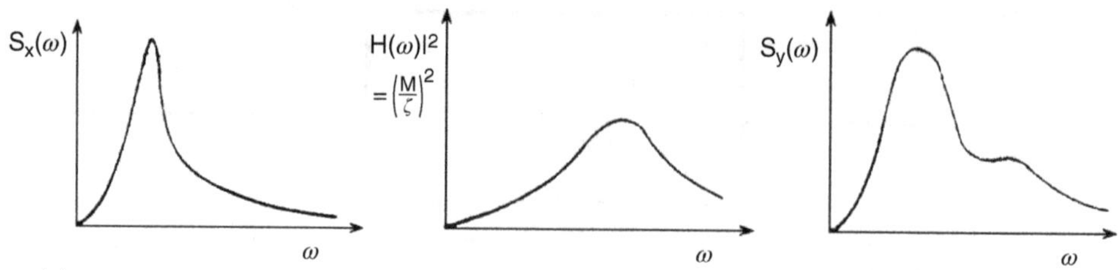

(a) (b) (c)

Figure 9.20 Relationship between input (wave) spectrum and output spectrum [29]. (a) Input spectrum. (b) Transfer function. (c) Output spectrum. Reprinted courtesy of the Ship Structure Committee, Washington, DC.

where

$$\begin{aligned} Q(t) &= \text{excitation (input)} \\ q(t) &= \text{output} \\ C &= \text{damping factor} \\ \omega_1 &= \text{natural frequency (circular)} \end{aligned}$$

For a sinusoidal excitation $Q(t) = Qe^{i\omega t}$, the response will also be sinusoidal $(q_0 e^{i\omega t})$ and will be given by

$$q(t) = q_0 e^{i\omega t} = |H(\omega)| Q_0 e^{i\omega t}$$

where

$$|H(\omega)| = \frac{1}{\omega_1{}^2 - \omega^2 + 2iC\omega_1\omega} \tag{9.56}$$

The output response is

$$q(t) = |H(\omega)| Q(t) \tag{9.57}$$

Thus a sinusoidal input produces a linearly related sinusoidal output for a SDOF system. As mentioned previously, the linearity assumption was introduced for the entire sea state, in which case the results obtained from the theory of linear random processes apply to the excitation produced by a sea state. For linear random processes [7, 49],

1. If the excitation (wave input) is a stationary random process, then the response (output) is also a stationary random process.
2. If the input is a normally distributed random process, the output is also a normally distributed random process.
3. If the mean of the random process is zero, then so is that of the output.
4. If the input is an ergodic process, then so is the output.

These are important conclusions with regard to the nature of the output, whether it is ship motions or loads. It has to be added, however, that in severe sea states certain responses are not linear, and thus it is necessary to perform nonlinear analysis.

In practice, RAOs are obtained for any type of response and are determined in one of two ways:

1. Analytically, by solving the general dynamic equations of motion of the ship;
2. Experimentally, using data obtained from towing tank tests.

In the first approach the ship is treated as a rigid body, and the forces acting on it at its centre of gravity are obtained by applying Newton's second law:

$$F = \frac{d}{dt}(mv) \; \mathrm{F} = \frac{\mathrm{d}}{\mathrm{dt}}(\mathrm{mv}) \tag{9.58a}$$

$$M = \frac{d}{dt}(I\omega) \; \mathrm{M} = \frac{\mathrm{d}}{\mathrm{dt}}(\mathrm{I}\omega) \tag{9.58b}$$

These equations constitute a system of six coupled nonlinear equations that correspond to the six-degree-of-freedom system of the ship (forces in the directions of the three coordinate axes and moments about the same axes). The resulting equations are difficult to solve, and solutions have been obtained by first decoupling some of the motions followed by a linearisation of the motions for which each solution is obtained.[7] Following the solution of the equations of motion, bending moments and shearing forces and indeed any other force acting on the hull girder can be calculated. This is performed using computer programs. Once the RAOs have been determined, Equations (9.58a–b) can be applied to obtain the corresponding response spectra for motions and loads.

Summarising, the procedure to obtain the short-term response consists of the following steps:

1. Measurements at sea in a particular location of wave elevations at specified angles (usually 30° intervals) 2. Statistical treatment of measurements in order to obtain parameters of wave spectrum density function 3. Solution of equations of motion (or towing tank tests) in order to obtain RAOs 4. Response spectrum (Equation (9.56)) using RAO and wave spectrum 5. Solution of equations of motion (Equations (9.58a–b)) in order to obtain loads and motions

Figure 9.21 includes experimental results that include a wave height spectrum, the resulting stress spectrum and the corresponding RAO spectrum.

Figure 9.22 includes results from theoretical (linear and nonlinear analysis) as well as experimental methods for containerships.

Structural Design Considerations

In structural design we need to determine the maximum load that is expected to act on the hull girder during the lifetime of a ship so as to be able to then select a suitable design load.

[7] *Linear strip theory,* which is based on decoupling of heave and pitch from other motions, is outlined in Section 5.7.

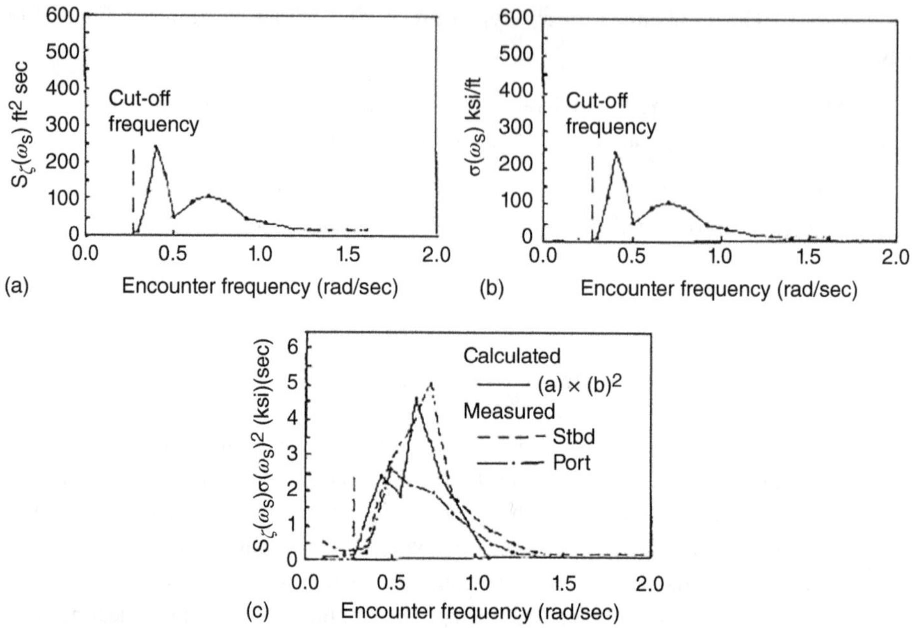

Figure 9.21 Wave height, RAO and stress data (measured and calculated) for SL-7 containership [15]. (a) Wave spectrum. (b) RAO. (c) Stress spectrum. Reprinted courtesy of the Ship Structure Committee, Washington, DC.

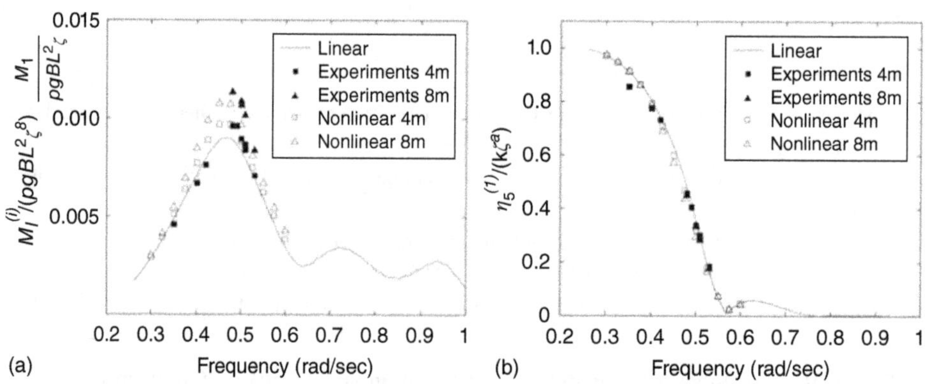

Figure 9.22 Containership RAOs using different methods (23 knots) [9]. (a) Vertical bending moment. (b) Pitch. Reprinted courtesy of Elsevier.

Over a 20- to 25-year period the expected number of waves encountered is of the order of 10^8. In statistical terms then, the probability of exceedance of a once-in-a-lifetime wave load will approach 10^{-8}. Whether or not a load of such a magnitude will actually be experienced in the case of any particular ship will depend on the properties of the probability distribution that describes the occurrence of these peak events. It is therefore necessary to study the probabilistic properties of *extreme loads*.

It ought to be made clear that in matters relating to the sea environment several different kinds of loads may be designated as extreme. Thus, we may speak of the maximum load applied to a particular ship during a storm that lasts a few hours or of the maximum loading experienced over one year, during which several storms will be encountered. We can also speak of the maximum load encountered during the lifetime of a particular ship. The discussion may be extended further, since a classification society will need to consider the distribution of peak loads that will be encountered by a fleet of ships. Fleets may consist of identical vessels operating in a particular environment or, as is the real case, a large variety of ships operating in many different environments. In the case of a fixed offshore or a coastal structure, meteorological data based on measurements in one location will provide a valid picture of the loading acting on it. For ships, however, this is not the case, since ships can navigate worldwide and also avoid storms. In this discussion we will limit ourselves to the case of a single ship.

Before considering long-term data, it is necessary to determine the loading that will arise in the short term. The reason for this is that, apart from the difficulties arising from the handling of large quantities of data, the characteristics of extreme loads in the short term differ from those in the long term, principally because they are the result of individual storms that arise in a particular region and have specific parametric characteristics. The most important feature of short-term data is its *stationarity*.

Over a (stationary) period that lasts several hours, a number of measurements are obtained. The wave height measurements conform to the Rayleigh distribution, and for each period one of these will be the peak value. The peak values of a series of individual storms as well as the annual peaks will also conform to some probability distribution, and it is the purpose of *statistics of extremes* to describe the properties of such distributions.

Basic Concepts in Extreme Value Theory: The Fisher–Tippett–Gnedenko Theorem

As mentioned previously, in order to provide a rational basis for the determination of primary design loads, it is necessary to estimate the maximum wave height that a ship (or fleet of ships) will encounter during a specific period (a few hours or the operating lifetime), as well as its probability of occurrence. Consideration therefore has to be given to the extreme values of certain variables. Such an analysis faces specific difficulties, since the probability distributions generally used are not suitable for the accurate estimation of extreme values, as the emphasis is on the distribution of the large bulk of the sample data and not of the outlier points. Extreme values are rare and may not even be included in the sample data. It is therefore necessary to employ methods that can overcome these problems. The first step involves formulating the data in a manner that is suitable for extreme value analysis. Three approaches can be used to achieve this.

In the first approach *block maxima* or *minima* series are derived. In many situations it is customary and convenient to extract annual maxima or minima, thereby generating a series of annual maxima (minima). In the marine environment these could be the maximum wave heights observed in a given location during each year for a specified number of years.

In the second approach peak values reached for any period, during which values exceed or fall below a certain threshold, termed the *peak-over-threshold* method, are extracted from a continuous record. In the marine environment this would involve specifying a particular wave height and then counting the number of times this is exceeded during a specified length of time.

The most direct way of determining the largest peak in a sequence of peaks is to use *order statistics*. Order statistics, in contrast to classical statistics that deal with measures of the overwhelming proportion of values of a distribution, such as the mean and the standard deviation, can be used to focus attention on the extremities (tail-ends) of a distribution. The measured values $(x_1, x_2, \ldots, x_n)$ (wave heights) that are random variables need to be rearranged in ascending order of magnitude $(y_1, y_2, \ldots, y_n)$, as was in fact done in the introductory example, in the last column of Table 9.1. If we now take a number of samples of n waves, the extreme value measured for each sample will differ each time. It is therefore seen that the extreme value is a random variable that follows its own probability law.

Let us then consider a sequence of independent and identically distributed (iid) variables $X_1, X_2, \ldots, X_n$ with cumulative distribution function F and let $M_n = max(X_1, X_2, \ldots, X_n)$ denote the maximum value of each of these. A relevant example is the distribution of maximum total bending moments of a ship measured every year over n years, where M_n is the corresponding maximum value of these.

The probability that an extreme (threshold) value z is not exceeded by the maximum value M_n is given by

$$P(M_n \leq z) = P(X_1 \leq z, X_2 \leq z, \ldots, X_n \leq z) \tag{9.59}$$

Since the variables $X_1, X_2, \ldots, X_n$ are independent,

$$P(M_n \leq z) = P(X_1 \leq z)P(X_2 \leq z), \ldots, P(X_n \leq z) \tag{9.60}$$

Furthermore, since variables $X_1, X_2, \ldots, X_n$ are identically distributed,

$$P(M_n \leq z) = (F(z))^n \tag{9.61}$$

This is the probability of *non-exceedance* of the value z and is the distribution function of the nth order statistic. The corresponding probability density function is obtained by differentiating Equation (9.61) with respect to z. This gives

$$f(z) = nf(z)[F(z)]^{n-1} \tag{9.62}$$

This is thus the initial probability density function of the wave (or load) peaks. Rice's distribution should be used as the initial distribution and thus Equation (9.39) and Equation (9.40) hold for $F(z)$ and $f(z)$, respectively.

The probability of *exceedance* of z is, from Equation (9.61),

$$P(M_n > z) = 1 - (F(z))^n \tag{9.63}$$

In each case it is necessary to determine the manner in which the maximum value is reached, especially as the distribution function is not generally known. An important theorem can be invoked in order to provide a practical solution to this problem. This is the *Fisher–Tippett–Gnedenko theorem* that is also known as the *extreme value theorem of statistics* [54]. This is a general result in extreme value theory that describes the asymptotic distribution of extreme-order statistics. The theorem states that the maximum of a sample of independent, identically distributed (iid) random variables after normalisation can only *converge in distribution* to one of three possible distributions:

1. The Gumbel distribution (asymptotic type I)
2. The Fréchet distribution (asymptotic type II)
3. The Weibull distribution (asymptotic type III)

Therefore, if the distribution of a normalised maximum converges, then the limit has to be one of a particular family of distributions.[8] A more formal statement of the theorem follows:

Let $X_1, X_2, \ldots, X_n$ be a sequence of iid random variables, and $M_n = max(X_1, X_2, \ldots, X_n)$. If a sequence of pairs of real numbers (a_n, b_n) exists such that

$$a_n > 0 \qquad \text{and} \qquad \lim_{n\to\infty} P\left(\frac{M_n - b_n}{a_n}\right) \leq x = F(x)$$

where F is a non-degenerate distribution function, then

$$F(x) = \exp\left\{-\left(1 + \xi\left(\frac{x-\mu}{\sigma}\right)\right)^{-1/\xi}\right\}$$

where

$$\xi = \text{shape parameter}$$
$$\sigma = \text{scale parameter } (> 0)$$
$$\mu = \text{location parameter}$$

The real numbers (a_n, b_n) are used to obtain a normalised (standardised) form of the distribution of M_n.

The shape parameter ξ governs the distribution type (Figure 9.23):

$\xi = 0$ Type 1 (Gumbel, light-tailed)
$\xi > 0$ Type 2 (Fréchet, heavy-tailed)
$\xi < 0$ Type 3 (Weibull, bounded)

If the distribution of M_n has a light tail with a finite upper (lower) bound (Figure 9.23), which is the case for ocean wave heights, the *Weibull law* holds, as has been found by plotting wave height data on Weibull probability paper [29]. This states that

$$F(x) = 1 - \exp\left(-\left(\frac{x}{k}\right)^l\right) \tag{9.64a}$$

[8] For details on the conditions for convergence, please consult Ref. [10].

where k and l are scale and shape parameters to be determined from data such as wave amplitude or wave-bending moment amplitude. The corresponding probability density function is

$$f(x) = \left(\frac{l}{k}\right)\left(\frac{x}{k}\right)^{l-1} \exp\left(-\left(\frac{x}{k}\right)^{l}\right) \tag{9.64b}$$

We thus have a suitable representation of maximum wave height long-term data for ocean waves. An example fit of a Weibull distribution to maximum stress data used in fatigue calculations is shown in Figure 9.24. The relevant Weibull parameters are also given in this figure.

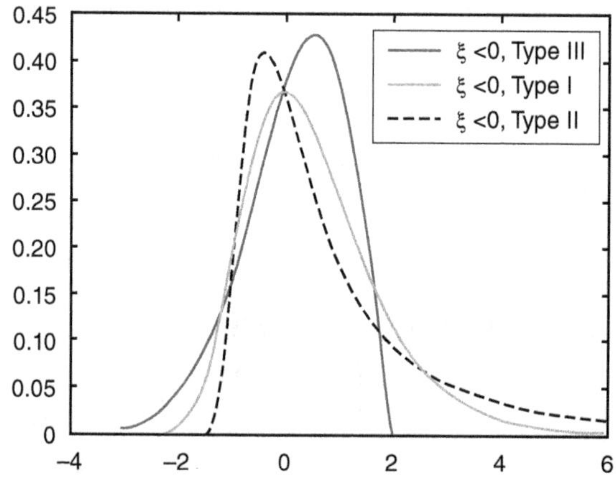

Figure 9.23 The three asymptotic extreme-type distributions.

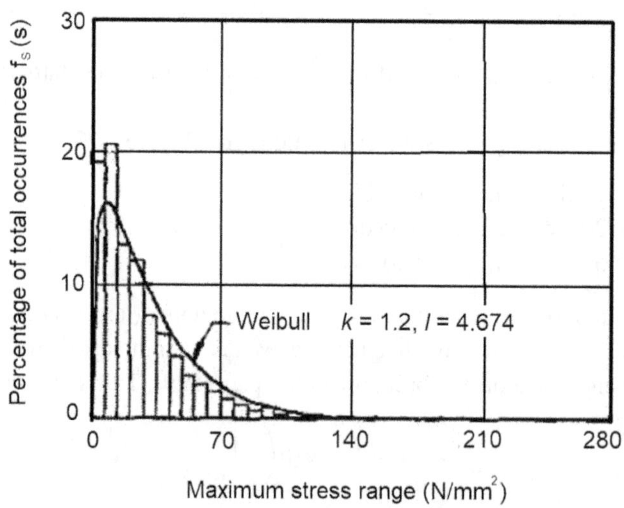

Figure 9.24 The Weibull distribution fitted to maximum stress range data measured at sea used in fatigue analysis.

The Weibull distribution reduces to two important special cases:

(a) The *exponential distribution* (when $l = 1$, $k = \lambda$)
(b) The *Rayleigh distribution* (when $l = 2$, $k = (\sqrt{2})H_{rms}$)

The Weibull parameters can be estimated using several methods [29]:

1. Method of moments
2. Weibull probability paper
3. Maximum likelihood method
4. Order statistics

The time interval between peak events is the *return period.* When measured in years, the return period of exceedance of a particular wave load is equal to the inverse of the probability that this wave load is exceeded in any one year. The *m*-year return level is associated with a return period of *m* years. We may thus speak of a 50- or a 100-year wave load. A derivation of the expression for the return period is included in Section 9.2.3.1.

In ship structural design, however, it is necessary to allow for ship operations as well, given that ships follow particular routes and may take avoiding action in the case of severe storms. Of more relevance then is the *probability of encounter*, since return periods reflect wave statistics alone [31].

Extreme Loads in Short-Term Seas

Before considering long-term data it is necessary to determine the short-term extreme values. At a practical level, wave data are obtained by measuring wave height at fixed time intervals. Thus over a (stationary) period lasting several hours, a number of measurements will be obtained. These will have some form of distribution, and for each period one of these will be the peak value. As discussed in the previous section, the peak values conform to a particular type of asymptotic probability distribution, and it is necessary to determine the properties of such a distribution.

In the case of short-term data analysis, we know that the *initial distribution* is the Rayleigh distribution, which describes wave height measurements. In this case the form of the exceedance probability density function $g(y_n)$ is shown in Figure 9.25, which also

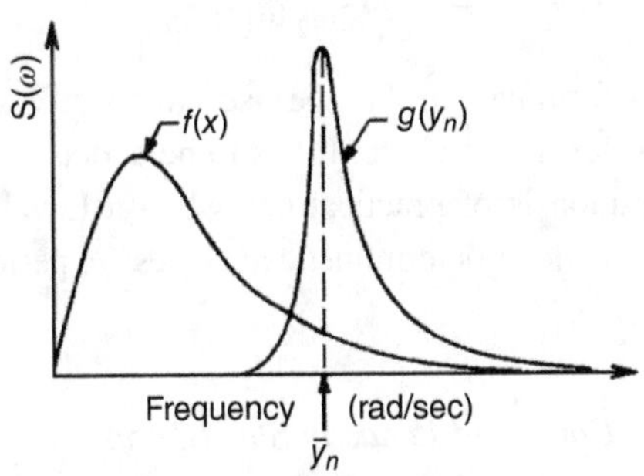

Figure 9.25 Rayleigh and corresponding extreme value distribution [40]. Reprinted courtesy of Cambridge University Press.

shows the initial distribution, $f(x)$. The exceedance probability density function is seen to be much more concentrated about its mode than the initial distribution, so that its characteristic measure can be taken to be the *modal* or *probable extreme value* [40]. The peak value of the extreme distribution is obtained by setting

$$\frac{dg(y_n)}{dy} = 0 \tag{9.65}$$

that gives, from Equation (9.62),

$$f'(y_n)F(y_n) + (n-1)\{f(y_n)\}^2 = 0 \tag{9.66}$$

If the initial distribution is a Rayleigh distribution of wave heights, its probability density function is given by Equation (9.34). Then, Equation (9.66) becomes

$$\left\{\frac{1}{4m_0} - \left(\frac{y_n}{4m_0}\right)^2\right\}\left(1 - e^{-\frac{y_n^2}{8m_0}}\right) + (n-1)\left(\frac{y_n}{4m_0}\right)e^{-\frac{y_n^2}{8m_0}} = 0$$

Since y_n is large, $(y_n/4m_0)^2 \gg (1/4m_0)$, so that the term $1/4m_0$ can be discarded. For large values of n, being the number of waves, the most probable extreme value is

$$\overline{y_n} = 2\sqrt{2m_0 \ln n} \tag{9.67}$$

This can then be used as a basis for the design load. Ochi [40] has shown that there is a 63.2 per cent chance that the largest wave will exceed $\overline{y_n}$. Since this is a high probability, it is necessary to consider a wave height of sufficient magnitude that has an acceptably low probability of exceedance. A risk parameter, α, is defined as follows [40]:

$$\lim_{n\to\infty} P[extremewave > \overline{y_n}] = \lim_{n\to\infty}\{1 - G(\overline{y_n})\} = \lim_{n\to\infty}\{1 - (F(\overline{y_n}))^n\} = \alpha \tag{9.68}$$

For $\alpha \ll 1$, large n and a narrow-band spectrum, Equation (9.67) becomes

$$\overline{y_n} = \sqrt{2m_0 \ln(n/\alpha)} \tag{9.69}$$

Ochi has suggested $\alpha = 0.01$. Substituting in Equation (9.69), we obtain

$$\overline{y_n} = \sqrt{2m_0 \ln(100n)} \tag{9.70}$$

for wave amplitude. This formula can then be used to specify a design load for ultimate strength considering short-term seas alone. It has to be added that information concerning the extreme value distribution is of practical use when a Level III reliability analysis is performed. If semi-probabilistic or deterministic analyses are performed, then only the peak value is required [31].

Combined Loads in Short-Term Seas

Up to this point in this chapter we have considered the motion of the sea surface and the manner in which individual loads and load effects that act on the hull girder can be

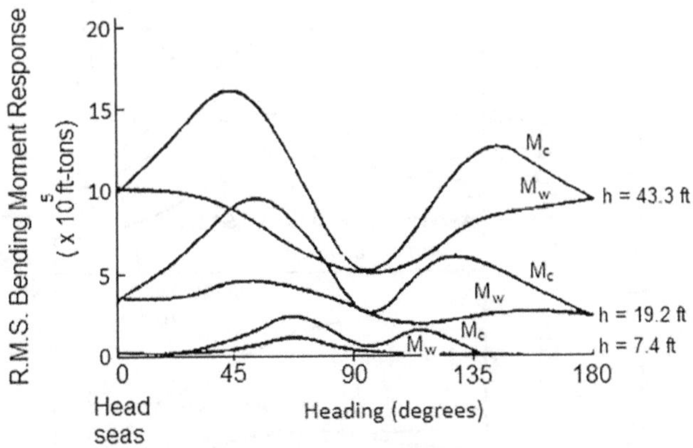

Figure 9.26 Vertical and combined wave-induced bending moments (M_v, M_c) in long-crested seas based on theoretical calculations [29]. Reprinted courtesy of the Ship Structure Committee, Washington, DC.

calculated. However, the actual loading acting in any given situation is the outcome of the concurrent action of several loads. In order to develop a rational basis for design, and particularly for ultimate strength design, it is necessary to consider the *combined effect* of loads, and in particular those that occur under extreme conditions. The task faced by the designer is therefore that of computing the effect of joint action, which is the resultant of a finite number of individual responses. Examples of individual loads are vertical bending, horizontal bending and torsion, and different computational procedures are used to calculate each of these. The procedures used are in general incompatible with one another. In addition to this difficulty, there arise two further related complications: (1) the phase difference between individual time-varying periodic components and (2) the differing random nature of the load components (Figures 9.26 and 9.27). Both of these aspects also must be allowed for. With regard to the latter, it is necessary to determine the *degree of correlation* between the different components and to develop a method of combining them.

In previous work, stress measurements in the time domain have been analysed and high- and low-frequency components have been separated out. Mansour reasoned that it would be possible to follow a reverse procedure, so that the combined response is determined from analytically determined components.

The proposed procedure, described in [28], [30] and [31], consists of the following steps:

1. Calculate the mean value of the still-water bending moment that results from the net longitudinal load curve and thermal gradients.
2. Combine low-frequency (wave-induced) rigid body response with high-frequency response (slamming or springing).
3. Add the results from step 2 to those of 1.

Step 1

In Step 1, the stresses resulting from bending in still water as well as thermal gradients should be combined to obtain the mean value for rigid body motion and high-frequency

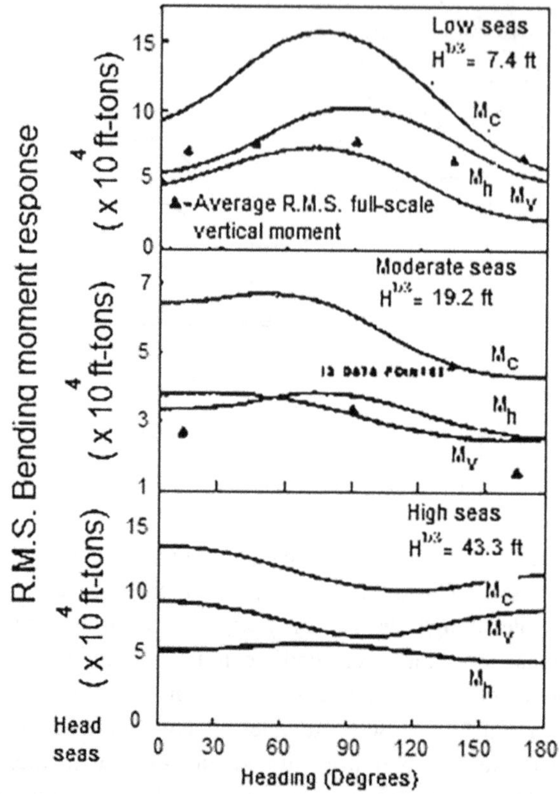

Figure 9.27 Vertical, horizontal and combined moments (M_c, M_h, M_v) in short-crested seas based on theoretical calculations [29]. Reprinted courtesy of the Ship Structure Committee, Washington, DC.

response. Still-water bending moments differ for each loading condition but essentially vary little for the duration of each voyage. Details are given in Chapter 12, and it is seen that if a statistical description of still-water bending moments is adopted, the normal distribution fits the data best. Using mean values and standard deviations, extreme values can be determined.

The question of environment-related thermal gradients has been discussed in Sections 3.1 and 6.2.4. The most important thermal stress fields are due to the temperature difference between air and sea-water as well as due to diurnal variations. For the North Atlantic, average diurnal changes of air temperature are approximately 6°C whereas those of deck plating lie in the range 5°C–25°C, depending on cloud cover and colour of plating. Thermal stresses can be calculated using either a two-dimensional approach or a finite element model of a ship, once thermal gradients have been determined along the length of a ship. In adverse weather conditions thermal gradients are not expected to be large, but in clear weather a heavy swell may occur, so it is necessary to consider several representative conditions [28].

Once the thermal stress field is determined, this should to be added to the stress fields resulting from several still-water bending moment distributions.

Step 2

A description of the environment–ship system, which for moderate sea states behaves as a linear system, is included in Section 9.2.1. For such a system the outcome of the concurrent action of a number of loads is obtained by summing these using the principle of superposition. Therefore, in the case of multiple load action, Figure 9.19 can be replaced by Figure 9.28, which describes the behaviour of a linear system in the time domain.

In the time domain the input and output are described by $\tilde{x}(t)$ and $\tilde{y}(t)$, respectively, where the tilde designates a random quantity. Individual loads are represented by separate branches enclosed by the dotted lines. In lieu of response amplitude operators that are used in the frequency domain, in the time domain the *impulse response functions* $h_i(.)$ relate input to output. These represent the response of each linear system to unit excitation multiplied by time (.) (response to the Dirac delta function). The a_i terms are introduced as linear multipliers that may be required when considering combined effects and may be set to unity when considering individual effects. Since the system is linear, the combined effect is obtained by adding individual effects, hence the summation sign.

In systems in which the input is defined by a simple expression such as a constant force, it is possible to solve the equation of motion and trace out the response. In the case of wave loading, however, this cannot be done, and it is necessary to resort to a method that is capable of providing solutions in the case of generalised input. To this end the *convolution integral method*, discussed in Appendix G, is used. For an individual input the output is given by

$$y(t) = \int_0^\infty h_i(\tau)x(t-\tau)d\tau \tag{9.71}$$

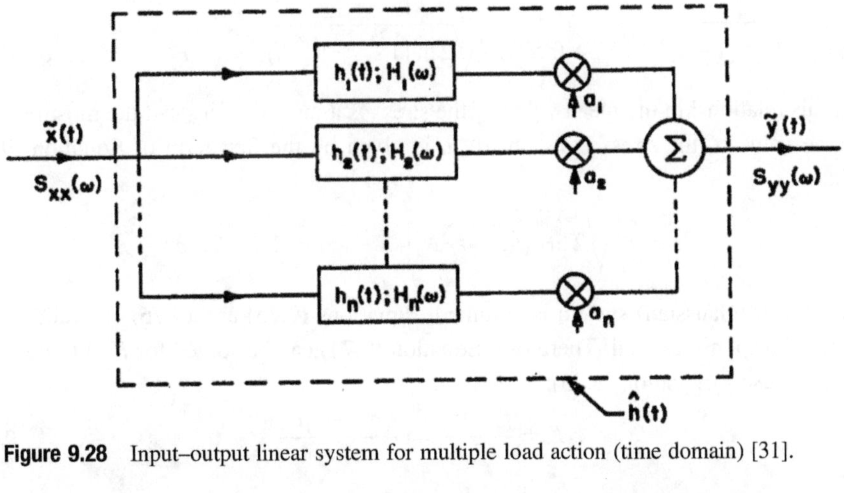

Figure 9.28 Input–output linear system for multiple load action (time domain) [31].

In Equation (9.71) $h_i(\tau)$ are *impulse response functions* of individual components and $x(t-\tau)$ is the common input, which is the time-varying history of wave elevation.

For n different linear inputs,

$$y(t) = \sum_{i=1}^{n} a_i \int_0^{\infty} h_i(\tau)x(t-\tau)d\tau \tag{9.72}$$

where a_i are constants to ensure uniformity of units and direction. They can be set equal to unity if not needed.

We consider now the combined response when two loads act (e.g. vertical bending and horizontal bending) so that when one of these loads reaches its peak value resulting in a peak stress, the probability that the other load is also at its peak value is equal to K – the *probabilistic load combination factor*. Therefore, for a linear system involving two branches, the combined response is

$$f_c = f_1 + Kf_2 \qquad f_1 > f_2 \tag{9.73}$$

where f_c is the combined stress and f_1 and f_2 are the peak stresses arising from individual loads. The value of K depends on the degree of correlation between the two loads. A characteristic feature of loads acting on ship structures is that they all originate from ocean waves – that is, they have a common source. On this basis, an approach assuming a Gaussian wave process with the ship acting as a linear filter was developed [30]. The values of f_1 and f_2 are obtained from the extreme value theory of Cartwright and Longuet-Higgins, who determined the following expression for the expected extreme stress peak f_i in N peaks in a Gaussian sea state [6]:

$$f_i = E[f_{i\max}] = \alpha_i \sigma_i \tag{9.74}$$

where σ_i^2 is the variance of the ith stress process and α_i is a multiplier that depends on the number of peaks N and the bandwidth parameter ε via the following relationship:

$$\alpha_i = \sqrt{2\ln(1-\varepsilon_i^2)^{1/2} N_i} + 0.5772/\sqrt{2\ln(1-\varepsilon_i^2)^{1/2} N_i} \qquad i = 1, 2, \ldots, c \tag{9.75}$$

In this relationship the mean value of the stress is assumed. If instead the most probable extreme value of the stress is assumed, α_i is given by the first term of Equation (9.75), so that

$$\alpha_i = \sqrt{2\ln\left[\sqrt{1-\varepsilon_i^2 N_i}\right]} \qquad i = 1, 2, \ldots, c \tag{9.76}$$

If a linear (Gaussian) system is assumed, Equations (9.75) and (9.76) are valid for the combined response as well. Therefore, Equation (9.74) can be solved for K in terms of the variances, using Equation (9.75):

$$K = \frac{f_c - f_1}{f_2} = \frac{\alpha_c \sigma_c - \alpha_1 \sigma_1}{\alpha_2 \sigma_2} \tag{9.77}$$

where σ_c, σ_1 and σ_2 are the root mean square values of the peak stresses and α_c, α_1 and α_2 are the corresponding multipliers determined from Equation (9.78). The variance ${\sigma_c}^2$ of the combined output is given as the 0th moment, m_0, of the output spectrum [30]:

$$ {\sigma_c}^2 = \sum_{i=1}^{n} {a_i}^2 {\sigma_i}^2 + \sum_{i=1}^{n} \sum_{j=1}^{n} a_i a_j \rho_{ij} \sigma_i \sigma_j \tag{9.78} $$

where ${\sigma_i}^2$ are the variances of each load component. In the frequency domain (note Equation (9.23)),

$$ {\sigma_i}^2 = \int_0^\infty |H_i(\omega)|^2 S_x(\omega) d\omega \tag{9.79} $$

and ρ_{ij} are correlation coefficients between individual load components,

$$ \rho_{ij} = \frac{1}{\sigma_i \sigma_j} \int_0^\infty \mathrm{Re}\left[|H_i(\omega)|^2 H_j^*(\omega) \right] S_x(\omega) d\omega \tag{9.80} $$

Substituting for σ_c in Equation (9.77) from Equation (9.78) and noting that $\alpha_c = \alpha_1 = \alpha_2 = 1$, since Equations (9.74) and (9.76) involve stresses rather than bending moments, the following expression is obtained [31]:

$$ K = \frac{m_r}{r} \left[m_c \sqrt{1 + r^2 + 2\rho r} - 1 \right] \tag{9.81} $$

where

$$ r = \frac{\sigma_2}{\sigma_1} \tag{9.82a} $$

$$ m_r = \sqrt{\frac{\ln\left[(1 - {\varepsilon_1}^2)^{1/2} N_1\right]}{\ln\left[(1 - {\varepsilon_2}^2)^{1/2} N_2\right]}} \tag{9.82b} $$

$$ m_c = \sqrt{\frac{\ln\left[(1 - {\varepsilon_c}^2)^{1/2} N_c\right]}{\ln\left[(1 - {\varepsilon_1}^2)^{1/2} N_1\right]}} \tag{9.82c} $$

The stresses are fully correlated when $\rho = 1$, in which case, from Figure 9.29, $K = 1$, independently of the stress ratio. When $\rho = 0$ (uncorrelated stresses), $K = 0.10$ for $r = 0.2$ and $K = 0.41$ for $r = 1$. If the load processes (both individual and combined) are narrow-banded and have a similar central frequency, $m_r = m_c = 1$. Examples are vertical and horizontal bending moments.

In 1973, the ISSC recommended a value of $\rho = 0.32$ for stresses due to primary bending moments. In a study on large tankers it was found that ρ depends on sea state and heading

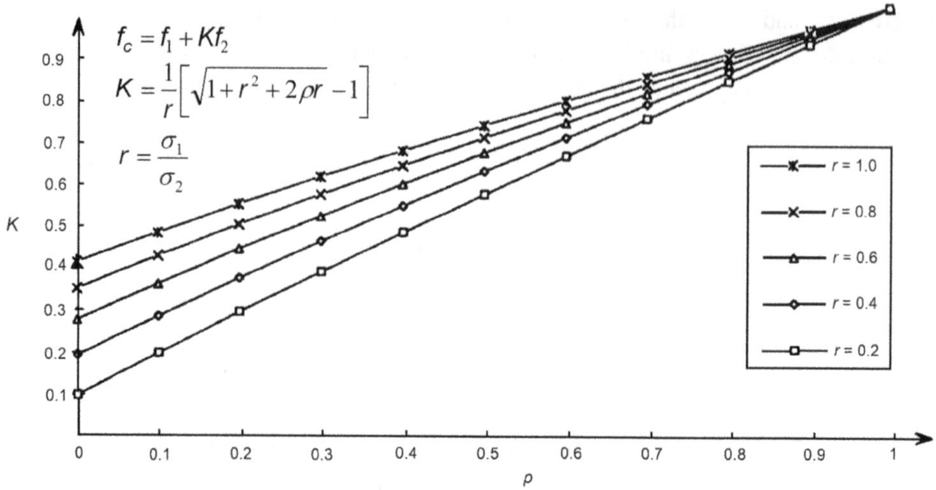

Figure 9.29 Correlation factor K for two correlated loads [30]. Reprinted courtesy of the Society of Naval Architects and Marine Engineers.

with values close to 0.45. Application of Equations (9.81)–(9.82) leads to the conclusion that about 60 per cent of the stresses due to the horizontal moment should be added to those due to the vertical bending moment.

9.3 Long-Term Analysis of Sea Loads and Ship Response

The approach described previously is valid only for short-term load combinations in a single sea state when stationarity holds. Should it be necessary to consider the joint action of more than two loads, this approach can readily be extended, and the case of three correlated load combinations is described in [31].

In Section 2.2.1, in which a single sea state was studied (short-term analysis), the properties of the Rayleigh distribution were used to determine the probability that a particular wave height level (H_1) is exceeded (Equation (9.24)).

If it is desired to perform such a calculation for the entire operating life of a ship, it is necessary to consider all the sea states that the ship will encounter and the frequency of encounter of each. These are described by particular wave height spectra, and it is thus necessary to include all possible wave spectra in the analysis.

The probability of exceedance of a particular wave height during the lifetime of a ship will then be obtained by integrating the product of the probability of exceedance of the wave height for each and every sea state encountered and the frequency of occurrence of this sea state. That is,

$$P[H > H_1]_{longterm} = \int_0^{\infty} [P[H > H_1]|m_0][P[m_0]]dm_0 \tag{9.83}$$

where $P[H > H_1]$ is the probability of exceedance of wave height H_1 during a sea state described by m_0. For N sea states,

$$P[H > H_1]_{longterm} = \sum_{i=1}^{N} [[P[H > H_1]|m_0][P[m_0]]] \tag{9.84}$$

In principle, this is the correct formulation of the long-term probability of exceedance of a wave height H_1. However, this needs to be modified for reasons that are given in the next section.

9.3.1 Differences with Short-term Analysis: Full-scale Measurements at Sea

In ship structural design it is necessary to consider (a) failure of the hull girder due to overloading, (b) fatigue failure of individual structural members and (c) the fact that the world fleet consists of a large number of ships that are at sea at any point in time. Each of these questions raises separate issues, the latter being of particular importance for regulatory bodies.

In preceding sections we outlined the approach used to determine the loads that act on the hull girder (bending moments, shearing forces, inertial forces) as well as motion response (linear and angular). These quantities have been calculated on the basis of wave height spectra that represent sea states whose duration extends to a few hours, that is, for periods necessary to capture the effects of individual storms. The analysis was based on the following assumptions:

1. Ocean waves are a *stationary* random process, that is, the statistical parameters are independent of the point in time at which measurements are initiated.
2. A single sample of measurements is representative of all other samples that may be obtained (*ergodicity*).
3. Ship response is linear.

The theory based on these assumptions has been found to give satisfactory results provided that:

(a) The duration of the period considered does not extend beyond a few hours.
(b) Sea states up to moderate level are considered.
(c) The ship is of conventional, wall-sided hull form and advances at a moderate speed.

Long-term measurements of wave heights have shown that conditions at sea are such that relatively long periods of moderate conditions are interrupted by storms that have a comparatively short duration. During storms, extreme conditions may arise on occasion. Corresponding to each storm there will be a maximum value, one of which will be the peak experienced during the ship's operating life. This peak value behaves as a random variable in much the same way short-term peaks do.

Long-term response is affected by a number of other factors as well that differentiate it from short-term response. These include:

1. The region of the ocean, the season of the year and the ship's loading condition.
2. In long-term analysis, independence of relative heading from the sea state cannot be assumed for ships that operate on a single route. Hence, the joint probability

density function of H_s, T_0, V, α cannot be expressed analytically for all seasons and areas.

3. The combined response is not narrow-banded. One reason is that when springing occurs, a second peak arises in the vertical bending moment spectrum (Figure 3.8).
4. As a result of (1), (2) and (3), the rigorous analysis of long-term response must be carried out in the time domain. Seasonal effects would necessitate this to extend to one year's duration. Since stationarity does not extend beyond a two- to three-hour period, such an analysis is not a feasible proposition.

It transpires that since the response is not narrow-banded (item (3) above), it is necessary to take into account the number of waves for each short-term sea state, so that Equation (9.84) needs to be modified.

9.3.2 Statistical Description of a Critical Wave Height

If we consider the lifetime loading of a ship, we may have available long-term data over, say, a 10-year period. We may therefore extract a set of annual maximum values for this period. The question then arises as to what ought to be the value of the maximum load that we need to consider for a ship whose lifetime is 25 years. Linked to this are questions concerning the probability of occurrence of such a load and, furthermore, what ought to be the acceptable probability of failure for a ship. We are therefore led to having to generate the statistical properties of what we consider to be an unacceptable event, this being the *exceedance* of a particular wave load that is related to the exceedance of a particular wave height. The statistical properties of the occurrence of a particular wave height can be determined in the following two ways:

1. In terms of the magnitude of the critical wave height and frequency of occurrence
2. In terms of the period between instances of exceedance of the critical wave height

Return Period, Waiting Time and Probability of Occurrence

The latter approach is used when long periods are involved, as in the case of offshore structures. These are termed *return periods* and are usually expressed in years. In formal terms, a return period T_R of a random event of magnitude Q is defined as the average length of time in which, statistically, Q will be equalled or exceeded once, taking a very long-term record [11]. If the probability of non-exceedance in any one year is $P[H < H_1]$, the probability of exceedance will be $1 - P[H < H_1]$. The return period, T_R, in years, is then

$$T_R = \frac{1}{1 - P[H < H_1]} \tag{9.85}$$

Clearly, occurrences are not spaced out at regular intervals so that the return period calculated in this manner cannot be used as an event predictor, and for this reason it may be called an *estimated return period*. It is simply a statistical average over the long term that is based on available data. In principle, the return period can be used to specify design values in certain situations. In the case of fixed offshore or coastal structures, if the return period exceeds the design life of a structure, we may speculate that the probability of failure due to overloading is satisfactory. The *m*-year return level is associated with a return period of *m* years. We may thus speak of a 50- or a 100-year wave. As noted previously, this does not in any way preclude the occurrence of this wave load during the lifetime of the structure, and possibly more than once.

If, however, we had data extending to 70 years, during which there were 7 years in which exceedance occurred, not necessarily at regular intervals, the *average return period* would be 70/7 = 10 years on the basis of this data.

Of more practical use is the *waiting time* for a first occurrence, and the corresponding *probability of non-exceedance* of a specified wave height for a specific length of time (years). We can interpret this sequence as a set of *Bernoulli trials* in which non-exceedance is assigned the value of unity and exceedance the value of zero. A sequence consisting of eight annual non-exceedances followed by one exceedance would be 111111110.

We let T be a random variable that measures the number of trials (years) it takes for the first exceedance to occur. If during the 70-year period, at one stage the waiting time between consecutive occurrences was 5 years, then the corresponding probability of occurrence for $T = 5$ trials would be the product of the probabilities of occurrence for each of the Bernoulli trials, since these are independent events. These consist of four non-occurrences (1111) for each of which the probability is (1-p), and one year of occurrence (0) for which the probability is p. In general, therefore, the probability that the waiting time is T trials, $P[T]$, is given by

$$P[T] = (1-p)^{T-1}p \qquad T = 1, 2, 3, \ldots, \; \mathrm{P[T]} = (1-\mathrm{p})^{\mathrm{T}-1}\mathrm{p} \tag{9.86}$$

A plot of $P[T]$ against the waiting time T shows that, as expected, there is a progressive reduction in $P(T)$ with increasing T (Figure 9.30a). This is in fact the *geometric distribution*. Figure 9.30b shows this for $p = 0.9$.

Since consecutive occurrences do not occur at regular intervals, it is necessary to calculate the probability of occurrence for all other intervals as well. The average probability of all the

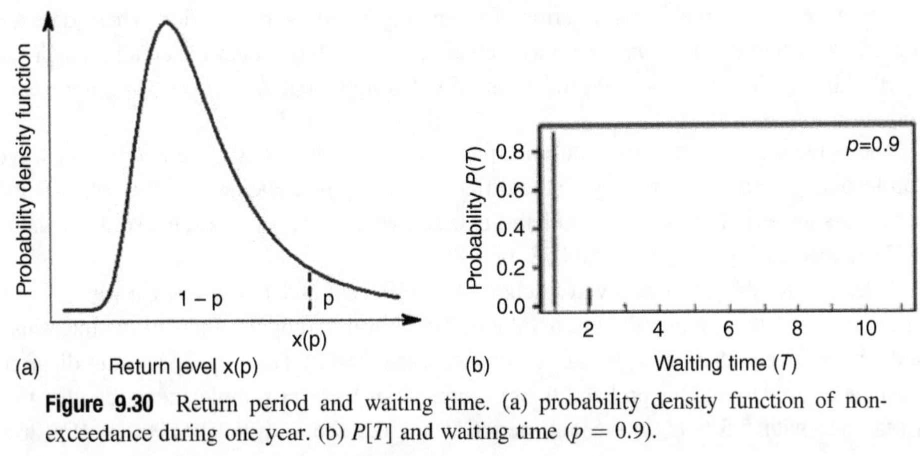

Figure 9.30 Return period and waiting time. (a) probability density function of non-exceedance during one year. (b) $P[T]$ and waiting time ($p = 0.9$).

possible waiting times is given by the *expected value* of T. The expected value $E[X]$ of a discrete random variable X with a probability distribution function $f(x)$ is given by

$$E[X] = \sum_i x_i f(x_i) \tag{9.87}$$

In this case,

$$\begin{aligned} E[T] &= \sum_{T=1}^{\infty} T\left((1-p)^{T-1}p\right) = \\ &= p + 2p(1-p) + 3p(1-p)^2 + \ldots + np(1-p)^{n-1} + \ldots \\ &= p\left(1 + 2(1-p) + 3(1-p)^2 + \ldots + n(1-p)^{n-1} + \ldots\right) \\ &= \frac{1}{p} \end{aligned} \tag{9.88}$$

from convergent power series theory. That is, the expected value of a waiting time that is geometrically distributed is the inverse of the probability of the event. Returning to the case of the fixed offshore structure mentioned previously, we may say that the *theoretical return period* is the inverse of the probability that a wave height will be exceeded in any one year. For a 100-year wave, therefore, the probability of occurrence is $1/100 = 0.01$.

In ship structural design, however, it is necessary to allow for ship operations as well, given that ships follow particular routes and may take avoiding action in the case of severe storms. Of more relevance is the *probability of encounter*, since return periods reflect wave statistics alone [31]. Furthermore, as mentioned in the previous section, the question of *non-stationarity* arises over the long term.

Description of Waves in Terms of Wave Height and Period: The Scatter Diagram

We have seen that the Rayleigh distribution is used in the short-term description of a sea state, and since in this case linearity is assumed, corresponding bending moments can be calculated. In the case of long-term conditions, descriptions of sea states are available for particular regions of the ocean in terms of wave height and average period. These data are presented in *scatter diagrams*, such as Table 9.3 that includes data for conditions in the North Atlantic. The oceans of the world are divided into what are termed *Marsden zones*, shown in Figure 9.31 [13].

The wave scatter table gives the joint probability of significant wave height and wave zero-crossing period. The rows give the different wave height classes, and the columns the zero-crossing period classes. The probability is shown as parts per thousand, so that a value of 78 represents a probability of 0.078, or 7.8%.

If H_i denotes the significant wave height of an interval i and T_j the average wave period of an interval j, these being the mean values of the lower and upper limits of the intervals i and j, respectively, then let S_{ij} be the sea state characterised by $\left(H_i, T_j\right)$. The scatter diagram can then be used to find the probability of occurrence of S_{ij}. For example, the probability that a sea state with $5.5 > H_s > 6.5$ and period $T = 10.5$ sec is $1{,}269/100{,}000 = 0.01269$.

Table 9.3 *Scatter diagram of sea states in the North Atlantic described as an occurrence per 100,000 observations used in long term predictions*

Hs/Tz	1.5	2.5	3.5	4.5	5.5	6.5	7.5	8.5	9.5	10.5	11.5	12.5	13.5	14.5	15.5	16.5	17.5	18.5	Sum
0.5	0	0	1.3	133.7	865.6	1,186	634.2	186.3	36.9	5.6	0.7	0.1	0	0	0	0	0	0	3,050
1.5	0	0	0	29.3	986	4,976	7,738	5,569.7	2,375.7	703.5	160.7	30.5	5.1	0.8	0.1	0	0	0	22,575
2.5	0	0	0	2.2	197.5	2,158.8	6,230	7,449.5	4,860.4	2,066	644.5	160.2	33.7	6.3	1.1	0.2	0	0	23,810
3.5	0	0	0	0.2	34.9	695.5	3,226.5	5,675	5,099.1	2,838	1,114.1	337.7	84.3	18.2	3.5	0.6	0.1	0	19,128
4.5	0	0	0	0	6	196.1	1,354.3	3,288.5	3,857.5	2,685.5	1,275.2	455.1	130.9	31.9	6.9	1.3	0.2	0	13,289
5.5	0	0	0	0	1	51	498.4	1,602.9	2,372.7	2,008.3	1,126	463.6	150.9	41	9.7	2.1	0.4	0.1	8,328
6.5	0	0	0	0	0.2	12.6	167	690.3	1,257.9	1,268.6	825.9	386.8	140.8	42.2	10.9	2.5	0.5	0.1	4,806
7.5	0	0	0	0	0	3.0	52.1	270.1	594.4	703.2	524.9	276.7	111.7	36.7	10.2	2.5	0.6	0.1	25,86
8.5	0	0	0	0	0	0.7	15.4	97.9	255.9	350.6	296.9	174.6	77.6	27.7	8.4	2.2	0.5	0.1	1,309
9.5	0	0	0	0	0	0.2	4.3	33.2	101.9	159.9	152.2	99.2	48.3	18.7	6.1	1.7	0.4	0.1	626
10.5	0	0	0	0	0	0	1.2	10.7	37.9	67.5	71.7	51.5	27.3	11.4	4	1.2	0.3	0.1	285
11.5	0	0	0	0	0	0	0.3	3.3	13.3	26.6	31.4	24.7	14.2	6.4	2.4	0.7	0.2	0.1	124
12.5	0	0	0	0	0	0	0.1	1	4.4	9.9	12.8	11	6.8	3.3	1.3	0.4	0.1	0	51
13.5	0	0	0	0	0	0	0	0.3	1.4	3.5	5	4.6	3.1	1.6	0.7	0.2	0.1	0	21
14.5	0	0	0	0	0	0	0	0.1	0.4	1.2	1.8	1.8	1.3	0.7	0.3	0.1	0	0	8
15.5	0	0	0	0	0	0	0	0	0.1	0.4	0.6	0.7	0.5	0.3	0.1	0.1	0	0	3
16.5	0	0	0	0	0	0	0	0	0	0.1	0.2	0.2	0.2	0.1	0.1	0	0	0	1
Sum	0	0	1	165	2,091	9,280	19,922	24,879	20,870	12,898	6,245	2,479	837	247	66	16	3	1	100,000

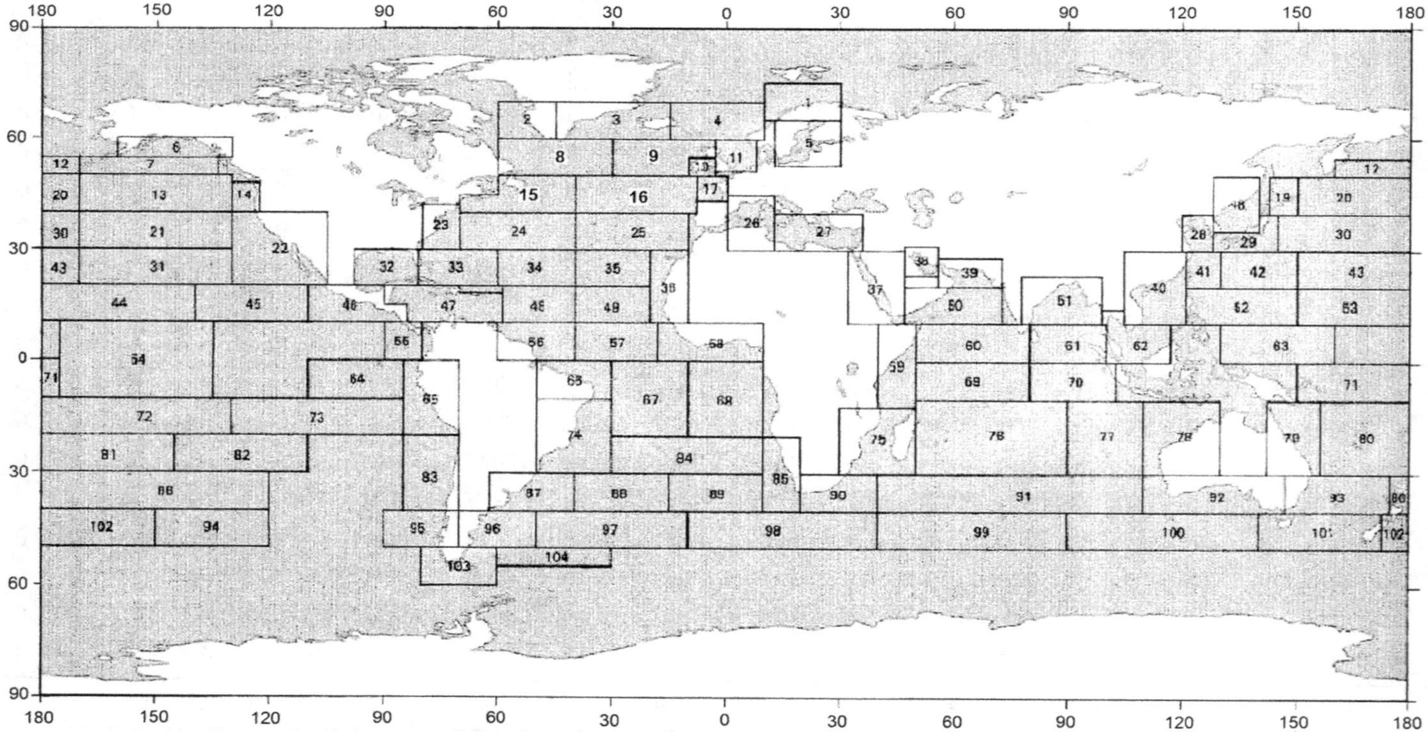

Figure 9.31 World ocean map subdivided into Marsden zones (the North Atlantic region consists of zones 8, 9, 15 and 16).

On the other hand, the probability that the significant wave height encountered in the North Atlantic exceeds 9.5 m is (285+124+51+21+8+3+1)/100,000 = 0.00493.

9.3.3 The Long-Term Cumulative Distribution (LTCD) Method

Long-term analysis was originally developed to analyse full-scale measurements obtained at sea on board ships over periods extending to three years. Figure 9.32 shows data for the oil tanker S.S. *Idemitsu Maru* obtained during a series of voyages in the North Atlantic (1961–1964).

These data have been plotted against Beaufort number (wind strength). The data obtained was analysed by calculating the root mean square value of the magnitude of peak-to-trough stress variation. As expected, as the Beaufort number increases, the average value of the root mean square stress range, denoted by an encircled cross in Figure 9.32, also increases. Also, for each weather condition, the data can be approximated by a normal distribution. Similar findings were obtained for other vessel types as well (containerships).

Data of the type mentioned in Section 9.3.1 was used to provide a basis for methods to predict response to long-term loading. The methods developed are:

1. The *extreme value* method, used for ultimate strength design [40]
2. The *long-term cumulative distribution* (LTCD) method, used for both ultimate strength and fatigue design [22]
3. The *simulation* method [43–45]

Of these, the extreme value method has been outlined in Section 9.2.1. In this section the long-term cumulative distribution (LTCD) method is described whereas the simulation method is described in the section that follows.

Basis of the LTCD Method

The LTCD method is based on the prediction of short-term probabilities that are then combined on the basis of assumed lifetime service profiles to determine long-term probabilities. In the case of a stationary Gaussian process, the statistical descriptors such as the mean value and the variance are calculated for all time and locations. In the sea environment this is not what occurs, because the spectra of ocean waves vary over time and distance so that the process is non-stationary. There is no probabilistic description available of the response of a ship to a non-stationary sea environment. This makes it necessary to make use of the assumption that within the vicinity of a ship conditions change gradually enough for it to be possible to model them as a series of stationary processes. These processes are of short duration and near term; as discussed in previous sections, the wave heights of a storm whose duration is or the order of three hours and involves approximately 1,000 waves is modelled using the Rayleigh distribution.

The first step involves the selection of suitable wave height spectra that cover the range of significant wave height (severity) and spectral shape.

The second step is to obtain the *RAO*s for the loads. This, as discussed previously, can be done using either model tests or theoretical means (strip theory programs). Because of the

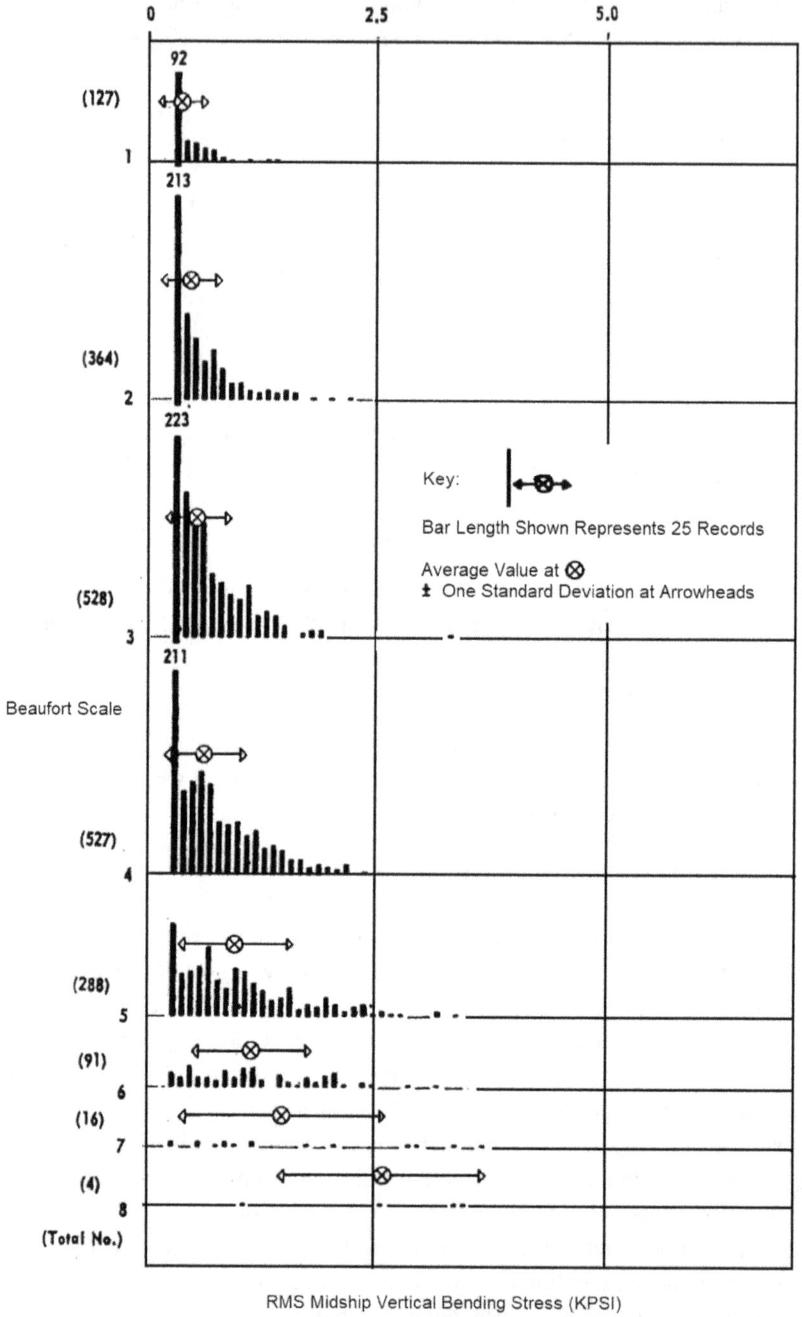

Figure 9.32 Distributions of root mean square values of peak to trough stresses grouped according to wind force (Beaufort Number) for S.S. *Idemitsu Maru* (adapted from [23]). Reprinted courtesy of the Ship Structure Committee, Washington, DC.

extensive testing required (number of wavelengths, headings, speeds and loading conditions), it is preferable in the case of conventional hull forms to use computational techniques that, however, require verification by testing.

Subsequent to this, the load response spectra are generated by superposition of the wave spectra. The spectra used may correspond to long-crested seas (regular wave pattern propagating in one direction) or short-crested seas (in which case the waves propagate over a range of directions). In the latter case, which is a more accurate representation of real conditions, wave energy is diffused, leading to lower bending moments. The spreading function, $f(\mu)$, given by Equation (9.34), is used to describe confused (short-crested) single-storm seas. In order to simplify the analysis, wave systems that come from more than one direction are normally neglected.

In addition to the sea environment, it is necessary to introduce specific parameters that describe ship operations and corresponding simplifying assumptions:

- Ship loading conditions (reduced to two conditions)
- Speed (assumed constant for each sea state considered)
- Heading with respect to prevailing sea conditions (also assumed constant)

Individual short-term distributions are thus characterised by the following parameters:

- Significant wave height, $H_{1/3}$
- Spectral shape, S
- Directional characteristics of a single wave system (or of a combination of storms and/or swell)

Response spectra are then obtained for ranges of values of the above parameters. The spectra are then integrated to obtain the root mean square values, σ, for each short-term response.

From the short-term load spectra it is necessary to obtain the long-term cumulative distribution. This is done by combining the short-term (Rayleigh) distributions, weighted by the frequency of occurrence of the different spectrum shapes, ship headings and significant wave heights. It is reasonable to assume that the ship, whilst operating in a particular short-term sea state, will have a given loading condition and will maintain a constant speed and heading for the duration of that sea condition. On this basis, the short-term functions are conditional probabilities, since each applies while the ship is operating in a particular steady sea condition.[9]

The variable under consideration, x, is the peak-to-mean amplitude of the bending moment. In a short-term analysis, the short-term load spectrum is assumed to be Rayleigh-distributed and is thus defined by its Rayleigh parameter, $\sigma^2(\equiv m_0)$. The value of m_0 in turn depends on significant wave height $H_{1/3}$, spectral shape S, and heading ψ with respect to prevailing sea conditions. These are assumed to be mutually independent. The long-term formulation is expressed as a joint probability of x and σ as follows [22]:

$$q(x, \sigma) = P[x|\sigma]P[\sigma] \tag{9.89}$$

[9] Appendix B.

where

$P[x|\sigma]$ = probability of occurrence of x for a given value of σ
$P[\sigma]$ = group of functions that define the variation of σ with combinations of sign, wave height $H_{1/3}$ spectral shape S and heading ψ

For a Rayleigh distribution,

$$P[x|\sigma] = \left(\frac{x}{\sigma^2}\right)e^{-x^2/2\sigma^2} \tag{9.90}$$

The density function of x is thus obtained by integrating over all possible sea states described by σ, where $0 < \sigma < \infty$:

$$q(x) = \int_0^\infty q(x, \sigma)d\sigma \tag{9.91}$$

We are interested in the probability of exceedance of a threshold value of x, say x_i, and thus we require the cumulative distribution function. This is obtained by integrating a second time:

$$Q(x > x_i) = \int_{x_i}^\infty \int_0^\infty q(x, \sigma)d\sigma dx = \int_{x_i}^\infty \int_0^\infty P[x|\sigma]p[\sigma]d\sigma dx \tag{9.92}$$

Substituting from Equation (9.90),

$$Q(x > x_i) = \int_0^\infty e^{-x_i^2/2\sigma^2} P[\sigma]d\sigma \tag{9.93}$$

where $\sigma = \sigma\left(H_{1/3}, S, \psi\right)$.

If the joint probability density function for these parameters is $P\left[H_{1/3}, S, \psi\right]$, then the fraction of time during which $H_{1/3}$, S, ψ simultaneously lie in the ranges $\left(H_{1/3}, H_{1/3} + dh\right)$, $(S, S + dS)$, $(\psi, \psi + d\psi)$, respectively, is given by $P\left[H_{1/3}, S, \psi\right]dHdSd\psi$.

If the three parameters are independent, so that

$$P\left[H_{1/3}, S, \psi\right] = P\left[H_{1/3}\right]P[S]P[\psi] \tag{9.94}$$

there arise three possible combinations of pairs of values of these parameters, and thus it is necessary to decide on how this procedure is to be implemented. However, it is not possible to determine a priori the manner in which σ varies with these parameters, and closed-form mathematical expressions for the probability density and the joint density functions for long-term response cannot be obtained.

Studies of full-scale measurements at sea have indicated that if data are grouped into different sea (or wind) groups, the Rayleigh parameters of the response within each group follow a truncated normal distribution. This is clear from the data included in Figure 9.32.

On this basis the function $p[c]$ can be replaced by the product of the probability density function of σ for each significant wave height, $p\left[H_{1/3}\right]$, and the probability of encounter of each wave group by the ship, $p^*[\sigma]$. This is normally distributed as noted in the full-scale measurements above, and σ in $p^*[\sigma]$ depends on S and ψ. Thus,

$$p[c] = p[H_{1/3}]p^*[\sigma] \tag{9.95}$$

Weather Group	Beaufort No.	Wind velocity (knots)
I	0–3	1–10
II	4–5	11–21
III	6–7	22–33
IV	8–9	34–47
V	10–12	48–71

The properties of the spectra used are described by Lewis and Zubaly [22] whereas equal probability of headings is assumed. Then, for the assumed normal distribution for σ in each weather condition and heading,

$$p^*[c] = 1/\sqrt{2\pi s^2} e^{-(\sigma-m)^2/2s^2} \tag{9.96}$$

where m and s are the mean value and standard deviation of each σ for each combination of S and ψ. Substituting Equation (9.96) in Equation (9.92) and performing the numerical integration for the selected combinations of $H_{1/3}$, S and ψ, the long-term cumulative distribution function is obtained:

$$Q(x > x_i) = \sum_{H_{1/3}} \sum_{S,\psi} e^{-x_i^2/2\sigma^2} p[H_{1/3}] \left(1/\sqrt{2\pi s^2}\right) e^{-(\sigma-m)^2/2s^2} \tag{9.97}$$

It is seen from Equation (9.97) that summation with respect to $H_{1/3}$ is carried out last so that a series of long-term curves for different wave groups are obtained first (Figure 9.33). From these the final long-term result can be presented graphically as shown in Figure 9.33, in this case for a 20-year operating life. The Weibull three-parameter distribution can be fitted to it although this is not necessary.

It is important to consider Figure 9.33 in a little more detail. Curves such as this are used to estimate the probability that a given bending moment (or stress) X_j is exceeded in one half-reversal. The probability of exceedance increases from 10^{-10} to 1.0, as the load level that is being considered is reduced. The value obtained for the probability of exceedance has to be interpreted in statistical terms, that is, it is the limit of the ratio of the total number of exceedances to the number of observations, when many readings are obtained [12]. The second scale along the abscissa gives the number of reversals n for which the ratio of the number of exceedances of X_j to the observed average number of exceedances equals unity. Of practical interest is the probability that actual failure will occur. In order to calculate this, the response of the structure needs to be properly modelled, including uncertainties relating to material properties, fabrication, strength calculation, etc. The ultimate strength of the hull girder is a property that is independent of the applied load and has to be represented in probabilistic terms.

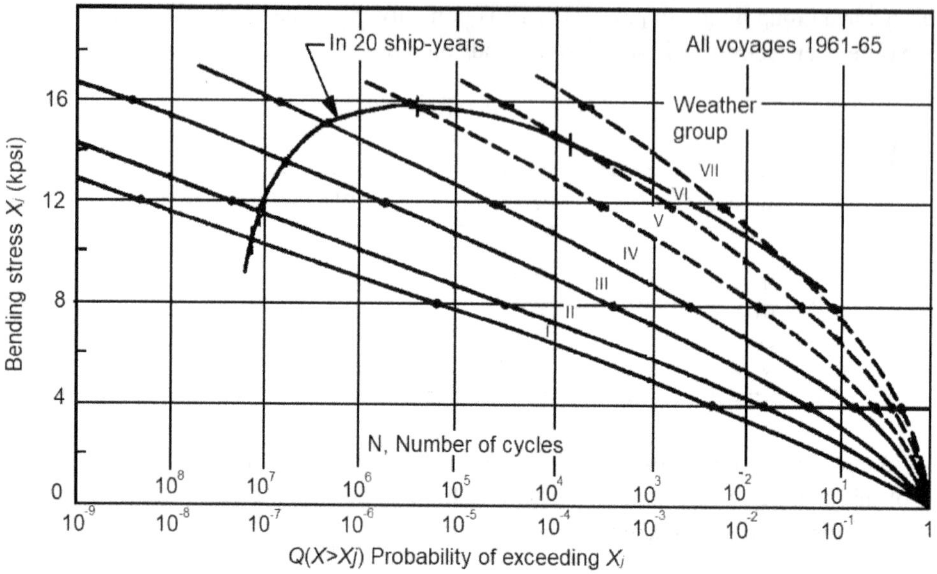

Weather Group	Beaufort No.	Wind velocity (knots)
I	0–3	1–10
II	4–5	11–21
III	6–7	22–33
IV	8–9	34–47
V	10–12	48–71

Figure 9.33 Probability that X_j will be exceeded in one half-cycle for different weather groups (SS *Wolverine State & Hoosier State*) [12]. Reprinted courtesy of the Ship Structure Committee, Washington, DC.

A Fleet of Ships

From the perspective of a regulatory body such as a classification society, it is necessary to consider a large number of ships for each of which the strength is modelled using an appropriate distribution. In each case the expected loading and the resulting probability of failure will differ.

On average, the bending moment value that will be exceeded once in a lifetime can be obtained directly from the cumulative distribution, once the number of cycles (duration) has been selected. When a large number of similar ships are considered, the lifetime exceedance value will be higher for some and lower for others. The probability that the average value will be exceeded is equal to 0.67, which is significant, and therefore this cannot be used as a basis for design.

For design purposes, it is necessary to calculate the *lifetime probability*. This is the probability that a given bending moment will be exceeded in any one or more of a large number of ships, say $\overline{N}$. In order to calculate this probability, let y be the number of ships of the total number being considered that experience at least one exceedance.[10] The value of y lies in the range $0 \leq y \leq \overline{N}$. The binomial distribution[11] can then be used to describe behaviour, since in each case either the load is exceeded at least once or else it is not exceeded [17]. If the probability of exceedance for one ship is P, the frequency function of y is

$$f(y) = \binom{\overline{N}}{y} P^y (1-P)^{\overline{N}-y} \qquad y = 0, 1, 2, \ldots, \overline{N} \tag{9.98}$$

For a large fleet in which a small number of failures arise (such that $\overline{N}P < 5$), the binomial function is approximated by the Poisson frequency function,

$$f(y) = \frac{e^{-\overline{N}P}\left(\overline{N}P\right)^y}{y!} \qquad y = 0, 1, 2, \ldots, \overline{N} \tag{9.99}$$

Thus the probability that exactly r of the $\overline{N}$ ships will experience an exceedance of X_j in their lifetimes is $f(r)$.

Single-Ship Case

We shall now consider the case of a single ship in order to determine a stress or bending moment that has a very low probability of exceedance. In principle this is modelled using the binomial distribution, for which the density function is given by

$$f(x) = \binom{n}{x} Q^x (1-Q)^{n-x}$$

where Q = probability of exceedance of X_j and n = number of half-oscillations in the lifetime of the ship ($=10^8$). Substituting,

$$f(x) = \binom{10^8}{x} Q^x (1-Q)^{10^8-x}$$

In this problem $Q \ll 1$, $n \gg 1$ and $nQ < 5$. For these figures it is possible to use the Poisson density function to approximate the binomial model. The Poisson density function is a one-parameter discrete function:

$$f(x) = \frac{m^x e^{-m}}{x!}$$

It can be shown that m is both the mean and the variance, and since the Poisson density function approximates the binomial density function,

$$\overline{N} = nQ \tag{9.100}$$

Hence,

$$f(x) = \frac{(nQ)^x e^{-nQ}}{x!} \tag{9.101}$$

is the density function for x, where x is the number of exceedances of a given X_j.

[10] The ships must all be of the same type so that the cumulative probability curve is valid for all of them.
[11] The *binomial distribution* is the probability distribution of the number of successes in a sequence of n independent yes/no experiments, each of which yields success with probability p.

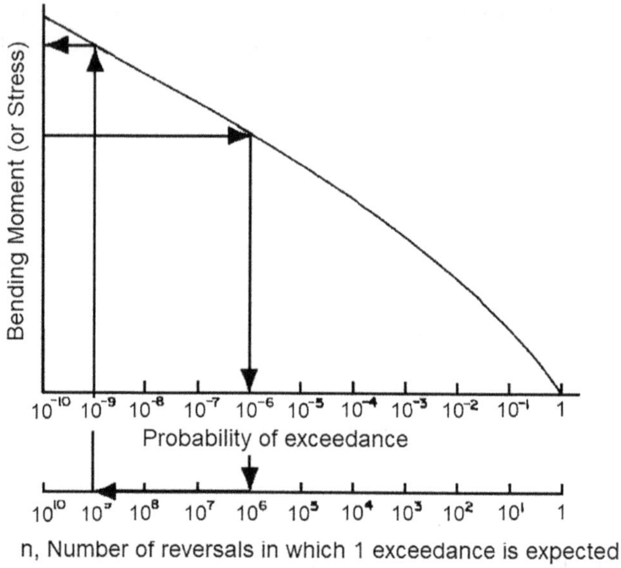

Figure 9.34 Probability of exceedance of bending moment (or stress) ($Q(X > X_j)$) during one reversal [12]. Reprinted courtesy of the Ship Structure Committee, Washington, DC.

As an example of use of this result, assume that during its lifetime a ship is subjected to $n = 10^8$ stress reversals. Equation (9.101) can then be used to calculate the stress X_j such that the probability that it will not be exceeded equals 0.99. In this case, $f(0) = 0.99$, since $x = 0$ for no exceedances. Hence,

$$f(0) = \frac{(nQ)^0 e^{-nQ}}{0!} = 0.99$$

Substituting for n the given value, we obtain $e^{-10^8 Q} = 0.99$. Hence, $Q = 10^{-10}$. The meaning of this is that if a ship is designed using this bending moment value (X_j) as reference, the probability that this bending moment will not be exceeded during the lifetime of the ship is equal to 0.99. In the general case,

$$e^{-nQ} = 0.99$$

Substituting $n = 10^8$ and using tables for exponential functions, a value of $Q = 10^{-10}$ is obtained. This value is then entered in a probability curve such as Figure 9.34, and the value of the bending moment or stress (X_j) is read off on the left vertical scale. If we follow the above procedure for a general n, we have

$$e^{-nQ} = 0.99$$

Hence

$$Q = \frac{10^{-2}}{n}$$

Consequently the probability of exceedance diagram can be used to determine the critical load (bending moment or stress) for a specified number of reversals n by entering the diagram along the reciprocal scale along the abscissa, translating by an appropriate amount to the left (two units for a probability of 0.01) and then reading off the critical load from the curve (Figure 9.34). This means that for a lifetime of $n = 10^8$ cycles, the stress that one expects to be exceeded just once is the value that corresponds to $n = 10^8$.

9.3.4 The Simulation Method

This section describes a method proposed in [43]–[45] to perform long-term analysis that is capable of allowing for the following factors:

- Sea areas
- Seasons
- Short-crestedness of waves
- Type of ship
- Speed and heading of ship
- Loading condition
- Route
- Spectral density bandwidth (input and output)

It involves the simultaneous simulation of multiple realisations of individual load time series and is rendered practical by a significant reduction in the computational effort required when compared to time-series analysis. This is achieved through the application of pertinent statistical theory and techniques

The output consists of

- Long-term CDFs of individual load peaks
- Long-term CDFs of combined load peaks
- Long-term correlation coefficients
- Nonlinear load combinations

The data required to represent the conditions to which a ship may be exposed to during its lifetime is such that it is not possible for it to be fully processed and analysed. In the simulation method, the effects listed above are allowed for in a single-voyage simulation, thereby rendering the analysis practicable. The procedure followed requires that peak loads and load combinations are introduced in a suitable manner. The development of the simulation method is described in the following section and is summarised in the flow chart of Figure 9.40.

Stages in the Simulation Method

1. Simulation of wave loads. Global Wave Statistics (GWS) data are used for the regions in which each ship navigates, and routes are then plotted using appropriate software. Wave data for each region are given in the form of scatter diagrams that represent normalised

observations of each combination of H_s and T_0 at discrete time intervals. Data for each season of the year are introduced corresponding to eight different wind/wave directions.

2. The conditional JPDF of wave height and period, $f(H_s, \overline{T_0})$, is obtained and fitted to each scatter diagram. It has to be noted that sea severity depends on geographic location as well as on the stage of growth and decay of a storm. As a result, there is no scientific basis for selecting a particular probability distribution to represent the statistical properties of a sea state [35]. For present purposes, the first task is to select an appropriate distribution function for the significant wave height, H_s. For a certain ocean area (A), season (S) and wave direction (β), Ochi [40] has proposed that the conditional distribution of the significant wave height, $f(H_S)$, given a certain ocean area, season and wave direction, can be represented by the *generalised gamma distribution*, whose probability density function is given by

$$f(H_S) = \frac{c}{\Gamma(m)} \lambda^{cm} H_s^{cm-1} \exp\{-(\lambda H_s)^c\} \tag{9.102}$$

where c, m and λ are distribution parameters that are calculated from scatter diagram data.

The conditional distribution of period, T_0, given H_S, for specific A, S and β, can be represented by the conditional log-normal distribution:

$$f(\overline{T_0}|H_S) = \frac{1}{\sqrt{2\pi}\sigma(H_S)\overline{T_0}} \exp\left\{-\frac{\left[\ln \overline{T_0} - \mu(H_S)\right]^2}{2\{\sigma(H_S)\}^2}\right\} \tag{9.103}$$

Because the JPDF cannot be derived directly from data, the method used is to find the marginal probability density function $f(H_s)$ and the conditional probability density function $f(\overline{T_0}|H_s)$. The parameters involved in the latter distribution function are evaluated from analysis of each conditional distribution for $\overline{T_0}$ for a specified H_s, and presented as a function of H_s. Then the JPDF can be represented as the product of these, as follows:

$$f(H_S, \overline{T_0}) = f(H_s) f(\overline{T_0}|H_s) =$$

$$= \frac{c}{\Gamma(m)} \lambda^{cm} H_s^{cm-1} \exp\{-(\lambda H_s)^c\} \frac{1}{\sqrt{2\pi}\sigma(H_s)\overline{T_0}} \exp\left\{-\frac{\left[\ln \overline{T_0} - \mu(H_s)\right]^2}{2\{\sigma(H_s)\}^2}\right\} \tag{9.104}$$

Expressions for the parameters $\mu(H_s)$ and $\sigma(H_s)$ of the conditional log-normal distribution were obtained by Mathiesen and Bitner-Gregersen [34] as follows:

$$\mu(H_S) = a_1 + a_2 H_S^{a_3} \tag{9.105a}$$

$$\sigma(H_S) = b_1 + b_2 \exp\{b_3 H_S\} \tag{9.105b}$$

where a_1, a_2, a_3, b_1, b_2 and b_3 are constants determined from scatter data.

The JPDF of wave height and period is thus specified by a total of nine (unknown) constants. These are calculated by fitting the JPDF to the scatter data using the *maximum likelihood estimation* (*MLE*)[12] that can be used, provided the JPDF has the form given in Equation (9.104). The *likelihood function* $\mathcal{L}(\theta)$ required is

$$\mathcal{L}(\theta) = \prod_{\substack{H_s,\, T_s combinations \\ in scatter diagrams}} f_{H_s,T_0|A,S,\beta}(h_s, t_0|\alpha, s, \beta; \theta) \qquad \theta \in \Theta \tag{9.106}$$

where $\theta = [c, m, \lambda, a_1, a_2, a_3, b_1, b_2, b_3]$ is the vector of unknown parameters and Θ is the parameter space $(c, m, \lambda, \sigma(H_S) > 0)$. $\mathcal{L}(\theta)$ is expected to be larger for values of θ near the ones that generated the data so that the parameters are found by maximising $\mathcal{L}(\theta)$. The parameters were obtained for eight different wind/wave directions for each area and season.

3. The explicit dependence of the conditional JPDF in Equation (9.104) on wind/wave direction, β, is obtained by fitting to each parameter a piecewise cubic Hermite polynomial.
4. The conditional JPDF of H_s, T_0 and β for a given area and season is then

$$f\left(H_S, \overline{T_0}, \beta|A, S\right) = f\left(H_S, \overline{T_0}|A, S, \beta\right) f(\beta|A, S) \tag{9.107}$$

The conditional probability of wind/wave direction, β, given area and season (*A*,*S*), *f*(β| *A*,*S*), was estimated from wind direction data in Global Wave Statistics.

At this stage the modelling of the sea environment is complete and it becomes feasible to reduce the extent of the data for computational purposes.

For a given area and season, a triplet (H_s, T_0, β) is sampled from the conditional joint probability density function using the *rejection sampling technique*.[13] Each sample is assigned the probability that is proportional to the value of the conditional JPDF in Equation (9.107) for that sample. These values have been sampled based on the current position and the randomly determined season (Figure 9.35). Equal probability is assigned to all four seasons. The black crosses in Figure 9.35 are acceptable sample points, since they lie within the bounds of the joint probability density function for (H_s, T_0) whereas points that lie beyond the bounds are rejected.

The relative heading α of the ship to the wind/wave direction, β, is given by $\alpha = \gamma - \beta$ (Figure 9.36). Ship's speed is dependent on wave height and relative wave direction, for each ship type, and is obtained from stored data. In this way the long-term operation profile is modelled (simulated) using data representative of the conditions the ship is expected to encounter.

Load Combinations in Long-term Analysis

Up to this point we have considered peak loading conditions. A closely related question is that of *extreme load combinations*, that is, the magnitude of a second load (or possibly a

[12] The *MLE* method is described in statistics textbooks such as [8]. For ease of reference, details are included in Appendix F. The method was implemented in this case using a standard computer-based routine (*MATLAB*).

[13] The purpose of the rejection sampling technique is to create a population of sample points for a particular probability density function for which data generation is difficult or not possible.

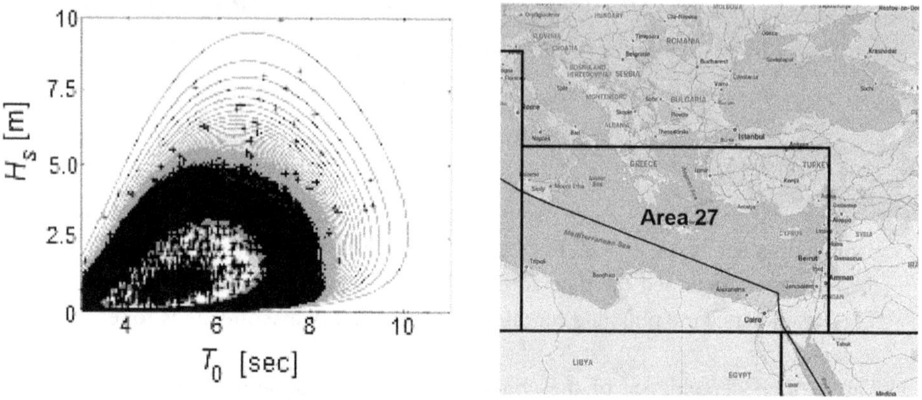

Figure 9.35 Joint probability density function for (H_s, T_0) for Area 27 (Europe–Asia route) in winter and 1,000-sample sea states superimposed (black crosses) [43]. Reprinted courtesy of Elsevier.

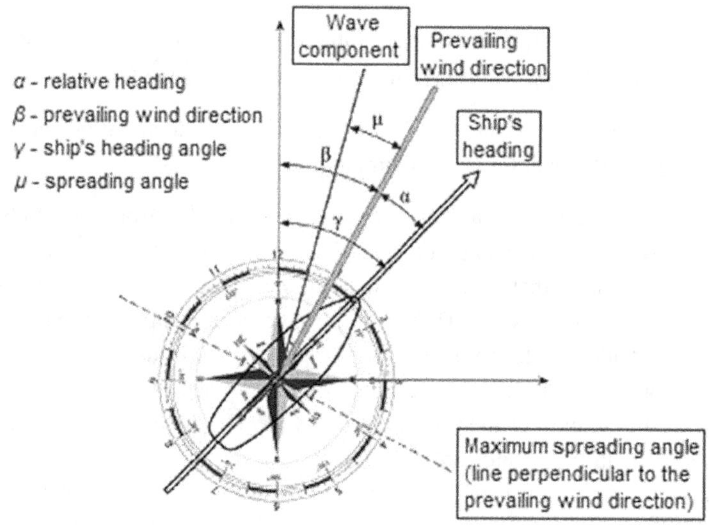

Figure 9.36 Ship's heading and related parameters [43]. Reprinted courtesy of Elsevier.

third one as well) when one particular load is at its peak value. In statistical terms this requires the determination of correlation coefficients of peak loads. In the case of short-term analysis, solutions are available and are presented in Section 9.2. It is, however, necessary to consider long-term analysis in the light of the differences between the two in order to develop a complete model of long-term loading of the hull girder (the methods and approaches described in Sections 9.3.2 and 9.3.3 are approximate).

In order to be able to make use of the reduced data sample so that the lifetime profile is simulated in a single voyage, it is necessary to use appropriate correlation coefficients. The following formula, drawn from standard statistical analysis, gives the

(Pearson) *sample correlation coefficient* r between two individual random variables X, Y (Appendix F):

$$r = \frac{\sum_{i=1}^{n} (X_i - \overline{X})(Y_i - \overline{Y})}{\sqrt{\sum_{j=1}^{n} (X_j - \overline{X})^2} \sqrt{\sum_{k=1}^{n} (Y_k - \overline{Y})^2}} \tag{9.108}$$

where X_i, Y_i are loads, $\overline{X, Y}$ are their corresponding mean values and n is the number of simulated load values. This expression thus gives an unbiased sample estimate of the true correlation coefficient for a single voyage. It is required, however, that the correlation coefficients using data for a single voyage approach the true correlation coefficient ρ that holds for the lifetime of the ship, which extends to 25 years.

The proof that the unbiased estimates of the correlation coefficient $R_{1voy}, R_{2voy}, \ldots, R_{1year}, R_{2year}, \ldots, R_{25year}$ all approach the true correlation coefficient ρ in probability as n increases is given in [44].[14] The proof relies on the fact that non-stationary time series of individual loads can be transformed to stationary series by randomly permuting the simulated load values within each time series. If the load values x_i remains paired with y_i, then the correlation estimate remains unchanged.

This is possible by making reference to *Slutsky's lemma,* which is a result of probability theory on the modes of convergence of random variables.[15] It states that if $X_n \xrightarrow{D} X$ (that is, if X_n converges to X *in probability*) and $Y_n \xrightarrow{D} y_0$ (that is, if Y_n, a second random variable converges in probability to y_0, a constant), then

$$\text{(a)} \ X_n + Y_n \xrightarrow{D} X + y_0 \tag{9.109a}$$

$$\text{(b)} \ X_n Y_n \xrightarrow{D} X y_o \tag{9.109b}$$

That is, the sum (a) and product (b) of the two random variables converge to the random variable that is the sum (product) of the two original variables.

It should be borne in mind that even for a single voyage the number of simulated load values is large (of the order of 10^6). This number is sufficient to ensure convergence of the unbiased sample correlation coefficient r to the long-term correlation coefficient ρ.

Ship Response

The next stage of the analysis concerns the definition of ship structural response. The hull girder is treated as a rigid body, and all six low-frequency wave-induced loads are calculated (one force and one moment about each axis). The wave environment is assumed to be modelled by the ISSC spectrum, and short-crestedness is allowed for through the use

[14] Modes of convergence of random variables are discussed briefly in Appendix C. The proof of stationarity of the correlation coefficient is also included.

[15] Slutsky's lemma is presented in statistics and probability texts such as [8] and [3]. A proof is given in [3].

of the spreading function (Equation (9.48)). The response amplitude operators (load transfer functions) are calculated using a suitable strip theory program.[16] The output spectrum for each load is then found from

$$S_y(\omega_e) = \int_{-\pi/2}^{+\pi/2} S_y(\omega_e, \mu)d\mu = \int_{-\pi/2}^{+\pi/2} \left|H_y(\omega_e, \alpha - \mu)\right|^2 S_x(\omega_e, \mu)d\mu = \\ = S_x(\omega_e) \int_{-\pi/2}^{+\pi/2} \frac{2}{\pi} \cos^2\mu \left|H_y(\omega_e, \alpha - \mu)\right|^2 d\mu \tag{9.110}$$

where

$x, y =$ indices denoting input (wave elevation) and output (force or moment)
$\omega_e =$ wave encounter frequency (Equation (9.50))
$S(\omega_e) =$ one-sided encounter spectral density
$\mu, \alpha =$ angles indicated in Figure 9.36

The integration in Equation (9.110) is performed numerically. At this point the *spectral representation theorem* is invoked, according to which any stationary time series of a random process can be treated as a random superposition of independent sine and cosines that oscillate at different frequencies and have amplitudes whose magnitudes are proportional to the corresponding variances. In turn, these are calculated using the spectral density of the random process. The time record of the wave elevation obtained on board a ship can thus be expressed as a Fourier integral, and according to the central limit theorem, it is approximately a Gaussian stochastic process. It can be approximated by a sum consisting of a finite number of independent periodic components as follows:

$$x(t) = \int_{\omega_{e\min}}^{\omega_{e\max}} X(\omega_e)\cos(\omega_e t + \phi)d\omega_e \approx \sum_{i=1}^{N} X_i \cos(\omega_{e,i}t + \phi_i) \tag{9.111}$$

where the summation is performed over N equally spaced encounter frequencies in the range $\omega_{emin} < \omega_e < \omega_{emax}$, X_i is the amplitude of the component oscillating at frequency f, and ω is the random phase angle. Since the time average of the wave (potential) energy for a range of encounter frequencies, $\Delta\omega_e$, is

$$S_x(\omega_{e,i})\Delta\omega_{e,i} = \frac{1}{2}\rho g X_i^2 \tag{9.112}$$

the amplitude of a periodic component is given by (ignoring ρg)

$$X_i = \sqrt{2S_x(\omega_{e,i})\Delta\omega_{e,i}} \tag{9.113}$$

[16] The frequency-domain modified strip theory program *SEAWAY* was used, that allows for forward speed effects (Chapter 4).

where

$$\Delta\omega_e = (\omega_{e\max} - \omega_{e\min})/N \tag{9.114}$$

Hence, the simulated output time series can be obtained from

$$T(t) = \sum_{i=1}^{N} \sqrt{2S_T(\omega_{e,i})\Delta\omega_{e,i}}\cos(\omega_{e,i}t + \phi_i + \phi_T) \tag{9.115}$$

The magnitudes of the rigid body loads can then be obtained using the general formulation (Equation (9.110)), where the input is wave height $x(t)$ as given in Equation (9.111) and the appropriate RAO is introduced for each load. Prior to evaluating individual loads, it is necessary to consider the effect that a linear filter has on each parameter (frequency, amplitude and phase) for each load. For a linear system, the frequency of the output of a periodic component is the same as that of the input. The phase angle of the output, however, will be modified by the phase part of the RAO, H, and the amplitude will be changed by a factor $|H|$.

In the case of the vertical bending moment, *VBM*, the magnitude is given by

$$VBM(t) = \sum_{i=1}^{N} \sqrt{2S_{VBM}(\omega_{e,i})\Delta\omega_{e,i}}\cos\left(\omega_{e,i}t + \phi_i + \phi_{VBM,i}\right) \tag{9.116}$$

Analogous expressions are obtained for longitudinal, shear and horizontal forces, horizontal bending and torsion, where appropriate values of spectral density $S_x(\omega_e, i)$ and phase angle ϕ are used.

Results

An important question that arises in connection with the simulation method concerns the length of load time series required to achieve convergence of the long-term correlation coefficients to the true value in probability, as the length of the series is increased. A measure of convergence is the COV, and it was found that this reduces as the length of the series is increased and that it converges fast to the true coefficient (50 consecutive simulations of one voyage were found to be required).[17]

A single voyage (Figure 9.37) involves 130 individual short time series, thus a total of 6,500 sea-state operational profiles were required for convergence. The total number of profiles, taking into account different heading angles, season area, speed and loading condition, is of the order of 10^8, but the use of rejection sampling reduces this to 6,500.

A comparison of the probability of exceedance of VBM peaks in the sagging condition of the results of the simulation method with linear and nonlinear theory [16] that allows for slamming is shown in Figure 9.38. It is evident that the results lie between those of linear and nonlinear theory.

Results from comparative studies are included in Figures 9.39a–d. Figure 9.39a shows the stress induced by the combined action of vertical and horizontal bending moments. This is obtained by adding Equation (9.118) to the corresponding one for the horizontal bending

[17] Table 2 in [44].

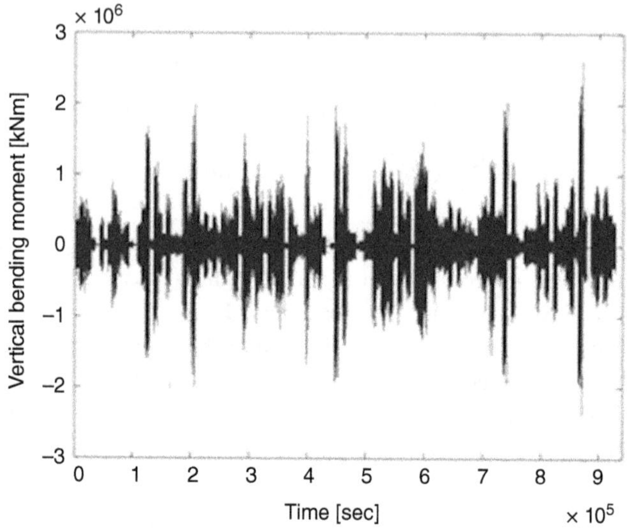

Figure 9.37 Time series of VBM for a single voyage (containership on North Atlantic route) [43]. Reprinted courtesy of Elsevier.

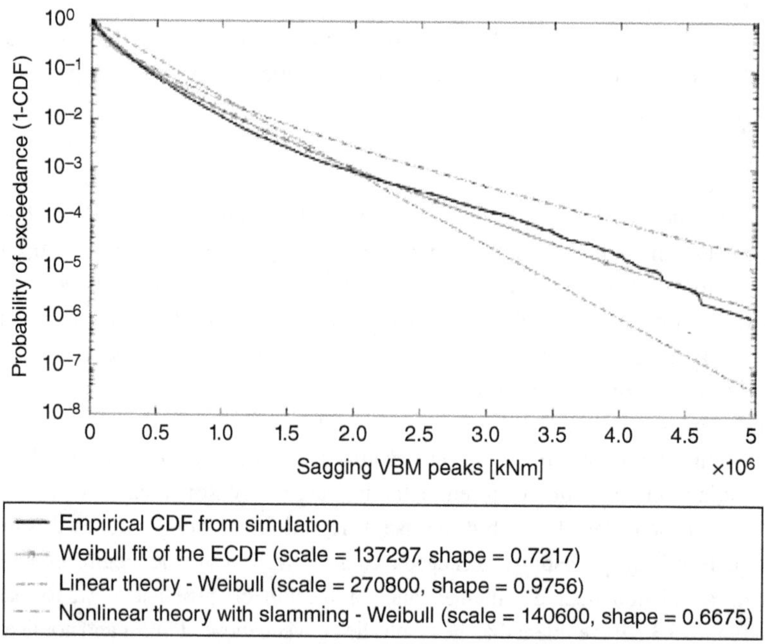

Figure 9.38 Comparison of the simulation method with linear and nonlinear theory [44].

moment, that is, by the addition of load time series [44]. In this way long-term exceedance probabilities can be obtained for combined stress peaks.

Figure 9.39b shows the effect that sea states encountered in different regions of the world have on the probability of exceedance of sagging vertical bending moments. It is

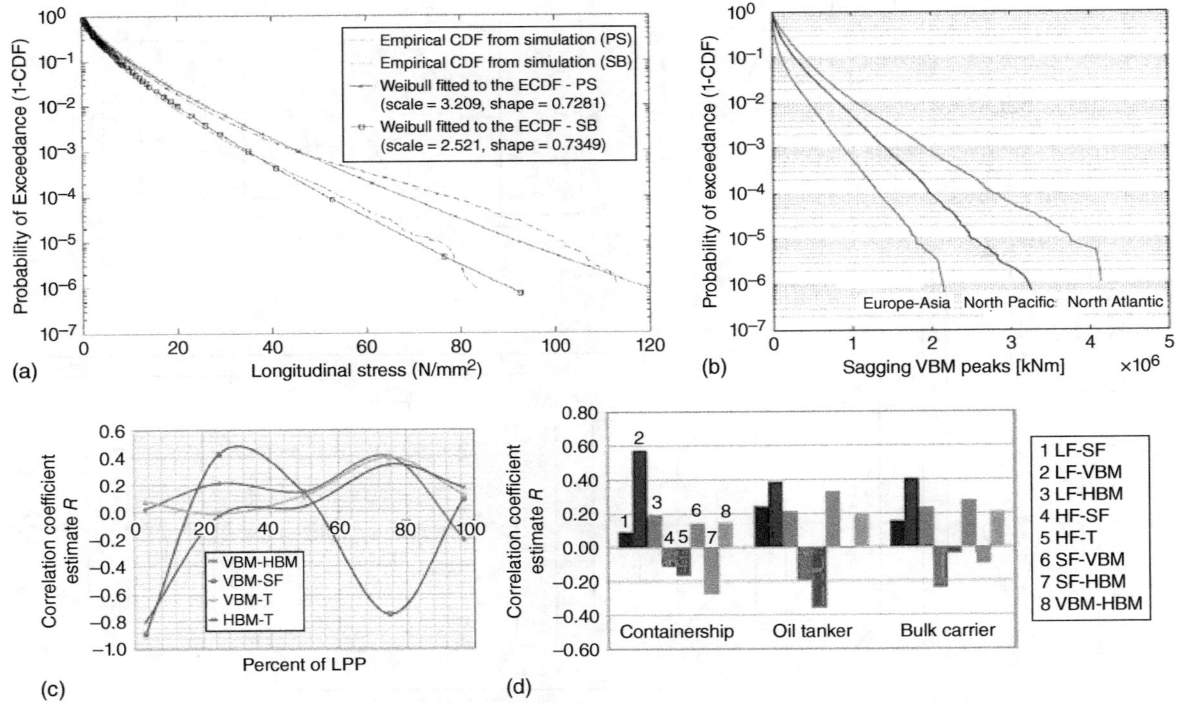

Figure 9.39 The effect of various parameters on long-term exceedance probability [43]. (a) Combined stress in sheer strake due to VBM and HBM. (b) Route effects. (c) Longitudinal position effect. (d) Correlation coefficients R for different ship types. Reprinted courtesy of Elsevier.

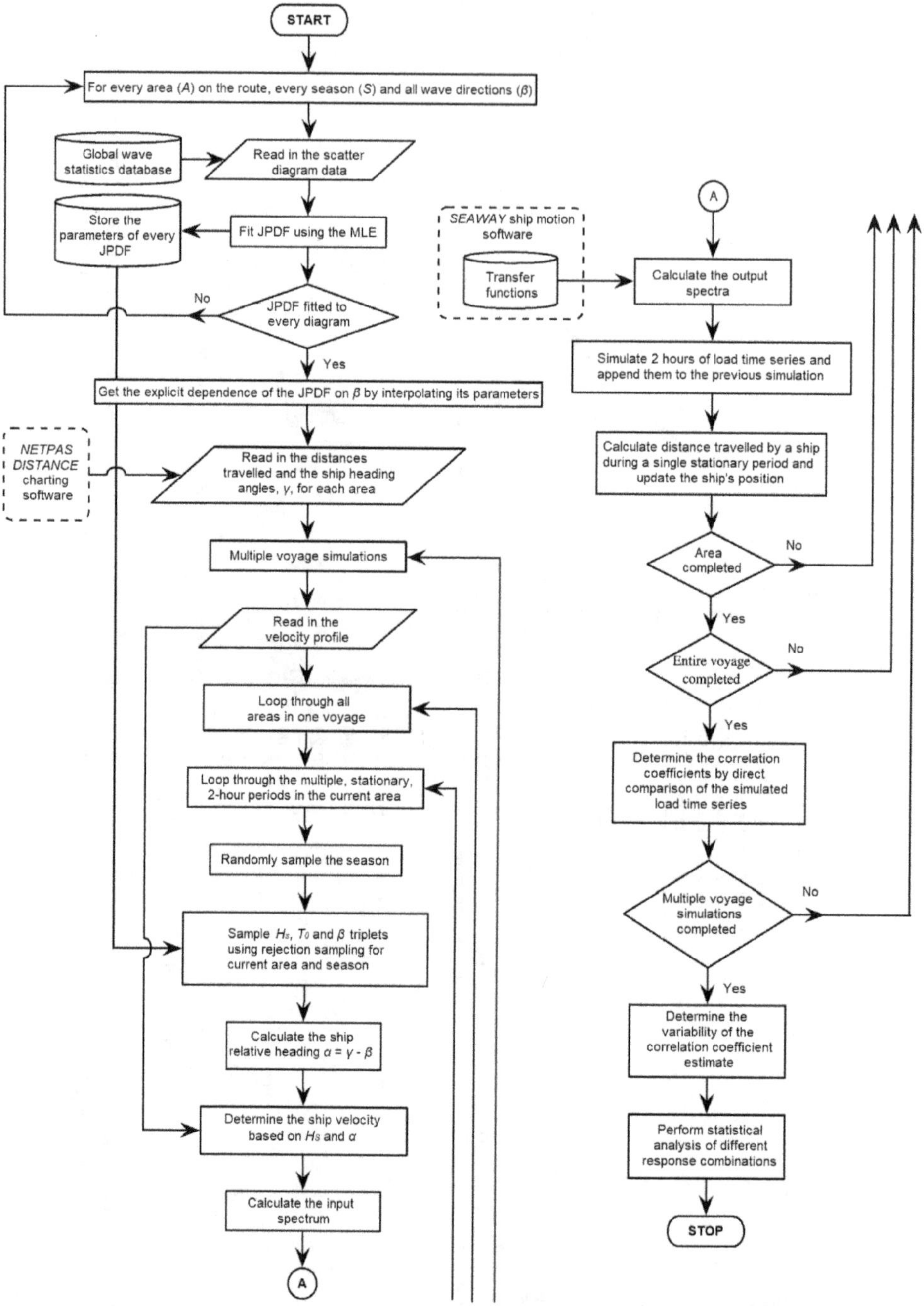

Figure 9.40 Flow chart for the simulation method [43]. Reprinted courtesy of Elsevier.

seen that a significant effect is observed, as the probability of exceedance may increase by several orders-of-magnitude as a result of a sea state of a particular region, especially for large loads. It is therefore important to allow for the range of operation that individual ships are designed for. Figure 9.39c illustrates the effect that position along the length of the hull girder may have on individual loads. Correlation matrices for pairs of loads indicate correlation changes with position (Figure 9.39d). In particular, it was found that at the quarter-length positions at which the vertical shear forces are at their peak values, the correlation between the vertical bending moment and the shearing force becomes very high.

References

1. American Bureau of Shipping (2018) *Rules for Building and Classing Steel Vessels*.
2. Bhattacharyya R. (1978) *Dynamics of Marine Vehicles*. New York: John Wiley & Sons.
3. Bickel P. J. Doksum K. A. (2001) *Mathematical Statistics: Basic Ideas and Selected Topics*. 2nd ed. Vol. 1. Upper Saddle River, NJ: Prentice Hall.
4. Bretschneider C. L. (1964) Generation of Waves by Wind: State of the Art. NESCO Report SN-134-6.
5. Burnside O. H. Hudak Jr. S. J. Oelkers E. Chan K. Dexter R. J. (1984) Long-term corrosion fatigue of welded marine steels. Ship Str. Comm. Rep. SSC-326, Washington, DC.
6. Cartwright D. E. Longuet-Higgins M. S. (1956) The statistical distribution of the maxima of a random function. *Proc Roy. Soc. London. Ser. A Math. Phys. Sc.* Vol. 237 No. 1209 212–232.
7. Crandall S. H. Mark W. D. (1963) *Random Vibration in Mechanical Systems*. 2nd ed. New York: Academic Press.
8. Davison A. C. (2011) *Statistical Models*. Cambridge: Cambridge University Press.
9. Drummen I. Wu M.-K. Moan T. (2009) Experimental and numerical study of containership responses in severe head seas, *Mar. Struct.* Vol. 22 172–193.
10. Fisher R. A. (1930) *The Genetical Theory of Natural Selection*. Oxford: Oxford University Press.
11. Hallam M. G. Heaf N. J. Wootton L. R. (1978) *Dynamics of Marine Structures: Methods of Calculating the Dynamic Response of Fixed Structures Subject to Wave and Current Action*. Rept. UR 8 2nd ed. London: CIRIA Underwater Eng. Group.
12. Hoffman D. Lewis E. V. (1969) Analysis and Interpretation of Full-Scale Data on Midship Bending Stresses of Dry Cargo Ships. Ship Str. Comm. Rep. SSC-196 Washington, DC.
13. Hogben N. da Cunha N. M. C. Olliver G. F (1986) *Global Wave Statistics* Old Woking: Unwin Brothers Ltd.
14. International Association of Classification Societies (2001) Recommendation No. 34 Standard Wave Data.
15. Jan H. Y. Chang K. T. Wojnarowski M. (1978) Comparison of Stresses Calculated Using the Daisy System to those Measured on the SL-7 Containership Program. Ship Struct. Comm. Rept SSC-282 Washington, DC.
16. Jensen J. J. Pedersen P. T. Shi B. Wang S. Petricić M. Mansour A. E. (2008) Wave Induced Extreme Hull Girder Loads on Containerships. *SNAME Ann. Mtg* 128–152.

17. Karst O. J. (1969) Statistical Techniques. Appendix A of Hoffman D. Lewis E. V. Analysis and Interpretation of Full-Scale Data on Midship Bending Stresses of Dry Cargo Ships. Ship Str. Comm. Rep. SSC-196 Washington, DC.
18. Kinsman B. (2012) *Wind Waves: Their Generation and Propagation on the Ocean Surface*. New York: Dover Publications.
19. Kitaigorodskii S. A. (1961) Applications of the theory of similarity to the analysis of wind-generated wave motion as a stochastic process. *Izv. Geophys. Ser. Acad. Sci. USSR*. I, 105–117.
20. Krogstad H. Arntsen Ø. A. (2000) Linear Wave Theory Part A. Norw. Univ. Sc. & Techn. Trondheim.
21. Lewis E. V. (1967) Predicting Long-Term Distributions of Wave-Induced Bending Moments on Ship Hulls. *Proc. SNAME Spring Meeting*.
22. Lewis E. V. Zubaly R. B. (1981) Predicting Hull Bending Moments for Design. *Proc. SSC/SNAME Extreme Loads Response Symposium.* Arlington, VA 31–62.
23. Little R. S. Lewis E. V. (1971) A Statistical Study of Wave-induced Bending Moments on Large Ocean-going Tankers and Bulk Carriers. *Trans SNAME* Vol. 79 117–168.
24. Lloyd A. R. J. M. (1989) *Seakeeping: Ship Behaviour in Rough Weather*. Chichester: Ellis Horwood.
25. Longuet-Higgins M. S. (1952). On the statistical distribution of the height of sea waves. *J. Mar. Res.* Vol. 11 245–266.
26. Longuet-Higgins M. S. (1969) A nonlinear mechanism for the generation of sea waves. *Proc Roy. Soc.* Vol. A311 371–389.
27. Mansour A. E. (1972) Methods of computing the probability of failure under extreme values of bending moment. *J. Ship Res.* Vol. 16 113–123.
28. Mansour A. E. (1981) Combining Extreme Environmental Loads for Reliability-Based Designs. *Proceedings SSC/SNAME Extreme Loads Response Symposium*, Arlington, VA 63–74.
29. Mansour A. E. (1990) *An Introduction to Structural Reliability Theory*. Ship Struct. Comm. Rep. SSC-351, Washington, DC.
30. Mansour A. E. (1995) Extreme loads and load combinations. *J. Ship Res.* Vol. 39 53–61.
31. Mansour A. Liu D. (2008) Strength of Ships and Ocean Structures, Paulling J. R. (Ed) *Principles of Naval Architecture Series*, Jersey City, NJ: SNAME.
32. Mansour A. Thayamballi A. (1993) Probability-based Ship Design. Loads and Load Combinations. Ship Str. Comm. Rep. SSC-373, Washington, DC.
33. Massel S. R. (1996) *Ocean Surface Waves: Their Physics and Prediction*. Singapore: World Scientific.
34. Mathiesen J. Bitner-Gregersen E. (1990) Joint distribution for significant wave height and zero-crossing period. *Appl. Ocean Res.* Vol. 12 93–103.
35. McCormick M. E. (2010) *Ocean Engineering Mechanics with Applications*. Cambridge: Cambridge University Press.
36. Miles J. W. (1957) On the generation of surface waves by shear flow. *Jnl Fluid Mech.* Vol. 3 185–204.
37. Miles J. W. (1960) On the generation of surface waves by turbulent shear flow. *Jnl Fluid Mech.* Vol. 7 469–478.
38. Miller N. S. (1975) *Ocean Engineering*. Lecture Notes, Dept of Naval Arch & Ocean Eng. Univ Glasgow, Scotland.
39. Mitsuyasu H. (2002) A historical note on the study of ocean waves. *J. Oceanography* Vol. 58 109–120.

40. Ochi M. E. (1999) *Ocean Waves: The Stochastic Approach*. Cambridge: Cambridge University Press.
41. Ochi M. E. (1981) Principles of Extreme Value Statistics and their Application. *SSC/SNAME Extreme Loads Response Symposium*, Arlington, VA 15–30.
42. Papoulis A. (1990) *Probability & Statistics*. Englewood Cliffs, NJ: Prentice-Hall International.
43. Petricić M. Mansour A. E. (2011) Long-term correlation structure of wave loads using simulation. *Mar. Struct.* Vol. 24 97–116.
44. Petricić M. Mansour A. E. (2010) Estimation of the long-term correlation coefficients by simulation. Proc. ASME 2010 29th Int. Conf. Ocean Offshore & Arctic Eng. *OMAE2010* 6–11 June 2010, Shanghai 1–11.
45. Petricić M. Mansour A. E. (2011) Extreme Loads and Load Combinations. Keynote Address, *IX HSMV Conference* Naples 1–40.
46. Pierson W. J. Moskowitz L. (1964) A proposed spectral form for fully-developed wind seas based on the similarity theory of S.A. Kitaigorodskii. *J. Geophysical Res*. Vol. 69 No. 24 5181–5190.
47. Pierson W. J. Neumann G. Janes R. W. (1958) Practical methods for observing and forecasting ocean waves by means of wave spectra and statistics. Hydrogr. Off. Publ. No. 603, U.S. Navy Hydrogr. Off. Washington, DC.
48. Phillips O. M. (1966) *Dynamics of the Upper Ocean*. Cambridge: Cambridge University Press.
49. Price W. G. Bishop R. E. D. (1974) *Probabilistic Theory of Ship Dynamics*. London: Chapman Hall.
50. Rice S. O. (1954) Mathematical Analysis of Random Noise, *Bell Sys. Tech. J.* Vol. 23 1944; Vol. 24, 1945. Reprinted in Wax N (Ed): *Selected Papers on Noise and Stochastic Processes*, Dover.
51. St. Denis M. Pierson W. J. (1953) On the Motion of Ships in Confused Seas. *SNAME Annual Meeting*, 281–357.
52. Ucar.edu. [Accessed 8 January 2021]. www.isse.ucar.edu/staff/katz/docs/pdf/ncdckatz.pdf
53. Wikipedia contributors. Extreme value theory. Wikipedia, The Free Encyclopedia. 31 December 2020. [Accessed 8 January 2021]. https://en.wikipedia.org/w/index.php?title=Extreme_value_theory&oldid=997338933
54. Wikipedia contributors. Fisher–Tippett–Gnedenko theorem. Wikipedia, The Free Encyclopedia. 17 December 2020. [Accessed 8 January 2021]. https://en.wikipedia.org/w/index.php?title=Fisher%E2%80%93Tippett%E2%80%93Gnedenko_theorem&oldid=994787391

10

Design of Hull Girder for Strength

> [D]esign synthesis has suffered from altogether too gross neglect; it has rarely been treated in professional papers. This is unfortunate, for the synthetic work is far more important than the analytic – a point that may be argued by analogy. In music a complete knowledge of harmony (almost an analytic science) will not insure the composition of even the simplest melody (essentially a synthetic art). The same is true also in engineering.
>
> *Manley St. Denis*

10.1 Engineering Design and Ship Structures

Before discussing ship structural design, it is useful to go back one step and remind ourselves of some basic ideas relating to design, in particular engineering design, which is our main concern. The subject is in certain ways elusive, and it is necessary to define in a clear-cut way the specific problem we will be dealing with and the way in which it will be approached. According to [7], 'the key element of design is *representation*' where the term 'representation' refers to the artefact as well as to the process by which the design is completed.

> Engineering design consists of three tasks: *synthesis*, *analysis* and *evaluation*. Synthesis is the task of assembling a set of primitive design elements or partial designs into one or more configurations that clearly and obviously satisfy a few key objectives and constraints. Analysis is the task of performing those calculations needed to assess the behaviour of the current synthesis. The evaluation task concerns a comparison of the attributes and behaviour of the current design with the stated design specifications and constraints in order to see whether this synthesis is acceptable. [7]

It follows that a multiplicity of design solutions can be reached. This general statement is applicable to the ship structural design problem.

10.2 Design of Ship Structures for Strength: A Definition of the Task

In the context of ship structures we understand the term *design* to refer to the definition and representation of the artefact prior to its construction as well as the process being followed.

The definition of the artefact concerns all functional requirements that are required to satisfy the interested parties, to a greater or lesser extent, such as range of application, durability and cost. One reason for which ship structural design is not as clear-cut a process as the design of other artefacts is that it is intertwined with other aspects of the behaviour and of the design of the artefact, such as arrangement of spaces, hull form, propulsion equipment, environmental issues, etc., as indicated in Figures 10.1a and 10.1b. It is

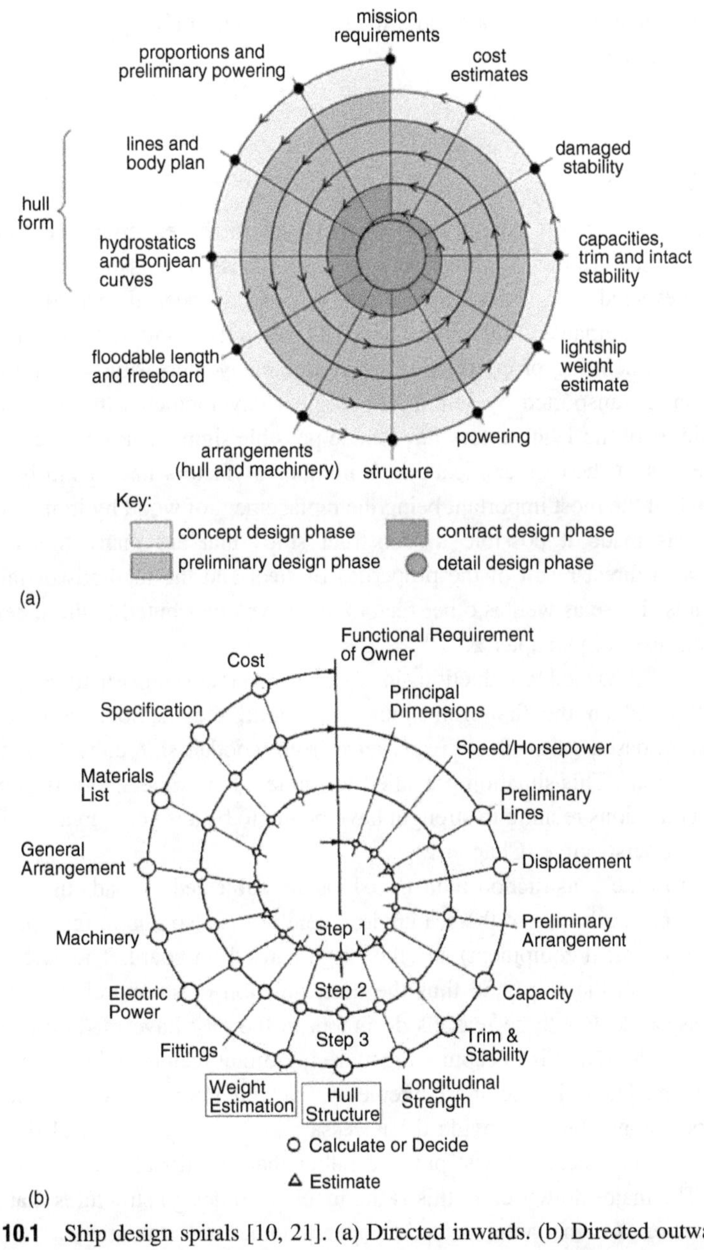

Figure 10.1 Ship design spirals [10, 21]. (a) Directed inwards. (b) Directed outwards.

dependent upon and in turn influences these other aspects. For this reason the end-product is based on considerations that are not purely structural (*objective needs).* Furthermore, the parties involved have competing (*subjective*) needs so that the end-product of the design process is seen to be a compromise that takes place at different levels.

Ship structural design, as indeed all other facets of ship design, is practiced in a competitive commercial environment and has to produce a viable solution using finite resources of manpower, infrastructure and time. For this reason it cannot afford to be more complicated or require more man-hours than necessary. It broadly aims at satisfying the requirements and needs of:

- Owners
- Shipyards
- Regulatory bodies

Until relatively recently, ships were much closer to the experience of long-distance travellers than they are now, as they were the only means of crossing the earth's oceans. With the widespread use of aircraft, the ship has largely been displaced as a means of transportation for humans. On the other hand, its role in international trade has increased, so that an overwhelming proportion of the continuously increasing volume of goods is nowadays being transported by sea. Technological developments that have taken place since the middle of the 19th century have made possible significant increases in the size of ships as well as in their diversification. A number of factors have contributed to these advances, one of the most important being the replacement of wood by iron and later on by steel. This has made it possible to construct ships that are many times larger than previously, as a direct result of the properties of steel and the methods of fabrication of steel structures. These as well as other factors that have contributed to these developments have been discussed in Chapter 2.

The use of steel has led to reductions in scantlings so that such structures can be classed as *thin-walled.* When the first iron ships were built, the methods used to determine scantlings were based on available experience from wooden ship construction and a few simple calculations. This situation prevailed for quite a long time, and it is only relatively recently that questions relating to strength have begun to be answered in a satisfactory way from a rational perspective (Chapter 2).

In contrast to the construction industry, ships are subjected to loads that vary continuously. This is primarily due to the sea environment but is also due to the operation of the ship itself (mechanical equipment) and the cargo carried on board. The external environment interacts with the ship, and thus the determination of structural response is by no means an easy task. For these reasons designers in the past have made use of relatively simple models whose aim is to capture the more important features of real conditions. Past rule requirements led to the use of arrangements that were not in any way optimum or even on certain occasions able to provide the necessary strength. Recommended practice was generally based on successful past practice rather than a rational appraisal of prevailing conditions. The major drawback to this is the inability to design structures that have novel configurations or ones that are expected to operate under new conditions.

With regard to structural response, the use of experimental results and later on of theoretical calculations and more recently of computer-based numerical calculations, it has become possible to define problems and obtain solutions using *rational methodologies*. *Rational structural design* aims at the systematic use of mathematical methods and the understanding of structural behaviour so as to be able to allow for external conditions and the properties of the structure itself. In this way economic considerations are introduced, albeit in an indirect manner.

As a consequence, in recent years a number of methods have been introduced to ship structural design that are based on the mathematical analysis of the behaviour of the various structural components that are encountered in ship structures. Classification society rules include sections on *buckling*, *fatigue* and *structural reliability*.

In recent years the situation has changed, since the transition from the *permissible stress approach* to the *limit state design* approach[1] is now established in the marine field as well. At this stage we ought to point out the differences between structures used in civil engineering and those in naval architecture. The former consists in large part of beams, columns and rods that are loaded at their supported ends as well as at intermediate positions. If one member fails and if this is a critical member, then it is possible that the whole structure may collapse. This will happen when the load in the failed member is not redistributed in a successful way so that adjacent members fail as well. In such circumstances we say that the structure has low *redundancy*.

Ship structures, on the other hand, are *extended structures* that consist of thin stiffened sheets of plating that are used throughout the ship. In most cases they are rectangular and flat, and are supported along their boundaries and possibly at intermediate positions as well. There are also cases in which the plating is curved with the curvature being either single or double, such as in way of the fore and aft ends of the ship. The stiffening is used to provide sufficient strength to the plating, and thus the aim of the designer is to select materials, arrangements and scantlings in order that the structure (1) can withstand the maximum load that it may be subjected to during its lifetime, (2) does not deflect more than the requisite amount, and (3) continues to fulfil its purpose for its expected lifetime. Because of their form, ship-type structures are highly redundant – that is, they can continue to fulfil their purpose even though one section of them has failed.

Commercially exploitable structures are also directly bound by economic constraints. Thus the *cost of construction* is of interest to both owner and shipyard. Operating expenses include *repair and maintenance costs,* which together with the cost of construction constitute the *life-cycle cost*, whose minimisation is in the interest of the owner. Such questions should be considered at early stages in the design process and indeed that of the overall design, so that suitable structural arrangements can be selected. For example, the reduction of weld lengths reduces the number of stress concentrations and other imperfections that reduce the fatigue life of the structure. In order to reduce the cost of construction, shipyards have in certain cases introduced novel arrangements such as the

[1] Also known as *load and resistance factor design* (*LRFD*).

unidirectional stiffening system (Figure 1.6). Another approach has been to reduce as much as possible regions of the outer shell structure that have double curvature.

Environmental considerations have become one of the most important issues that concern human activities. In the maritime field, they have been considered with respect to *pollution* and, more recently, *sustainability*. The former relates to the measures already taken following a number of serious pollution incidents, whereas the latter has resulted from a realisation of the impact of human activities in general on the environment, which bring about changes on a different time scale and at a broader level. The structural designer has to be aware of the implications of these issues and to understand how they affect ship design and operation, in order to incorporate the lessons that can be drawn in an effective way.

Examples of measures that have been implemented are, with respect to oil pollution prevention, the gradual replacement of all single-skin oil tankers by double-skin tankers. Analogous measures apply also to other related ship types, such as chemical tankers. With respect to sustainability, examples of ongoing initiatives are:

- Ballast water management, in order to limit the transfer of micro-organisms from one part of the globe to another
- Control of greenhouse gas emissions (carbon dioxide) and noxious gas emissions such as nitrogen and sulphur oxides that are present in engine exhaust gases

Clearly then ship structural design is a complex process that needs to cater to technical, economic, operational as well as environmental considerations. We can summarise matters as follows:

Ship structural design involves the selection and specification of form, materials and dimensions of the hull structure and individual structural members such that a ship can operate under conditions of external loading, economic exploitation and environmental protection constraints that are in accordance with the requirements of the owner, the builder and the prevailing rules and regulations.

For a design to be able to fulfil all of the above requirements, which generally compete with one another, compromises have to be made. In other words, the end result of the design procedure has to be a compromise (and hopefully an 'optimised' in some sense) design. The question then arises as to what importance or weighting is to be given to each of the parameters involved in such an optimisation process. This process does not necessarily lead to an 'objective selection' but will depend to a certain extent on which party is directly involved in the design. If it is a shipyard, the cost of construction will understandably be of paramount importance, whereas if the design is carried out by a consultant on behalf of the owner or by the owners design team, the life-cycle cost ought to feature largely in the end result. Economic considerations play an important role in the case of non-commercial designs, such as naval vessels, as well. The question of optimum ship structural design is considered in Chapter 14.

10.3 Stages in Structural Design

The three stages that comprise the ship structural design process are *concept design*, *preliminary* (or *basic*) *structural design* and *detail structural design*. Shipyards also need to plan the production process, so that additional information and sets of production drawings are required. In this chapter we shall discuss concept design and preliminary structural design. Detail structural design is considered in the recently published book by the author (*Local Strength of Ship Structures*). Matters relating to the production process lie outside the field of ship structural design and are not considered.[2]

There are two distinct approaches currently followed by shipyards involved in merchant ship design. In the majority of cases – that is, those in which the new design does not involve radical departures from current practice – the design work is assigned to an experienced engineer who performs parametric studies on the existing ship structure. If the new design involves technically complex components, or else if a degree of novelty is introduced, a design team is formed. The team, which consists of members that have different perspectives and expertise, defines the basic structural configuration (*synthesis*), following which a number of concept designs are produced. The best of these is then selected. This stage is termed *concept design*, which is followed by the *preliminary design stage*.

10.3.1 Concept Design

A scrutiny of the orderbooks of the world's shipyards shows that the overwhelming majority of new builds concern ships that are of a similar design to existing vessels and are ordered either for fleet renewal or expansion. These vessels trade on well-established routes, transport the specific commodities for which they have been conceived and built, and have operating expenses that generally do not change very much over time (e.g. oil tankers, chemical tankers, gas carriers, containerships and large bulk carriers). Specifications for new ships incorporate the latest requirements of the rules for construction and other regulations, and in certain cases they may anticipate future requirements. Owners are most able to contribute to innovations in ship design when these are linked to commercial interests, as their experience in this aspect is invaluable. Examples are the arrangement of cargo tanks and holds, the performance of loading and discharging equipment, the operation of propulsion machinery and other equipment and, last but not least, maintenance and repair issues.

For all of these reasons, the concept design stage is in many cases a question of selecting a particular design of a ship type in preference to a competing one. In this, economic considerations feature highly. Shipyards and design consultancies market portfolios of designs that can be adapted to suit owners' specific requirements.

Having described the current 'state of practice', it has to be said that in recent years efforts by research groups have been made to harness the capabilities of computer-based structural design with operations research at the concept design stage. In this way computer models can be used to compare alternative arrangements and ranked on the basis of

[2] It has to be mentioned, however, that if a concurrent engineering approach is used, questions relating to the production process will be faced early on in the design process.

(a) (b)

Figure 10.2 Containerships. (a) Classic design. (b) Innovative design.[3]

(a) (b)

Figure 10.3 Oil tankers. (a) Classic design. (b) Innovative design.[4]

stakeholders' requirements and priorities. The method, which is the current state-of-the-art, is described in Chapter 14 with reference to the structural design of a RoPax ship.

The concept design stage comes into its own when a new mission profile is being considered or a new technology that will affect the ship's operation is being introduced. This may result in the development of a new type of ship, such as an offshore support vessel or an Arctic oil tanker, or the introduction of new gas storage technology in gas carriers, among others. Categories of vessels for which the concept design stage is also important include fast ferries, cruise ships and large luxury yachts. In such cases the mission statement will be the starting point for the first thoughts as to what the new vessel will look like and what its innovative features may be. Over time the trend has been for more specialised types to be introduced as the volume of seaborne trade increases, bringing with it economies of scale. Technological solutions have then enabled market requirements to be realised.

Figures 10.2–10.4 show three pairs of designs, the first of which is the classic or traditional one that has evolved since the ship type first originated and the second is an innovative design. In Figure 10.2 a new hull form for a container feedership has been used, which features a bow form that incorporates the superstructure. The new bow ensures soft

[3] Ulstein DX541 1200 TEU container feedership fitted with X-BOW®.

[4] Det Norske Veritas *Triality* VLCC design.

(a) (b)

Figure 10.4 Passenger (Ropax) ferries. (a) Classic design. (b) Innovative design (trimaran).

entry into waves, thus reducing speed loss, pitch and heave accelerations. It also contributes to the elimination of slamming and vibration problems that are associated with conventional bow flare. The use of the forward superstructure overcomes the problem of reduced visibility that arises in the traditional design due to the storage of containers on the main deck (Figure 10.2a). Figure 10.3b shows a concept design for a VLCC that has environmentally friendly features. One of these is that the vessel is designed not to carry ballast water; to achieve this objective, the new hull form is not wall-sided. Figure 10.4b shows a trimaran passenger ferry whose design has been based on a navy prototype. In all of these designs important modifications to the hull form have been introduced. These have corresponding repercussions in structural design.

In times of economic growth, increased volumes of trade lead owners and indeed the whole shipping sector to consider the construction of ships that are significantly larger than any that have been built in the past. This occurred in the 1970s when oil tankers of up to half a million deadweight tonnes were built and even larger ones conceived (million-tonne tankers). A similar situation occurred in 2008 when ultra-large containerships, LNG carriers and other ship types were designed and built and even larger ones anticipated. Once a concept design has been finalised, a preliminary (or basic) design procedure is performed in which the most important features of the ship are established. The preliminary design forms the basis of the detail design in which attention is paid to local strength considerations that include compressive strength and fatigue.

10.3.2 Preliminary (Basic) Structural Design

The fundamental difficulty of design procedures in general stems from the fact that in the final product, a number of objectives have to be satisfied simultaneously. However, when starting out, the available information cannot lead to an end-solution for any one individual requirement. For this reason it is necessary to satisfy each requirement in a sequential manner and an iterative approach has to be used. Furthermore, there are available at the disposal of the designer a variety of techniques of varying complexity and accuracy. This methodology has been formalised by the use of *design spirals* to represent the ship design and the ship structural design procedure (Figures 10.1 and 10.6).

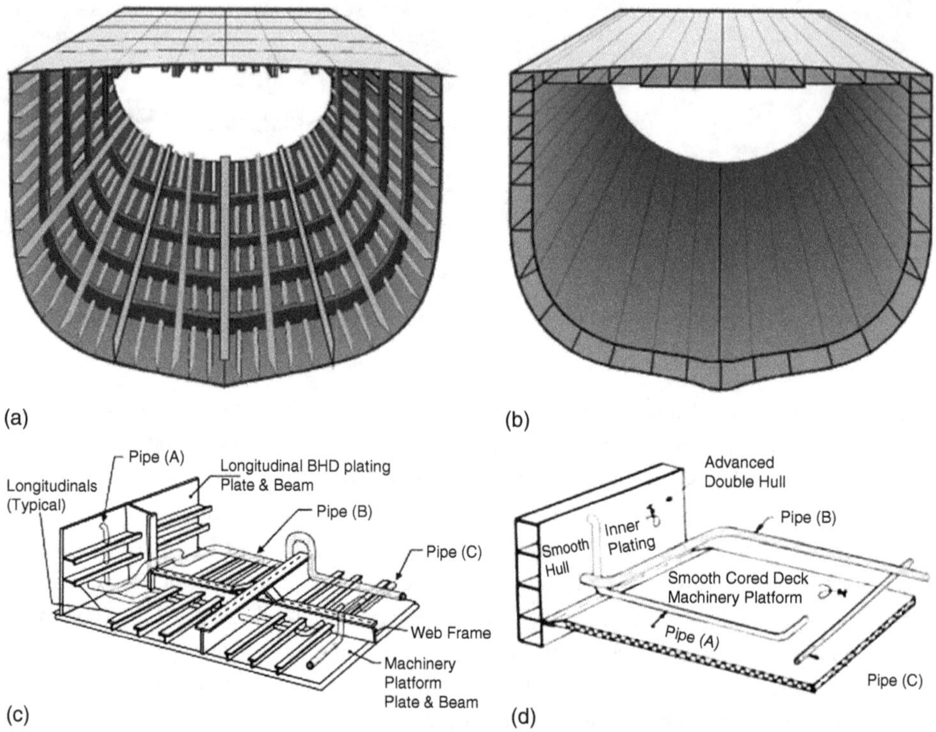

Figure 10.5 Warship construction and impact of advanced double hull on outfitting [4]. (a) Conventional single hull. (b) Advanced double hull (ADH). (c) Layout with conventional framing system. (d) Layout with ADH framing system.

The use of design spirals is based on the premise that when starting out, results will be rough estimates; it therefore pays to make use of simple, approximate methods that can be applied without excessive effort. At later stages techniques that are much more accurate can be used, as the margin of error gradually reduces. Starting from the owners' requirements, the design spiral is laid out so that all stages are in an ordered and logical sequence. Following completion of each loop of the spiral, a more accurate set of results is obtained. In the inner loop (first cycle) of Figure 10.1b a rough estimate of the structural arrangement is obtained (synthesis) whereas in the second loop (second cycle) more accurate calculations are carried including strength and vibration particulars. In loop 3 (third cycle) midship section scantlings are obtained and drawings prepared.

Each step has a direct or an indirect effect on the other steps in the design process (Figure 10.1). Thus structural design influences overall design in that weights and centres of gravity, position and geometry of openings for cargo loading and ease of construction are specified by it. Weights and centres are used to estimate the positions of the centres of gravity for the vessel lightship and the outfit items. The weight distribution can affect overall hull deflections. It also determines displacement and is used to obtain an estimate of construction costs. Accurate weight estimates are needed to produce meaningful stability data. Structural arrangements are in turn affected by extraneous requirements; for example,

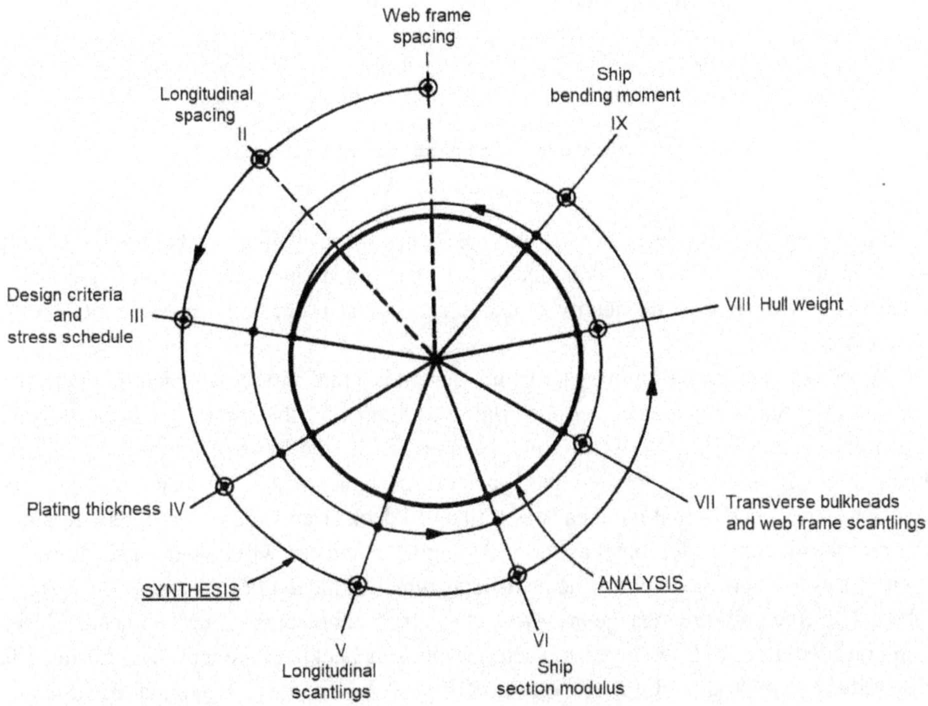

Figure 10.6 Design spiral for a longitudinally stiffened midship section [8].

the longitudinal positioning of bulkheads depends on subdivision requirements, and the type of stiffening (longitudinal or transverse) may be affected by internal arrangements of cargo or mechanical systems.

We shall now concentrate on structural design; it is important to plan structural calculations in an orderly manner, and to this end a structural design spiral can be prepared, as shown in Figure 10.6 [8]. This particular design spiral is specific to a longitudinally stiffened hull girder, but in principle a design spiral for any stiffening system can be prepared.

During the early stages the process is denoted as *synthesis* whereas in later stages it is denoted as *analysis*. In the synthesis stage, selection of form, materials and estimates of scantlings are made and are based on existing designs and intelligent judgement. During later stages the use of analysis comes into its own, as structural mechanics and optimisation software can be applied using techniques of increasing complexity and accuracy. As a result, calculations including scantlings estimates will converge and a design in which all external requirements have been taken into account will be produced. The interaction of synthesis and analysis is seen to be an important feature of ship structural design.

In preparing a design spiral the following matters have to be decided upon:

- Initial data (such as an existing design)
- The stages to be included in the design spiral
- The calculation methods to be used during each stage

Table 10.1 *Preliminary ship structural design*

First cycle	Synthesis
Second cycle	Classification society rules
Third and later cycles	Rational design methods

When preparing the design spiral, a balance has to be maintained between the order of accuracy of the results and the effort expended in each of the calculation methods. A measure of the quality of the design spiral is the rapid convergence to the final result.

The outcome of the preliminary structural design is a midship section plan that indicates the layout of the structure, selected materials and figures for the scantlings. A preliminary general arrangement plan (GA) can then be prepared on the basis of the midship section plan. This can also be compared with existing designs. In the GA plan the layout of transverse and longitudinal bulkheads as well as all deck arrangements, double bottom and double sides if applicable, forepeak, aftpeak, engine room, etc. will be indicated. This plan then acts as the starting point for the next stage, which is the *detail structural design* stage. The detail structural design stage involves selecting structural arrangements at a local level, and consideration has to be given to such questions as buckling strength, fatigue strength, accessibility and producibility. The question of producibility lies outside that of structural design and for this reason will not be addressed in this text.

We have seen that the methods used in each cycle become more accurate as the final design is approached. Table 10.1 lists the general approaches that have been used in preliminary ship structural design

The term 'rational design' has been used in the field of ship structures to denote a variety of ideas and methods. Invariably it implies that during the design process, engineering fundamentals and possibly the economic interests of one or more of the parties involved are being taken into consideration. We should not forget that engineering structures, and ships in particular, represent serious investments that must provide a satisfactory return. For this reason a sound economic basis is a prerequisite for any proposed design. A series of questions arise that need to be addressed, and these are discussed in Chapter 14 that deals with optimum structural design. In recent years computer-based techniques are promoting the development of new approaches to structural design, and especially ship design, such that the design spiral approach discussed in this chapter is being gradually superseded. These are touched upon in Chapter 1 with reference to warship design techniques, and in particular systems engineering and concurrent engineering. An example of innovative structural design is shown in Figure 10.5. It is expected that these approaches will find their way into merchant shipbuilding as well, initially at least in the case of large, well-organised shipyards with a forward-looking approach.

At this point, however, we shall postpone our discussion of rational design methods and turn to a subject that has to be addressed before that. That subject is the 'design philosophy' of ship structures.

10.4 Design Principles, Design Criteria and Design Philosophy

Up to this point in this chapter we have addressed the task of ship structural design in general terms. In doing so we have discussed the breadth of scope of ship structural design as well as the stages involved. We now need to extend our discussion to the concepts of *engineering principles, design principles, design criteria* and, finally, *design philosophy*. This will allow us to clearly define and distinguish between these terms. We will see how each of these is related to the others, and this will enable us to gain a deeper understanding of the key components of the ship structural design process.

Engineering principles are the fundamental concepts and physical laws that need to be adhered to when solving a specific engineering design problem. Clearly, the concepts and physical laws that are applicable in each case differ, depending on the nature of the problem.

Design principles, on the other hand, are the overall guiding intent for judging the worthiness of a design. They represent the accumulated wisdom of researchers and practitioners that are involved in design. In the marine field, the most explicit articulation of a general design principle is that of the IMO Goal-Based Standard (GBS). The Tier 1 GBS states:

Ships shall be designed and constructed for a specified design life to be safe and environmentally friendly, when properly operated and maintained under the specified operating and environmental conditions, in intact and specified damage conditions, throughout their life.

This principle encompasses issues within the narrow sphere of the ship, but also at a broader level, thereby acknowledging the wider consequences that marine operations have. Focusing on ship structural design, the clearest statement of design principles at a formal level is given in the IACS Harmonised Common Structural Rules (2019), quoted below verbatim:

4. DESIGN PRINCIPLES

4.1.2 General

The Rules are based on the following overall principles:

- *The safety of the structure can be assessed by addressing the potential structural failure mode(s) when the ship is subjected to operational loads and environmental loads/conditions.*
- *The structural requirements are based on consistent design load sets which cover the appropriate operating modes of a bulk carrier or oil tanker.*

The ship's structure is designed such that:

- *It has a degree of redundancy. The ship's structure should work in a hierarchical manner and, in principle, failure of structural elements lower down in the hierarchy do not result in immediate consequential failure of elements higher up in the hierarchy.*
- *It has sufficient reserve strength to withstand the wave and internal loads in damaged conditions that are reasonably foreseeable e.g. collision, grounding or flooding scenarios. Residual strength calculations are to take into account the ultimate reserve capacity of the hull girder, considering permanent deformation and post-buckling behaviour.*
- *The incidence of in-service cracking is minimised, particularly in locations which affect the structural integrity or containment integrity, affect the performance of structural or other systems or are difficult to inspect and repair.*

- *It has adequate structural redundancy to survive in the event that the structure is accidentally damaged by a minor impact leading to flooding of any compartment.*

4.1.3 Limit state design principles

The rules are based on the principles of limit state design. Limit state design is a systematic approach where each structural element is evaluated with respect to possible failure modes related to the design scenarios identified. For each retained failure mode, one or more limit states may be relevant. By consideration of all relevant limit states, the limit load for the structural element is found as the minimum limit load resulting from all the relevant limit states.

The limit states [. . .] are divided into the four categories: Serviceability Limit State (SLS), Ultimate Limit State (ULS), Fatigue Limit State (FLS) and Accidental Limit State (ALS). The Rules include requirements to cover the relevant limit states for the various parts of the structure.

At a less formal level, practitioners implement design principles such as the ones listed below.[5] These are of a general nature, in the sense that they are applicable to other branches of engineering as well.

1. Keep it simple – everything in general.
2. Incorporate feedback from in-service experience.
3. Ensure that loads are readily transferred, thereby ensuring that the structure is highly redundant.

It is seen that a design principle needs to be supported by one or more measures of the extent to which the principle is fulfilled. This is achieved by the formulation and use of *design criteria*.

Design criteria are thus the explicit measures used to monitor the degree of success of a particular design in meeting one or more design principles. They are in effect the goals that a project must achieve for it to be judged acceptable.[6] They can be divided into *primary* (or *essential*) and *secondary* (or *optional*) criteria. Primary criteria are those that need to be satisfied by statutory and/or other requirements. A project will not proceed if it does not meet these goals. Secondary criteria are those that are desirable but not essential. They may be specified by any one or more of the stakeholders involved (e.g. owners, builders, charterers, funding bodies).

Differentiating between primary and secondary criteria establishes a hierarchy in design choices. In marine structural design, the primary design criteria concern safety and serviceability. The former commonly includes the ultimate and the fatigue limit state.[7] Service experience shows that accidental loads and insufficient strength due to gross errors or omissions made during design, fabrication or operation contribute to the risk of failure. Other examples of secondary criteria are inspectability and repairability.

A *design philosophy* is a fundamental guiding principle that dictates how a designer approaches his/her practice with regard to a particular task. Within the realm of engineering, a design philosophy is founded upon the twin pillars of engineering principles and design principles, and its role is to provide a framework for the implementation of a design process. It is thus seen to be the synthesis of, on the one hand, fundamental scientific laws

[5] A more detailed list is included at the end of Chapter 1.
[6] http://web.mit.edu/course/21/21.guide/designcr.htm.
[7] Limit states are discussed in Section 10.6.

and, on the other, the accumulated wisdom based on engineering judgement and past practice.

As such, a design philosophy remains the choice of the designer, insofar as it satisfies particular design criteria, both primary and secondary. As a result, different groups of designers will adopt different design philosophies with regard to the same task, even though they all adhere to the same engineering principles. Design philosophies evolve with the passage of time as a result of scientific and technological innovations as well as inventions.

A relevant example of a design philosophy that has evolved through time concerns structural design as well as ship structural design. During the latter part of the 19th century the understanding of the importance of the yield point of metals as a failure criterion led to the elastic design philosophy, which led to the use of the working stress method. Later on, the understanding of the role of plastic strength led to the plastic design philosophy and the use of plastic design methods. More recently, the use of much more slender sections in ship structures has made it necessary to calculate buckling strength, and thus ultimate load methods were developed. The most recent development in the marine field is the *limit state design* approach, which deals with requirements on several fronts (ultimate strength, fatigue strength, serviceability and accidental limit state). Particular techniques have been developed for each of these limit states.

With regard to ships, safety, reliability and post-damage (residual) strength are fundamental prerequisites for their continued operation given the environment within which they operate and the consequences of failure. The environment ensures that the condition of ship and other marine structures is slowly but continuously degraded throughout their life. They therefore need to be able to operate despite the presence of flaws and local damage that lead to reductions in strength, although a significant proportion of these will be detected and repaired at some point in time. We need to contrast this with the case of aircraft, for which any loss of strength will have catastrophic consequences. For this reason aircraft are built so that no such eventuality arises. The design philosophies followed in these two cases are thus seen to differ at a fundamental level. In the case of ships a *fail-safe* approach is followed whereas in the case of aircraft a *safe-life* approach is followed.

In the next sections we shall discuss the implementation of the fail-safe design philosophy in ship structural design. We will do so having in mind sea-going ships that are built of steel and are subjected to the loading normally expected in the sea environment. If other materials are used or if vessels operate under different conditions, they may need to be designed using other approaches.

10.4.1 Elastic Design of Ship Structures

When a ship's structure is subjected to the loads that arise under normal operating conditions, upon removal of these loads, it has to revert to its initial intact condition so that it can continue to carry loads throughout its expected lifetime. In order for the structure to be able to fulfil this role in a satisfactory way, it is necessary that the stresses that arise as a result of normal loading are elastic, otherwise permanent plastic deformations will arise. These may result in a reduction of the load-carrying capacity of the structure, and continuous application of normal range loads may possibly lead to eventual collapse. It is for this reason that we generally make use of *elastic design* when we consider normal loading conditions.

Classification societies recognised this early on and introduced rules to ensure that structures behave elastically by imposing an upper limit to the *permissible stress,* which is less that the yield stress of the material ($\sigma_a = 175/k$ N/mm^2 where k is a material factor[8]). When this requirement was initially imposed, ship structures were generally proportioned so that other local failure modes such as buckling and fatigue did not arise and it was not necessary to consider them. This situation arose from the use of empirical sizing of scantlings in the absence of knowledge regarding loads as well as of analytical tools necessary to assess structural strength. In this way it was ensured that the structure did not fail when the maximum load was applied to it at some point during its operating lifetime. As a result, the operating life of ships was limited by matters other than structural.

However, with the increase in size of merchant ships during the second half of the 20th century, buckling problems arose in certain ship types,[9] and it became clear that it is necessary to rely less on empirical formulas and instead to make use of a first-principles approach to be able to check buckling strength. As a result, classification societies introduced buckling manuals, and it became necessary to check the elastic and, more importantly, elasto-plastic buckling strength of structural members. One other practice introduced with the construction of large ships was the use of high-strength steel. This facilitates fabrication and reduces weight. However, a new unforeseen problem arose – fatigue. Since thinner plating could be used, the magnitude of the alternating stress range increased, and this brought about a reduction in fatigue life. It therefore became necessary to introduce fatigue strength assessment as part of the ship design procedure. Indeed, this is a relatively recent development, and classification society rules today display a wide variety of approaches to this question. The important issue is that when a ship structure is subjected to normal operating loads, elastic design is required to ensure continued operation; the techniques introduced are there to ensure that this is the case.

10.4.2 The Role of Plastic Design

There arise circumstances in which the applied loads exceed those that are expected to act on a structure under normal operating conditions. We can conceive of such scenarios with the help of the available experience of the operation of such structures and thus define one or more types of *extreme loads* or loading conditions. Since these are improbable events, their frequency of occurrence may be one or more orders-of-magnitude lower than that of normal loads. Having defined such a series of scenarios, we can design the structure to withstand these loads without suffering the consequences of catastrophic failure. In doing

[8] $k = 1$ for mild steel. For higher-strength steels k decreases accordingly (Table 6.2).
[9] Shear buckling of the webs in transverse frames of large tankers.

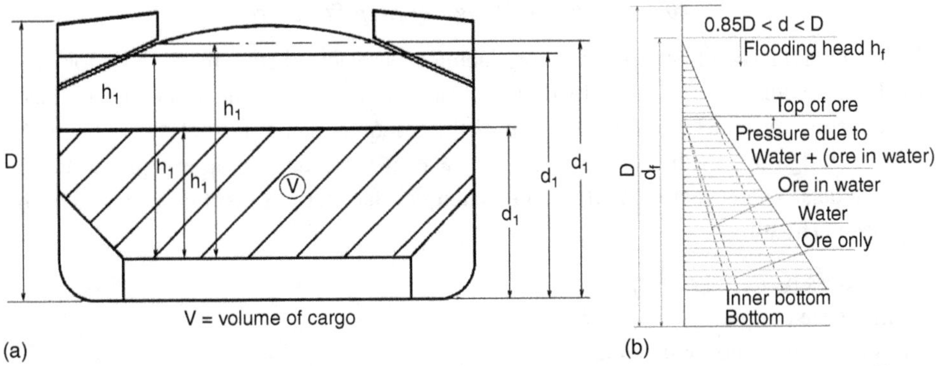

Figure 10.7 Flooded condition of bulk carrier hold. (a) IACS UR S20 loading scenario.[10] (b) Loads acting on transverse bulkhead.

so, we acknowledge the extreme but at the same time improbable nature of these loads, and allow permanent plastic deformations to arise. The purpose of this is to avoid the serious consequences that will result from collapse but at the same time not to impose an unduly severe weight penalty on the structure.

A classic example of such a situation arises when, following damage to the structure of a bulk carrier, ingress of sea-water takes place in a loaded cargo hold (Figure 10.7). If the density of the cargo carried is high, as in the case of mineral cargoes, only a small percentage of the volume of the hold will be occupied. As a result, the hold will be flooded so that the net loading on any bulkhead will be the sum of the normal and the extraordinary load. If the bulkhead and bottom structure are designed to withstand an abnormal load situation such as this, the damage will not propagate to adjoining compartments, and the integrity of the hull girder will be preserved. Otherwise, failure will arise in an uncontrolled manner in other parts of the structure with possibly dire consequences.

Another example concerns warships and in particular the hull girder loading that results from atmospheric or underwater explosions during wartime action. The structure can be designed to withstand specific levels of explosive loading in such a way that permanent plastic deformations result without, however, leading to rupture or global collapse. The purpose of this is to safeguard human life and equipment under such extreme conditions.

There arise other situations in which the magnitude and direction of the applied loading is such as to warrant the direct application of plastic design methods. One such example is the structure in the bottom forward region of the hull girder of polar, ice-breaking ships. In order to withstand the large local loads, the plating is designed to be stocky (low *b/t*) so that failure is not elasto-plastic, as is usual in ship structures, but plastic (material yielding). A similar situation arises in the case of the pressure hull of deep-sea submersibles.

[10] h_1 = cargo level (m), h_f = flooding head (m), V = volume of cargo.

10.4.3 Rational Design of Ship Structures

Rational (*engineering*) *design* aims at producing an artefact suited to the purpose for which it is built. Requirements are specified in a precise and accurate manner, and the artefact is subsequently designed and created to fulfil those requirements. In order to implement a rational design approach it is necessary to have in place an appropriate procedure and relevant data. There exist many different techniques that can form part of a rational design approach. These include

- The stress hierarchy model
- Limit state design
- Engineering economy methods
- Strip theory, used to obtain hydrodynamic loads acting on the hull girder (Chapter 4)
- Structural reliability-based and risk-based design (Chapter 12)
- Finite element procedures (Chapter 13)
- Optimisation methods (Chapter 14)

The stress hierarchy model is a valuable tool in understanding the response of ship hull girder components to quasi-static loading, described in Chapters 4 and 5. It underlies the approach that is followed in the IACS Common Structural Rules for oil tankers and bulk carriers and is discussed in the companion volume, *Local Strength of Ship Structures*.

Limit state design is discussed in the next section whereas, with the exception of engineering economy methods, the other approaches to rational ship structural design listed above are treated in other chapters of this book.

10.5 Limit State Design of Ship Structures

We have seen in the previous section that elastic and plastic design fulfil important roles in ship structural design. In many cases the application of one criterion alone is sufficient, whereas in others it is necessary to apply both. We therefore have to first identify and investigate the possible failure modes for the applied loading, following which we need to establish the criterion (-a) that will be implemented. In doing so, we need to calculate the failure load for each of these. We can then identify the critical failure mode, which is that which corresponds to the lowest induced stress for a given structural arrangement. If the anticipated load exceeds the critical load, the structure has to be redesigned.

10.5.1 The Need for a More Comprehensive Approach

Let us go back to elastic design. The aim of elastic design is to ensure that the structure does not fail by yielding, because if this did occur, it would result in permanent deformations or possibly collapse. To succeed in this, a maximum permissible stress is specified, and in this way an empirical safety factor is introduced. The permissible stress range lies then below the yield stress level by a preordained margin. This situation is satisfactory provided other failure modes such as buckling do not arise. However, buckling is possible when in-plane compressive loading is applied. In the case of ship-type structures, which are of intermediate slenderness, inelastic buckling arises at an average stress that is below the yield stress of the material. Ensuring that the stresses remain below the yield stress does not therefore guarantee that buckling will not occur. To ensure structural integrity it is therefore necessary to take into account this failure mode as well. In order to do so, the buckling strength of each and every component of the hull girder that is subjected to in-plane compression has to be calculated. It therefore becomes apparent that implementing the working stress approach[11] is not sufficient to ensure structural integrity.

The calculation of buckling strength is then a step in the right direction, as it ensures the integrity of individual structural members. To this end, several years ago classification societies issued buckling design manuals for use in parallel with rules. However, this still left open the calculation of the ultimate strength of the hull girder. The collapse of hull girders involves certain members yielding in tension whereas others buckle in compression; the latter occurs at lower loads than the former. It is therefore necessary to calculate the collapse load of the entire structure, and only by comparing it with the expected maximum value of the applied load it is then possible to estimate the margin of safety. If the true margins of safety are calculated for all ships, they can all be designed to the desired level of safety. In other words, rules can be drafted such that the reliability of an entire fleet of ships of comparable type is equal to or above a stipulated minimum standard.

If, however, we pause to reflect on the possible events a ship's structure may be subjected to throughout its lifetime, it becomes clear that overloading is not the only possible source of failure. Indeed, it is not only failure considered in the narrow sense that needs to be considered. During its lifetime a structure suffers from (1) the presence of inherent defects that are the result of production, fabrication procedures, etc.; (2) normal wear and tear, in particular that resulting from exposure to the marine environment; and (3) unexpected loads for which it has not been designed, such as abnormal waves, grounding, fire, explosions, etc.

For these reasons, it becomes clear that it is necessary to adopt an alternative, much more comprehensive approach to ship structural design. Such an approach will have to be based on the following:

1. Calculate the strength and structural properties such as stiffness, of the hull girder.
2. Assess the effects of production and fabrication on structural properties and strength.
3. Gain an understanding of the effects that the marine environment has on individual structural members in representative regions of ship structures, for each ship type.
4. Acquire an understanding of the consequences of failure in a wide variety of circumstances (scenarios).
5. Collect data on the frequency of failures, and in general all specified undesirable outcomes.

[11] Up to the middle of the 20th century ship structures had large reserves of strength against buckling so that the working stress approach ensured that designs were safe. When larger vessels started to be built, buckling became an important consideration, and classification societies introduced buckling calculations to ensure structural integrity.

10.5.2 Limit States for Ship Structures according to IACS

The approach that has to be followed in order to cater for failure as well as for eventualities connected with (1)–(3) above is to introduce what is termed the *limit state* approach that distinguishes an acceptable from an unacceptable state in general. Examples of unacceptable states are structural collapse, excessive deformations, fatigue cracking and the outcome of an accident. The concept of the limit state is portrayed graphically in Figure 11.7 as the boundary that separates the safe from the unsafe domain. For present purposes the concept is extended to denote any undesirable state. One of the most pertinent features is the variability of the factors that determine the performance of a ship throughout its working life. This has proven to be a serious challenge, and to this end reliability-based techniques have been developed and applied. These are described in Chapters 11 and 12. Chapter 11 provides an introduction to structural reliability theory whereas Chapter 12 focuses on ship-related applications.

In the IACS rules, the following limit states are prescribed:

- Serviceability limit state
- Ultimate limit state
- Fatigue limit state
- Accidental limit state

The Serviceability Limit State

The *serviceability limit state* (SLS) concerns firstly the ability of the structure to act as a satisfactory working platform under normal working conditions and secondly for it to be inspected and repaired without undue difficulty. More specifically, for these aims to be fulfilled,

(a) At a global level,
  - The hull girder has to have adequate global longitudinal stiffness.
  - The hull girder has to have acceptable vibration characteristics.
  - For passenger ships, noise and vibrations are to be within prescribed limits.

(b) At a local level,
  - Structural members and substructures are to have adequate stiffness.
  - Structural members and substructures are to have acceptable vibration characteristics.
  - Unacceptable levels of deformations that affect the efficient use of structural or non-structural elements or the functioning of equipment ought not arise.
  - Damages due to normal wear and tear, such as corrosion and fatigue cracks, have to be limited by proper maintenance and careful local design.
  - Local damages such as buckling and yielding that may reduce the durability of the structure or affect the efficiency of structural or non-structural elements must be avoided.
  - Deformations and deflections that may affect the aesthetic appearance of the structure (and/or the ship) are to be avoided (this is not an IACS requirement).
  - The means of access and compartment layout are to be arranged bearing in mind the need for accessibility and ease of repair.

Table 10.2 *Classification society hull girder stiffness requirements*

Class society	Hull Girder Moment of Inertia required minimum value
ABS	$I = L\,\frac{SM}{33.3}$ (cm^{2}m^2) where SM = required section modulus
	$(SM)_{min} = 0.01C_WL^2B(C_B+0.7)$ cm^2 – m (C_W = wave coefft)
DNV-GL	$I_{yR-gr} = 3C_WL^3B(C_B+0.7)10^{-8}(\text{m}^4)$

We shall focus on the aims that need to be fulfilled that concern the behaviour of the hull girder at a global level.

The need to ensure sufficient global stiffness arises for the following reasons:

(a) To limit hull girder deflections in order to retain cargo carrying capability
(b) To measure hull girder vibrations so as to be able to avoid resonance with wave loads and equipment operation
(c) In the case of large, flexible ships such as ULCSs, to maintain wave bending moments and vertical deflections at acceptable levels

This is achieved by ensuring a minimum value of the second moment of area, I, of the hull girder amidships. Examples of formulas included in recent rules are given in Table 10.2.

In the case of cargo ships, minimum hull girder stiffness is required to ensure that the longitudinal deflections of the hull girder do not exceed permissible levels under any condition of loading. This is necessary for several reasons. Firstly, excess flexibility will affect the cargo-carrying capability of the ship. In a ballasted condition, a cargo ship will most likely be in a hogging or a near-hogging condition. As loading progresses, the weights increase in the midships region and the degree of hogging gradually reduces, so that hull girder changes to a sagging condition. If the hull girder is too flexible, the draught amidships will increase excessively and will reduce in way of the fore and aft ends. The result is that limitations to the quantity of cargo that can be loaded need to be imposed in order to maintain the required freeboard amidships. Secondly, equipment and systems that extend over a large proportion of the ship's length, such as piping, are subjected to excessive bending and cannot function effectively. Supports are thus overloaded and may suffer damage over time as a result of alternate loading.

At a local level, excessive longitudinal hull girder deflections will result in increases in the amplitudes of alternating stresses, resulting in a reduction of the fatigue life of structural details.

The wave-induced bending moment is currently calculated on the basis of length, breadth, the block coefficient and the wave coefficient (Equations (4.49a–b)). This is tantamount to assuming that the hull girder behaves as a rigid body. This is acceptable for the majority of ships, but for certain types of large, open-deck cargo ships that have a low bending stiffness, such as large containerships, it has been found that the actual wave-induced bending moment exceeds that estimated using rigid body assumptions. Estimates of this exceedance range from 7 per cent in the general case up to 22 per cent in selected

Table 10.3 *Sound pressure limits for passenger accommodation expressed in dB(A) [5]*

	DNV			BV			LR	
Location	1	2	3	1	2	3	1	2
Top grade cabins	44	47	50	45	50	50	45	50
Standard cabins	49	52	55	50	55	55	45	50
Public rooms	55	58	62	55	60	65	55	60
Open recreation decks	65	65	70	65	70	75	65	70

cases. This means that such vessels, as built according to classification society rules, do not have sufficient strength. Clearly this represents a matter for further investigation; this departure has been attributed to the effects of hydroelasticity [2].[12]

Longitudinal stiffness is one of the parameters that affect global hull girder vibration, the others being displacement and ship length. If it is judged necessary to bring about changes to the hull girder natural frequencies, it will be necessary to add material to the hull girder, as for every ship a minimum value of the moment of inertia is specified by classification society rules. Clearly this is not a desirable path to follow because it leads to an increase in lightship, a loss of cargo-carrying capability and a cost increase. Another option to explore would be to modify the longitudinal distribution of weights. This is possible in the case of cargo ships such as bulk carriers and oil tankers that sail in part-loaded or ballasted conditions for part of the time. In these circumstances, it is possible to select loading arrangements such that resonance with propeller excitation frequencies is avoided. However, such a solution would meet with limited success [24].

In the case of warships, hull girder resonant frequencies due to wave action may affect equipment operation. It has been proposed that these frequencies be modified by suitable changes to distribution of masses on board and stiffness [6].

The most successful way of avoiding vibration problems is to reduce or, if possible, remove the sources of excitation. Some of these, such as hull excitation due to propeller blade action, cannot be removed totally, but accurate estimation of frequencies, redesign and proper mounting of equipment contribute towards this goal. Classification societies require that specific calculations be carried out such as for the entire propeller-shaft system, and appropriate measures need to be implemented. In the case of the main engine vibration, characteristics are provided by the manufacturer.

Linked to vibration is *noise* generation. Sources of noise and its propagation are shown in Figure 10.8, and it is seen that these are also sources of vibration.

Naval submarines are a type of vessel for which noise reduction and/or elimination are of vital importance. Special technologies have been developed to this end, but this is not a topic that will be pursued further at this point. With regard to merchant and passenger ships, the International Maritime Organisation (IMO) has issued a noise code that specifies

[12] A case in point is the containership *MSC Napoli*, whose hull girder collapsed during a voyage in the English Channel in heavy weather conditions.

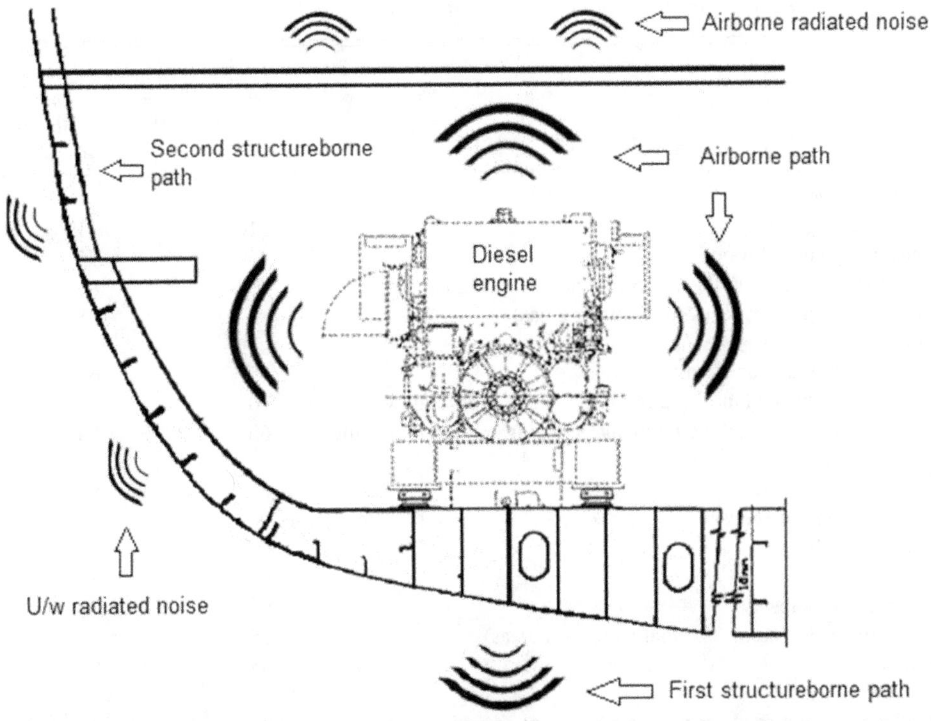

Figure 10.8 Noise transmission [1].

acceptable noise levels for different regions of a ship in order to ensure acceptable environments for crew and passengers (Table 10.4) [19]. The International Standards Organisation (ISO) has subsequently issued a series of documents (ISO 20282-3 series) that provide more details on measurement, methods of attenuation and related matters and are used as a basis by classification societies. With regard to global hull girder vibration, ISO 20283-2 recommends methods to obtain an overall picture of the vibration behaviour of the ship by measurement of natural frequencies and vibration responses related to the hull operating deflection shapes.

A sound pressure level L_p (SPL) of a sound or noise, expressed in decibel (dB), is given by the following expression [19]:

$$L_p = 10 \log \frac{p^2}{{p_0}^2} \tag{10.1}$$

where

p = sound pressure, in Pascal

p_o = 20 μPa (reference level)

In addition to the above requirements, classification societies have introduced notations that rank habitability standards for crew and passengers. Such notations are provided for

Table 10.4 *Sound pressure limits in dB(A)*[13] *for various spaces [19]*

Designation of rooms and spaces	Ship size 1,600–10,000 GT	≥10,000 GT
Work spaces		
Machinery spaces	110	110
Machinery control rooms	75	75
Workshops other than those forming part of machinery spaces	85	85
Non-specified work paces (other work areas)	85	85
Navigation spaces		
Navigating bridge and chartrooms	65	65
Look-out posts, incl. navigating bridge wings and windows	70	70
Radio rooms (with radio equipment operating but not producing audio signals)	60	60
Radar rooms	65	65
Accommodation spaces		
Cabins and hospitals	60	55
Messrooms	65	60
Recreation rooms	65	60
Open recreation areas (external recreation areas)	75	75
Offices	65	60
Service spaces		
Galleys, without food processing equipment operating	75	75
Serveries and pantries	75	75
Normally unoccupied spaces		
Spaces referred to in section 3.14 of the rules	90	90

passenger and cruiseships that aim at providing a superior level of comfort. Table 10.3 includes the acceptable sound pressure levels in various regions for use by passengers [5].

The differences between the acceptable sound pressure levels that are included in Tables 10.3 and 10.4 are evident.

The Ultimate Limit State

The need for an *ultimate limit state* (also called ultimate limit strength (ULS)) is based on strength considerations. In the case of a ship structural member it is necessary to estimate the expected load profile for a given design life. The ULS corresponds to the collapse of the structure due to loss of structural stiffness and strength related to:

- Attainment of the maximum resistance capacity of sections, members or connections by gross yielding, rupture or fracture

[13] *Decibel (dB):* A dimensionless unit of measure of the ratio of two quantities each of which is equal to or proportional to power. *dB(A)* are decibel scale readings that have been adjusted to attempt to take into account the varying sensitivity of the human ear to different frequencies of sound.

- Instability in part or of the entire structure resulting from buckling or plastic collapse of plating, stiffened panels and support members

The vertical hull girder bending moment, M, in hogging and sagging conditions, to be considered in the ultimate strength check, is to be taken as:

$$M = \gamma_S M_{sw-U} + \gamma_W f_\beta M_{wv} \tag{10.2}$$

where

M_{sw-U}	Permissible still-water bending moment, in kNm, in hogging and sagging conditions at the hull transverse section considered as defined in Table 10.5
M_{wv}	Vertical wave-bending moment, in kNm, in hogging and sagging conditions at the hull transverse section
γ_S	Partial safety factor for the still-water bending moment, as defined in Table 10.6
γ_W	Partial safety factor for the vertical wave-bending moment, as defined in Table 10.6
f_β	Heading correction factor, as defined in section 1 of the rules

The ultimate bending moment capacities of a hull girder transverse section in hogging and sagging are defined as the maximum values of the curve of bending moment capacity versus the curvature χ of the transverse section considered (Figure 10.9). The curvature χ is positive for the hogging condition and negative for the sagging condition.

The hull girder ultimate bending capacity, M_U, is calculated as described in Appendix D.

The Fatigue Limit State

The *fatigue limit state* (FLS) is introduced in response to the need to prevent fatigue cracks from occurring as a result of repeated loading. If these occur in the outer shell plating of the hull girder, the watertightness of the hull envelope is jeopardized. Less serious, although

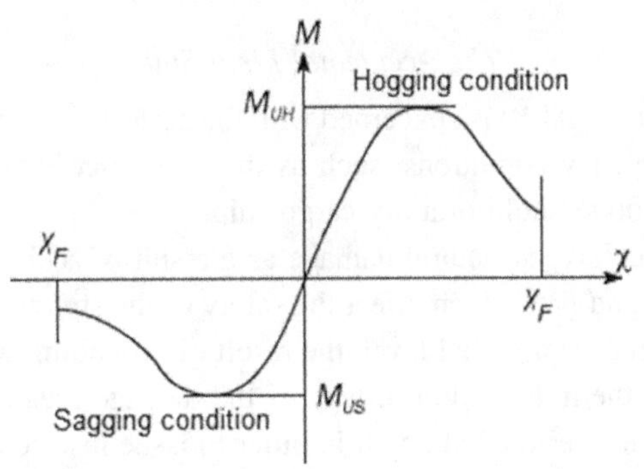

Figure 10.9 Bending moment capacity M versus curvature χ. Reprinted courtesy of IACS, London.

Table 10.5 *Design load scenarios*

Design load scenarios		Permissible still-water bending moment, M_{sw-U}
	S + D	M_{sw-h} or M_{SW-s}
A	S + D	M_{sw-p-h} or M_{SW-p-s}
	A: S + D	M_{sw-f}
B	S + D	Maximum sagging still-water bending moment for operational sea-going homogeneous full-load condition*

* The maximum still-water bending moment is to be taken from the departure condition with the ship homogeneously loaded at maximum draught and corresponding arrival and any mid-voyage conditions.

Table 10.6 *Partial safety factors*

Design load scenarios	γ_S	γ_W
A	1.0	1.2
B	1.0	1.3

important from an operational point of view, is the loss of watertightness of internal compartments, with adjacent compartments being possibly flooded. This may lead to cargo damage and general deterioration in vessel performance. In order to pre-empt such events it is necessary to pay attention to the design of structural details.

Fatigue cracks act as stress-release mechanisms at a local level, and thus loads carried by individual members seek alternate routes. It is thus possible that apparently innocuous fatigue failures may lead to overloading of adjacent structural members, thereby leading to their failure (compressive buckling, tensile yielding).

In addition, the marine environment brings about a deterioration of the condition of structural members through the mechanism of electrochemical corrosion. As a result, the fatigue of individual members is accelerated by *corrosion-fatigue*.

The Accidental Limit State

The *accidental limit state* (ALS) is concerned with the performance of the hull girder or of a part of it under emergency conditions, such as the one described earlier in this chapter, which concerns the flooded hold of a dry cargo ship.

ALS represents excessive structural damage as a result of accidents such as collisions, grounding, explosion and fire, which affect the safety of the structure, persons on board as well as the environment. At a global level, the result of grounding would be partial loss of structural integrity of the hull section in way of the damage location so that it would be necessary to estimate the residual strength in order to issue instructions to the crew.

The vertical hull girder ultimate bending capacity in the damaged condition is to be checked for the damage conditions specified in the rules (Table 10.7).

Table 10.7 *Design load scenarios*

	Design load scenario	Permissible still water bending moment in damage, M_{sw-D}
Collision	A: S+D	M_{sw-h} or M_{sw-s}
Grounding	A: S+D	M_{sw-h} or M_{sw-s}

The vertical hull girder ultimate bending capacity in the damaged condition at any hull transverse section is to satisfy the following criterion:

$$M_D \leq \frac{M_{UD}}{\gamma_{RD} C_{NA}} \tag{10.3}$$

where

- M_D Vertical bending moment in the damaged condition, in kNm, to be obtained as specified in the rules
- M_{UD} Vertical hull girder ultimate bending capacity in the damaged condition, in kNm
- γ_{RD} Partial safety factor for the vertical hull girder ultimate bending capacity in the damaged condition, to be taken equal to $\gamma_{RD} = 1.0$
- C_{NA} Neutral axis coefficient taken as
  $= 1.0$ for grounding
  $= 1.1$ for collision

The calculation of the residual strength of the hull girder in case of grounding or collision is discussed in Section 7.3.4.

Interaction of Limit States

In addition to the above limit states, it is necessary to consider the interaction of two or more of these. For example, the occurrence of a grounding incident (ALS) may be followed by deep plastic collapse and rupture of the bottom structure (ULS), ingress of water in the cargo compartment resulting in cargo damage (SLS), overloading and subsequent rupture of bulkheads that may have been weakened by extensive corrosion (ULS), and finally flooding of an adjacent cargo compartment (SLS).

In the well-known case of the OBO *Derbyshire*, the sequence of events was initiated by abnormal waves striking and rupturing the hatch cover of cargo hold no 1 (SLS and ULS) and concurrent flooding of the forepeak tank because a manhole had been left open (SLS). The events that followed have been reported extensively in the literature (ULS).

In order to establish a rational approach to such a sequence of events it is necessary to establish benchmark studies for the most important scenarios. This would permit comparisons to be made regarding the probability of occurrence of each of these for different ship types, sizes, routes, cargoes transported, etc.

Table 10.8 *Limit states to be considered when assessing hull girder global strength*

Failure mode	Description	Limit states
Yielding	Yielding results from the action of lateral and/or in-plane loads and is avoided by imposing limits to induced stresses. When hull girder collapse occurs, one section is subjected to tensile loading that leads to yielding failure.	SLS, ULS, ALS
Buckling (elasto-plastic)	Buckling is the result of in-plane compressive loading or of a combination of loads and leads to collapse at a local level. Progressive collapse leading to hull girder collapse may occur if intact members are unable to sustain additional loading	ULS, ALS

Failure Modes

The failure modes to be considered when assessing limit states in relation to the global strength of the hull girder are listed and discussed in Table 10.8. For both types of failure modes and for all listed limit states, IACS specifies a probability of exceedance of the critical load equal to 10^8.

10.6 Design Loads in the IACS Common Structural Rules

The improved understanding of the marine environment and marine structures has contributed to the design of novel structures, notably in the offshore industry. The regulatory regime that persisted until the end of the 20th century involved a number of classification societies that competed with one another, with differences in structural requirements being one important criterion. For reasons discussed in Section 1.4.3, it was decided that competition between societies on the grounds of technical requirements should be eliminated, and to this end two groups of societies were formed, with a view to bringing together their collective experience in order to draft rules intended for universal application.

10.6.1 Background

Operating in accordance with the Goal-Based Standards introduced by the IMO, the International Association of Classification Societies introduced rules for oil tankers and bulk carriers that were later unified as the Harmonised Common Structural Rules.

An important prerequisite was that a ship built in accordance with the new rules would be as safe and robust as would have been required by any of the pre-existing rules. It ought to be added at this point that a study reported in Section 12.6.2 has showed that in the case of a CSR-compliant Aframax oil tanker the strength is marginally enhanced when compared with the pre-CSR design.

Design Basis for Oil Tankers

- Design Life – 25 years
- External Environment
  - North Atlantic wave environment
  - Air temperature: –15°C
  - Sea temperature: 0°C
- Internal Environment
  - Specific gravity cargo: Min 1,025 kg/m^3 (900 for fatigue)
  - Max cargo temperature: 80°C
  - Min cargo temperature: 0°C
  - Corrosion margins based on a 25-year design life

Before considering design loads it is necessary to specify the design basis according to which a ship is expected to operate. For oil tankers this is given above [12].

10.6.2 Requirements for the Determination of Design Loads for CSR

In this section we shall describe the methodology introduced and followed with regard to the definition and estimation of design loads, as reported in IACS documents and other publications. The proposed methodology is intended for use by classification societies in local strength analysis and also in direct (finite element-based) design.

An important priority has been the implementation of an approach that enables the designer to move away from the traditional, quasi-static wave load model described in Chapter 4 to the use of a more rational way of modelling structural loading. In order to achieve this, it has been necessary to define design loads in such a way that:

1. They are based on statistical data that represent actual conditions.
2. Use is made of properties of statistics theory that facilitate the handling of data and does not require the processing of large amounts of information by the designer, which may introduce subjectivity of interpretation and possible errors.
3. They allow for the global operation of ships throughout their expected lifetime.
4. They are applicable using existing methods of structural analysis.

More specifically:

- A clear description of load definition and load application is required.
- Load combination factors are used.
- Rule design wave loads are based on current IACS practice.
- Load formulations are validated by direct calculations.
- Sloshing, slamming and bow impact use available current methods.

As is evident from the Design Basis for oil tankers given above, the design loads are meant to account for the full operating lifespan of the structure, taking into account

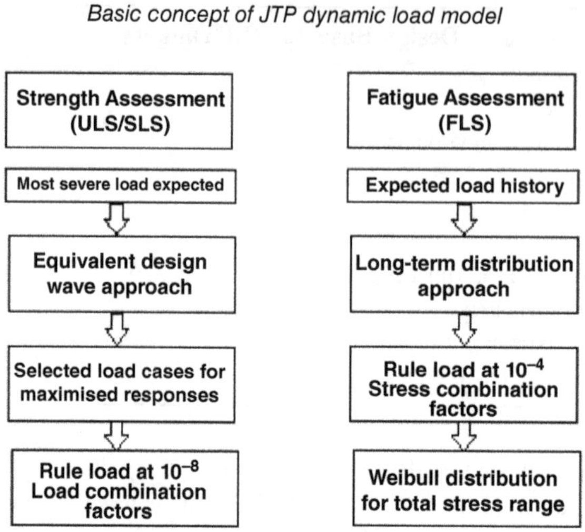

Figure 10.10 The basic concept of the JTP dynamic load model [12].

environmental conditions and allowing also for the probability of occurrence of extreme conditions. In order to incorporate into the proposed methodology the requirements listed above, the JTP group[14] proposed the sequences shown in Figure 10.10 for strength and fatigue assessment of oil tankers.

In discussing this dynamic load model we shall focus on strength rather than fatigue assessment. It is seen that the model introduces the estimation of the severest load expected during the operating life of a ship and thus satisfies item 1 listed above. We shall later see how this estimation is carried out. The 'equivalent design wave approach' is the means by which statistics theory is used to 'facilitate the handling of data and does not require the processing of large amounts of information by the designer' as listed above in item 2. Subsequent to this are the selected load cases for maximised responses. These, as we shall see, are short-term sea states that generate loads that are very close to the long-term maximum loads. In the final step, rule loads are given on the basis of a probability of encounter of 10^{-8} during the lifetime of the ship. This is the maximum load a ship is expected to encounter during its 25-year operating life. Finally mention is made of load combination factors, which are introduced to allow for the concurrent action of more than one load type (vertical bending moment, horizontal bending moment, etc). A methodology to introduce load combinations was developed specifically to this end by the IACS classification societies involved.

The steps followed to determine design loads in the IACS procedure are listed below [14].

[14] Joint Tanker Project: American Bureau of Shipping, Det Norske Veritas and Lloyds Register.

The IACS procedure for design load calculation

1. Detailed direct structural analyses of global FE models of three bulk carriers and one double hull VLCC are carried out in order to determine the dominant load components that are most critical for the strength of ship structures.
2. Equivalent design waves are proposed together with the regular design wave conditions such as wave encountering angles, wave lengths and wave heights.
3. Hydrodynamic pressure distributions under each EDW are developed.
4. Load combination factors for hull girder loads, ship motions and accelerations due to ship motions under each EDW are identified.
5. The proposed extreme design loads are verified by structural analysis

10.6.3 Design Sea States

In a rational ship structural design environment, the design loads that will be used are based on *design sea states*, which ought to generate the highest load each ship is expected to encounter during its operating life. They therefore have to include the severest conditions that the ship is expected to encounter. Once the appropriate design loads have been specified, structural calculations can be carried out that will lead to scantlings that will ensure that the ship will perform in a satisfactory way for at least the duration of its specified design life, provided proper maintenance is carried out. The determination of design sea states is therefore of fundamental importance to structural design and ship operation.

Ships are expected to operate for at least a 25-year period during which they will encounter 10^8 waves of differing periods and heights. In order to determine the load profile the ship will be subjected to during this period it is necessary to know the voyages to be performed and the sea states expected to prevail during each voyage (*long-term analysis*). In a linear system (moderate sea states) it is possible to obtain the response for each individual sea state in the frequency domain. If, however, the response is nonlinear, as in the case of severe storms, time-domain simulation involving lengthy calculations is required. For this reason long-term analysis is a complicated and time-consuming procedure, and it becomes necessary to seek alternate ways of defining design sea states.

It has been found that long-term extremes can be estimated by considering only a few short-term sea states,[15] and to this end several authors, such as [3], have proposed procedures. On this basis, a *design sea state* can be defined as the short-term sea state that generates response values that are equivalent to the long-term predictions of stress (exceedance probability $Q = 10^{-8}$, all headings) for the primary structural members of a ship.

In order therefore to develop a benchmark for critical short-term sea states it was necessary to also perform a series of long-term analyses. A series of short-term analyses was also carried out, and following the selection of the severest, or *dominant*, sea states, a

[15] A *short-term sea state* refers to conditions that last for a few hours. Typically, for a three-hour period, 1,000 waves are expected to be encountered.

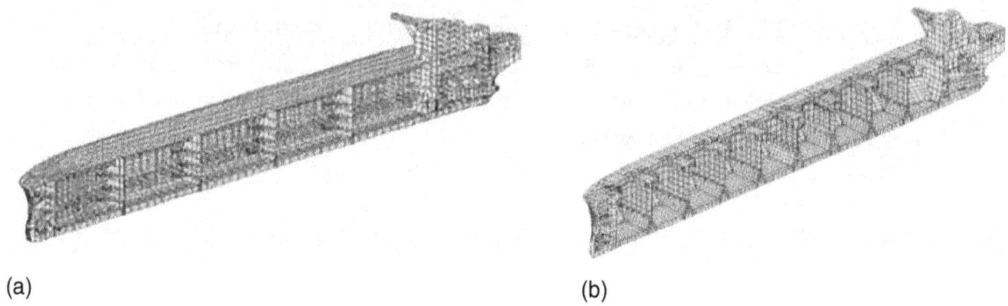

Figure 10.11 Finite element models used in direct stress analysis. (a) Double hull VLCC model. (b) Capesize bulk carrier model. Reprinted courtesy of Elsevier.

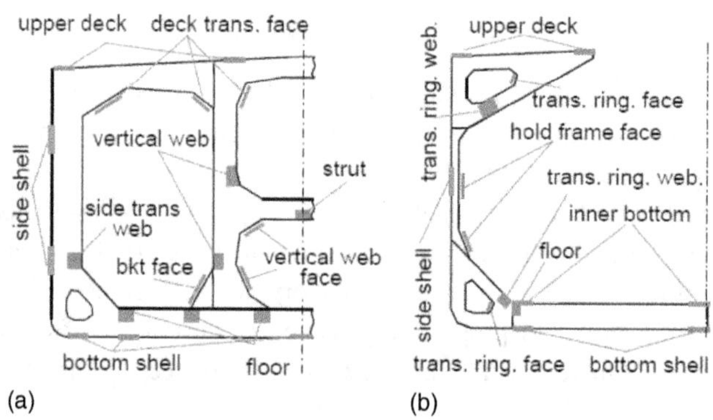

Figure 10.12 Selected critical structural members [23]. (a) Double-hull VLCC. (b) Capesize bulk carrier. Reprinted courtesy of Elsevier.

number of these were selected as design sea states. In order to select the design sea states from the dominant sea states, dominant load components were identified, as discussed in Section 10.6.3.2.

Short-Term Analysis

Finite element models of three bulk carriers and one double hull VLCC (Figure 10.11) were developed and used to obtain stress RAOs for selected primary structural members, some of which are indicated in Figures 10.12a and 10.12b. These were obtained for short-term sea states for a constant wave height. These are described by the ISSC spectrum shown in Figure 9.14 in conjunction with the spreading function given by Equation (9.48). Calculations were carried out for series of (1) loading conditions, (2) wavelengths and (3) wave encounter angles. Details are included in Table 10.9.

Ship speed was varied according to sea state severity (Figure 10.13). Thus, speed was reduced to 5 knots in extreme conditions, this being the minimum speed required to maintain manoeuvring and other operations. For fatigue load calculations the average speed over a 25-year period was assumed to be $0.75V_{des}$.

Table 10.9 *Analysis conditions of ship models*

	Tanker	Full loading condition	Partial loading condition		Ballast condition
		Full loading		Ballast condition	
Loading condition	Bulker	Homogeneous	Alternate	Normal Ballast	Heavy ballast
Wave height		$H = 5.0$ m			
Wave length		$\lambda/L_{pp} = 0.2, 0.4, 0.6, 0.8, 1.0, 1.2, 1.4, 1{,}7, 2.0, 2.3$			
Wave encounter angle		$\chi = 0°, 300, 60°, 90°, 120°, 150°, 180°, 210°, 240°, 270°, 300°, 330°$			
Ship speed		$V_{ref} = 0.75\,V$ (V = max. ahead service speed in knots)			
Short-term prediction		Wave spectrum: ISSC-1964 Directionality function: $\cos^2\theta$			
Long-term prediction		Wave scatter diagram (IACS Recommendation No. 34 as amended)			

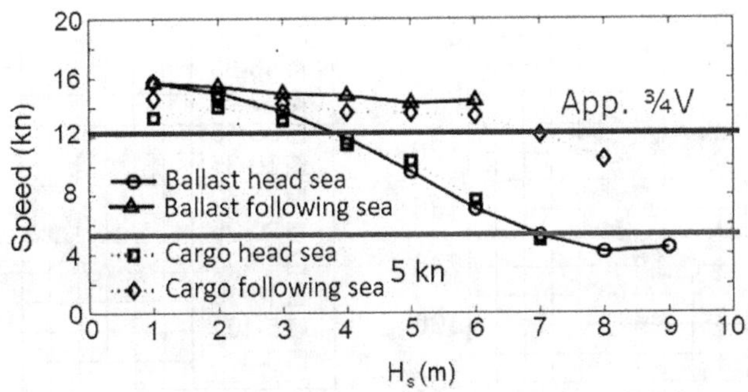

Figure 10.13 Effect of sea state severity on ship speed [16]. Reprinted courtesy of IACS, London.

Selection of Dominant Short-Term Sea States

It has been shown [20] that the maximum expected response under the most severe short-term sea states almost coincides with the long-term predicted value. On this basis, sea states for which the standard deviation of the response becomes a maximum were proposed as dominant short-term sea states. A sea state is specified by its mean wave period T_V and the wave encounter angle χ. The procedure followed to identify dominant sea states consists of the following steps:

1. Conduct structural analysis of selected examined members.
2. Determine RAO for each load condition.
3. Perform short-term predictions of stresses using RAOs and wave spectra.
4. Select sea states (identified by combinations of mean wave period T_V and wave encounter angle χ) for which the standard deviation of the response per significant wave height becomes a maximum.

Consider, for example, Figure 10.14, which shows the RAO for hydrodynamic pressure along the waterline plotted against mean wave period, T_V. It is seen that in the case of Ship

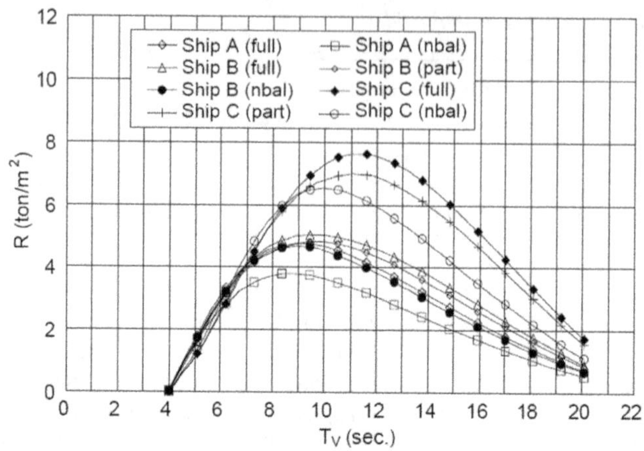

Figure 10.14 Response amplitude operator (RAO) for hydrodynamic pressure along waterline. Reprinted courtesy of Elsevier.

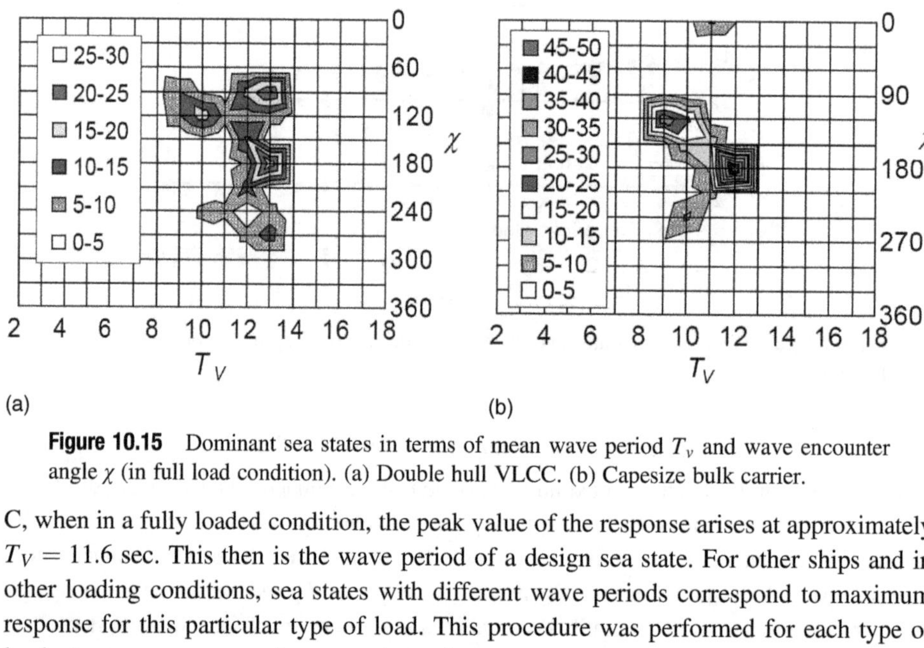

Figure 10.15 Dominant sea states in terms of mean wave period T_v and wave encounter angle χ (in full load condition). (a) Double hull VLCC. (b) Capesize bulk carrier.

C, when in a fully loaded condition, the peak value of the response arises at approximately $T_V = 11.6$ sec. This then is the wave period of a design sea state. For other ships and in other loading conditions, sea states with different wave periods correspond to maximum response for this particular type of load. This procedure was performed for each type of load, stress component and structural member.

Figure 10.15 illustrates the properties of the dominant sea states selected on the basis of the load response amplitude operators. The sea states are described in terms of mean wave period T_v and encounter angle χ, and it is seen that the dominant short-term sea states lie in a narrow range of values.

Identification of Dominant Load Components

In the next step, the load components that reproduce short-term sea states equivalent to the selected short-term sea states are identified. These are termed *dominant load components.*

The dominant load components are identified by comparing the structural responses of the examined members in the selected short-term sea states with the long-term stress predictions, based on the procedure described below.

Identification of dominant load components

1. The standard deviation $[R_{STRESS}/H_W(T_{Vi},\chi_i)]$ of the stress corresponding to the selected short-term sea state for each load component is determined.
2. The structural response (a_i) of the examined member in the selected short-term sea state for each load component is determined from the expression given below:

$$a_i = 2H_{Wi}[R_{STRESS}/H_W(T_{Vi,}\chi_i)1.9 \quad (10.4)$$

3. The dominant load component for the examined member is selected by comparing the determined structural response (a_i) with the long-term prediction of stress (b_i).

   The comparison is performed by calculating the percentage ratio of a_i and b_i (ratio Ra), from:

$$Ra = \frac{a_i}{b_i} \text{X } 100\% \quad (10.5)$$

The load components considered when selecting dominant load components are grouped as follows:

1. Ship motions (surge, sway, heave, roll, pitch and yaw)
2. Accelerations (vertical, transverse and longitudinal accelerations at the centreline and side of deck at midship, bow and stern sections)
3. Wave-induced sectional forces and moments (vertical bending moment, horizontal bending moment, torsional moment, vertical and horizontal shear forces)
4. Hydrodynamic pressures (along the waterline, bilge keel and bottom centreline at midship, bow and stern sections)

The above procedure was implemented for each load condition and each stress component. Results showed that in the case of the load components that are included in Table 10.10, the maximum response values in the corresponding short-term sea states can be used to estimate the long-term stress, with an error of ±10 per cent. These were thus selected as dominant load components.

On this basis, the short-term sea states corresponding to these load components are proposed as *design sea states* and are included in Table 10.10. In the case of design sea state (DSS) *L*-180, vertical acceleration is the result of pitching and heaving motions, and the vertical bending moment reaches its maximum value. In the case of DSS *L*-0, the vertical acceleration reaches its minimum value. In the case of DSS *R*, the transverse acceleration due to roll motion and the asymmetrically distributed hydrodynamic pressure are at their maximum values. In the case of DSS *P*, the hydrodynamic pressure along the waterline in way of the weather side and the vertical acceleration due to heaving reach their maximum values (Figure 10.16).

Table 10.10 *Dominant load components, their wave encounter angles and corresponding design sea states*

No	Load component	Wave encounter angle	Design Sea State (DSS)	Design Condition
1	Vertical Bending Moment	0° (head sea)	*L*-180	*H*
2	Vertical Bending Moment	180° (foll. sea)	*L*-0	*F*
3	Roll	90° (beam sea)	*R*	*R*
4	Hydrodynamic pressure	90° (beam sea)	*R*	*P*
5	Wave Torsional Moment	60° (oblique sea)	*P*	*T*

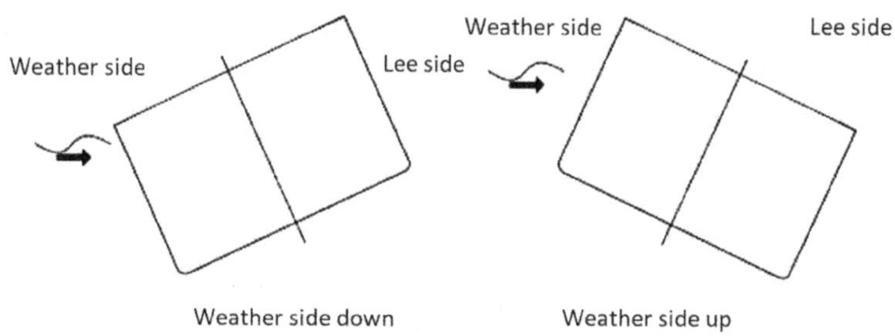

Figure 10.16 Definition of direction of beam sea for design conditions *R* and *P*. Reprinted courtesy of IACS, London.

The long-term sea states considered were based on the wave scatter diagram for the North Atlantic region included in [11]. The world map is shown in Figure 9.31, and the relevant scatter diagram is included in Table 9.3. In order to obtain the long-term prediction of the maximum load, Fukuda's approach was followed [9]. This is essentially the LTCD method, which is described in Section 9.3.3.

10.6.4 The Equivalent Design Wave

The idea of an equivalent design wave (EDW) stems from past practice, in which a quasi-static wave of a particular geometry (such as the trochoid) was used to represent a critical wave loading condition of the hull girder (Chapter 4). In the case of the IACS CSR, the design wave is determined on the basis of quite different principles, as described below.

The *equivalent design wave* used in the CSR is a regular (sinusoidal) wave which, when applied to the hull girder, gives rise to a maximum short-term value of a sinusoidal load parameter that corresponds to worst short-term conditions, assuming this approximates with sufficient accuracy the long-term value of the parameter. The load parameter may, for example, be the bottom pressure or the transverse acceleration at a given location.

The IACS definition is given below [16]:

An EDW is defined as a regular wave that reproduces the same response value as the reference design value, hence the envelope value at a certain probability level of a maximised response.

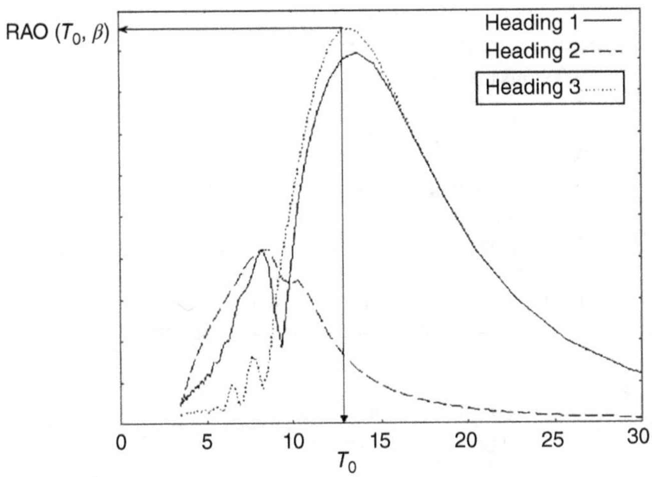

Figure 10.17 Determination of EDW heading and wave period [16]. Reprinted courtesy of IACS, London.

The EDW is defined by its amplitude, its period and its heading with respect to the ship (A_0, T_0, β_0). In order to obtain these, the load RAO is plotted against wave period for all headings considered. For the heading for which the peak value of the RAO is a maximum, the corresponding period is found (Figure 10.17).

The amplitude of the equivalent design wave, A_0, is equal to the ratio of the long-term value of the response to the peak value of the RAO determined using short-term analysis, at a given period T_0 and heading β_0:

$$A_0 = \frac{\text{Long term value}}{RAO(T_0, \beta_0)} \tag{10.4}$$

The long-term value of the response is calculated at a specified probability of exceedance that depends on the nature of the response. Thus for ultimate strength a probability of 10^{-8} is used, whereas for fatigue the corresponding probability is 10^{-6}.

Results obtained from the short-term analysis are compared with those obtained from long-term analysis for the Capesize bulk carrier in Figure 10.18a. These results correspond to the structural members indicated in Figure 10.18b.

It is seen that the differences between the short-term and long-term predictions do not exceed 25 N/mm^2, with differences being less than 10 N/mm^2 in most cases. If still-water loading is taken into account as well, these discrepancies will reduce proportionately.

The seven EDWs that have been proposed for use in the determination of extreme design loads in yielding, buckling and ultimate strength assessment of structural members of bulk carriers and oil tankers are included in Table 10.11.

The encounter angles of the above EDWs are shown in Figure 10.19.

Table 10.11 *Equivalent design waves (EDWs) for the determination of extreme loads*

No	Maximisation of	EDW
1	Vertical bending moment in head sea	HSM
2	Vertical acceleration at FP in head sea	HSA
3	Vertical bending moment in following sea	FSM
4	Roll motion in beam sea	BSR
5	Hydrodynamic pressure along waterline	BSP
6	Wave torsional moment at $x = 0.25L$ in oblique seas	OST
7	Pitching acceleration in oblique sea	OSA

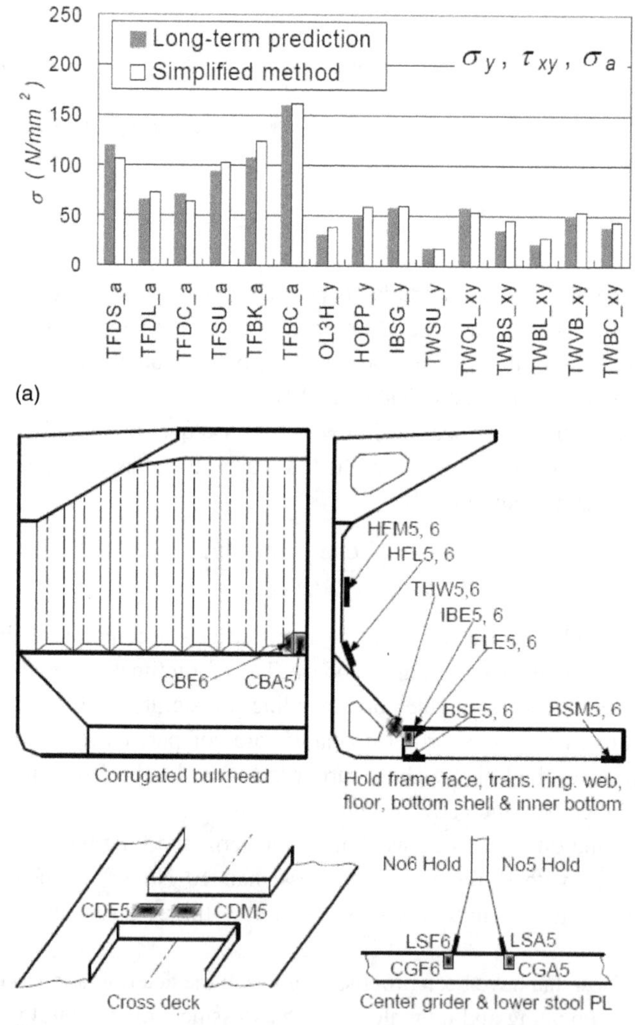

Figure 10.18 Validation of results obtained using short-term analyses (capsize bulk carrier). (a) Comparison of predicted stresses. (b) Location of members where stress comparisons were carried out. Reprinted courtesy of Elsevier.

Table 10.12 *Design wavelength for each EDW*

EDW	Regular design wavelength, λ (m)
HSM	$0.6(1+f_T)L$
HSA	$0.6(1+f_T)L$
FSM	$0.6(1+2f_T/3)L$
BSR	$gT_\theta^2/(2\pi)$
BSP	$(0.2+0.4f_T)L$
OST	$0.45L$
OSA	$0.70L$

$f_T = T_{LC}/T_{SC}$
T_{SC}: Design moulded draught (m)
T_{LC}: Draught amidships for the relevant loading condition (m)

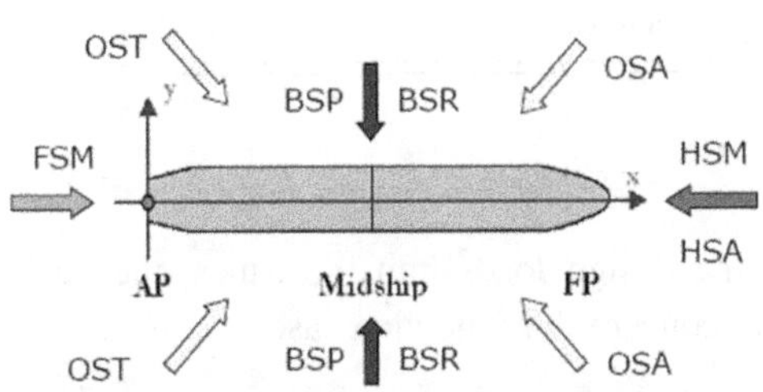

Figure 10.19 Encounter angles of EDWs. Reprinted courtesy of IACS, London.

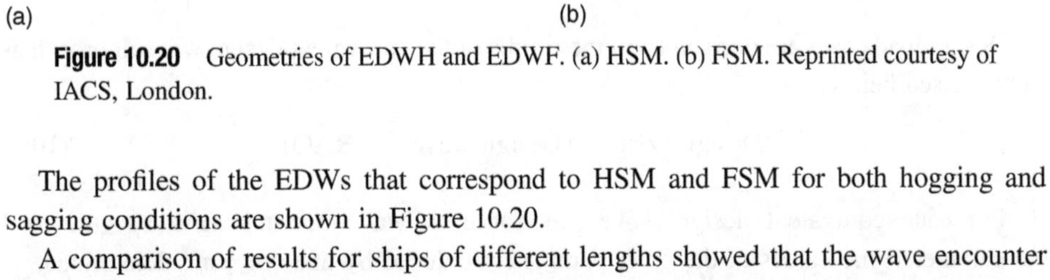

Figure 10.20 Geometries of EDWH and EDWF. (a) HSM. (b) FSM. Reprinted courtesy of IACS, London.

The profiles of the EDWs that correspond to HSM and FSM for both hogging and sagging conditions are shown in Figure 10.20.

A comparison of results for ships of different lengths showed that the wave encounter angles that correspond to the maximum value of the RAO remain the same throughout, so that these encounter angles can be used for the design regular wave. With regard to wavelength and wave height of the EDWs, a series of expressions are proposed for use. These are included in Tables 10.12 and 10.13. Comparison with direct analysis indicated that these expressions retain sufficient accuracy for design purposes.

Table 10.13 *Design wave height for each EDW*

EDW	Regular wave height, H (m)
HSM	$1.31C_w\sqrt{\frac{L+\lambda-125}{L}}$
has	$1.09C_w\sqrt{\frac{L+\lambda-125}{L}}$
FSM	$1.31C_w\sqrt{\frac{L+\lambda-125}{L}}$
BSR	$0.88C_w\sqrt{\frac{L+\lambda-125}{L}}$
BSP	$1.5C_w\sqrt{\frac{L+\lambda-125}{L}}$
OST	$1.38C_w\sqrt{\frac{L+\lambda-125}{L}}$
OSA	$1.38C_w\sqrt{\frac{L+\lambda-125}{L}}$

λ: regular design wave length for each EDW (m)
C_w: wave coefficient (m)

10.6.5 Design Loads

In order to introduce the design loads into the Rules, they are expressed in terms of simplified formulae and concern the following cases:

1. Ship motions, accelerations and hydrodynamic pressures, as follows:
   a. Ship motions (surge, sway, heave, roll, pitch and yaw)
   b. Accelerations at the centre of gravity of the ship (vertical, transverse)
   c. Accelerations at any point on the ship
   d. External and internal hydrodynamic pressures
   e. Vertical bending moment and shearing force
   f. Horizontal bending moment
2. Internal pressure due to dry cargo
3. Static pressure
4. Dynamic pressure

Design load cases are derived from the results of the equivalent design wave approach as summarised below.

$$\text{Design Load} = (\text{Design wave}) \times (\text{RAO}) \tag{10.5}$$

1. For each equivalent design wave (i.e. for each wave maximising a specific load parameter) the values of the other parameters induced by that wave are determined.
2. The value of each load parameter for the equivalent wave is expressed as a fraction of its long-term value, so that, for each wave, a combination is obtained in which the maximised parameter appeared with a factor equal to 1.0 and the other with lower combination values.
3. Each combination identifies a load case.

4. Analysing the load cases so obtained for all the considered ships, it is possible to identify some recurrences in the combinations between the load parameters. Finally, the design load cases are identified.

References

1. American Bureau of Shipping (2017) *Noise and Vibration Control for Inhabited Spaces* Guidance Notes.
2. Andersen I. M. V. Jensen J. J. (2012) On the Effect of Hull Girder Flexibility on the Vertical Wave Bending Moment for Ultra Large Container Vessels. *ASME 31st Int. Conf. Ocean Offshore & Arctic Eng. OMAE 2012* 259-266.
3. Baarholm G. S. Moan T. (2000) Estimation of nonlinear long-term extremes of hull girder loads in ships *Mar. Struct.* Vol. 13 495–516.
4. Beach J. E. (2012) 40+ years of ships structures advances: Opportunities gained opportunities lost. *ASNE & Columbia Group Breakfast Seminar* Washington, DC.
5. Blanchet A. (2000) Comfort Class and Passenger Ships. *Inter. noise 2000 The 29th Int. Congress and Exhibition on Noise Control Engineering* Nice.
6. Chalmers D. W. (1988) Hull structural design using stiffness as a criterion. *RINA Annual Spring Meeting, London.*
7. Dym C. L. (1994) *Engineering Design: A Synthesis of Views.* Cambridge: Cambridge University Press.
8. Evans J. H. (Ed) (1975) *Ship Structural Design Concepts.* Centreville, MD: Cornell Maritime Press.
9. Fukuda J.-I. (1967) Theoretical determination of design wave bending moments. *Japan Shipb. & Mar. Eng.* Vol. 2 13–22.
10. 'Home - AccessScience from McGraw-hill education' *Accessscience.com.* Available: www.accessscience.com (accessed 19 December 2020).
11. International Association of Classification Societies (2001) Standard Wave Data. Recommendation No. 34.
12. International Association of Classification Societies (2004) Joint Tanker Project. Development of Common Structural Rules for Oil Tankers. Presentation by A. Johnston.
13. International Association of Classification Societies (2006) Common Structural Rules for Bulk Carriers. Background Document Chapter 4 – Design Loads.
14. International Association of Classification Societies (2012) EDW Definition of Extreme Loads. TB Report No Pt1, Ch 4, Sec 2 (Rev 1 July 2012).
15. International Association of Classification Societies (2020) *Common Structural Rules for Bulk Carriers and Oil Tankers.*
16. International Association of Classification Societies (2020) *Common Structural Rules for Bulk Carriers and Oil Tankers. Technical Background Rule Reference.*
17. International Association of Classification Societies (2020) Unified Requirement: Longitudinal Strength Standard (S11) (Rev. 10).
18. International Maritime Organisation (2002) Bulk carrier safety. Rept. on FSA Study on Bulk Carrier Safety. MSC 75/5/2.
19. International Maritime Organisation (2012) Annex 1 Resolution MSC.337 (91) Adoption Of The Code On Noise Levels On Board Ships.
20. Mano H. Ueno H. (1972) A simplified estimation of long-term distributions of various random variables induced by ocean waves, and a study on the influences of operating conditions on the extremes. *J. Soc. Nav. Arch. Japan* 235–247 (in Japanese).

21. Okumoto Y. Takeda Y. Mano M. Okada T. (2008) *Design of Ship Hull Structures* Berlin: Springer Verlag.
22. Paik J. K. Thayamballi A. K. (2007) *Ship-shaped Offshore Installations. Design Building and Operation*. Cambridge: Cambridge University Press.
23. Shigemi T. Zhu T. (2003) Practical estimation methods of the design loads for primary structural members of tankers. *Mar. Struct.* Vol. 16 275–321.
24. Todd F. H. (1962) *Ship Hull Vibration*. London: Edward Arnold (Pub.) Ltd.
25. 'Welcome to the MPI Group – marine publishing training & marketing' *Mpigroup.co.uk*. Available: www.mpigroup.co.uk (accessed 19 December 2020).
26. Wikipedia contributors 'Design' *Wikipedia The Free Encyclopedia* 30 November 2020. Available: http://en.wikipedia.org/w/index.php?title=Design&oldid=991521297 (accessed 19 December 2020).

11

Aspects of Uncertainty

Structural Reliability Theory and Fuzzy Logic

Hope is not a strategy.

Betsy Beyer

11.1 Types of Uncertainty

A dictionary definition of *uncertainty* is:

The lack of certainty. A state of having limited knowledge where it is impossible to describe exactly the existing state, a future outcome, or more than one possible outcome.

As such, the causes of uncertainty are several. In the case of engineering and scientific problems, the issue is having sufficient information in order to assess the behaviour of structures or equipment as well as to make practical decisions that relate primarily to design issues. Information may be imprecise, fragmentary, unreliable, vague, ambiguous, incomplete or deficient in some other way. The most well-known type of uncertainty that characterises engineering problems is due to *random processes*, such as the wave loading of ships. These lie outside the control of the analyst, and in order to model them, probability theory and statistics are used. In this chapter, probability theory will be used to establish the basis of structural reliability theory, and in the next chapter *marine structural reliability theory* and its use are discussed.

As intimated above, uncertainty arises for a variety of other reasons, in addition to random processes:

- A lack of clarity or precision, or the use of an approximate description when specifying a particular requirement. In this case, the term *fuzziness* is used.
- Vagueness, in which case the information provided is not specific. With information that is vague, it is difficult to form any interpretation at the desired level of specificity.
- Ambiguity, in which case the information is liable to misinterpretation.
- Ignorance, meaning lack of sufficient information about a particular issue (incompleteness)

These additional sources of uncertainty need to be considered in appropriate ways, as they are of importance in engineering and ship design. We shall focus on the first of these, as this characterises the opinions of individuals or bodies that are involved in design decisions in which the grading of particular features takes place and may be all-important. As such, it is not a type of uncertainty that can be catered for in the same way as vagueness and ambiguity. For example, when a designer is asked to express an opinion on how important the cost of construction of a ship is, their opinion is not given in quantitative (numerical) terms but in linguistic terms. They will not say, 'We feel construction cost is 20 per cent important' but that 'Construction cost is not particularly important.' Such communication needs to be assessed in a rational manner so that a practical decision can be reached. One way of doing so is to express a qualitative assessment in quantitative terms. Solutions to problems of this kind can be obtained by turning to *fuzzy logic*. Fuzzy logic and fuzzy set theory, which have seen rapid growth in many engineering and non-engineering fields since their inception in 1965 [32], are discussed in the last sections of this chapter. The application of fuzzy logic in a ship structural design problem is discussed in Chapter 14.

11.2 Risk and Reliability

A survey of incidents and accidents in the marine field reveals features that are common to other industries such as air transportation, land transportation and the nuclear power industry. During the lifetime of a ship, there exist a variety of undesirable events that may or may not occur, some of which may have serious consequences. For example, if the outer shell plating that forms part of the bow structure of a ship is subjected to wave impact, permanent dishing is the most likely outcome. If, on the other hand, a ship whose hull girder has been weakened as a result of corrosion is discharged in an improper manner, the ship may possibly break in half. Clearly, the consequences of the events in these two cases differ significantly. Additionally, the probabilities of occurrence of these events differ, as is clear from practical experience and recorded data. Nevertheless, it is necessary to design structures that have to operate in the real world, in which both of these events are possible. A design approach therefore has to take into account both the consequences of an event as well as the probability of that event occurring. It is possible to do this by introducing the concept of *risk*. In this context,[1] risk is 'a measure of the expected loss that results from a random event' [5] and can be quantified as the expected value of an undesirable outcome. If discussion is confined to a single outcome, risk can be determined in quantitative terms from the following expression:[2]

$$\textit{Risk} = (\text{Consequences of an event}) \times (\text{Probability of that event occurring}) \tag{11.1}$$

[1] In finance and investment analysis, the concept of risk is used in the valuation of all kinds of investments (real estate, stocks, etc.) and investment regimes (creditworthiness of an individual, a firm or a country) and is expressed in quantitative terms. A "risk premium" is added to the nominal cost of the investment. Risk is also viewed as a measure of volatility, the *volatility index* being used as a quantitative measure of this.

[2] This definition has been attributed to Blaise Pascal and dates from the 17th century [7].

Two or more events whose consequences and probabilities of occurrence differ can then be compared directly. In the marine field, the sources of risk are many and can be broadly subdivided into the following categories:

- Failure of equipment
- External events (*force majeure*), that is, random events such as freak weather conditions
- The human factor, involving possibly the crew, officers, company management, regulatory bodies, existing rules and regulations

A study of marine accidents reveals that serious incidents may arise as a result of failure at several levels concurrently or sequentially (crew negligence, inadequate reporting, poor company organisation, inadequate design, poor maintenance, etc.). It thus becomes necessary to analyse the manner of progression of failure or malfunction throughout the system.[3] Furthermore, when considering the operation of a complex system, it is necessary to include the risk associated with more than one event, so that it becomes necessary to introduce the concept of *total risk*:

$$\textit{Total Risk} \equiv [(p_1, C_1), (p_2, C_2), \ldots, (p_n, C_n)] \tag{11.2}$$

where p_n and C_n are, respectively, the probability of occurrence and the consequences of a particular event. It is necessary to specify the duration for which the risk is calculated (e.g. per annum) as well as to specify the units in which the consequences are measured. Generally speaking, the consequences are most easily quantified in monetary terms, although this may result in oversimplification and loss of objectivity when interpreted at a broader, social level. The presence of risk has long been acknowledged in the maritime industry,[4] and insurance premiums reflect in part the risk involved in individual operations, such as the transportation of a quantity of cargo that has a certain value from a particular loading port to a particular destination using a particular vessel owned by a particular company.

The first explicit application of the risk concept was carried out in the nuclear industry, and Farmer produced curves in which consequences (number of fatalities) were plotted against annual frequency of occurrence (Figure 11.1). These curves show that as the number of fatalities increases (that is, the consequences become more serious), the frequency of occurrence of the corresponding event decreases. The curves also show the relative fatality risk in various industries.

The implementation of the risk approach has followed two separate paths: (1) *risk management,* which includes *risk assessment* and *risk control*; and (2) *risk communication*. Risk management begins with actively identifying possible *hazards* leading to the ongoing management of those risks deemed to be acceptable. Risk assessment and risk control are indicated in more detail in Figure 11.2 [4]. Risk assessment includes *risk analysis* and *risk*

[3] In the case of complex structures, it is possible for failure of an individual member (such as a stiffened plate) to lead to failure (collapse) of another member, such as a transverse frame or an adjoining stiffened plate. The manner in which individual members interact within a structure and the corresponding probabilistic analysis of the various failure mechanisms is conducted using techniques that come under *systems reliability theory*. This is a distinct field of the probabilistic analysis of structures that will not be discussed in this chapter.

[4] As mentioned in Section 1.4.3 risk was recognised and quantified in the maritime sector during the late 18th century when underwriters at Lloyd's instructed surveyors to assess the seaworthiness of ships in order to classify them accordingly, and estimate risk and, hence, insurance premiums. This was an important step forward at that time.

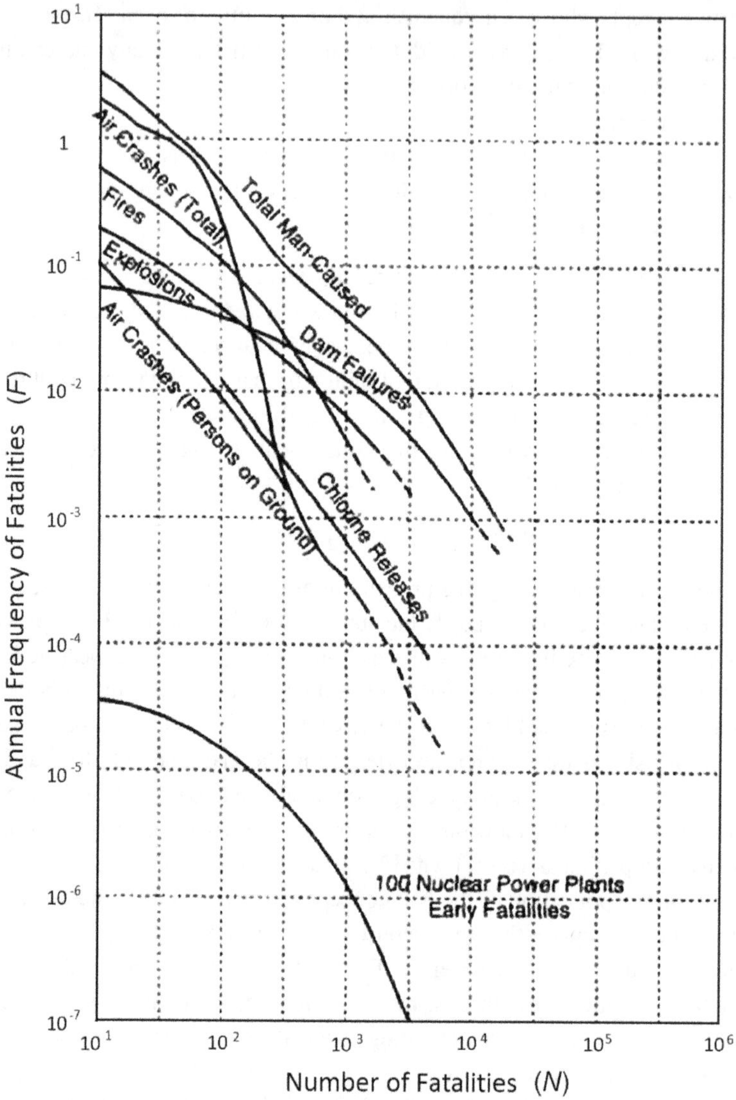

Figure 11.1 Fatal accidents and frequency of occurrence [2].

evaluation. It aims at risk identification, event probability assessment and consequence assessment. Its purpose is to provide answers to the following questions:

1. What can go wrong?
2. What is the probability that it will go wrong?
3. What are the consequences if it does go wrong?

Techniques that have been developed to perform risk assessment include:

1. *Preliminary Hazard Analysis (PrHA).* Used to identify and prioritise hazards leading to undesirable consequences early on in the life of a system. Also used to determine

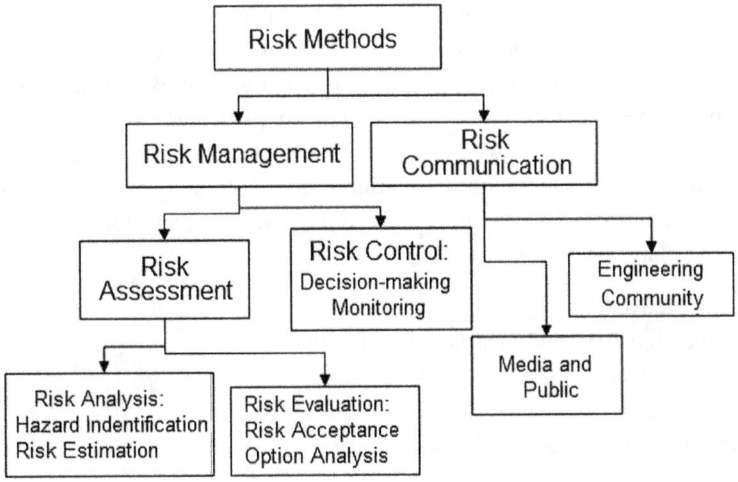

Figure 11.2 The implementation of the risk concept in engineering [5].

recommended actions to reduce the frequency and/or consequences of prioritised hazards.

2. *Hazard and Operability Studies (HAZOP).* These are used to identify system deviations and their causes that can lead to undesirable consequences. They are also used to determine recommended actions in order to reduce the frequency and/or consequences of the deviations.
3. *Failure Modes and Effects Analysis (FMEA).* Used to identify the components (equipment) failure modes and the impacts on the surrounding components and the system
4. *Failure Modes Effects and Criticality Analysis (FMECA).* Identifies the components (equipment) failure modes and the impacts on the surrounding components and the system, and criticality of failures.
5. *Fault Tree Analysis (FTA).* Identifies combinations of equipment failures and human errors that can result in an accident.
6. *Event Tree Analysis (ETA).* Identifies various sequences of events, both failures and successes, that can lead to an accident

The first two of these are based on expert opinions and are thus qualitative risk methods, whereas the remainder are quantitative risk methods.

As defined previously, the evaluation of risk involves the assessment of the effect of occurrence of an event, which is carried out using *consequence analysis*, as well as the calculation of the probability of occurrence of the event. Consequence analysis is an evaluation of the predicted outcome of an incident and how it affects the surrounding equipment and people. It involves the use of theoretical models that may be analytical or numerical as well as experimental techniques. When considering a new design, it can be used to optimise a structural arrangement, reduce the risk from an unacceptable level by improving design, develop an emergency preparedness plan and assess possible mitigation systems.

The calculation of the probability of occurrence of an event, on the other hand, is carried out using *reliability analysis*, which we will consider further in this chapter.

Before turning to reliability analysis, it is necessary to stress the importance of the question concerning the *acceptable level of risk* in particular circumstances. This is a difficult question to answer, as the perception of acceptable risk by society at large is not based on informed, rational criteria. Furthermore, the acceptable level of risk may change with the passage of time. In order to determine the level of acceptable risk in general, the following steps have been proposed [17]:

1. Define alternatives
2. Specify the objectives and measures for effectiveness
3. Identify consequences of alternative
4. Quantify values for consequences
5. Analyze alternatives to select the best choice.

Having specified risk in quantitative terms for a particular case, it is possible to consider the related or desired values for the two components of risk, which are the probability of failure and the consequences of failure. In order to provide a rational basis for a design value for the probability of failure, when a risk approach is implemented, a *consequence analysis* must be performed. Following this, the desired or *target probability of failure* can be specified. We see therefore that in order to implement classical structural reliability methods in conjunction with a risk-based approach, prior calculations are required. For this reason we designate these calculations, which involve risk assessment and consequence analysis, as *Level IV* reliability methods.[5] The reliability calculations that are directly concerned with the structure in question and aim at providing design data are termed *Level III*, *Level II* and *Level I* methods for reasons that will become clear in later sections of this chapter.

It is useful at this stage to consider the contents of Table 11.1 that includes target annual probabilities of failure for failures whose consequences have different levels of severity. This data have been proposed by the International Maritime Organisation for use by classification societies when drafting rules for ship construction [15].

11.3 Basic Probability Ideas and a Definition of Reliability

Before considering probabilistic ideas in ship structural analysis and design, it will be useful to remind ourselves of the corresponding concepts and their role in traditional structural design. Let us therefore consider a simple beam that is subjected to a varying external load. In traditional design, fixed (deterministic) values are assumed for the *design stress* and for the beam properties that are associated with strength (elasticity modulus, yield stress, ultimate stress).

[5] The target probability of failure (or reliability) is a function not only of the consequences of failure but also of model uncertainty, structural redundancy and the possibility of early warning. In Level IV reliability analysis, a wider perspective has to be adopted.

Table 11.1 *Target annual failure probabilities* (p_f) *and corresponding reliability indices* (β) *for different failure consequences [11]*

Class of failure	Not serious p_f	Not serious β	Serious p_f	Serious β
I – Redundant structure	10^{-3}	3.09	10^{-4}	3.71
II – Significant warning before the occurrence of failure in a non-redundant structure	10^{-4}	3.71	10^{-5}	4.26
III – No warning before the occurrence of failure in a non-redundant structure	10^{-5}	4.26	10^{-6}	4.75

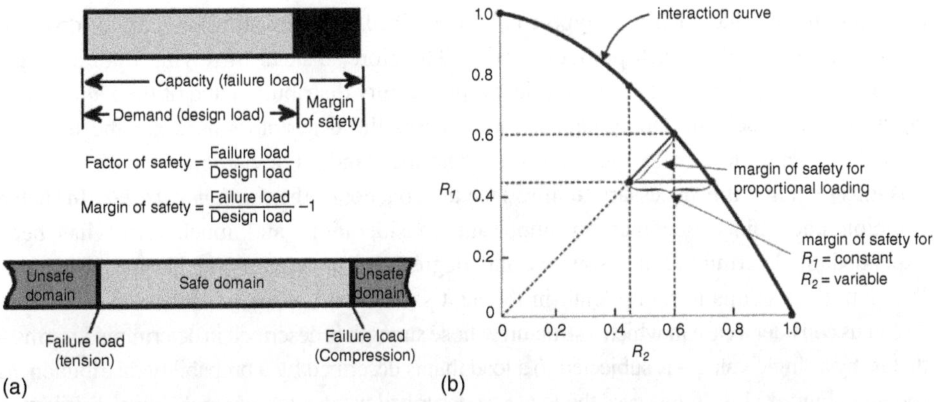

Figure 11.3 Margins of safety in deterministic formulations. (a) Single load acting. (b) Two concurrent loads (R_1, R_2 are the ratios of actual to failure loads).

For present purposes we can consider failure to occur at the onset of yielding. If the design stress is less than the yield (failure) stress, safety is ensured provided the design stress is never exceeded. In this manner safety is ensured throughout the lifetime of the structure.

The difference between the design stress and the failure stress, which in this case corresponds to initial yielding, represents a margin of safety[6] (Figure 11.3a). Regardless of the type of failure, safety is ensured by introducing a *factor of safety,* defined as the ratio of the strength to the permissible load. In order to ensure safety, the factor of safety, when defined as the ratio of failure load/design load, must exceed the value of unity.

In the general case, more that one load acts, so that it is necessary to allow for their interaction and the consequences this has on structural safety. An example is presented in Figure 11.3b, which shows the effect of bending and axial loading on a beam-type structure. The margin of safety is here represented in geometrical terms, and can be defined

[6] The margin of safety can be defined in two different ways: (1) as a measure of *structural capacity* and (2) as a means of satisfying *design requirements* (requirement verification). The margin of safety can be a treated as a measure of the excess capacity of a structure, which is held "in reserve" during loading.

in different ways, depending on the manner in which the loads are applied. In Figure 11.3b two different ways are indicated. In the first case proportional loading is applied whereas in the second one load is maintained constant whilst the other load is increased. Such an approach can be extended to include additional load types, but in that case a geometrical representation is not possible.

This is the *permissible stress* approach, and it is clear that the issue of *uncertainty* with regard to safety does not arise, since safety is perceived as being present or absent.

Let us now consider a ship at sea. In Chapter 3 we noted that the environmental loads can vary significantly (distributions in Figures 3.2 and 3.5 (bending, thermal loads), Figure 3.11 (slamming loads) and Figure 3.8 (springing loads)). Furthermore, several different types of loads, in addition to the environmental loads, generally act concurrently (Table 5.12). Following extensive measurements as well as theoretical investigations, it has been determined that each load type can be described using existing probability distributions that are suited to each particular case.[7] Therefore, instead of relying upon a single value, the description consists of a particular probability distribution and of the values of its descriptors for each individual case. In most cases the descriptors used are the first and second moment, these being the mean value and the standard deviation of the distribution. It is thus seen that it is necessary to introduce the concept of the *random variable*. In such a situation uncertainty becomes an important consideration, and much effort has been expended in determining the sources and degrees of uncertainty in marine structures. We shall discuss this in more detail in the next section and in the next chapter.

Let us consider a case in which a structure whose strength is described in deterministic terms – that is, by a single value – is subjected to a load that is described by a probability distribution, as shown in Figure 11.4. In this case the load L is described using a known probability distribution (F_L, μ_L), whereas the strength has a constant value s. It is seen that, although a large proportion of load parameter values L is less than the strength parameter s (*domain of safe operation*), there are some that do exceed it, indicated by the shaded region (*failure domain*). There arises, therefore, the possibility of failure, and the structure is not in a safe state for all values of the load parameter L. The shaded region in fact represents the *probability of failure* p_f of the structure, and if the ordinate of the intersection is known, this probability can be readily evaluated with respect to a given load distribution. In this particular case,

$$p_f = P[s - L \leq 0] = F_L(s) = P[s/L \leq 1] \tag{11.3}$$

where F_L is the probability distribution function of the load and $F_L(s)$ is the probability that F_L exceeds s. From Figure 11.4 we observe that the probability of failure can be reduced by (1) increasing the strength s, (2) reducing the mean value of the load L and (3) reducing the spread of the load, expressed in quantitative terms by the standard deviation of the distribution.

Closely related to the probability of failure of a structure is the concept of *reliability*, which is a measure of the ability of the structure to continue functioning in a satisfactory

[7] Strictly speaking, frequency data obtained from measurements are not identical to the probability of occurrence of an event, which is axiomatic in nature [3]. That is, the difference between statistical frequency and probability of occurrence does not reduce to zero as the number of data points $n \rightarrow \infty$. For practical purposes, however, it is considered that probability of occurrence can be deduced from measurements.

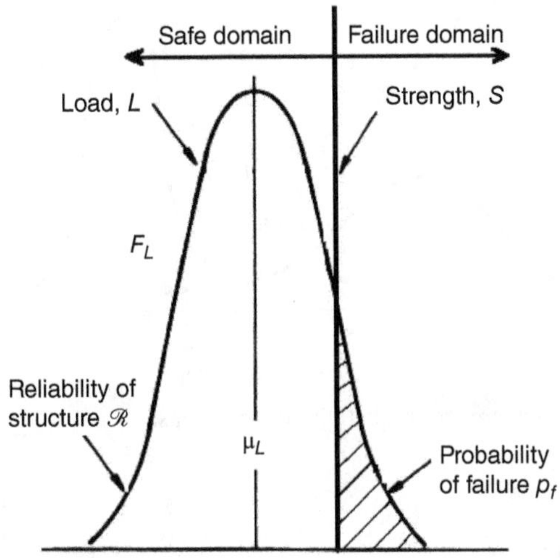

Figure 11.4 Load L and strength s (random variable and scalar respectively).

manner ('non-failure'). In quantitative terms, the reliability of a structure is defined to be equal to the probability that it does not fail, that is,

$$\boldsymbol{R} = 1 - p_f \tag{11.4}$$

where $\boldsymbol{R}$ is the reliability of the structure.

It has been found that the properties that influence structural response are characterised by uncertainty, as are the applied loads. These include material properties (elastic modulus, yield stress, ultimate stress), structural properties (mean thickness, out-of-plane imperfections) as well as those introduced during fabrication (geometrical imperfections and residual stresses). Clearly then it is necessary to allow for *variability* in both the applied loads and the structural response. This will involve the interaction of the corresponding probability distributions, and as in the case of loading, there exist particular distributions that are suited to represent structural response. The distributions that represent load and response are *independent* of each another, since changing the parameters of one does not affect those of the other. Figure 11.5 shows two probability distributions that overlap and may represent the load and response parameters for a particular structure.

Since the two distributions overlap, the probability of structural failure is greater than zero $(p_f > 0)$. We therefore need to determine this probability and the associated margin of safety. This is no longer represented by an area but by a distribution, whose nature and properties are not known.

In this case, the probability of structural failure may be restated as

$$p_f = P[S < L] = P[(S - L) \leq 0] \tag{11.5}$$

where the load L and strength S are now both random variables.[8]

[8] Capital letters are used as symbols for random variables whereas small letters are used for deterministic variables.

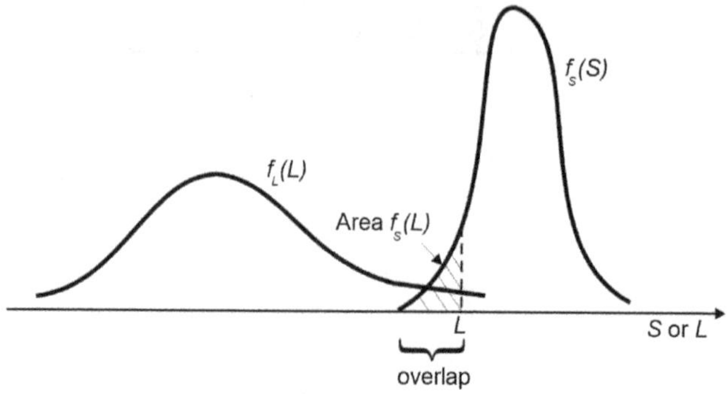

Figure 11.5 Load and strength probability distributions.

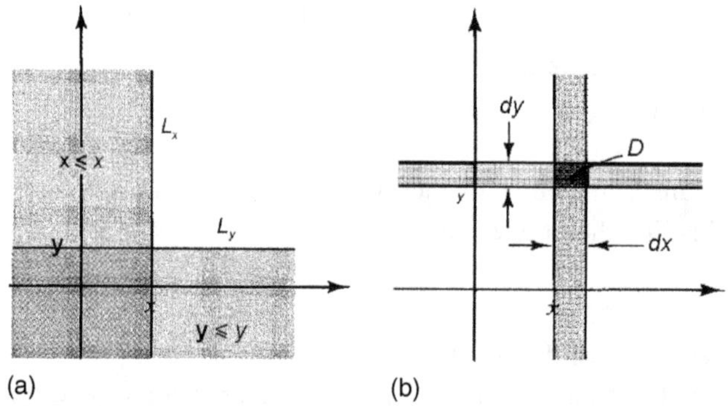

Figure 11.6 Geometrical interpretation of event realisations [1]. (a) Domain defined by $(\boldsymbol{x} < x) \cap (\boldsymbol{y} < y)$. (b) Domain defined by $(x < \boldsymbol{x} < x{+}dx) \cap (y < \boldsymbol{y} < y{+}dy)$.

In general, strength and loading are completely specified by, respectively, S and L. In the case of ship structures, they are often modelled using the log-normal distribution and the Weibull distribution, respectively. Since they are independent variables, the cumulative distributions $F_L(x)$ and $F_S(y)$ denote the probabilities that S and L are less than x, y, respectively. Thus, $F_S(x) = P[S \leq x]$ and $F_L(y) = P[L \leq y]$. However, their joint properties – that is, the probability that a point (x, y) is included in an arbitrary region of the plane, D – cannot be determined directly from $F_L(x)$ and $F_S(y)$. In order to determine this probability the *joint cumulative distribution function* or *joint distribution* $F_{xy}(x, y)$ of the random variables x and y has to be evaluated. This is defined to be the probability of the event,

$$[\boldsymbol{x} \leq x, \boldsymbol{y} \leq y] = [\boldsymbol{x} \leq x] \cap [\boldsymbol{y} \leq y] \tag{11.6}$$

This consists of the outcomes ζ such that $x(\zeta) \leq \boldsymbol{x}, y(\zeta) \leq \boldsymbol{y}$. Thus,

$$F_{xy}(x, y) = P[\boldsymbol{x} < x, \boldsymbol{y} \leq y] \tag{11.7}$$

This is the probability that the point (x, y) lies in the shaded region (domain) D of Figure 11.6a. For continuous random variables $\boldsymbol{x}$, $\boldsymbol{y}$, the probability that a point with coordinates $(\boldsymbol{x}, \boldsymbol{y})$ lies in the rectangle defined by $(x, x + dx)$ $(y, y + dy)$ is given by

$$P[x < \boldsymbol{x} \leq x + dx, \;\; y < \boldsymbol{y} \leq y + dy] = f(x, y)dxdy \tag{11.8}$$

where the function *f*(*x, y*) is the *joint density function* of the random variables ***x, y*** (Figure 11.6b).

For an arbitrary domain *D* the probability that (*x, y*) lies in it is given by

$$P[(\boldsymbol{x},\boldsymbol{y}) \subset D] = \iint_D f(x,y)dxdy$$

from which

$$f(x,y) = \frac{\partial^2 F(x,y)}{\partial x \partial y} \tag{11.9}$$

if the derivatives exist. For independent events,

$$P(A \cap B) = P(A)P(B) \tag{11.10}$$

Since strength and load are independent variables, that is, the events $[\boldsymbol{x} \le x]$ and $[\boldsymbol{y} \le y]$ are independent,

$$P[\boldsymbol{x} \le x, \boldsymbol{y} \le y] = P[\boldsymbol{x} \le x]P[\boldsymbol{y} \le y] \tag{11.11}$$

for any (*x, y*). Therefore,

$$F_{xy}(x,y) = F_x(x)F_y(y) \tag{11.12}$$

Differentiating,

$$f_{xy}(x,y) = f_x(x)f_y(y) \tag{11.13}$$

where $f_{xy}(x,y)$ is the joint density function of random variables *x, y*.

Thus if two random variables are independent, their joint distribution function is given by Equation (11.12) and their joint density function is given by Equation (11.13).

In order therefore to determine the probability of failure, p_f, of a structure for which the strength and load are represented by random variables *S* and *L*, respectively, we need to evaluate (Figure 11.8):

1. The probability that the strength *S* is less than or equal to *x*. This is given by the cumulative distribution function $F_S(x)$ and is represented by the shaded area to the left of *x* in density distribution *S*.
2. The probability that the load *L* lies in the range *x* to $x + dx$. This is given by $f_L(x)dx$, where f_L is the probability density function of the load density distribution *L*.

The probability of failure, p_f, is then the integral of the product of these probabilities over the range $-\infty$ to $+\infty$, given by [30][9]

$$p_f = P[S - L \le 0] = \int_{-\infty}^{+\infty} F_S(x)f_L(x)dx \tag{11.14}$$

[9] A proof of the relevant theorem is included in [13].

Equation (11.14) gives the total probability of failure p_f in the form of a convolution integral and is in effect the product of the probabilities of occurrence of two independent events summed over all possible occurrences. The numerical result obtained thereby does not have an absolute meaning and has to be treated as an intermediate result in a decision-making process.

We see that the probability of failure is not the area of overlap of the two density functions $f_S(x), f_L(x)$, shown in Figure 11.8.

The reliability of the structure $\boldsymbol{R}$ as defined in Equation (11.4) can then be applied in the case in which the strength and load are random variables so that

$$\boldsymbol{R} = 1 - p_f = 1 - \int_{-\infty}^{+\infty} F_S(x) f_L(x) dx \tag{11.15}$$

Up to this point it was assumed on an intuitive basis that the margin of safety, *M*, is the difference between the strength *S* of a structure and the applied load, *L*. There arise cases in which it is required to introduce components of each of these distributions; also, the relationship between them is not necessarily of the form adopted previously. It therefore becomes necessary to consider the relationship between load and strength in much more general terms. This is achieved by introducing the concept of a *limit state function* $g(S, L)$ that can assume a variety of forms. For present purposes,

$$g(S, L) = M = S - L \tag{11.16}$$

The value of the limit state function $g(.)$ then represents different states of equilibrium of a structure. Thus a *safe state* is denoted by $M > 0$, an *unsafe or failure state* by $M < 0$ and a *limit state* by $M = 0$. In Figure 11.7, which illustrates the simple two-dimensional case, the probability of failure is represented by the volume below the surface of the joint density function in the failure domain ($S < L$).

11.4 Structural Reliability Methods

In this section we will describe the methods that have been developed to enable reliability theory to be used in practical engineering applications. Classical reliability theory was developed and used in the study of the mass production of identically manufactured products such as electric light bulbs. For problems of this type reliability is measured in terms of the number of failures and the *mean time to failure* (*MTTF*). In structural engineering a different approach is needed because structures are unique in nature and failure takes place when excess loading of some kind is applied. *Structural reliability* is therefore concerned with the probability of failure of an individual structure rather than the failure frequency of a large number of items. The use of probability-based methods in structural engineering was pioneered during the first half of the 20th century in the former Soviet Union [16, 28] and in Germany [24]. During the Second World War the question of reliability of complex technical installations generated some interest. The approach became much more widely considered later, and progress ensued in several

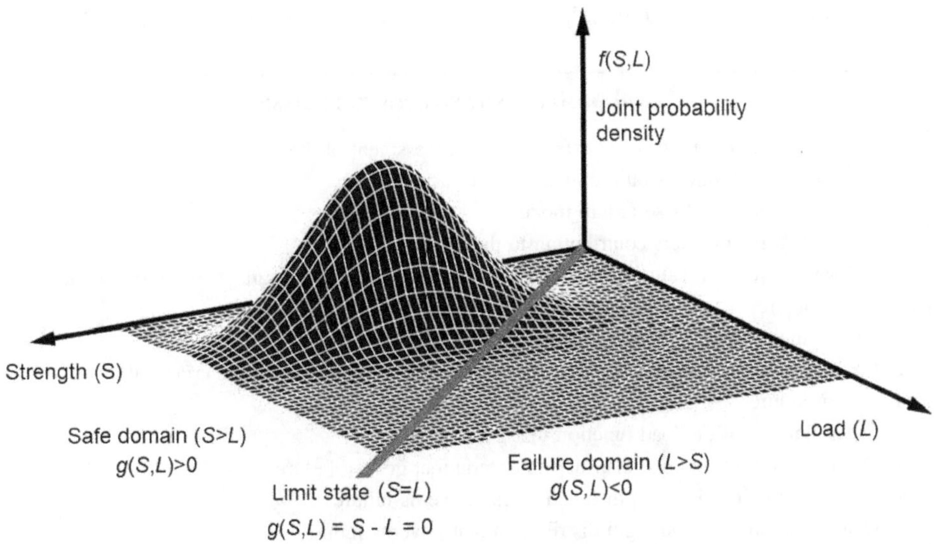

Figure 11.7 Joint probability distribution function of two independent variables (S, L) and state domains [18].

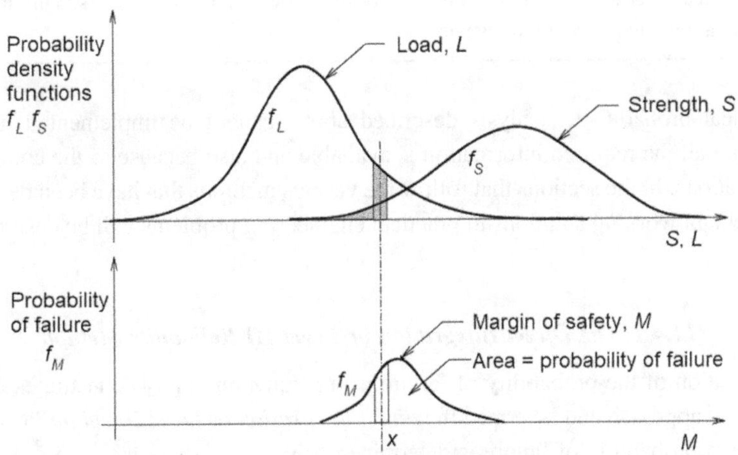

Figure 11.8 Strength, load and margin of safety distributions [30].

fields (military, electrical, civil, mechanical, aerospace, nuclear engineering as well as naval architecture).

In structural engineering the principal aim of a probabilistic approach is to define the required or existing margin of safety in rational terms, and to obtain a quantitative measure of performance of a structure. The measure of performance used is the probability of failure or reliability of the structure, as defined by Equation (11.15). When implementing such an approach, account has to taken of the uncertainties that characterise the problem in a systematic manner.

An ideal probabilistic structural analysis would include the following stages [10]:

Ideal probabilistic structural analysis

1. Analysis of failure modes and effects and an assessment of their criticality
   - Identify all significant failure modes of the structure
   - List the causes of these failure modes
   - Identify all parameters contributing to these causes
   - Determine the criticality of all significant failure modes to the satisfactory performance of structures
   - Rank the most critical failure modes
2. Formulation of the relationship between the critical parameters and the criteria that govern failure
3. Determination of the load function that governs failure
4. Determination of the load distribution criteria that govern failure
5. Determination of the strength function that governs failure
6. Determination of the strength distribution that governs failure
7. Determination of the probability of failure or reliability associated with load that governs failure and the strength distribution for each critical failure mode
8. The upper bound of the total probability of failure (lower bound of the reliability) is then the sum of the individual probabilities of each of the critical failure modes, assuming that these modes are mutually exclusive events.

The ideal probabilistic analysis described above cannot be implemented in practice because not all the required information is available and also because of the complexity of the calculations. In the sections that follow the various methods that have been developed in order to obtain working solutions to practical engineering problems will be described.

11.4.1 The Direct Integration or Level III Reliability Method

The calculation of the probability of failure using Equation (11.14) constitutes the direct, probabilistic approach and is termed the *direct integration* or *Level III reliability method.*[10] In general, a probability of failure is determined from

$$p_f = \int \cdots \int f_X(x_1, x_2, \ldots, x_n) dx_1 dx_2, \ldots, dx_n \tag{11.17}$$

where $f_X(.)$ is the joint probability distribution function of the design variables $x_1, x_2, \ldots, x_n$. The domain of integration is over the unsafe region of operation of a structure. The limit state function may be represented as $g(x_1, x_2 \ldots, x_n)$, and the corresponding unsafe region (integration domain of Equation (11.16)) is given by

$$g(x_1, x_2, \ldots, x_n) \leq 0 \tag{11.18}$$

[10] The procedure described in the previous section, used to determine risk level, is often termed the *Level IV* procedure.

If the variables $x_1, x_2, \ldots, x_n$ are the strength and load, failure is defined by $S < L$ so that the unsafe region is defined by

$$g(x_1, x_2) = g(s, l) = S - L \leq 0 \tag{11.19}$$

Applying Equation (11.17), the probability of failure is

$$p_f = \int_{[(s,l):s-l\leq 0]}\int f_{S,L}(s, l)dsdl \tag{11.20}$$

and since *S, L* are statistically independent,

$$p_f = \int_{[(s,l):s-l\leq 0]}\int f_S(s)f_L(l)dsdl \tag{11.21}$$

Therefore,

$$p_f = P[S - L \leq 0] = \int_0^{+\infty} F_S(l)f_L(l)dl \tag{11.22a}$$

Also,

$$p_f = 1 - \int_0^{+\infty} F_L(l)f_S(l)dl \tag{11.22b}$$

Note that in Equations (11.22a) and (11.22b) the integration is performed over the range 0 to $+\infty$. This is because strength and load do not assume negative values.

11.4.2 Level II Reliability Methods

The direct integration method requires the determination and use of information that is in most cases not available and/or difficult to obtain, and for this reason is not used in practice. The difficulties stem from the fact that data are not available for the full range of values of the joint probability distribution function (Figure 11.7). Furthermore, difficulties arise when evaluating the convolution integral.[11]

In order to circumvent these difficulties it is necessary to proceed with information that is available and also avoid having to carry out the integration of the convolution integral. In this and the subsequent section we shall describe methods that have been developed that can be used in practical circumstances. In the case of ship structural design, they have found application in the preparation of classification society rules and are also recommended in design office practice, especially in the case of novel structures.

[11] The direct probabilistic approach has been used to calculate the probability of failure of a hull girder under extreme bending moment [19].

The Mean-Value First-Order Second Moment (MVFOSM) Method

We proceed by considering the first part of Equation (11.14):

$$p_f = P[S - L \leq 0]$$

Since $M = S - L$, the probability of failure, p_f, can be expressed as

$$p_f = P[M \leq 0] = \int_{-\infty}^{0} f_M(m)dm = F_M(0) \tag{11.23}$$

The margin of safety is represented by a probability distribution whose form will depend upon the distributions of the load and strength and their overlap. The integration is performed over the range $-\infty \leq M \leq 0$, since the probability of failure is geometrically represented by the area of the region below the curve to the left of the *y*-axis (Figure 11.9).

Let us assume that both the strength S and load L are normally distributed. In this case the margin of safety M is, in accordance with basic probability theory, also normally distributed. The mean (expected) value μ_M and the standard deviation σ_M of M are given by [1]

$$\mu_M = \mu_S - \mu_L \tag{11.24a}$$

and

$$\sigma_M{}^2 = \sigma_S{}^2 + \sigma_L{}^2 \tag{11.24b}$$

where $\mu_S, \mu_L, \sigma_S, \sigma_L$ are the mean values and standard deviations of the strength and load distributions *S, L*, respectively.

From Equation (11.23), the probability of failure p_f is given by

$$p_f = F_M(0) = \Phi\left(\frac{-\mu_M}{\sigma_M}\right) = 1 - \Phi\left(\frac{\mu_M}{\sigma_M}\right) \tag{11.25}$$

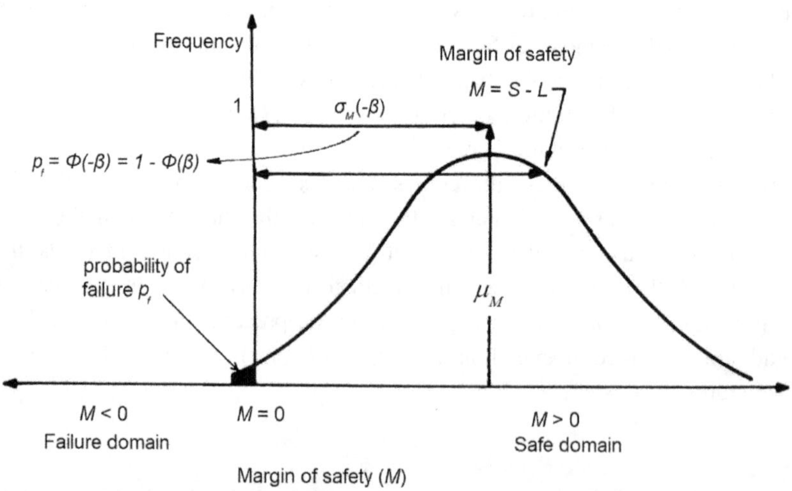

Figure 11.9 The probability density function of the margin of safety, *M*.

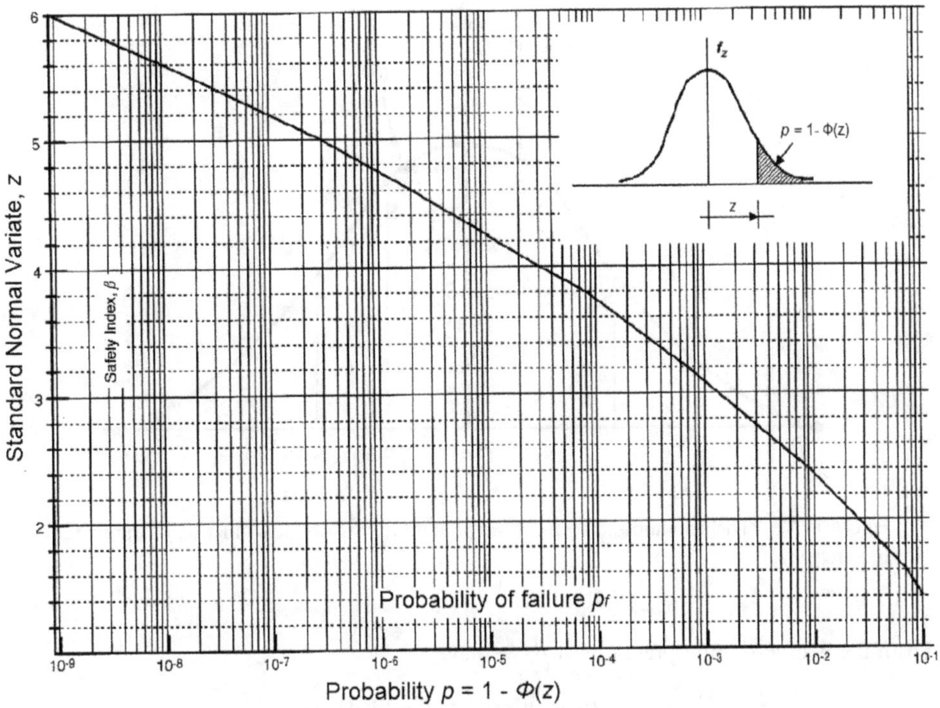

β	1.3	2.3	3.1	3.7	4.2	4.7	5.2
$p_f(=\Phi(-\beta))$	10^{-1}	10^{-2}	10^{-3}	10^{-4}	10^{-5}	10^{-6}	10^{-7}

Figure 11.10 Reliability index β and probability of failure p_f for a normally distributed margin of safety [22]. Reprinted courtesy of the Society of Naval Architects and Marine Engineers.

The ratio μ_M/σ_M can then be set to be identically equal to β, which is termed the *safety* or *reliability index* [9]. As the safety index β increases, the probability of failure decreases. Therefore, the safety index is a measure of the probability of failure. Since $\beta\sigma_M = \mu_M$, it is seen in Figure 11.9 that the failure state defined by $M \leq 0$ lies β times the standard deviation σ_M from the mean value μ_M. We thus have a geometric interpretation of the safety index in the case of normally distributed variables. In this case, the probability of failure can be exactly determined from the safety index (Figure 11.10).

β	1.3	2.3	3.1	3.7	4.2	4.7	5.2
$p_f(=\Phi(-\beta))$	10^{-1}	10^{-2}	10^{-3}	10^{-4}	10^{-5}	10^{-6}	10^{-7}

Substituting from Equations (11.24a–b),

$$\beta = \frac{\mu_M}{\sigma_M} = \frac{\mu_S - \mu_L}{\sqrt{{\sigma_S}^2 + {\sigma_L}^2}} \tag{11.26}$$

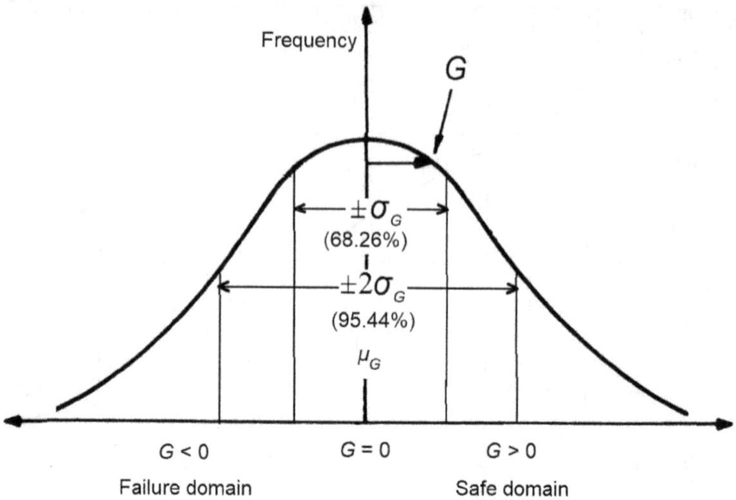

Figure 11.11 The standardised form of the probability distribution function of the margin of safety, *G*.

From the geometric interpretation of β, we have

$$p_f = \Phi(-\beta) = 1 - \Phi(\beta) \tag{11.27}$$

In order to use the table for the standard normal distribution, the distribution for the margin of safety, *M*, has to be converted to a standard normal distribution for which $\mu = 0$ and $\sigma = 1$. This is achieved by a linear transformation that renders the *standardised margin distribution, G* (Figure 11.11). This transformation consists of subtracting from *M* the mean value μ_M and dividing by the standard deviation, σ_M, as follows:

$$G = \frac{M - \mu_M}{\sigma_M} \tag{11.28}$$

The probability of failure p_f can then be obtained from standard normal probability tables. The relationship between reliability index and probability of failure is shown in Figure 11.10. For other distributions, the variation of probability of failure with safety index is given in Figure 11.12 [22].

Let us return to Equation (11.18), which is valid in the general case with *n* parameters. If the limit state function *g*(.) is nonlinear, then approximate values for μ and σ for the margin of safety can be obtained by using a linearised limit state function. This can be done by expanding the limit state function in a Taylor series about a selected point and by retaining only the first order terms:

$$g(x_1, x_2, \ldots, x_n) \approx g(x_1{}^*, x_2{}^*, \ldots, x_n{}^*) + \sum_{i=1}^{n} \left(\frac{\partial g}{\partial x_i}\right)_{x^*} (x_i - x_i{}^*) \tag{11.29}$$

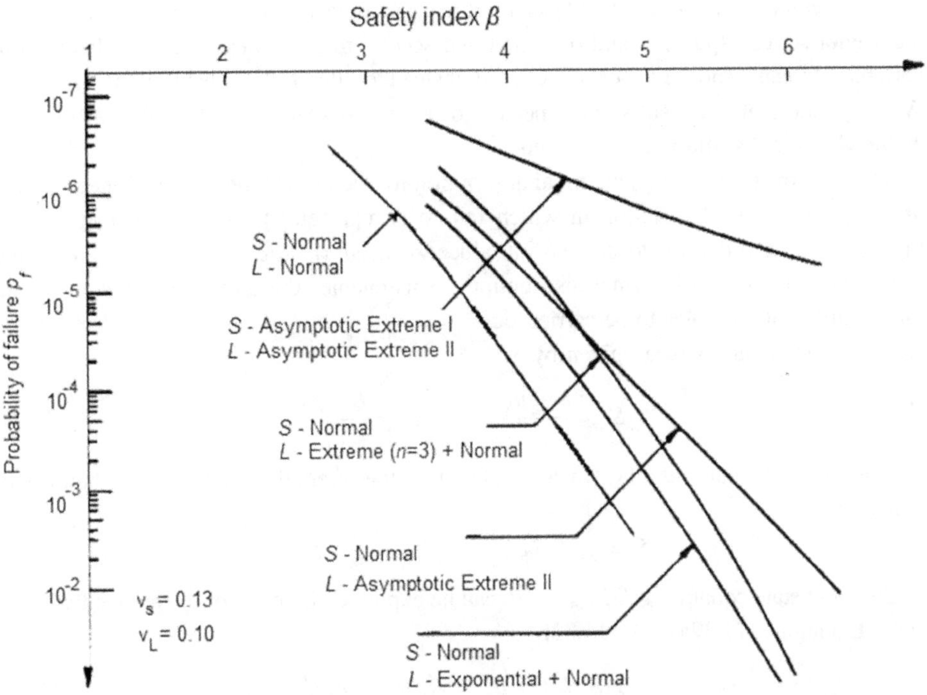

Figure 11.12 Probabilities of failure and safety indices for different combinations of strength and load distributions (log-normal plot) [21]. Reprinted courtesy of the Ship Structure Committee, Washington, DC.

where

x_i are the applied load and strength parameters, considered as random variables

$x_i{}^*$ are the values that x_i assume at the linearisation point

$g(.)$ is the limit state function that links these variables.

A first choice for the location of the linearisation point is for it to be at the mean values of the x_i parameters. Denoting mean values by $\overline{x_i}$, the mean and variance of M are given by

$$\mu_M \approx g(\overline{x}_1, \overline{x}_2, \ldots, \overline{x}_n) \tag{11.30a}$$

$$\sigma_M{}^2 \approx \sum_{i=1}^{n} \sum_{j=1}^{n} \left(\frac{\partial g}{\partial x_i}\right)_{\overline{x_i}} \left(\frac{\partial g}{\partial x_j}\right)_{\overline{x_j}} \rho_{x_i x_j} \sigma_i \sigma_j \tag{11.30b}$$

where $\rho_{x_i x_j}$ is the correlation coefficient and subscripts $\overline{x_i}$, $\overline{x_j}$ indicate that the gradients are evaluated at the mean point. In the case of independent (uncorrelated) variables,

$$\sigma_M{}^2 \approx \sum_{i=1}^{n} \left(\frac{\partial g}{\partial x_i}\right)_{\overline{x_i}}^2 \sigma_{x_i}^2 \tag{11.31}$$

This method is called the *FOSM method* because only first-order terms are retained in the Taylor series expansion and only first and second moment descriptors of the random variables (means and variances) are used instead of the full probability distributions. We conclude that the results are expected to be approximate, as not all the information included in the distributions is retained.

We will now provide a geometrical description for the FOSM method, analogous to that of the probabilistic description in which failure is represented by the region $M < 0$ in Figure 11.9. In order to do so, we introduce *reduced variates* S', L' to transform the distributions to standard normal distributions. This enables the direct comparison of data for the different variables to be carried out.

The reduced variables are given by

$$S' = \frac{S - \mu_S}{\sigma_S} \qquad L' = \frac{L - \mu_L}{\sigma_L} \tag{11.32a–b}$$

Rearranging Equations (11.31a) and (11.31b), the strength and load are obtained in terms of S', L'

$$S = \mu_S + S'\sigma_S \qquad L = \mu_L + L'\sigma_L \tag{11.33a–b}$$

The limit state equation $g(S,L) = S - L$ can be expressed in terms of the reduced variates using Equations (11.32a) and (11.32b):

$$g(S', L') = \mu_S{}'\sigma_S S' - \mu_L - \sigma_L L' \tag{11.34}$$

Equation (11.33) represents a straight line in the space of reduced variates and for $g(.) = 0$ the line separates the safe and unsafe (failure) states. That is,

$$\mu_S + \sigma_S S' - \mu_L - \sigma_L L' = 0 \tag{11.35}$$

This is the straight line shown in Figure 11.13. The distance d from the origin to the line is then a measure of structural reliability. From analytic geometry,

$$d = \frac{\mu_S - \mu_L}{\sqrt{\sigma_S{}^2 + \sigma_L{}^2}} \equiv \beta \tag{11.36}$$

We note that d is identically equal to the safety index, β, defined by Equation (11.27). If the density distributions of strength and load are not normal, corresponding normal distributions need to be obtained. For equivalent normal S and L, the probability of failure is given by

$$p_f = \Phi(-\beta) \tag{11.37}$$

where $\Phi(.)$ is the standard normal distribution function.

The Hasofer-Lind Reliability Index

If instead of the limit state function $M = S - L$ a function that includes nonlinear terms[12] was considered, and linearisation took place at the mean values of the random variables,

[12] For example, $M = \sigma_0 Z - L$, where σ_0 and Z are both treated as random variables. The presence of the product $\sigma_0 Z$ renders M nonlinear.

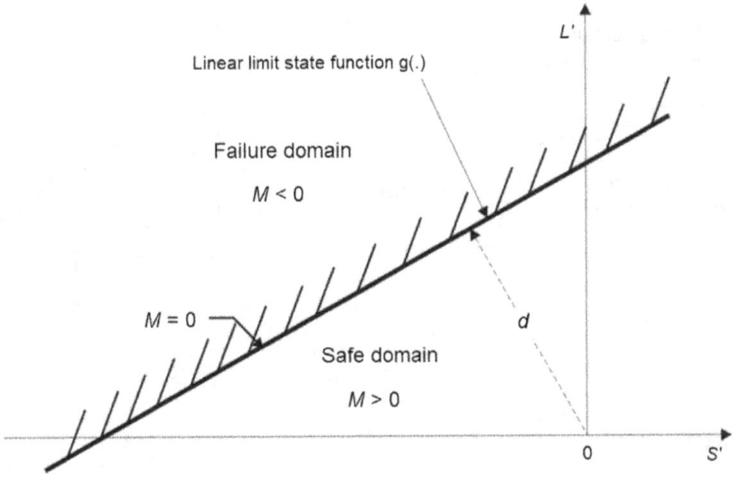

Figure 11.13 Reduced variate space (S', L') for strength and load [1].

errors would be introduced at increasing distances from the linearisation point if the higher-order terms that appear in Equation (11.29) are disregarded. Furthermore, the value of the safety index depends on the limit state function, despite the fact that $p_f = F_M(0)$. As a result, if the limit state function changes, the reliability index will change.

These two problems can be resolved by use of the *Hasofer-Lind* procedure. In this, instead of selecting a linearisation point whose coordinates are the mean values of the random variables, linearisation is carried out at a suitable location on the failure surface. In this way, the results no longer depend on the properties of the random variable distributions. On the failure surface the limit state function g(.) = 0 and its gradients are independent of the properties of the random variables.

It is first necessary to convert to standard form the random variables x_i using the following transformation:

$$y_i = \frac{x_i - \bar{x}_i}{\sigma_{x_i}} \tag{11.38}$$

The result of this transformation in the case of a linear limit state function is shown in Figure 11.14. It is seen that in the normalised variate space there exists rotational symmetry with respect to the standard deviations. The *Hasofer-Lind reliability index* is then defined as the shortest distance from the origin to the failure surface (Figure 11.15) [12]. By defining the index in this manner it becomes invariant with respect to the failure function.

In mathematical terms, the Hasofer-Lind index is defined as follows:

$$\beta_{HL} = \min_{y} \sqrt{\sum_{i=1}^{n} y_i^2} \tag{11.39}$$

If the limit state function is linear, the reliability index as defined by Equation (11.36) (Cornell index) will coincide with the Hasofer-Lind index. It is seen from Equation (11.39)

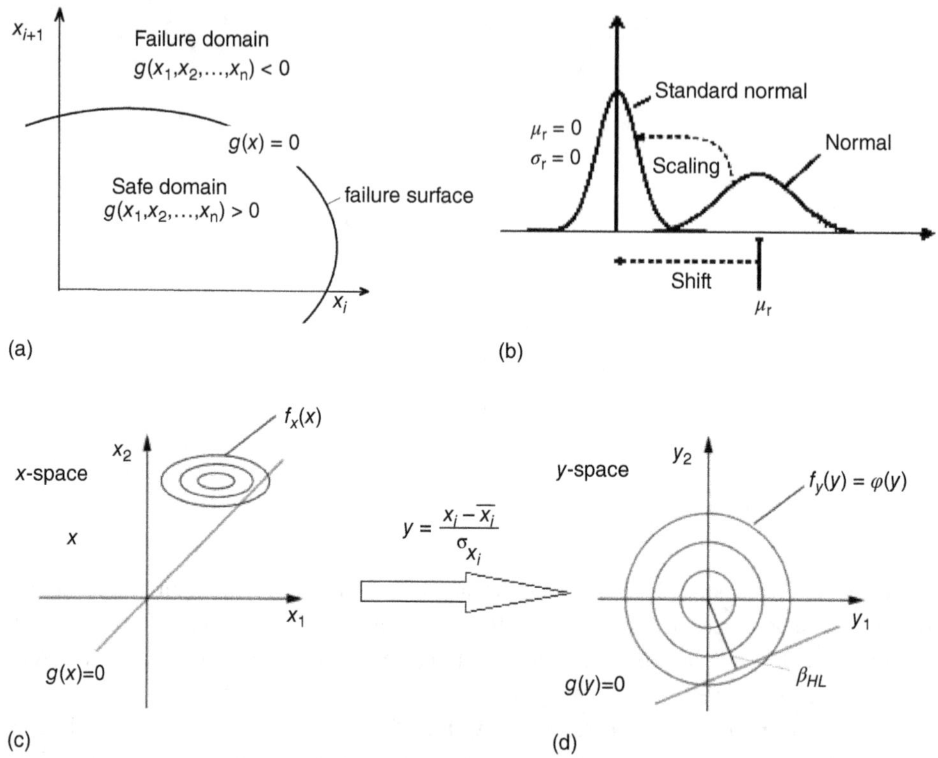

Figure 11.14 The failure surface concept and transformation to normalised variate space y [18]. (a) General case. (b) Transformation to standard normalised variate. (c) Two-dimensional case. (d) Normalised variate space (y).

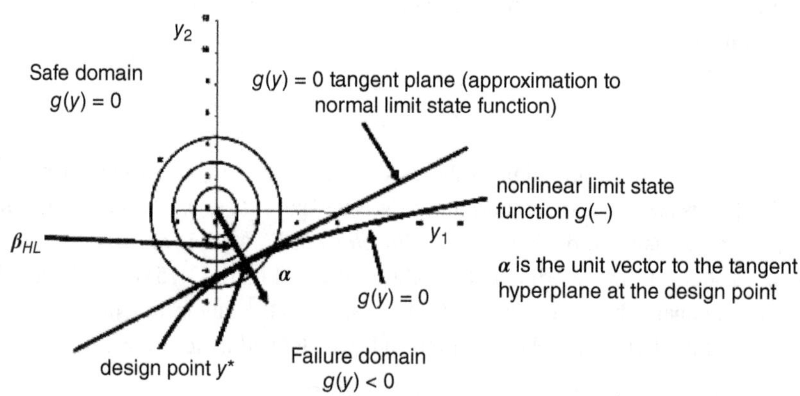

Figure 11.15 Nonlinear limit state function and the Hasofer-Lind reliability index [18].

that it is necessary to determine the coordinates of the design point y^* on the failure surface in order to calculate the index β_{HL}. These, however, cannot be obtained directly, and an iterative approach is required. This is given below.

Determination of the Hasofer-Lind reliability index, β_{HL}

1. Select design point

$$y_i^* = -a_i^* \beta_{HL} \tag{11.40}$$

2. Calculate unit normal vector $\boldsymbol{a}^*$ from

$$\alpha_i^* = \frac{\left[\frac{\partial G}{\partial y_i}\right]_{y_i^*}}{\sqrt{\sum_i \left[\frac{\partial G}{\partial y_i}\right]_{y_i^*}^2}} i = 1, 2, \ldots, n \tag{11.41}$$

3. Find the reliability index β_{HL} from

$$G(\beta_{HL}\alpha_1, \beta_{HL}\alpha_2, \ldots, \beta_{HL}\alpha_n) = 0 \tag{11.42}$$

4. The co-ordinates of the design point are then

$$y^* = (\beta_{HL}\alpha_1, \beta_{HL}\alpha_2, \ldots, \beta_{HL}\alpha_n)^T \tag{11.43}$$

5. Check for convergence in y^*. If convergence has not been achieved, go back to step 1.

In order to initiate the iterative procedure, it is necessary to select a suitable starting point. The coordinates of the most likely design point in the original variate space are given by

$$x_i^* = \bar{x}_i + \sigma_{x_i} y_i^* = \bar{x}_i - \sigma_{x_i} a_i \beta \tag{11.44}$$

Non-Gaussian Distributions: The Rackwitz–Fiessler Procedure

The reliability index has been defined as a function of the probability of failure when the variables, x_i, are normally distributed. In the case of marine structures, however, the loading follows an extreme-type distribution such as the Weibull distribution, whereas strength variables such as the yield stress are log-normally distributed. Therefore, in order to be able to use the method described in the previous section, this has to be modified so that it is applicable for non-normal distributions as well.

In the case of normally distributed variables Equation (11.27) is used. In order to be able to extend the use of this relationship, it is necessary to obtain the moments of the normal distribution that is equivalent to the true distribution. These values can then be used in the iterative procedure described in the previous section. In most cases the regions of the distribution that are most likely to contribute to failure are the extremes. It therefore makes sense to approximate the true distribution at the point of linearisation, x^*, at which in any case the reliability index is minimised and the failure probability is at its maximum value.

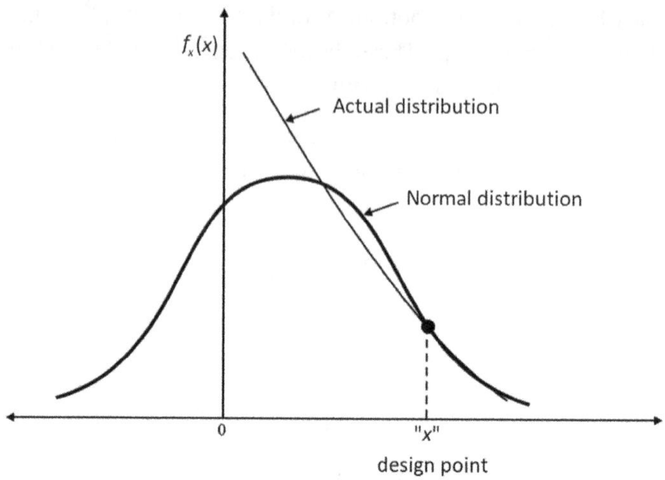

Figure 11.16 Actual and equivalent normal density distribution [21]. Reprinted courtesy of the Ship Structure Committee, Washington, DC.

If the values of the corresponding normal distribution $\Phi(.)$ are set equal to those of the true distribution $F_x(x^*)$ and density function $f_x(x^*)$ at the linearisation point, the mean value $\mu_x{}'$ and standard deviation $\sigma_x{}'$ of this distribution can be obtained. Thus,

$$\Phi\left[\frac{x^* - \mu'_x}{\sigma'_x}\right] = F_x(x^*) = \frac{1}{\sqrt{2\pi}\sigma'_x}\exp\left[-\frac{1}{2}\left[\frac{x^* - \mu'_x}{\sigma'_x}\right]^2\right] \tag{11.45a}$$

$$f_x(x^*) = \frac{1}{\sigma'_x}\phi\left[\Phi^{-1}[F_x(x^*)]\right] \tag{11.45b}$$

where $\Phi(.)$ is the required normal distribution function. Solving the above two expressions for μ'_x and σ'_x, the following expressions are obtained for the first and second moment of the equivalent normal distribution:

$$\mu'_x = x^* - \Phi^{-1}[F_x(x^*)]\sigma'_x \tag{11.46a}$$

$$\sigma'_x = \frac{\phi\left[\Phi^{-1}[F_x(x^*)]\right]}{f_x(x^*)} \tag{11.46b}$$

The matching of the original distribution with the equivalent normal one is shown in Figure 11.16. The two distributions have the same value and slope at the linearisation point, x^*.

The above procedure was introduced in [26] and is known as the *Rackwitz–Fiessler procedure*. Equations (11.46a–b) need to be introduced to the iterative procedure, and if the iteration is carried out in normalised space, the moments of the equivalent distribution have to be transformed into normalised space in each step of the iteration procedure.

Determination of Safety Index for Nonlinear Limit State Functions and Non-Normal Distributions

The solution methodology outlined in this section can be used to determine the safety index β when a limit state function is nonlinear and non-normal distributions are involved [21]. Consider, for example, the ultimate strength limit state, defined when the stress in the extreme fibres in a hull girder section, located at a distance y_{max}, from the neutral axis of bending reaches yield ($\sigma_{max} = \sigma_0$). Applying simple beam theory,

$$\frac{\sigma_0}{y_{max}} = \frac{M_{SW} + M_W}{I}$$

where M_{SW} and M_W are the still-water and wave-induced bending moments.

$$\text{Hence, } \sigma_0 = \frac{M_{SW} + M_W}{I/y_{max}} \qquad \text{and} \qquad \sigma_0 Z = M_{SW} + M_W$$

$$\text{Thus, } \sigma_0 Z - M_{SW} - M_W = g(x) \tag{11.47}$$

This then is the limit state function, $g(.)$, for ultimate strength. It is nonlinear, since the first term is a product of two random variables: the yield stress, σ_0, and the section modulus, Z. In the reduced space, applying the procedure defined by Equations (11.32)–(11.34) gives the following expression for the limit state function:

$$g(x') = (\mu_{\sigma_0} + \sigma_{\sigma_0}\sigma_0{}')(\mu_Z + \sigma_Z Z') - (\mu_{SW} + \sigma_{SW} M_{SW}{}') - (\mu_W + \sigma_W M_W{}') = 0 \tag{11.48}$$

where the prime superscript denotes a reduced variable and μ, σ are the mean value and standard deviation of each variable.

The gradients required for the calculation of the direction cosines Equation (11.41) are

$$\frac{\partial g}{\partial Z'} = \sigma_Z(\mu_{\sigma_0} + \sigma_{\sigma_0}\sigma_0{}') = \sigma_Z \sigma_0 \tag{11.49a}$$

$$\frac{\partial g}{\partial \sigma_0{}'} = \sigma_{\sigma_0} Z \tag{11.49b}$$

$$\frac{\partial g}{\partial M_{SW}{}'} = \sigma_{SW} \tag{11.49c}$$

$$\frac{\partial g}{\partial M_W{}'} = -\sigma_W \tag{11.49d}$$

The coordinates of the most likely failure point are then found from Equation (11.40). Substituting these in Equation (11.48), the following expression is obtained:

$$(\mu_Z - \sigma_Z \alpha *_Z \beta)(\mu_{\sigma_0} - \sigma_{\sigma_0}\alpha *_{\sigma_0}\beta) - (\mu_{SW} - \sigma_{SW}\alpha *_{SW}\beta) - (\mu_W - \sigma_W \alpha *_W \beta) = 0 \tag{11.50}$$

This equation is solved for β. In each iteration, the direction cosines $\alpha_i{}^*$ are evaluated at the current expected failure point for which corresponding values for σ_0 and Z are used.

The iteration procedure requires initial values, and to this end the mean values for these variables can be used. The procedure is repeated until convergence in the value of the reliability index is achieved.

This is the solution procedure followed when all random variables are normally distributed. If one or more random variables are not normally distributed, it is necessary to use the mean value and standard deviation of the corresponding normal distribution. Assume that in this case the wave-induced bending moment M_w follows the exponential distribution. The descriptors of the corresponding normal distribution are given by Equations (11.46a–b). The values obtained are substituted in Equation (11.50), following which the reliability index can be evaluated.

The exponential (Gumbel) distribution function and density function are

$$F(x) = 1 - e^{-(x/\lambda)} \qquad x \geq 0 \tag{11.51a}$$

$$f(x) = \frac{1}{\lambda} e^{-(x/\lambda)} \qquad x \geq 0 \tag{11.51b}$$

where λ = mean or expected value of the wave load amplitude.

For an exponential distribution with a parameter λ, Equations (11.46a–b) give

$$\mu_{mw}^{n} = m_{W}^{*} - \Phi^{-1}\left[1 - e^{-m_{W}^{*}}\right]\sigma_{mw}^{n} \tag{11.52a}$$

$$\sigma_{mw}^{n} = \lambda e^{m_{W}^{*}/\lambda}\Phi\left\{\Phi^{-1}\left[1 - e^{-m_{W}^{*}/\lambda}\right]\right\} \tag{11.52b}$$

where μ^n and σ^n are, respectively, the mean value and standard deviation of the corresponding normal distribution for the wave-induced bending moment.

It is of interest at this point to turn to existing ships and to determine their reliability indices for the ultimate strength limit state. Results for a series of 18 ships are plotted against ship length in Figure 11.17 [21]. It is evident that that the reliability indices differ; they lie in the range $6 > \beta > 4$ and are found to increase with ship's length. The notional probabilities of failure that correspond to the above values of reliability index are indicated in Table 11.1.

11.4.3 The Level I (Partial Safety Factor) Method

A key objective of engineering design is to ensure that the structure performs in a satisfactory manner. In the case of ship structures this has to be achieved under conditions of uncertainty, and thus the assurance of performance is achievable only in probabilistic terms. It is therefore necessary to base a design on probabilistic considerations.

On the other hand, ship structural design requires a simple and clearly outlined format. Traditional codes were based on the idea of a permissible (design) stress so as to ensure that the stresses do not extend beyond the elastic range. Implicit in these were factors of safety that were selected on an empirical basis. In order to avoid changing code formats, it would be appropriate to introduce corresponding factors of safety based on probabilistic considerations.

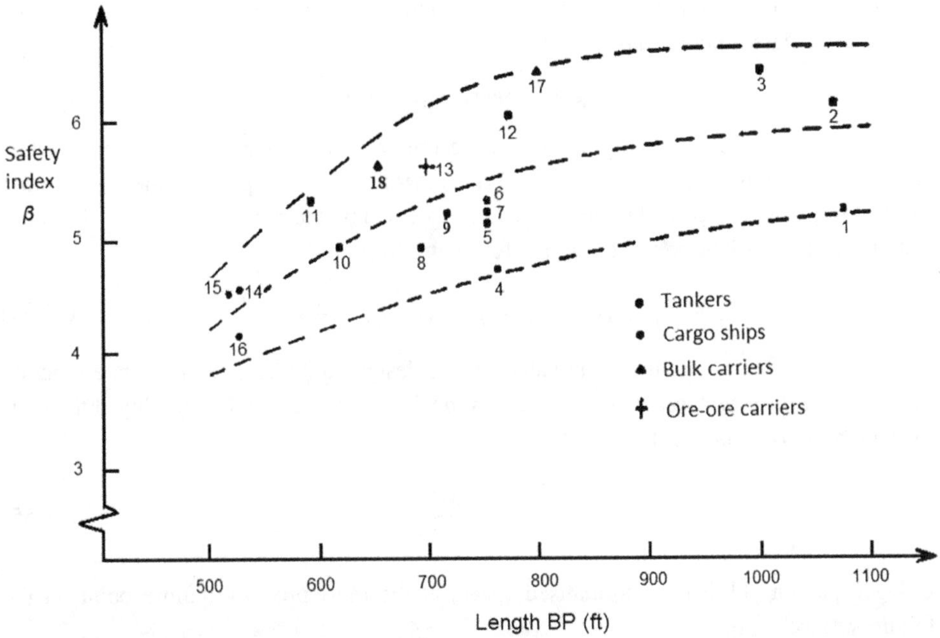

Figure 11.17 Reliability indices for 18 ships plotted against length (L_{BP}) [21]. Reprinted courtesy of the Ship Structure Committee, Washington, DC.

The results obtained from any calculation of the reliability of a structure can only be used for comparison purposes. Therefore, in order to determine the level of safety that will be enshrined in a new code, it is necessary to ensure that the resulting designs will have a reliability that is at least equal to that of structures that have performed satisfactorily in the past. It thus becomes necessary to calculate the reliability of existing designs. This can be done using the Level II methods described in the previous section, leading to reliability indices β, which are then used to calculate the *partial safety factors* that can be used in a code, or indeed in a new design. The use of partial safety factors is termed the *Level I reliability method* or *partial safety factor method*. Strictly speaking, it cannot be classed as a reliability method, since it does not make use of distribution information.

It is seen that at each stage, as a simpler method is sought, information is discarded. Level IV methods include full probabilistic data as well as results from consequence analysis. Their scope therefore extends beyond that of structural performance, with broader issues considered. Level III methods involve a fully probabilistic analysis of a structure. In Level II methods, distribution parameters such as the second moment and joint probability function data are discarded. For this reason, these methods are often called *semi-probabilistic*. The Level I method is deterministic, since it does not include distribution data. It makes use of the mean value and partial safety factors, which are based on the reliability index.

In this section we will outline how partial safety factors are obtained from the reliability index for a given limit state function.

We return to the limit state function *g(S, L)*, defined by Equation (11.16). In the general case, a limit state function has the form

$$g(\boldsymbol{x}) = g(x_1, x_2, \ldots, x_n) \tag{11.53}$$

where $\boldsymbol{x} = (x_1, x_2, \ldots, x_n)$ is a vector of design variables. The partial safety factors $\gamma_1, \gamma_2, \ldots, \gamma_n$ can then be introduced as safety factors that are multiplied by the mean values in order to ensure a specified reliability level, expressed by the reliability index β. The limit state function can then be written in the following form:

$$g(\gamma_1\mu_{x_1}, \gamma_2\mu_{x_2}, \ldots, \gamma_n\mu_{x_n}) = 0 \tag{11.54}$$

where $\mu_{x1}, \mu_{x1}, \ldots, \mu_{xn}$ are the mean values of the design variables. The limit state function describes a *failure surface,* so that $\gamma_i\mu_{xi}$ lie on the failure surface, and preferably at the *most probable failure point.* At this point, $x_i = x_i{}^*$,

$$\gamma_i = \frac{x_i{}^*}{\mu_{x_i}} \tag{11.55}$$

Using Equation (11.40), in normalised space, at the most probable failure point on the failure surface,

$$x_i'{*} = -\alpha_i{}^*\beta \qquad i = 1, 2, \ldots, n \tag{11.56}$$

where

$$\alpha_i^* = \frac{\left(\frac{\partial g}{\partial x_i'}\right)_*}{\sqrt{\sum_i \left(\frac{\partial g}{\partial x_i'}\right)_*^2}} \tag{11.57}$$

are the direction cosines along the reduced variate space axes x_i'. Thus the original variate is

$$\begin{aligned} x_i{}^* &= \mu_{x_i} + \sigma_{x_i} x_i'{*} = \\ &= \mu_{x_i} - \alpha_i{}^*\beta\sigma_{x_i} = \\ &= \mu_{x_i}(1 - \alpha_i{}^*\beta v_{x_i}) \end{aligned} \tag{11.58}$$

where v_{xi} is the coefficient of variation. Comparing Equation (11.56) with Equation (11.58) we obtain

$$\gamma_i = 1 - \alpha_i{}^*\beta v_{x_i} \tag{11.59}$$

For linear limit state functions, Equation (11.59) is used to obtain the values of the partial safety factors. In the case of nonlinear limit state functions, the iterative procedure proposed by Rackwitz given on the next page has to be used. Equation (11.59) can then be used to obtain the partial safety factors for a given reliability index, β. If the reliability index is given, step 4 in the above procedure is omitted [26]. For a linear limit state function, Equation (11.54) holds. This can be expressed as

$$g(\boldsymbol{x}) = a_0 + \sum_i a_i \boldsymbol{x}_i \tag{11.60}$$

where a_0 and a_i are constants. The limit state function is then

$$a_0 + \sum_i a_i \gamma_i \mu_{x_i} = 0 \tag{11.61}$$

Determination of Partial Safety Factors, γ_i, for Nonlinear Limit State Functions
Assume initial values of x_i^* $i = 1, 2, \ldots, n$ and obtain $x_i'^* = \frac{x_i^* - \mu_{x_i}}{\sigma_{x_i}}$
1. Evaluate $\left(\partial g/\partial x_i'\right)^*$ and α_i^* at x_i^*.
2. Form $x_i^* = \mu_{x_i}(1 - \alpha_i^* \beta v_{x_i})$
3. Substitute above x_i^* in $g(x_1^*, x_2^*, \ldots, x_n^*) = 0$ and solve for β
4. Using the value of β obtained above re-evaluate $x_i'^* = -\alpha_i^* \beta$
5. Repeat steps 2 through 5 until convergence is achieved

Because of linearity, the partial derivatives $\left(\partial g/\partial x_i'\right)^*$ are independent of x_i. Thus,

$$\left(\frac{\partial g}{\partial x_i}\right) = \alpha_i \sigma_{x_i} \tag{11.62}$$

Therefore, from Equation (11.59), the partial safety factors are

$$\gamma_i = 1 - \frac{\alpha_i \sigma_{x_i}}{\sqrt{\sum_i (\alpha_i \sigma_{x_i})^2}} \beta v_{x_i} \tag{11.63}$$

For the linear limit state function $g(S, L) = S - L$ with a given reliability index β, the partial safety factors are first written as reduced variables (Equation (11.26)). Differentiating, we obtain

$$\left(\frac{\partial g}{\partial S'}\right) = \sigma_S \qquad \left(\frac{\partial g}{\partial L'}\right) = -\sigma_L \tag{11.64}$$

Therefore,

$$\alpha_S = \frac{\sigma_S}{\sqrt{\sigma_S^2 + \sigma_L^2}} \qquad \alpha_L = \frac{\sigma_L}{\sqrt{\sigma_S^2 + \sigma_L^2}} \tag{11.65}$$

The partial safety factors are then

$$\gamma_S = 1 - \alpha_S \beta v_S = 1 - \beta v_S \frac{\sigma_S}{\sqrt{\sigma_S^2 + \sigma_L^2}} \tag{11.66a}$$

$$\gamma_L = 1 - \alpha_L \beta v_L = 1 + \beta v_L \frac{\sigma_L}{\sqrt{\sigma_S^2 + \sigma_L^2}} \tag{11.66b}$$

We note that $\gamma_S > 1$ and $\gamma_L < 1$, which is as expected. It is important to note that the values of the partial safety factors obtained are a function of the reliability index so that they do not have universal application.

This brings to a conclusion the introduction to structural reliability theory in this chapter. Its application to the marine field and examples of use are discussed in Chapter 12.

11.5 Fuzzy Logic and Fuzzy Set Theory

In the remainder of this chapter we shall describe a method that has been developed to deal with a type of uncertainty that differs from that which characterises physical processes, which has been discussed in this chapter up to this point. This type of uncertainty features in human decision-making in general; the techniques that will be described have found application in ship structural design.

Fuzzy logic and, by extension, *fuzzy set theory* are used to allow for graded answers rather than the yes/no answers that characterise classical (Boolean) logic. Fuzzy logic is thus an extension of classical logic. The need to extend classical logic arises because communication takes place in approximate, or fuzzy, terms rather than precise or *crisp* terms, and since computers cannot handle information in approximate terms, in order for them to be able to do so, a specific procedure has to be set up. In the next section we shall provide a brief outline of classical logic that will help in understanding the basis of fuzzy logic and fuzzy set theory.

11.5.1 Classical and Fuzzy Logic

In order to understand the connection between classical and fuzzy logic we need to refer briefly to certain basic ideas of classical logic. The first concept is that of the *proposition,* which is a single declarative statement or collection of declarative statements. In classical logic such statements may be either true or false – that is, they may have a *truth value* 'true' or a truth value 'false'.

A proposition affirms (or proposes) that something is the case – that is, it says something about something. More formally, it *predicates* (or attributes) the *subject* with some property or quality. As Aristotle discovered, this can be done in a limited number of ways. In logic, this is called the study of *categorical propositions* – that is, statements involving categories of things. A *categorical proposition* expresses a relation between two categories or groups of things by stating either that all, none or some of one category belong or do not belong to a second category of things. A *categorical argument* is defined as one composed of categorical propositions. The four standard form categorical propositions of classical logic thus are:

- All S is P
- No S is P
- Some S is P
- Some S is not P

For example,

- Man is mortal is a categorical proposition that returns the value 'true'.
- $12 + 9 = 3 - 2$ is a categorical proposition that returns the value 'false.

There are four elements to each categorical proposition:

1. A *subject* term, indicated by the letter 'S'. A subject term is an expression (a word or a phrase) that describes a category. The subject is the category about which something is being said.
2. The *quantifier*, which is a word or phrase that indicates whether the proposition concerns an entire category, part of a category or just a single member of the category. In symbolic language, two quantifiers are used to provide a measure of the extent of applicability of the proposition in question. The *universal quantifier*, denoted by the symbol $\forall$, states that the statements within its scope are true for every value of the specific variable. $\forall x P(x)$ is read as 'for every value of x, $P(x)$ is true'. The *existential quantifier*, denoted by the symbol $\exists$, states that the statements within its scope are true for some values of the specific variable. $\exists x P(x)$ is read as 'for some values of x, $P(x)$ is true'.
3. A *predicate term* indicated by the letter 'P'. The predicate is the quality or property attributed to the subject.
4. A form of the verb 'to be', which connects the subject term to the predicate term. This is known as the *copula*.

Two or more propositions may be joined together to form a *connective proposition*. This is done using *logical connectives*. Commonly used connectives include 'but', 'and', 'or', 'if... then', and 'if and only if'. The connectives that use the words AND and OR to join two propositions are termed *conjunction* and *disjunction*, respectively. The *negation* of a proposition, meaning that it is false when A is true and it is true when A is false, is written as $\neg$A. If a proposition B is a logical consequence of another (A), the connective in this case is termed *implication* and is denoted by $A \supset B$ or $A \rightarrow B$ (if A then B). Lastly, the *if-and-only-if* connective is true if both propositions are the same, that is, they are either true or false.

We now turn to the classical *truth tables* for connective propositions that involve the use of the connectives AND, OR and NOT.

Figure 11.18 includes the tables that include all possible combinations of Boolean truth values for propositions A and B, which may be zero or unity. In each case, A = 0 means that A is a false proposition, whereas A = 1 means that A is a true proposition. For the AND connective (first table), the third column returns a value of unity if the connective proposition A AND B is true and zero otherwise.

Our aim is to extend the range of allowable values for propositions A and B to include all real numbers that lie in the range zero to unity. In order to do so it is necessary to select functions that preserves the existing Boolean truth table in each case.

In the case of the AND connective, we succeed in this if we use the *min* (A, B) operation. This operation selects the lowest value of each pair of truth values (A, B). Implementing this, the truth table included in Figure 11.19 is returned. Using the same reasoning, the *max* (A, B)

AND

A	B	*min*(A,B) or $A \vee B$
0	0	0
0	1	0
1	0	0
1	1	1

OR

A	B	*max*(A,B) or $A \wedge B$
0	0	0
0	1	1
1	0	1
1	1	1

NOT

A	1–A
0	1
1	0

Figure 11.18 Classical (Boolean) logic truth tables for AND, OR, NOT connectives [23].

AND

A	B	*min*(A,B) or $A \vee B$
0	0	0
0	1	0
1	0	0
1	1	1

OR

A	B	*max*(A,B) or $A \wedge B$
0	0	0
0	1	1
1	0	1
1	1	1

NOT

A	1–A
0	1
1	0

Figure 11.19 Fuzzy logic truth tables for MIN, MAX and NOT operations [23].

IMPLICATION

A	B	$A \Rightarrow B$
0	0	1
0	1	1
1	0	0
1	1	1

IF and only IF

A	B	$A \Leftrightarrow B$
0	0	1
0	1	0
1	0	0
1	1	1

Figure 11.20 Fuzzy logic truth tables for IMPLICATION and IF-AND-ONLY-IF operations [23].

operation returns the truth table for the OR connective and the (1-A) operation returns the truth table for the NOT connective. Comparing the corresponding truth tables included in Figures 11.18 and 11.19, it is seen that they are identical. The important difference between the two is that in the case of the fuzzy logic truth tables, intermediate values are permitted rather than just zero or unity that are used in classical logic. In addition to these we introduce two further connectives: IMPLICATION, which returns a truth value "1" if B follows from A and 0 otherwise and the IF-AND-ONLY-IF connective (Figure 11.20).

Approximate reasoning is the process by which a possibly imprecise conclusion is deduced from a collection of imprecise premises [25]. The types of approximate reasoning are:

1. *Categorical reasoning*. In this mode of approximate reasoning, the subjects are assumed to be in canonical (simplest possible) form.

2. *Qualitative reasoning*: A means to express conceptual knowledge such as the physical system structure, causality, start and end of processes, assumptions and conditions under which facts are true, qualitatively distinct behaviour, etc.
3. *Syllogistic reasoning*. A syllogism is a logical argument in which a proposition (the conclusion) is inferred (deduced) from two others (the premises). An example of two premises (one major and one minor) follows:
   S1 As are Bs (all dogs are animals)
   S2 Cs are Ds (some dogs bite)
   The conclusion takes the following form:
   S3 Es are Fs (some animals bite)
   S1 and S2 are given quantifiers whereas S3 is deduced or inferred.
4. *Dispositional reasoning*. A disposition is a proposition that is preponderantly but not necessarily always true [33]. Most dispositions are of the form A is B or As are Bs where A and B are fuzzy predicates (for example, smoking is harmful). A disposition signifies that the conditional probability of B given A is high, where 'high' should be interpreted as a fuzzy probability. A disposition may be expressed in the following two canonical forms:
   Unconditional: usually (X is A)
   Conditional: usually (X is A if Y is B)
   where X and Y are variables, A and B are fuzzy predicates and usually is a fuzzy quantifier that may be represented as a fuzzy number in the range zero to unity. The quantifier usually links dispositional and syllogistic reasoning and is important for this reason.

11.5.2 Fuzzy Sets

We are now in a position to consider *fuzzy sets* bearing in mind classical sets and their operations, discussed previously. A fuzzy set is a class of elements (members) with a continuum of grades of membership. Examples are the set of all beautiful women and the set of all tall men, for which a degree of subjectivity arises when defining them. Furthermore, these sets are characterised by a degree of continuum as regards the extent to which each candidate is a member of the set or not. Some candidate members undisputedly are members of the set, whereas others undisputedly are not. There exist also candidate members that fulfil some but not all the requirements of a full member of the set. In this case, participation (membership) of the set is *partial*.

In order to express in quantitative terms the degree of membership of a particular element (member) x of fuzzy set A, a *membership function,* $\mu_{\mathrm{A}}(x)$ assigns to each element a grade whose value ranges from zero to unity. If X is the *universe of discourse* and its elements are denoted by x, then a fuzzy set A in X is defined as a set of ordered pairs:

$$\mathrm{A}(x),\ \mu_A(x) \qquad x \in \mathrm{X} \tag{11.67a}$$

or

$$\mathrm{A} = \{x, \mu_A(x) | x \in \mathrm{X}\} \tag{11.67b}$$

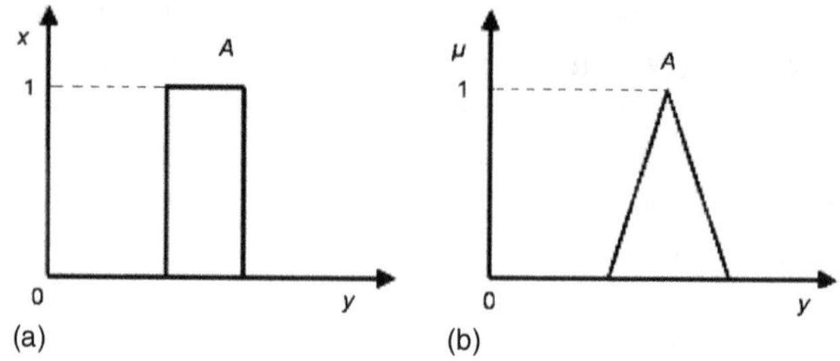

Figure 11.21 Membership functions χ_A, μ_A. *(a) Classical set A. (b) Fuzzy set A.*

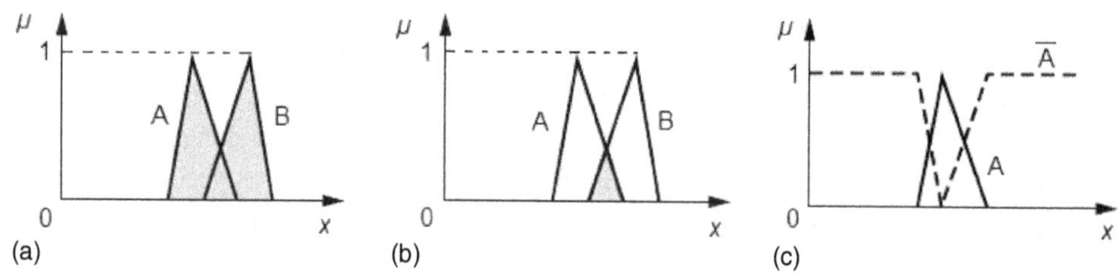

Figure 11.22 Fuzzy set operations. (a) Union. (b) Intersection. (c) Complement.

If the universe of discourse X is discrete and finite,

$$\mathrm{A} = \left\{ \frac{\mu_A(x_1)}{x_1} + \frac{\mu_A(x_2)}{x_2} + \cdots \right\} = \left\{ \sum_i \frac{\mu_A(x_i)}{x_i} \right\} \tag{11.67c}$$

In Equation (11.67c) the horizontal bars do not denote division, nor do the plus signs denote addition. The summation sign denotes the aggregation of each element. This notation was introduced in [32].

We can illustrate the differences in the membership status of a set in the case of classical logic, χ_A, with that of a fuzzy set, μ_A, using the sketches in Figures 11.21 and 11.22.

In each case, the degree of membership ranges from zero to unity. In the case of the classical set, the degree of membership of all members is equal to unity and there are no partial members. In the case of this particular fuzzy set, the degree of membership varies linearly from zero to unity and from unity to zero (triangular form). The form of the membership function can vary and is selected using expert knowledge of the problem considered.[13]

Having determined the operations that return the same values as in classical sets for zero and unity, the following *operations on fuzzy sets* A and B are defined:

Union $$\mu_{A\cup B}(x) = \mu_A(x) \vee \mu_B(x) \tag{11.68a}$$

Intersection $$\mu_{A\cap B}(x) = \mu_A(x) \wedge \mu_B(x) \tag{11.68b}$$

Complement $$\mu_{\bar{A}}(x) = 1 - \mu_A(x) \tag{11.68c}$$

[13] The height of the membership function may not necessarily reach unity, in which case the function is termed *subnormal*.

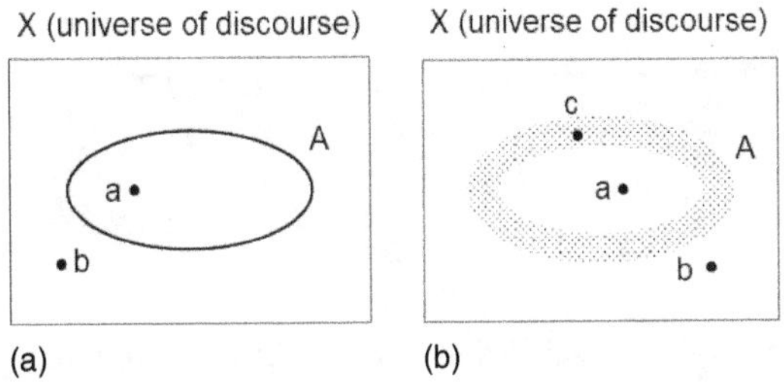

Figure 11.23 Classical and fuzzy sets. (a) Classical set. (b) Fuzzy set.

Having defined fuzzy set operations, it is necessary to consider the properties of fuzzy sets. It is useful at this point to illustrate the differences between a classical set and a fuzzy set before discussing the properties of fuzzy sets.

It is seen in Figure 11.23 that in the case of the classical set, elements are either full members or are not members of the set, whereas in the case of the fuzzy set there exist partial members such as c, as well. Because of this, classical sets are also termed *crisp* sets, in contradistinction to fuzzy sets. In the case of the classical set, $A \cup \overline{A} = X$ where $\overline{A}$ is the complement of A (Figure 11.24a). This is in effect the expression of the *law or principle of the excluded middle (or non-contradiction),* which is one of the cornerstones of Aristotelian logic.

In the case of the fuzzy set, $A \cup \overline{A} \neq X$, as seen in Figure 11.24b. Therefore, the law of the excluded middle does not hold for fuzzy sets. Apart from this, all other properties of crisp (classical) sets hold in the case of fuzzy sets as well. These are as follows [26]:

Commutativity	$A \cup B = B \cup A$
	$A \cap B = B \cap A$
Associativity	$A \cup (B \cup C) = (A \cup B) \cup C$
	$A \cap (B \cap C) = (A \cap B) \cap C$
Distributivity	$A \cup (B \cap C) = (A \cup B) \cap (A \cup C)$
	$A \cap (B \cup C) = (A \cup B) \cup (A \cup C)$
Idempotency	$A \cup A = A$
	$A \cap A = A$
Identity	$A \cup \emptyset = A$
	$A \cap X = A$
	$A \cup \emptyset = A$
	$A \cup X = X$
Transitivity	If $A \subseteq B$ and $B \subseteq C$, then $A \subseteq C$
Involution	$\overline{\overline{A}} = A$
De Morgan's principles	$\overline{A \cap B} = \overline{A} \cup \overline{B}$ and $\overline{A \cup B} = \overline{A} \cap \overline{B}$

Logical IF-THEN rules

Fuzzy sets and fuzzy operators are the subjects and verbs of fuzzy logic. In order to formulate a *conditional statement* in fuzzy logic an IF-THEN rule is used. This assumes the form:

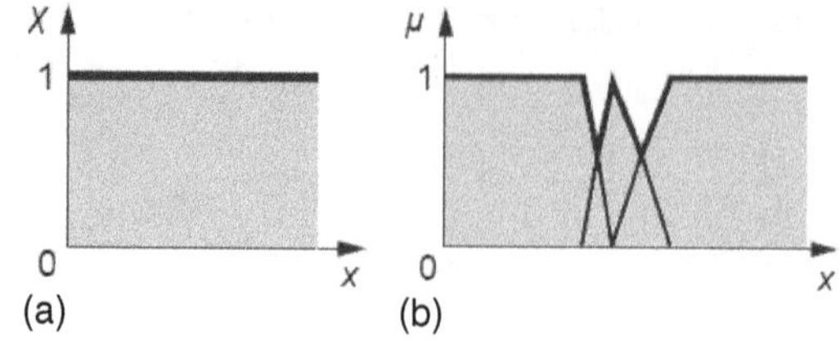

Figure 11.24 The law of the excluded middle. (a) Crisp set. (b) Fuzzy set.

IF (X is A) THEN (Y is B)

where A and B are linguistic values defined by fuzzy sets. The IF-part of the rule is termed the *antecedent* and the THEN-part is termed the *consequent* and thus, in general,

IF (antecedent) THEN (consequent)

Fuzzy IF-THEN rules act in the following ways:

Assignment statements that use '=' (equal to sign) for the purpose of assignment. They are of the following form:

a = hello
climate = summer

Conditional statements that use the 'IF-THEN' rule base as follows:

IF temperature is high THEN climate is hot
IF food is fresh THEN eat

Unconditional statements, of the following form:

GOTO 3
Switch off the engine

Interpreting an IF-THEN rule involves the following steps:

- Evaluation of the antecedent. This involves *fuzzifying* the inputs and applying any necessary *fuzzy operators*. Fuzzification involves allocating a degree of membership to each antecedent, in the range zero to unity. If there is only one part to the antecedent, then this is the degree of support for the rule.
- Application of fuzzy operators to multiple-part antecedents. If this is the case, the antecedent has to be resolved to a single number in the range zero to unity and this is the support for the rule.
- The third step is the application of the *implication method*. For an IF-THEN rule, the antecedent, p, *implies* the consequent, q. In binary (classical) logic, if p is true, then q is also true ($p \rightarrow q$). In fuzzy logic, if p is true to some degree of membership, then q is also true to the same degree. That is, if p is assigned a value less than unity, the output fuzzy set q is truncated according to the implication method.

In most practical problems more than one rule is required. The output of each is then a fuzzy set. The output fuzzy sets are then *aggregated* into a single output fuzzy set. This is then *de-fuzzified*, so that a crisp number is the final outcome.

11.5.3 Fuzzy Inference

The entire procedure is termed *fuzzy inference* and consists of the same steps in all types of problems. The steps need to be set up to solve each particular problem individually and are summarised below:

1. *Fuzzification* of the input variables
2. Application of the fuzzy operator (AND or OR) in the antecedent
3. Implication from the antecedent to the consequent
4. Aggregation of the consequents across the rules
5. *Defuzzification*

The block diagram included in Figure 11.25 illustrates the connection of the various components of a fuzzy inference system.

The components of a FIS are:

- Rule Base contains fuzzy IF-THEN rules
- Database defines the membership functions of fuzzy sets used in fuzzy rules
- Decision-making Unit performs operations on rules
- Fuzzification Inference unit converts crisp quantities into fuzzy quantities
- Defuzzification Inference unit converts fuzzy quantities into crisp quantities

It is now possible to describe how fuzzy logic is used, given that the information that is available is in verbal form, and that this needs to be manipulated by a computer that traditionally works with binary (yes/no) logic. The procedure followed consists of a number of steps, in all cases. The steps need to be set up to solve each particular problem individually. What in fact occurs is a mapping from a given input to an output. This is termed *fuzzy inference*. The fuzzy inference procedure consists of the following five steps [23]:

1. *Fuzzification* of the input variables
2. Application of the fuzzy operator (AND or OR) in the antecedent
3. Implication from the antecedent to the consequent
4. Aggregation of the consequents across the rules
5. *Defuzzification*

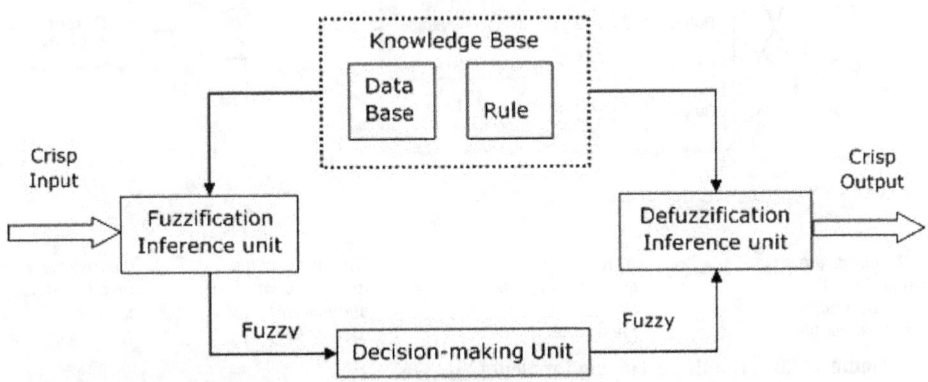

Figure 11.25 A fuzzy inference system (FIS).

11.5.4 Steps in Fuzzy Inference System

In this section we shall consider the steps followed in the operation of a fuzzy inference system. In order to illustrate how these are implemented we shall refer to an example used in [23]. The problem in question is the 'Basic Tipping Problem', that in fuzzy terms is defined to be a two-input, one-output, three-rule system (Figure 11.26). We note that the operands are problem dependent; thus in this case the Input comprises of the service and food and the Output is the level of tipping. The rules that are set up are also problem dependent.

For this problem three Rules are specified as indicated in Figure 11.27c. We see that they are all of the IF-THEN type. For Rule 1 the antecedent is 'service is poor or food is rancid', and the consequent is 'tip is cheap'. It may also be necessary to select *weights* for each rule. In the simplest case, all rules carry the same weight.

We are now in a position to implement the fuzzy inference process, following the steps listed above.

Step 1

In Step 1 the input has to be fuzzified. This is done by allocating fuzzy sets to each of the relevant linguistic variables (Figure 11.27). In this example the linguistic variables for the Input variable service are 'poor', 'good' and 'excellent'. For the Input variable food they are 'rancid' and 'delicious'. The fuzzified input obtains a value on the vertical axis in the interval zero to unity, depending on the form of the membership function. It is also necessary to select membership functions for the linguistic variables of the consequent ('cheap', 'average', 'generous').

On this basis, we apply the appropriate membership functions for service and food from the above. If we judge the food to be delicious (8), the membership function $\mu = 0.7$. The

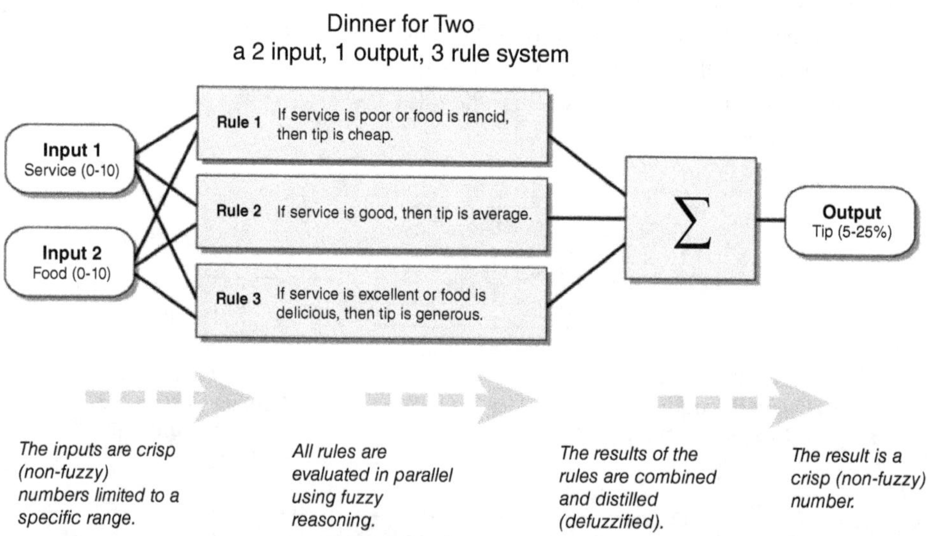

Figure 11.26 The Basic Tipping Problem [23].

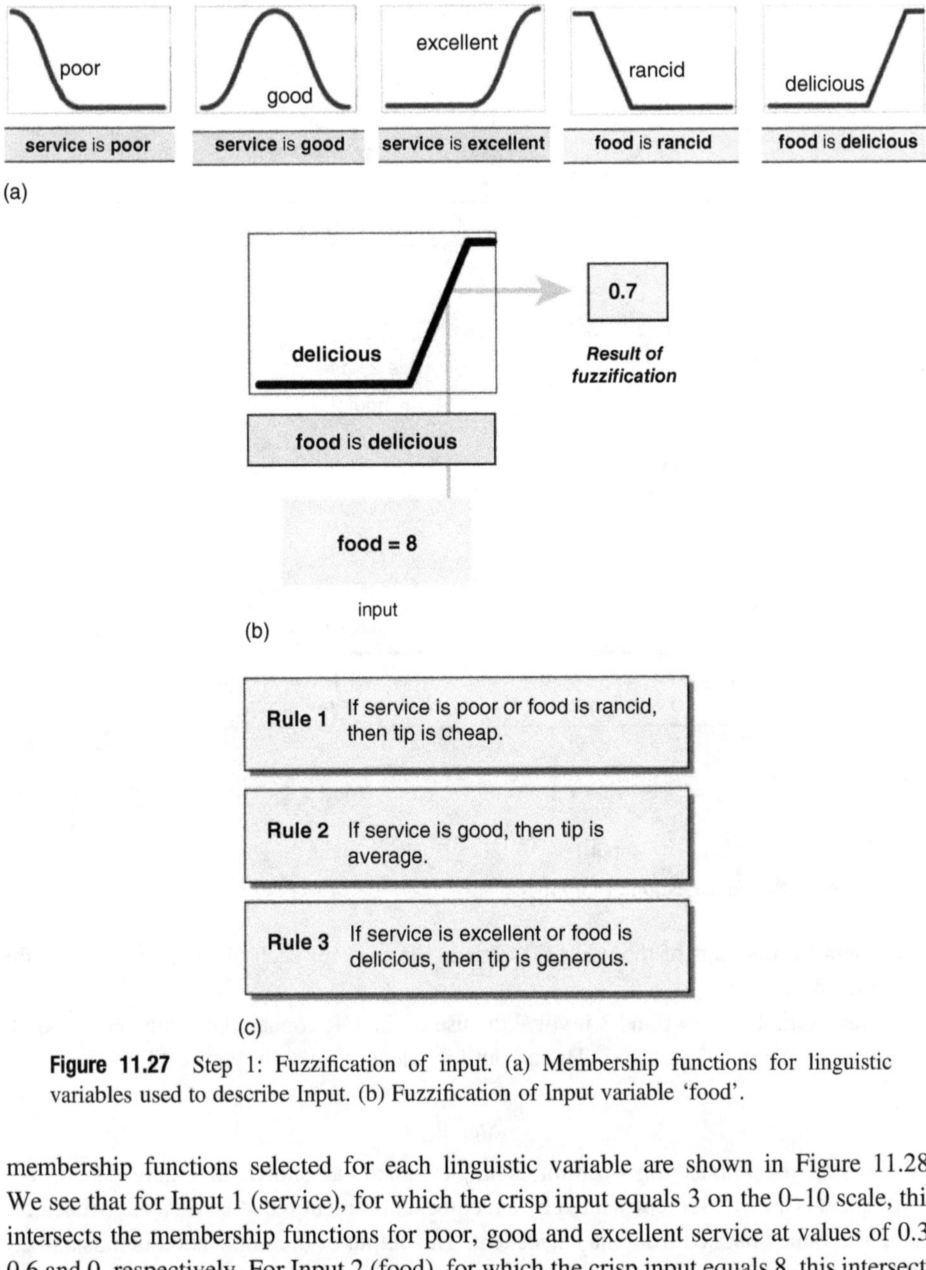

Figure 11.27 Step 1: Fuzzification of input. (a) Membership functions for linguistic variables used to describe Input. (b) Fuzzification of Input variable 'food'.

membership functions selected for each linguistic variable are shown in Figure 11.28. We see that for Input 1 (service), for which the crisp input equals 3 on the 0–10 scale, this intersects the membership functions for poor, good and excellent service at values of 0.3, 0.6 and 0, respectively. For Input 2 (food), for which the crisp input equals 8, this intersects the membership functions for 'rancid' and 'delicious' at values of 0.3 and 0.7, respectively.

Step 2

We then apply each rule (fuzzy operator). If the antecedent of a rule has more than one part, the fuzzy operator is applied to obtain the one number that represents the result of the

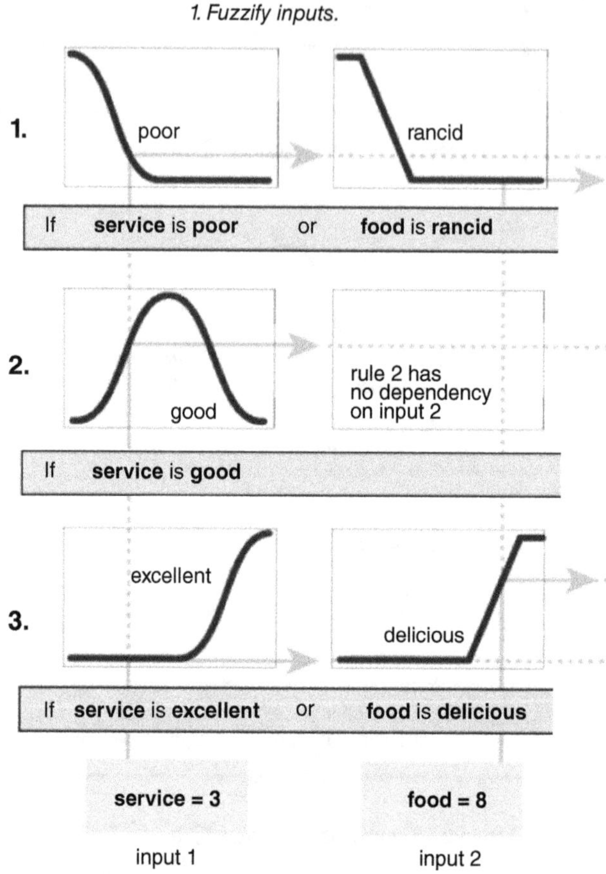

Figure 11.28 Fuzzification of all inputs.

antecedent for that part of the rule [23]. This is the case for each of the three rules of the example problem.

In this example, rules 1 and 3 involve the use of the OR connective, which corresponds to the use of the max() function. The output for rule 3 is given in Figure 11.29.

Step 3

In this step the implication operator is implemented, as shown in Figure 11.30. The consequent is a fuzzy set represented by a membership function that is reshaped (truncated) using the value obtained from the antecedent. The output of the implication is thus a fuzzy set. This process is applied for each rule.

Step 4

In the next step, since in this problem more than one rule has been implemented, the results of all rules need to be combined, or *aggregated*, into a single fuzzy set. It is necessary to produce a meaningful output based on the results of the use of more than one rule. When

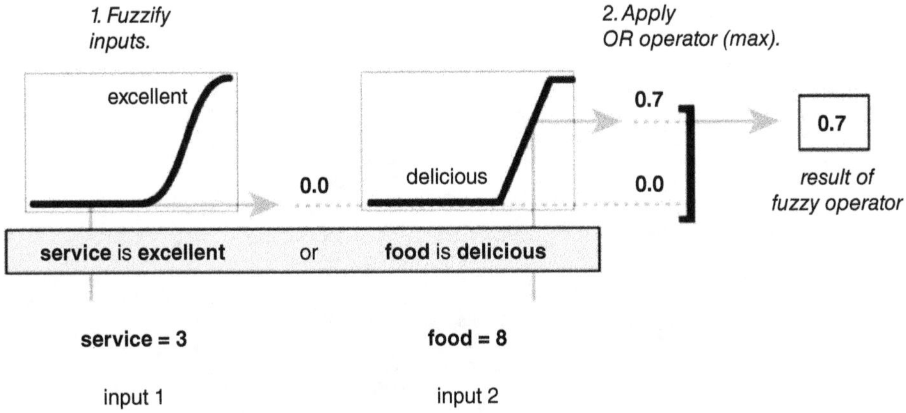

Figure 11.29 Sample output for rule application (rule 3).

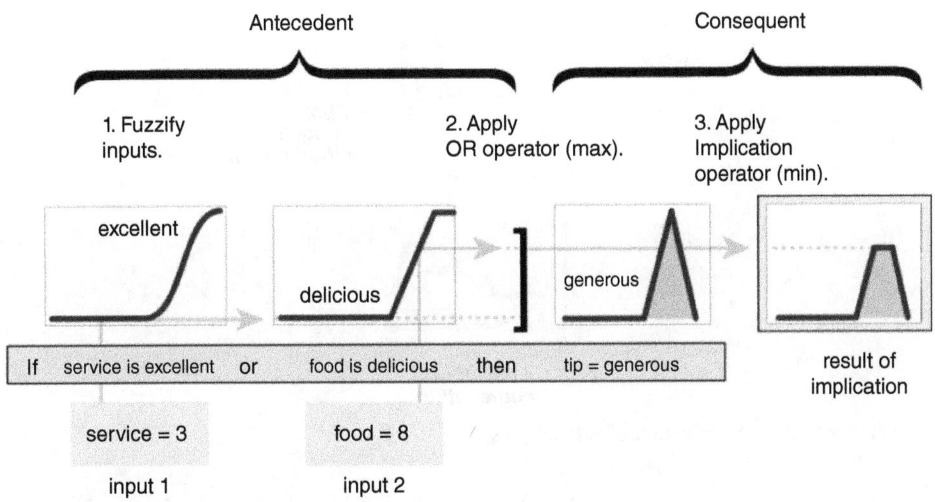

Figure 11.30 Application of the implication function for rule 3.

aggregation is carried out, fuzzy sets are combined in a suitable way to produce a single fuzzy set. Aggregation can be carried out using the standard fuzzy intersections and unions operations (Figure 11.31). In determining an aggregation strategy, two possible cases arise. In the first case a set of rules must be *jointly satisfied*, in which case the rules are connected by 'AND' connectives and the output is found by the fuzzy intersection of all rule consequents:

$$y = y^1 \cap y^2 \cap \cdots \cap y^r$$

This is defined by the membership function,

$$\mu_y(y) = \min\left(\mu_{y1}(y), \mu_{y2}(y), \ldots, \mu_{yr}(y)\right) \qquad \text{for } y \in \mathrm{Y} \tag{11.69}$$

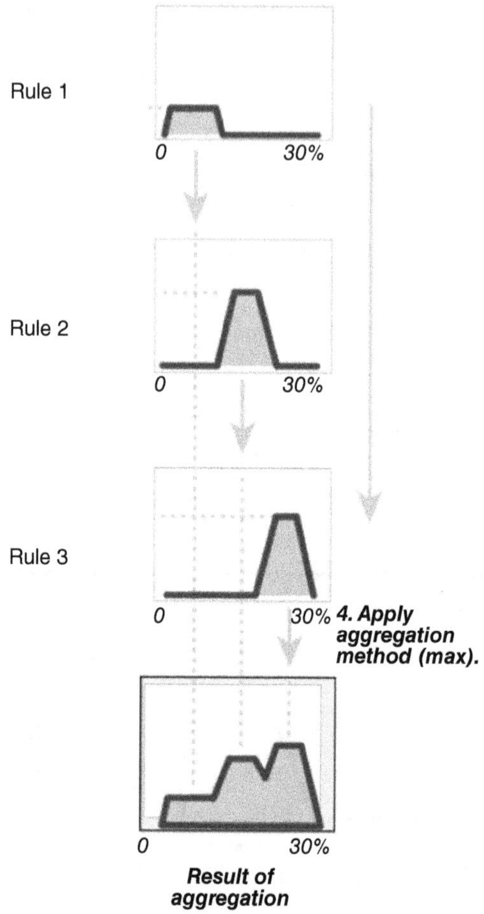

Figure 11.31 Aggregation of rule outputs.

In the second case, one rule at least has to be satisfied in which case the rules are connected by 'OR' connectives. In this case the aggregated output is found by the fuzzy union of all individual rule contributions, as

$$y = y^1 \cup y^2 \cup \cdots \cup y^r \tag{11.70}$$

This is defined by the membership function,

$$\mu_y(y) = \max\left(\mu_{y1}\ (y), \mu_{y2}\ (y), \ldots, \mu_{yr}\ (y)\right) \qquad \text{for } y \in Y \tag{11.71}$$

The input used in aggregation in each case is the list of the truncated output functions that have resulted from the implication process. The output is a fuzzy set for each output variable. Since in this example it is required that one rule be satisfied, aggregation involves the use of the max() function.

Step 5

The last step in the FIS consists of transforming the fuzzy set that is the result of aggregation into a single (crisp) number. This is called *defuzzification*, and there are several ways in which it can be accomplished. Two of the most frequently used ones are the *centroid method* and the *weighted average* method (Figure 11.32). In the first of these, also known as the *centre of gravity method*, the output is obtained from the following algebraic expression:

$$z^* = \frac{\int \mu_{\underset{\sim}{C}}(z) \cdot z \, dz}{\int \mu_{\underset{\sim}{C}}(z) \, dz}$$

The second method is computationally efficient but is restricted to sym-metrical output membership functions. The output is given by the following expression:

$$z^* = \frac{\sum \mu_{\underset{\sim}{C}}(\bar{z}) \cdot \bar{z}}{\sum \mu_{\underset{\sim}{C}}(\bar{z})} \tag{11.72}$$

where Σ denotes the algebraic sum and where $\bar{z}$ is the centroid of each symmetric membership function. The method involves the weighting of each membership function in the output by its respective maximum membership value.

The entire process is summarised in the block diagram included in Figure 11.33. Calculations commence in the lower left-hand part of the figure and proceed in a clockwise manner, ending with a crisp output obtained in the last step, following defuzzification.

In the FIS that has been described up to this point the output of each rule is a fuzzy set derived from the output membership function and the implication method used. This type of FIS is called the *Mamdani system* after the person who introduced it. It is not the only type of FIS in common use but it is intuitively clear and well suited to expert system

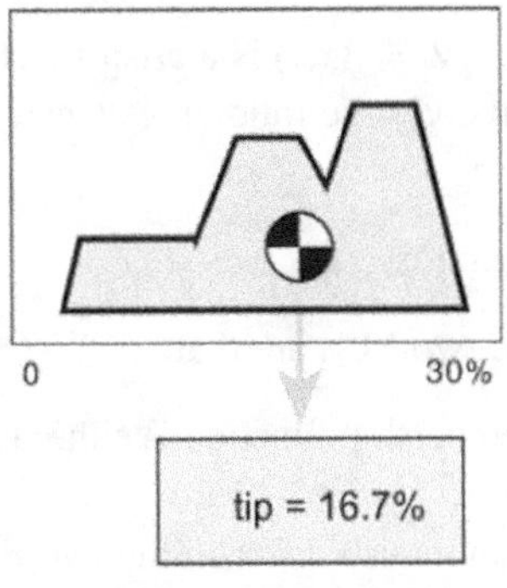

Figure 11.32 Defuzzification using the centroidal method.

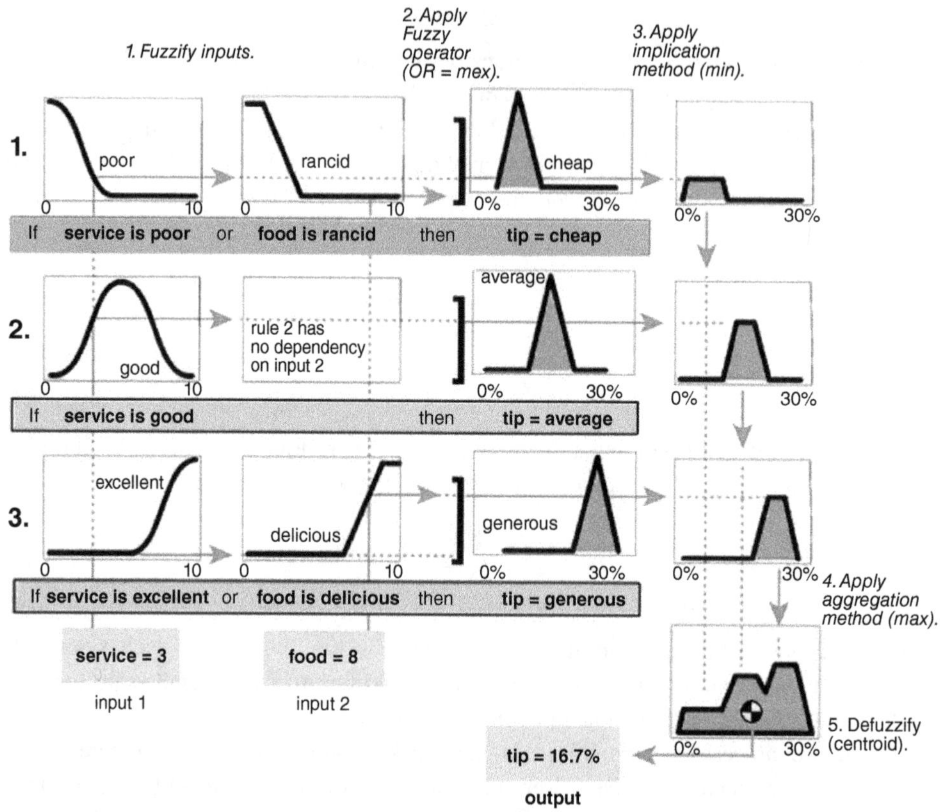

Figure 11.33 Example of a fuzzy interference system (Mamdani system) [23].

applications in which the rules are created from human knowledge. The block diagram included in Figure 11.34 illustrates the essential operations of the Mamdani system.

Mention should be made of one other commonly used FIS, the Takagi-Sugeno FIS. In this, the rule format is

$$\text{IF (X is A) AND (Y is B) THEN Z} = f(x, y)$$

were A and B are fuzzy sets and $Z = f(x,y)$ is a crisp function in the consequent.

The FIS in the TS model involves the following steps:

1. Fuzzification of inputs
2. Application of the fuzzy operator

The major differences of the two FISs are that:

1. In the TS FIS the output membership function is either linear or constant whereas in the Mamdani FIS it is a fuzzy set.
2. Aggregation and defuzzification also differ in the two FISs.
3. There exist more mathematical rules for the TS than for the Mamdani FIS.

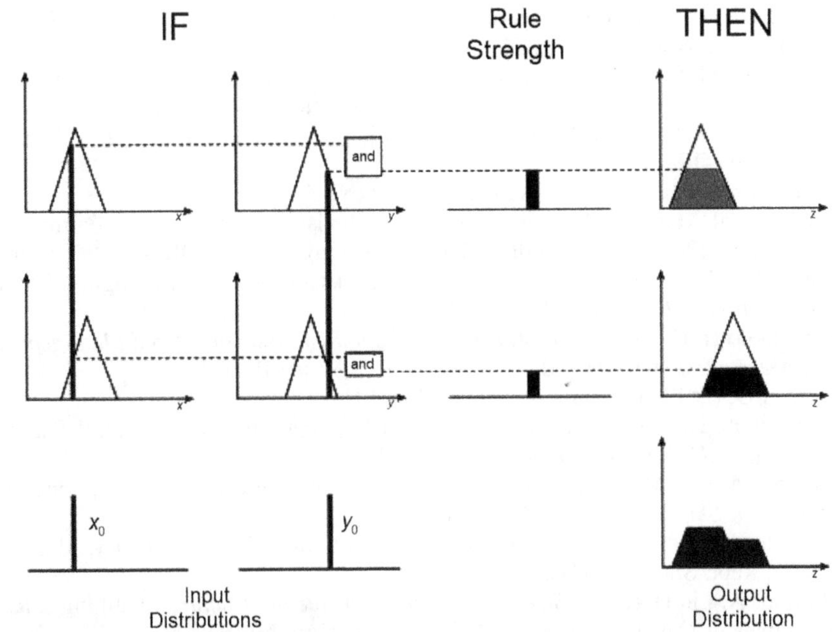

Figure 11.34 The Mamdani system.

References

1. Ang A. H.-S. Tang W. H. (1990) *Probability Concepts in Engineering Planning and Design. Vol. I – Basic Principles*. New York: John Wiley & Sons.
2. van Asselt, M. B. A. (2000). Risk. In *Perspectives on Uncertainty and Risk.* Dordrecht: Springer Netherlands 147–204.
3. Augusti G. Baratta A. Casciati F. (1984) *Probabilistic Methods in Structural Engineering*. London: Chapman & Hall.
4. Ayyub B. M. Assakkaf I. Atua K. Engle A. Hess P. Karaszewski Z. Kihl D. Melton W. Sielski R. A. Sieve M. Walsman J. White G. J. (1998) Reliability-based Design of Ship Structures: Current Practice and Emerging Technologies *Research Rept.* to the US Coast Guard SNAME T&R Rept. R-53.
5. Ayyub B. M. Assakkaf I. A. Beach J. E. Melton W. M. Nappi Jr N. Conley J. (2002) Methodology for Developing Reliability-Based Load Resistance Factor Design (LRFD) Guidelines for Ship Structures. *Naval Eng. J.* 23–41.
6. Beaumariage Jr. G. N. (1960) Syllogistic Reasoning. *J. Exp. Educ.* Vol. 29 No. 1 61–71.
7. Bernstein P. L. (1996) *Against the Gods: The Remarkable Story of Risk.* New York: John Wiley & Sons.
8. Brünner E. et al. (2012) Design Principles and Criteria. Committee IV.1 Rept. in Fricke W. Bronsart R. (Eds) *Proc. 18th Int. Ship & Offsh. Struct. Cong*. Vol. 1 Schiffahrts-Verlag Hansa GmbH & Co KG Hamburg 435–506.
9. Cornell C. A. (1969) A Probability-based Structural Code *ACI J.* Vol. 66 974–985.
10. Daidola J. C. Basar N. C. (1980) Probabilistic Structural Analysis of Ship Hull Longitudinal Strength. Ship Struct. Comm. Rept. SSC-301 Washington, DC.

11. Det Norske Veritas (1992) Structural Reliability Analysis of Marine Structures. Classification Notes No. 30.6.
12. Hasofer A. M. Lind N. C. (1974) Exact and invariant second moment format. *J. of the Engineering Mechanics Division ASCE* Vol. 100 No. EMI 111–121.
13. Huntingdon E. V. (1939) Frequency distribution of sum and quotient. *Ann. Math. Stat.* Vol. 10 195–198.
14. International Maritime Organisation MSC 79/INF.5.
15. International Maritime Organisation (2013) Revised Guidelines for Formal Safety Assessment (FSA) for Use in the IMO Rule-making Process. MSC-MEPC.2/Circ. 12.
16. Khotsialov N. F. (1929) 'Строительная промышленность (Strength of stocks), *Construction Industry*. № 10 840-844.
17. Kumamoto H. Henley E. J. (2000) *Probabilistic Risk Assessment and Management for Engineers and Scientists* 2nd ed. Hoboken, NJ: Wiley-IEEE Press.
18. Larsson O. (2019) Reliability Analysis, Lund Univ.
19. Mansour A. E. (1972) Probabilistic Design Concepts in Ship Structural Safety and Reliability. *SNAME Transactions* 64–94.
20. Mansour A. (1974) Approximate probabilistic method of calculating ship longitudinal strength. *J. Ship Res.* Vol. 18 No. 3 203–213.
21. Mansour A. E. (1989) *An Introduction to Structural Reliability Theory*. Ship Struct. Comm. Rept. SSC-351 Washington, DC.
22. Mansour A. Liu D. (2008) *Strength of Ships and Ocean Structures*. Paulling J. R. (Ed) (Principles of Naval Architecture Series). Jersey City, NJ: SNAME.
23. Mathworks Inc (2008) *Fuzzy Logic Toolbox™ – User's Guide*. Natick, MA.
24. Mayer M. (1926) *Die Sicherkeit der Bauwerke und ihre Berechnung nach Grenzkräften anstatt nach zulässigen Spannungen* (The safety of structures and its calculation by border forces rather than by permissible stresses). Berlin: Springer Verlag.
25. McGraw-Hill *Dictionary of Scientific & Technical Terms*, 6th ed. McGraw-Hill.
26. Rackwitz R. Fiessler B. (1978) Structural reliability under combined random load sequences. *Comp. & Struct.* Vol. 9 489–494.
27. Ross T. J. (2010) *Fuzzy Logic with Engineering Applications*. 3rd ed. Chichester: John Wiley & Sons.
28. Streletskii N. S. (1947) 'Основы статистического учета коэффициента запаса прочности сооружений' (Fundamentals of statistical safety factors of structural strength). Stroyzdat.
29. 'The world standard in knowledge since 1768' *Encyclopædia Britannica*.
30. Thoft-Christensen P. Baker M. J. (1982) *Structural Reliability Theory and Its Applications*. Berlin: Springer-Verlag.
31. Wikipedia contributors 'Fuzzy control system' *Wikipedia The Free Encyclopedia* 1 December 2020. Available: https://en.wikipedia.org/w/index.php?title=Fuzzy_control_system&oldid=991692032 (accessed 19 December 2020).
32. Zadeh L. A. (1965) Fuzzy sets. *Information and Control* Vol. 8 338–353.
33. Zadeh L. A. (1988) Dispositional logic. *Appl. Math. Lett.* Vol. 1 95–99.

12

Ship Structural Reliability Theory and Applications

> The ability to correctly and consistently provide the necessary safety margins while meeting the twin requirements of structural safety and economy is key to the design of successful structures.
>
> *J. K. Paik & A. K. Thayamballi*

12.1 Introduction: Applications of Ship Structural Reliability Theory

The overall aim of structural reliability theory when applied to a specific field of engineering is to take into account in a systematic manner the uncertainties that feature in a new design as well as those that influence the safety of an existing structure. A major task of structural reliability theory is to determine and subsequently allow for the properties of the distributions that describe the parameters in question, so that meaningful results are obtained.

More specifically, *ship structural reliability theory* can be used to address the following:

1. To develop probability-based codes
2. To calculate the reliability and hence the inherent safety of existing or future designs. This is particularly valuable in the case of novel designs or novel applications of existing designs. This applies not only to the entire hull girder but also to major and minor components.
3. To perform failure analysis. This a posteriori analysis aims at understanding the cause and sequence of complex failure modes that may have had serious consequences for the structure itself, humans and the environment.
4. To evaluate proposed designs by comparing safety levels with those of successful existing designs.
5. To contribute to risk evaluation studies. Reliability analysis is one of the two components of risk analysis. When complemented by well-established consequence studies, risk can be quantified. Such studies may prove beneficial when assessing the lifecycle cost of a structure, allowing for repairs, maintenance and off-hire periods. In the case of warships, this would be inactive periods, a cause for concern for any naval command. The relevant parameters would include detail structural design and related fabrication

costs on the one hand and inspection, repair and maintenance costs on the other. In the case of single ships, this may not appear to be an important issue, but when a fleet of vessels is considered, the potential savings that may result from a rational approach could be significant.

6. To extend the operating life of ship hull structures. At this point in time, the operating life is assumed to be 25 years, and rules for ship construction are currently based on this assumption. It is possible, however, to extend the design life significantly, provided the structural design of details and their fabrication are carried out following the proper procedures. There exist rules of certain classification societies that may be applied to extend the operating life of surface ships to more than 40 years. There is no technical reason this cannot be extended much further. The need to do so is not only of interest to shipowners and especially navies but also to society at large, for a variety of reasons. In this, reliability studies can play an important role.

In order to implement ship structural reliability theory it is necessary to have an understanding of the random processes specific to ships at sea. It is therefore necessary to consider the uncertainties that characterise the loading and strength of marine structures. In doing so, relevant probabilistic descriptions need to be identified and their descriptors calculated. These include the first- and second-order moment or the coefficient of variation (COV).[1] This data can subsequently be used in determining target reliabilities so that subsequently design calculations can be carried out.

Following this, selected examples of the application of structural reliability theory in the marine field will be considered. The first one will describe the calculation of partial safety factors for use in the IACS Common Structural Rules for oil tankers. The provision of a rational basis for ship construction rules issued by classification societies is arguably the most important contribution that structural reliability theory has made in the field of marine structures. The discussion is limited to hull girder ultimate strength, a topic that has been treated in an earlier chapter of this text.

Reliability-based analysis will be discussed next. This is a wide-ranging subject that touches upon a number of topics that have not been considered in this text, in particular local strength and fatigue. Reliability-based analysis is a necessary prerequisite to any other application of structural reliability theory, and it can prove to be of immense value in specialised studies that deal with problems that involve an appraisal of safety standards of an existing structure. The problem considered here is the assessment of the effect of the introduction of the CSR for oil tankers, and this is performed by comparing the designs of a pre-CSR with a CSR-compliant design of the same ship.

The last example concerns the use of risk analysis in the structural design of an oil tanker.

[1] The *coefficient of variation (COV)* is a measure of dispersion of a probability distribution and is defined as the ratio of the standard deviation σ to the mean $|\mu|$. It is independent of the unit in which measurements have been taken, and is thus a dimensionless number. Therefore, it should be used for comparison between data sets with different units or widely differing means, instead of the standard deviation. In reliability analysis it fulfils an important role by acting as a quantitative measure of uncertainty.

12.2 Uncertainty and the Safety Concept in the Marine Environment

Uncertainty pervades all physical processes and can lead to difficulties, especially when designing structures that are innovative or that are expected to encounter conditions that have never been experienced in the past.[2] This type of uncertainty can be classed as *objective* (or *random*, or *physical*) or *subjective* (or *systematic*). Objective uncertainties are innate to the nature of the structure or the applied loading, whereas subjective uncertainties include the non-measurable inaccuracies of engineering analyses, any non-measured variances of construction and fabrication, as well as the unavoidable and non-determinable errors associated with the prediction of future conditions [1]. The latter are therefore human-related and can thus be reduced with advances in scientific understanding, technology, etc. There exist other classifications in the literature, but in the case of marine structures this classification serves our purpose. Typical representations of the response, loading and strength in ship structures are shown in Figure 12.1. The first of these is a plot of the wave elevation, this being a measure of wave loading. The actual wave elevation itself is characterised by objective uncertainty, whereas the plot obtained and traced has, in addition, subjective uncertainty. The second graph shows the long-term loading of a hull girder, represented by a Weibull distribution, whereas the last graph shows the variability in yield stress of steel, which is a measure of structural strength.

The elimination of subjective uncertainties may permit them to be reclassified as objective. Clearly, in the case of one particular variable, it is possible for uncertainties of one or the other kind to arise, as well combinations of both. The need therefore arises to develop techniques to combine uncertainties, whatever their cause or origin may be.

When calculations are performed that involve a quantity that is characterised by uncertainty, it is necessary to allow for that uncertainty, and preferably to err on the side of safety but not excessively so. Thus, when estimating structural properties, such as, for example, the yield stress of a series of sheets of steel plating, a finite number of tensile tests are carried out and the results plotted (Figure 12.1c). These results provide an estimate of the true mean value of the yield stress and of the variation about this mean value (*sample mean*). Since the actual true value cannot be obtained, a difference arises between the estimated value and the true mean value. This difference is termed *bias*. In addition to this, the measurements are seen to be characterised by *spread*. Statistical analysis shows that they can be modelled using the normal or the lognormal distribution.

In structural calculations it is necessary to allow for the uncertainty that arises as a result of these variations, and this is done by specifying *nominal* values. In this case it would be a nominal value for the yield stress. In order to ensure that the material will conform to requirements, the nominal value for the yield stress of this particular series of sheets of steel plating is set equal to two standard deviations less than the sample mean for these sheets of plating. This ensures that the probability that the yield stress of any given sample will exceed the nominal value is greater than approximately 95 per cent, this being an acceptable *confidence interval*.

[2] Examples are vehicles for space exploration, deep sea exploration and nuclear reactors.

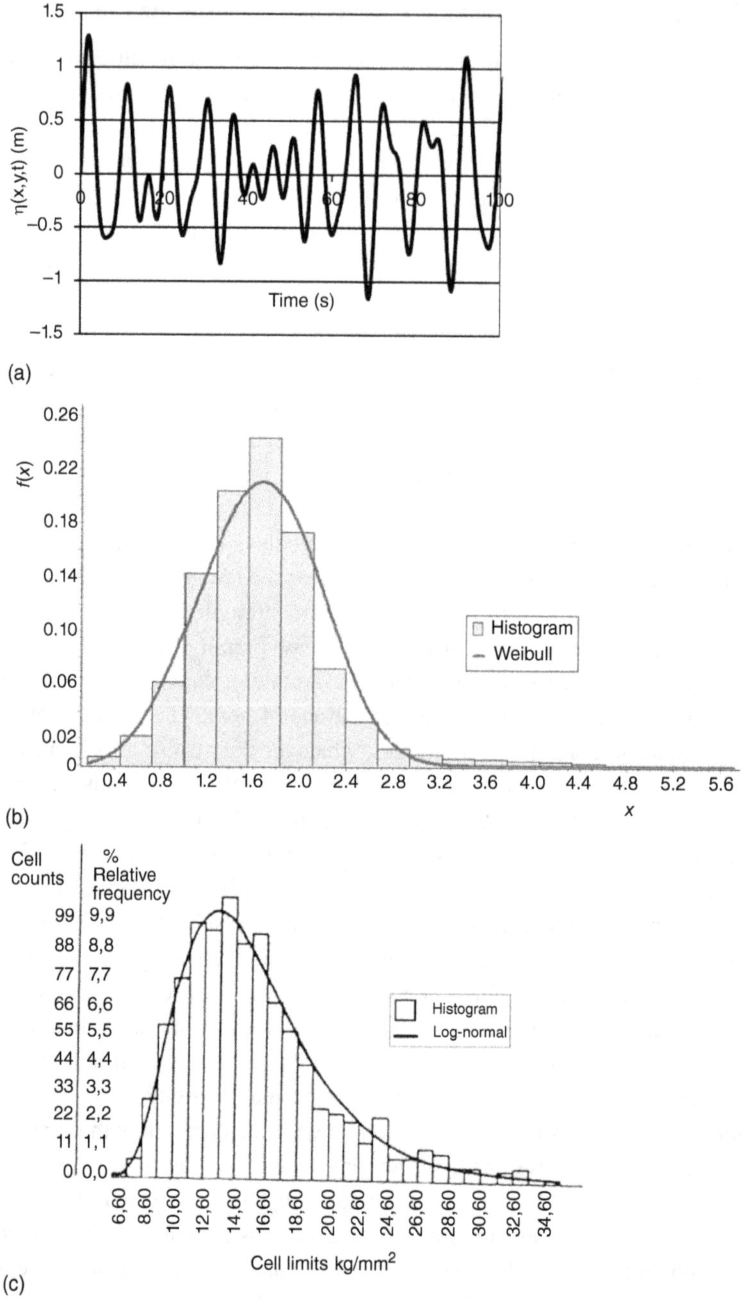

Figure 12.1 Statistical representation of loading and strength. (a) Realisation of wave elevation variation with time. (b) Long-term significant wave height distribution. (c) Yield stress distribution of steel (*Weibull*) [10] (lognormal) [11]. Figure 12.1b courtesy of the Ship Structure Committee, Washington, DC. Figure 12.1c reprinted courtesy of Det Norske Veritas, Høvik.

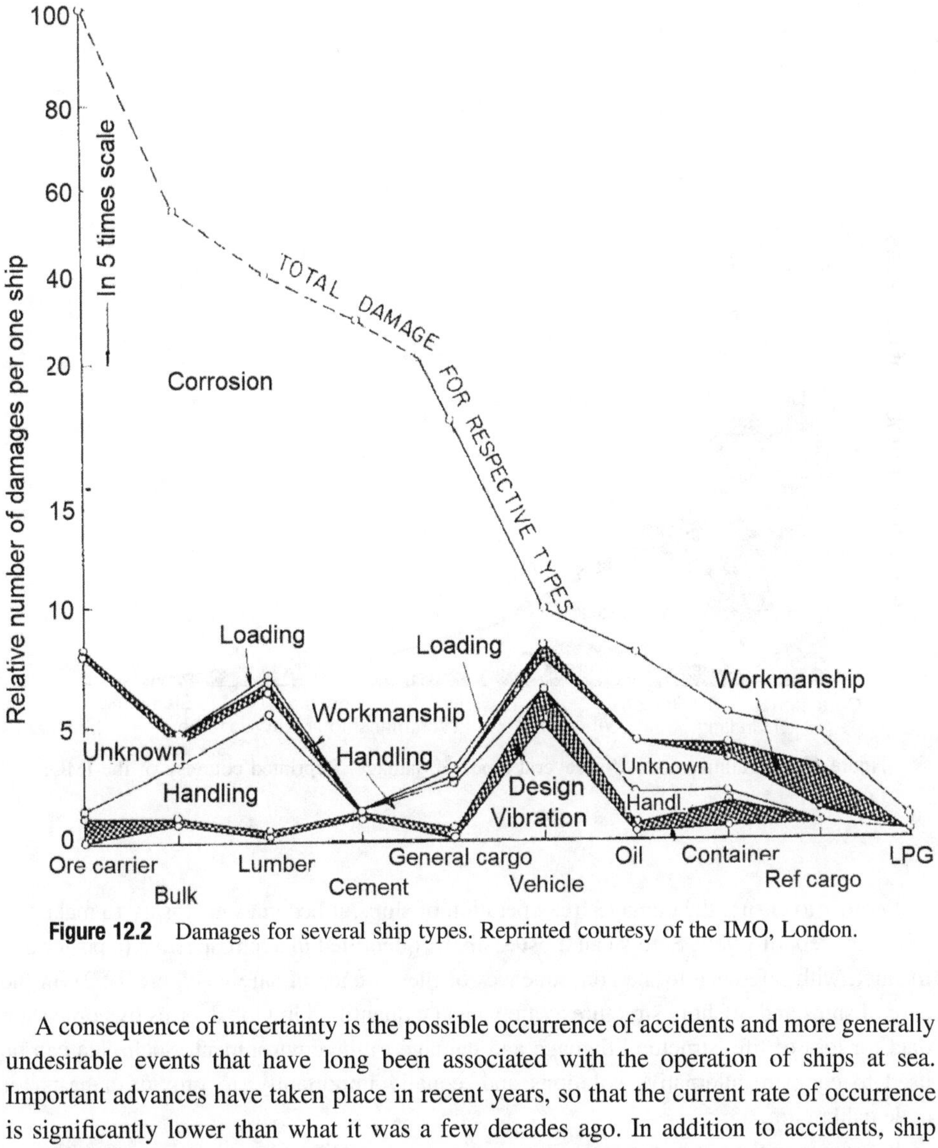

Figure 12.2 Damages for several ship types. Reprinted courtesy of the IMO, London.

A consequence of uncertainty is the possible occurrence of accidents and more generally undesirable events that have long been associated with the operation of ships at sea. Important advances have taken place in recent years, so that the current rate of occurrence is significantly lower than what it was a few decades ago. In addition to accidents, ship structures suffer a whole range of minor effects that do not imperil overall structural integrity[3] and may collectively be termed *damages*. Figure 12.2 shows the relative number of damages per ship for a number of ship types. Although the data used in this figure relate to the 1980s, it is clear that for each ship type there are a number of different types of causes. Figure 12.3 shows the contribution of different causes to the total damage sustained by the ship structure.

[3] *Structural integrity* is understood to be the preservation of both individual parts of a structure as well as their mutual relation in an intact state.

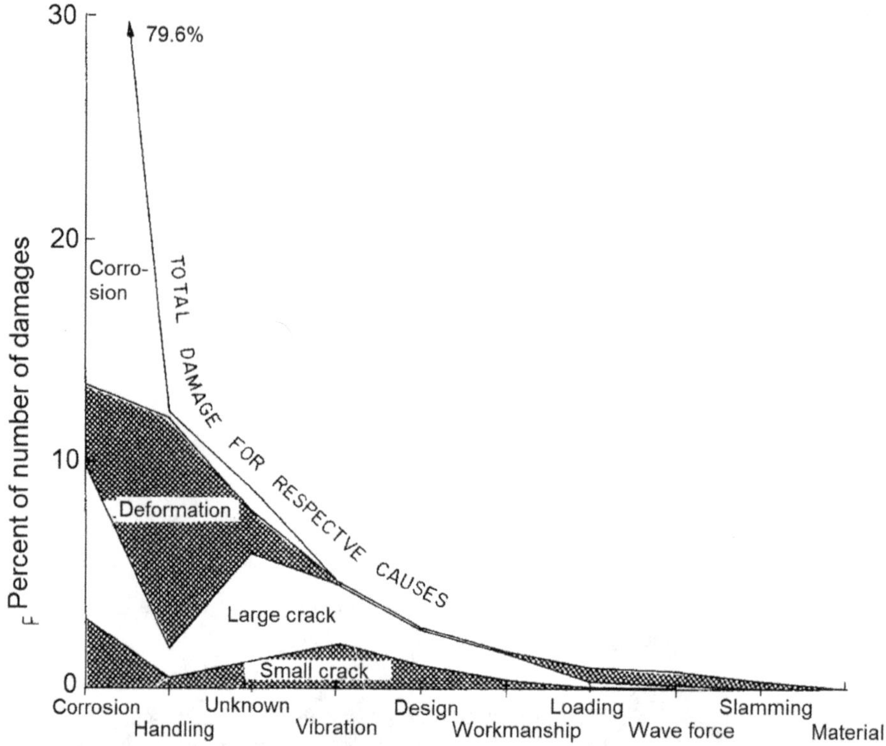

Figure 12.3 Failure modes for several types of damage. Reprinted courtesy of the IMO, London.

In order to ensure the damage-free operation of ships, it becomes necessary to make use of the concept of *safety*. Safety can most easily be quantified in a retrospective (a posteriori) manner, with reference to the consequences of the absence of safety (Figure 12.2). In the case of ships and offshore structures, safety can be quantified in global terms by comparing data for loss of life, structural damage and damage to the environment. Such data can be used to compare alternative solutions and, equally importantly, to provide a basis for code calibration.

In order to make use of the safety concept in a constructive way, a precise definition and rational means of quantifying it are needed. In writing about ship operations, Goodman (1983) defined safety as 'the freedom from danger or from unacceptable risks of harm to the person, the environment or property'. More recently, the following indirect definition was provided: 'A structural system is safe if it does not impair sustainability through an excessive expected loss due to accidents (that is, to random events)' [3].

In many cases, the failure of an engineering component will not necessarily endanger the safety of a human operator or the operation of another component. Hence in order to use the term in a precise manner it is necessary to specify the entity whose safety is being considered. Strictly speaking, therefore, safety is a relative concept.

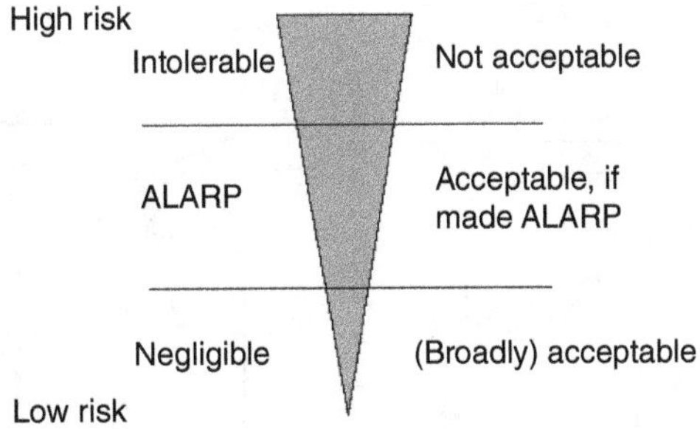

Figure 12.4 The ALARP principle [26]. Reprinted courtesy of IACS, London.

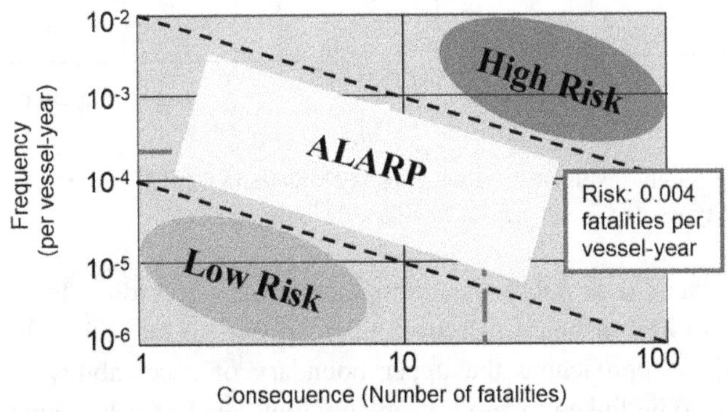

Figure 12.5 Risk acceptance and the ALARP concept [22]. Reprinted courtesy of IACS, London.

The manner that has evolved in the maritime sector is to link safety with risk via the ALARP principle. The ALARP (*A*s *L*ow *A*s *R*easonably *P*racticable) principle refers to a level of risk that is neither negligibly low nor intolerably high (Figure 12.4).

Figure 12.4 shows that above an upper bound there is a risk level that cannot be tolerated. In this region, the risk level is not acceptable and has to be reduced, irrespective of the costs involved. Below a lower bound there is a risk level that is broadly acceptable, and since in this region risk is negligible, no risk reduction measures are required. If, however, the risk level lies between the two bounds, termed the ALARP region, risk should be reduced to a level that is as low as is reasonably practicable by making use of cost-effective measures. The term 'cost-effective' is used to indicate that the costs involved in reducing the level of risk should not be disproportionate to the benefits gained. To assess this, cost-benefit studies need to be carried out.

A graphical representation of these ideas is obtained by making use of *FN* plots, such as Figure 11.1 and Figures 12.6a and 12.6b. In these plots, the frequency of accidents, (F), is plotted against consequences, expressed in terms of the number of fatalities, (N). The graph is divided into three regions, designated as Low Risk, ALARP and High Risk, as indicated in Figure 12.5. In the maritime sector, the acceptable level of risk involves 0.004 fatalities per vessel per year, and this defines the ALARP region [22].

Data for existing ships are plotted in Figure 12.6a that refers to the loss of life in different ship types whereas Figure 12.6b shows corresponding data for different types of bulk

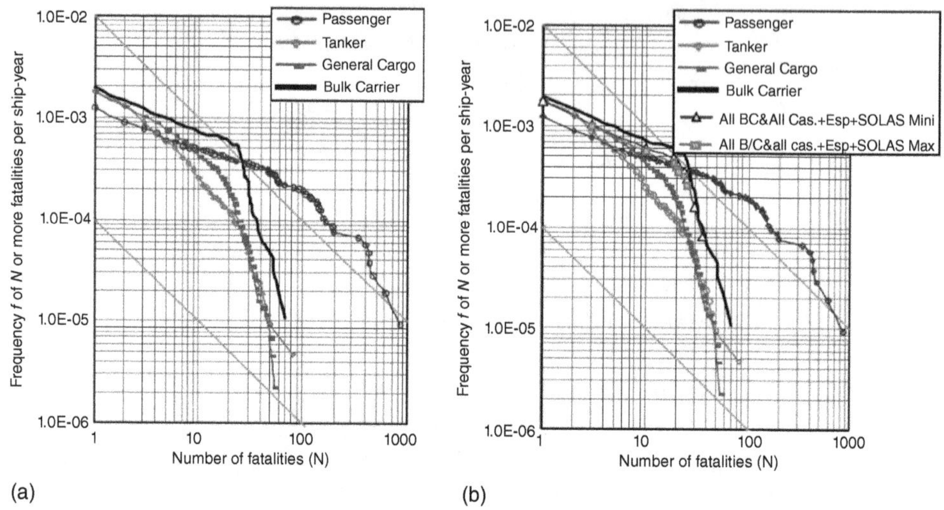

Figure 12.6 Safety levels of existing ships (FN diagrams). (a) Frequency of fatalities for various ship types. (b) Frequency of fatalities for bulk carriers.

carriers. Each diagram is subdivided into three zones according to the level of risk tolerance. Risk to life on board merchant ships appears to lie within the ALARP zone, with bulk carriers approaching the upper boundary of acceptability. Passenger ships, however, appear to be linked to an unacceptably high level of risk to human life.

12.3 Risk Mitigation in the Maritime Sector: Formal Safety Assessment

In the maritime sector, the use of such methods has been incorporated in a procedure recommended by the International Maritime Organisation (IMO) in order to evaluate the success or otherwise of measures that have been proposed to improve safety at sea.

The procedure is termed *formal safety assessment* (FSA) and is described in detail in various IMO documents [26–28]. Important features are described in this section. FSA is intended for use in the development of rules by classification societies and is aimed at enhancing safety, including the protection of life, health, the marine environment and property. It involves the use of risk analysis tools such as those mentioned in Chapter 11, as well as cost-benefit assessment.

In the past, revisions and changes to regulations and safety procedures in the maritime sector have in general been introduced following accidents. Formal safety assessment can assist because of its proactive nature, in that it may anticipate risk and thus alert regulatory bodies. By its nature, it addresses issues that extend beyond purely technical aspects, and in order to be successful, judicious application of the available techniques is required. These may include analytical modelling, historical data and expert judgement. The steps involved are summarised below with some brief accompanying comments. In the last section of this chapter the implementation of parts of FSA is demonstrated in the risk-based structural design of a Suezmax oil tanker.

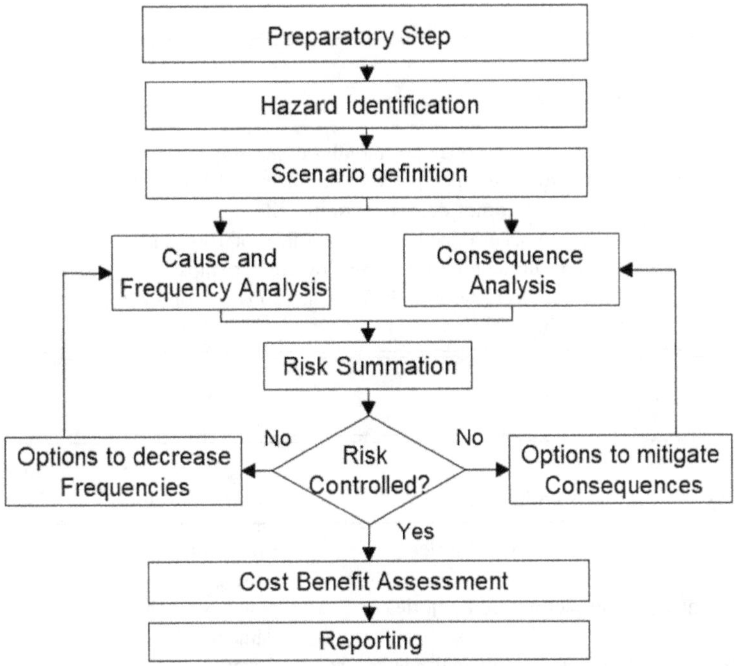

Figure 12.7 The FSA procedure according to IACS. Reprinted courtesy of IACS, London.

Figure 12.7 summarises the steps in the FSA procedure as proposed by IACS. A FSA is carried out in the following steps:

Preparatory Step

1. Identification of hazards
2. Risk analysis
3. Risk control options
4. Cost-benefit assessment
5. Recommendations for decision-making

In the preparatory step, the problem being considered is carefully defined in relation to the regulations under review or to be developed. In Step 1, which involves *hazard* identification and accident scenario definition, both creative and analytical techniques may be used, in an effort to identify all relevant hazards. Hazards are understood to be sources of potential risk that may or may not come to the forefront in any particular scenario. For example, corrosion within ballast tanks as stipulated in rules may, under normal operating conditions, prove to be acceptable. However, should grounding or contact with a berth occur, the weakened structure and adjoining members will suffer much more serious damage than otherwise; other side effects are also possible, such as ingress of water in adjoining spaces. Clearly, the identification of hazards requires a deep understanding of ship operations, as well as the underlying engineering principles upon which codes are based. Historic data are often used, but probabilistic modelling of failures and development

Table 12.1 *Frequency index,* FI *[26]*

FI	Frequency	Definition	F (per ship-year)
7	Frequent	Likely to occur once per month on one ship	10
5	Reasonably probable	Likely to occur once per year in a fleet of 10 ships, i.e. likely to occur a few times during the ship's life	0.1
3	Remote	Likely to occur once per year in a fleet of 1000 ships, i.e. likely to occur in the total life of several similar ships	10^{-3}
1	Extremely remote	Likely to occur once in the lifetime (20 years) of a world fleet of 5,000 ships	10^{-5}

Table 12.2 *Severity index,* SI *[26]*

SI	Severity	Effects on human safety	Effects on ship	S (equiv. fatalities)
1	Minor	Single or minor injuries	Local equipment damage	0.01
2	Significant	Multiple or severe injuries	Non-severe ship damage	0.1
3	Severe	Single fatality or multiple severe injuries	Severe damage	1
4	Catastrophic	Multiple fatalities	Total loss	10

of scenarios are strongly recommended. Hazards and scenarios have to be ranked in order of importance using a risk index. To facilitate ranking the consequence and probability indices are defined on a logarithmic scale. From the definition of risk, given by Equation (11.1), taking logarithms we obtain

$$Log(risk) = Log(probability) + Log(consequences) \tag{12.1}$$

For the needs of the maritime sector Tables 12.1 and 12.2 can be used to assess frequency and severity.

We may hence define a *risk index* as follows:

$$\text{Risk index} = \text{Frequency index} + \text{Severity index} \tag{12.2}$$

A risk index table may thus be constructed (Table 12.3):

Step 2 relates to *risk analysis*, which is expected to be a detailed investigation of the causes and initiating events and consequences of the most important accident scenarios identified in step 1. Methods such as those mentioned in the previous section can be used. Frequency of occurrence is simply the ratio of the number of casualties to the number of ship-years. An example of a cause-and-frequency analysis is included in Figure 12.8 [22].

The consequences are defined as the ratio of the number of fatalities to the number of ship-years. An example of a consequence analysis for the case of water ingress in the forepeak is included in Figure 12.9.

Table 12.3 *Risk index table [26]*

		Severity (SI)			
		1	2	3	4
FI	Frequency	Minor	Significant	Severe	Catastrophic
7	Frequent	8	9	10	11
6		7	8	9	10
5	Reasonably probable	6	7	8	9
4		5	6	7	8
3	Remote	4	5	6	7
2		3	4	5	6
1	Extremely remote	2	3	4	5

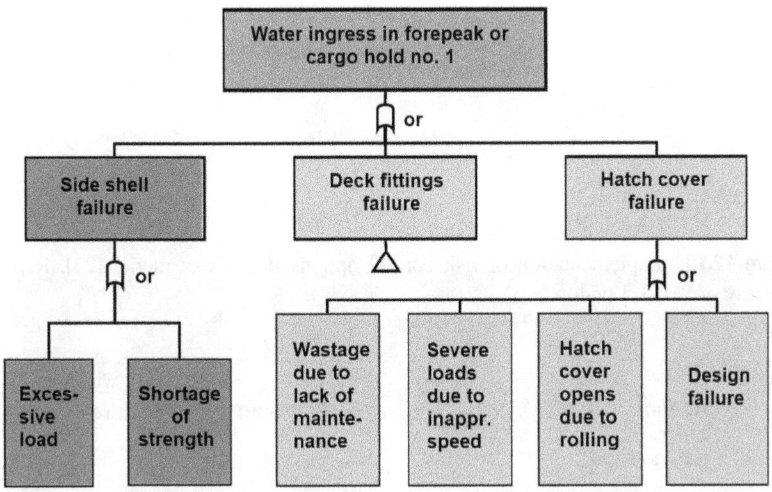

Figure 12.8 Cause and frequency analysis [22]. Reprinted courtesy of IACS, London.

Step 3 relates to the identification of *risk control measures* (RCMs) and then to potential *risk control options* (RCOs) (Figure 12.10). The following four stages constitute step 3:

1. Focussing on risk areas needing control
2. Identifying potential RCMs
3. Evaluating the effectiveness of the RCMs in reducing risk by re-evaluating step 2
4. Grouping RCMs into practical regulatory options

The outcome of this step is a list of RCOs. In cases where two or more elementary RCOs are introduced simultaneously, the calculation of risk reduction and of the cost-benefit effectiveness is not that simple. This step strongly relies on expert opinion.

Step 4 involves carrying out a cost-benefit assessment of the RCOs proposed in step 3 (Figure 12.11).

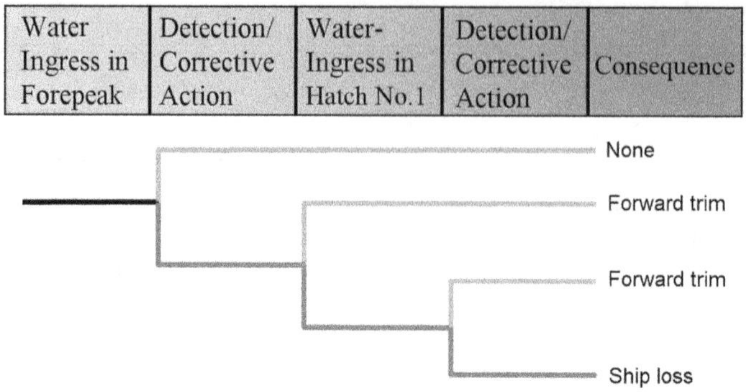

Figure 12.9 Consequence analysis (example) [22]. Reprinted courtesy of IACS, London.

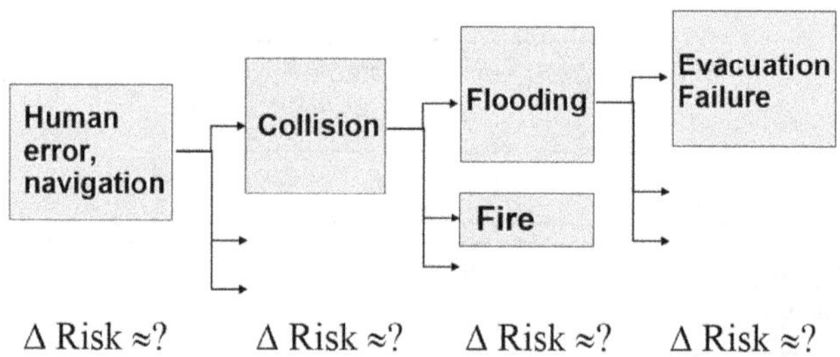

Figure 12.10 Implementation of risk control options in the risk model [22]. Reprinted courtesy of IACS, London.

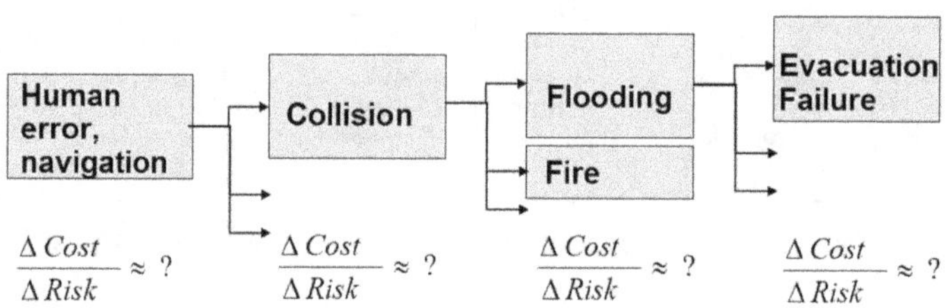

Figure 12.11 Example structure for cost-benefit assessment [22]. Reprinted courtesy of IACS, London.

1. Consider the frequency and consequence of risks assessed in step 2 in order to define the base case in terms of risk levels of the situation under consideration.
2. Rank the RCOs, defined in step 3, so as to facilitate the understanding of the costs and benefits resulting from the adoption of each RCO.
3. Estimate the costs and benefits for all RCOs.
4. Estimate and compare the cost effectiveness of each option, in terms of the cost per unit risk reduction, by dividing the net cost by the risk reduction achieved as a result of implementing the option.
5. Rank the RCOs from a cost-benefit perspective in order to facilitate the decision-making recommendations in step 5 (e.g. to screen those which are not cost effective or

impractical). The types of costs that may be included are investment costs, costs related to operation, training, inspection, maintenance. The types of benefits include reductions in fatalities and injuries as well as damage to the environment.

The output of step 5 includes figures for the costs and benefits for each RCO, the costs and benefits for parties that are directly affected and finally the cost effectiveness expressed in terms of indices.

Step 5 concerns recommendations for decision-making. These ought to include the comparison and ranking of hazards and RCOs. The following are expected:

1. A comparison of the alternative options, based on the potential reduction of risks and cost-effectiveness where new rules are needed
2. Information to enable the review of results obtained in previous steps
3. Recommended RCO(s) and related information

It is clear from the above that FSA is a process fraught with difficulties and possible misinterpretations. If used properly, it is a powerful tool in the hands of regulatory authorities and supervisory bodies such as the IMO. There exist difficulties in the execution of FSAs, which may be categorised as both objective and subjective, in the sense that these terms were used in the discussion of uncertainties. Objective difficulties stem from the definition of risk. Risk has two components, both of which are difficult to evaluate: (1) the probability of failure of an extreme event, which in the case of a novel structure or a structure that is expected to operate in novel conditions or in a novel manner may have never been experienced; and (2) the consequences of failure. The quantification of consequences has serious innate difficulties, since consequences extend well beyond the purely technical sphere and touch upon social and environmental issues. Subjective difficulties arise because the results of a FSA study will most likely have important and wide-ranging repercussions, and since it is possible to conduct the study so as to favour a particular result, the temptation to do so may arise. The use of FSAs is a new field, and further work is necessary to improve their robustness and stakeholders' trust in them.

12.4 Sources of Uncertainty

In the following sections we will consider the most important sources of uncertainty that influence the response of marine structures.

12.4.1 The Environment

The environment within which surface ships operate was discussed in Chapter 9, where it was shown that wave energy spectra are used to describe individual sea states. The uncertainties that arise in the description of the sea environment relate to:

- Wave spectra, and in particular the variability in spectrum shape. Figure 12.12a shows that the error in wave-induced bending moment is large for seas with small average wave

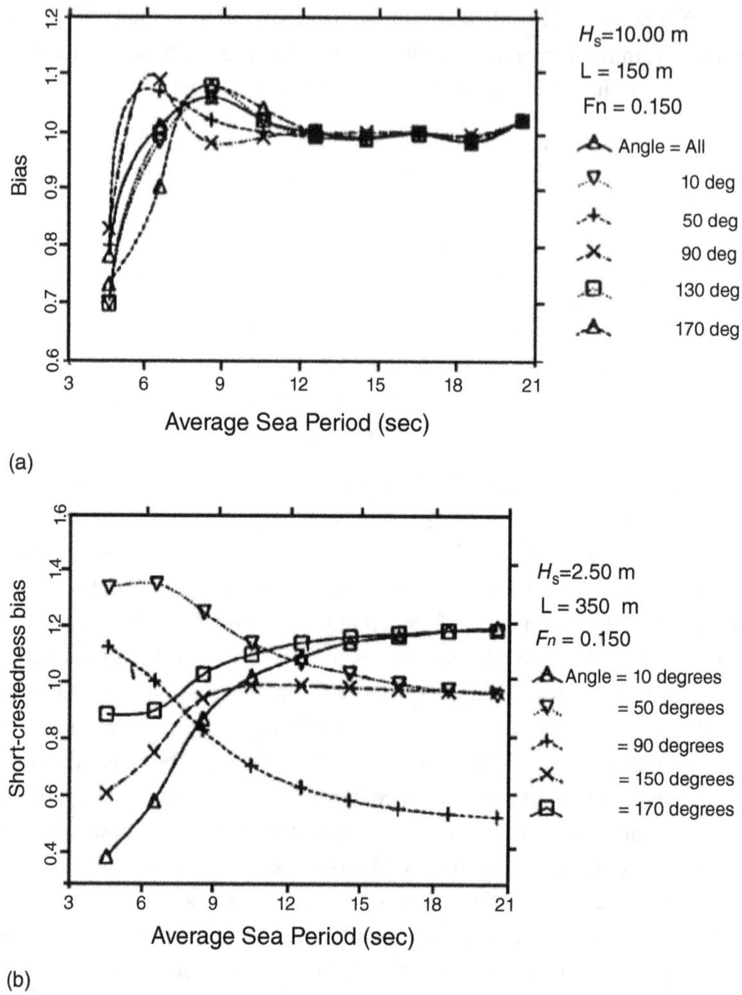

Figure 12.12 Effect of spectral shape variability (Series 60 $C_B = 0.80$) [36]. (a) Bias in mean-square wave-induced bending moment ($L = 150$ m). (b) Bias in short-crestedness for various angles of dispersion ($L = 350$ m). Reprinted courtesy of the Ship Structure Committee, Washington, DC.

periods ($T < 9$ sec) whereas for seas with large average wave periods ($T > 9$ sec) the error is lower [36].

- Directionality of wave systems. The use of spreading functions has been described in Chapter 9. If it is assumed that the energy distribution is independent of the significant wave height and frequency, as is done in practice, errors are introduced, since as the significant wave height increases, directional spreading decreases.
- Shortcrestedness. This is related to directionality and is a measure of the effect of directionality on energy transfer (Figure 12.12b).

- Visual observations of wave heights. It has been found that visual observations tend to underestimate small waves and overestimate large ones.
- Wave height estimates, due to correlation of subsequent wave peaks. It has been shown that assuming independent peaks leads to conservative estimates of the maximum wave height.

12.4.2 Primary Load Effects

In Section 3.1 a fictitious scenario of a voyage of a bulk carrier was described. It is seen that during the voyage the longitudinal primary loading of the hull girder changes, as do the local loads. We shall focus on the time-domain variations of the primary loading, which includes the longitudinal vertical bending moment and shear force. The primary loading is conventionally subdivided into still-water loads and wave-induced loads. When in port, the wave-induced loads are zero and any changes in still-water loads are ignored – that is, the still-water bending moment is assumed to remain constant. For present purposes, the assumed still-water loads are those prior to departure and depend on the selected loading condition for the subsequent voyage. In the case of cargo ships a number of such conditions arise, the most important being 'fully loaded departure' and 'ballast departure'. All typical permitted departure conditions are included and described in a ship's loading manual. At sea, both still-water and wave-induced bending moment and shear components act. The still-water components change gradually during the voyage as fuel and other consumables are used and possibly ballast transfers take place so that upon arrival, the still-water condition will be either 'loaded arrival' or 'ballast arrival'. During the subsequent voyage, the loading conditions will most probably differ, and it becomes evident that when the entire operating life of the ship is considered, the still-water loads need to be treated in probabilistic terms.

The wave-induced components, on the other hand, vary continuously whilst the ship is at sea and are the larger of the two components. Since they vary in a random manner, they are treated as stochastic processes and can be described either by short-term or long-term statistics. The latter are used when determining design loads used in classification society rules.

The above primary loads are in the low-frequency range. Other low-frequency primary loads are the horizontal bending and torsional moments of the hull girder. In addition to low-frequency loads, high-frequency wave-induced loads also act; these include slamming, bow flare impact, deck wetting and springing. The first three of these are transient loads whereas the fourth is a steady-state load. When determining design loads, it is necessary to calculate the long-term combined action of low-frequency and high-frequency loads. In order to achieve this, the long-term extreme values of both low-frequency and high-frequency loads are required. In the case of wave-induced loads a long-term prediction is based initially on a linear analysis, which is then corrected for nonlinear effects using data obtained from experiments and/or full scale measurements [45].

The most important primary load combinations are:

- Extreme vertical bending loads, with co-existing horizontal bending and torsional loads. These are important amidships, where the vertical bending moment is at its peak.

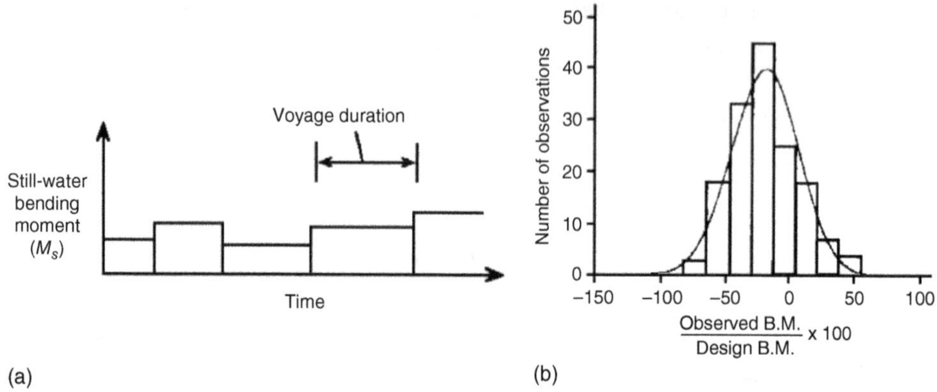

Figure 12.13 Still-water bending moment (M_S). (a) Time-domain variation. (b) Bending moment distribution of oil tankers at $x = 0.46L$ [15]. Figure 12.13b reprinted courtesy of the Society of Naval Architects and Marine Engineers.

- Extreme torsional loads, with co-existing vertical and horizontal bending loads. These are important in the region approximately a quarter-length from the bow of the ship where the torsional rigidity is lowest.

In addition to determining the long-term extreme values of these loads it is necessary to estimate the degree of correlation between them. However, the first step is to provide an adequate probabilistic description of the individual loads being considered. In the next section we shall consider the long-term probability distributions of the still-water and wave-induced vertical bending loads. In order to understand the sources of uncertainty in primary loads it is useful to consider the single voyage of the bulk carrier that was described in Section 3.1. It is easiest to do so by treating the still-water and wave-induced bending moments separately.

Still-Water Bending Moment

With regard to the still-water bending moment it is seen that it remains essentially constant for the duration of one voyage. A time-domain plot of the still-water bending moment therefore has the form shown in Figure 12.13a. This is seen to be a *Poisson rectangular pulse process*. Frequency analysis has shown that in the case of cargo ships the amplitude of the still-water bending moment follows the normal distribution and differs according to ship type (Figure 12.13b). Thus, in the case of oil tankers there is much greater variability due to the large number of ballast voyages. In a study on still-water load parameters of a number of ships of differing types [15], the total variability was considered to depend on the variability (a) from voyage to voyage for a particular ship, (b) from one ship to another, for ships of a particular type, and (c) from one ship type to another.

The value of a load effect X (bending moment or shear force) was thus modelled as

$$X_{ijk} = m_o + m_k + m_j + \varepsilon_i \tag{12.3}$$

Table 12.4 *Variability in still-water bending moments* (M_S) and shear forces (V_S)

	Mean values for all ships of a single type		Standard deviation for a particular ship		Standard deviation for all ships (particular type)	
Type of ship	$(m_o + m_k)$		σ_ε		$\sqrt{\sigma_\varepsilon^2 + \sigma_j^2}$	
	M_S	V_S	M_S	V_S	M_S	V_S
Cargo	0.50	–0.17	0.28	0.20	0.30	0.31
Containership	0.72	–0.32	0.16	0.10	0.20	0.29
Bulk Carrier	–0.008	0.04	0.30	0.34	0.38	0.52
OBO	0.80	0.03	0.30	0.25	0.41	0.49
Chemical carrier	–0.005	–0.06	0.22	0.20	0.36	0.30
Ore/oil carrier	–0.44	0.12	0.22	0.25	0.37	0.75
Tanker	–0.12	0.07	0.21	0.20	0.44	0.55

where

X_{ijk} = load effect arising during the ith voyage, applied to the jth ship, of type k
m_o = mean value of load effect in all ships
$m_o + m_k$ = mean value of load effect of ship of type k
$m_o + m_k + m_j$ = mean value of load effect for the jth ship of type k
ε_i = variation of load effect from voyage-to-voyage

The following statistical parameters are defined:
Variance of load effect for a particular ship = σ_ε^2
Variance of load effect for all ships of a particular type = $\sigma_\varepsilon^2 + \sigma_j^2$
Variance of load effect of all ships = $\sigma_\varepsilon^2 + \sigma_j^2 + \sigma_k^2$

As the population expands, the variance increases. Data for the distribution of M_S and still-water shear force, V_S for various types of cargo ships are included in Table 12.4 [30], based on data for ships included in [15]. In Table 12.4 positive bending moments correspond to hogging. The bending moment values have been normalized by dividing by the values prescribed by classification societies.

Wave-Induced Bending Moment

It has been found that when the peaks of the wave-induced vertical bending moment are plotted on Weibull paper, they fit a two-parameter Weibull distribution (Figure 12.14). The distribution function of the Weibull distribution is

$$F(M_w) = 1 - \exp\left[-\left[\frac{M_w}{k}\right]^l\right] \qquad (12.4)$$

where

l = shape parameter of the distribution
k = scale parameter (characteristic value of stress)

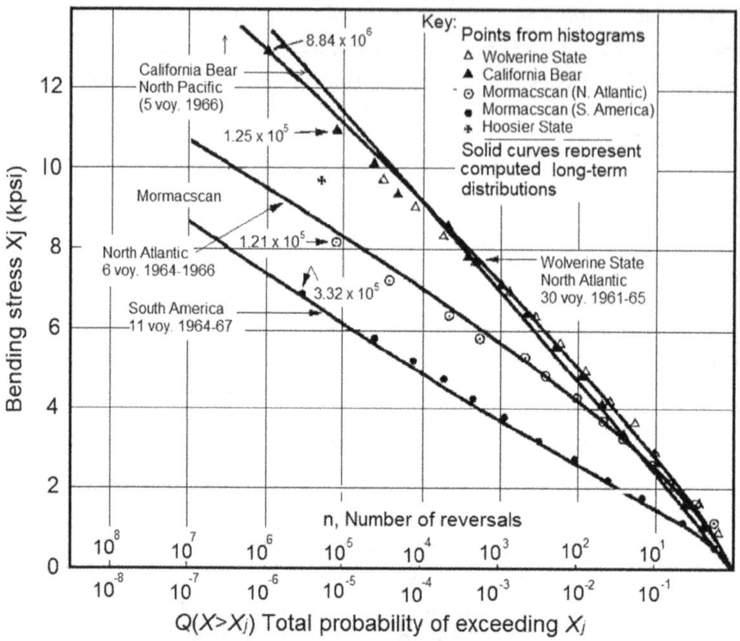

Figure 12.14 Long-term distributions of bending stresses (Weibull) [19]. Reprinted courtesy of the Ship Structure Committee, Washington, DC.

For a wave-induced bending moment M_w, the distribution function gives the probability that the bending moment is less than this value. The corresponding probability density function is obtained by differentiating Equation (12.4) to give

$$f(M_w) = \frac{l}{k}\left(\frac{M_w}{k}\right)^{l-1} \exp\left[-\left(\frac{M_w}{k}\right)^{l}\right] \tag{12.5}$$

The area under the probability density function between the y-axis and a particular value of bending moment corresponds to the probability of occurrence of this bending moment for a ship with a shape parameter l. The area on the right-hand side of a particular bending moment represents the probability of exceedance of this range of bending moments in the lifetime of the structure.

The *short-term response* can be determined using theoretical means (strip theory programs) whose results can then be compared with test data and full scale data, in order to assess modelling uncertainties as well as natural uncertainties such as nonlinear behaviour in way of the fore and aft ends, whether the ship is subjected to a hogging or sagging moment, and the magnitude of the significant wave height. The results are:

1. The Salvesen–Tuck–Faltinsen (*STF*) linear strip theory method underestimates sagging wave bending moments.
2. Errors in the *STF* method are larger in beam seas than in head or following seas.

Table 12.5 *Relative contribution of various types of uncertainty to total uncertainty in extreme bending moment*

Type of uncertainty	Contribution (%)
Spectral shape variability	61
Uncertainty in transfer function	27
Random uncertainty	12

Table 12.6 *Uncertainty in long-term vertical wave bending moment [13]*

Case	Modelling uncertainty Exp. Bias/COV	Random uncertainty	Total uncertainty Exp. Bias/COV
Tankers (hogging)	1.13/0.04	0.07	1.13/0.08
Tankers (sagging)	1.13/0.04	0.07	1.13/0.08
Containerships (hogging)	0.88/0.05	0.07	0.88/0.09
Containerships (sagging)	1.28/0.04	0.07	1.28/0.08
Any ship in hogging	1.0/0.15	0.07	1.0/0.17
Any ship in sagging	1.2/0.88	0.07	1.2/0.11
Any ship hogging or sagging	1.1/0.15	0.07	1.1/0.17

3. The *STF* method underestimates sagging and overestimates hogging wave bending moments because it does not distinguish between them.
4. The error of the *STF* method due to nonlinearities reduces as the block coefficient is increased, because in this case the wall-sided assumption is approximated much more accurately.

Of greater importance is the estimation of the uncertainties in estimating the *maximum long-term vertical wave bending moment*, M_W. The most important contributory factors to uncertainty in the estimation of M_W are included in Table 12.5 [30]. The *contribution* is defined as the square of the ratio of the particular uncertainty to the total uncertainty. The recommended characteristics of the distributions describing other load types are included in Table 12.7.

Results for individual ship types are summarised in Table 12.6, for which a modified version of Ochi's method was used [13].

An important conclusion of this study is that uncertainty reduces when ship type(s) and/or the type of moment are specified. Therefore, if a design code distinguishes between different ship types and thus specifies appropriate load and strength factors in each case, it will permit the design of more efficient ships. A second conclusion was that theoretical predictions underestimate loading effects such as the sagging bending moment, especially in the case of ships with low block coefficients. In this case nonlinearity effects become important because of the reduction of the wall-sided effect, as mentioned earlier. It is

Table 12.7 *Recommended probabilistic characteristics of load random variables [18]*

Random variable	Distribution type	Mean to nominal ratio	COV
Still-water bending moment, M_{SW}	Normal	Merchant ships: 0.4–0.6 Naval ships: 0.7	Merchant ships: 0.3–0.9 Naval ships: 0.15
Lifetime extreme wave bending moment, M_W	Gumbel	1.0	0.1–0.2
Whipping bending moment, M_{wh}	Gumbel	Mean value determined using spectral analysis	0.2–0.3
Springing bending moment, M_{sp}	Gumbel	1.0	0.3
Hydrostatic pressure due to still-water, P_{sw}	Normal	Merchant ships: 0.4–0.6 Naval ships: 0.7	0.15
Hydrostatic pressure due to waves, P_W	Gumbel	1.0	0.15
Hydrostatic pressure due to dynamic effects, P_D	Gumbel	1.0	0.25
Hydrostatic pressure due to loads, P_{WD}	Weibull	1.0	0.25

important to have an accurate estimate of the peak sagging bending moment, as this type of loading can lead to buckling of the deck plating, which is an important failure mode.

To summarise, different effects lead to underestimation or overestimation of the magnitudes of loads or load effects that act on ship hull girders. Thus, if nonlinearities, wave directionality and heavy weather countermeasures are neglected, the load effects will be underestimated. On the other hand, if errors in visual observations of wave heights and the effect of short-crestedness on sea severity are disregarded, the load effects will be overestimated.

12.4.3 Structural Behaviour

We shall now turn to the behaviour of structures of the type used in the marine field, ignoring for the moment load effects and the corresponding structural response. Structural types of interest include thin plating and extruded beams, which have a variety of shapes and sizes. The degree to which variability in properties and fabrication has an effect on response has been extensively documented for marine structures as well as civil, aerospace and mechanical engineering structures. As regards the nature of uncertainties of structures, these concern properties at a microscopic level as well as at the level of the entire structure. We shall consider briefly the variability and its effects on primary strength estimation in:

1. Material properties (elasticity modulus, yield stress, ultimate tensile stress, Poisson's ratio)
2. Structural properties, that is, ship overall dimensions and structural arrangement (length, breadth, depth, section modulus)
3. Mathematical modelling (analytical and numerical solutions)

Table 12.8 *Yield stress of steel – statistical properties assuming a lognormal distribution*

Magnitude	Mean value (MPa)	COV	Ratio
Ordinary-strength steel			
Average	259	0.068	1.11
Minimum	235	0.026	1.00
Maximum	305	0.116	1.22
High-strength steel			
Average	344	0.089	1.22
Minimum	275	0.066	1.10
Maximum	458	0.100	1.35

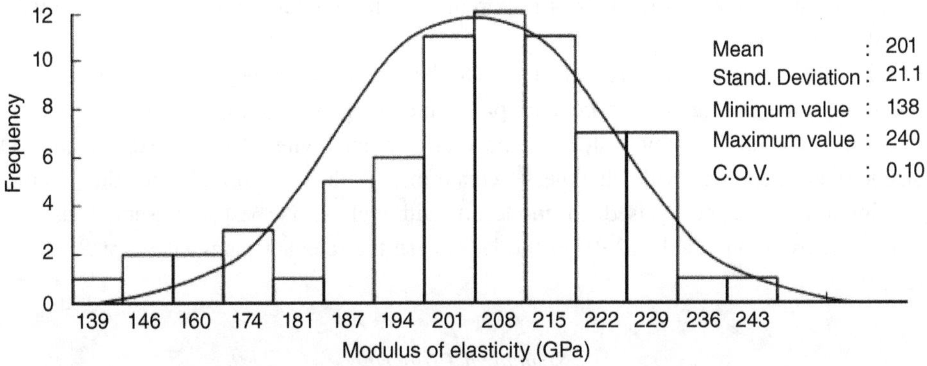

Figure 12.15 Variability in modulus of elasticity of steel (normal distribution) [17].

The first two are essentially natural uncertainties whereas the third is a modelling uncertainty. We shall consider each in turn.

Material Properties

Variations in the modulus of elasticity reflect the degree of material anisotropy (grain structure orientation), and this depends on the grade of steel. Increases in the modulus of elasticity result in proportional increases in buckling strength and bending stiffness. The one environmental parameter that has an effect on the elastic modulus is temperature, but for the practical range of temperatures experienced, the effect is negligible (temperature increases result in a quasi-linear reduction in elastic modulus). Figure 12.15 shows data to which a normal distribution has been fitted, although in other cases the lognormal distribution has been used.

Yield stress is the most important property with regard to strength, and for this reason numerous studies have been published on its variability and the causes for this. It is affected by loading direction, roll direction, temperature, steel type, thickness and year of production. Of these the most important were found to be type and, to a lesser extent, plate thickness, the others being of negligible importance. The lognormal distribution is usually fitted to data obtained for individual steel types (Table 12.8).

Table 12.9 *Average and ranges for the statistics of ultimate strength of shipbuilding steel*

	$\overline{\sigma_u}$ (MPa)	COV of σ_u	$(\overline{\sigma_u}/\sigma_u)$	Distribution
Average	428	0.04774	1.046	Normal
Minimum	412	0.02300	1.007	
Maximum	446	0.09100	1.090	

Ultimate Tensile Stress

In the case of high tensile steel, the ratio of ultimate tensile stress to yield stress decreases, and when this occurs, the ductility has been observed to reduce. As a result, if yielding occurs, rupture may follow. This is important in fatigue-related problems but also when tensile loading is applied.

Ultimate tensile stress is affected by the same variables as yield stress, the most important being steel type and plate thickness. The ultimate strength ratio for both mild and high-strength steels is generally much higher than for HSLA and HY steels, and increases with thickness reductions. Both the normal and the Weibull distributions have been used to model it, although it is best represented using a normal distribution (Table 12.9). In the latter case the relevant parameters are $\beta = 4.13$, $\alpha = 0.29$.

Structural Properties

The important variables in this respect are depth and section modulus. With regard to principal dimensions, variability is quantified with respect to tolerance. Table 12.10 includes data on variation of depth based on average tolerance whereas Table 12.11 provides data on ship breadth.

Depth

Table 12.10 *Uncertainty in ship depth (D) based on tolerance*

Data Point	Tolerance (mm)	Standard Deviation (mm)	Mean (mm)	COV
1	6.35	2.11	6,096	0.000347
2	12.70	4.24	10,973	0.000386
3	0.10	3.66	10,973	0.000333
4	12.70	4.24	7,925	0.000535
5	9.52	3.17	27,737	0.000114
6	12.70	4.24	15,240	0.000278

Breadth

Table 12.11 *Averages and ranges for the statistics of ship breadth (B)*

	Mean value (mm)	Standard Deviation (mm)	COV
Average	28.727	3.33	0.0001833
Minimum	1.829	0.61	0.0000695
Maximum	60.960	4.24	0.0003333

Table 12.12 *Ratio of actual to rule section modulus* (Z_a/Z_r)

Ship	Z_a/Z_r	Ship	Z_a/Z_r	Ship	Z_a/Z_r	Ship	Z_a/Z_r
CS 3	1.04	OBO 3	1.00	TK 7	1.00	TK 31	1.02
CT 2	1.00	OBO 4	1.00	TK 8	1.00	TK 32	1.02
CT 2	1.00	OBO 5	1.06	TK 18	1.12	TK 33	1.02
BC 5	1.00	OBO 6	1.00	TK 19	1.12	TK 34	1.02
BC 9	1.01	CH 1	1.00	TK 20	1.12	TK 35	1.02
BC 10	1.00	CH 2	1.15	TK 21	1.12	TK 36	1.02
BC 14	1.01	CH 3	1.15	TK 22	1.00	TK 37	1.02
BC 15	1.01	OO 2	1.02	TK 23	1.00	TK 38	1.04
		OO 3	1.02	TK 24	1.07		

BC = bulk carrier
CS = cargo ship
CT = containership
CH = chemical carrier
OBO = ore/bulk/oil carrier
OO = ore/oil carrier
TK = tanker

Section Modulus

Table 12.12 below provides data on the ratio of the actual to the rule section modulus. It is observed that the actual section modulus is equal to or exceeds the rule requirement in all cases. The exceedance lies in the range 0–15%.

Table 12.13 gives probabilistic data on the strength of mild steel and high strength steel used in shipbuilding.

12.5 Combination of Uncertainties

In representing the response of structural members it is necessary to allow for the presence of objective and subjective (modelling) uncertainties. A method is required to allow for the presence of uncertainties that characterise several random variables. Structural properties

Table 12.13 *Recommended statistical properties of primary strength random variables*

Variable	Nominal value	Statistical information		
		Mean	COV	Distribution type
Yield stress (mild steel)	σ_0	1.30 σ_0	0.124	Lognormal
Yield stress (HSS)	σ_0	1.19 σ_0	0.083	Lognormal
Ultimate stress	σ_u	1.05 σ_u	0.075	Normal
Modulus of elasticity	E	$0.987E$	0.076	Normal
Poisson's ratio	ν	ν (=0.3 mild steel)	0.00	Deterministic
Section Modulus	Z_R	1.04 Z_R	0.05	Lognormal
First yield moment, M_O	$\sigma_0 Z$	$\overline{\sigma_0 Z}$	0.15	Lognormal
Plastic collapse moment, M_P	$\sigma_0 Z_P$	$\overline{\sigma_0 Z_P}$ or	0.18	Lognormal
		$c\overline{\sigma_0 Z}$	0.18	Lognormal

remain constant whereas loads vary continuously throughout the lifetime of the structure. We shall first consider uncertainties in the properties of structural members. *Linear error theory* has been successfully applied to model them. In the following section the basic equations of linear error theory are given and their use is demonstrated in an example [9].

Following this, the uncertainties characterising the combined action of the loads will be considered.

12.5.1 Strength Uncertainties Using Linear Error Theory

Let us consider the case of yield stress of steel, σ_0. A statistical survey of data for a particular grade has provided the following information:

Descriptor	Unit	Value
Mean $\bar{x}$	MPa	285
Standard deviation σ	MPa	20
Coefficient of variation ν	%	7.0

The guaranteed minimum or *nominal* value, $\sigma_{0(nom)}$, is obtained by deducting 2σ from the mean value, so that in this case, $\sigma_{0(nom)} = 285 - (2)(20) = 245$ MPa.

The spread in values arises partly because of variations in material properties at the micro-structural level (objective uncertainties) and also as a result of measurement techniques (subjective uncertainties). One source of subjective uncertainty is the strain rate effect,[4] which is dependent on experimental technique. If this were to be removed,

[4] When a steadily increasing tensile load is applied to a specimen at a particular strain rate, the yield stress has been found to depend on the strain rate, and is higher than that experienced under constant strain, this being the *static yield stress*. For this reason different testing methods lead to different results.

a different sample mean value would be obtained although the standard deviation would not be expected to change. Allowing for strain rate effect in this case leads to a new mean value of 270 MPa. The nominal value is now no longer 2σ from the mean but 1.25σ.

The difference between the mean value obtained from statistical analysis and the nominal value, which is used in structural calculations, is termed *bias*, ζ. Since bias will always be present, it is a systematic error and its value in percentage terms is obtained from

$$\zeta = \left(\frac{\overline{x} - x_{nom}}{x_{nom}}\right)(100) \tag{12.6}$$

In the previous case, $\zeta = 100(270 - 245)/245 = 10\%$. It is therefore necessary to combine these uncertainties. In the case strength is assumed to be a function f of several variables such as yield stress, length, breadth. etc.,

$$P = f(x_1, x_2, \ldots, x_n) \tag{12.7}$$

The mean, standard deviation and COV of each of these variables are, respectively, $\overline{x_i}$, σ_i, v_i. Taylor's theorem is used to expand the function about the mean values, so that a first order approximation is

$$P = f(\overline{x_1}, \overline{x}_2, \ldots, \overline{x}_n) + \sum_{i=1}^{n}(x_i - \overline{x}_i)\frac{\partial P}{\partial x_i}\bigg|_{x=\overline{x}_i} + H(0) \tag{12.8}$$

If variables $x_1,\ x_2, \ldots, x_n$ are normally distributed and if higher-order terms $H(0)$ are neglected, we obtain

$$P = f(\overline{x_1}, \overline{x}_2, \ldots, \overline{x}_n) \tag{12.9a}$$

$$\sigma_P{}^2 \cong \sum_{i=1}^{n}\left[\frac{\partial P}{\partial x_i}\bigg|_{x=\overline{x}_i} \sigma_i\right]^2 \tag{12.9b}$$

Table 12.14 provides data on the uncertainties that characterize ship scantlings and the fabrication procedure.

$$v_P{}^2 = \left(\frac{\sigma_P}{\overline{P}}\right)^2 = \sum_{i=1}^{n}\left[\frac{\partial P}{\partial x_i}\bigg|_{x=\overline{x}_i} \frac{x_i}{\overline{P}}\right]^2_{x_i=\overline{x}_i} v_i{}^2 \tag{12.9c}$$

Systematic errors are allowed for using

$$v_T{}^2 = v_P{}^2 + v_S{}^2 \tag{12.10}$$

where v_T is the total COV and v_S is the COV of the systematic errors.

Table 12.14 *Typical random and systematic uncertainties [9]*

Item	COV (%)	Bias ζ (%)
Strength		
Yield stress	6–9	5–15
Plate thickness	1–4	–2 to +2
Cross-section of scantlings	2–4	–2 to +2
Welding stresses	10–15	0–20
Plate distortions	30–50	10–40
Stiffener distortions	10–15	
Compression panel strength	10–15	
Ultimate moment	15–20	
Loads		
Dead loads	5–10	–10 to –50
Live loads	10–20	
Wave loads – strength	15–25	
Fatigue	<40	
Wind	20–30	

Example

A solid round bar is subjected to axial tension. Given mean values of diameter of 30 mm and yield stress of 290 N/mm^2, and standard deviation of 3 mm and 25 N/mm^2, respectively, find the COV of P and the lower 5% strength.

$$P = \frac{\pi}{4} d^2 \sigma_0$$

$$\overline{P} = \frac{\pi}{4} \overline{d}^2 \overline{\sigma_0} = \frac{\pi}{4}(30^2)(290) = 205 \text{ kN}$$

$$v_P{}^2 = \sum_{i=1}^{n} \left[\frac{\partial P}{\partial x_i}\bigg|_{x=\overline{x}_i} \frac{x_i}{\overline{P}} \right]^2_{x_i=\overline{x}_i} v_i{}^2$$

$$\frac{\partial P}{\partial d} = \frac{\pi}{4} 2\overline{d}\overline{\sigma_0} \qquad \frac{\partial P}{\partial d}\frac{\overline{d}}{\overline{P}} = \frac{\pi}{4}\frac{2\overline{d}\overline{\sigma_0}\overline{d}}{\frac{\pi}{4}\overline{d}^2\overline{\sigma_0}} = 2$$

$$\frac{\partial P}{\partial \sigma_0} = \frac{\pi}{4}\overline{d}^2 \qquad \frac{\partial P}{\partial \sigma_0}\frac{\overline{\sigma_0}}{\overline{P}} = \frac{\pi}{4}\frac{\overline{d}^2\overline{\sigma_0}}{\frac{\pi}{4}\overline{d}^2\overline{\sigma_0}} = 1$$

Hence,

$$v_P{}^2 = 2^2\left(\frac{3}{30}\right)^2 + 1^2\left(\frac{25}{290}\right)^2 = 0.0474$$

and

$$v_P = 0.218$$

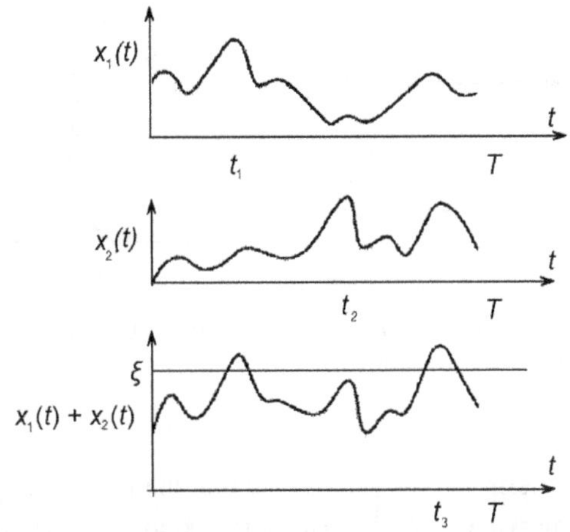

Figure 12.16 Realisations of two random processes and their sum during an interval *T*. (The number of upcrossings of the sum of the random variables is found for level ξ during *T*.)

12.5.2 Combination of Primary Loads

In Section 12.4 the uncertainties surrounding the still-water and wave-induced bending moments were discussed, and it was seen that the probability distributions of the two differ, as do their respective statistical parameters. In order to obtain the parameters of the distribution of the total bending moment on a hull girder it is necessary to combine these distributions. The general problem that is faced is shown in Figure 12.16 in which the amplitudes of two mutually independent, stationary processses $x_1(t)$, $x_2(t)$ are added to obtain the sum $x_1(t) + x_2(t)$. It is seen that the peak values of the three processes do not coincide during a specific time interval, so that it would be inappropriate to add the expected peak values of the component distributions in order to estimate that of the resultant. This has been confirmed in the case of primary loading of ship hull girders. An alternative approach is therefore needed.

Barrier-Crossing Theory

Before considering the more complex problem of the primary loading of a ship, we shall provide some results on the general problem of combining two loads. We shall discuss the case of the single load that can subsequently be extended to the case of multiple loads.

In Figure 12.16, a particular time-domain distribution of a random variable is shown that in this case is the sum of two independent random processes. For present purposes it is immaterial whether this trace refers to a single variable or a combination of variables. Our aim at present is to interpret the (safe or unsafe) state of a structure or component in an alternative way to that presented up to this point in this chapter. In Figure 12.16 a level x-ξ is included, and this can be thought of as being a boundary between safe and unsafe operation (failure). That is, if the process $x_1(t) + x_2(t)$ is perceived as being a margin of safety, the condition $x = \xi$ represents a limit state condition. We can then formulate the load

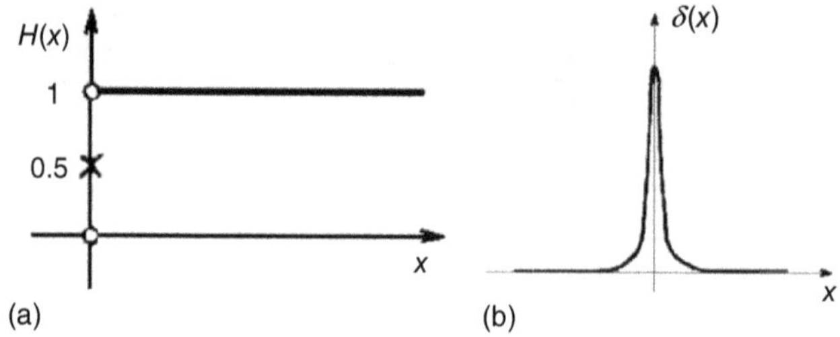

Figure 12.17 The Heaviside (step) function and its derivative [42]. (a) The Heaviside function. (b) The Dirac delta function.

combination problem in terms of the probability that the process $x_1(t) + x_2(t)$ exceeds the value ξ during a time interval $0 \leq t \leq T$. The corresponding distribution function will then be equal to the probability of failure of the structure. This probability can be expressed either in terms of the duration of exceedance or in terms of the number of *upcrossings*, that is, the number of times the process exceeds the value ξ during a time interval $0 \leq t \leq T$. We therefore need to refer to *barrier crossing theory*, which will be outlined following [42].

In Figure 12.16 the number of upcrossings during the time interval $0 \leq t \leq T$ is equal to 2. The determination of the probability of exceedance of a specified level is facilitated by transforming the original problem to a simpler one, in which the relevant information concerning the number of upcrossings and their properties is retained, whereas the exact time-domain response is discarded. We shall refer to the salient points of the method.

The first step is to introduce the *Heaviside function*, which is defined by (Figure 12.17a):

$$H(x) = \begin{cases} 0 & for \quad x < 0 \\ 1/2 & for \quad x = 0 \\ 1 & for \quad x > 0 \end{cases} \tag{12.11}$$

The Heaviside function can also be defined as the integral of the Dirac delta function:

$$H(x) = \int_{-\infty}^{x} \delta(s)ds \tag{12.12}$$

The Dirac delta function can be expressed as the limit of the sequence of standard normal distributions as the standard deviation approaches zero (Figure 12.17b):

$$\lim_{\sigma \to 0} \left[\frac{1}{\sigma\sqrt{2\pi}} e^{-x^2/2\sigma^2} \right] = \delta_\sigma(x) \tag{12.13}$$

As $\sigma \to 0$, the standard normal distribution becomes more peaked, and when $\sigma = 0$ it has zero area and unit height. It is then identical to the Dirac delta function. A new random process, $y(t)$, that may be termed the *upcrossing function* of $x(t)$ at ξ can then be defined by

$$y(t) = H(x(t) - \xi) \tag{12.14}$$

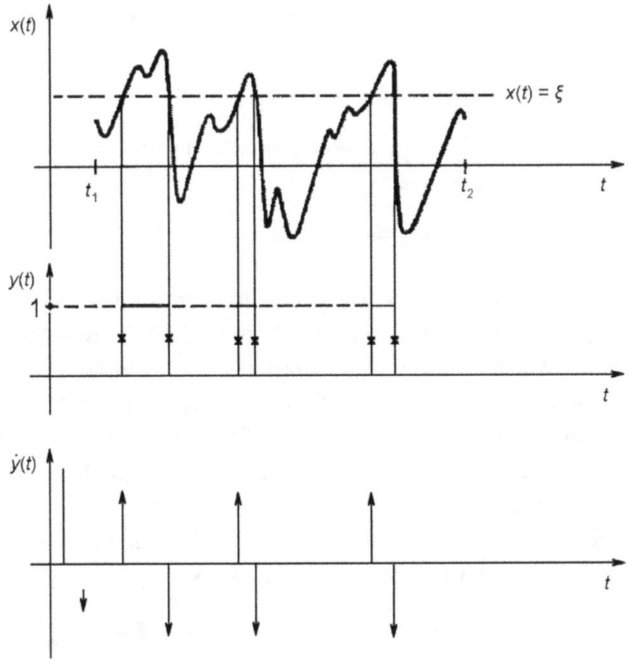

Figure 12.18 Time-domain response of $x(t)$ and $y(t)$, the upcrossing function of $x(t)$ [42].

or

$$y(t) = \begin{cases} 0 & for \quad x(t) < \xi \\ 1/2 & for \quad x(t) = 0 \\ 1 & for \quad x(t) > \xi \end{cases} \tag{12.15}$$

The derivative of $y(t)$, which is $y'(t)$, is then given by (using formal differentiation)

$$y'(t) = x'(t)\delta(x(t) - \xi) \tag{12.16}$$

The plots of $x(t)$, $y(t)$ and $y'(t)$ are included in Figure 12.18. It is seen that the true time-domain response of the process $x(t)$ is replaced by a process that consists of a sequence of pulses that have unit magnitude and duration equal to the upcrossing period. The plot of $y'(t)$ consists of a series of instantaneous pulses whose sign depends on whether upcrossing or downcrossing takes place.

It can be shown [42] that the average number of upcrossings per unit time, $v(\xi,\ x')$, is given by

$$v_x(\xi) = \int_0^\infty x' f_{xx'}(\xi, x')dx' \tag{12.17}$$

where $f_{xx'}$ is the joint probability density function for the process $x(t)$ and its derivative $x'(t)$. This is known as *Rice's formula*.[5] If ξ is defined to be the most probable maximum

[5] Rice S. O. (1944) Mathematical analysis of random noise. *Bell System Tech. J.* Vol. 23 282–332.

value of $x'(t)$,[6] the *probability of exceedance* (*POE*) of ξ in a specified time interval T is given by

$$POE(\xi, x', T) = 1/v(\xi, x')T \tag{12.18}$$

The distribution function F_x for the most probable maximum value $x_{max}T$ during the specified time interval T is thus given by

$$F_x(x_T) = 1 - POE = 1 - \frac{1}{v_x T} \tag{12.19}$$

Let us now return to the question of determining the probability of failure of the resultant of two independent random variables, shown in Figure 12.16. More precisely, the problem is that of determining the probability that the process $x_1(t) + x_2(t)$ has a value larger than $x(t) = \xi$ during a time interval $0 \le t \le T$. That is, $P[x_{max}(t) > \xi]$ for $0 \le t \le T$.

This is equal to the probability that at $t = 0$, $x > \xi$ plus the probability that for $0 \le t \le T$ more than one upcrossing occurs. The latter is equal to the sum of the probabilities of all possible numbers n of upcrossings during the interval T, that is, $\infty > n \ge 1$. If, therefore, the number of expected upcrossings per unit time, $v_x(\xi)$, is known, the total expected number of upcrossings during an interval T will be $v_x(\xi)T$. Thus,

$$P[x_{max}(t) > \xi] \le P[x(0) > \xi] + v_x(\xi)T \tag{12.20}$$

In general,

$$P[x_{max}(t) > \xi] \cong v_x(\xi)T \tag{12.21}$$

and

$$P[x_{max}(t) > \xi] = 1 - F_\Xi(\xi) \tag{12.22}$$

where $F_\Xi(\xi)$ is the distribution function of the maximum value of the random process $x(t)$ in the time interval T. Therefore,

$$F_\Xi(\xi) = 1 - v_x(\xi)T \tag{12.23}$$

Thus the problem of determining the distribution function F_Ξ for the maximum value of $x_1(t) + x_2(t)$ is that of calculating the expected number of upcrossings per unit time, which is given by Rice's formula (Equation (12.17)). In the case of the process $x_1(t) + x_2(t)$ the joint density function is given by the following expression, which is a generalised form of the convolution integral of probability theory [42]:

$$f_{xx'} = \int_{x_1=-\infty}^{+\infty} \int_{x_1'=-\infty}^{+\infty} f_{x_1x_1'}(x_1, x_1') f_{x_2x_2'}(x - x_1, x' - x_1') dx_1' dx_2 \tag{12.24}$$

In this expression $f_{x_1x_1'}(x_1, x_1')$ and $f_{x_2x_2'}(x - x_1, x' - x_1')$ are the joint density functions for x_1, x_1' and x_2, x_2', respectively. However, because the calculation of the joint density functions is difficult, a number of alternative approaches have been developed to determine

[6] The *most probable maximum* is the value for which the probability density function of the maxima of the variable has its peak, and is also called the *mode* of the distribution.

the probability of failure of a structure subjected to loads represented by several random variables.

Extreme Value of the Bending Moment during the Operating Life of a Ship

As mentioned earlier, the two most important primary loads that act on ship hull girders are the still-water bending moment, M_S, and the wave-induced bending moment, M_W. Since the extreme value of M_S is associated with the ship-loading procedure whereas the extreme value of M_W is affected by the random environment, they are not correlated with respect to the extreme value of the combined bending moment. As a result, the peak values of the loads are not expected to occur at the same instant in time (Figure 12.16). Furthermore, the actual peak value of the combined load will not necessarily occur at the same time as any of the peak values of the individual distributions (Figure 12.16).

Therefore, if the total bending moment is obtained by summing the peak values of the two distributions, this will result in an overestimation of the peak value of the combined load distribution. It is thus seen that if a rule uses the simple addition rule of the peak values, this may penalise the structure unduly. In order to avoid this, it becomes necessary to determine the most probable peak value of the combined load distribution.

To achieve this, it is necessary to treat the random loads as stochastic processes in the time-domain and to combine them as vector or scalar processes. The following two types of approaches can be used [33, 45]:

1. Treat them as stochastic processes and combine them directly. Such methods are the *general formulation*, the *point-crossing* method, the *load coincidence* method and the *Ferry Borges-Castanheta* method, which is described below. Alternatively,
2. Use a deterministic approach in which the characteristic values of the distributions are combined. Examples are the *peak coincidence* method, *Turkstra's rule* and the *square root of the sums rule*.

The stochastic process approaches have all been implemented in the problem of combining still-water and wave-induced bending and all lead to acceptable results. The deterministic methods on the other hand are less accurate and generally underestimate the total bending moment. In a recent comparative study it was concluded that the Ferry Borges-Castanheta method, which offers a simple treatment of the loads, is the most accurate in determining the most probable maximum total bending moment.

We shall now describe the Ferry Borges-Castanheta method following which we shall describe the approach proposed in [14].

The Ferry Borges-Castanheta Method

In this method the time period T considered for the variable(s) in question is subdivided into intervals of constant duration τ, and the value assigned to the variable is constant for the duration of each interval. When the action of a second variable (load) is considered during the same time period as the first, the time period is subdivided in such a way that the subintervals used for each variable are multiples or submultiples of each other. The intensities of each variable during each subinterval are mutually independent and

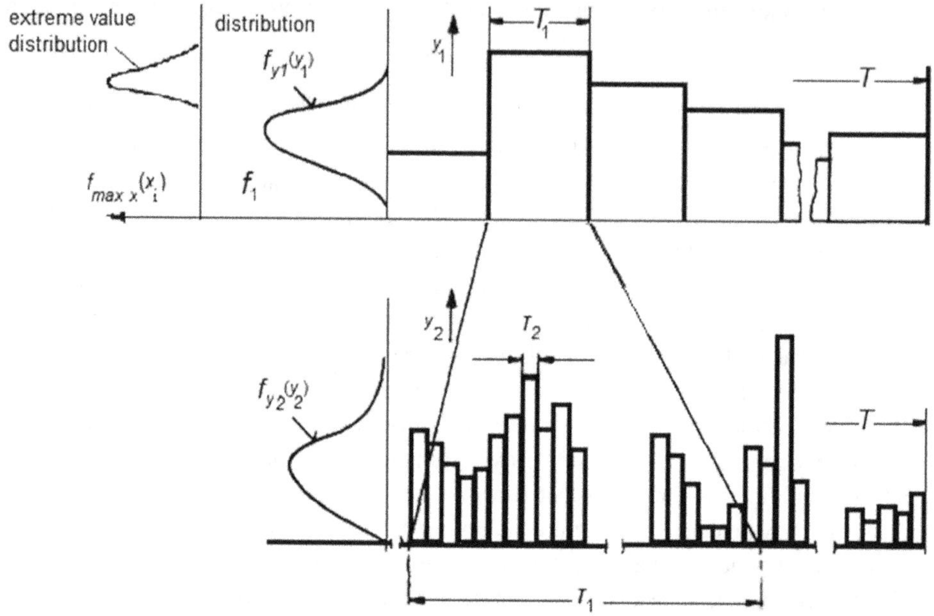

Figure 12.19 The Ferry Borges-Castanheta process.

identically distributed. They can therefore be represented by a single probability distribution (Figure 12.19). In Figure 12.19, variable 1 may represent the still-water bending moment M_S, and this is treated as being constant during an interval τ_1. This interval may represent the duration of a complete voyage, including the time at port and at sea. Variable 2, on the other hand, may represent the wave-induced bending moment M_W. The time interval in this case is treated as being much shorter in order to accurately model the fact that it varies at a much higher frequency. The time interval τ_2 is, however, chosen to be a submultiple of τ_1, as shown in Figure 12.19. We note that in this case there arise subintervals during which variable 2 can assume zero value (in this case, whilst the vessel is in port).

We need to consider the maximum or *extreme* value of each variable during the time period *T*. Let

$$M_n = \max\{X_1, X_2, X_3, \ldots, X_n\} \tag{12.25}$$

where $\{X_1, X_2, \ldots, X_n\}$ is a sequence of values that are independent random variables having a common distribution function *F*. For present purposes, the variables X_i may represent the average values of the bending moment (still-water or wave-induced) measured over a specified interval τ.[7] Then, M_n is the maximum or *extreme value* of an individual loading process over *n* time intervals. It was shown in Chapter 9 (Equation (9.60)) that the probability that the extreme value M_n is less than a specified value *z* is given by

$$P[M_n \leq z] = (F(z))^n \tag{12.26}$$

[7] In the case of still-water bending moments, the duration of the time interval ranges from 2 days for container ships to 28 days for oil tankers [13].

where $F(z)$ is the distribution function of the load intensity (bending moment). This is then the distribution function of the extreme values, shown in Figure 12.19 for variable 1.

If a second variable is introduced (wave-induced bending moment), it becomes necessary to determine the probability distribution of the maximum of the distribution of the combined processes during the time period T. If the time intervals of the two processes are such that the ratios τ_2/τ_1 and T/τ_1 are integers, the extreme value distribution of the combined variables, $F(x)$, is given by the convolution integral,

$$F(x) = \left\{ \int_{-\infty}^{x} f_{X_1}(z)[F_{X_2}(x-z)]^m dz \right\}^n \tag{12.27}$$

The wave-induced bending moment can be represented by variable X_2 in Equation (12.27). If a third load variable, X_3, needs to be introduced, this is done recursively. Once the probability distribution of the maximum value of the combined event is obtained, this is used in combination with statistical data and strength parameters to find the reliability (safety) index, β.

The accuracy of the Ferry Borges-Castanheta method improves as the number of voyages increases, and provides satisfactory results for a ship's lifetime. The method proves to be more accurate than other methods because it takes into consideration the rate of occurrence of the loads as well as their occurrence.

Equation (12.7) gives the distribution of the combined loads. It is then necessary to determine the expected value of the peak load of the combined load distribution. This can then be used in combination with strength data to determine the reliability index β.

The Load Combination Factor Method

In order to combine the extreme values of the still-water and wave-induced bending moments, the use of *load combination factors* was proposed, whose values are determined by performing reliability analysis. The following two equations were proposed [14]:

$$M_T = \varphi(M_S + M_W) \tag{12.28a}$$

$$M_T = M_S + \psi M_W \tag{12.28b}$$

where M_T is the total bending moment and (φ, ψ) are load combination factors. The load combination factor φ has the character of a bias and reduces both load components equally. The load combination factor ψ is a load reduction factor that is applied only to the component with the highest variability, M_W. Calculations for the load combination factors for different values of *truncation factor*[8], T_R, are included in Table 12.15.

[8] The *truncation factor* is introduced to allow for the effect of human control on the load distribution. This arises because in certain cases the maximum permitted load by the rules is exceeded for certain distributions of weight [14]. In the absence of human control $T_R = 1$ and with absolute control $T_R = 0$ (for tankers $T_R = 0.50$ and for containerships $T_R = 0.25$).

Table 12.15 *Results of the load combination calculations using the Ferry Borges-Castanheta model, in terms of total load* $(M_S + M_W)$ and the corresponding load combination factors (T_R= truncation factor) [14]

	T_R	20 years		
		$M_S + M_W$	φ	ψ
Tankers	0.10	259	0.95	0.91
(sagging)	0.50	311	0.94	0.90
Containerships	0.90	337	0.91	0.78
(hogging)	0.10	256	0.96	0.92
	0.50	302	0.90	0.79
	0.90	342	0.79	0.51

In order to determine the numerical values of the load combination factors it is necessary to perform a rigorous analysis of the individual time domain distributions of the loads. The probability of failure of the structure is then analysed by assessing the probability that the combined load process exceeds a threshold value, which corresponds to the limit state being considered. In the case of still-water and wave-induced bending of the hull girder this would be the ultimate limit state or the fatigue limit state, for each of which different acceptable probabilities of failure (target reliability levels) would need to be stipulated.

When performing time domain analysis, a number of methods have been proposed so as to avoid having to consider the entire lifetime of an individual ship, which involves 10^8 load variations over a 20- to 25-year period. In doing this it is necessary to use data relating to the rate of occurrence in time (frequency), its time duration and the intensity of each load. A method that has proven to be successful in modelling the action of combined loads in general is that proposed by Ferry Borges and Castanheta [8]. This method can be used in combination with a reliability method in order to calculate the reliability (safety) index, β, of the structure. The results of the reliability analysis are used to select target reliability levels and to verify that the load combination factors obtained produce designs that satisfy these target reliability levels.

The accuracy of the model obtained using the Ferry Borges-Castanheta method can be checked using a much more accurate, computer-based approach that performs an analysis of a large number of load variations, the *Monte Carlo simulation method* [29]. This can also be used in lieu of any of the more approximate methods.

12.6 Applications

In the sections that follow, a number of applications of structural reliability theory in the marine field will be considered. The first of these concerns the provision of a rational basis for ship construction rules issued by classification societies. The case of hull girder ultimate strength is selected, a topic that has also been treated in earlier sections of this chapter. Arguably, the provision of a rational basis for class society rules is the most important contribution that structural reliability theory has made in the marine field until now. Two further cases are described, the first being an assessment of the effect that the introduction of the CSR has had on the reliability of an Aframax oil tanker, the second being the risk-based structural design of an oil tanker.

12.6.1 Development of Rules for Ship Construction

As already mentioned, a new code has to provide an overall level of safety at least equal to that which characterises existing structures that have performed satisfactorily. The overriding aim of a code is to ensure that structures will have the necessary safety, robustness and performance qualities that are expected and considered acceptable for the duration of their operating life. The question of acceptability raises a number of issues that are complex and extend beyond purely technical issues. This topic will be considered in Section 12.6.2 in which the safety levels of the CSR for oil tankers are discussed.

In order to propose an acceptable level of safety for a new code, it is necessary to make an evaluation of the standards of safety that are inbuilt in present-day codes. This can be done by:

1. Determining the probability of failure that is inherent in current rules for new designs,
2. Extending this by assessing the inherent safety of existing ships, both new and aged,
3. Comparing safety standards with those inbuilt in codes of other industries,
4. Conducting consequence analyses, that will lead to a rigorous appraisal of risk,
5. Making professional judgements with regard to the safety standards required for ship structures.

In the case of the IACS Common Structural Rules, a number of cases of a particular ship type were selected and, using relevant data on loading and strength, the reliability, expressed in terms of the reliability index β, was calculated for each of these. The target reliability was then specified, with past performance acting as a guide. If a risk-based approach is desired, the consequences of failure have also to be allowed for when determining the reliability index. Using this approach the variability in target reliability reduces and becomes closer to the required safety level. Following this, the partial safety factors and the nominal loads are calculated. Inclusion of information on the consequences of failure leads to the necessity to consider different levels of required safety, that is, different reliability indices for different components of the hull structure.

The calibration of a ship structural code generally includes the following steps:

1. *Definition of scope*. This involves specifying the type(s) of ships for which the code is intended, including the range of principal dimensions in each case.
2. *Definition of code objective*. Selection of a particular limit state, such as ultimate strength with respect to yielding, plastic collapse of the entire hull girder, or collapse that is associated with one of the local failure modes. Other limit states are fatigue, serviceability and accident. Ship structures, as opposed to aircraft structures, are designed to be *fail-safe* rather than *safe-life* (Section 10.4). This means that should a

failure occur during their lifetime, they ought to be able to continue operating without placing human life and the environment at risk. Aircraft structures, on the other hand, are designed to ensure safety throughout their working life, because failure (e.g. airframe fatigue) would lead to loss of life and of the aircraft itself. Offshore structures consist of certain components that may not lead to loss of life and of the entire structure and of certain other components that may not be easily repairable during their working life, primarily due to inaccessibility and/or difficulties in repair and maintenance work. For these, a safe-life approach would be appropriate. In order to mitigate the consequences of a safe-life approach, which may result in additional costs and possibly overdesign, a *defect-tolerant* approach has been developed and implemented.

3. *Selection of code format.* Definition of the form of the limit state function, G, in conjunction with the relevant variables, which may be scalar or random. Scalar values are introduced directly, whereas for each random variable a *characteristic value,* x_c, is used. In order to allow for the uncertainty of each random variable, corresponding partial safety factors, γ_x, are introduced. In the case of the loads these are termed *load factors*, and in the case of strength, *strength* or *resistance factors*.[9]
4. *Characteristic values.* It should be underlined that there is no hard-and-fast rule governing the determination of characteristic values. Their purpose is to provide a basis upon which the code format may be formulated allowing for the various types of uncertainty that arise. Since uncertainty is involved, it is important to use realistic and somewhat conservative estimates of the actual values for each random variable. It is wise to err on the side of caution, and for this reason, when considering load random variables, these are considered to define limitations of operational levels that should rarely be exceeded. If a probabilistic description of a particular load type exists, the characteristic value is expressed with respect to a given exceedance probability. In the case of the still-water bending loads that act on a hull girder, the loading condition that results in the maximum bending moment is selected. In the case of environmental loads, the most probable extreme value for a given return period is used. If the relevant information is available, joint probability data of various parameters is incorporated. With regard to the characteristic value of strength, this defines a lower limit below which the strength parameter should rarely lie. Thus in the case of steel, the characteristic strength is based on the extremities of the distribution (5th percentile of the test results). For yield strength, the lower value of (a) the minimum upper yield stress, (b) the yield strength at 0.2% offset, (c) 0.85 times the minimum tensile strength of the material, is used.
5. *Selection of the target reliability index level.* In principle, the target reliability aims at ensuring safety in ship operations. It is necessary, however, to take into account current ship construction practice as well as performance, and this is done by analysing a number of existing ships that are representative of those for which the code is to be drafted. In an earlier study, about 300 ships of different types that were classed with ABS were analysed and the reliability index for each calculated [33]. It was found that

[9] In the United States the terms *load factor* and *resistance factor* are used.

that the range of values within which the reliability index lay was narrow, which indicated a high degree of consistency within the rules. On this basis, the reliability index β for the ultimate limit state was set equal to 3.20, this being the average value. This corresponds to a notional probability of failure $p_f = 0.7 \times 10^{-3}$. As to classification with respect to failure consequences (Table 11.1), it comes under the category of 'not serious', provided the structure will fail in a ductile manner and exhibit some degree of reserve strength.

6. *Calculation of partial safety factors*. The set of partial safety factors replaces the set of safety factors used in deterministic codes. In the case of loading, they link the characteristic value of each random variable to the corresponding design value, x^*, by expressions of the type $x^* = \gamma_x x_c$. In the case of strength, following past practice in deterministic codes, the expression used is of the form $x_c = \gamma_x x^*$. The procedure followed and the relevant equations are included in Section 11.4.3.
7. *Verification of the system of partial safety factors*. Following calculation of the set of partial safety factors, it is necessary (1) to assess whether the target safety level is realised for the range of structural types the code is intended for and (2) to reduce the spread in resulting safety of the test cases. The first requirement is most easily accomplished by recalculating the partial safety factors in an iterative manner, and this is done for a selected number of test cases. With regard to the second step, one way to minimise the spread of the resulting reliability indices is to use an optimisation technique such as constrained minimisation. An objective function to be minimised is defined. This includes the partial safety factors, the corresponding probabilities of failure and corresponding frequency weighting factors. Minimisation of this function can be done using the Lagrange multiplier method, with the target failure probability fulfilling the role of an active constraint.

The next section will describe the use of structural reliability to obtain the Level I format (deterministic) partial safety factors used in the IACS Common Structural Rules for oil tankers (Figure 12.20). The particular case discussed is hull girder ultimate strength, which is one of the most important types of failure. The text is primarily based on the IACS background document [21] and the Common Structural Rules for oil tankers [23].

The strength formulation for ultimate hull girder strength involves the calculation of the hull girder strength of a selected transverse section amidships, M_U, and the maximum load applied to the hull girder. This consists of the still-water bending moment M_{SW} and the wave-induced bending moment M_W (Figure 12.21). The ultimate strength limit state function is therefore

$$g(x) = M_U - M_{WV} - M_{SW} \tag{12.29}$$

The code format, as specified by IACS, is

$$\gamma_S M_{SW} + \gamma_W M_{WV} \leq \frac{M_U}{\gamma_R} \tag{12.30}$$

where the characteristic values of the bending moments M_{SW}, M_{WV} and M_U are used and γ_S, γ_W and γ_R are corresponding partial safety factors. The characteristic values are those stipulated in the rules and the respective formulas for M_{SW} and M_{WV} are given below.

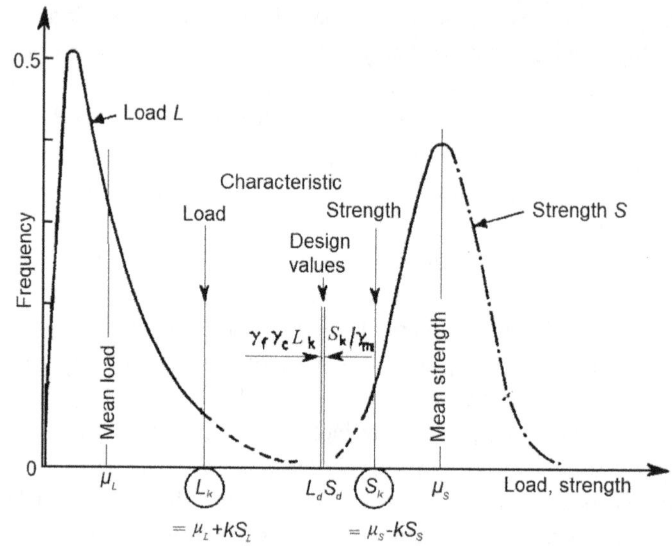

Figure 12.20 Partial safety factors in a Level I code [7].

Sagging bending moments only are considered, and the critical collapse mode involves interframe elasto-plastic buckling in compression of the strength deck. The ultimate strength is calculated using the *single-step method* that is a simple but accurate approach described in Appendix D and in the Rules [23].

It is seen that three partial safety factors are introduced in the code format. This is because the corresponding random variables each have their own probabilistic characteristics and uncertainties and therefore need to be treated separately. Thus, γ_W covers environmental, wave load and prediction uncertainties whereas γ_R covers material, geometric and prediction uncertainties. In order to allow for uncertainties in the bending moments, two combinations are considered. In the first the minimum still-water bending moment is used, and in the second the maximum still-water sagging moment in a homogeneous full load condition is used. It was found that the rule minimum M_{SW} (2006) was larger than the value for the homogeneous loading condition for all test cases and lower than the rule minimum for the ships when built.

The characteristic value of M_{SW} is the rule minimum value, which is obtained from the following expressions [23]:

(a) For hogging,

$$M_{sw-h-\min} = f_{sw}\left(171 C_w L^2 B (C_B + 0.7) 10^{-3} - M_{wv-h-mid}\right) \tag{12.31}$$

(b) For sagging,

$$M_{sw-s-\min} = -0.85 f_{sw}\left(171 C_w L^2 B (C_B + 0.7) 10^{-3} - M_{wv-s-mid}\right) \tag{12.32}$$

where $M_{wv-h-mid}$ and $M_{wv-s-mid}$ are the envelope values of the vertical wave bending moment in hogging and sagging conditions, respectively. The coefficient f_{sw} is equal to unity in way of the midship section.

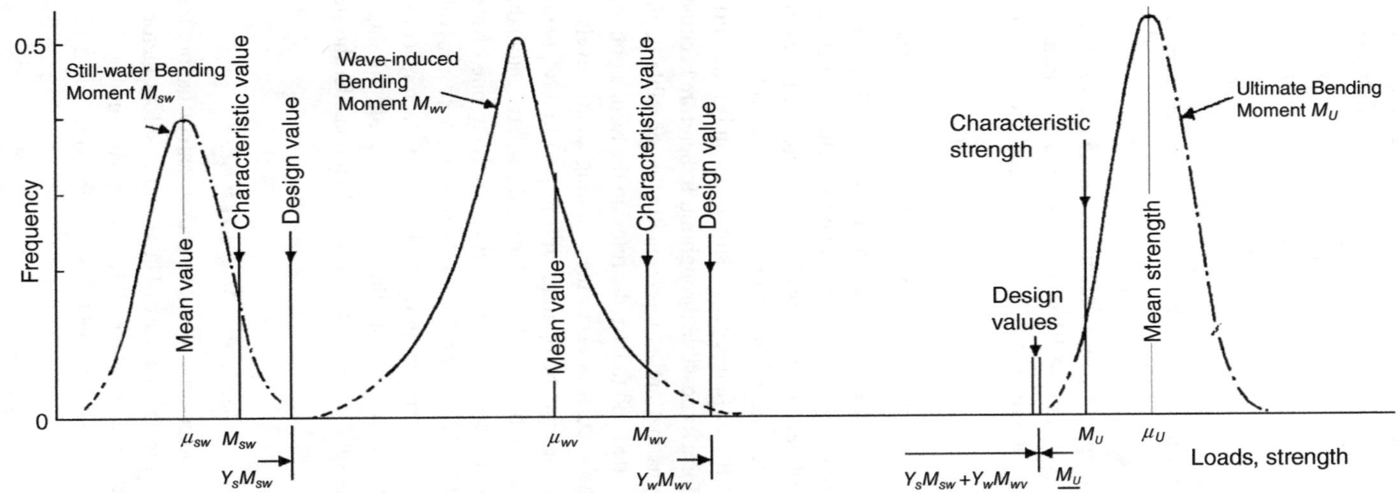

Figure 12.21 Applied bending moments and ultimate moment distributions.

Table 12.16 *Main dimensions of test ships used in calibration of partial safety factors*[10]

Case	Ship type	Lpp (m)	Breadth (m)	Depth (m)
1	Suezmax	263	48	22.4
2	Product/Chem. carrier	174	27	17.6
3	VLCC1	320	58	31.0
4	VLCC2	316	60	29.7
5	Aframax	234	42	21.0

The characteristic strength, M_U, was calculated using the *single-step* method, described in Appendix D and in Appendix A2/1 of the Common Structural Rules. In this, sagging loading is applied until compressive failure in the deck and then failure in tension in the bottom structure.

Calibration of Partial Safety Factors γ_S, γ_W and γ_R

The rules cover a range of ship arrangements and sizes, for all of which $L_{BP} \geq 150$ m. Both product tankers and crude oil carriers are included. Their structural arrangements vary, primarily as a function of deadweight capacity. It is therefore necessary, since a single rule is used, for the partial safety factors to ensure that the probability of failure is less than the stipulated target probability of failure for all ship types and sizes. Furthermore, the level of safety should be as uniform as possible throughout. It therefore becomes necessary to select a number of ship designs that are as representative as possible of the range of ships for which the rules are intended and to base the rules on the results obtained from all these test cases (Table 12.16). As a consequence, the resulting safety levels will not be identical throughout, nor will the designs be optimal with respect to any given criterion.

If only one design were to be considered, the partial safety factors that satisfy a particular target probability of failure could be obtained by applying one of the reliability methods described in Chapter 11, such as *FORM*. This would result in a value of the reliability index, β, from which the corresponding partial safety factors could then be obtained. However, because of the need to apply the rules to a range of ship types and sizes, the procedure followed was split into two stages. We shall consider the steps followed in each of these two stages.

Stage 1: Single Test Case

The purpose of the first stage was to obtain partial safety factors that would satisfy the required safety level, β, for one test case. These partial safety factors were used as initial estimates for the second stage, in which all test cases were analysed. The steps followed in stage 1 are listed below and the approach is summarised in Table 12.17 [21].

[10] An additional three ships were included in the analysis for which details are not given in [21].

Table 12.17 *Probabilistic description of hull girder loading and strength [21]*

Variable name	Distribution	Parameters
	(a) Environment and hull girder loading variables	
Significant wave height (3-hour sea state)	Weibull (3-parameter) $f_{H_S}(h_S) = \frac{1}{k}\left(\frac{h_S - m}{k}\right)^{1-1} \exp\left\{-\left(\frac{h_S - m}{k}\right)^{1}\right\}$	$k = 2.721$ m $l = 1.401$ $m = 0.866$ m (location parameter)
Zero-crossing wave period	Lognormal $f_{T_z\|H_s}(t_z\|h_s) = \frac{1}{\sqrt{2\pi}\sigma t_z} \exp\left[\frac{(\ln t_z - \mu)^2}{2\sigma^2}\right]$	$\mu = E(\ln T_z) = 0.623 + 1.356 \cdot {h_s}^{0.123}$ $\sigma = \sqrt{Var(\ln T_z)} = 0.146 + 0.044 \cdot e^{-1.712 \cdot h_s}$
Short-term extreme wave-induced bending moment	Gumbel	
Annual extreme wave-induced bending moment	Numerically computed	North Atlantic conditions are assumed. The net thickness approach is used allowing for 50% reduction in corrosion allowance. Sagging condition
Still-water bending moment	Normal	Mean = 0.7x (max of manual) Standard deviation = 0.2x (max of manual) (+ sensitivity)
	(b) Strength variables	
Yield strength (normal grade)	Lognormal	Mean = 269 N/mm^2, COV = 0.08 Lower limit = 0.0
Yield strength (high strength)	Lognormal	Mean = 348 N/mm^2 COV = 0.06 Lower limit = 0.0

Table 12.17 (*cont.*)

Variable name	Distribution	Parameters
	(c) Model uncertainties	
Wave-induced bending moment, uncertainties in linear result	Normal $\phi_\tau(t) = \frac{1}{\sigma_S\sqrt{2\pi}} \exp\left[-\frac{1}{2}\left(\frac{t-m}{\sigma_S}\right)^2\right]$	Mean = 1.0 Std. dev = 0.1
Wave-induced bending moment, nonlinear effects	Normal $\phi_\tau(t) = \frac{1}{\sigma_S\sqrt{2\pi}} \exp\left[-\frac{1}{2}\left(\frac{t-m}{\sigma_S}\right)^2\right]$	Mean = 1.0 Std. dev = 0.1 (+ sensitivity)
Still-water bending moment	Normal $\phi_\tau(t) = \frac{1}{\sigma_S\sqrt{2\pi}} \exp\left[-\frac{1}{2}\left(\frac{t-m}{\sigma_S}\right)^2\right]$	Mean = 1.0 Std. dev = 0.1
Capacity calculation	Normal $\phi_\tau(t) = \frac{1}{\sigma_S\sqrt{2\pi}} \exp\left[-\frac{1}{2}\left(\frac{t-m}{\sigma_S}\right)^2\right]$	Mean = 1.05 Std. dev = 0.1

The partial safety factors are thus equal to the ratios of the design point values, which correspond to the most probable values at failure, to the characteristic (rule) values, referred to in the previous section.

In step 1, the characteristic values are obtained using the formulas given in the previous section. For steps 2–4, the assumptions made regarding the assumed distributions and related uncertainties are listed in Table 12.17.

The uncertainties concerning the still-water bending moment M_{SW} (step 2) are listed in Table 12.17. They relate to the bending moment itself as well as modelling uncertainties.

In order to model the wave-induced bending moment it is first necessary to represent the state of the sea surface (Chapter 9). The region that is considered, because of the severity of the prevailing conditions and the number of ships trading in this area, is the North Atlantic (Figure 9.31). This is represented by a probabilistic model that includes significant wave height and zero upcrossing period data, obtained over the long term (*joint environmental model*).

Stage 1 – Single Test Case to Determine Partial Safety Factors

1. Calculate the characteristic values for M_W, M_{SW} using Equations (12.12) and (12.13) and the characteristic value for M_U, using the single step method.
2. Calculate the first and second distribution moments for M_S on the basis of the assumed distribution, the actual loading and the assumed model uncertainty.
3. Calculate the first and second distribution moments for M_W using the joint environmental model, hydrodynamic analysis and model uncertainty.
4. Combine the two bending moments to obtain the total bending moment M, using Turkstra's combination rule.
5. Determine the type of distribution for M_U and its first and second moments using statistical material properties, geometrical uncertainty and model uncertainty assumptions.
6. Apply a Level II reliability method (*FORM*) to obtain the coordinates of the most probable point of failure (design point).
7. Obtain $M_{W,DP}, M_{SW,DP}, M_{U,DP}$.
8. Calculate the partial safety factors $\gamma_S, \gamma_W, \gamma_R$ from $\gamma = \frac{M_{DP}}{M}$.
9. If $p_f > p_f$ (target), recalculate.
10. If $p_f \leq p_f$ (target), END.

Joint environmental models can be developed in several different ways (Maximum Likelihood Model, Conditional Modelling Approach, Nataf Model).[11] This particular model has been based on Global Wave Statistics (GWS) (H_s, T_z) data (Table 9.3) and was developed using the Conditional Modelling Approach [2].

A three-parameter Weibull distribution for the significant wave height, H_S, is adopted:

$$f_{H_S}(h_S) = \frac{l}{k}\left(\frac{h_S - m}{k}\right)^{l-1} \exp\left\{-\left(\frac{h_S - m}{k}\right)^{l}\right\} \tag{12.33}$$

[11] *Multivariate extremes theory* provides the theoretical basis for joint environmental models [5].

Table 12.18 *Coefficients* $a_i, b_i (i = 1-3)$ [21]

Coefficient	Value	Coefficient	Value
a_1	0.623	b_1	0.146
a_2	1.356	b_2	0.044
a_3	0.123	b_3	-1.712

The zero upcrossing wave period, T_Z, for a given significant wave height H_S is given by the conditional lognormal distribution:

$$f_{T_z|H_S}(t_z|h_s) = \frac{1}{\sigma(h)t_z\sqrt{2\pi}} \exp\left\{-\frac{(\ln t_z - \mu(h))^2}{2\sigma(h)^2}\right\} \tag{12.34}$$

where μ is the *logarithmic mean*, given by

$$\mu = E(\ln T_Z) = a_1 + a_2 h_S{}^{a_3} \tag{12.35}$$

and σ is the corresponding standard deviation:

$$\sigma = Std(\ln T_Z) = b_1 + b_2 e^{b_3 h_S} \tag{12.36}$$

The coefficients $a_i, b_i, (i = 1-3)$ were originally determined using the *GWS* data and the actual figures are included in Table 12.18. The logarithmic mean μ has since been adjusted in order to obtain a more accurate description of conditions prevailing in the North Atlantic as well as worldwide trading.

In order to find the wave-induced bending moment for the ultimate limit state of hull girder collapse, the most probable extreme sea condition (storm) that a ship is likely to encounter during its lifetime has to be determined. The ship is assumed to remain in this peak sea condition for the duration of a storm lasting 3 hours ($N = 1{,}000$ wave peaks). In a short-term sea condition, the distribution of the wave heights above the mean value (and hence that of the wave loads) is Gaussian and the process is stationary. The sea state is represented by the *significant wave height*, H_s together with either the *peak period*, T_p, or the *zero up-crossing period*, T_z. The peak period is the wave period determined by the inverse of the frequency at which a wave energy spectrum has its maximum value. The zero upcrossing period is the average time interval between two successive upcrossings of the mean sea level (Figure 9.4).

The distribution of extreme values for this sea state can then be determined using any one of the following methods [32]:

1. Use order statistics to determine the extreme value for a set of N peaks.
2. Determine the asymptotic type-1 distribution.
3. Perform an upcrossing analysis.
4. Derive the Vanmarke extreme distribution on the basis of a two-stage description of the random process.

The method recommended and used in the IACS rules for the long-term (annual) extreme value of an environmental variable such as the significant wave height is the asymptotic type-1 (Gumbel) distribution:

$$F_C(x) = \exp\left\{-\exp\left[-\frac{(x-U)}{A}\right]\right\} \tag{12.37}$$

where A and U are distribution parameters related to the mean μ (= U+0.557A) and standard deviation σ (= 1.283A) of the Gumbel variable. The resulting estimate should be compared to results provided by other methods, such as those listed above.

The derivation of the transfer functions was carried out using the PROBAN program [35, 44]. The wave-induced bending moments were subsequently obtained using 3D linear theory and the effect of forward speed included. It is acknowledged, however, that linear theory is not suited to modelling the response to extreme weather conditions. Nonlinearities are due to (a) large amplitude wave fields, (b) the variable geometry of the ship's hull as it emerges and enters the water, and (c) the effect of wave impacts such as bottom slamming, bow waves and deck wetting. These issues have been discussed in Chapter 3. The use of nonlinear methods in order to allow for nonlinear effects indicated a wide spread of results. In order to cater for this, a modelling uncertainty factor with a mean value of 1.0 and a COV of 0.1 was applied to the total bending moment in the IACS rules.

In step 4 of stage 1 the still-water bending moment, M_{SW}, and the wave-induced bending moment, M_W, are combined using Turkstra's rule.

Following this, the reliability analysis was carried out in PROBAN and is summarised in the flow chart of Figure 12.22.

The reliability analysis used (*FORM*) produces a value for the reliability index and a corresponding probability of failure, p_f. This is compared to the target probability of failure

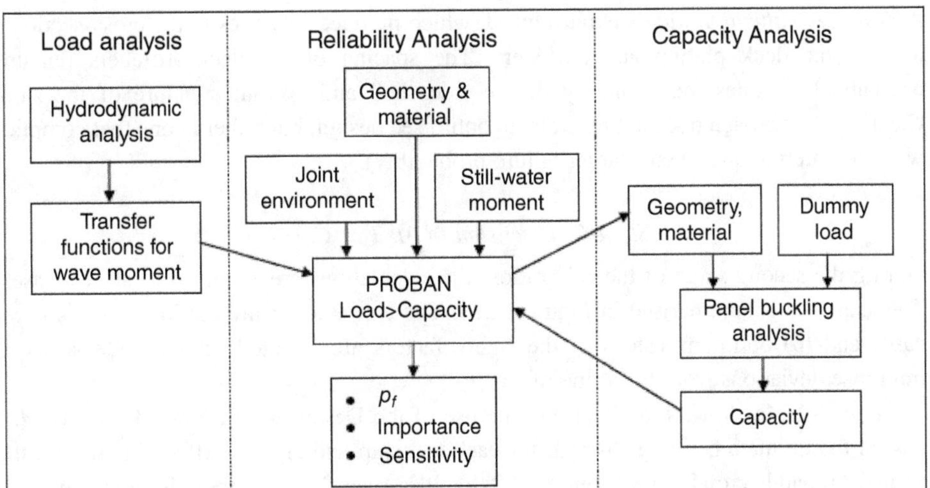

Figure 12.22 Reliability analysis and related calculations for IACS hull girder collapse. Reprinted courtesy of IACS, London.

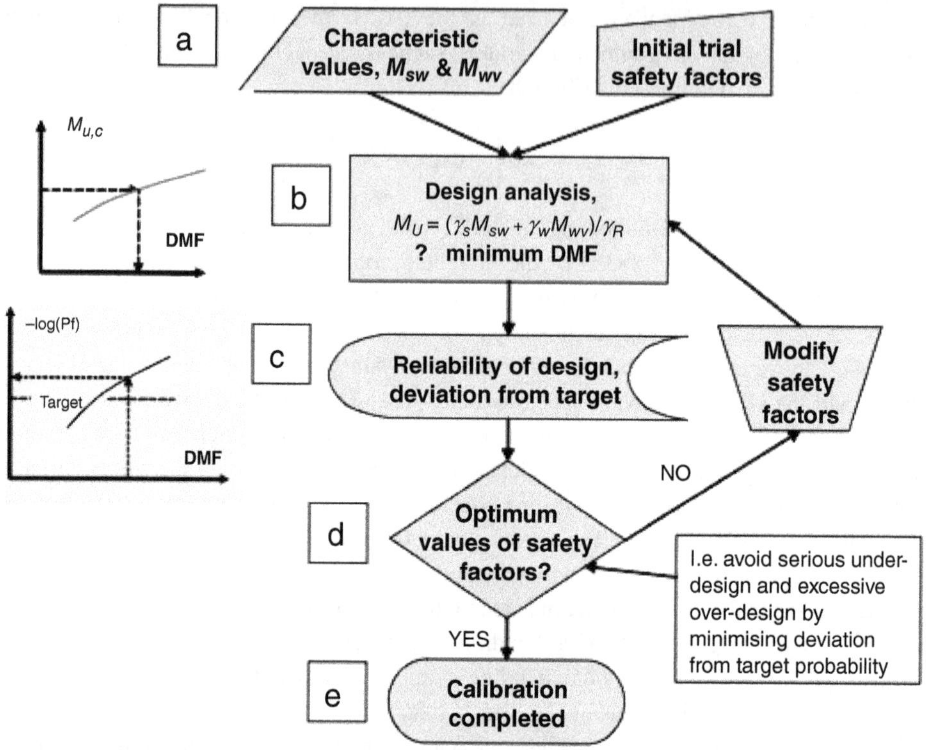

Figure 12.23 Calibration of partial safety factors [21]. Reprinted courtesy of IACS, London.

p_{ft}. If $p_f > p_{ft}$, the hull girder strength needs to be modified (Figure 12.23). This is achieved by changing the deck compressive strength, since this is the critical region of the section. A *design modification factor* is introduced, which produces changes to the cross-sectional area of the deck plating and stiffeners. The spacing of the deck stiffeners remains unchanged, as does the geometry of the side shell and bottom structure of the ship. Clearly this approach does not result in an optimised design, but rather in one that complies with the safety requirement (target failure probability).

Stage 2: Extension to All Test Cases

During the second stage of the calibration, the calculations are extended to all test cases. The sequence is summarised in Figure 12.23. The aim is to ensure that (a) $p_f < p_{ft}$ in all cases and (b) optimum values of the safety factors are selected for the code so as to minimise deviations from the value of p_{ft}.

Figure 12.24a shows the effect that the use of the Design Modification Factor (*DMF*) has on the ultimate bending strength for each test ship and also the effect this has on the annual probability of failure (Figure 12.24b). It is seen that the strength varies approximately linearly with *DMF*, which is proportional to the deck area. Figure 12.24b is a lognormal plot of probability of failure versus *DMF*. It is seen that there is a spread in the

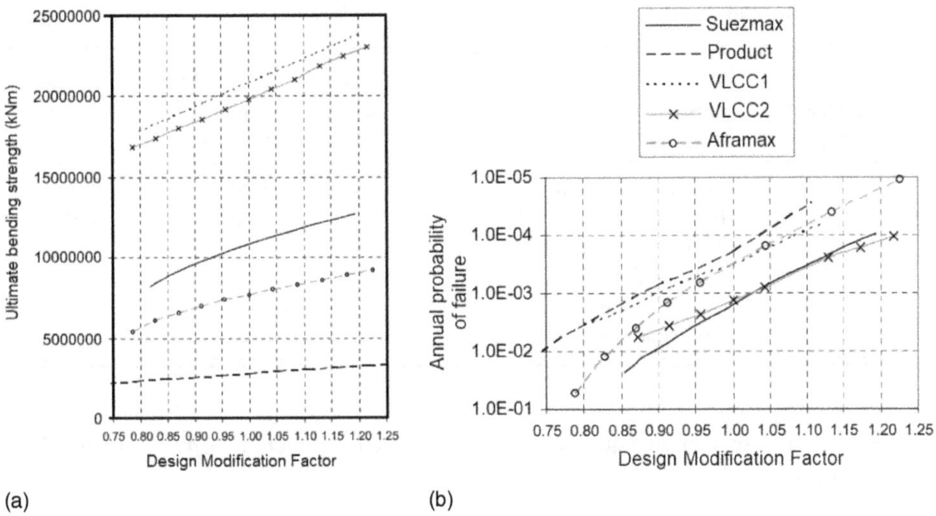

Figure 12.24 Use of design modification factor. (a) Characteristic ultimate bending strength. (b) Logarithm of annual p_f.

required value of the DMF for a given value of p_{ft} for the test cases considered. Strength calculations were carried out assuming (a) rule thicknesses, (b) a 50 per cent reduction in corrosion allowance and (c) a 100 per cent reduction in corrosion allowance. The calibration of the partial safety factors was done assuming a 50 per cent reduction in corrosion allowance.

The steps followed in stage 2 of the calibration process are summarised below:

Stage 2 – Calibration and Optimisation of Partial Safety Factors

1. Calculate the characteristic values for the bending moments for each test case.
2. Set the partial safety factors equal to the values obtained for the test case in stage 1.
3. Apply rule criteria to determine ultimate bending moment for each test case (as in stage 1).
4. Apply *FORM* and calculate p_f.
5. If necessary apply or modify the *DMF* and recalculate the ultimate bending moment.
6. Check the value of p_f. If $p_f < p_{ft}$ continue. Otherwise, repeat steps 3–5.
7. Calculate the deviation of p_f from p_{ft}.
8. Minimise deviations of the probability of failure using constrained optimisation procedure to find optimum values of partial safety factors.
9. Recalculate if necessary.
10. Calibration complete.

In step 8, the reduction of the difference of the value of the probability of failure in each test case with the target failure probability was formulated as a constrained minimisation of the squares of the differences of the individual probabilities of failure, p_{ft}, and the target probability of failure, p_{ft}. The weighting factors were all selected to be equal to unity, so

Table 12.19 *Calibrated partial safety factors for hull girder ultimate strength based on 50%*t_{corr}

Target probability of failure p_f	Design load combination	Still-water bending moment γ_{sw}	Wave-induced bending moment γ_{wv}	Ultimate bending moment γ_R
$1.5(10)^{-3}$	Permissible still-water bend mom	1.00	1.20	1.10
	Maximum bending moment for homogeneous full-load condition	1.00	1.30	1.10

that each test case carried equal weight. The objective function, $\Delta p^2{}_{min}$, assumed the following form:

$$\Delta p^2{}_{min} = \sum_{i=1}^{N} \left(p_{ft} - p_{fi}(\gamma)\right)^2 \tag{12.38}$$

where

p_{ft} is the target annual probability of failure
γ is the vector of partial safety factors subjected to calibration
p_{ft} is the annual probability of failure for the *i*th case associated with the partial safety factors
N is the number of cases included in the calibration analysis

This is a classical nonlinear optimisation problem that can be solved using Lagrange multipliers. It is evident that the final choice of partial safety factors does not correspond to an optimum design for any individual test case. The primary aim of a reliability-based code is to ensure a minimum required safety level for the range of sizes of the types of ships for which it is intended.

The calibrated partial safety factors that are used in the CSR for oil tankers are included in Table 12.19. It should be noted that these correspond to plating that has a thickness reduction that is equal to 50 per cent of the corrosion allowance. If the net thickness is used (zero corrosion allowance), the probability of failure increases by a factor of 10. Correspondingly, for plating with a full corrosion allowance, the probability of failure is much reduced. Furthermore, North Atlantic conditions have been used, and no weather routing is allowed for. Lastly, a rigorous consequence analysis using, for example, formal safety assessment, as stipulated in the IMO requirements, has not been performed, so that, strictly speaking, the code is not risk compliant.

However, the most important piece of information contained in Table 12.19 is the target probability of failure, which has been set equal to $(1.5)(10)^{-3}$. This classifies hull girder collapse as 'Not Serious'. Furthermore, it assumes that the structure behaves in a ductile manner and has reserve strength capacity.

Table 12.20 *Proposed safety standards for ship structures [34]*

Failure Mode	Consequences	Merchant ships	Warships
Primary (initial yield)	-	5.0 (2.9×10^{-7})	6.0 (1.0×10^{-9})
Primary (ultimate)	Very serious	3.5 (2.3×10^{-4})	4.0 (3.2×10^{-5})
Secondary	Serious	2.5 (6.2×10^{-3})	3.0 (1.4×10^{-3})
Tertiary	Not serious	2.0 (2.3×10^{-2})	2.5 (6.2×10^{-3})

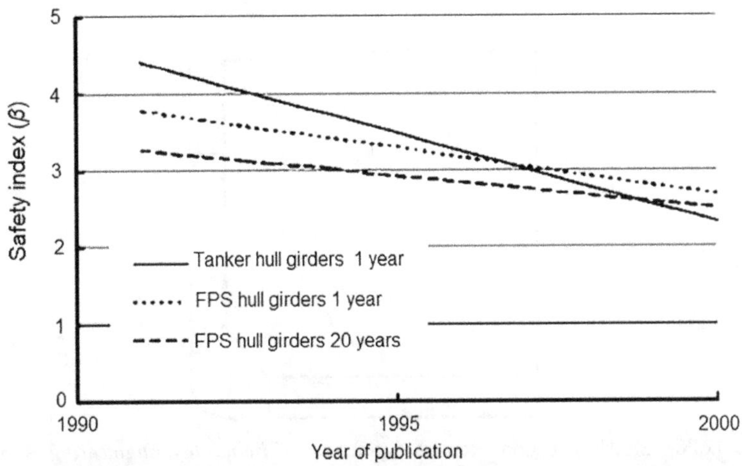

Figure 12.25 Trends in reliability indices for oil tankers and FPSs (1991–2000).

It has been shown, however, that with the passage of time, the reliability of ship structures has been diminishing (Figure 12.25) [38]. The authors of this study expressed concern, and the reasons for this lowering of safety standards were not clear at the time. It has to be mentioned that with the drafting of the IACS Common Structural Rules in 2006, the trend observed in Figure 12.25 is not expected to continue downwards. However, this question remains unanswered and needs further study.

In another study, an assessment of safety standards over a wide range of industries was carried out, and detailed proposals were made with regard to ship structures [34]. These are included in Table 12.20, which gives the reliability index, β, and the corresponding notional probability of failure.

A comparison of the data contained in Tables 12.19 and 12.20 shows that current safety standards lie below those that have been proposed by industry experts. It remains to be seen whether in the future there will be a need to raise safety standards. This will depend on data availability and processing, accident occurrence as well as the economic factors that affect the maritime industry.

The recording and availability of safety data will enable target failure probabilities to be established on a firmer foundation in the future, and for rules to be rewritten as necessary. The need to conduct consequence assessment studies in a rigorous manner and to draft rules for different ship types, configurations and sizes will also have to be addressed.

Table 12.21 *Main particulars of Aframax oil tanker*

Description	*Magnitude*
Length between perp. (L_{pp})	236 m
Moulded breadth (B)	42.0 m
Moulded depth (D)	21.0 m
Scantling draught (T)	15.6 m
Deadweight (DWT)	114,000 mt

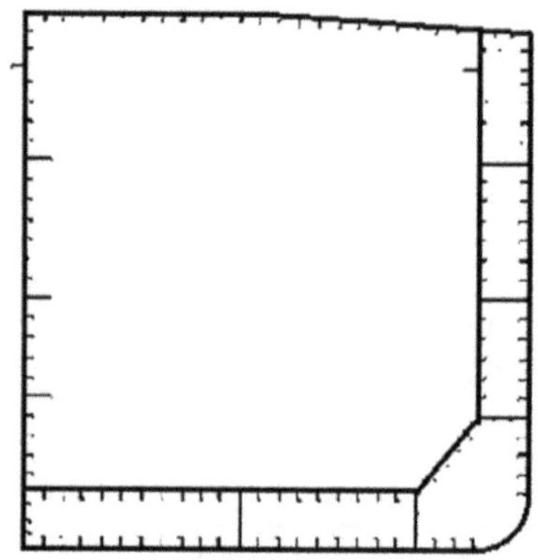

Figure 12.26 Midship section. Reprinted courtesy of *Reliability Engineering & System Safety.*

12.6.2 A Study on the Effect of the Introduction of the Common Structural Rules

In this second example, structural reliability theory will be used to calculate the probability of failure of a ship that was designed prior to the introduction of the Common Structural Rules, and this will be compared with the design of the same ship modified so that it is CSR compliant. The ship in question is an Aframax oil tanker, and this study is reported in [39]. The methodology used follows the procedure recommended by the International Association of Classification Societies (IACS); this is summarised in Figure 12.22.

Description of the Ship

The ship considered was an Aframax oil tanker whose design complied with the pre-CSR rules. The particulars of the vessel are included in Table 12.21, and the midship section is shown in Figure 12.26.

Pre-CSR Design

The deck and bottom area of the ship were constructed of high tensile (HT) steel (AH32), while in the region around the neutral axis mild steel (ST235) was used. Since this tanker was awarded an ICE-1C class notation, HT steel AH36 was used in the side shell in the ice belt region as well. In addition, HT steel was used for the centreline bulkhead in order to comply with shear strength requirements.

CSR-compliant Design

The modifications necessary to comply with CSR requirements were:

1. An increase in the cross-sectional area of the deck longitudinals of 15 per cent
2. An increase in the thickness of the main deck plating of 1 mm
3. An increase in the thickness of the uppermost strakes of the centreline bulkhead and of the inner hull plating of 1.0–1.5 mm.

The above were sufficient to satisfy the criteria for the ultimate longitudinal strength. However, additional strengthening was introduced in order to satisfy local strength requirements as well. This consisted of:

1. An increase of the inner bottom and the inner hull plating thickness by 1.5 mm
2. An increase of the area of the bottom longitudinals by 17 per cent.

The net result of these modifications was an increase of the cross-sectional area of the midship section of 5 per cent. Calculations showed that the ultimate bending capacity of the modified section required by CSR was thereby satisfied.

The Procedure

Before embarking on the detailed calculation of the necessary variables required to assess the reliability of the two designs, it is necessary to make some preliminary comments on the basis upon which this study will be carried out. This is because reliability studies can be based on different assumptions, as well as because different techniques can be used. In such circumstances it is not possible to compare the results from different sources; probabilities of failure are notional probabilities, and their value lies in them being able to be compared directly with the results from other studies. To overcome this problem a concerted effort was made by a number of classification societies to lay the foundations for a common basis for the reliability studies that are performed in the marine industry. These are reported in [16] and are listed below:

Ship operational service. Ships are generally designed and built to operate worldwide with unrestricted service. It makes sense to consider the worse possible conditions encountered at sea. In order to err on the side of caution, service on the North Atlantic is considered for the lifetime of the ship.

Ship types. There exist standard designs for the most important types of merchant ships.

Use of wave data. In compliance with the operational service requirements mentioned above, wave scatter diagrams for the North Atlantic are recommended.

Reliability formulation. It is recognised that with the passage of time, the probability of failure of a ship increases as a result of corrosion and other reasons. On the basis of existing studies, a time-independent first-order reliability formulation corresponding to one year of operation is considered to be the most adequate and practically applicable method for the reliability assessment of a ship.

Design code format. The reliability code format involves the use of one partial safety factor for hull girder strength and one partial safety factor for each of the applied loads (still-water and wave-induced bending moment).

Load Analysis

The first part of the procedure involves calculating the external loads applied to the ship as well as establishing their probabilistic description. These loads are the still-water bending moment and the wave-induced bending moment.

Still-water Bending Moment

The probabilistic nature of the still-water bending moment has been discussed in Section 12.4.2, and in this study a Gaussian distribution was used for the probabilistic description of the still-water bending moment for one voyage. The descriptors of the distribution (mean value and standard deviation) were calculated using the information contained in the ship's loading manual for three loading conditions (full load (FL), ballast (BL) and partial loading (PL)) taking into account both departure and arrival conditions. Once the mean value (μ_{sw}) and the standard deviation (σ_{sw}) of a normal distribution are known, the corresponding extreme value is obtained from the corresponding Gumbel distribution.[12] The parameters of this distribution are obtained from the normal distribution using the following relations:

$$x_e^* = F_{sw}^{-1}\left(1 - \frac{1}{n_{sw}}\right) \qquad \alpha = \frac{1 - F_{sw}}{f_{sw}} \tag{12.39}$$

where

n_{sw} = number of occurrences of a load condition in the reference period (1 year)
F_{sw} = cumulative distribution function
${F_{sw}}^{-1}$ = inverse of the cumulative distribution function
f_{sw} = probability density function of a normal distribution with parameters μ_{sw} and σ_{sw}

According to the operating scenario given in Table 12.22, $n_{sw} = 5.4$ for full load and ballast and $n_{sw} = 27.4$ for the partial loading condition. The mean value μ_{se} and standard deviation σ_{se} of the Gumbel distribution are thereby obtained from:

$$\mu_{se} = x_e^* + 0.5772\alpha \qquad \sigma_{se} = \frac{\pi}{\sqrt{6}}\alpha \tag{12.40a–b}$$

Table 12.23 includes the parameters of the stochastic model of the still-water bending moment for one voyage and for one year, calculated according to this procedure.

[12] Section 10.2.3.

Table 12.22 *Operational profile of tankers*

Loading condition	Harbour	Full	Ballast	Partial
Percentage of spent time (%)	15	35	35	15
Voyage duration (days)	-	23.5	23.5	2.0

Table 12.23 *Parameters of still-water bending moment distributions (MNm)*

	One voyage		One year	
Loading condition	μ_{sw}	s_w	μ_{se}	σ_{se}
FL	−1,229	456	−1,819	405
PL	1,444	387	2,240	226
BL	2,555	46	2,615	41

Wave-induced Bending Moment

The environmental conditions that should be considered when estimating the extreme wave-induced bending moment for ships that comply with the IACS longitudinal strength standard (UR S11) are specified in [25]. This document requires that:

- The scatter diagram for the North Atlantic (areas 8, 9, 15 and 16) should be used (Table 9.3). The data from the GWS are further modified by IACS in order to take into account the limited wave steepness more properly.
- A ship speed equal to zero is to be assumed.
- The two-parameter Pierson–Moskowitz wave height spectrum is recommended (Equation (9.45)).
- Short-crested waves with the wave energy spreading function proportional to $\cos^2\mu$ have to be used (Equation (9.48)).
- All heading angles should have equal probability of occurrence and maximally 30° spacing between headings should be applied.
- Proper corrections for nonlinear effects are to be applied.

Hydrodynamic loading and the calculation of the transfer functions of the wave-induced load effects was carried out using the WADAM program,[13] based on the sink–source 3D method. The panel model of the Aframax tanker is shown in Figure 12.27. The calculated transfer functions for the full load condition and different headings are included in Figure 12.28.

In accordance with the IACS recommended procedure, the long-term analysis is performed for three loading conditions: full load (FL), ship in ballast (BL) and partial loading condition (PL). As seen in Figure 12.29 and from a detailed comparison of results, it is

[13] The program forms part of the SESAM suite and is made available by DNV-GL.

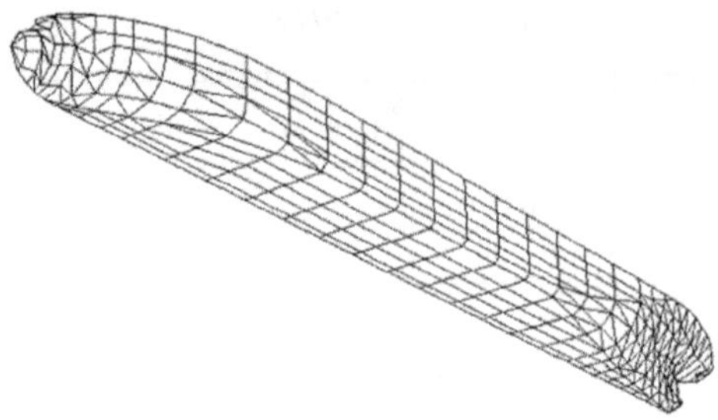

Figure 12.27 Panel model (underside view). Reprinted courtesy of *Reliability Engineering & System Safety.*

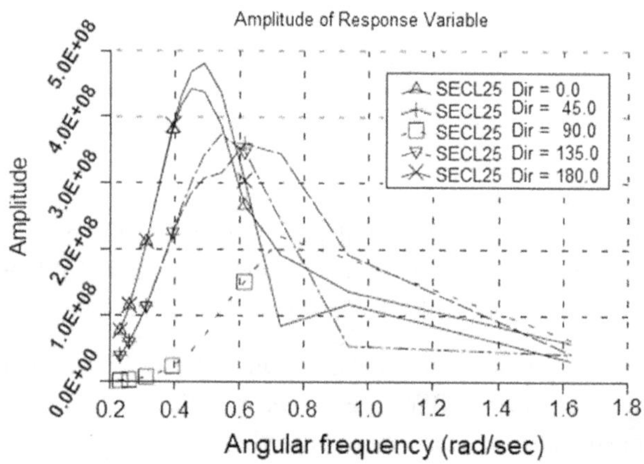

Figure 12.28 Transfer functions of VWBM calculated using WADAM. Reprinted courtesy of *Reliability Engineering & System Safety.*

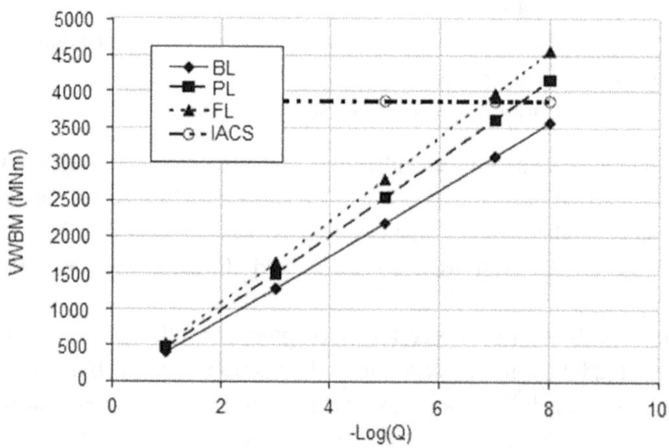

Figure 12.29 Long-term distributions of VWBM for three different loading conditions. Reprinted courtesy of *Reliability Engineering & System Safety.*

concluded that the long-term extreme value obtained by direct 3D hydrodynamic analysis is, on average, 6 per cent larger than the mean value of the IACS rule hogging and sagging wave-induced bending moments. The highest extreme VWBM is obtained for the fully loaded condition and is found to be 18 per cent larger than the mean rule value.

Statistical Models of Vertical Wave Bending Moment for Reliability Analysis

A Weibull two-parameter model is usually used to approximate the long-term probability distribution of the vertical wave-induced bending moment (Figure 12.14):

$$F(x_p) = 1 - e^{(-x_p/k)^l} \tag{12.41}$$

where

$$k = \text{Weibull scale parameter}$$
$$l = \text{Weibull shape parameter}$$
$$F(x_p) = P\left[\text{amplitude of the wave-induced bending moment} < x_p\right]$$

The probability that the response amplitude does not exceed a given (extreme) value, x_e, over a longer time period, such as for one full voyage, one year or more, is given by the Gumbel law:

$$F(x_e) = \exp(-\exp(-(x_e - x_e^*/\alpha))) \tag{12.42}$$

The parameters $x_e{}^*$ and α are derived from the parameters of the Weibull distribution Equation (12.41) using the following relationships:

$$\alpha = \frac{k}{l}(\ln n)^{(1-l)/l} \qquad x_e^* = k(\ln n)^{1/l} \tag{12.43}$$

where n represents the number of response cycles in a given long-term period. The mean value and standard deviation of the Gumbel distribution are given by Equations (12.40a–b).

Modelling uncertainty in determining the vertical wave-induced bending moment is allowed for by introducing two parameters, χ_w and χ_{nl} (linear and nonlinear wave loading modelling uncertainty). The first of these is used to allow for simplifications, assumptions and inaccuracies of the linear engineering model used to predict extreme wave load on the ship's hull. Following IACS recommendations [28], this is treated as being normally distributed with a mean value of unity and a COV of 0.10. The second of these is of importance when there are significant differences in the hogging and sagging bending moments. In this study the following nonlinear correction factors were used:

$$\Phi_S = \frac{M_S}{M_L} = \frac{2R}{1+R} \qquad \Phi_H = \frac{M_H}{M_L} = \frac{2}{1+R} \tag{12.44a–b}$$

where R is the ratio of the sagging to the hogging moment given by IACS UR S11 (Equation (4.55)). The uncertainty due to nonlinear effects was assumed to be normally distributed with mean value equal to the nonlinear correction factors given by Equations (12.44a–b) and a COV of 0.15, on the basis of earlier studies.

Table 12.24 *Load combination factors, ψ*

Loading condition	ψ
FL	0.92
BL	0.91
PL	0.80

Combination of Still-water and Wave Bending Moment Treated as Stochastic Variables

As discussed in Section 12.5.2, it is necessary to take into account the fact that peak still-water and wave-induced bending moments do not coincide. Therefore, in order to allow for this when calculating the total bending moment, load combination factors are introduced. The factors used in this study are included in Table 12.24.

These factors were calculated in an earlier study using the Ferry Borges-Castanheta method (Section 12.5.2.4). It ought to be added that the IACS recommendation is for the *Turkstra* rule to be applied when calculating load combination factors. A comparison of results obtained by the two methods carried out as part of the present study indicated minor differences.

Strength Capacity of the Hull Girder

For present purposes the strength capacity was taken to be the ultimate vertical bending moment of the hull girder. This was calculated using the two methods that are described in this text. These are:

1. Progressive collapse analysis (PCA) based on column-type interframe elasto-plastic buckling. The stress–strain relationships used were those proposed in the CSR for oil tankers. This method is described in Section 7.4.3.
2. The single-step procedure proposed by IACS (HULS-1), which is described in Appendix D. In this case the finite element program PULS, provided by DNV-GL, was used.

Calculations were performed for each ship for the intact case (using gross scantlings) and the corroded case (using net scantlings). The results are included in Table 12.25.

In this study the results of the PCA were used as reference with regard to reliability calculations. A single uncertainty variable χ_U was used to account for yield strength variability as well as model uncertainty, and a lognormal distribution with a mean value of 1.1 and a coefficient of variation equal to 0.12 were used.

Reliability Formulation

The limit state equation that corresponds to the ultimate strength limit state was used in the reliability analysis. The equation incorporates the variables and their corresponding uncertainties discussed previously, as follows:

Table 12.25 *Ultimate bending moments (MNm)*

Hull state	Method	Initial design		CSR-compliant design	
As-built	PCA	8,246	10,737	9,078	11,503
	HULS-1	7,983	-	8,709	-
Corroded	PCA	6,813	9,265	7,587	10,008
	HULS	6,787	-	7,476	-

$$\chi_u M_u - M_{sw} - \psi \chi_w \chi_{nl} M_w < 0 \tag{12.45}$$

where

$$M_u = \text{ultimate hull girder bending moment (deterministic)}$$
$$M_{sw} = \text{extreme vertical still-water bending moment (rv)}$$
$$M_w = \text{extreme vertical wave-induced bending moment (rv)}$$

The reliability analysis was carried out for both sagging and hogging failure modes, and for each of these modes the safety index was calculated for the three stipulated loading conditions: full load (FL), ship in ballast (BL) and partial loading (PL). It is necessary to allow for the duration of each of these periods during the year of operation, and to this end the data included in Table 12.22 were used.

Annual safety indices $\beta_{FL}, \beta_{PL}, \beta_{BL}$ and associated failure probabilities $P_{f,FL}, P_{f,PL}, P_{f,BL}$ were calculated for each of the loading conditions separately for failures in sagging and hogging. The global annual safety index for each of two failure modes β_t was then determined from

$$\beta_t = -\Phi^{-1}\left(P_{f,FL} + P_{f,PL} + P_{f,BL}\right) \tag{12.46}$$

The reliability method implemented in this study was the Level II method, described in Section 11.4.2. The general purpose reliability analysis program PROBAN was used to this end [44].

Results and Conclusions

Results consisting of global annual failure probabilities and safety indices are included in Table 12.26. It is seen that the highest probability of failure arises in the sagging mode in the fully loaded condition. The probability of failure in the hogging condition is significantly lower than in sagging. Comparing results for the initial design with the CSR-compliant design shows that the lowest safety indices are 2.04 and 2.60, respectively, both in the case of the corroded condition in sagging. These figures correspond to a 5.7 times lower failure probability in sagging than the tanker designed in accordance with the old rules. In the corroded condition, the ratio of failure probabilities equals 4.4.

The failure probabilities of the CSR-compliant design are somewhat lower than the target reliabilities proposed by the IMO. The reasons for this are:

Table 12.26 *Safety indices in sagging and hogging for as-built and corroded ships*

Ship	Hull state	Condition	β_{FL}	β_{BL}	β_{PL}	P_{ft}	β_t
Initial design	As-built	Sag	3.03	6.71	6.52	1.223E-03	3.03
	Corroded	Hog	6.32	5.18	5.43	1.393E-07	5.14
		Sag	2.04	6.15	5.94	2.068E-02	2.04
		Hog	5.82	4.38	4.68	7.371E-06	4.33
CSR-compliant design	As-built	Sag	3.52	7.00	6.81	2.158E-04	3.52
	Corroded	Hog	6.55	5.54	5.77	1.912E-08	5.50
		Sag	2.60	6.46	6.26	4.661E-03	2.60
		Hog	6.08	4.80	5.07	9.928E-07	4.76

1. In the present study, reliability methods have been applied to an individual case, in contrast to the 'rule approach' that caters to the whole population of ships.
2. Slightly different statistical models have been used in the two analyses.
3. Different reliability formulations and methodologies have been used in the computation of long-term distributions of wave-induced bending moments.

12.6.3 Risk-Based Approach for Structural Design of a Double-Hull Suezmax Tanker

In this final example demonstrating the use of structural reliability analysis, the structural design of a Suezmax oil tanker is performed using a risk-based approach. The study is reported in [40], and the methodology followed is based on IMO documents [26, 28] and a similar case documented in [20].

In order to understand the procedure that was followed it is useful to refer to Figure 12.7 that summarises the steps of FSA, since in the present study this is the methodology that has been followed.

The problem was specified to be the structural design of a double hull Suezmax oil tanker using risk-based criteria. The single hazard that was considered in this case was the ultimate collapse of the hull girder in vertical bending (sagging condition) when the ship is in a fully loaded condition at sea. Risk can be modified by changes to the probability of failure, to the consequences of an accident or both. In the present case the risk-control option that was considered was the strengthening of the main deck by varying the strength of the main deck, the reliability and thereby the probability of failure of the hull girder in vertical bending changes. The extent to which this change in deck strength from the initial value is needed is determined from a cost–benefit analysis. The consequences, being the second component of risk, were assessed by taking into account loss of the ship, loss of cargo, fatalities and damage to the environment.

The purpose of the cost–benefit analysis was to determine an *optimum safety index*, β_{opt}, for which the total expected costs $C_t^{\beta_t}$ would be minimised, these being the sum of the *actual* cost of strengthening the main deck, $C_{me}^{\beta_t}$, and the *potential* cost of structural failure, $C_{Tf}^{\beta_t}$.

Table 12.27 *Main particulars of Suezmax double-hull oil tanker*

Description	Magnitude
Length overall (L_{OA})	280.0 m
Length between perp (L_{pp})	270.0 m
Moulded breadth (B)	48.20 m
Moulded Depth (D)	23.00 m
Design draught (T)	16.00 m
Scantling draught (T_{sc})	17.10 m
Block Coefficient (C_B)	0.83
Deadweight (at T)	152,300 mt
Deadweight (at T_{sc})	166,300 nit

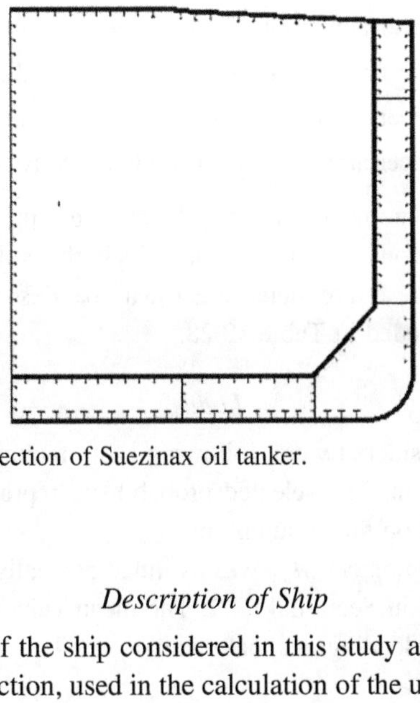

Figure 12.30 Midship section of Suezinax oil tanker.

Description of Ship

The principal particulars of the ship considered in this study are included in Table 12.27. The discretised midship section, used in the calculation of the ultimate bending moment, is shown in Figure 12.30.

Structural Reliability Analysis

In this study, as in the preceding study, vertical bending is considered alone, as for this type of ship the other longitudinal bending moments (horizontal, torsional) are assumed small and do not contribute to the loss of strength. The limit state function, $g(X)$, used thus corresponds to ultimate collapse of the ship in longitudinal vertical bending and is expressed as follows:

$$g(X) = \chi_u M_u - (\chi_{sw} M_{sw} + \chi_{wv} \chi_{nl} M_{wv}) \tag{12.47}$$

Failure is defined by $g(X) < 0$.

Table 12.28 *Statistical description of model uncertainty factors*

Model uncertainty factor	Mean	COV
Ultimate strength (χ_u)		
Still-water bending (χ_{sw})	1.0	0.1
Wave-induced (linear) (χ_{wv})	1.0	0.1
Wave-induced (nonlinear) (χ_{nl})	1.0	0.1

In this equation,

M_u = ultimate bending moment capacity

M_{sw} = still-water bending moment

M_{wv} = wave-induced bending moment

χ_u = model uncertaintyfactor for ultimate capacity

χ_{sw} = model uncertaintyfactor for still-water bending

χ_{wv} = model uncertainty factor for linear wave-induced bending

χ_{nl} = model uncertainty factor for nonlinear wave-induced bending

The ultimate bending moment, M_u, is obtained by progressive collapse analysis (PCA).[14] In the next sections the manner in which the still-water and wave-induced bending moments (M_{sw}, M_{wv}) are determined will be described. Data for the model uncertainty factors are included in Table 12.28.

Loads

The bending moments considered were, following past practice, the peak expected values during one year of operation. The selected probabilistic representations described in the following have been based on this assumption.

The still-water bending moment M_{sw} was assumed normally distributed, in accordance with [15] and as discussed in Section 12.4.2. The mean value is 0.7 times the maximum value given in the ship's loading manual, with a standard deviation of 0.2 times the maximum value.

The loading manual gives the value $M_{sw,sag} = -2{,}610.1$ MNm (sagging condition).

The extreme value of the vertical wave-induced bending moment is obtained from a two-parameter Weibull distribution:

$$F_{M_{wv}}(M_{wv}) = 1 - \exp\left[-\left(\frac{M_{wv}}{k}\right)^l\right] \tag{12.48}$$

where $l = 1.0$, which is equivalent to assuming an exponential distribution for M_{wv}. The scale parameter w is determined such that

[14] Section 7.4.

$$P[M_{sw} > M_{wv,\max}] = 10^{-8} \tag{12.49}$$

that is, the probability of a still-water bending moment exceeding the maximum wave-induced bending moment is less than 10^{-8}. The scale parameter k is thus given by

$$k = -\frac{M_{wv,\max}}{\ln(10^{-8})} \tag{12.50}$$

The maximum value of the wave-induced bending moment amidships in the sagging condition was obtained using the IACS formula included in the Common Structural Rules (2010 edition):

$$M_{wv-sag} = -0.11 C_w L^2 B (C_B + 0.7) \tag{12.51}$$

where L, B, C_B and C_w are the length, breadth, block coefficient and wave coefficient, respectively.

The distribution of annual maximum values of the wave-induced bending moment is described by a Gumbel model, whose parameters are defined from the initial Weibull distribution:

$$F_{M_{wve}}(M_{wve}) = 1 - \exp\left[-\left(\frac{M_{wve} - l}{k}\right)\right] \tag{12.52}$$

where M_{wve} is a random variable representing the extreme values of the wave-induced bending moment. The parameters of the Gumbel distribution are given by

$$\lambda = k(\ln n_c)^{1/k} \tag{12.53a}$$

$$\theta = \frac{k}{l}(\ln n_c)^{(1-l)/l} \tag{12.53b}$$

where n_c is the mean number of wave load cycles expected over the reference period, T_R, for a given operational profile of the ship and mean wave period $\overline{T}$ of the sea state considered. In the present study $T_R = 1$ year and $\overline{T} = 8$ sec. The period of operation of the ship in a fully loaded condition was $0.35T_R$. An estimate for n_c is therefore given by

$$n_c = 0.35\frac{T_R}{\overline{T}} \tag{12.54}$$

Combination of Still-water and Wave-induced Bending Moments

The Turkstra rule for load combinations was used, as recommended by IACS. According to this, for two random variables two load combinations are required, in each of which the peak value of one random variable is added to a randomly selected value of the other. It has been found [18] that for the conditions prevailing in a study such as the present one, the critical combination is that of the annual extreme value of the wave-induced bending moment combined with a randomly selected value of the still-water bending moment. The data used are included in Table 12.29.

Table 12.29 *Data used in combining the still-water and wave-induced bending moments*

Random Variable	Distribution type	Mean value [MNm]	Stand. dev. [MNm]	COV
M_{SW}	Normal	1,834.1	524.0	0.29
M_{WV}	Gumbel	5,000.8	435.9	0.09

Ultimate Strength Assessment and Design Modification

In the present study the variability of strength was modelled via distributions for the yield stress of steel and scantlings. In the former case the yield stress was assumed to be lognormally distributed and the characteristic yield stress was defined as the 5th percentile of the distribution [18]. In the latter case, variability in scantlings was described by a normal distribution with a mean value equal to the design value and a COV of 0.05. A deterministic value was assumed for the elastic modulus of steel.

As mentioned in Section 12.6.1.3, in this study the strength of the hull girder is varied by modifying the cross-sectional area of the strength deck, the reason being that when the ship is in a sagging condition, it is the deck that is critical to survival, since failure occurs by interframe buckling. Increasing the deck cross-sectional area increases the ultimate bending moment and the reliability index, which is tantamount to reducing the probability of failure. It therefore becomes possible to monitor the change in failure probability as a function of deck area change.

Following the IACS procedure described in [21], a *design modification factor* (DMF) is introduced. This reflects the change in deck cross-sectional area with respect to the initial design. The design modification was achieved by increasing or decreasing the plate and stiffener thickness in increments of 1 mm. The spacing of stiffeners and other members was kept constant. Detailed calculations concerning the actual elasto-plastic collapse strength of stiffened plating were not carried out, and this simplification adds a degree of conservatism to the calculation. Table 12.30 gives the resulting deck area and DMF for changes in thickness. This was done for gross scantlings (intact structure) and net scantlings (corroded structure).

The results of the progressive collapse analysis are included in Figure 12.31 in which the ultimate capacity is obtained for a range of values of the DMF.

Reliability Calculation

The reliability of the hull girder was calculated using the semi-probabilistic Level II first-order second moment method described in Section 11.4.2. The program COMREL was used, and the output is included in Table 12.31, which includes the reliability index (β) and the probability of failure (p_f) as a function of the DMF.

It is seen that there is a proportionate increase in the reliability index with increase in DMF. This is as expected, since as the thickness increases, so does the ultimate capacity and the reliability index.

Table 12.30 *Thickness change in scantlings, corresponding deck cross-section area and Design Modification Factor (DMF)*

	Intact condition		Corroded condition	
Change in thickness (mm)	Deck area (mm^2)(× 106)	DMF	Deck area (mm^2)(× 106)	DMF
−5	1.19	0.75	1.03	0.72
−4	1.27	0.80	1.11	0.78
−3	1.35	0.85	1.19	0.83
−2	1.43	0.90	1.27	0.89
−1	1.51	0.95	1.35	0.94
0	1.59	1.00	1.43	1.00
+1	1.67	1.05	1.51	1.06
+2	1.75	1.10	1.59	1.11
+3	1.83	1.15	1.67	1.17
+4	1.91	1.20	1.75	1.22
+5	1.99	1.25	1.83	1.28
+6	2.07	1.30	1.91	1.33
+7	2.15	1.35	1.99	1.39

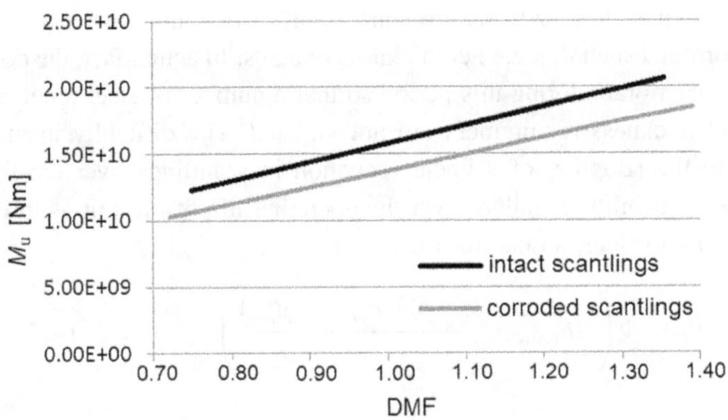

Figure 12.31 Effect of modifying deck cross-section on ultimate bending capacity.

Cost–Benefit Analysis

As mentioned in Section 12.6.3, the aim of the study is to calculate the optimum reliability index for which the total expected costs $C_t^{\beta_t}$ are minimised, these being the sum of the *actual* cost of strengthening the main deck, $C_{me}^{\beta_t}$, and the *potential* cost of structural failure, $C_{Tf}^{\beta_t}$. Both of these vary with the reliability index. It is therefore necessary to quantify, in monetary terms, the changes to the cost of modifying the deck structure and the corresponding changes to the cost of failure. The final year of operation of the ship (year 25) was used as a basis for a range of values of $1.5 < \beta < 5$ with a spacing equal to 0.5. The annual probability of failure (P_f) is given by

Table 12.31 *Reliability index (β) and probability of failure* (p_f) as a function of the DMF

Intact condition			Corroded condition		
DMF	β	p_f	*DMF*	β	p_f
0.75	3.42	3.05E-04	0.72	2.41	7.89E-03
0.80	3.73	9.29E-05	0.78	2.77	2.73E-03
0.85	4.01	2.92E-05	0.83	3.12	8.79E-04
0.90	4.28	9.25E-06	0.89	3.44	2.86E-04
0.95	4.53	2.92E-06	0.94	3.73	9.53E-05
1.00	4.76	9.30E-07	1.00	4.01	2.96E-05
1.05	4.99	2.99E-07	1.06	4.28	9.23E-06
1.10	5.20	9.81E-08	1.11	4.53	2.90E-06
1.15	5.40	3.26E-08	1.17	4.77	9.24E-07
1.20	5.59	1.11E-08	1.22	4.99	2.99E-07
1.25	5.76	4.20E-09	1.28	5.19	1.01E-07
1.30	5.91	1.65E-09	1.33	5.37	3.78E-08
1.35	6.06	6.71E-10	1.39	5.54	1.45E-08

$$P_f = \Phi(-\beta_t) \tag{12.55}$$

where Φ is the cumulative normal distribution.

It is assumed that in year 0 the structure is intact (as-built), whereas in year 25 the structure is corroded such that the net thickness remains. In actual fact, the condition of the structure will deteriorate during this period so that a number of steel renewals are carried out so the net thickness requirement are not violated. The difficulty in modelling such changes led to the adoption of a linear reduction in scantlings over the lifetime of the structure. The probability of failure over the operating life of the ship is thus interpolated between these two values, so that for the *i*th year (t_i),

$$P_{fi} = \Phi\left(-\beta_{gross} + \frac{(t_i - 1)\left(\beta_{gross} - \beta_{net}\right)}{24}\right) \qquad i \in [1, 25] \tag{12.56}$$

In the present study the actual cost considered concerns the steelwork modifications to the main deck.[15] Based on steelwork cost, the following expression is used:

$$C_{me}^{\beta_t} = \left(DMF^{\beta_t} - 1\right) W_{steel} C_{steelwork/ton} \tag{12.57}$$

where W_{steel} is the weight of the hull girder spanning 40 per cent of the length of the ship about the midship section, $C_{steelwork/ton}$ is the cost of steelwork per ton of steel, and $DMF^{\beta t}$ is the design modification factor associated with a reliability index β_t.

The potential cost of failure is calculated considering the operational life of the ship, which was assumed to be 25 years. For this period, following the recommended procedure

[15] Clearly this is not the only real cost that would arise, as there is opportunity cost involved related to the additional time off-hire for the ship.

by the IMO, an annual discount rate γ of 5 per cent was assumed. The costs that contribute to the total cost of failure are:

1. The ship's resale price, $C_n(t)$, being a function of age, t
2. The costs associated with the loss of cargo, C_c
3. The costs related to an accidental oil spill at sea, C_d
4. The costs associated with the loss of human life, C_v

The above are incorporated in the following expression:

$$C_{Tf}^{\beta_t} = \sum_{t=1}^{T} P_f(\beta_t)[C_n(t) + (C_c + C_d + C_v)]e^{-\gamma t} \tag{12.58}$$

It is seen that the cost of failure is a function of the probability of failure.

For year t, the resale price of the ship, $C_n(t)$, is given by the following expression, assuming linear depreciation:

$$C_n(t) = C_{n0} - \left(C_{n0} - C_{scrap}\right)\left(\frac{t}{25}\right) \tag{12.59}$$

where

C_{n0} = purchase cost of the ship

C_{scrap} = scrap price of the ship

The purchase cost was estimated using data compiled by UNCTAD for the period 2003–2010 and inflation rates included made available by the OECD, leading to a figure of USD70.5 million. The scrap price was calculated on a per ton of steel-weight basis using a figure of USD277 per ton.

The costs related to the loss of cargo were calculated following past practice, by assuming that in the case of structural failure, 20 per cent of the deadweight capacity is spilt. The following expression was used:

$$C_c = \left(C_{crude/ton}\right)(DWT)\left(P_{spill}\right) \tag{12.60}$$

where the price of crude oil, $C_{crude/ton}$, was, using data for 2014, USD108/ton and P_{spill} is the percentage of cargo spilt. Of the quantity of oil spilt, it is assumed that 10 per cent will reach the shoreline, so that the costs related to clean-up operations can be obtained from the following expression:

$$C_d = \left(P_{spill}\right)(P_{sl})(CATS)(DWT) \tag{12.61}$$

where P_{sl} is the proportion of oil reaching the shoreline and CATS is the *C*ost to *A*vert a *T*on of oil *S*pilt. The CATS criterion was based on past data, giving a figure of USD60,000 per ton. On the basis of past data, when the ship is lost at sea, up to one-quarter of the crew on board may be assumed to lose their lives (P_{crew}), so that for a Suezmax oil tanker with a crew of 25 (n_{crew}) the following expression is used:

$$C_v = n_{crew}P_{crew}\text{ICAF} \tag{12.62}$$

where ICAF is the Implied Cost to Avert a Fatality. A figure for the ICAF was obtained using data for the Life Quality Index (LQI), which is defined to be a function of the gross domestic product per capita, g, life expectancy at birth, e and the percentage of time spent in an economic activity, w. An estimate for the latter is $w = 1/8$. The following expression for LQI is used:

$$\mathrm{LQI} = g^w e^{1-w} \tag{12.63}$$

A safety measure is implemented if the marginal LQI is positive, that is, $\Delta LQI > 0$:

$$\frac{\Delta e}{e} > -\frac{\Delta g}{g}\frac{w}{1-w} \tag{12.64}$$

where Δe and Δg are the change in life expectancy at birth and the GDP per capita, respectively, that result from the application of a safety measure. Assuming that saving a human life corresponds to saving one-half of the life expectancy, the optimal acceptable cost for saving a human life is given by

$$|\Delta g| = \frac{g}{2}\frac{1-w}{w} \tag{12.65}$$

The optimal cost that should be invested in order to avoid a fatality is then

$$ICAF = \frac{ge}{4}\frac{1-w}{w} \tag{12.66}$$

Using the above expression, a figure of USD3.85 million was calculated for the ICAF. The steelwork cost and the total cost of failure are calculated for a range of values of the reliability index β_t, and are subsequently plotted on the same chart as shown in Figure 12.32. The total cost is obtained from

$$C_{\mathrm{t}}^{\beta_{\mathrm{t}}} = C_{\mathrm{me}}^{\beta_{\mathrm{t}}} + C_{Tf}^{\beta_t} \tag{12.67}$$

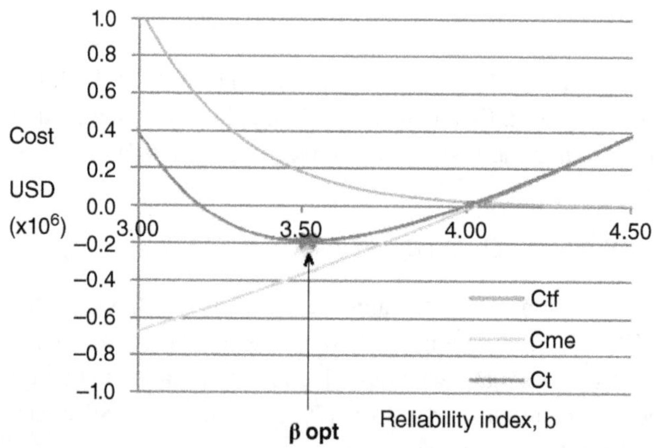

Figure 12.32 Variation of cost of structural modification (C_{me}), cost of failure (C_{tf}) and total cost (C_t) with reliability index β.

The minimum value of the total cost, C_t, is seen to arise for a reliability index equal to 3.52, which is thus the value of β_{opt}

12.7 A Note on Probabilistic and Deterministic Structural Design

Probabilistic structural design differs from deterministic structural design in one or two possible ways. Firstly, the code used has a probabilistic basis, that is, the safety factors have been determined using probabilistic (Level III) or, more usually, semi-probabilistic (Level II) methods. The code itself, however, is a deterministic (Level I) procedure and is implemented in a manner identical to that of the codes that are not reliability-based. Secondly, the structure is designed using a direct probabilistic or a semi-probabilistic approach such as *FORM* or *SORM*. The difference between Level III and Level II methods is that the former are fully probabilistic whereas the latter are semi-probabilistic, since part of the probabilistic description of the random variables has been discarded. In general, the type of distribution, the mean and the standard deviation are the information that is retained in Level II methods. In Level I (deterministic) methods, only the mean value is retained.

If structural optimisation is carried out, code requirements fulfil the role of active constraints, regardless of whether the code is deterministic or reliability-based. Reliability-based codes do not lead to an optimum design with respect to any particular criterion. Their role is to ensure a minimum safety level, expressed in terms of a target probability of failure or a target reliability index. Therefore, optimisation procedures are conducted as usual in the case of a reliability-based code. In the case of a probabilistic design, the target reliability or maximum probability of failure is an active constraint in the optimisation procedure.

References

1. Ang A. H.-S. (1970) Extended Reliability Basis of Structural Design Under Uncertainties. *Proc. SAE/ALAA/ ASME, 9th Rel. & Maintain. Conf.*
2. Bitner-Gregersen E. M. Haver S. (1991) Joint Environmental Model for Reliability Calculations. *Proc. 1st Int. Offsh. & Polar Eng. Conf.* (*ISOPE'91*). Edinburgh 11–15 August, 246–254.
3. Brünner E. et al. (2012) Design Principles and Criteria. Committee IV.1 Rept. In Fricke W. Bronsart R. (Eds) *Proc. 18th Int. Ship & Offsh. Struct. Cong.* Vol. 1 Schiffahrts-Verlag Hansa GmbH & Co KG Hamburg 435–506.
4. Card J. Haugland B. J. Pomeroy V. (2004) Developing the next generation of Classification Society Rules for Oil Tankers. *Design and Operation of Double Hull Tankers RINA Conf.* London 1–10.
5. Coles S. (2001) *An Introduction to Statistical Modelling of Extreme Values.* Berlin: Springer-Verlag.
6. Daidola J. C. Basar N. C. (1980) Probabilistic Structural Analysis of Ship Hull Longitudinal Strength. Ship Struct. Comm. Rept. SSC-301 Washington, DC.
7. Faulkner D. (1976) Some Thoughts on the Nature of Hazards affecting Structures in Deeper Water *Offshore Int. Conference* Birmingham.

8. Ferry Borges J. Castanheta M. (1972) *Structural Safety*. 2nd ed. Lisbon: Lab. Nac. Eng. Civil.
9. Frieze P. A. (1984) *Offshore Structural Design* Summer Course Dept. Naval Arch. Ocean Eng. Univ. Glasgow, Scotland.
10. Galanis G. Chu P. C. Kallos G. Kuo Y.-H. Dodson C. T. J. (2012) Wave height characteristics in the north Atlantic Ocean: a new approach based on statistical and geometrical techniques. *Stoch. Environ. Res. Risk Assess.* Vol. 26 83–103.
11. Goodman R. A. Mowatt G. A. (1977) Application of Strength Research to Ship Design. In Dowling P. J. Harding J. E. Frieze P. A. (Eds) *Steel Plated Structures*. London: Crosby Lockwood Staples 676–712.
12. 'Global Wave Statistics Online' *Globalwavestatisticsonline.com*. Available: www.globalwavestatisticsonline.com/Help/wave_scatter.htm (accessed 19 December 2020).
13. Guedes Soares C. (1984) Probabilistic Models for Load Effects in Ship Structures. Dept. Mar. Tech. Norw. Inst. Tech. Trondheim. Rept. No. UR-84-38.
14. Guedes Soares C. (1992) Combination of primary load effects in ship structures. *Prob. Eng. Mech.* Vol. 7 103–111.
15. Guedes Soares C. Moan T. (1988) Statistical analysis of still-water load effects in ship structures. *Trans SNAME* Vol. 96 129–156.
16. Guedes Soares C. Dogliani M. Ostergaard C. Parmentier G. Pedersen P. T. (1996) Reliability Based Ship Structural Design. *SNAME Annual Meeting*.
17. Hess P. E. Ayyub B. M. (1997) Variability in the Geometry and Imperfections of Surface Ship Structural Scantlings. Draft Paper, Carderock Division Naval Surface Warfare Center, Bethesda, Maryland.
18. Hess P. E. Bruchman D. Assakkaf I. A. Ayyubb B.M. (2002) Uncertainties in material strength geometric and load variables. *Naval Engineers J.* Vol. 114 139–166.
19. Hoffman D. Lewis E. V. Analysis and Interpretation of Full-Scale Data on Midship Bending Stresses of Dry Cargo Ships. Ship. Struct. Comm. Rept SSC-196 Washington, DC.
20. Hørte T. Wang G. White N. (2007) Calibration of the Hull Girder Ultimate Capacity Criterion for Double Hull Tankers. *Proc. 10th Int. Symp. Prac. Des. Ships & Float. Struct. (PRADS 2007)* Vol. 1 Houston, TX 553–565.
21. International Association of Classification Societies (2006) Common Structural Rules for Double Hull Oil Tankers Background Document Sect. 9/1 – Design Verification. Hull Girder Ultimate Strength.
22. International Association of Classification Societies (2002) Formal Safety Assessment. Overview and International Association of Classification Societies experience. Presentation to IMO MSC 75.
23. International Association of Classification Societies (2020) *Common Structural Rules for Bulk Carriers and Oil Tankers*.
24. International Association of Classification Societies (2012) Re-examination of the Environmental Data for use in Harmonised CSR. Rept. No Pt 1 Ch 4 Sec 6.
25. International Association of Classification Societies (2001) Standard Wave Data. Recommendation No. 34.
26. International Maritime Organisation (2013) Revised Guidelines for Formal Safety Assessment (FSA) for Use in the IMO Rule-making Process. MSC-MEPC.2/Circ. 12.
27. International Maritime Organisation (2007) Formal Safety Assessment MSC 83/INF.2.
28. International Maritime Organisation (2006) Goal-based New Ship Construction Standards. Linkage between FSA and GBS. MSC 81/INF.6.

29. Ivanov L. Wang G. Foster J. (2007) Probabilistic analysis of hull girder stresses using an analytical approach verified by the Monte Carlo simulation. Proc. 26th Intl Conf. Offshore Mech. & Arctic Eng. OMAE2007 San Diego, CA 1–11.
30. Kaplan P. (1984) Analysis and Assessment of Major Uncertainties Associated with Ship Hull Ultimate Failure. Ship Struct. Comm. Rept. SSC-332 Washington, DC.
31. Kaplan P. Benatar M. Bentson J. Achtarides T. A. (1984) Analysis and Assessment of Major Uncertainties Associated with Ship Hull Ultimate Failure. Ship Struct. Comm. Rept. SSC-322 Washington, DC.
32. Mansour A. E. (1989) An Introduction to Structural Reliability Theory. Ship Struct. Comm. Rept. SSC-351 Washington, DC.
33. Mansour A. Lin M. Høvem L. Thayamballi A. (1993) Probability-based Ship Design (Phase 1): A Demonstration. Ship Struct. Comm. Rept. SSC-368 Washington, DC.
34. Mansour A. Wirshing P. Luckett M. Plumpton A. et al. (1997) Assessment of Reliability of Existing Ship Structures. Ship Struct. Comm. Rept. SSC-398 Washington, DC.
35. Mathisen J. Birknes J. (2003) Statistics of Short Term Response to Waves First and Second Order Modules for Use with PROBAN. *DNV Report No 2003-0051* Rev. No. 02.
36. Nikolaidis E. Kaplan P. (1991) Uncertainties in Stress Analysis on Marine Structures. Ship Struct. Comm. Rept. SSC-363 Washington, DC.
37. Ochi M. K. (1978) Wave statistics for the design of ships and ocean structures. *Trans SNAME*, Vol. 86 47–69.
38. Paik J. K. Frieze P. A. (2001) Ship structural safety and reliability. *Prog Struct. Eng. Mater.* Vol. 3 198–201.
39. Parunov J. Guedes Soares C. (2008) Effects of Common Structural Rules on hull-girder reliability of an Aframax oil tanker. *Rel. Eng. & Sys. Saf.* Vol. 93 1317–1327.
40. Pires Guia J. (2014) Risk-based structural design of double hull tankers. *M.Sc. Thesis* Tech. Univ. Lisbon.
41. Sørensen J. D. (2004) Notes in Structural Reliability Theory and Risk Analysis. Inst. Build. Tech. & Struct. Eng. Aalborg Univ. Denmark.
42. Thoft-Christensen P. Baker M. J. (1982) *Structural Reliability Theory and Its Applications*. Berlin: Springer-Verlag.
43. Transportation Research Board (2003) Design of highway bridges for extreme events NCHPR Rept. 489 Washington, DC.
44. Tvedt L. (2006) PROBAN – probabilistic analysis. *Struct. Saf.* Vol. 28 150–163.
45. Wang X. Moan T. (1996) Stochastic and deterministic combinations of still-water and wave bending moments in ships. *Mar. Struct.* Vol. 9 787–810.
46. Creative commons – attribution-NonCommercial-NoDerivs 3.0 unported – CC BY-NC-ND 3.0. (n.d.). Creativecommons.org. Available: http://creativecommons.org/licenses/by-nc-nd/3.0/ (accessed 25 February 2021).

13

Hull Girder Strength Assessment Using the Finite Element Method

> It has often been said that a person doesn't really understand something until he teaches it to someone else. Actually a person doesn't really understand something until he can teach it to a computer, i.e. express it as an algorithm... The attempt to formalize things as algorithms leads to a much deeper understanding than if we simply try to understand things in the traditional way.
>
> *Donald Knuth*

13.1 The Introduction of the Finite Element Method to the Marine Field

The introduction and establishment of a new technique rarely take place for its own sake. Rather, it is the successful application to specific problems that cannot be solved using existing methods that will bring it to prominence. Such is the case of the finite element (FE) method and its use in ship structural analysis and design. In the marine field the earliest application of numerical software was used in the pressure hull analysis of submarines (BOSOR IV), as a natural extension of curved grillage analysis that had hitherto been used. During the early 1970s, large tankers were being built and designed in accordance with classification society rules that were based on the extrapolation of rules applicable to smaller ships. When it became apparent that this led to inadequate designs, the FE method was used to investigate this issue. An early application of the FE method in the marine field was in the strength and stress analysis of passenger ship structures, which have a complex geometry that does not allow the resulting stress fields to be predicted using linear mathematical theory. It was acknowledged that complex geometries and nonlinear structural response cannot be modelled using analytical techniques without introducing a significant degree of mathematical complexity, which cannot be justified in routine design work.

At the same time, the offshore industry expanded as a result of the need to extract oil in more hostile environments and deeper waters, which led to the development of novel platform designs that necessitated the use of powerful analysis and design tools.

To address both classes of problems, general-purpose FE software applications were used, such as NASTRAN, ANSYS and ASAS.

With the development of the FE method, the growth in computational power and the parallel development of specialised software (e.g. ABAQUS, ASAS/NL, MAESTRO, interactive graphic pre- and post-processors), the field has matured to the point that it can be applied routinely to the analysis and design of ship and submarine structures. Indeed, classification societies require its use when designing particular ship types such as bulk carriers, oil tankers, containerships and multi-deck ships. In the case of bulk carriers and oil tankers the types of analysis required are described in the IACS Harmonised Common Structural Rules, whereas in the case of multi-deck ships in specific rules for each type of vessel.

More recently, FE codes have been used in conjunction with computational fluid mechanics and multiphysics software[1] as part of a virtual process development (VPD) environment [25, 32]. This involves the simulation of a new product such as a ship within an environment that brings together structural response and fluid mechanics, using sophisticated tools for each of these.

13.2 Analysis of the Hull Girder

We can differentiate between the different types of analysis according to criteria used in analytical methods, that is, as:

1. Static or dynamic
2. Linear or nonlinear
3. Deterministic or probabilistic

In the case of (1) and (2), it is structural response that is being characterised and not the nature of the loading. In (3) it is both the loading and the response that can be modelled using deterministic or probabilistic criteria.

As in the case of the analytical methods described in earlier chapters, FE models are used to calculate hull girder stiffness and strength. A stress-oriented FE model of a hull girder provides a wealth of information on stress distributions, deflections, etc. These can be exploited in order to assess global response (e.g. longitudinal bending strength) as well as local response. In detailed local structural analysis a finer mesh than that used in global strength assessment is used. The latter uses input from the former in the form of boundary conditions that are used in the local strength model. A dynamic analysis of hull girder response is required when assessing vibration response and, more rarely, springing response, in which case the output consists of natural frequencies and mode shapes.

The question as to whether the analysis needs to be in the linear or the nonlinear range may depend on either the nature of the applied loading or the required output. Thus, a quasi-static analysis is generally performed in the linear range provided (a) sea states are moderate, (b) the hydrodynamic loading can be obtained by applying linear strip theory,

[1] *Multiphysics* treats *simulations* that involve multiple physical models or multiple simultaneous physical phenomena. In the case of ships it concerns the coupling of structural mechanics, computer-aided design (CAD) and fluid mechanics codes. It may additionally involve computer-aided manufacturing (CAM). The nonlinearly solved rigid-body ship motions and a simple FE Timoshenko beam model may be coupled with the Reynolds-averaged Navier–Stokes (RANS) equations [31].

and (c) the wave force acting on each section is proportional to the difference between the wave height and the still-water plane. A nonlinear analysis is required in the case of extreme sea states, in which case the preconditions for linearity are no longer valid. Additionally, residual strength and ultimate strength assessments are performed in the nonlinear (elasto-plastic) range.

A deterministic analysis is performed in the case of conventional hull forms that are subjected to normal loads whereas a probabilistic analysis is required in the case of novel hull forms or else if a vessel is exposed to unusual types of wave action.

13.2.1 Quasi-Static Analysis

The quasi-static analysis of hull girder strength involves the determination of the longitudinal distributions of vertical bending moment, shear force and torque, following which hull girder stiffness, deflections and primary stresses can be found.

A difficulty that arises in all ship-type structures is that both the hull transverse sections and the weight distribution vary in the longitudinal direction. As a result, the longitudinal net load distribution and hence the external loading distributions (bending moment, shear force and torque) cannot be calculated explicitly for each position along the length of the hull girder using a single calculation as is done in the case of a uniform beam. The weight distribution varies also in time, which is a further complication, since each voyage involves different loading arrangements. To allow for these factors, approximate methods are used in analytical approaches; these have been described in Sections 4.3 and 4.4. To account for the resulting discrepancies from exact solutions, empirical factors are introduced.

Longitudinal Strength Using Linear Elastic Analysis

The first departure from approximate analytical approaches involves the use of *matrix theory* (beam elements), which, provided certain conditions are fulfilled,[2] leads to accurate first-order approximations of global response. When a structure is modelled as a series of beam elements, its dimensions are reduced from three to one (+one) (Figure 13.1).

This enables variations in weight in the longitudinal direction to be allowed for. The vertical buoyancy forces are applied at the nodes and a reasonably accurate approximation to the load curve is obtained, depending on the number of elements used. This may be subsequently integrated numerically and the shear force and bending moment distributions obtained.

A second, more accurate approximation is to develop a 3D model of the entire hull girder or a sufficiently large part of it. In the former case this involves an accurate representation of the entire structure. Weights such as those of machinery equipment and outfitting may be added, so that the weight distribution is modelled much more accurately. In this manner longitudinal strength can be determined with sufficient accuracy under quasi-static loading conditions.

[2] This is the case if only bending response is involved. If other, more complex modes such as coupled bending and torsion arise, as in the case of vessels with large deck openings, it is necessary to resort to a more advanced method.

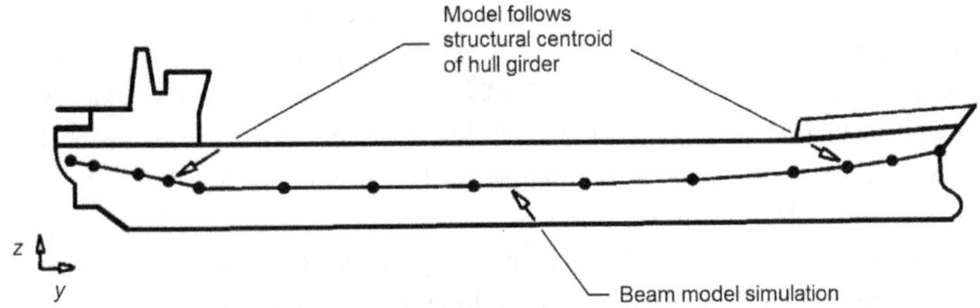

Figure 13.1 Simplified analysis of hull girder for longitudinal strength study [11].

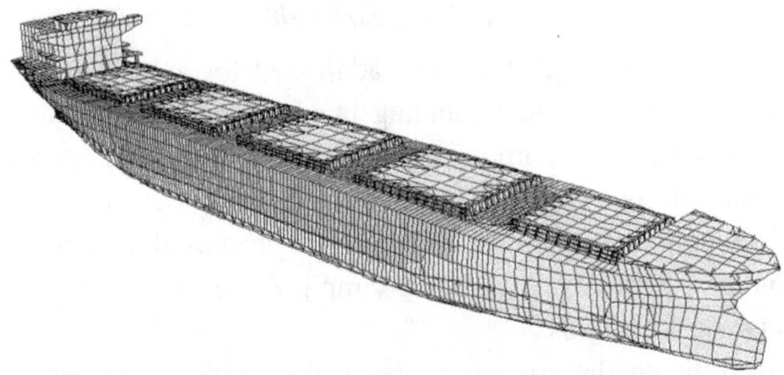

Figure 13.2 3D FE global model of a Handysize bulk carrier.[3]

A global structural model is used to determine:

1. The overall stiffness of the hull girder (vertical deflections in the case of vertical bending and rotations about the longitudinal axis when torsional loading is applied)
2. The stresses that arise in primary strength members (primary stresses). These are needed to check that strength requirements are met. They are also used when checking the yield and buckling strength of primary support members such as girders and frames.
3. In addition, the nature of the required boundary conditions for use in any subsequent local strength analysis can be extracted. These consist of the relevant deflections and rotations at the boundary nodes of the local strength model.

In practice it is not always necessary to model the entire hull girder. This is because the required information can be obtained by focussing on a portion of the hull girder amidships, extending to three holds (cargo tanks) in the longitudinal direction. To this end the IACS Common Structural Rules provide detailed guidance on the modelling and analysis of this region of the hull girder, although classification societies stipulate their own requirements.

A global strength analysis may be required in the case of special-type ships such as those used for heavy lift operations. It is generally not required in the case of bulk carriers of conventional form. The same is true in the case of general cargo ships.

Direct strength analysis is also required in the case of passenger ships. This is because the superstructure extends over most of the ship's length and the internal structure

[3] This model was developed using Bureau Veritas software (MAESTRO).

comprises a large number of decks that are supported by pillars and longitudinal and transverse bulkheads. The structure has many discontinuities such as knuckle points, recesses, non-continuous bulkheads and decks as well as geometric irregularities such as openings in decks and bulkheads, side shell doors and portholes. To evaluate the global stiffness of the structure the load paths have to be established and the global and local strength assessed [13].

In the case of containerships it may be necessary to perform a dynamic analysis in order to assess the possibility of springing. This is done by comparing wave encounter frequencies with those of the hull girder and of critical structural members.

Residual Strength

The question of residual strength has been addressed for a long time, albeit indirectly through the study of collision and grounding incidents. In these the emphasis has been placed on damage assessment and improved structural design to limit the effects of external impacts. Finite element codes have been used extensively in such studies. In the case of explicit residual strength assessments, FE codes have been used in various forms including conventional nonlinear FE analysis as well as simplified codes such as ISUM and ISFEM,[4] in addition to analytical methods.[5]

A characteristic difficulty that arises in the residual strength assessment of a transverse hull section is the need to consider damages that are not symmetrical with respect to the centreline of the hull girder. This introduces asymmetrical bending of the section. If a hull section is not symmetrical, regardless of whether there is damage present or not, rotation as well as translation of the neutral axis will arise during bending. Furthermore, the nature of the asymmetry will change as the loading is increased and additional members of the cross-section fail. A FE-based approach that has been developed in the case of symmetrical hull damage is described in [35] whereas in the more general case alternative approaches have been followed, such as [9], [33] and others. The application of a simplified FE method is described in [37].

Ultimate Strength of the Hull Girder

In addition to conventional quasi-static analysis, FE codes can be used to determine the ultimate strength of the hull girder. This is a true measure of strength, in contrast to the working strength. The ultimate strength calculation of the hull girder necessitates the use of nonlinear analysis methods, since for the majority of ship geometries, interframe elasto-plastic buckling failure arises in the individual stiffened plate elements that comprise the cross-section. The full-range elasto-plastic curve includes the linear elastic range, inelastic plate buckling, the post-buckling response of the plate, stiffened plate collapse as well as the post-collapse range for the stiffened plate. In the post-collapse range unloading occurs, and thus knowledge of the collapse strength of the stiffened plate is not sufficient.[6] This is because when the hull girder global bending strain is increased, the strain that is applied to

[4] ISUM: Idealised Structural Unit Method; ISFEM: Intelligent Supersize Finite Element Method (such as ALPS/HULL).
[5] The use of simplified analytical methods is discussed in Section 6.8.
[6] The collapse strength of a stiffened plate can be calculated using the *effective width* approach.

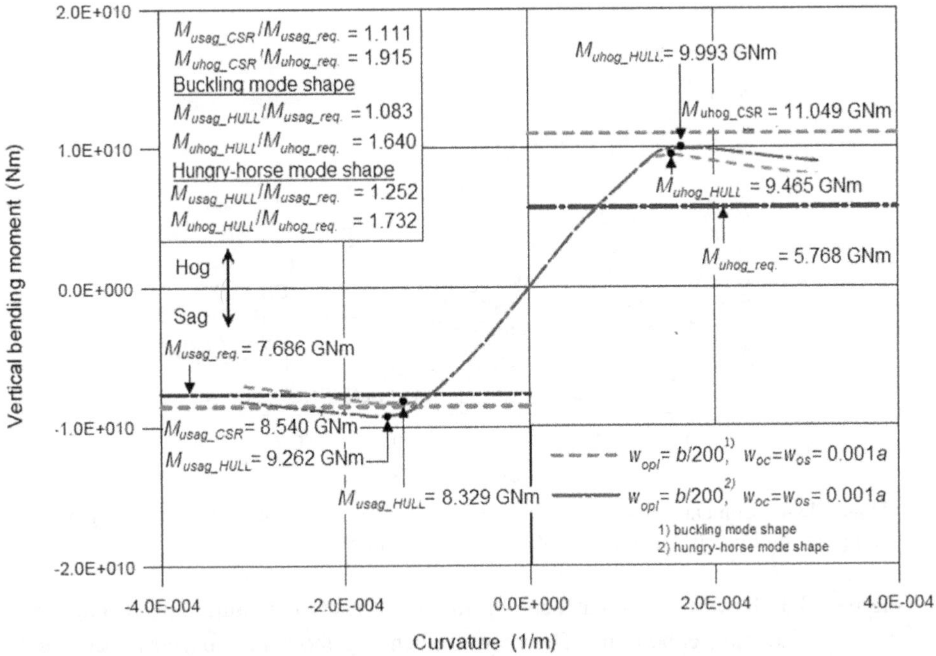

Figure 13.3 Comparison of hull girder bending moment curves between IACS-CSR and ALPS/HULL for a double hull oil tanker section [21].

each stiffened plate element will differ, since it depends on the distance of the element in question from the neutral axis of the entire cross-section.

However, nonlinear FE analysis is time-consuming due to the large number of elements required to represent the hull structure, requires a high level of expertise and necessitates making approximations with regard to geometrical and material imperfections, since little information is available regarding these [18]. In order to circumvent these difficulties, simplified FE codes have been developed, an early example of which was ISUM (the Idealized Structural Unit Method) [34]. In this approach the structure is modelled using a number of specially devised types of mid-scale elements, thereby reducing the extent of discretisation needed as well as computing time. In other software (MAESTRO) purpose-designed finite elements for use in ship structural design have been introduced.

This was developed further at a later stage, and a series of codes have evolved. On the other hand, IACS does not require nonlinear analysis to be used in routine design work and instead advocates the use of a simpler and easier approach, which has a semi-analytic basis. This is the incremental-iterative method, described in Appendix D. It is of interest to compare the results obtained using a FE procedure with that recommended by IACS.[7]

In Figure 13.3 results obtained using ALPS/HULL, a program based on ISUM, are compared with those of the IACS iterative-incremental method. ALPS/HULL is a special program for the progressive hull collapse analysis of ships using ISUM [20].

[7] Results from a second comparison are included in Appendix D.

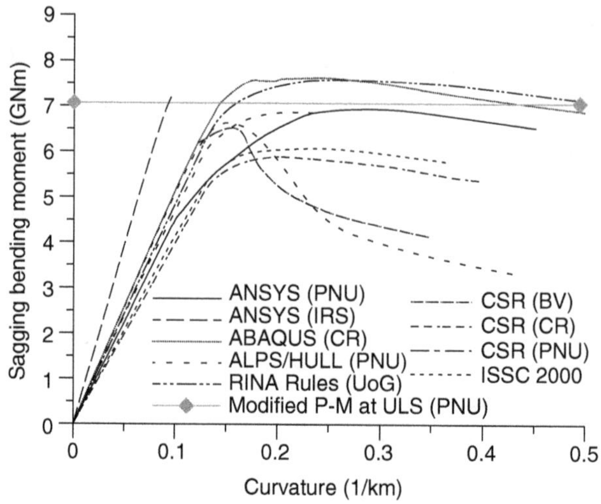

Figure 13.4 Comparison of progressive collapse analyses of containership hull girder using several FE codes and the modified Paik–Mansour formula [29].

Figure 13.3 shows the vertical bending moment versus curvature curves under both sagging and hogging conditions. The required bending moment capacities specified by IACS CSR as well as the IACS CSR method predictions of the ultimate bending moments are also shown in the figure. Results for two different initial imperfection shapes for the plating are included in the diagram. These are the 'hungry horse' shape, which is the form resulting from the action of residual welding stresses, and the 'buckling mode' shape, which corresponds to the lowest buckling stress of the plating. The former of the two is closer to what is observed in practice.

As expected, the ultimate hull girder bending moments are lower when the plate buckling mode is chosen. It is observed that IACS CSR predictions of ultimate sagging bending capacity correspond well with more refined ALPS/HULL progressive collapse analysis results. Both approaches show that the actual strength of this particular vessel is higher than the CSR required value. The difference is larger in the case of hogging than in the case of sagging. Comparing the results of the two approaches shows that in the case of hogging the IACS approach estimate is approximately 10 per cent higher than that of the ALPS/HULL program. In the case of sagging, the IACS approach underestimates the ultimate bending moment by 8 per cent. The general conclusion is that the CSR approach is satisfactory given that the effort and expertise required to generate results using it are substantially less than in the case of nonlinear FE analysis.

Figure 13.4 includes data on the ultimate bending moment of a containership in vertical bending using the modified Paik–Mansour formula[8] and several finite element codes. It is

[8] The modified Paik–Mansour formula is used to determine the ultimate bending moment of a hull girder subjected to vertical bending. It is an improvement on the original formula in that it allows for the spreading of tensile yield in vertical members. In contrast to the original formula, it requires an iterative approach because of the need to calculate two variables whereas equilibrium is defined by one equation alone [29].

seen that a satisfactory assessment of ultimate strength is obtained using the formula, the use of which requires much less effort than nonlinear finite element analysis. Furthermore, it is noted that there is limited spread in the prediction of ultimate strength using the finite element codes although the spread in post-collapse strength response is much more significant.

13.2.2 Vibration Analysis

In Chapter 8, which deals with hull girder vibration, empirical formulas and their underlying theory were discussed. Empirical formulas are of use in preliminary design in order to check for resonance with the various sources of steady-state excitation. If the lower vertical and horizontal hull vibration modes alone are of interest, a beam model can give satisfactory results. Classification societies recommend the use of empirical formulas and in the past suggested the use of non-uniform beam models. The use of beam theory in general terms is satisfactory provided it is known in advance that the hull girder will follow this type of response. However, in the case of ships with large deck openings a more detailed model is required, since coupling arises between horizontal bending and torsion. When calculating the frequencies of higher modes and in the later stages of design of all types of ships it is necessary to use the FEM as well as when assessing the vibration response at a regional or a local level.

In vibration analysis the aim is to determine the natural frequencies, and so it is not necessary to use as fine a model as is used in stress analysis. This is because in dynamic analysis the mesh has to model the vibration mode shapes with sufficient accuracy and not stress distributions that require fine mesh representation. It is, however, necessary to include all significant structural sections, since deflections, velocities and accelerations need to be predicted accurately. The mesh size has to be able to reproduce the actual vibration modes, and thus for the overall structure, a mesh size of three to four stiffener bays is acceptable (Figure 13.5). In critical areas such as the stern region and the superstructures a finer mesh is required. Suitable elements are bending plates or shells and beam elements, which enable the bending response to be reproduced accurately. Primary strength members as well as web frames are modelled using plane stress elements. The model is

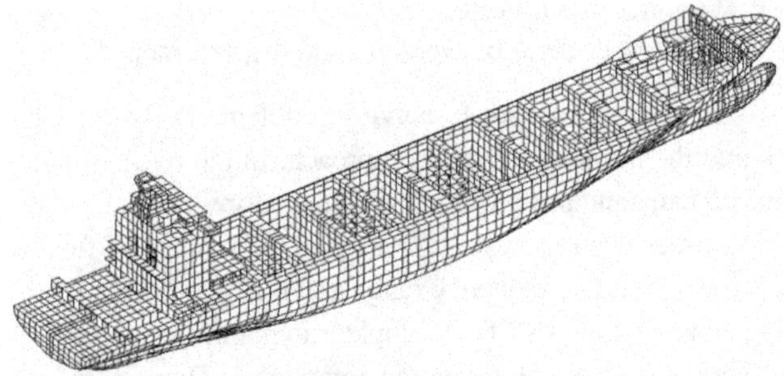

Figure 13.5 A global hull girder model suitable for vibration analysis [22].

oriented relative to deck planes and to main longitudinal and transverse structures. The number of degrees of freedom of a model of the hull girder for purposes of vibration calculations lies in the range of 20,000–40,000, corresponding to 50–150 natural vibration modes in the range of 0–20 Hz [30].

Two-dimensional FE models using membrane elements have been used to analyse deckhouse and aft body vibration and the vertical hull vibration of tankers. The results have been satisfactory in the case of superstructures but not in the case of deckhouses, where structural discontinuities arise. Furthermore, coupling effects with vibrations of the side shell or double bottom are not allowed for. Anti-symmetric vibrations, which are of particular interest in the case of containerships and are excited by the main engine or the propeller, cannot be analysed and a three-dimensional model is needed.

Three-dimensional FE models used to calculate hull girder natural frequencies need to include superstructures in all cases, as these have an important effect. When 3D models were used to calculate deckhouse vibration but without proper representation of the supporting structure, the results were found to be inaccurate. If, however, a 3D representation of the stern structure including the deckhouse is coupled with a beam model of the forward part of the hull, sufficiently accurate results can be obtained.

The Equations of Dynamic Equilibrium

The solution of the equations of dynamic equilibrium enables the extraction of mode shapes and frequencies of the hull girder. The hull girder is modelled using a large number of elements, and thus the solution of a multi-degree-of-freedom problem is sought. The generalised equation of motion for hull girder vibration has the following form:

$$\mathbf{M}\ddot{\mathbf{U}}(t) + \mathbf{C}\mathbf{U}(t) + \mathbf{K}\mathbf{U}(t) = \mathbf{F}(t) \tag{13.1}$$

where

$\mathbf{M}$ = mass matrix
$\mathbf{C}$ = damping matrix
$\mathbf{K}$ = stiffness matrix
$\mathbf{F}$ = applied force vector
$\ddot{\mathbf{U}}, \dot{\mathbf{U}}, \mathbf{U}$ = acceleration, velocity and displacement vectors

There exist two distinct approaches to solving problems in dynamic analysis: the *time-domain* approach and the *frequency-domain* approach. In the former the forcing function is defined in the time domain, and this approach is suitable for vibration resulting from transient loading of the hull girder such as wave impact and slamming. In frequency analysis the forcing function is defined in the frequency range and is discretised into a series of harmonic components using Fourier analysis. The complete dynamic response is then obtained by superposing the response due to each harmonic component. This type of analysis is suited to loads that are of a continuous sinusoidal (steady state) nature, such as the cyclic loads

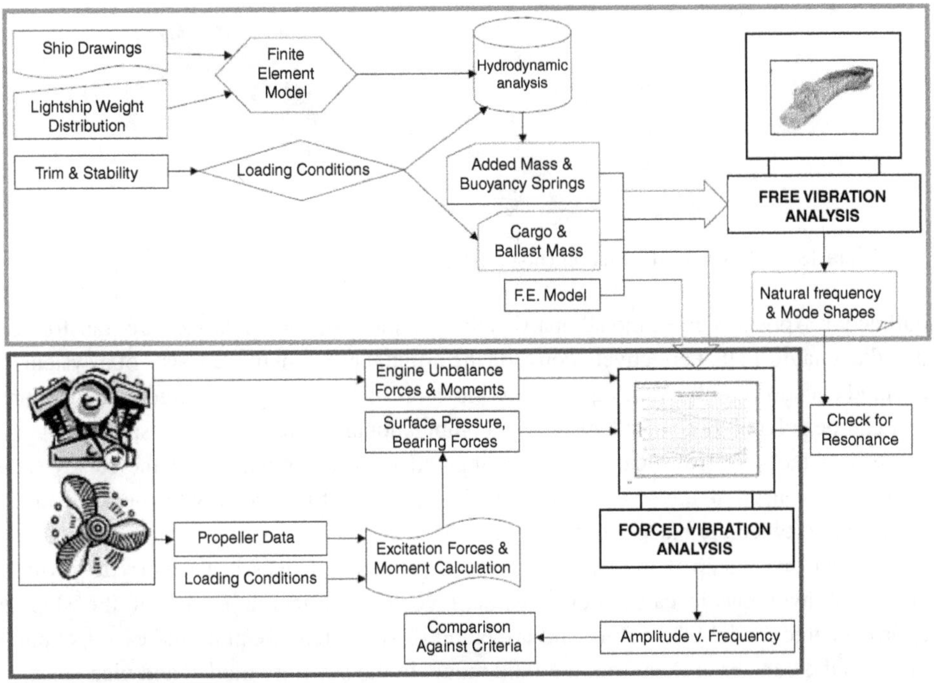

Figure 13.6 Vibration analysis procedure according to ABS.

resulting from the propeller and the various types of machinery.[9] From a computational point of view, frequency response analysis is much less demanding than time domain analysis.

The vibration analysis procedure as described by the American Bureau of Shipping is shown in Figure 13.6 [2]. It consists of two sets of calculations: the free vibration analysis and the forced vibration analysis. The finite element model developed for the free vibration analysis is also used in the forced vibration analysis, which is used to assess structural response to excitation by the main engine and/or the propeller. The resulting forced response frequencies are then compared with the corresponding natural frequencies obtained from the free vibration analysis to see whether resonance occurs.

Free Vibration Response

The free vibration response is determined in order to estimate the natural frequencies and modes of vibration of the hull girder. In this case, $\mathbf{C} = \mathbf{F}(\mathrm{t}) = \mathbf{0}$. The equation of motion, Equation (13.1), then becomes

$$\mathbf{M}\ddot{\mathbf{U}}(t) + \mathbf{K}\mathbf{U}(t) = \mathbf{0} \tag{13.2}$$

Since the equations of motion include a very large number of degrees of freedom, it is desirable to reduce the number of operations that need to be performed when solving the above system. This is achieved by implementing the *modal analysis* method, which is

[9] Equation (4.9).

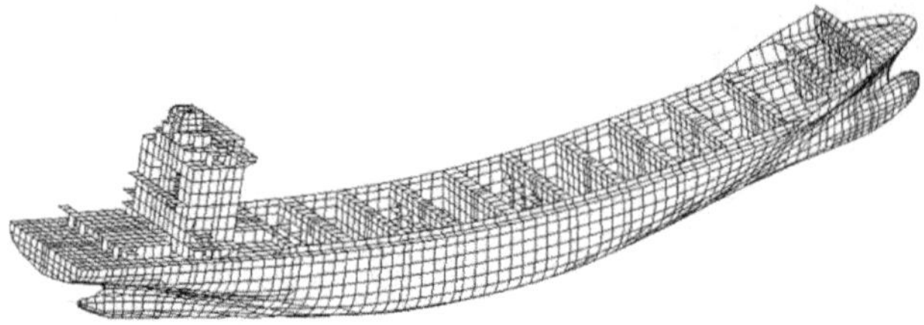

Figure 13.7 Two-noded vibration (first mode) [22].

outlined in Appendix E. In modal analysis the mass and stiffness matrices are transformed into diagonal form that effectively converts the original problem that consists of a system of coupled equations of dynamic equilibrium into a set of single-degree-of-freedom equations that can be solved with considerably less computational effort. The free response characteristics of the hull girder can then be compared with the frequency of the steady-state excitation to ensure that resonance is avoided. In practice it is necessary to check the mode shapes up to twice the blade-rate frequency.

The fact that a ship is immersed in water (wet mode) makes it necessary to perform numerical hydroelastic calculations. These involve a structural model of the ship, a hydrodynamic model of the fluid and a method of accurately coupling the two. Coupling involves the transfer of data from the fluid model to the structure model, and vice versa, at each time step. To model fluid behaviour the 3D Rankine panel method is used, whereas coupling is most efficiently carried out using the modal analysis method.

The structural model is a 3D FE model or, in the case of monohulls, a beam model. As discussed in Chapter 8, a Timoshenko beam approach is required, although it is also possible to introduce the effect of shear on torsion as well as the presence of closed compartments such as the engine room and fore and aft ends.

It has to be mentioned also that in the case of cargo ships it is necessary to allow for the different loading conditions, since these have an important effect on natural frequencies. The following items need to be included in such an analysis [7]:

1. Ship structure
2. Outfitting and equipment
3. Tank filling
4. Cargo[10]
5. Hydrodynamic masses (Chapter 8)

The two-node vertical vibration mode is shown in Figure 13.7 [22].

Forced Vibration Response

Forced vibration is the result of either steady-state or transient excitation forces, and thus all terms of Equation (13.1) need to be included in the analysis. The steady-state excitation forces are the result of propeller and main engine operation and are thus harmonic in nature.

[10] See Chapter 8 regarding cargo effectiveness, and in particular the case of vehicle cargoes.

Equation (13.1) is solved using the modal analysis method in which the loading is of the following form:

$$\mathbf{F}(t) = Fe^{i\omega t} \tag{13.3}$$

leading to an output of the form:

$$\mathbf{u}(t) = ue^{i\omega t} \tag{13.4}$$

Substituting in Equation (13.1), the equation of dynamic equilibrium becomes

$$[\mathbf{K} + i\omega\mathbf{C} - \omega^2\mathbf{M}]\mathbf{U} = \mathbf{F} \tag{13.5}$$

It is seen that damping, **C**, is present in Equation (13.5). In the case of ship structures, damping is a collective term that refers to the following:

1. Structural damping
2. Cargo damping
3. Hydrodynamic damping

The first relates to internal structural damping, which is the most important for vibration purposes, since the material is continually under stress. The mechanism that arises is energy absorption by *material hysteresis*. Factors that affect this are the quality of workmanship and the presence of residual stresses. Better-quality workmanship results in lower structural damping [22]. However, this is not necessarily desirable, as the number of cycles increases, thereby reducing fatigue life. Cargo damping also contributes to internal damping and will depend heavily on the nature of the cargo. Hydrodynamic damping is important for ship motions but is usually neglected in the case of vibration response as having a minor contribution.

Damping is expressed as a percentage of *critical damping*, which is defined as the value of damping such that the response of a given mode to an impulse returns to equilibrium without oscillating. It has been found that damping grows with increasing frequency, and both Lloyds Register and Germanischer Lloyd acknowledge this in their rules. Satisfactory results were obtained in the calculation of the damping coefficient as a function of frequency using the data included in Figure 13.8 [7].

The inclusion of damping presents difficulties with regard to its quantification as well as from a computational perspective. At present there is no method available that will provide a universally reliable estimate of damping, and it has been found that sister ships may exhibit markedly different levels. From a computational viewpoint it has to be mentioned that the damping matrix in general cannot be diagonalised because off-diagonal terms remain, indicating coupling. This problem is avoided by simply ignoring the coupling effects, as damping is in any case low; the American Bureau of Shipping gives a constant value of 1.5 per cent for the damping coefficient. Damping is of importance, however, in the case of transient loading because it affects the number of oscillations; as mentioned previously, these have a direct effect on the accumulated fatigue damage of the structure.

Figure 13.9 illustrates the response of the hull girder to forced excitation resulting from the operation of the main engine (Chapter 5) [7]. From this set of diagrams it is possible to assess the interaction of the excitation forces and the natural vibration modes of the hull

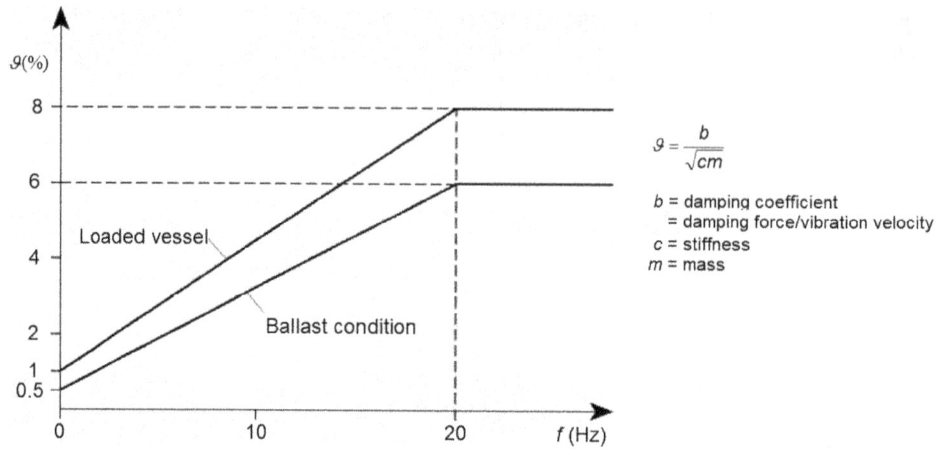

Figure 13.8 Modal damping in hull girder vibration [7].

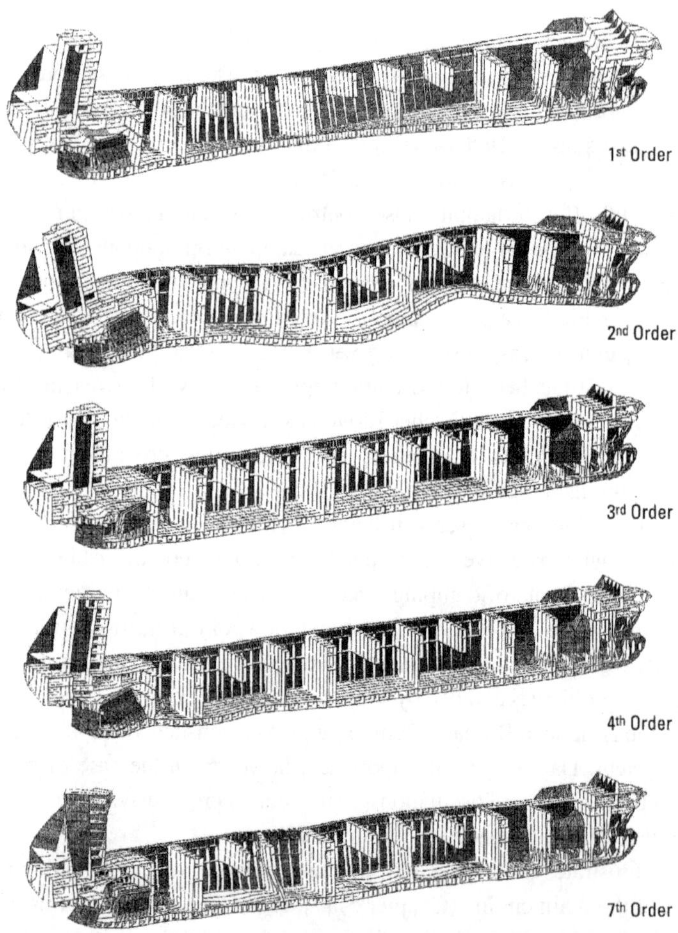

Figure 13.9 Vibration modes resulting from main engine excitation modes (first–seventh order) [7].

girder. This particular case concerns a seven-cylinder engine mounted on board a containership. It is seen that different orders excite different members of the ship structure, which may involve either the entire hull girder or substructures such as the deckhouse that is located directly above the engine room.

The first order (1.5 Hz) excites the fundamental torsional vibration of the hull girder. The vertical second order (3 Hz) results in four-noded vibration of the entire hull girder. The third order (4.5 Hz) is due to the X-type guiding moment of the main engine (Figure 5.36c) and is not transmitted to the hull girder, since the engine housing is stiff and absorbs the vibration. The fourth order (6 Hz) excites longitudinal vibration of the engine that is not, however, transmitted to the hull girder. The fifth and sixth orders have similar results to the third order and are not shown. The seventh order (10.5 Hz) causes H-type transverse vibration of the main engine (Figure 5.36c). The vibration frequency is close to the fundamental torsional vibration frequency of the deckhouse, and high-amplitude vibration is noted in the bridge deck.

13.3 Rule Requirements and Implementation of the FE Method

FE codes can be used to perform a wide range of studies when assessing ship structures. The IACS Common Structural Rules for bulk carriers and oil tankers (2018) refer to the following two categories:

1. Global strength assessment of the hull girder. This involves determining the linear elastic static response to the still-water (port) loading condition and to wave-induced loading (vessel at sea). The critical parameters are the induced primary stresses and deflections of the hull girder. The natural frequency of vibration of the hull girder can also be determined.
2. Local strength assessments that are performed using fine-mesh models for buckling strength purposes and very fine-mesh models for fatigue strength purposes.

The relationships between these types of studies are shown in Figure 13.10 [19]. We shall confine our discussion to large-scale types of analysis, which take place at the global and the cargo hold level. The hull structure can be modelled either in its entirety or else part of it, in which case a representative portion (cargo hold model) is considered.

A global model (Figure 13.2) provides the most accurate estimate of the longitudinal stiffness of the hull girder; in addition, estimates of the primary stresses are obtained. It also acts as a basis for (a) vibration studies (determination of natural frequencies), (b) strength estimates, following the onset of corrosion (use of a *digital twin*), and (c) the accurate assessment of boundary conditions when performing local strength analyses of structural members. These are performed to determine local buckling, yielding and fatigue strength.[11]

[11] A global structural analysis can also act as the basis for a *full stochastic analysis* that is carried out to perform fatigue limit state and ultimate limit state calculations of details with complex stress patterns and loading [13]. This is a local strength problem.

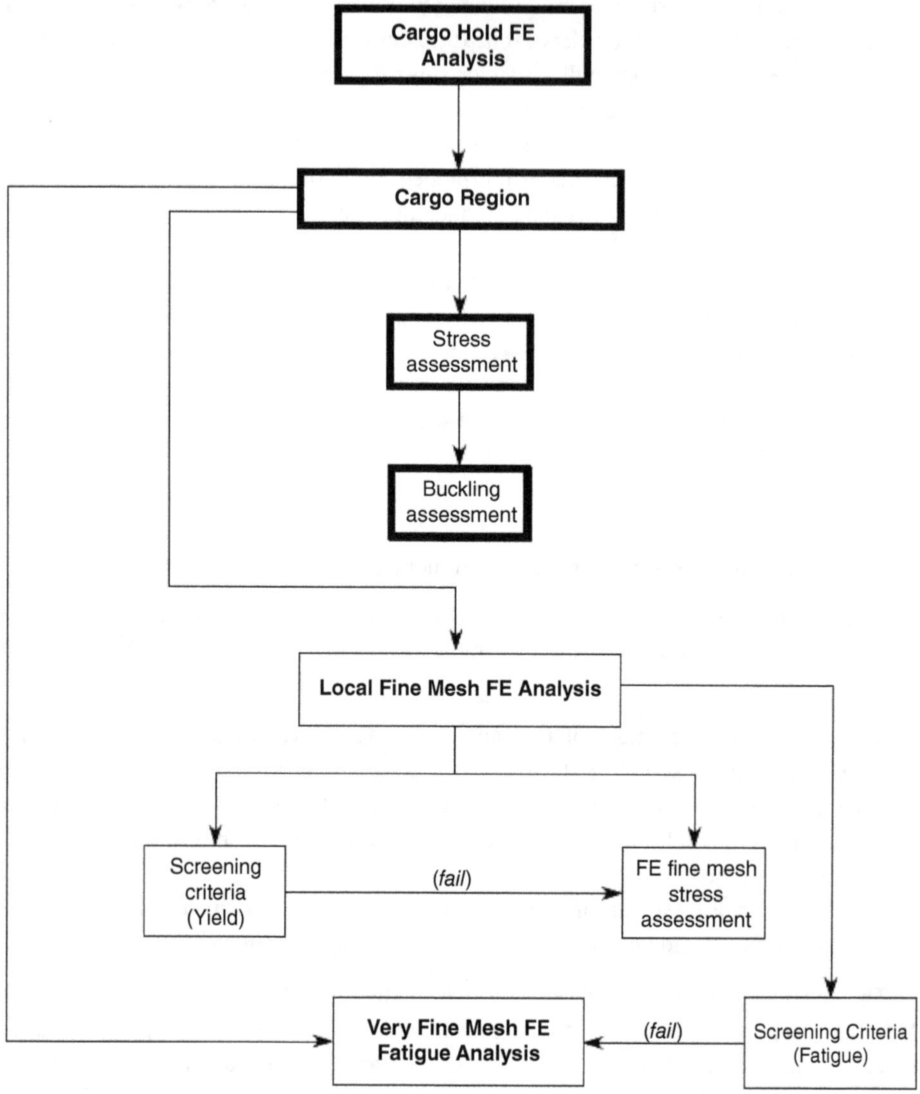

Figure 13.10 IACS analyses and their relationships [19]. Reprinted courtesy of IACS, London.

13.3.1 Support Conditions

A structural model needs to be supported so that rigid body translation does not take place and the correct type of loading is applied. The application of the external loads results in distortions and/or vibrations. An under-restrained model may exhibit higher deflections than the actual ones, whereas an over-restrained model may show high stresses at points where these do not arise. The points of support should be located away from areas of interest, as the applied loads may lead to imbalance in the model and will affect the local stress field. Points

	Location	Direction of support
Aft End	WL, CL (point A)	*X, Y, Z*
	CL, Upper deck (point B)	*Y*
Forward End	WL, CL (point C)	*Y, Z*

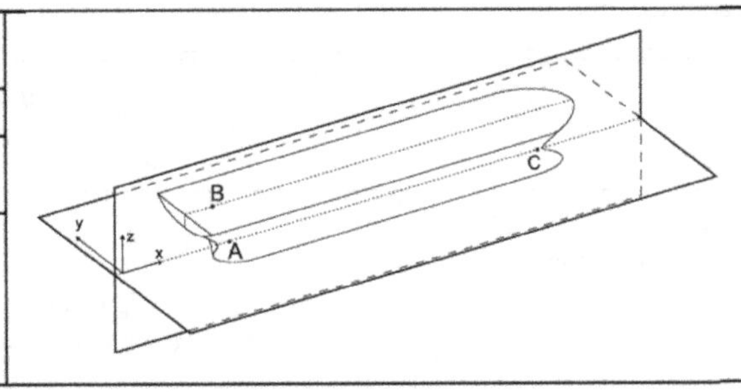

Figure 13.11 Example application of boundary conditions for global hull girder model [13].

of support of the model are generally located in way of the centreline, close to the fore and aft ends of the vessel (Figure 13.11). In general, three points are used, the first of which is restrained in three directions, the second in two and the third in one [13].

In the case shown in Figure 13.11, support points are located along the centreline close to the AP and the FP. Point A, which is located in way of the waterline and centreline at a transverse bulkhead in the aft region of the ship, is restrained from translating along all three coordinate axes. Point B, located in way of the uppermost continuous deck, is fixed in the transverse direction. Point C, which is located at the waterline level in way of the junction of the centreline with the forward collision bulkhead, is restrained in the vertical and the transverse directions.

Cargo Hold Model

Cargo hold models are used to assess the strength of primary strength members and bulkheads in the midships region. The loads acting are longitudinal bending in addition to normal loads that act on the outer shell plating as well as internal members. IACS specifies that 'the structural strength analysis of cargo holds is mandatory within the cargo hold region including the aft bulkhead of the aftmost cargo hold and the collision bulkhead'. Classification societies specify the extent of the structure to be analysed in the longitudinal direction. In Figure 13.12 an example of a three–cargo hold model is shown [19].

Since the fore and aft portions of the hull girder are ignored, it is necessary to apply appropriate conditions at the ends of the model that are equivalent to the actual loads that are imposed by these portions of the hull girder (Table 13.1, Figure 13.13).

The term *rigid link* refers to an element or a part of a structure that does not distort when subjected to loads of any sort. Rigid links are used along the boundary transverse sections of a cargo hold model in order to facilitate the application of the external bending moment, which results in in-plane translations of the transverse section in the longitudinal direction.

13.3.2 Loads Applied to a FE Model

The loads acting on the structural model consist firstly of *external loads*, that is, sea pressure acting on the outer hull surface. This includes hydrostatic pressure, hydrodynamic loading of

Table 13.1 *Boundary constraints at model ends for cargo hold models [19]*

	Translation			*Rotation*		
Location	δ_x	δ_y	δ_z	θ_x	θ_y	θ_z
Aft end						
Independent point	-	Fix	Fix	$M_{T\text{-}end}$ [12]	-	-
Cross-section	-	Rigid link	Rigid link	Rigid link	-	-
	End constraint beam[13]					
Fore end						
Independent point	-	Fix	Fix	Fix	-	-
Intersection of CL and inner bottom	Fix	-	-	-	-	-
Cross-section	-	Rigid link	Rigid link	Rigid link	-	-
	End constraint beam					

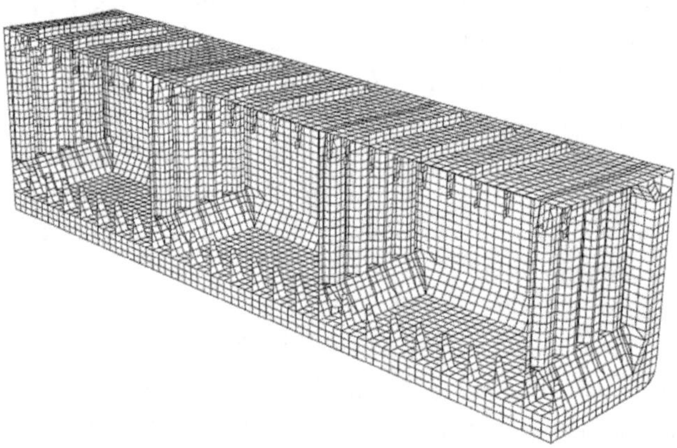

Figure 13.12 Example of a three–cargo hold model of a product tanker (portside half) [19]. Reprinted courtesy of IACS, London.

the hull girder and wave impacts on bottom, side and foredeck structure and of *internal loads*. The internal loads are the result of (a) gravitational forces exerted by dry and liquid cargoes, ballast seawater and other liquids, (b) setting pressure on relief valve and (c) inertial forces of dry and liquid cargoes, seawater ballast and other liquids. Loads can be calculated in a number of ways, of varying degrees of sophistication. These include linear (strip theory), nonlinear hydrodynamic load methods, the rule equivalent design wave (EDW) method, and component stochastic analysis. The basis of the EDW load method is discussed in Chapter 10.

13.3.3 Finite Elements for Ship Structural Analysis and Design

The number of types of finite elements that have been developed is quite large, and when modelling an entire hull girder, it is important to select the correct types and combinations

[12] Torsional moment acting at model end. [13] Figure 13.13b and part 1/chap. 7/sect. 2/para. 2.5.4 in [19].

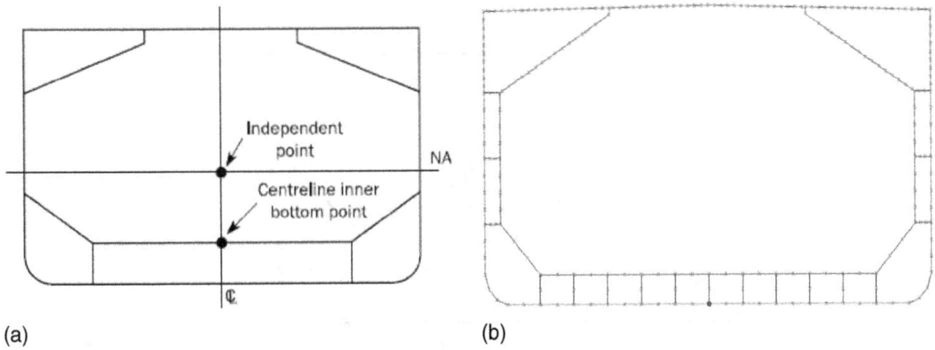

Figure 13.13 Means of supporting the model at its ends (support points and rigid links). (a) Support points at model end sections (single skin bulk carrier). (b) End constraint beams at extremities (double skin bulk carrier).

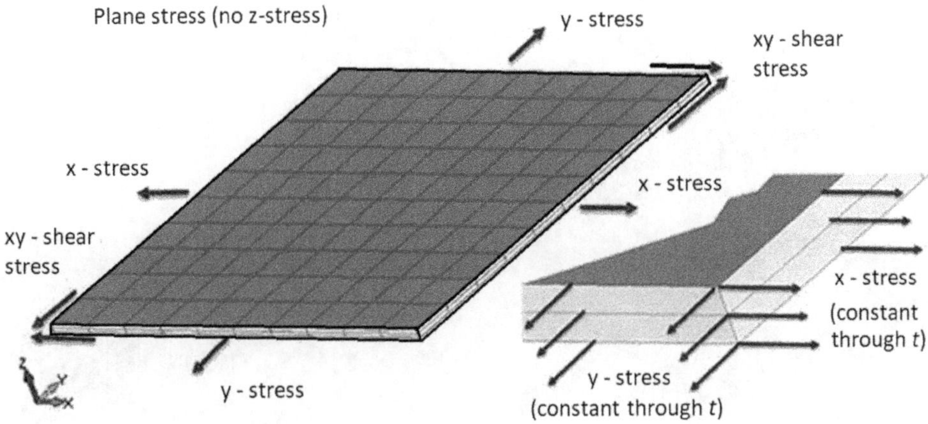

Figure 13.14 Plane stress state in a thin plate resulting from in-plane loading.

of elements for the study in question. In other words, it is necessary to ask not only what is the configuration of the structure being analysed but also how it is loaded and what will be the form of the induced stresses and deflections. For a thin-walled structure such as a ship, plane stress states prevail in practically all members of the structure. That is, the stresses normal to plane surfaces are assumed to be zero (Figure 13.14).

When a ship is immersed in water, the applied loading induces both in-plane and bending stresses, as seen in Chapter 10. However, if we are concerned with the response of the hull girder in primary bending only, the longitudinally effective structural members are loaded in the in-plane direction and the effects of loads that act normally to the longitudinally effective members, such as hydrostatic pressure, are ignored. The bending response of individual members at a local level is thus not part of the primary hull girder analysis. That is, we are seeking to reproduce the σ_1 stresses only and not the σ_2 or σ_3 stresses, described in Chapter 10.

The structural members that constitute the primary structure are the longitudinally effective plating and the attached stiffeners. These include the side shell structure, the

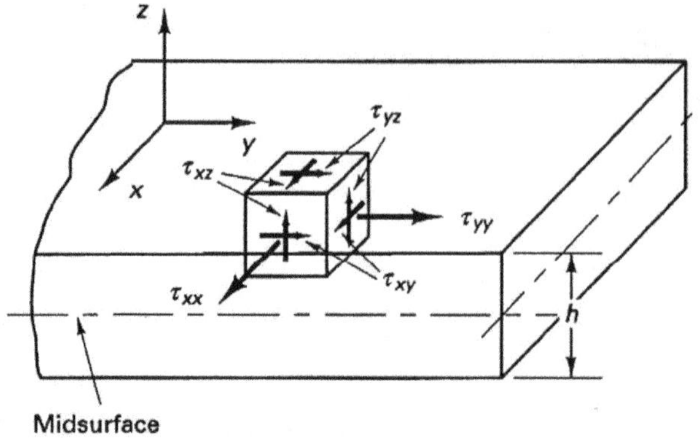

Figure 13.15 Membrane plate element.

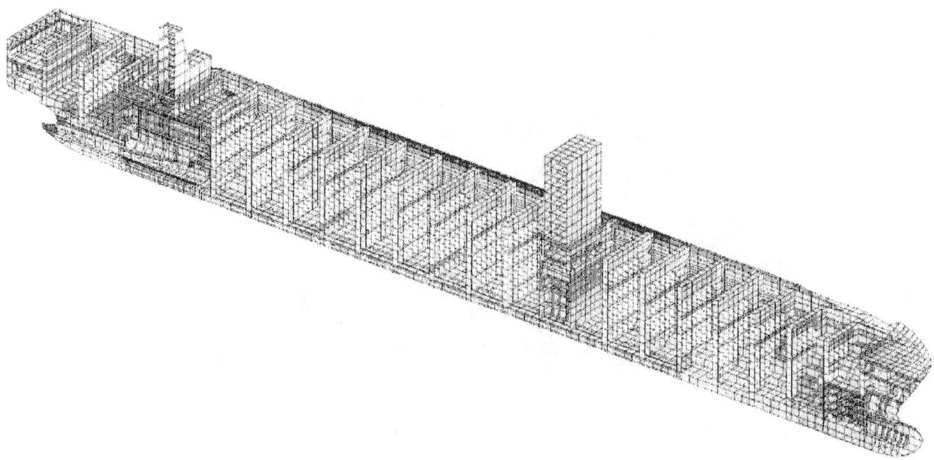

Figure 13.16 Finite element model of containership used for primary strength assessment [31].

decks and the outer and inner bottom structure. When modelling in-plane behaviour, the plating is represented by *membrane* plate elements (Figure 13.15) that behave according to plane stress theory (Figure 13.14). These would need to be combined with suitably attached *beam elements* that are used to represent stiffener action. The longitudinally effective members are supported by transverse bulkheads and other members that form part of the secondary structure. Since these also largely consist of stiffened plates, they can also be modelled using membrane plate elements in combination with beam elements [31]. Alternatively, the secondary structure may also be represented by beam elements, since their role in this case is to provide support to the primary structure.

In the case of the finite element model shown in Figure 13.16, in the cargo hold region the structure was discretised into elements whose length extended to approximately 3 metres (10 feet). Therefore, to model the length of structure corresponding to a 40-foot container, four elements were used. In total, 156 cross-sections were modelled for the entire hull girder [31].

If, in addition to primary bending, the effect of local loads that act normal to the plated surfaces is included, as is done in a quasi-static analysis (Chapter 10), then membrane elements are no longer suitable and *plate bending* or *shell elements* are needed. In principle, plate bending elements ought to be able to model bending response. However, the first plate bending elements that were based on Kirchhoff (thin) plate theory encountered difficulties in their use because of the simultaneous need to satisfy convergence requirements and to be effective in their applications [8]. It was found that more effective plate elements are formulated by using Reissner–Mindlin (thick) plate theory. In this theory, shear deformations are included, and therefore the line originally normal to the mid-surface does not remain perpendicular during deformation, this being analogous to Timoshenko beam theory.

Since in ship structures the plating that contributes to primary bending strength is generally loaded in the lateral direction as well, it is necessary to represent both in-plane and bending response. This is achieved in the most satisfactory manner by using shell elements that may be flat quadrilaterals. In such a case the finite element model of the hull girder shown in Figure 13.16 would require a significantly greater number of elements. There exist a wide variety of shell elements to choose from, and in doing so it is necessary to balance the accuracy of the element versus the computational time required. Elements with a small number of nodes are much more efficient computationally than are elements with a large number of nodes. The latter are much more accurate, and thus when using them, fewer elements are required. In the case of ship structures the discretisation is defined to a large extent by the topology of the structure, and thus it is not possible to reduce the number of elements beyond a certain point.[14] Therefore, the most efficient means is to make use of simple shell elements (four-noded) that represent both in-plane and bending response and are not computationally expensive. It ought to be borne in mind that an entire hull girder will require a very large number of elements in any case. For example, in the case of the finite element model shown in Figure 13.16 that is used to assess primary bending strength alone, the total number of elements used (including membrane and trusses) was close to 100,000. A model used to perform an accurate stress analysis would require a much greater number, and this is one good reason for which classification societies require only a representative section of the hull girder to be modelled.

13.4 Rational Design of the Hull Girder (Classification Society Approach)

Ship structural design involves three distinct stages, namely synthesis, analysis and detail structural design, as discussed in Chapter 10. Of these the most amenable to mathematical treatment are the second and third stages, although there has been progress in the development of computer-assisted decision support tools in concept design as well [6].[15] The

[14] Unless a different approach is used, such as ALPS/ISUM modelling.

[15] An example of use of an advanced structural design methodology that includes the concept design phase as well as the preliminary design phase is described in Chapter 14.

design process involves the systematic application of the different tools of structural analysis so that a number of requirements are fulfilled in the end-product. These include:

- Global strength requirements (primary normal and shear stress checks, stiffness requirements, ultimate strength). These can be studied by modelling either an entire ship or else by modelling a representative section of the hull girder, such as a cargo hold and its adjoining structure.
- Local strength requirements (yielding in tension, compression and shear, compressive buckling, fatigue). In this case one or more structural members such as a plate-beam combination are studied in isolation.
- Optimum structural design.

The first two concern issues of structural adequacy and are very much at the centre of classification society requirements, whereas the third concerns the satisfaction of the competing requirements of the stakeholders that are involved in a particular design.

When compared with other types of engineering structures, a number of features are seen to characterise marine structures. These are:

1. Marine structures are large and complex.
2. Marine structures are immersed in water, which exerts hydrostatic pressure on the outer shell plating. Also, certain internal compartments are subjected to lateral pressure loads. The types of external loads that act are many, but in general, wave loading has to be described in probabilistic terms.
3. Marine structures consist of thin sheets of plating to which orthogonal systems of stiffeners are attached.
4. Stiffening arrangements of an identical form are employed in a repetitive manner over large regions of the internal structure.
5. Global strength and local strength assessments are required and need to be integrated with one another.
6. Structural design is but one element of marine design, which includes general arrangement and outfitting, hydrodynamic design, seakeeping, propulsion, etc. These interact with one another in the ship design cycle.
7. Last but not least, merchant ships are built and operate in a competitive commercial environment that is subject to market forces unrelated to the economics of the ship construction industry. As a result, quality of ship construction may vary depending on shipyard management practice as well as the negotiating power of the stakeholders involved.

It is seen that the development of a procedure or a series of procedures that cater to all of the above is a demanding task. There currently exists one suite of software programs that cater to the first six of the points raised above (PARAMARINE). This is a series of programs that can be used to design all types of surface ships as well as submarines, and is thus unique in this respect. Beyond this, there are available special-purpose programs for marine structural assessment such as the software developed by the major classification societies (DNV-GL's NAUTICUS HULL, LR's SHIPRIGHT, Bureau Veritas'

Table 13.2 *Use of FE codes in ship structural assessment as required by DNV-GL*

Ship type	Extent of use of FEA
Bulk carriers	For ships equipped with cranes for heavy lifting operations and with a long centre cargo hold, a global strength analysis may be required
Container ships	Cargo hold FEA is to be carried out for the midship region and applied for one hold with typical hold arrangement
Multi-deck ro-ro ships and car carriers	The primary supporting members are to be assessed from a partial ship model as required
Passenger ships	The use of direct analysis methods is required
General cargo and MPP vessels	Supplementary analysis required for novel design and/or complex structural arrangements. Global FE analysis with equivalent design waves from direct wave load analysis

VERISTAR, etc.) as well as commercial programs such as MAESTRO, which differs from the above because it is expressly design oriented [18]. These address points (1)–(5) above.

In addition to these special purpose programs, there are available general-purpose FE programs such as ABAQUS, ANSYS, LS-DYNA and NASTRAN that can be used to study ship structures and in particular local strength problems.

In order to reduce computational time and complexity of calculations, special-purpose programs have been developed, especially taking into account points (1) and (3)–(4) above. In the earliest of these, ISUM (*I*dealised *S*tructural *U*nit *M*ethod), a portion of the structure that would normally be considered as consisting of a number of elements was treated as a single unit whose buckling and yielding properties were identical to those of the actual structure [34]. This approach was further developed for use in nonlinear structural analysis, thus resulting in more efficient and faster analysis without sacrificing accuracy [21].

As an illustration of the requirements of a classification society concerning the types of analysis needed for each ship type, Table 13.2 summarises DNV-GL's current requirements (as of 2019).

Of real value to the designer, however, is the combined use of the finite element method with structural optimisation methods. Recent advances in several fields now permit the application of several different kinds of optimisation methods in ship structural design [40], a task that until recently has not been successful for a variety of reasons. The complexity of the hull structure and that of structural response necessitates the use of efficient algorithms, and to this end *sequential linear programming* has been used in existing software [17].

One of the most recent approaches that have proven to be successful are *evolutionary algorithms*, of which *genetic algorithms* have been used in ship structural design with success [10].[17] The background theory to genetic algorithms is summarised in Chapter 14 and example applications are included.

The successful linking of a finite element procedure with optimisation algorithm(s) is but one, albeit important, step in the formulation of a *computer-based, rational structural*

[17] The ISSC 2018 Report by Committee IV.2 'Design Methods' includes a list of applications of ship structural optimisation techniques using mostly evolutionary algorithms [10].

Table 13.3 *Rationally based design of ship structures*

No	Step	Present volume
1	Definition of environmental loads	Chapters 3, 9
2	Global response analysis	Chapters 4–8
3	Substructure response analysis[16]	-
4	Limit state analysis	Chapter 10
5	Reliability-based structural constraints	Chapter 12
6	Optimisation problem	Chapter 14

ship design procedure. Such a procedure consists of a number of steps [17]; these are listed in Table 13.3. The background theory for each of these steps that is covered in the present volume is referenced in the third column of the table.

13.5 Case Studies

This section describes two cases in which the finite element method has been used, the aim being to illustrate some of the applications of the method in the marine sector. At the same time, ideas and concepts discussed in earlier chapters are utilised and the manner in which they can be used in conjunction with the finite element method becomes clearer.

The fundamentals of the FE method are not discussed, as the aim is to underline the nature of the problems and the difficulties encountered when assessing ship structural response. The finite element method is of universal application, and the cases discussed here should be treated only as a sample, given the limitations of a single chapter. Not all results or cases studied in the papers are discussed, and further information is obtained from the original references. In a previous section of this chapter reference has been made to the rational design of a containership, as understood by a classification society [31]. This case is supplemented forthwith with the following two which are structural strength assessments. One further example of the use of the FE method is included in Chapter 14, where the numerical model is coupled with optimisation techniques.

1. Nonlinear elasto-plastic finite element analysis to determine the ultimate strength of a bulk carrier in the alternate hold loading (AHL) condition [4, 5].
2. Strength comparison of single skin with double skin bulk carrier following collision damage [26].
3. Structural optimisation of a RoPax ship, including the concept design phase (Chapter 14).

The first case is an illustration of the calculation of the ultimate strength of the hull girder with the ship floating in water, also allowing for initial distortions in the plating that arise during fabrication. A number of questions concerning the generation of the finite element

[16] See the sister volume by this author: *Local Strength of Ship Structures*.

model as well as the state of the structure need to be resolved, and these are discussed in some detail. The second case deals with the ship collision problem, which requires a dynamic analysis to be performed. Such a study may be viewed as being one component of a risk analysis (the technical part of a consequence analysis) (Chapter 11). The text also deals with the estimation of the residual strength of the struck ship following each collision incident studied (Chapter 7).

13.5.1 Nonlinear Elasto-Plastic Finite Element Analysis to Determine the Ultimate Strength of a Bulk Carrier in the Alternate Hold Loading Condition

There arises one particular loading condition of bulk carriers that induces high stresses in the structure of the double bottom and is critical with regard to collapse of the entire hull girder. This is the *alternate hold loading condition* (AHL) that may be implemented when a ship is carrying high-density cargo. The AHL condition is used to raise the vertical centre of gravity of the hull girder in order to ease the ship's rolling motion, by reducing the roll frequency. An additional benefit, of a commercial nature, is that during loading and discharging the labour and equipment costs are reduced, as is the time to load and discharge the ship. Additionally, the cleaning time of the cargo holds that takes place following discharging is also reduced. There is thus an economic benefit to the owner or charterer. For a ship to be able to be loaded in this way additional strengthening of the bottom structure is required.

Figure 13.17a shows a nine-hold capesize bulk carrier in the AHL condition whereas structural particulars of a transverse section are included in Figure 13.18. It is seen that the net load acting on the bottom structure of any one cargo hold depends on whether the hold is loaded or not. The net uniformly distributed load acting on the bottom structure of the loaded holds is the difference between the weight of the cargo and the buoyancy force. The net effect is that the bottom structure of these holds bends downwards. In the case of the empty holds, on the other hand, the loading consists solely of the hydrostatic pressure acting on the outer shell plating, and these holds thus bend upwards. The magnitude of the hydrostatic pressure depends on the draught, and thus restrictions may need to be enforced regarding the loading arrangement of a ship as a function of the draught [19].

The main purposes of this study are to calculate the ultimate strength of a bulk carrier in this loading condition and to assess the effect this condition has on the structure at a local level on the basis of the IACS CSR-BC rules.[18] The longitudinal strength of bulk carriers depends to a large extent on the strength of the bottom structure, since the deck has hatch openings and the side shell is transversely framed (apart from the topside and hopper tanks). When the ship is in a hogging condition, the bottom structure is compressed, and thus collapse will involve elasto-plastic buckling.

The most vulnerable section is the double bottom below an empty hold, which in the AHL condition is induced to bend upwards. Of these, the one in way of the midship section is the critical one (Figure 13.17b).

[18] A comparison between CSR-BC and DNV requirements and results is also performed in the paper, but this is not discussed here.

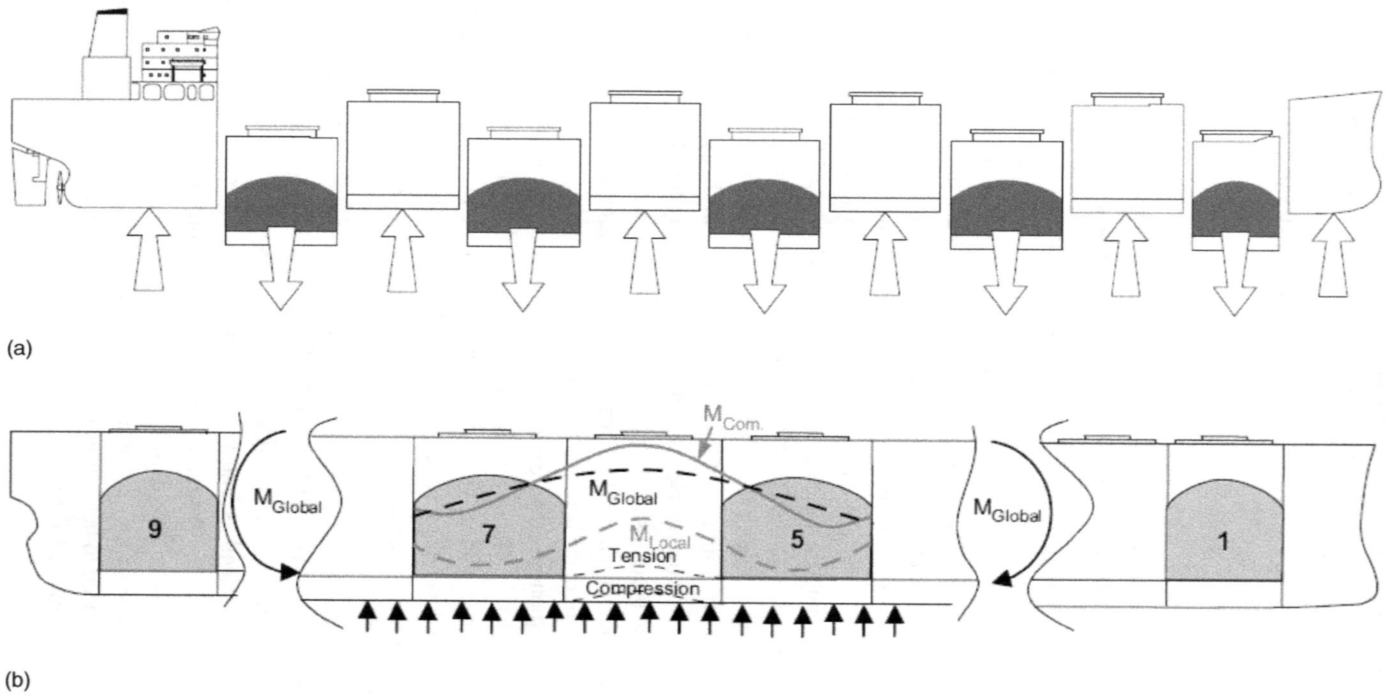

Figure 13.17 Capesize bulk carrier in AHL condition [4]. (a) Shearing action of the hull girder and net loads in cargo holds. (b) Longitudinal global and local bending moments acting on ½ + 3 + ½ holds. Reprinted courtesy of Elsevier.

No.	Dimensions	Type	σY(MPa)
1	390x27	Flat-bar	392
2	333x9+100x16	Tee-bar	352.8
3	283x9+100x14	Tee-bar	352.8
4	283x9+100x18	Tee-bar	352.8
5	333x9+100x17	Tee-bar	352.8
6	283x9+100x16	Tee-bar	352.8
7	180x32.5x9.5	Bulb-bar	235.2
8	283x9+100x17	Tee-bar	352.8
9	333x9+100x18	Tee-bar	352.8
10	333x9+100x19	Tee-bar	352.8
11	383x9+100x17	Tee-bar	352.8
12	383x10+100x18	Tee-bar	352.8
13	383x10+100x21	Tee-bar	352.8
14	300x27	Flat-bar	392

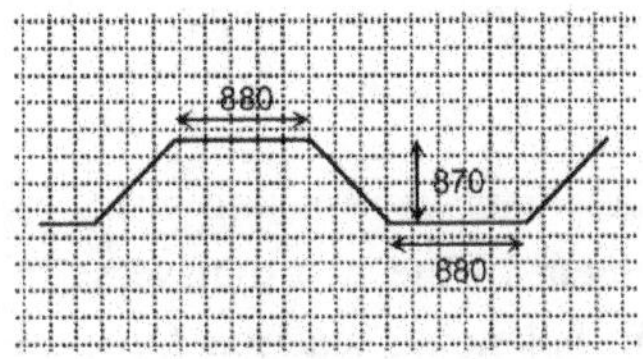

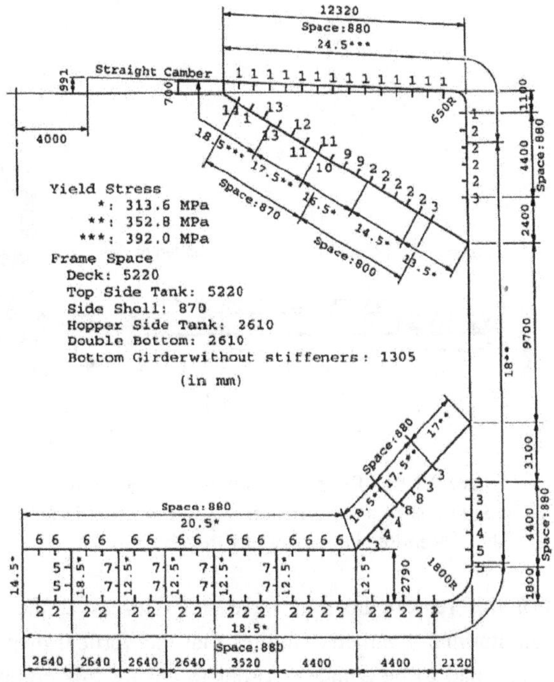

Figure 13.18 Transverse section of capesize bulk carrier being studied [4]. Reprinted courtesy of Elsevier.

In order to determine the ultimate strength, a finite element model of ½ + 1 + ½ holds is developed. The mesh has to be fine enough to capture both global and local effects and also for the boundary conditions to be transmitted accurately. Because of the extent and complexity of the structure, only the region where failure is expected is modelled using a fine mesh. The adjoining structure is modelled using a coarse mesh, leading to a total of more than 1 million degrees of freedom for the whole model. The ABAQUS program was used and the plating was modelled using the four-node shell element (S4) that has been shown to be satisfactory for such an application. Element performance comparisons and mesh convergence studies were carried out in order to achieve a balance between accuracy and computational time.

Since only one hold and two adjacent half-holds were modelled, it was necessary to apply appropriate boundary conditions at the ends of the model. In principle these are accurately defined by running a model of the entire hull girder and then excerpting the relevant conditions in way of the sections of interest. This, however, is not a practical solution in this case, and certain simplifying assumptions need to be made. Symmetry about the centreline plane exists, and thus it was possible to perform an analysis of only one-half of the structure. At the model ends the nodes were restricted to move in a single plane as the cross-section rotated under loading (in ABAQUS this is termed *surface-based kinematic coupling*) (Figure 13.19).

From diagrams of the longitudinal distribution of the transverse shear force it was noted that at the mid-length position of each hold the shear force is approximately equal to zero.

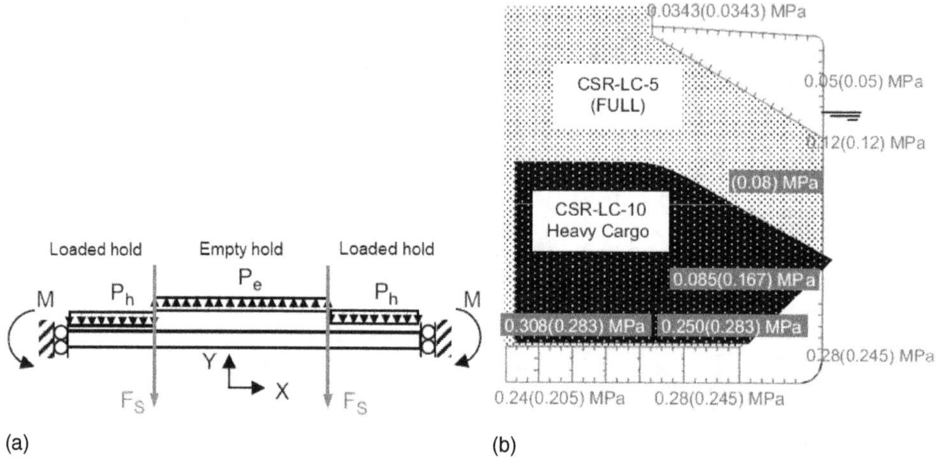

Figure 13.19 Boundary conditions and pressure loading of model [4]. (a) Boundary conditions of model in AHL condition. (b) Lateral pressure profiles according to CSR-BC.[19] Reprinted courtesy of Elsevier.

For this reason the transverse shear forces at the model ends were set to zero. Also, longitudinal symmetry implies that the vertical displacement of the two transverse bulkheads had to be equal. In order to ensure this condition, balancing reaction forces were imposed. In the still-water condition the AHL condition determines structural response. The rules specify a number of different loading conditions that include partially heavy cargo and fully loaded heavy cargo. In the analysis of these still-water conditions corresponding cases involving wave loads were added as well. Wave loading is specified in the rules using the equivalent design wave concept (Section 10.7). Loading is applied by performing incremental rotation of the model ends and the concurrent increase in the lateral pressure loading. The latter reaches its peak value at 10 per cent of the global bending load and is subsequently maintained constant.

The ultimate strength of the hull girder depends on the strength of the stiffened plate panels it consists of, since the collapse mode that triggers collapse of the hull girder is interframe buckling (Chapter 7). However, because stiffened plates are imperfection-sensitive, it is necessary to represent the initial imperfections as accurately as possible. These include initial distortions as well as residual stresses, both of which arise during the fabrication process. In the present study only initial distortions were considered. There are several ways in which this can be done, and in this case a linear superposition of buckling eigenmodes obtained from an eigenvalue buckling analysis was carried out. Since the buckling (eigen)modes of the plate and the stiffener differ and the stiffener eigenmodes are coupled with higher rather than lower plate eigenmodes, it is necessary to extract suitable initial distortion shapes from higher rather than lower eigenmodes, for which coupling arises. In other words, the actual buckled form of the plate has a larger number of half-sine

[19] Two load cases are included: CSR-LC-5 Full Loaded Cargo and CSR-LC-10 Heavy Cargo.

Table 13.4 *Load cases*[20]

Case No	Description	Imperfection	Load case
C1	PB-No-Imp	No	Pure bending
C2	CSR-LC-10-No-Imp (heavy cargo)	No	CSR-LC-10
C3	PB-Imp	Yes	Pure bending
C4	CSR-LC-10-Imp (heavy cargo)	Yes	CSR-LC-10
C5	CSR-LC-5-Imp (fully loaded cargo)	Yes	CSR-LC-5

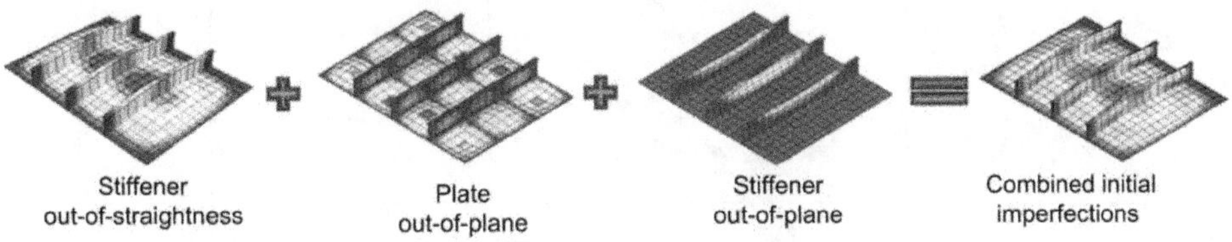

Figure 13.20 Definition of initial distortions at the level of the stiffened plate [4]. Reprinted courtesy of Elsevier.

waves in the longitudinal direction than those predicted by isolated plate buckling theory. This arises because the plate is welded to the stiffener, and its buckled form is thereby forced to be compliant with the buckled form of the stiffener. The procedure to obtain the initial imperfection forms is initiated by introducing the stiffener out-of-straightness deformation, as indicated in Figure 13.20. This is done by changing the thicknesses of the plate and stiffener so that the stiffened plate buckles as a whole. The same procedure is applied in the case of the plate and stiffener out-of-plane distortions. In each step the magnitude of the initial imperfection is introduced to the panel before starting the next step.

The effect of initial distortions was assessed by performing runs with and without these and by comparing results. Global response was assessed by comparing bending moment – end-rotation curves for a number of cases and also with results included in [39]. The response of the bottom structure that is critical to overall failure was described by stress distributions in the FE model at each stage of loading. The cases discussed here are listed in Table 13.4.

Figure 13.21 shows the results for the cases that are included in Table 13.4. It is first necessary to compare the cases that involve pure bending (without and with imperfections, C1 and C3, respectively) with that of [39]. The lower prediction for the ultimate strength provided by the latter is due to the inclusion of the residual stress effect. It is seen that in the post-collapse range, in cases C1 and C3 sharp unloading takes place. This arises because the compression of the bottom structure is uniform, and thus failure occurs throughout at the same load level. In cases C2, C4 and C5 the lateral pressure loading is allowed for, and comparing these with the pure bending cases (C1 and C3) provides an assessment of the effect of lateral pressure loading. From the curves it is seen that significant reductions, of

[20] Additional cases are discussed in [5].

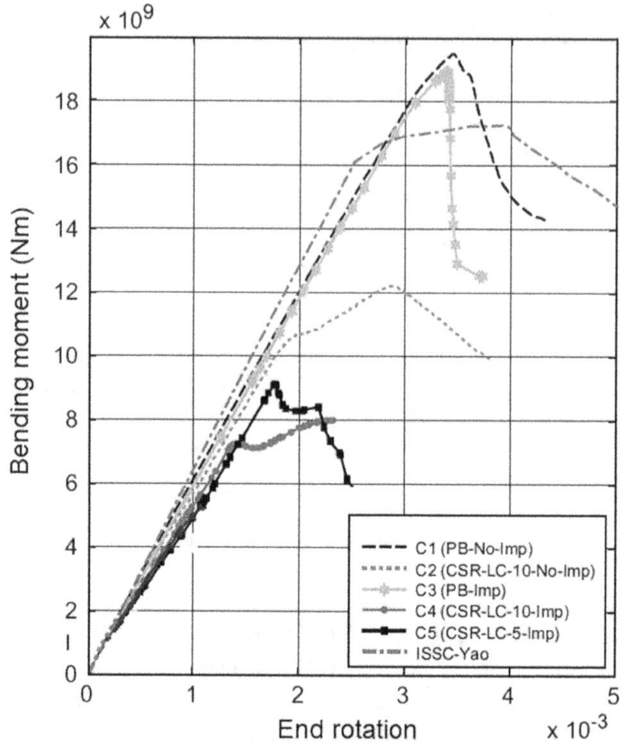

Figure 13.21 Ultimate strength of the hull girder predictions for various assumptions [4]. Reprinted courtesy of Elsevier.

the order of 50 per cent, are observed in both first failure as well as ultimate bending moment. It is also seen that the effect of initial distortions is more pronounced in the presence of lateral pressure, and the combined effect of lateral pressure and initial distortions results in the largest reductions in ultimate strength. This is attributed to the fact that transverse compression in way of the inner bottom and bottom shell in the empty hold in conjunction with high lateral pressure brings about a significant decrease in the stiffness and strength of the double-bottom structure.

The results plotted in Figure 13.21 show that the ultimate strength of a bulk carrier hull girder under the AHL condition may be reduced quite significantly when compared to other loading arrangements. This is the result of global stress redistribution between the outer bottom and the inner bottom structure. The stress redistribution arises because the outer shell plating is subjected to higher compressive loads and fails first. Following this, the load is borne by the inner bottom structure that fails at a later stage. This type of failure gives rise to the 'double-maxima' form of the curves for the combined loading cases. A comparison of curves C4–C5 (combined loading cases) with C1 and C3 (pure bending cases) in Figure 13.21 shows that one of the consequences of local loads is to convert post-collapse behaviour from an unstable to a stable form, which at the same time results in a significant reduction of the ultimate strength.

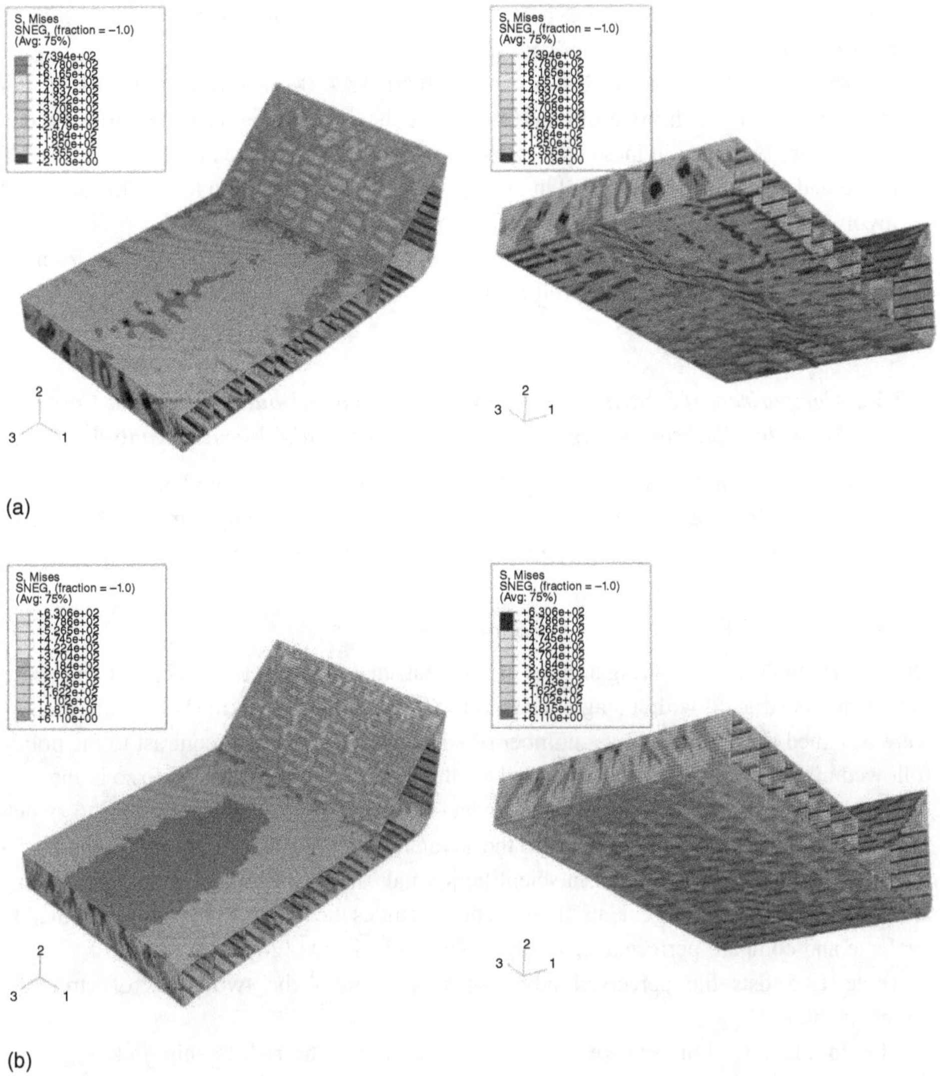

Figure 13.22 Von Mises stress contour plots at the ultimate failure load [5]. (a) Case C4 (CSR-LC-10-Imp). (b) Case C5 (CSR-LC-5-Imp). Reprinted courtesy of Elsevier.

Further information may be obtained by studying detailed stress distributions in the double-bottom structure. The following observations are pertinent:

1. The bending of the double bottom at the ultimate load level is visible for case C4 (CSR-LC-10-Imp), as shown in Figure 13.22a. The same was observed in case C2. The shear buckling of the centre girder and the girder at 18,480 mm from the centreline is also observed in both of these cases. Although collapse arises in the mid-span region of the empty hold in both cases, the collapse in case C4 involves the adjacent frame span

(Figure 13.22). It is therefore expected that the ultimate strength of the hull girder is reduced substantially, in comparison to case C2.

2. A comparison of Figure 13.22a for case C4 (CSR-LC-10-Imp) with Figure 13.22b for case C5 (CSR-LC-5-Imp) at ultimate load level shows that there is a clear difference in the collapse sequence in these two cases. The collapse load of the hull girder in case C5 is attained when the critical panel in the outboard part of the bottom fails, whilst case C4 involves buckling of two frame spans, as noted previously.
3. Shear buckling of the centre girder and of the side girder located at 18,480 mm from the centreline arises in case C5 as well as in case C4.

13.5.2 Comparison of Structural Response of Single and Double Skin Bulk Carriers Following Collision Using Nonlinear Dynamic Finite Element Analysis

In this section we shall describe the application of nonlinear dynamic FE analysis to the modelling of collision events involving bulk carriers that have different structural configurations, following which a comparison of structural response is carried out.

Background

During the 1990s and following a series of accidents that occurred to bulk carriers that until then were constructed with a single side skin (SSS), double side skin (DSS) bulk carriers were designed and marketed by a number of shipyards. The IMO, in contrast to the policy followed in the case of double skin oil tankers that it had endorsed, did not do so in the case of DSS bulkers. As a result, owners had to decide for themselves whether or not to switch from SSS to DSS ships. Yards promoted the advantages of the DSS design, and in parallel a number of publications appeared in scientific journals and conferences comparing the two alternative structural arrangements. This section includes the results of a study performed to evaluate and compare performance during collision incidents [26].

Table 13.5 lists the perceived advantages of each of the two types of structural arrangement.

The double side skin concept is one way of reducing the risk to ship and cargo in a collision incident, even though it does not cater for all possible scenarios. Thus, if the inner shell of a DSS ship is penetrated in addition to the outer shell, ingress of seawater will occur as in the SSS design. If the inner plating remains intact, there will be no damage to or loss of cargo, even though the outer shell plating may have been seriously ruptured. The possibility remains, however, for a ship to be lost, despite the fact that the inner shell structure remains intact following a collision incident. This is because following collision the struck ship may remain in contact with the striking ship or else it may continue to be grounded, as the case may be. As a result, the struck ship may temporarily be in a comparatively safe state, since its structure receives some form of external support. During salvage attempts the two vessels may be dislodged from one another, leaving the struck ship vulnerable. Its condition may deteriorate, given that it may not have sufficient residual strength to withstand the external loads. An analogous

Table 13.5 *Comparison of DSS bulk carriers with SSS bulk carriers*

a) Advantages of DSS design

(a) Structural strength

1. Increased shear strength compared to SSS design, especially when loaded in an AHL condition (previous case study). It has been found that in the case of a SSS bulker the shear strength dominates the structural design of the side shell structure, whereas a margin of safety of up to 40 per cent against failure in shear is inherent in the DSS design.
2. Increased transverse strength. In single skin bulk carriers, the topside tank and the lower hopper tank, being enclosed structures possess high torsional rigidity. The vertical frames and the outer shell plating they support are thus the most flexible part of the entire side structure. Excess loading of the side shell structure occurs during wave impact, in conjunction with hydrostatic pressure with the cargo hold being empty and the vessel being in a deep draught condition. In these circumstances the frames and outer shell plating bend about the lines of support provided at their lower and upper edges. These supports, being part of the torsionally stiff topside and lower hopper side tanks, do not bend. The alternate bending of the frames and side shell plating about these rigid supports is the reason for which fatigue cracks arise. These propagate and can lead to serious damage to the side shell structure. This mechanism has led to many failures of the side shell structure in single skin bulk carriers that have resulted in the loss of ships and lives. In DSS bulk carriers the side shell structure is cellular and has a significantly higher torsional rigidity and for this reason the failures that occurred in single skin ships do not arise.
3. The outer shell plating and its stiffening are not in direct contact with cargoes that may create a corrosive environment, such as coal. This leads to significant thickness reductions and a consequent reduction in strength.
4. Enhanced safety of the ship, crew and cargo due to increased collision resistance. If as a result of a low-to-moderate energy impact penetration of the outer shell structure occurs, the side structure is seen to have a greater degree of redundancy since the inner structure remains intact.

(b) Ship operation

1. Operational cost savings for owner and charterer. Double-skin structures have flat inner sides and as a result, the final stages of cargo discharging need not involve the use of grabs and pipes that have to manoeuvre around structural members such as frames. The holds can thus be discharged and cleaned in less time and at a lower cost. Additionally, the side structure can be inspected from within the double skin without the need for staging.
2. Reduced repair and maintenance cost throughout the lifetime of the ship
3. The need to designate a hold for ballast water containment does not arise, making ballasting and ballast water exchange at sea a simpler and safer operation. In SSS bulk carriers one cargo hold is designated to be watertight and with sufficient strength to withstand being fully flooded.
4. Enhanced protection of the environment due to greater collision resistance.

b) Advantages of SSS design

(a) Ship construction and operation

1. Reduced labour and material costs for fabrication and erection of the cargo holds and hence lower construction cost for the ship.
2. Increased cargo carrying capacity. For a bulk carrier of 75,000 dwt tonnes, the increase ranges from zero to 8 per cent, depending on the DSS arrangement that is used for purposes of comparison.

situation may arise when a ship is refloated following grounding. The situation may be complicated further with the ingress of seawater and the imposition of additional loads on the structure. Clearly circumstances differ in each particular case, and for this reason it is necessary to compare the strength of DSS and SSS bulk carriers in both the intact and the damaged state.

Aim of Study

The aim of this study was first to quantify the structural response of DSS bulk carriers following collision as well as that of SSS bulk carriers. Second, the residual strength in each case following the incident was determined and the results used to assess the safety margin (factor) in the damaged state. The study therefore comprises two parts, the first being the modelling of the collision event, for which a nonlinear dynamic finite element procedure was used, and the second being the calculation of the residual strength, and from this the safety margin, for each case. In addition, the effect of corrosion leading to plate and stiffener thickness reductions was assessed, using as a basis the Joint Bulker Project (Common Structural) Rules.[21]

The collision event was described and quantified by evaluating resistance forces, energy absorption and depth of penetration for a number of collision scenarios.

Description of Ships Involved

The ships that were studied are Capesize bulk carriers whose dimensions differ very slightly. Their midship sections are shown in Figure 13.23 and their principal dimensions are included in Table 13.6.

Additional details regarding scantlings are included in [26].

Modelling of the Collision Incident and Results

The nonlinear explicit FE code ANSYS LS-DYNA was used to develop the FE models of the struck ships and following that of the collision incident in each case. The struck ships were assumed to be still, and the bow sections of the striking ships were modelled as rigid bodies.

The analysis of the collision of two ships is split into two separate parts that are linked to one another: the *external dynamics* and the *internal mechanics*, linked by the common contact force. The external dynamics calculations are used to determine the motion of the two ships before and after impact and also to quantify the energy dissipated in causing damage to the struck ship. The internal mechanics focus on structural response and define the strength or resistance of the structure of a ship involved in an accident. The internal mechanics calculations can be carried out, apart from using nonlinear FEM modelling, in three more ways: use of empirical formulae, simplified analytical methods and model experiments.

[21] This topic is not discussed here. Details are included in [26].

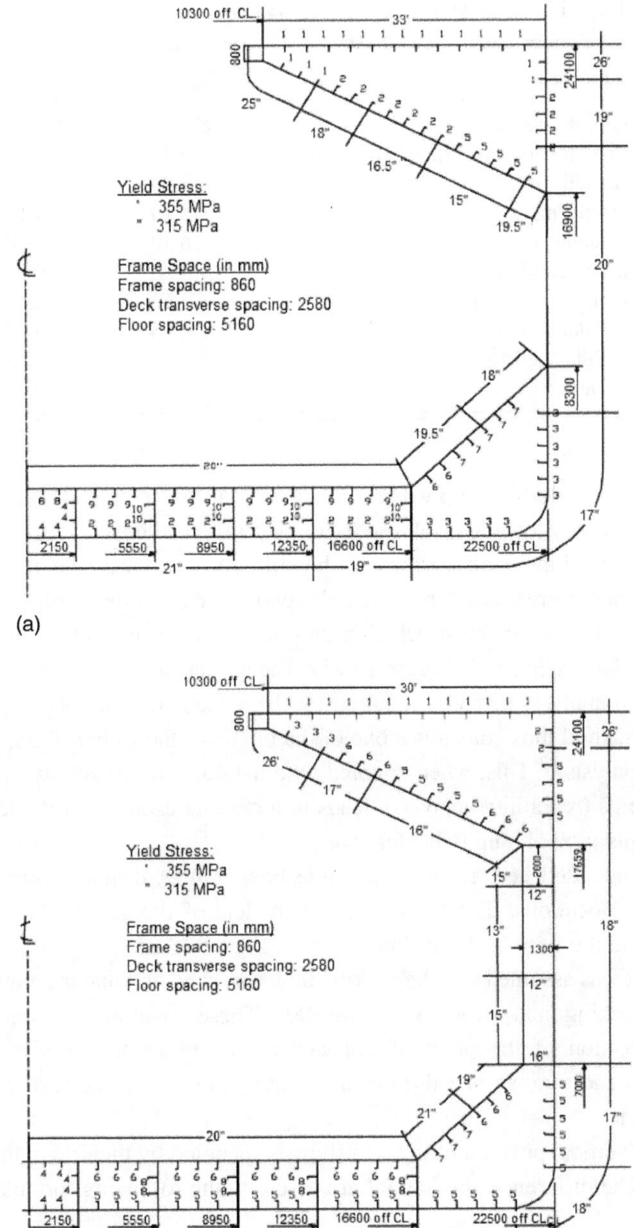

(b)

Figure 13.23 Midship sections of SSS and DSS bulk carriers [26]. (a) Single side skin bulker (SSS). (b) Double side skin bulker (DSS). Reprinted courtesy of Elsevier.

Table 13.6 *Principal dimensions of SSS and DSS bulk carriers*

Item	SSS	DSS
Length overall (m)	279.91	279.29
Length between perpendiculars (m)	279.33	279.00
Breadth (m)	45.00	45.00
Depth (m)	24.10	24.10
Draught (m)	16.50	17.70
Block coefficient	0.86	0.86
Frame spacing (m)	2.68	2.68
Bottom structure	0.86	0.86
Side shell	5.16	5.16
Deck		

In this study the colliding ship was assumed to be an Aframax-class oil tanker that was proceeding in all cases considered at a speed of 10.5 m/s (20 knots) at the instant of impact, whereas the struck ships were assumed to be still (zero forward velocity). A number of collision scenarios were considered, namely two different bow profiles, a number of different draught levels for the struck ship and two different impact locations (in way of a web frame and away from it) (Figure 13.25). The scenarios considered were required for the consequence analysis that was required by the stakeholders involved. The use of the finite element method thus constitutes one of the pillars of the technical part of a complete consequence analysis.[22] This, when coupled with data on the probability of occurrence of collision incidents (reliability analysis), leads to a rational estimate of the level of risk for collisions for this type of ship (Chapter 10).

The results included here relate to a single bow profile, different draught levels and different impact locations. The finite element models of the struck ships are shown in Figure 13.24 and the bow of the striking ships in Figure 13.25.

In all cases it was assumed that the striking bow was rigid and that the angle of incidence between the striking and struck ship was 90°. These assumptions together with the longitudinal position of the point of impact (in way of or far from a web frame of the struck ship) and a range of values of the relative draught of the two ships, define the collision scenario.

The relative vertical position of the two ships is specified by their draughts, as shown in Figure 13.25. The differences in draught are given by the following formulae [26]:

$$\Delta T_1 = T_{2\,\mathrm{max}} - \frac{3T_{1\,\mathrm{max}} + T_{1\,\mathrm{max}}}{4} \qquad \Delta T_2 = T_{2\,\mathrm{max}} - \frac{3T_{1\,\mathrm{max}} + T_{1\,\mathrm{max}}}{4} \tag{13.6a}$$

$$\Delta T_3 = \frac{T_{2\,\mathrm{min}} + 3T_{2\,\mathrm{max}}}{4} - T_{1\,\mathrm{max}} \qquad \Delta T_4 = \frac{T_{2\,\mathrm{max}} + 3T_{2\,\mathrm{min}}}{4} - T_{1\,\mathrm{max}} \tag{13.6b}$$

[22] Other issues that may be addressed in a consequence analysis may relate to loss of life, damage to cargo, damage to the environment, loss of earnings, social upheaval and others.

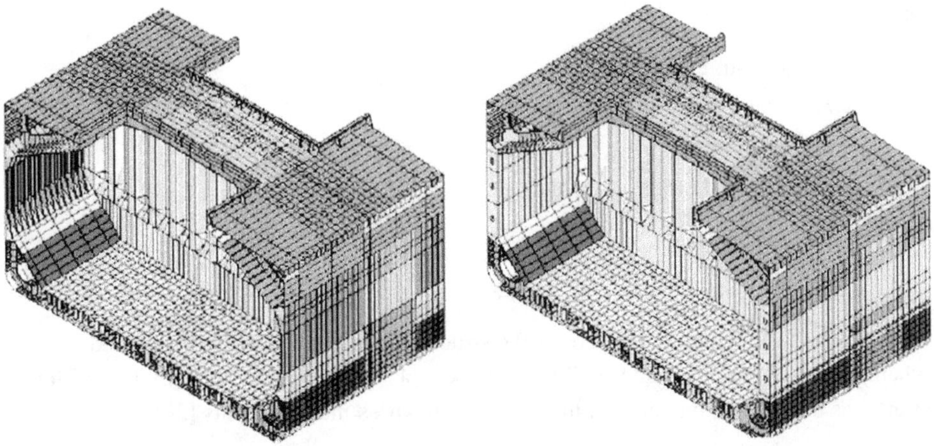

(a) (b)

Figure 13.24 Finite element models of struck ships [26]. (a) SSS design. (b) DSS design. Reprinted courtesy of Elsevier.

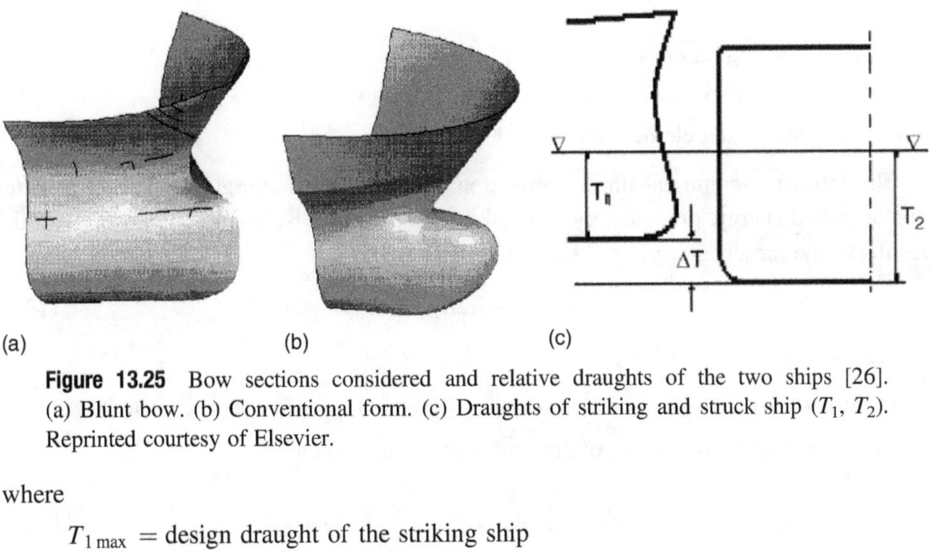

(a) (b) (c)

Figure 13.25 Bow sections considered and relative draughts of the two ships [26]. (a) Blunt bow. (b) Conventional form. (c) Draughts of striking and struck ship (T_1, T_2). Reprinted courtesy of Elsevier.

where

$T_{1\,\max}$ = design draught of the striking ship

$T_{1\,\min}$ = ballast draught of the striking ship

$T_{2\,\max}$ = design draught of the struck ship

$T_{2\,\min}$ = ballast draught of the struck ship

The total duration of the simulated collision scenarios is 0.3 seconds. Since the problem is inherently of a dynamic nature, it is necessary to allow for the effect of high plastic strain rates in the struck ship, since the value of the static yield stress no longer holds. This is achieved by using the following Cowper–Symonds type formula:

$$\sigma_y = \left[1 + \left(\frac{\dot{\varepsilon}}{C}\right)^{1/P}\right]\left(\sigma_0 + \beta E_p \varepsilon_p{}^{\mathit{eff}}\right) \tag{13.7}$$

where

$$\sigma_y = \text{dynamic yield stress}$$
$$\sigma_0 = \text{static yield stress}$$
$$\dot{\varepsilon} = \text{strain rate}$$
$$C, p = \text{material strain rate constants} = (3{,}200/\text{s}, 5)$$
$$E_p = \frac{E_{tan}E}{E - E_{tan}}$$
$$E_{tan} = \text{tangent modulus} = 625 \text{ MPa}$$

In the case of a collision, failure of the structure of the struck ship is specified to occur when tearing of the outer side shell plating takes place. The following definition of failure strain has been recommended on the basis of thickness measurements [26]:

$$\varepsilon_f(l_e) = \varepsilon_g + \varepsilon_e\left(\frac{t}{l_e}\right) \tag{13.8}$$

where

$$e_g = \text{uniform strain}$$
$$e_e = \text{necking strain}$$
$$t = \text{plate thickness}$$
$$l_e = \text{individual element length}$$

The following empirical linear correlation between the structural resistance parameter and absorbed energy, proposed by Minorsky, was used as a basis for comparison with the results of the present study:

$$W_C = 47.2R_T + 32.7 \tag{13.9}$$

where

$$W_C = \text{total energy absorbed by damage and striking and struck vessels (MJ)}$$
$$R_T = \text{damaged volume of structural steel } \left(\text{m}^3\right) \text{ given by}$$

$$R_T = \sum_{N=1} P_N L_N t_N + \sum_{n=1} P_n L_n t_n \tag{13.10}$$

The first summation represents the permanently (plastically) damaged volume of material of the striking ship, whereas the second concerns that of the struck ship and

$P =$ penetration for the *N*th (*n*th) member
$L =$ length of damage of *N*th (*n*th) member
$t =$ thickness of *N*th (*n*th) member

Tables 13.7 and 13.8 provide comparative data for the estimation of the internal energy for impact in way of a web frame and at midbay, respectively, for the two types of internal structural arrangement. The data for a given penetration depth are obtained using the finite element dynamic analysis program LS DYNA and the Minorsky formula (13.9).

Table 13.7 *Internal energy for impact i.w.o. web frame* (t = *0.1 s)*

	SSS			DSS		
		Internal energy (MJ)			Internal energy (MJ)	
Case No	Penetration (m)	LS DYNA	Minorsky	Penetration (m) t=0.1s	LS DYNA	Minorsky
1	2.36	50.2	54.1	1.10	61.5	65.4
2	2.48	83.7	81.3	2.22	73.8	59.1
3	2.73	107.2	93.6	1.96	90.1	75.8
4	2.79	79.3	89.9	2.83	69.7	76.3

Table 13.8 *Internal energy for impact between web frames* (t = *0.3 s)*

	SSS			DSS		
		Internal energy (MJ)			Internal energy (MJ)	
Case No	Penetration (m)	LS DYNA	Minorsky	Penetration (m)	LS DYNA	Minorsky
5	2.71	73.0	66.5	2.38	63.1	74.7
6	2.85	121.2	54.8	2.41	78.2	80.6
7	2.80	120.1	83.8	2.62	84.7	76.1
8	3.07	104.1	65.7	2.72	69.4	81.3

Table 13.9 *Rupture energies (in MJ) for DSS and SSS bulk carrier designs*

	DSS		SSS		
Case No	Outer shell	Inner shell	Outer shell	$\frac{DSS_{outer}}{SSS_{outer}}$	$\frac{DSS_{outer}}{SSS_{outer}}$
1	22.23	51.08	13.84	1.61	3.69
2	24.61	50.29	14.76	1.67	3.41
3	30.48	69.25	37.35	0.81	1.85
4	16.45	40.61	17.17	0.96	2.36
5	16.83	47.19	32.12	0.52	1.47
6	18.79	60.35	59.20	0.32	1.02
7	16.17	61.18	43.45	0.37	1.41
8	7.88	43.43	26.75	0.29	1.62

The results included in Table 13.9 (Column 6) show that in all cases the rupture energy for the DSS design exceeds that for the SSS design. For the cases listed the ratio varies quite significantly (from 1.02 to 3.69), depending on the relative draught. The results included in Column 5 show that for low-energy impacts, in the majority of cases the SSS design is more effective than the DSS design in resisting penetration.

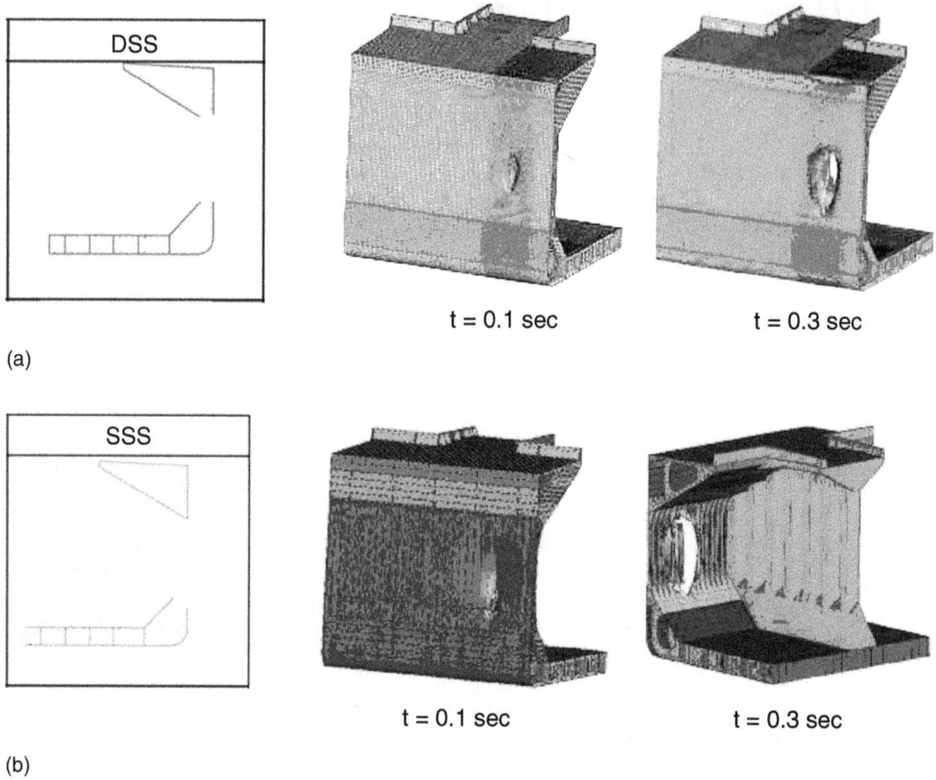

Figure 13.26 Deformation of the side shell structure of the struck ships [26]. Case 1: blunt bow striking in way of web frame. (a) DSS. (b) SSS. Reprinted courtesy of Elsevier.

Figure 13.26 shows the progress of damage of the struck ships at two different times following impact. It is noted that penetration of both the outer and the inner shell plating occurred.

Residual Strength Following Collision and Safety Against Hull Girder Collapse

Following impact and the loss of effectiveness of part of the side shell structure, the struck ship is subjected to combined vertical and horizontal bending (Chapter 6). It is thus necessary to calculate the residual strength in such a condition.

Residual strength calculations were carried out for both designs and were based on the use of the incremental-iterative method incorporated in IACS CSR. The method is described in Appendix D. In this study, the stiffened plate collapse curves were determined semi-analytically, as described in [27]. The ultimate hull girder bending moment was calculated for each damaged case and was compared to that in the intact condition. The total longitudinal bending moment, M_{total}, comprising of the still-water bending moment and the wave-induced bending moment is given by the following formula [3]:

$$M_{total} = K_{us}M_{sw} + K_{uw}M_{wv} \tag{13.11}$$

Table 13.10 *Residual strength of SSS and DSS bulk carriers in intact and damaged states*

		SSS		DSS	
		Hogging	Sagging	Hogging	Sagging
M_{Design} (collision)	M_{sw}	0.406	0.375	0.406	0.375
$M_{Plastic}$ (intact)	M_{wv}	0.605	0.635	0.605	0.605
$M_{vert\ ult}$ (intact)	M_{total}	0.830	0.820	0.830	0.820
$M_{comb\ ult}$ (intact)		1.779	1.779	1.823	1.823
$M_{comb\ ult}$ (collision)		1.606	1.289	1.570	1.312
Case No. 1		1.499	1.304	1.493	1.314
2		1.392	1.221	1.345	1.241
3		1.384	1.239	1.386	1.244
4		1.313	1.094	1.341	1.113
5		1.407	1.191	1.394	1.197
6		1.374	1.225	1.356	1.249
7		1.354	1.154	1.386	1.241
8		1.367	1.151	1.420	1.244
		1.364	1.126	1.393	1.201

Table 13.11 *Safety factors for SSS and DSS bulk carriers*

	SSS		DSS	
Case No	Hogging	Sagging	Hogging	Sagging
1	1.677	1.489	1.620	1.513
2	1.667	1.511	1.669	1.517
3	1.582	1.334	1.615	1.357
4	1.695	1.452	1.679	1.459
5	1.655	1.493	1.633	1.523
6	1.631	1.407	1.669	1.513
7	1.646	1.403	1.710	1.517
8	1.595	1.373	1.678	1.464

where K_{us} and K_{uw} are load combination factors. In the case of a collision incident, $K_{us} = 1.0$ and $K_{uw} = 0.7$ [3]. The still-water and wave-bending moments were evaluated using formulas included in the JBP (now CSR) rules.

Table 13.11 includes the safety factors determined from the following formula:

$$Safety\ factor = M_{Comb\ ult}/M_{total} \tag{13.12}$$

where $M_{Comb\ ult}$ is the combined vertical and horizontal bending moment ($\times 10^7$ kNm). $M_{Comb\ ult}$ and M_{total} are obtained from Table 13.10.

The following observations were made on the basis of the residual strength calculations that were carried out for the two bulk carrier designs:

1. The ultimate bending moment of an intact DSS bulk carrier in the hogging condition is a little less than that of a SSS bulk carrier. This is attributed to the minor differences in the scantlings of the primary members of the two designs.

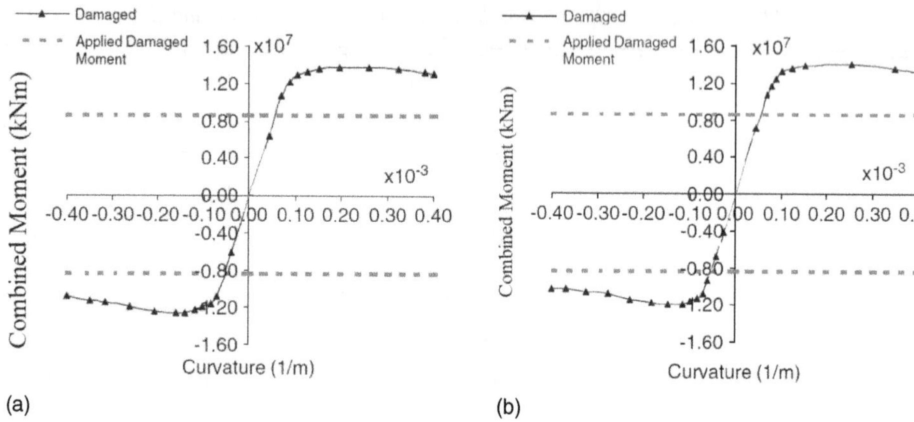

Figure 13.27 Bending moment – curvature relationships for Case 3.

2. The ultimate bending moments of both the intact and damaged SSS and DSS bulk carriers in sagging are considerably less than the corresponding ultimate bending moments in hogging. This is because the reduction in strength due to elasto-plastic buckling in way of the upper region of the side shell and deck structure that are both subjected to in-plane compressive loading in the sagging condition results in a substantial reduction of the residual strength. The largest reduction in residual strength is observed in case 3, for both SSS and DSS designs (Figure 13.27).
3. A thickness reduction in the bottom plating largely affects the hogging strength, whilst a thickness reduction in the deck plating largely reduces the sagging strength. This is because the bottom structure is in compression when the ship is in a hogging condition and the deck structure is in compression when the ship is in a sagging condition.
4. In the case of damaged DSS bulk carriers the reduction in strength is largest when the ship is in a sagging condition. This is because the deck stiffeners have low torsional rigidity and fail by tripping at a much lower load than the beam–column type failure of the stiffened plate that is generally expected to occur. This leads to a more severe reduction of the ultimate bending moment in sagging compared with to that in hogging, in which the bottom structure is in tension.
5. Hull girder collapse is more likely to occur in the sagging condition than in the hogging condition for both SSS and DSS designs, when involved in a collision.

The safety margin (factor) of SSS and DSS bulk carriers is used to assess damage tolerance and survivability in each case.

1. For the collision damage scenarios considered, the safety factor is greater in the hogging condition than in the sagging condition. This is because damage to the upper region of the structure as a result of a collision has a greater effect on the decrease in ultimate bending moment when the ship is in a sagging condition.
2. It is seen that the DSS design has a larger safety factor than the SSS design for nine of the cases considered, in both hogging and sagging conditions.

Conclusions

The following conclusions are reached.

With regard to the collision incident:

1. Ship structural design has an important influence on collision resistance. The energy absorption capability during a collision event depends on the structural geometry and scantlings of the side shell structure. These include the thicknesses of the outer shell plating, the inner shell plating, the side stringers, the transverse webs as well as the width of the side ballast tank and the lower and upper wing tanks.
2. When rupture occurs to the outer shell structure of the DSS ship, the energy absorbed is approximately 10 per cent less than the energy absorbed by the outer shell structure of the SSS ship. However, the total energy absorbed, which is the energy absorbed when the cargo hold skin (the inner shell in the case of the DSS design and the outer single shell in the case of the SSS design) ruptures, is 2.2 times larger for the DSS than for the SSS design. This shows that the outer shell structure of the DSS design is weaker than that of the SSS design, but when the impact is severe enough, the inner shell adds a significant degree of strength to the structure that exceeds that of the SSS design.

With regard to the residual strength of the damaged ships:

1. For the ships considered, the ultimate bending moment of the intact DSS bulk carrier in the hogging condition is slightly less than that of the intact SSS bulk carrier. This is because there exist minor differences in the scantlings of the primary members of the two designs. The reduction in ultimate strength of the hull girder is considered to be a function of the extent of damage, the mode of failure and the location of the damage.
2. The ultimate strength in sagging in both intact and damaged SSS and DSS bulk carriers is considerably less than the ultimate strength in hogging.
3. The damage that occurs in the upper side shell and deck structures that are subjected to in-plane compression as a result of the collision incident brings about a substantial reduction in residual strength in the sagging condition.
4. The largest reduction in residual strength arises in case 3, for both SSS and DSS bulk carriers. This is because the damage inflicted upon the upper side shell and deck structures leads to significant decrease in the ultimate bending moment in the sagging condition.

With regard to the safety margin:

The safety margin of both SSS and DSS bulk carriers is used to assess damage tolerance. Survivability can be calculated by comparing the ultimate strength with the expected maximum long-term hull girder bending moment.

1. For the collision scenarios the safety margin is greater in the hogging than in the sagging condition.
2. Of the cases studied, bulk carriers built to the DSS design possess a larger safety factor than SSS bulk carriers in 9 of the 16 cases.

The mean values of the safety factors calculated may be used to provide a quick estimate for practical ship applications, such as rescue and salvage operations of ships following collision. The calculations indicate that following a collision incident, hull girder collapse is a possibility when the ship is in a sagging condition. This holds true for both SSS and DSS bulk carriers.

References

1. Aksu S. et al. (2012) Quasi-Static Response. Committee II.1 Rept. In Fricke W. Bronsart R. (Eds) *Proc. 18th Int. Ship & Offsh. Struct. Cong*. Vol. 1 Schiffahrts-Verlag 'Hansa' GmbH & Co KG Hamburg 151–216.
2. American Bureau of Shipping (2015) *Guidance Notes on Ship Vibration*.
3. American Bureau of Shipping (1995) *Guide for Assessing Hull Girder Residual Strength*.
4. Amlashi H. K. K. Moan T. (2008) Ultimate strength analysis of a bulk carrier hull girder under alternate hold loading condition: a case study. Part 1: Nonlinear finite element modelling and ultimate hull girder capacity. *Mar. Struct.* Vol. 21 327–352.
5. Amlashi H. K. K. Moan T. (2009) Ultimate strength analysis of a bulk carrier hull girder under alternate hold loading condition: a case study. Part 2: double bottom bending and stress redistribution. *Mar. Struct.* Vol. 22 522–544.
6. Andrić J. Prebeg P. Žanić V. (2019) Multi-level Pareto supported design methodology-application to RO-PAX structural design *Mar. Struct.* 67 102638.
7. Asmussen I. Menzel W. Mumm H. (2001) *GL-Technology. Ship Vibration*. Germanischer Lloyd Rept.
8. Bathe K.-J. (1996) *Finite Element Procedures*. Upper Saddle River, NJ: Prentice-Hall.
9. Choung J. Nam J.-M. Ha T.-B. (2012) Assessment of residual ultimate strength of an asymmetrically damaged tanker considering rotational and translational shifts of neutral axis plane. *Mar. Struct.* Vol. 25 171–184.
10. Collette M. et al. (2015) Design Methods. Committee IV.2 Rept. In Guedes Soares C. Garbatov Y. (Ed) *Proc. 19th Int. Ship & Offsh. Struct. Cong*. Vol. 1 CRC Press 459–518.
11. Dambra R. Iaccarino R. Porcari R. (2000) The role of finite element technique in ship structural design. *Proc. First South European Technological Meeting Int. Conference: The role of simulation in improving products development process*. Monaco: Princip.
12. DNV GL (2018) Wave Loads DNVG-CG-0130.
13. DNV-GL (2016) Finite Element Analysis DNVGL-CG-0127.
14. Dow R. S. Hughes O. F. McNatt T. R. (1997) Unified first-principles ship structural design based on the MAESTRO methodology. In Johansson K. Koyama T. (Ed) *9th ICCAS* Yokohama.
15. Germanischer Lloyd (2003) Development of explanatory notes for harmonized SOLAS Chap. II-1. International Maritime Organization (IMO).
16. Hsu S.-H. Chang B.-C. Chen H.-C. (2009) Why Double Side Skin Bulk Carrier? – Study from a Strength Viewpoint. *TEAM 2009* Kaohsiung.
17. Hughes O. F. (1983) *Ship Structural Design: A Rationally-Based Computer-Aided Optimization Approach*. New York: John Wiley & Sons.
18. Hughes O. F. Paik J.-K. (2010) *Ship Structural Analysis and Design*. Jersey City, NJ: SNAME.
19. International Association of Classification Societies (2020) Chap. 7: Direct Strength Analysis in: *Common Structural Rules for Bulk Carriers and Oil Tankers Part 1*.
20. Jang B. S. Suh Y. S. Paik J. K. Seo J. K. Ju B. K. (2007) Ultimate Limit State Assessment of Tankers considering Common Structural Rules. *TSCF Shipb. Mtg*.

21. Kim D. K. Kim H. B. Mohd M. H. Paik J. K. (2013) Comparison of residual strength-grounding damage index diagrams for tankers produced by the ALPS/HULL ISFEM and design formula method. *Int. J. Naval Archit. Ocean Eng.* Vol. 5 47–61.
22. Lloyds Register (2015) *General Overview of Ship Structural Vibration Problems.*
23. Mansour A. Liu D. (2008) *Strength of Ships and Ocean Structures.* In Paulling J. R. (Ed) *Principles of Naval Architecture Series.* Jersey City, NJ: SNAME.
24. Mansour A. E. Lin Y. H. Paik J.-K. (1995) Ultimate strength of ships under combined vertical and horizontal moments. *Proc. 6th Int. Symp. Pract. Des. Ships & Mob. Units Seoul (PRADS 95)* 17–22.
25. McVee J. D. Reijmers J. (2005) A review of FEA technology issues confronting the marine & offshore industry sector. *Proc. FENet Meeting Malta* 111–118.
26. Özguc O. Das P. K. Barltrop N. (2005) A comparative study on the structural integrity of single and double side skin bulk carriers under collision damage. *Mar. Struct.* Vol. 18 511–547.
27. Özguc O. Das P. K. Barltrop N. (2006) A proposed method to evaluate hull girder ultimate strength *Ships & Offsh. Struct.* Vol. 1 335–345.
28. Paik J. K. et al. (2012) Ultimate Strength. Committee III.1 Rept. In Fricke W. Bronsart R. (Eds) *Proc. 18th Int. Ship & Offsh. Struct. Cong.* Vol. 1 Schiffahrts-Verlag Hansa GmbH & Co KG Hamburg 285–363.
29. Paik J. K. Kim D. K. Park D. H. Kim H. B. Mansour A. E. and Caldwell J. B. (2013). Modified Paik-Mansour formula for ultimate strength calculations of ship hulls. *Ships & Offsh. Struct.* 245–260.
30. Payer H. G. Asmussen I. (1985) Vibration response on propulsion-efficient container vessels. *Trans SNAME* Vol. 93 147–164.
31. Payer H. G. Schellin T. S. (2013) A class society's view on rationally-based ship structural design. *Ships & Offsh. Struct.* Vol. 8 319–336.
32. Ringsberg J. et al. (2015) Quasi-Static Response Committee II.1 Rept. In Guedes Soares C. Garbatov Y. (Eds) *Proc. 19th Int. Ship & Offsh. Struct. Cong.* Vol. 1 CRC Press 141–207.
33. Samuelidis M. Koukounas D. Pollalis C. (2013) Residual Strength of Damaged Ship's Hull. *Proc. Intl. Symp. Pract. Des. Ships (PRADS2013).* Changwon City Korea 1003–1010.
34. Ueda Y. Rashed S. M. H. (1991) ISUM (idealized structural unit method) applied to marine structures (mechanics strength & structural design). *Trans. JWRI.* Vol. 20 123–136.
35. Villavicencio R. Liu Z. J. Amdahl J. Guedes Soares C. (2013) Influence of the neutral axis displacement on the residual strength of a damaged tanker double-bottom structure *Ships & Offsh. Struct.* Vol. 8 663–674.
36. Wang E. D. Bone J. S. Ma M. Dinovitzer A. (2019) Guidelines for Evaluation of Marine Finite Element Analyses. Ship Struct. Comm. Rept. SSC-475 Washington, DC.
37. Wang G. Chen Y. Zhang H. Shin S. (2000) Residual Strength of Damaged Ship Hull. In *Ship Structures for the New Millennium: Supporting Quality in Shipbuilding* Vol. 13 Arlington, VA.
38. Williams C. (2002) Double hulls double benefits *Graig Horizons.* Graig Shipping Plc Newsletter No. 2 Cardiff.
39. Yao T. et al. (2000) Ultimate hull girder strength. Rept. of Special Task Committee VI.2 In Ohtsubo H. Sumi Y. (Eds) *Proc. 14th Int. Ship & Offsh. Struct. Cong.* Vol. 2. Amsterdam: Elsevier 21–91.
40. Žanić V. Andrić J. Prebeg P. (2013) Design synthesis of complex ship structures. *Ships & Offsh. Struct.* Vol. 8 383–403.

14

Optimum Design of Ship Structures

Design is a funny word. Some people think design means how it looks.
But of course, if you dig deeper, it's really how it works.

Steve Jobs

14.1 Introduction

Up to this point our aim has been to ensure that the structure can withstand the loads to which it will be subjected so that it will perform satisfactorily throughout its operating life. That is, we have been concerned with *structural adequacy*. The purpose has been to ensure that the structure will fulfil its role with regard to the normal as well as the extreme conditions to which it may be subjected during its operating life. In the past, structural adequacy was ensured by limiting stress levels and deflections (*permissible stress* and *adequate stiffness criterion)*. Permissible stress is a useful means of ensuring adequate strength at the preliminary design stage of a ship, but it does not provide a measure of its true ultimate strength, nor does it provide an idea of the safety margin against overall collapse. Structures exhibit types of response for which permissible stress is an adequate measure of strength (fatigue, thermal stresses), but there are others for which it is not, such as nonlinear elastic and elasto-plastic behaviour, buckling and fracture. More recently, limit state design has been introduced, but this too is concerned with structural adequacy. On the other hand, adequate stiffness of the hull girder and individual structural members is required in order to avoid:

- Reduction of cargo-carrying capacity resulting from excessive hogging in the still-water condition[1]
- Damage to mechanical systems such as the propeller shaft, piping systems, machinery, etc.
- Reduction in fatigue life
- Excessive vibration amplitudes, with corresponding adverse effects

[1] The weight of cargo loaded is estimated using the fore and aft draught marks.

It becomes apparent, however, that when proposing an arrangement for a ship structure,

1. There are a number of alternatives that may be followed.
2. Decisions have to be made at different levels.
3. These decisions have important economic consequences.

The challenge then is to develop and make use of approaches that will offer a genuine capability to find the 'best' and not just an adequate structure, thereby maximising benefits for the parties concerned. The term *optimisation* is used to denote any process that operates in a systematic manner and leads to the best possible use of available resources in order to fulfil one or more aims, against a background of predefined conditions.

In structural design, optimisation can be implemented in various ways and at different levels. At an abstract level, *structural optimisation* aims at the best definition of a structure that is used to transmit a particular load or combination of loads from one region in space to another. By 'best' we understand to be the structure that fulfils one or more simultaneous requirements and that operates under specific conditions. Such requirements may be minimum construction cost, weight restrictions, spatial distribution, minimum maintenance cost, etc. Structural optimisation procedures lead to a number of alternative solutions that need to be assessed and compared. In order to do this, *measures of merit* are required.

The primary aim of ship structures is to ensure the integrity of the vessel and to provide a watertight envelope for the cargo and persons on board. The structure thus has to withstand the loads that act as well as provide a desired distribution and size of cargo spaces for the particular ship type. The designer then has to ensure that the structure does this in the best possible manner. In the process of design, it may be possible to select *topology* (structural arrangement), *shape* and the *sizing* of structural members. When considering ship structures, it is helpful to keep in mind examples of these. With regard to structural arrangement, examples are:

- The layout and arrangement of cargo tanks in oil tankers, especially large ones
- The structural arrangement used to support a dry cargo ship tweendeck (cantilever, pillars, combination)
- The type of stiffening arrangement, whether single or double skin, used in the side shell of dry cargo ships
- The stiffening arrangement (longitudinal or transverse) in each region of a transverse section and along the length of the hull girder
- The form of the upper and lower stools that support transverse bulkheads in bulk carriers
- The number of cross-ties used in oil tanker swash bulkheads

With regard to shape, examples are

- Bulkhead form (flat stiffened or corrugated)
- Type of cross-section of pillar used to support a deck (hollow circular, rectangular hollow, solid section, I beam)
- Sandwich-type structure versus conventional arrangement
- Primary stiffener cross-section (flatbar, T-bar, angle bar, bulb section)

- Frame cross-section
- Types of brackets used to connect primary members to major members (such as transverse bulkheads)

With regard to size, the following are generally included:

- Plate thickness
- Web height
- Web thickness
- Stiffener spacing
- Transverse member dimensions
- Material properties (yield stress)

The selection and extent of use of different materials, on the other hand, are not normally included in optimisation studies but are done following the experience and capability of a shipyard, although a possible exception to this is the use of composites.

Despite the growth of hardware capabilities and the development of optimisation methods, progress in the optimisation of ship structures has been slow. Until recently only small-scale optimisation problems have been studied. The application of optimisation to ship structure problems has gone through a number of phases, and it is useful to outline developments.

The application of formal structural optimisation concepts in ship structural design was first described in Europe by Moe and Lund [33] and in the United States by Evans [14] and Adamchak [1].[2] Initially optimisation was confined to strictly structural questions, that is, structural form and scantlings. A brief description of some of the methods used is included in Sections 14.2–14.8. One of the problems that were encountered of the approach followed was that the algorithm had to be adjusted to each particular structure.

At a later stage, when optimisation was combined with finite element procedures, the computational effort and complexity increased significantly. To overcome these difficulties two avenues were followed: either more efficient search algorithms were introduced [16], or the optimisation problem was divided into more than one level,[3] thus effectively simplifying it. Despite these advances, difficulties remained in that it was not possible to bring together optimisation of topology, shape scantlings and materials. Furthermore, structural optimisation has to be seen as part of the overall ship design problem so that other questions such as hydrodynamics, powering, propeller, equipment layout and size are considered as well. The problem thus concerns the successful introduction of *multi-criteria* and *polyvalent* design optimisation. Because of these difficulties, optimisation procedures have not been widely used in ship design practice.

Recently, however, more generic approaches that address more than one objective have begun to be used. These methods rely on computer-based models of evolutionary processes, and a variety of approaches have been explored.

In this chapter we shall trace developments and provide examples of some of the earlier methods. These examples will help to draw some useful conclusions with regard to the ship

[2] A review of early work on ship structural optimisation is included in [25]. [3] For example, [22] and [36].

structure optimisation problem in general. We shall then focus on the latest methods that are being used, of which genetic algorithms is one, and present examples of results reached in recent years.

14.2 Simultaneous Mode Design

When a prismatic section such as a column is subjected to a compressive load, in order to select the dimensions of the section, the expression that gives the buckling load is used. It is seen that this depends on the length of the column as well as the distribution of the material within each section. The dimensioning of the cross-section remains indeterminate, however, and clearly there exist many different geometries that can be used to fulfil a particular strength requirement. How then do we decide on a particular type of cross-section and how do we select dimensions for it? One answer to this question is suggested by elementary strength of materials theory and the fact that any one type of cross-section can fail in a variety of modes. It would make sense to proportion the selected cross-section so that the failure stresses for each mode coincide; in this way, the structure would not be over-designed with respect to any one failure mode.

Let us therefore consider a column that transmits a load P over a distance L. The column may fail either by yielding or instability. For a given cross-sectional shape and area, the buckling load is equal to P_{cr}. If every dimension of the column is multiplied by a factor N, linear dimensions are factored by N whereas areas are factored by N^2. For constant axial stress, the axial strength increases in proportion to the cross-sectional area. Therefore,

$$P_{cr_2} = N^2 P_{cr_1} \tag{14.1}$$

N can be expressed in terms of L, and thus,

$$N = \frac{L_2}{L_1} \tag{14.2}$$

Substituting this in Equation (14.1) gives

$$\frac{P_2}{{L_2}^2} = \frac{P_1}{{L_1}^2} \tag{14.3}$$

The quantity P/L^2 will be the same for all columns in which the same axial stress develops, provided their cross-sections are of the same form. It is therefore possible to compare columns that have the same type of cross-section using this quantity, termed the *structural index*.[4] Of the columns being compared it is then possible to select one on the basis of weight or other criteria.

The allowable stress is given by the Euler formula:

$$\sigma = \frac{P}{A} = \frac{\pi^2 E \rho^2}{L^2} \tag{14.4}$$

[4] Originally derived by Wagner and subsequently used by Shanley [42].

The weight will be a minimum when the stress is a maximum. Since the maximum stress is determined by instability (buckling strength), it will depend on the proportions of the cross-section. This suggests that we should replace ρ^2 by a non-dimensional factor that depends on the proportions of the cross-section. A *shape parameter* c can thus be defined as follows:

$$c = \frac{\rho^2}{A} \tag{14.5}$$

where ρ is the radius of gyration $(= I/A)$. For a hollow cylindrical tube of diameter D and wall thickness t, $\rho^2 = D^2/8$ and $A = \pi Dt$. Thus, $c = D/8\pi t$.

Substituting in Equation (14.4) and since $A = P/\sigma$, we obtain

$$\sigma_1 = \frac{\pi^2 EcP}{L^2 \sigma_1}$$

Hence,

$$\sigma_1 = \pi \sqrt{Ec\left(\frac{P}{L^2}\right)} \tag{14.6}$$

Rearranging Equation (14.6) and substituting for c we obtain

$$\frac{P}{L^2} = \frac{8{\sigma_1}^2}{\pi E\left(\frac{D}{t}\right)} \tag{14.7}$$

The *structural index* is seen to be inversely proportional to the diameter to thickness ratio (D/t). For overall buckling, an upper limit to D/t exists, but at a critical load, local buckling of the tube wall will take place. Timoshenko obtained the following expression for the local symmetrical buckling stress of a thin-walled tube:

$$\sigma_2 = kE\left(\frac{t}{D}\right) \qquad \text{where} \qquad k = \frac{2}{\sqrt{3(1-\nu^2)}} \tag{14.8}$$

although test data indicate a value $k = 0.4$. For optimum design, we assume that overall buckling at σ_1 will occur simultaneously with local buckling of the wall tube at σ_2. By equating Equations (14.6) and (14.8) we obtain an expression for the optimum value of D/t for which the allowable stress is a maximum [15, 42]:

$$\sigma_1 = \pi \sqrt{Ec\frac{P}{L^2}} = \sigma_2$$

Thus,

$$\pi \sqrt{\frac{E}{8\pi}\left(\frac{D}{t}\right)\frac{P}{L^2}} = kE\left(\frac{t}{D}\right)$$

$$\pi \frac{E}{8}\left(\frac{D}{t}\right)\frac{P}{L^2} = kE^2\left(\frac{t}{D}\right)^2$$

Hence,

$$opt\left(\frac{D}{t}\right) = 2\left\{\frac{k^2 E}{\pi \frac{P}{L^2}}\right\}^{\frac{1}{3}} \tag{14.9}$$

If inelastic buckling takes place, the elastic modulus has to be replaced by an inelastic modulus.

The process described above is termed *simultaneous mode design optimisation.* As mentioned earlier, its rationale is based on the premise that material is used most efficiently if the failure load is the same for all modes.[5]

This particular problem concerns the design of an individual structural member. However, ship structures consist of a large number of members, and therefore any attempt at optimisation needs to address a much wider range of questions. In the next section we shall describe one of the most common approaches used early on in the optimum design of ship structures.

14.3 A Heuristic Approach to Form Optimisation

Having considered the optimisation of the dimensions of a single structural member, we will now consider the optimisation of the form of a member, albeit using a heuristic rather than a formal mathematical approach. This problem was considered in [9] and concerns a beam whose task is to transfer a central point load to two supports; the problem statement is given in Figure 14.1a [9].

Figure 14.1 includes a number of alternative designs whose minimum scantlings have been set to ensure adequate robustness and corrosion resistance. These designs are compared in Figure 14.1f, in which cost efficiency (C/C_0) is plotted against weight efficiency (W/W_0). Reference values[6] for both weight and cost are given in the figure. The following general conclusions were drawn. Firstly, form is the most influential parameter on the quality of the design (when compared to beams, trusses are seen to be superior). Secondly, gains made by improved layout and scantlings are smaller than those obtained by selecting a suitable form. Thirdly, the minimum thickness constraint was found to have an important effect and is an active constraint. Finally, of the materials used, steel is seen to be most efficient (although this design parameter is affected by material and fabrication costs).

Clearly then, when we turn our attention to ship structures, it becomes necessary to take a closer look at the question of form optimisation and to investigate whether it is in fact possible to improve existing designs in this respect. The complexity of ship structures and the actions of the external loading lead us to consider three levels of form: *primary, secondary* and *tertiary*.

[5] It has been found, however, that when the failure loads of two or more buckling modes are very close to each other, nonlinear interactions take place that lead to a reduction in the actual failure load of the structure. It is therefore necessary to designate the mode and load for which the structure will be designed as well as to ensure that the failure loads that correspond to other failure modes will exceed this by a reasonable margin.

[6] Valid at the time of publication of this volume.

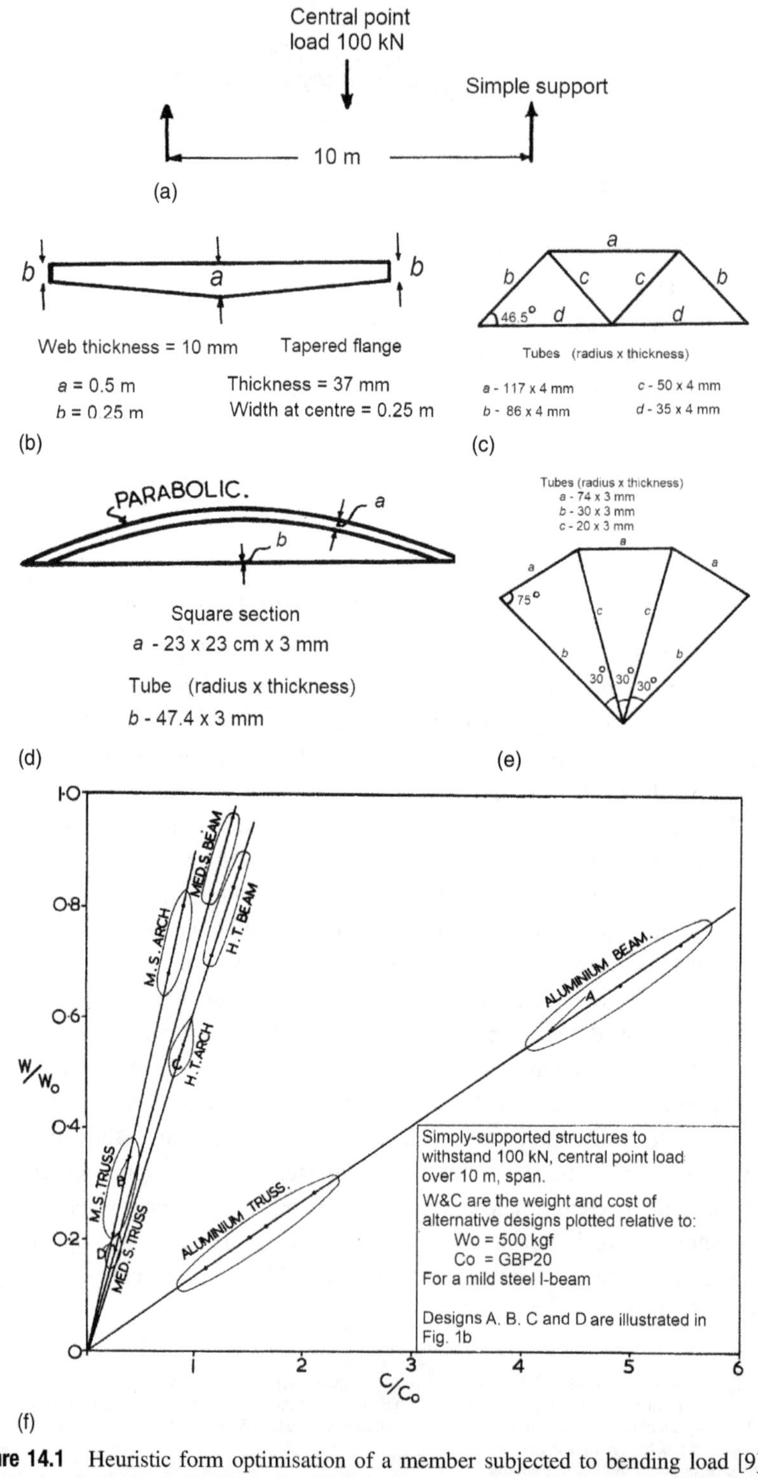

Figure 14.1 Heuristic form optimisation of a member subjected to bending load [9]. (a) Statement of problem. (b) Tapered aluminium beam. (c) Mild steel truss. (d) Arch (HT steel). (e) Medium steel truss. (f) Comparison of weight and cost for alternative designs.

Primary form refers to general structural layout, that is, the number, layout and proportions of cargo compartments, ballast tanks and fuel tanks. These variables depend on the quantity and the properties of the cargo transported, and in a number of cases, safety issue arise. The problem may be stated thus [9]:

Given a specified structural function and constraints imposed for safety reasons, can the presently adopted primary forms be improved upon?

The most important issues affecting the choice of primary form are operational and safety requirements, and these depend to a large extent on the type of ship in question. As a result, freedom of choice of primary form is restricted, and the most important variable affecting primary form becomes the relative importance of longitudinal versus transverse loading of the hull girder. In the case of an individual compartment, as ship size increases, the transverse loading to which it is subjected assumes more importance than the longitudinal loading. This is seen if we compare two ships that differ only by a scale factor S. In the case of longitudinal strength, if the effective wave height varies as (ship length)^m, then the mean deck thickness will vary approximately as $S^{(m+1)}$. The transverse loads on the structure will vary as S^3, so that the corresponding deck thickness will be proportional to S^2. In the case of large ships, $m = 0.3$, and thus it is clear that transverse strength requirements become more important and the need for form optimisation of the transverse stiffening system also becomes apparent. This concerns the distribution and position of transverse bulkheads and horizontal decks.

Secondary form has to do with the configuration of gross panel members once their position has been decided. The constraints to which these are subjected concern operational and production issues. Loading consists of (a) predominantly lateral pressure, (b) in-plane loading and (c) combined loading. The choice may be single or double skin and whereas the stiffening system may be unidirectional, orthogonal or unstiffened. In the case of bulk cargo ships, corrugated bulkheads have been extensively used in place of flat stiffened plating.

Finally, we consider the tertiary or elemental form. This essentially relates to selecting the appropriate type of stiffener cross-section, as form optimisation of plate members is meaningless (although in the case of composite materials or sandwich-type construction this requires special study). Caldwell and Woodhead [9] used dimensional analysis to compare different member forms and concluded that the choice of optimum form does not necessarily lead to conventional prismatic beams. It was also disclosed that trusses may have a role to play in ship structures in specific instances.

14.4 Basic Concepts of Mathematical Structural Optimisation

The aim of structural optimisation is to select *design variables* to achieve, within the *constraints* placed upon structural behaviour, topology and other factors, a goal of optimality that is defined by an *objective function,* for specified loading and operating conditions. The optimisation problem is thus seen to be based on three components:

design variables, constraints and the objective function. The optimisation problem is formulated in the so-called *design space* within which the design variables may assume any value. Optimisation theory includes a large number of methods that have been developed in order to solve particular categories and types of problems. In general, computer-based approaches are used, since real-life practical problems involve large numbers of design variables and constraints, so hand calculations or graphical methods cannot be used.

Design variables are the quantities that are allowed to vary such that an objective function is fulfilled within the imposed constraints. In structural optimisation these may include dimensions and structural properties of individual members such as thickness and section modulus, but they may also relate to the configuration of a structure as well as material properties. There are differences in each of these, as in the first case continuity in the values of design variables is possible, whereas in the latter two the situation is not so clear.

The sum total of all design variables may be designated by the vector $\boldsymbol{x}$, the ith design variable being $\boldsymbol{x}_i$. It is possible to describe the design variables using the design space, which is defined by axes each of which corresponds to one design variable. If therefore an optimisation problem involves three design variables, the corresponding design space is three-dimensional, and in this case it is possible to provide a pictorial representation. In general, however, the number of variables exceeds the number three, so it is not possible to use sketches, in which case an n-dimensional space is required, termed a *hyperspace*. A point k in the design space is a design with variables $\boldsymbol{x}^k$.

The *objective function* is the function whose greatest or least value is sought in an optimisation procedure. Within the design space a large number of designs can be selected, for each of which the objective function assumes a particular (scalar) value. This value corresponds to a selected property of the design that may be structural, such as weight, stiffness or strength, or else it may be non-structural, such as cost. For this reason the objective function is often called a *merit* or *cost function* in optimisation literature. In the case of the structural design of a weight-sensitive vessel such as a catamaran or a SWATH, weight will be the prime contender for minimisation. Structural weight, W, can be calculated using the following expression:

$$W = \sum_{i=1}^{N} \rho_i L_i A_i \tag{14.10}$$

where ρ_i, L_i and A_i are, respectively, the material density, length and cross-sectional area of the ith member. If density and length are maintained constant, then the only design variable will be the cross-sectional area. In vector notation,

$$W = \rho \boldsymbol{L}^T \boldsymbol{A} \tag{14.11}$$

It is seen that in this case the objective function is *linear*. If the length were also to vary, the objective function would be *nonlinear*. In the case of a linear problem that includes

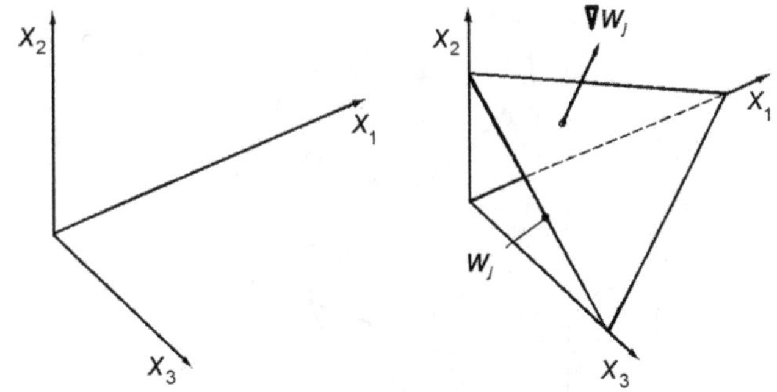

Figure 14.2 Optimisation parameters [17]. The *gradient* of the objective function, ∇W, is a vector that includes all derivatives of the objective function with respect to each of the design variables, specifically (a) design space (three variables) and (b) objective function W in design space.

three design variables (Figure 14.2a) it is possible to sketch the part of the objective function that lies within the design space (Figure 14.2b).

It is seen that the locus of the objective function W is a plane. This includes all feasible design points. In n-dimensional space the locus of a linear objective function is termed a *hyperplane*, and if the objective function is nonlinear, its locus is termed a *hypersurface*.

$$\nabla W = \left[\frac{\partial W}{\partial x_1} \cdots \frac{\partial W}{\partial x_n}\right]^T \tag{14.12}$$

The gradient vector is normal to the function in question at the point at which it is taken (Figure 14.2b). It defines the direction of design change in which the objective function increases most rapidly for a given increment in change.

Constraints are restrictions that need to be satisfied in order for the design to be acceptable. They may take the form of a limitation on a design variable or a group of variables, in which case they are termed *explicit* constraints, or else they may take the form of limitations on quantities whose dependence on design variables cannot be stated directly, such as stresses, in the case of structural design (*implicit* constraints). Such limitations may be expressed in terms of equalities or inequalities so that in the case of a equality constraint,

$$g_i(\boldsymbol{x}) = 0 \qquad i = 1, \ldots, E \tag{14.13a}$$

for E constraints. In the case of an inequality constraint,

$$g_i(\boldsymbol{x}) \leq 0 \qquad i = 1, \ldots, I \tag{14.13b}$$

Constraints are also classified as *side* and *behaviour* constraints (Figure 14.3). In the former case, a limitation is imposed on the maximum or minimum value a design variable may assume (e.g. minimum plating thickness). Side constraints may also be imposed on groups of design variables. Behaviour constraints are applied to impose limitations on

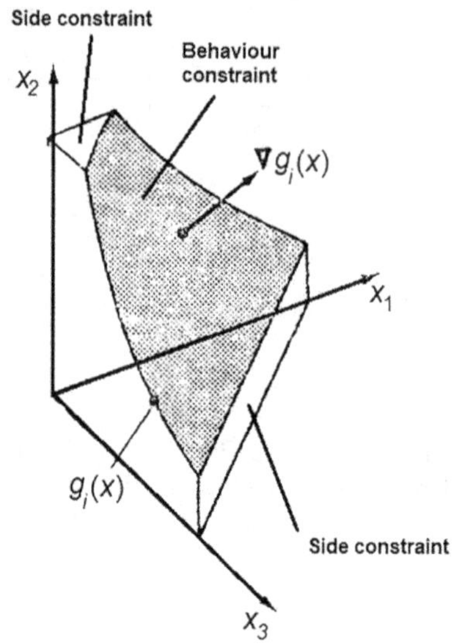

Figure 14.3 Side and behaviour constraints.

displacements and stresses, for example. As such they are generally implicit constraints, whereas side constraints are explicit.

Within the design space, the locus of each constraint condition is a surface upon which the constraint is satisfied as an equality constraint. For an indeterminate elastic structure, the constraint surface is continuous and curved (Figure 14.3).

14.5 An Example Involving Optimisation of Topology and Scantlings

We shall now describe the application of structural optimisation theory to a ship structures problem that was published in [32] and [33]. This problem concerns the optimisation of the topology and the scantlings of a hatch cover, the aim being to achieve a minimum weight design.

Problem Definition

A hatch opening of width 6 m (Figure 14.4a) is to be covered by aluminium box beams, whose cross-section is shown in Figure 14.4b. The design load is 10 kN/m^2. The flange thickness and the height are to be selected so that the weight of the box beam is a minimum.

The following requirements are to be satisfied

1. $h, t_f \geq 0$
2. Maximum shearing stress $\tau \leq \tau_{all}$
3. Maximum bending stress $\sigma \leq \sigma_{ball}$
4. Buckling of flanges must not occur $\sigma_b \leq \sigma_k$
5. Maximum deflection $w \leq Kl_0$

Table 14.1 *Expressions for constraints variables in terms of free variables*

No	Quantity	Mathematical expression
1	Load on box girder	$q_1 = 6kp/cm$
2	Maximum shearing force	$Q = q_1 l_0/2 = 1{,}800kp$
3	Maximum bending moment	$M = q_1 {l_0}^2/8 = 270{,}000kpcm$
4	Maximum shear stress in web	$\tau = \frac{Q}{2t_S h} = \frac{1{,}800}{(2)(0.5)(h)} = \frac{1{,}800}{h}$
5	Maximum bending stress in flange	$\sigma_b = \frac{M}{I}\frac{h}{2} = \frac{M}{bt_F h} = \frac{4{,}500}{t_F h}$
6	Buckling stress of flanges (σ_k)	$\sigma_k = \frac{4\pi^2 E}{12(1-\nu^2)}\left(\frac{t_F}{b}\right)^2 = \frac{E{t_F}^2}{1{,}000}$
7	Maximum deflection	$w = \frac{5}{384}\frac{q_1 {l_0}^4}{EI} = \frac{(56.2)(10^4)}{Et_F h^2} l_0$

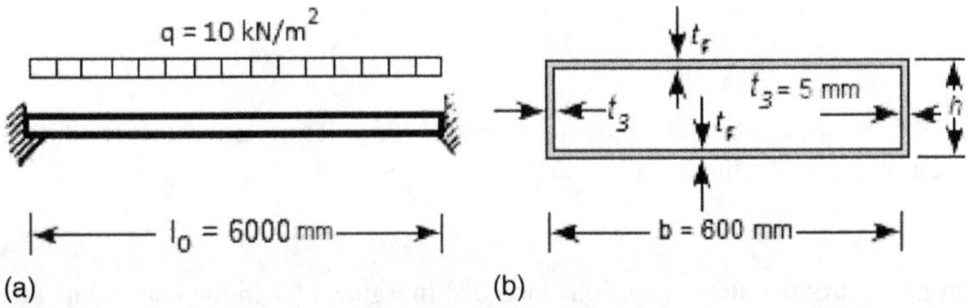

Figure 14.4 Design of hatch cover [33]. (a) Load application. (b) Cross-section. Reprinted courtesy of the Royal Institution of Naval Architects, London.

Solution

Following Equations (13a–b), these requirements can be written as

$$
\begin{aligned}
g_1 &= h \geq 0 \\
g_2 &= t_f \geq 0 \\
g_3 &= \frac{\tau_{all}}{\tau} - 1 \geq 0 \\
g_4 &= \frac{\sigma_{ball}}{\sigma_b} - 1 \geq 0 \\
g_5 &= \frac{\sigma_k}{\sigma_b} - 1 \geq 0 \\
g_6 &= \frac{Kl_0}{w} - 1 \geq 0
\end{aligned}
\tag{a}
$$

In the next step relations for τ, σ_b, σ_k and w in terms of the free variables t_f and h are introduced. These are listed in Table 14.1 [33].

The moment of inertia is assumed to be given by the following simplified expression:

$$I = \frac{1}{2} bt_F h^2 \tag{b}$$

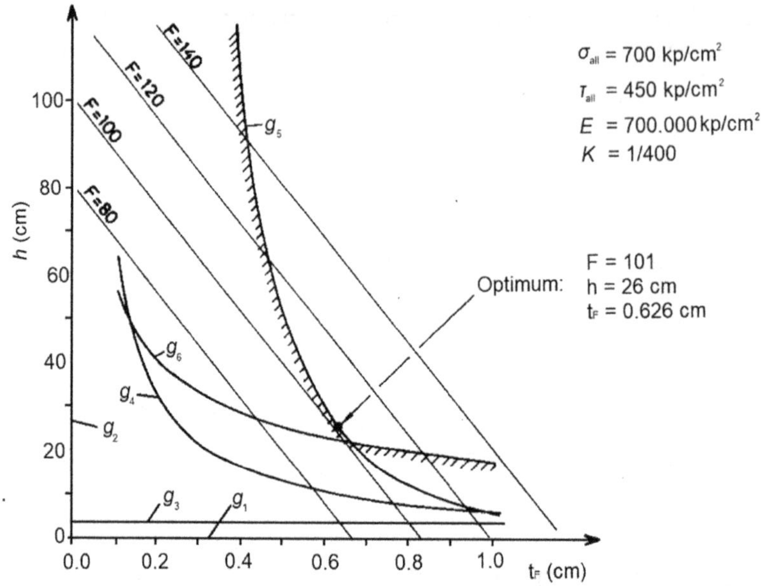

Figure 14.5 Constraints and objective function [33]. Reprinted courtesy of the Royal Institution of Naval Architects, London.

We can now introduce the expressions included in Table 14.1 in the constraints (a). The constraints are thereby expressed in terms of the free variables only and are plotted in Figure 14.5 for the values included in the figure. The portion of the plane that lies above and to the right of the hatched portions of the curves g_5 and g_6 represents the set of feasible or acceptable designs.

Since the stated aim is to minimise the weight of the box girder, the objective function assumes the following form:

$$F(t_F, h) = \frac{W}{\gamma l_0} = 2t_s h + 2t_F b = h + 120t_F \tag{c}$$

where W is the weight of the box girder. The weight function is included in Figure 14.5 and is plotted along an axis perpendicular to the t_F-h plane. An inclined plane passing through the origin is thus obtained. A series of lines of constant weight is shown in Figure 14.5. The optimum solution is indicated in this figure and is seen to lie on the tangent to g_5, at which point $F = 101$. We note that the active constraint is buckling, although the deflection constraint (g_6) is also close.

14.6 Nonlinear Programming

In general, the objective function as well as the relationships of the design variables to the behaviour variables (Equation (14.11)) contain nonlinear terms, which leads to a *nonlinear programming* problem. One approach to solving such problems that are however unconstrained is *Powell's method*, which is illustrated in Figure 14.6.

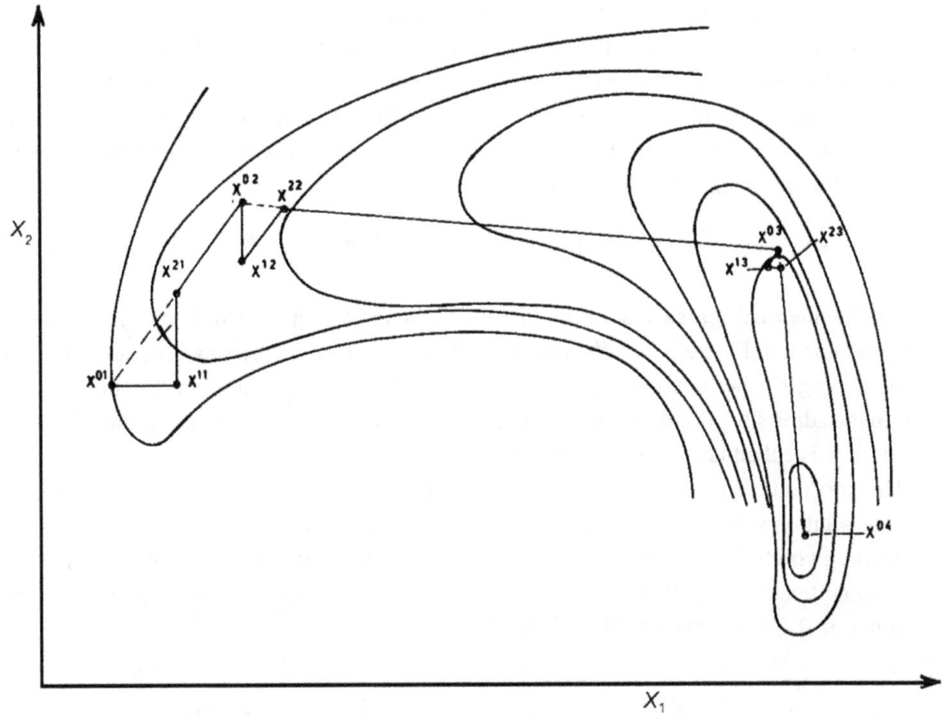

Figure 14.6 Powell's method for nonlinear unconstrained programming problems [33]. Reprinted courtesy of the Royal Institution of Naval Architects, London.

In this particular example two free variables exist, (x_1, x_2). Starting from an arbitrarily chosen point (x_{01}) the search proceeds along a line parallel to the x_1 axis until a minimum value is reached. The search then continues in a perpendicular direction until a new minimum is reached. The third stage consists of a search in the direction obtained by extrapolating the line segment joining x_{01} and x_{21}. In this way a new minimum is reached that in Figure 14.6 is represented by point x_{02}. This process is repeated so that a new point x_{03} is obtained, and so on. If the problem involves n free variables, it is formulated and solved in n-dimensional space.

As mentioned above, this method can be used when there are no constraints imposed. The presence of constraints may exclude minima from the feasible area, and therefore it is necessary to allow for their existence. This can be done by resorting to an alternative method,[7] or else the method described above has to be appropriately modified. Moe and Lund [33] introduced a transformation of the existing problem to a series of unconstrained minimisations (Sequential Unconstrained Minimisation Technique (SUMT)) (Figure 14.8). This approach is based on the use of a *penalty term,* which is introduced to compensate for any constraint violation.

The objective function $F(x)$ then becomes $P(x, r_k)$, where

$$P(x, r_k) = F(x) + r \sum_{i=1}^{m} G(g_i(x)) \tag{14.14}$$

[7] One method used to solve ship structural optimisation problems requiring constrained minimisation is the *Lagrange multiplier method* [9].

In this expression G is an operator on the ith constraint and r is a positive constant termed the *penalty parameter*. The second term is the penalty term, and our task is then to perform an unconstrained minimisation of this expression for a series of values of r_k ($k = 1, 2, \ldots$) so that the resulting minima of Equation (14.14) converge to the solution with $r_k \to \infty$. One form of the penalty function G that has proven to give satisfactory results is $1/g$ so that Equation (14.14) becomes

$$P(x, r_k) = F(x) + r \sum_{i=1}^{m} 1/g_i(x) \tag{14.15}$$

The geometrical representation of Equation (14.15) for the example discussed previously is shown in Figure 14.7. We observe that as subsequent approximations are obtained by increasing r_k, the function P approaches the optimum. In this case the solutions all lie within the admissible domain, which is bounded by F and g_5, the active constraint. For this reason this is called an *interior* penalty formulation.

This procedure does not necessarily lead to a global optimum of the objective function, and the solution will generally correspond to a local optimum within the neighbourhood of the starting point. One additional difficulty arises from the fact that certain design variables have to be treated as continuous whereas in fact some of them will be integer in nature (number of stiffeners, plate thickness, etc.).

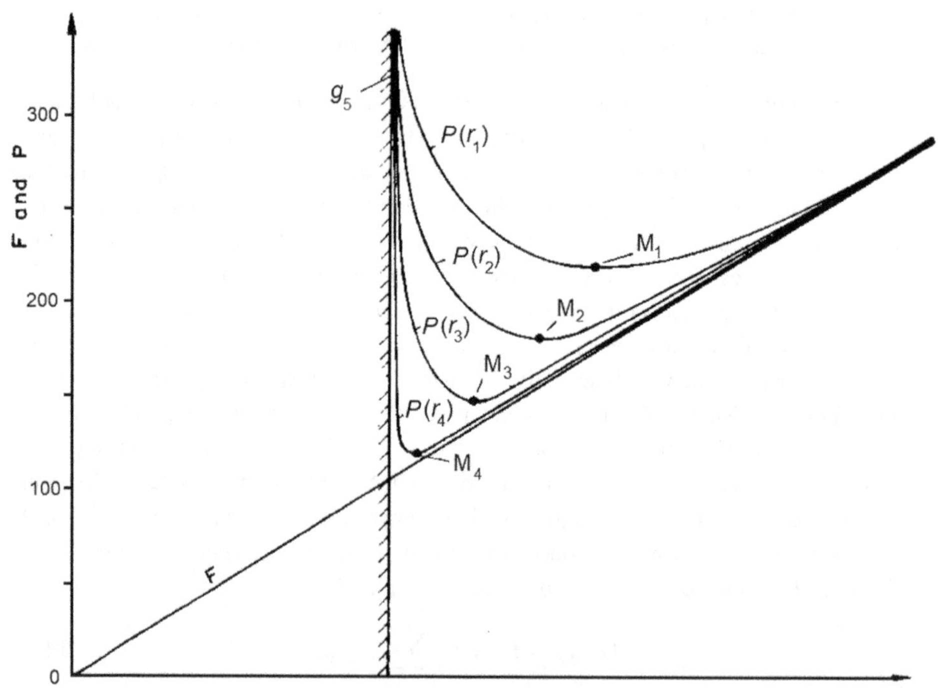

Figure 14.7 Objective function $F(h, t_p)$ and transformed function $P(h, t_p, r_k)$ [33] (section α-α in Figure 14.5). Reprinted courtesy of the Royal Institution of Naval Architects, London.

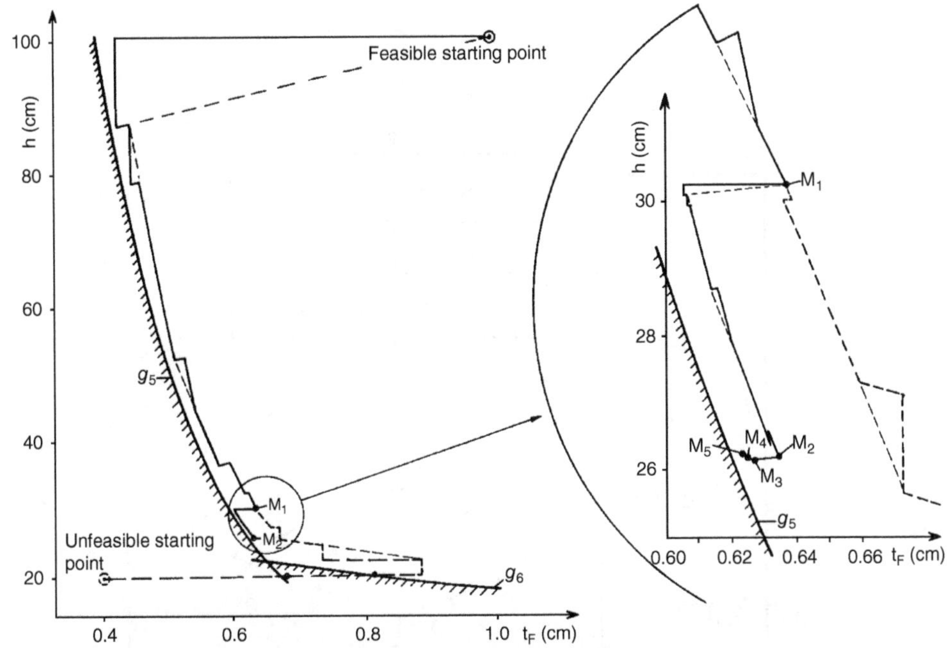

Figure 14.8 Application of SUMT starting from feasible and unfeasible starting points [33]. Reprinted courtesy of the Royal Institution of Naval Architects, London.

14.7 Rational Design and Optimisation of Ship Structures

In ship structural design, optimisation is an important part of the rational design process, as illustrated in Figure 14.9. Rational structural design requires that a large number of design cycles be executed so that it becomes necessary to use optimisation algorithms that are computationally efficient and sufficiently accurate for the design problem.

The *redesign* problem, which constitutes the second stage in the design cycle (Figure 14.9), can thus be formulated in terms of an objective function, design variables $\boldsymbol{X}$ and NC constraints, of which NI are inequality constraints. If $g_i(X)$ is the ith constraint function, the redesign problem is stipulated as follows:

Calculate the values of the design variables $\boldsymbol{X}$ that satisfy the following constraints:

$$g_i(\boldsymbol{X}) \geq 0 \qquad i = 1, \ldots, NI \tag{14.16a}$$

$$g_i(\boldsymbol{X}) = 0 \qquad i = NI + 1, \ldots, NC \tag{14.16b}$$

such that the objective function,

$$\sum_{j=1}^{N} V(K_j, X) \tag{14.17}$$

is optimised.

N is the number of design variables and K_j is a cost coefficient.

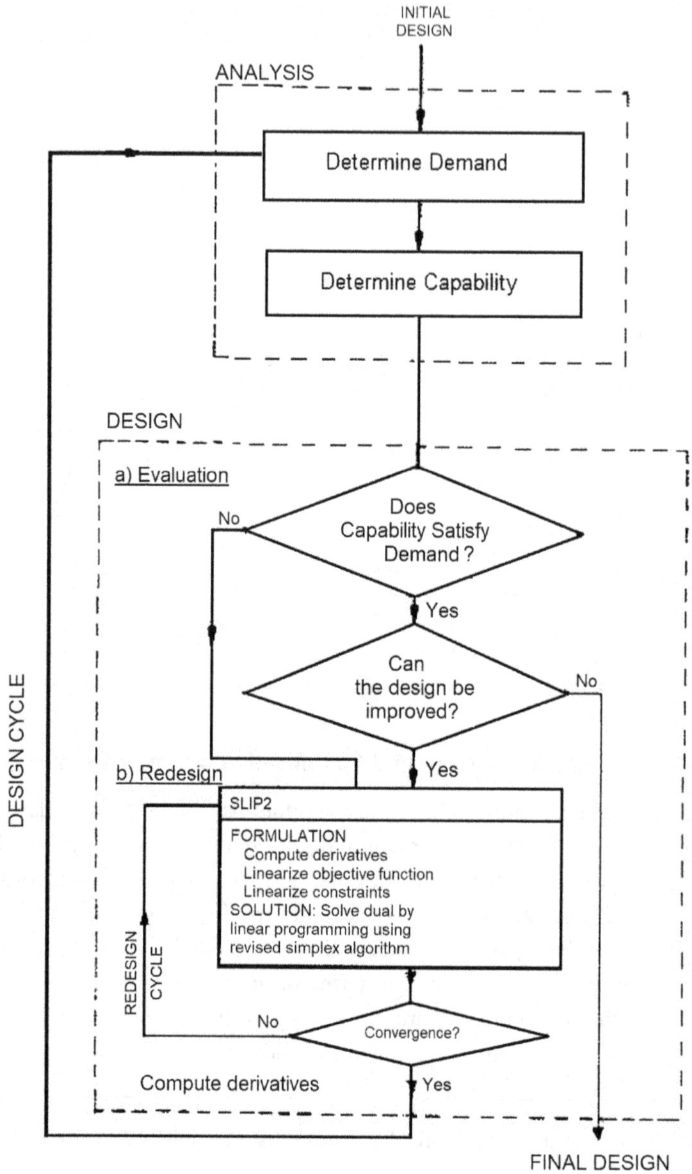

Figure 14.9 The rational design process [31].

14.8 Linear and Sequential Linear Programming: Implementation in the MAESTRO Program

If the objective function and the constraints are all linear, we have what is termed a *linear programming* (LP) problem. In this case the solution is to be found at a vertex of the multidimensional polyhedral *feasible space* formed by the inequality constraints. LP problems are encountered in many fields, although in most practical situations *nonlinear programming* (NLP) is required, because either the objective function or the constraints, or both, are nonlinear. In structural optimisation in general, both are nonlinear, since the formulas describing the geometry of a member include products of dimensions (objective function and constraints) and the strength formulation includes parameters such as the section modulus that also includes products (constraints). LP, however, retains its importance because one way of solving NLP problems is to convert them into a series of LP ones whose solution can be obtained much more efficiently. LP and other techniques to solve such problems are discussed in textbooks dealing specifically with optimisation theory and programming, such as [6]. The *simplex method* is the most well-known one used in LP, the basic idea being to proceed from one feasible solution to another in such a way that the value of the objective function decreases continually until the minimum is reached.

In the case of nonlinear problems that involve a very large number of design variables such as the structure of a ship, it has been found more efficient to use *sequential linear programming* (SLP) rather than NLP. The latter is used effectively when the number of design variables is limited. In contrast, in ship structural optimisation,

- There are many design variables.
- The number of constraints greatly exceeds the number of design variables.
- Most constraints are nonlinear and are thus computationally time-consuming.

To solve problems of this type, an improved version of SLP was developed for use in the computer program MAESTRO.[8] In SLP the nonlinear constraints and the objective function are linearised using first-order derivatives of the Taylor expansions of these functions. In problems that are characterised by the features listed above, this leads to a lack of convergence of the optimisation procedure. To overcome this, second-order terms were used to improve the linear representation of the nonlinear constraints in a modified version of SLP, termed SLIP2 [22, 31]. When a nonlinear problem is linearised, in theory it is not possible to determine the optimum if it does not coincide with a constraint – that is, when it is a *field optimum*. However, in complex, highly constrained systems the optimum is nearly always located at constraint intersections so that linearisation of second-degree terms does not lead to convergence problems when searching for the correct solution. The SLIP2 algorithm can be formulated in two ways:

1. A standard problem with one nonlinear objective function and a mix of linear and nonlinear constraints
2. A multi-objective ('goal') programming problem that includes a mix of multiple linear and nonlinear objective functions (goals) and a mix of linear and nonlinear goal constraints in addition to the ordinary linear and nonlinear constraints.

In this section the linearisation in the case of a single nonlinear objective function with both linear and nonlinear constraints will be described, following [31]. As mentioned previously, in SLP the objective function and the nonlinear constraints are linearised by

[8] MAESTRO is a design-oriented ship structural optimisation program [22].

using the first-order terms of a Taylor series expansion. In this case, for a constraint function $g(X)$ the first-order expansion about a design point X^0 is given by

$$\begin{aligned} g(\boldsymbol{X}) &\cong g(\boldsymbol{X}^0) + (X_1 - X_1{}^0)\left(\frac{\partial g}{\partial X_1}\right)_0 + (X_2 - X_2{}^0)\left(\frac{\partial g}{\partial X_2}\right)_0 + \ldots \\ &\quad \ldots + (X_N - X_N{}^0)\left(\frac{\partial g}{\partial X_N}\right)_0 \\ &\cong g(\boldsymbol{X}^0) + \left(\frac{\partial g}{\partial X_j}\right)_0 X_j - \left(\frac{\partial g}{\partial X_j}\right)_0 X_j{}^0 \quad j = 1, \ldots, N \end{aligned} \tag{14.18}$$

Hence, the linear approximation to the nonlinear constraint $g(\boldsymbol{X}) \geq 0$ for the design point ($\boldsymbol{X}$) is

$$\sum_{j=1}^{N} \left(\frac{\partial g}{\partial X_j}\right)_0 X_j \geq -g(\boldsymbol{X}^0) + \sum_{j=1}^{N} \left(\frac{\partial g}{\partial X_j}\right)_0 X_j{}^0 \tag{14.19}$$

Then, for NC constraints, Equation (14.19) becomes

$$\sum_{j=1}^{N} p_{i,j} X_j \geq q_i \qquad i = 1, \ldots, NC \tag{14.20}$$

where

$$p_{i,j} = \left(\frac{\partial g_i}{\partial X_j}\right)_0 \quad \text{and} \quad q_i \geq -g(\boldsymbol{X}^0) + \sum_{j=1}^{N} \left(\frac{\partial g}{\partial X_j}\right)_0 X_j{}^0 \tag{14.21}$$

In a similar manner, the linearised version of the objective function is

$$V(\boldsymbol{X}) = \sum_{j=1}^{N} \left(\frac{\partial V}{\partial X_j}\right)_0 X_j + V(\boldsymbol{X}^0) - \sum_{j=1}^{N} \left(\frac{\partial V}{\partial X_j}\right)_0 X_j{}^0 \tag{14.22}$$

Letting

$$m_j = \left(\frac{\partial V}{\partial X_j}\right)_0 \qquad \text{and} \qquad n = V(\boldsymbol{X}^0) - \sum_{j=1}^{N} \left(\frac{\partial V}{\partial X_j}\right)_0 X_j{}^0$$

We have

$$V(\boldsymbol{X}) = \sum_{j=1}^{N} (m_j X_j) + n \tag{14.23}$$

The first term in Equation (14.23) represents the increase in the objective function if the design is changed and can thus be referred to as the *incremental cost*. The second term represents the *base cost*, which remains constant. It is thus not necessary to include it in the redesign problem.

The linearised redesign problem can thus be restated as follows:

Find the values of the design variables $\boldsymbol{X}$ that minimise the incremental cost:

$$\sum_{j=1}^{N}(m_j X_j) \tag{14.24}$$

and satisfy the linear constraints:

$$\sum_{j=1}^{N} p_{i,j} X_j \geq q_i \qquad i = 1, \ldots, NI \tag{14.25a}$$

and

$$\sum_{j=1}^{N} p_{i,j} X_j = q_i \qquad i = NI + 1, \ldots, NC \tag{14.25b}$$

When using a first-order approximation, if the derivatives $(\partial g/\partial \boldsymbol{X})_0$ are small or if the constraint function $g(\boldsymbol{X})$ has a high degree of curvature (large values of the second derivatives near $\boldsymbol{X}^0$), then the tangent plane is a poor approximation to $g(\boldsymbol{X})$, and the linearised constraint differs substantially from the actual constraint $g(\boldsymbol{X}) = 0$ (Figure 14.10).

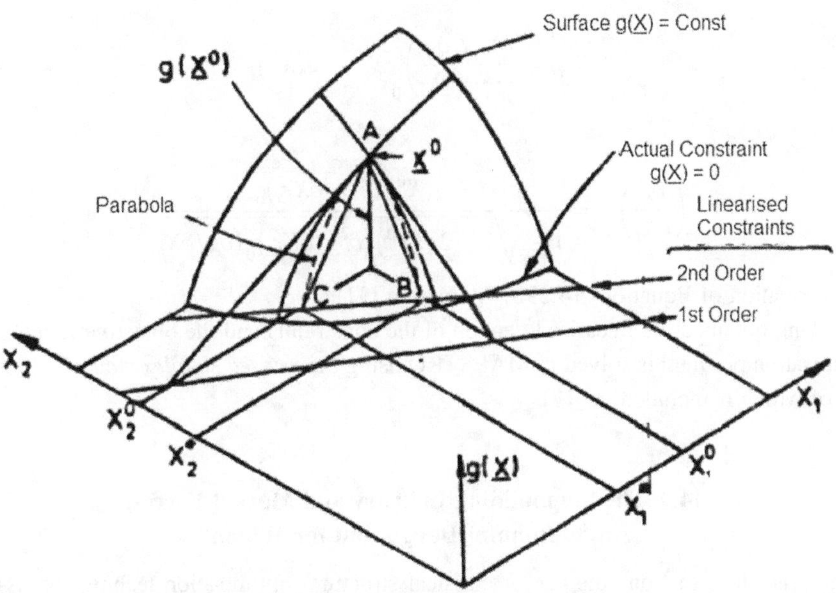

Figure 14.10 First- and second-order linearisation of constraints [31].

An improved result is obtained by including the second-order terms of the Taylor expansion, as follows:

$$g(\boldsymbol{X}) = g(\boldsymbol{X}^0) + \sum_{j=1}^{N}(X_j - X_j{}^0)\left(\frac{\partial g}{\partial X_j}\right)_0 + \\ +\frac{1}{2}\sum_{j=1}^{N}\sum_{k=1}^{N}(X_j - X_j{}^0)(X_k - X_k{}^0)\left(\frac{\partial^2 g}{\partial X_j \partial X_k}\right)_0 \tag{14.26}$$

Sufficient accuracy is achieved if only the uncoupled second derivatives are retained. In this case the previous equation becomes

$$g(\boldsymbol{X}) = g(\boldsymbol{X}^0) + \sum_{j=1}^{N}(X_j - X_j{}^0)\left(\frac{\partial g}{\partial X_j}\right)_0 + \\ +\frac{1}{2}\sum_{j=1}^{N}(X_j - X_j{}^0)^2\left(\frac{\partial^2 g}{\partial X_j{}^2}\right)_0 \tag{14.27}$$

Using this approach, the nonlinear constraints can be replaced with linear approximations that use second-order terms. The resulting system of linear constraints can be written as follows:

$$\sum_{j=1}^{N}\left(\frac{\partial g}{\partial X_j}\right)_0^* X_j \geq q_i \qquad i = 1,,, \ NI \tag{14.28a}$$

$$\sum_{j=1}^{N}\left(\frac{\partial g}{\partial X_j}\right)_0^* X_j \geq q_i \qquad i = NI + 1,,, \ NC \tag{14.28b}$$

where

$$q_i = \sum_{j=1}^{N}\left(\frac{\partial g}{\partial X_j}\right)_0^* X_j{}^0 - g(\boldsymbol{X}^0) \tag{14.29a}$$

and

$$\left(\frac{\partial g}{\partial X_j}\right)_0^* = \frac{g(\boldsymbol{X}^0)(\partial^2 g/\partial X_j{}^2)_0}{1 - \sqrt{1 - 2g(\boldsymbol{X}^0)(\partial^2 g/\partial X_j{}^2)_0(\partial g/\partial X_j)_0^2}} \tag{14.29b}$$

The derivation of Equation (14.29b) is given in [31].

Having obtained the linearised version of the constraints and the objective function, the optimisation problem is solved in MAESTRO using the *revised simplex method*, a description of which is included in [17].

14.9 The Shipbuilding Industry and Market Forces: An 'Optimum' Design, but for Whom?

In the preceding sections the use of classical structural optimisation techniques vis-á-vis ship structures was outlined. A number of measures of merit were mentioned, but it is seen that no attempt was made at assessing the quality of the structure with reference to non-structural issues that are of importance in the realm of engineering structures at a broader level. Such issues are related, on the one hand, to the investment potential of the structure and, on the other, to the industry within which it operates. It is therefore necessary to go beyond the narrow confines of structural considerations and to take a broader view of ship design, ship structural design as well as construction (production). In doing so we need to consider the market within which ships operate and the roles of key participants. This will enable us to gain a better understanding of how and why ship structures are designed and built the way they are.

Key participants (*stakeholders*) are bodies or groups of individuals that are directly or indirectly involved in structural design and thereby in the operation and performance of a particular ship. Stakeholders that are directly involved include shipowners (designers), shipyards, charterers and classification societies. Stakeholders that are indirectly involved include hull underwriters, financial institutions, government and/or flag institutions that represent passenger and crew interests, and international organisations such as the IMO. The maritime industry is very broad, and ships operate in different market sectors that abide by different practices. It is therefore necessary to distinguish between these in order to have a meaningful discussion with regard to stakeholders. Shipping markets are broadly classified as *charter* and *liner markets*, the former abiding by the rules of supply and demand for transportation services that define freight rates, the latter abiding by freight rates agreed upon by the carriers involved. Charter markets concern bulk cargo transportation whereas liner markets concern container shipping, passenger and cruise ship operations as well as other, more specialised sectors such as vehicle transportation. The latter comprise some of the more complex designs from the point of view of the structure of the ship.

In general, the most important stakeholders with regard to ship structural design issues are the owner/operator, whose interests may be represented by the designer, the builder, the classification society, repair yards and hull underwriters. Bulk cargo shipping operates on a global scale and is not restricted by national legislative practices with regard to the pricing of services. Freight rates depend on factors that affect the supply and demand of shipping services in the short term and as a result can exhibit a high degree of volatility. Such fluctuations are reflected in related markets such as the newbuilding market,[9] despite the fact that shipyard costs do not vary to the same extent.

The overwhelming majority of bulk cargo ships are built in the Far East, with three countries dominating this market: China, Japan and South Korea (Figure 14.11). There exist major shipyards in other parts of the world such as in Europe and the United States, but for the foreseeable future it appears that the dominant position of the major players with regard to these types of ships will remain unchallenged. It therefore pays to examine the shipbuilding market in that part of the world. In China, ship structural design is carried out by yard design offices and design bureaux such as MARIC (Marine Design and Research Institute) and SDARI (Shanghai Merchant Ship Design and Research Institute) in

[9] In addition to the newbuilding market, other markets affected by freight rates are the *second-hand* market and the *demolition* market.

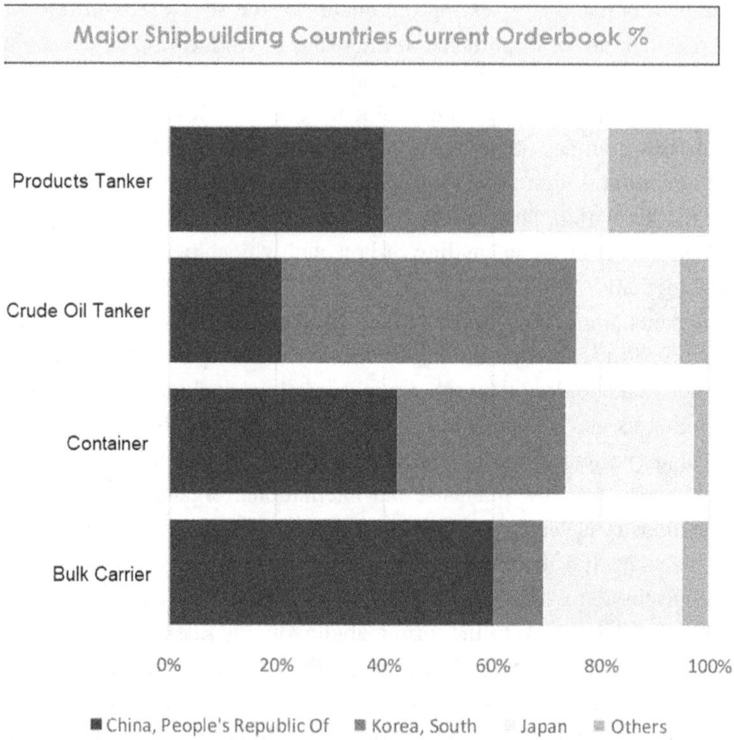

Figure 14.11 Orderbook for major shipbuilding countries (2018) [29].

co-operation with individual owner's consulting naval architects. Both of these organisations are controlled by CSSC, the China State Shipbuilding Corporation. In Japan and Korea the major shipyards market their own portfolios of designs. A characteristic of the shipbuilding sector for bulk cargo ships is thus the high degree of concentration in one geographical region.

With regard to passenger and cruise ships, shipyards located in Europe control a major share of the market. In this market sector ships are designed by specialist design offices in collaboration with the owner/operator as well as the yard.

The world shipowning market, on the other hand, is characterised by its high degree of fragmentation, as the number of owners that operate around the world is large and varied. The largest shipowning companies are in certain cases state-owned or form part of large industrial conglomerates, such as the major oil companies. There exist many independent owners that operate large fleets and that specialise in different ship types. The most established companies are the most active ones in the newbuilding market sector. In addition to these there exist a very large number of owners that operate ships acquired on the second-hand market and whose policy is to remain in this market. Owners' ship acquisition policies are understandably not an openly discussed topic. We shall therefore limit ourselves to some general comments in order to probe the motives and modes of thinking behind ship investment decisions.

Table 14.2 *Stakeholders*

Directly involved with ship structural design		
Stakeholder	*('Rational') desires*	*Negotiates with*
Owner/operator	Maximisation of profit via Minimisation of life cycle cost, incl. maintenance and repair of hull girder Freight market performance	Shipyard Class society Repair yards Underwriters
Shipyard	Competitive product via Minimum construction cost Compliance with requirements Technological advances Newbuilding market performance	Owner Class society Labour organisations Materials suppliers Subcontractors
Classification Society	Safe performance via compliance with ship construction and operation requirements	Owner Shipyard IACS
Indirectly involved with ship structural design		
Charterer	Compliance with charter party terms Market performance	Owner Underwriters P&I Club Cargo receivers Stevedores
Provider of finance (bank, private equity)	Loan security and adherence to loan repayment terms	Owner Shipyard Market performance
Underwriters/P&I Club	No claims (safe and incident-free performance)	Owner Charterer
Government/flag	Safety in general terms Acceptable working terms and conditions Crew living conditions and welfare	Owner P&I Insurer Shipyard
International organisations	Compliance with international regulations regarding safety of shipping (all aspects)	Owner Classification Society Flag state Labour organisations

With regard to the performance of the ship from the point of view of the ship structure as well as an investment vehicle, the perceived aspirations ('rational desires') of the major stakeholders are listed in Table 14.2. We shall discuss briefly the role of the major stakeholders, these being the owners, the yards and the funding institutions (banks, private funds, individual shareholders, etc).

On technical grounds, the operating life of a bulk cargo ship is expected to exceed 20–25 years, but for the owner who places an order for a newbuilding, important issues are the level and tenor of the loan on offer by the bank, the duration of the time charter required by the bank and the projected operating expenses over this period. The calculations carried out by the owner thus focus on a 10-year horizon, since any repayment of loans usually extends for 8–10 years. The owner expects to be able to repay the bank, pay for the operating

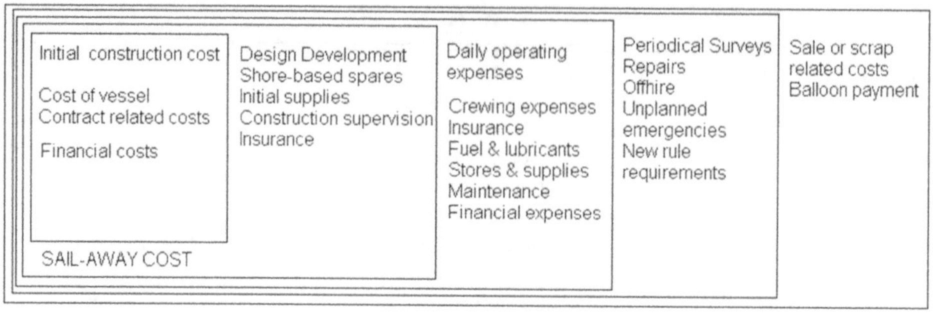

Figure 14.12 Life-cycle cost elements for merchant ships.

expenses of the ship (Figure 14.12) and retain a meaningful profit over this period. Profits for owners come from ship trading and also from the second-hand sale and purchase market. As a result, most owners do not retain control of their ships throughout their life, since market fluctuations provide opportunities for sales at profitable levels. This makes the second-hand market a very active and important one. As a result, owners have, in principle, no strong incentive to keep a high standard of upkeep and maintenance because the potential profits in the second-hand market may well dwarf any reduction in value resulting from poor maintenance. It is thus tempting to not invest in good upkeep, especially in the case of second-hand ships, and many owners do what is necessary to satisfy classification society and other requirements (such as oil tanker vetting) in order to ensure the continuous and unimpeded operation of their ships. Maintenance policies followed will in any case depend on ship type. For example, in the case of chemical tankers and LNG carriers that are highly capital intensive and involve the transportation of highly combustible/explosive cargoes, maintenance standards need to be above the levels deemed acceptable for other ship types.

There arise certain important differences in the case of specialised ships such as cruise-ships with regard to ownership and operation. In general, a cruiseship is designed and operated for a significantly larger proportion of its life by the owner who places the initial order with the yard. The ship is designed and built to serve a particular market, and the characteristics of this market play an important role in the design and thereby the structural design of the ship as well. In contrast, for bulk cargo ships, versatility is considered to be a prime asset, since it facilitates the transportation of different kinds of goods as well as the resale of the ship in the second-hand market. In the case of a cruiseship, maintenance and upkeep of a high standard are required because of statutory as well as market requirements.

One other stakeholder closely linked to the performance of the structure of a ship is the classification society. The role of class is to ensure that standards are adhered to during the design and construction as well as during the operating life of the ship. The original role of classification societies was to act as technical advisors to underwriters. Nowadays their role extends far beyond this, since having valid class certificates is a necessary prerequisite for chartering a ship as well as the issuance of statutory certificates. Class may also be called upon to act as technical representative/advisor to financial institutions, legal entities, governments and flag states.

Banks and other sources of finance provide capital and need to ensure that capital and interest will be repaid as stipulated in the loan agreement. Banks do not assume any active role in technical issues, although they retain consultants that provide advice on yard capabilities, other technical issues, terms agreed, etc.

On the other hand, hull structure requirements are but one part of ship design, others being arrangement of spaces, hydrodynamic and powering requirements, sea-keeping, vibration, maintenance and life cycle cost, etc. There exists a range of possible solutions that will satisfy each stakeholder to a greater or lesser degree. In practice, it is also other, non-technical factors that enter the picture and have an influence on the solution that will finally be selected. In the case of bulk cargo ships, shipping market conditions are important (Figure 14.13), as these determine the negotiating strength of each party. When shipping markets boom, owners are prepared to invest, which increases demand for ship construction. This results in competition for berths, and consequently newbuilding prices rise. Shipyards are then in a more powerful position to dictate terms on both economic and technical aspects. If the market is depressed, as it usually is, the shipyards seek out business from among the fewer owners who are prepared to invest and are more willing to negotiate in order to attract orders. The second important feature of this particular market stems from the fact that, because of the cyclical nature of shipping, owners do not retain staff dedicated exclusively to ship design. Shipyards, in contrast, do maintain ship design personnel that seek to provide solutions with yard priorities uppermost in mind. These have to do with the types of ship being built, the available infrastructure of the yard, the quality of manpower and the cost and financing arrangements. As a result, there exists an imbalance of expertise and negotiating power in favour of shipyards. One other reason for which market conditions generally favour shipyards is that because of the significant capital investment involved in ships, owners rarely place orders for a series of ships. As a result, they are generally compelled to accept the features of 'standardised' designs offered by yards. Designs are therefore carried out by yards, and inevitably their own priorities will feature prominently; as mentioned in Table 14.2, these are related to minimisation of production costs. Nonetheless, shipyards compete with each other in providing competitive products by introducing technology that will reduce operating costs for the owner, such as fuel consumption.

In order therefore to provide a solution, the problem has to be considered at a more general level. Table 14.2 includes stakeholders and their 'rational desires' that can be abstracted into the following categories:

- Production cost (involving design, manufacturing, assembly, etc.)
- Robustness (insensitivity to uncontrollable parameters)
- Minimum life-cycle cost (construction, operation, manning, maintenance and repair, etc.)
- Safety (crew/passenger safety as well as ship safety such as collision and grounding resistance)
- Retrofitting (adapting to new role, introduction of new equipment)
- Environmental protection (pollution avoidance)
- Comfort (passengers)

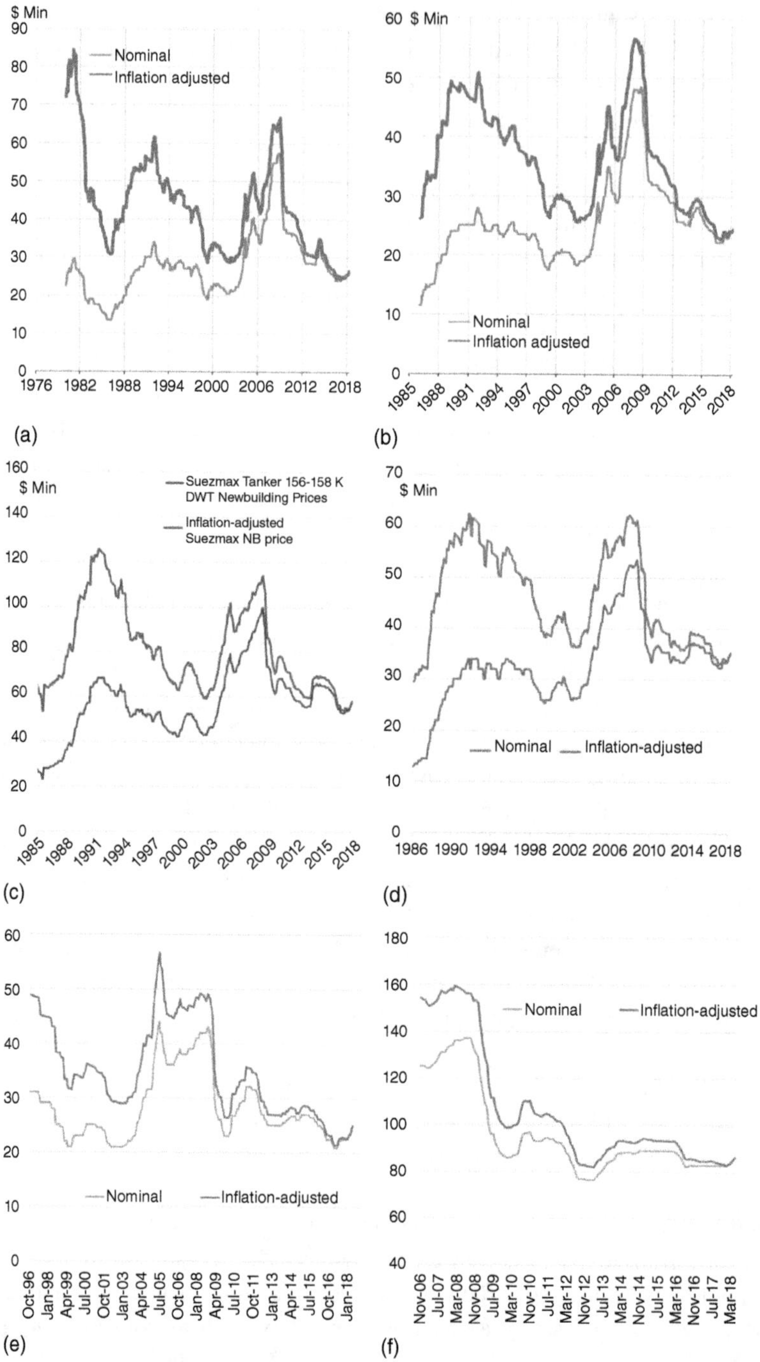

Figure 14.13 Newbuilding prices of cargo ships (in millions of USD) [29]. (a) 81,000 dwt Panamax bulk carriers. (b) 61,000–64,000 dwt Supramax bulk carriers. (c) 156,000–158,000 dwt Suezmax oil tankers. (d) 47,000–51,000 dwt Products carriers. (e) 1,700 TEU containerships. (f) 8,800 TEU containerships.

It is seen that the desires (or *objectives*) of stakeholders may be shared or may compete. For example, if in the case of a passenger ship comfort standards are upgraded, production costs will rise. There are other objectives that, when aimed at, will enhance one another in general terms (e.g. safety, robustness and environmental protection). The above list includes items that extend beyond structural design and encompass the entire design of a ship. Many efforts have been made to design and build ships with one or some of the above as an overriding objective [35]. In the next section we will discuss optimisation studies that aim at fulfilling a single objective, before considering the more complex problem of multi-objective optimisation.

14.10 Design for X

The use of mathematical optimisation techniques in ship-related applications including design, production and operation was initiated with a single objective and a single stakeholder in mind. In this section a number of objectives termed *Design for X* and the relevant means of reaching the desired goal will be discussed. It has to be borne in mind that the objectives do not necessarily change if multi-objective optimisation is considered. It is therefore important to understand what the designer is aiming at in each case. These include:

1. Production costs
2. Maintenance and repair costs of the hull girder
3. Robustness of the hull structure
4. Safety of the ship, cargo and crew
5. Protection of the environment
6. Life Cycle Costs
7. Passenger comfort

A brief discussion of items (1)–(5) follows. Issues concerning item (7) that are related to structural design (hull girder vibration) have been discussed in Chapter 10, whereas other issues such as ship motion characteristics and noise attenuation lie outside the scope of the present text.

14.10.1 Production Costs

'Design for production costs' aims at reaching a target cost by treating it as an independent design parameter. The objective is to make the design converge to an acceptable cost rather than letting the cost converge to the design. When designing for production cost, the manufacturing cost needs to be estimated during the concept and early design phases in order to identify high cost components and estimate the costs of alternative designs.

In ship production there are a number of components that need to be considered, including manufacturing, fabrication, materials and labour. The two main principles followed in design for ship production are:

1. Designs should strive for simplicity.
2. Designs should be suitable for the facilities of the shipyard in question.

On this basis, the following guidelines have been proposed, as distilled from the technical literature [35]:

1. Design the hull structure so that simple, fast and cost-effective fabricating methods can be implemented.
2. Standardise as much as possible, by reducing complexity.
3. Use modularity wherever possible. This involves fabricating smaller parts (modules) of the ship independently.

14.10.2 Maintenance and Repair Costs of the Hull Girder

It is the shipowners'/operators' concern that maintenance and repair involve low costs. This aspect of structural design has not received the attention it merits. Maintenance and repairs are carried out throughout the operating life of a ship and can affect its operation in many ways if not planned for at the design stage. In order to assess the magnitude of the problem and the factors that affect it, it is necessary to quantify costs for each category of damage that may occur under normal circumstances, such as corrosion, fatigue cracks and buckling throughout the lifetime of a ship. Related to damages and wear are the issues of *inspectability* and *accessibility*, which will affect the ease with which surveys and repairs are carried out. Examples of measures that may be taken to reduce maintenance and repair costs are, among others, (a) reduction of weld lengths; (b) ensuring that all ballast tank spaces are easily accessible and are coated with light-coloured protective coatings; (c) use of corrosion resistant steels in cargo tanks in the case of oil tankers; and (d) use of drones equipped with cameras for surveying inaccessible regions and of crawlers to measure plate thickness in inaccessible regions. Items (a), (b) and (c) can be planned at the design stage, which is when the experience of owners and class surveyors can contribute in a significant way.

14.10.3 Robustness of the Hull Structure

A robust structure is one in which the consequences of structural failure are not disproportionate to the effect causing the failure. That is, when taken to extremes, the failure of a single member of the structure will not lead to serious damage to and/or collapse of the entire structure. The degree of robustness is measured by comparing the seriousness of failure to the cause of that failure.

An example is the case of a tweendeck supported by a number of pillars. If the number of pillars is large enough, when some of them fail, the remaining pillars may be able to continue to support the deck. If the number of pillars used is insufficient, the deck may collapse.

Robustness is a measure of the predisposition of a structural system to loss of global equilibrium and global stability as a result of a failure scenario [3]. *Fragility* of a structure or element, on the other hand, may be expressed in quantitative terms by the conditional probability of failure for a given hazard event.

We recall the energy treatment of a system that comprises the structure in question and its immediate surroundings. In a conservative system the total energy content, ΔE, remains constant:

$$\Delta E_{Tot} = \Delta P + \Delta K = \Delta E + \Delta W + \Delta K = \text{const} \tag{14.30}$$

where

ΔP = change in potential energy

ΔK = change in kinetic energy

ΔE = change in internal energy, including strain and other types of energy

ΔW = change in external work done

The internal strain energy E_S consists of elastic strain energy, E_e, and the dissipated plastic work, E_p. If $E_p > 0$, the system is non-conservative.

A robustness index, I_R, can be based on the damage energy of a system for a given hazard scenario and is thus independent of structural resistance and risk [3]:

$$I_R\,(A_L|H) = \frac{\text{Damages up to unavoidable collapse state for hazard } h}{\text{Damages up to collapse state for hazard } h} \tag{14.31}$$

where $I_R(A_L|H)$ is the robustness index of a structure subjected to a leading action A_L and a set of hazard scenarios $H = \{h_1, h_2, \ldots h_n\}$. More specifically, I_R is calculated from

$$I_R(A_L|H) = \frac{D_{uc} - D_{1stfailure}}{D_c - D_{1stfailure}} \tag{14.32}$$

where

$D_{1stfailure}$ = damage energy of the structure when first failure occurs

D_{uc} = damage energy corresponding to the state of unavoidable collapse

D_c = damage energy corresponding to the collapse state

These quantities are evaluated for each hazard scenario considered. The robustness index assumes values $0 \leq I_R \leq 1$ and for $D_c - D_{1stfailure} = 0$, $I_R = 1$. Figure 14.14 [3] illustrates the progress of damage that the structure sustains until collapse occurs and the nomenclature used in Equation (14.32). Details of the procedure followed to determine I_R are included in [3].

Robustness is an issue of concern in the case of progressive collapse, and in the case of the hull girder this may arise when the ultimate bending moment of the hull girder is reached.

In the case of ship structural design it is necessary to understand how external loads are transferred throughout the structure and the important role of structural details such as connections, brackets, etc.

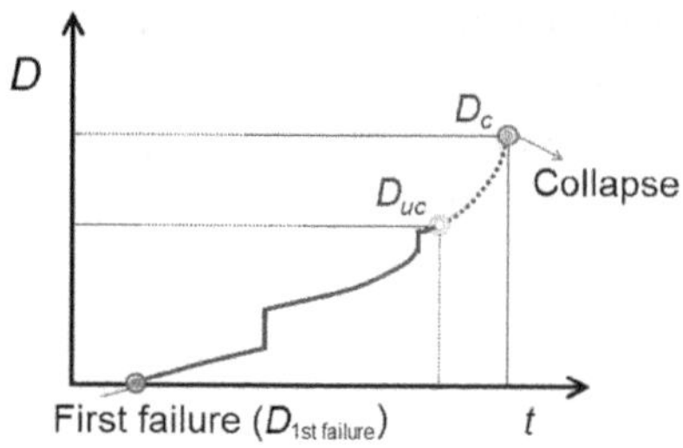

Figure 14.14 Structural robustness index nomenclature.

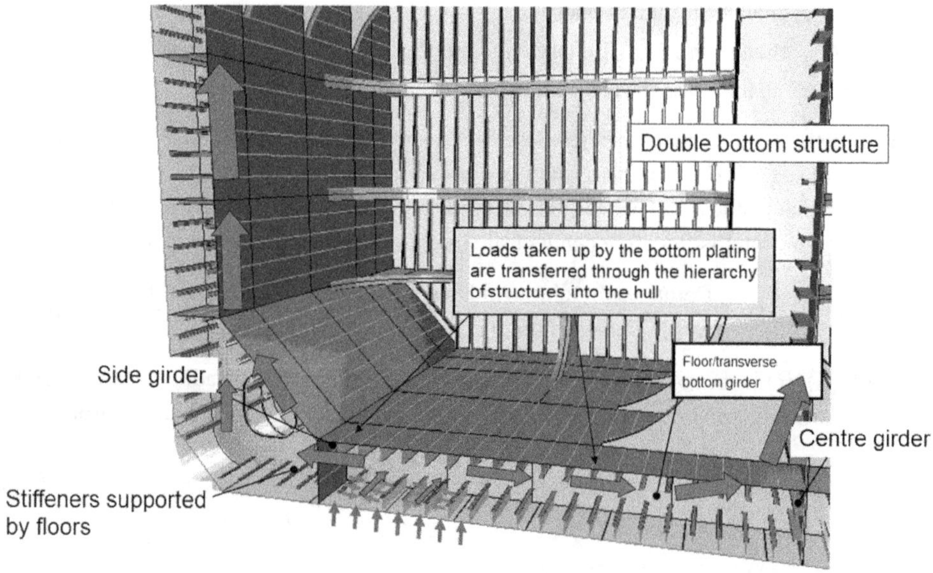

Figure 14.15 Load transfer mechanism of hydrostatic pressure acting on the bottom structure of a double skin bulk carrier [12]. Reprinted courtesy of Det Norske Veritas, Høvik.

The design options available to improve robustness are to:

- Modify or redesign the structure at a local level to ensure more effective load transfer (Figure 14.15 provides an example of the load transfer mechanisms in the case of hydrostatic pressure acting on the bottom plating of a single skin bulk carrier).
- Increase the redundancy of individual elements.
- Improve inspection and maintenance procedures to address damage due to wear and tear.

14.10.4 Safety of the Ship, Cargo and Crew

This section discusses measures to improve safety against extreme loads due to wave-induced damage, collision and grounding events. Fire hazards ought to be included as well, although these are important in specific types of ships such as passenger ships, chemical tankers and gas carriers as well as the offshore industry. Measures taken to improve safety are divided into two major groups: *active* and *passive safety measures* [13]. Of concern to the structural designer are the latter, and these are intended to come into play after an accident has occurred. These measures relate to structural integrity, crashworthiness and resistance to rupture. International organisations such as IACS have been active on this front, and rules for ship construction now include methods of calculating the residual strength of the hull girder following an accident. IMO requirements relate to subdivision rules that are probabilistically based.

The effects of wave-induced loads and the resulting damage have been discussed in Chapter 3. With regard to grounding events, these are categorised as *soft grounding* events that occur on soft sea bottoms and *hard grounding* events that occur on rocky ground. To study ship response in the case of soft grounding events, classical mechanics, small-scale and full-scale tests as well as computer software have been used. With regard to hard grounding events, it is necessary to take into account both ship motion and the response of the structure to the surface of the seabed. Theoretical methods are available as well as computer software.

Ship collisions have been studied extensively, and classical mechanics, computer-based approaches as well as tests have been used. At the design stage it is necessary to assess the crushing strength of bow structures as well as the effect the impact has on the struck ship. An example of such an analysis is included in Chapter 13.

Specific cases in which collision and grounding resistance considerations are of primary importance arise in the case of ships carrying dangerous cargoes, such as oil tankers, chemical carriers and gas carriers. Ships powered by nuclear reactors are another example.

14.10.5 Protection of the Environment

The concept of safety as discussed earlier in the book can be extended to the environment – an issue that traditionally was of importance in the case of oil tankers and other cargo ships that transport dangerous cargoes but more recently for all ocean-going ships with respect to flue gas emissions. International bodies have introduced an extensive list of regulations that pertain to each type of ship in order to protect the environment. Direct stakeholders such as the owners and indirect ones such as the shipyard will bear the costs and consequences of any new measures and understandably will proceed with designs that satisfy rule requirements and no more. Introducing safety measures in ship design and ship structural design can increase costs significantly, and it is important to assess the burden of risk and the financial consequences among the stakeholders concerned. The issue is one of multi-objective optimisation in which there is competition between stakeholders – that is, there are opposing objectives that need to be reconciled. A rational basis for reconciling technical and financial aspects is needed, and in [30] a game theory approach [20] was implemented in order to reach a *competitive optimum*. This is meant to be the fairest possible choice and represents the best compromise among stakeholder preferences. The method was applied to the case of a 40,000 dwt tonne chemical/products carrier on the question of how and to what extent crashworthiness can be increased above that of a standard design. It was concluded that it is possible to increase crashworthiness without any deleterious financial consequences for the stakeholders involved. It is therefore possible, in principle, to design ships with safety features that exceed regulation requirements without necessarily financially penalising the stakeholders concerned. The proposed approach needs to be applied on a case-by-case basis.

14.11 Multi-Criteria Decision-Making

Table 14.2 includes the stakeholders that are directly and indirectly related to ship structural design, as well as their 'rational desires', or primary objectives. It is clear that, because of the multiplicity of stakeholders, we are dealing with a *multi-stakeholder* problem, in which the objectives differ. When viewed from the perspective of an individual stakeholder, the ranking of the objectives that may be included in a list such as Table 14.2 will differ. We see therefore that we are dealing with a *multi-criteria decision-making* problem.

Multi-criteria decision-making (MCDM) is a field of operations research dedicated to the development and implementation of decision support tools and methodologies used to confront complex decision problems involving multiple criteria, goals or objectives of a conflicting nature [28]. The aim is to determine the 'best' from a number of alternatives, this being the one with the highest degree of desirability with respect to all relevant criteria. MCDM techniques have been employed in the posterior evaluation of Pareto front optimal solutions in order to choose the best one [34].

MCDM problems can be classified in several ways, depending on:

- The number of decision-makers involved (single decision-maker, group decision-making);
- The type of data (deterministic, stochastic, fuzzy);
- The type of information and the salient features of the information;
- Whether the solutions are *explicitly* or *implicitly* defined.

Solutions that are explicitly defined concern multi-criteria *evaluation* (*or selection*) problems. In these there exist a finite number of alternative solutions that are known in advance. An example is the case of a shipping company that is interested in purchasing a ship on the second-hand market, in which case a finite number of candidate ships are available. One approach to solving such a problem is included in the next section. Another example would be the selection of a main engine for a ship from a number of candidate solutions.

Solutions that are implicitly defined concern multi-criteria *design* problems (*multi-objective mathematical programming* problems). In these the alternatives are not explicitly known and a solution has to be found by solving a mathematical model. This is the case in ship structural design and ship design in general.

There are a large number of solution methods available for such problems, and these can be broadly classified as a priori and as a posteriori approaches. In the former, sufficient preference information is expressed before the solution process is initiated. Examples are the *utility function* method, the *lexicographic* method and *goal programming*.[10] In a posteriori solution methods the aim is to produce all Pareto optimal solutions. These methods include *genetic algorithms*, *particle swarm optimisation* and *simulated annealing*. A discussion on genetic algorithms with example applications is included in Section 14.13. There are also *hybrid methods* that combine multi-criteria decision-making with multi-objective decision-making.

Multi-criteria decision-making has seen rapid growth and expansion as its role in different application areas has increased significantly. We shall limit our discussion to two techniques, namely the *analytic hierarchy process* and *fuzzy set theory*, both of which have been applied in ship structural design optimisation. In addition to these we may note *case-based reasoning* (CBR), in which cases similar to a problem from an existing database of cases are retrieved and a solution to a decision-making problem based on the most similar cases is proposed. In the early stages of design, when proposals are still being conceived, designers often use older design and construction plans as part of the proposal to adequately convey the simple plan for the proposed design. For this reason CBR was applied to identify existing designs that best match with early design requirements in the case of engine room monitoring and automation systems design. The method was found to be successful not only at the early proposal stage but also during the actual detailed design.

In this chapter we are primarily interested in multi-criteria design problems, since ship structural design is one such problem. Before discussing these problems we shall illustrate the use of a MCDM technique in multi-criteria evaluation problems using a simple example.

14.11.1 Multi-Criteria Evaluation/Selection Problems

In a single decision-maker multi-criteria evaluation problem, there is one decision-maker (or entity), whom we can call the 'Manager', whose task is to select a single solution from amongst several; each of them has different *attributes*.[11] The Manager has in mind a number of *criteria* (or goals, or objectives) against which each of the candidate solutions is assessed. Thus, the selected attributes of the candidate solutions are necessary in assessing their performance against the specified criteria but are not necessarily identical to them because in many cases the information required is not explicitly available (see example that follows).

A multi-criteria evaluation problem can be expressed in matrix format. A decision matrix $\boldsymbol{A}$ is an $(M \times N)$ matrix in which element x_{ij} indicates the performance of alternative X_i when it is evaluated in terms of criterion C_j (for $i = 1,2,3,\ldots,M$, and $j = 1,2,3,\ldots,N$). It is also assumed that the decision-maker has determined the weights of relative performance of the criteria (denoted as w_j for $j = 1,2,3,\ldots,N$).

[10] See, e.g. [6].

[11] Attribute: a quality or characteristic inherent in or ascribed to someone or something (*The American Heritage Dictionary of the English Language*, 5th ed). In the literature on MCDA the terms 'criterion' and 'attribute' are used interchangeably, as are the acronyms MCDM and MCDA. In this text the meaning of these terms is treated as being distinct, so that the acronym MCDA is not used.

The general multi-criteria evaluation problem can be defined as follows [44, 52]:

Let X = {X_i, for i = 1,2,3,. . .,M} be a (finite) set of decision alternatives and C = {C_i, for j = 1,2,3,. . .,N} a (finite) set of criteria according to which the desirability of an action is judged. Determine the optimal alternative A^* with the highest degree of desirability with respect to all relevant criteria C_i.

	Criteria				
	C_1	C_2	C_3	...	C_N
Alt	w_1	w_2	w_3	...	w_N
X_1	x_{11}	x_{12}	x_{13}	...	x_{1N}
X_2	x_{21}	x_{22}	x_{23}	...	x_{2N}
X_3	x_{31}	x_{32}	x_{33}	...	x_{3N}
.	.	.	.	...	.
.	.	.	.	...	.
X_M	x_{M1}	x_{M2}	x_{M3}	...	x_{MN}

An example follows in which the simplest of techniques required to measure the importance of individual attributes is used (*weighted additive valuation*).

Example

Consider a shipping company that is interested in acquiring a ship on the second-hand market. The company operates cargo ships of a particular type, and following discussions with brokers, a list that includes currently available ships is prepared for Managers. In order to reach a decision on which of the candidate ships best fulfils the company's requirements the following steps need to be followed:

1. Clarify the problem.
2. Identify attributes of candidate solutions.
3. Identify criteria (objectives).
4. Implement a measurement of effectiveness.

1. Clarify the Problem

At the first stage, it is necessary to ask whether the purchase of a ship is necessary and to compare with alternative choices, such as the chartering of a ship from the market or the rescheduling of a ship that forms part of the owner's existing fleet. Another issue that may have to be considered is whether the ship to be purchased needs to comply with new regulations or with impending ones. Other issues will need to be addressed at this stage; these are particular to the circumstances of the Manager at the given point in time. Answers to such questions need to be soundly supported using data analysis.

2. Identify Attributes of Candidate Solutions

At the second stage it is necessary to identify and list the attributes of the ships that will assist in the decision-making process. These are included in the information contained in the report accompanying the list of candidate ships. On the basis of this report the Manager prepares the following list:

- Age
- Price
- Fuel consumption
- Hull and machinery condition
- Draught
- Manning requirements
- Location
- List of previous owners
- Yard of build

3. Identify Criteria (Objectives)

At the third stage it is necessary to identify the criteria according to which the candidate ships will be evaluated. If the company intends chartering the ship on a long-term time-charter, the charterer may well impose an age limit on the ship. This will act as an active constraint and may well exclude one or more candidate ships. If the remaining candidate ships are located in different parts of the globe and ships in specific regions are not acceptable for charter because of the distance and the imposition of a time limit for charter commencement, then one or more further ships may not be eligible. This process narrows down the choice to a short list.

The list of criteria that the company prepares (not in order of preference), follows:

1. Price C_1
2. Operating expenses C_2
3. Age C_3
4. Yard of build/previous owners C_4

4. Implement a Measurement of Effectiveness

Item 1 depends primarily on type of ship, size, age and market conditions. The price is negotiable and will affect the owner, with the ship's location having some role to play in this.

Item 2 is a function of ship type, size, age, condition and crewing costs. Data are available for the owner to have an accurate estimate of costs, barring unforeseen circumstances such as undisclosed damages, previous claims, impending legal action, etc.

Item 3 is an attribute of the ship, and a clear choice exists among the short-listed ships, although the condition of the ship, which is a function of the quality of build, the types of cargoes carried in the past and the maintenance standards of the previous owners, influences operating costs. The owner's judgement and knowledge of the market will aid in this.

Item 4 is an attribute of the ship and, same as for item 3, a clear choice exists among the short-listed ships. The owner's judgement and knowledge of the market will aid in this.

Additive Valuation

The simplest method that can be used is to set up a scale ranging, for example, from zero to unity, against which each of the candidate ships will receive a score for each of the four listed attributes.

1. In the first step the criteria are defined (C_1, C_2, C_3 and C_4) and for each of these, each ship receives a score on a scale from zero to unity.
2. In the second step the four criteria need to be weighted, namely:
   - How important is the first criterion (price) in the total decision? w_1
   - How important is the second (OPEX) in the total decision? w_2
   - How important is the third (age) in the total decision? w_3
   - How important is the fourth (yard of build) in the total decision? w_4

   The four weightings w_1, w_2, w_3 and w_4 should be normalised, such that

$$\sum_{i=1}^{4} w_i = 1$$

3. In step (1) each ship j has received a score for criterion i. Hence, x_{ij} is the score of ship j in criterion i. The total score for each ship is calculated from

$$\sum_{i=1}^{4} x_{ij} w_i$$

   This is the score for ship j.
4. The ship with the highest total score is selected.

The approach followed in this example is one, simple way of reaching a decision based on the subjective ranking of selected criteria. There are many other ways of selecting and ranking criteria in MCDM.

14.11.2 Multi-Objective Optimisation and Pareto Optimality

In mathematical terms, a *multi-objective optimisation* problem that involves K criteria is expressed as follows:

$$\min \boldsymbol{F}(x) = [f_1(\boldsymbol{x}), f_2(\boldsymbol{x}), f_3(\boldsymbol{x}), \ldots f_K(\boldsymbol{x})] \tag{14.33}$$

$\boldsymbol{x} = [x_1, x_2, \ldots x_N]^T$ subject to the following equality and inequality constraints,

$$h_i(\boldsymbol{x}) = 0 \qquad i = 1, \ldots, I \tag{14.34a}$$

$$g_j(\boldsymbol{x}) \geq 0 \qquad j = 1, \ldots, J \tag{14.34b}$$

The K criteria $f_1(x) - f_k(x)$ are individually dependent on the N unknown parameters in the vector $\boldsymbol{x}$. In order to obtain a solution to this problem it is necessary to introduce the

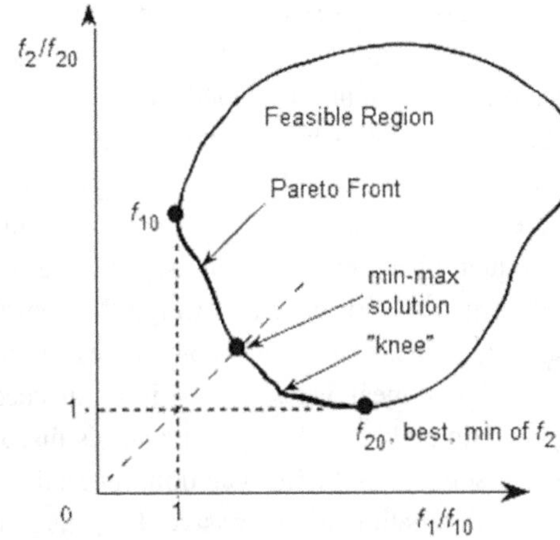

Figure 14.16 Pareto front between solutions f_{10} and f_{20} [34].

concept of *Pareto optimality*. This was proposed by Pareto in 1906 and is expressed as follows [34]:

A design is (Pareto) optimal if it satisfies the constraints and is such that no criterion can be further improved without causing at least one of the other criteria to decline.

Pareto optimality recognises the nature of co-existing conflicting or competitive criteria and results in a *set of solutions* that are considered equally satisfactory. A geometrical representation of the set of solutions for two criteria is included in Figure 14.16, in which the enclosed space represents the set of all possible solutions, in non-dimensional form. The subset of optimal solutions (Pareto solutions) has the form of a curve that joins points f_{10} and f_{20} that correspond to the two optima. A comparison of an enclosed solution and a Pareto one shows that the former can be improved upon; the Pareto set is therefore a set of *non-dominated* solutions.

We realise that a paradoxical situation arises because whilst on the one hand the solution consists of a series of optima, on the other a single product has to be designed and produced. In order to focus on a single design, it will be necessary to introduce additional criteria upon which a decision can be based. Examples of such criteria in the case of ship construction will be considered further in this chapter. A rational procedure that will lead to a single design has to be based on the ranking of each trade-off solution in accordance with the stated stakeholder objectives. Examples of these are maximum shipyard profit, minimum ship price, minimum maintenance cost and minimum life-cycle cost. Ideally, higher-level information is used to select one of the Pareto optimum solutions.

One way of tackling this problem is to rank the objectives by assigning *preference factors* to each of these. In this way a *composite objective function,* as the weighted sum of the objectives, can be formed [11]. The introduction of weighting factors converts the multi-objective optimisation problem to a single-objective optimisation problem. For this we can then obtain a single solution, as required. This approach is characterised by a degree of subjectivity and is not straightforward, since it may involve non-technical, qualitative and experience-driven information. We see that an important difference is that in ideal multi-objective optimisation an existing optimal solution is selected whereas in the preference-based optimisation a new solution is obtained.

An examination of the different objectives shows that in most cases they cannot necessarily be compared directly – for example, shipyard capability, freight market fluctuations and material costs are incommensurate. In the case of passenger ships, for example, we also have to be able to quantify objectives such as passenger comfort and aesthetics. We realise that we have to normalise, prioritise and weigh the contributions of the various objectives in arriving at a suitable measure. In addition, values can interact or conflict with each other, in both linear and nonlinear ways. Methods to succeed in reconciling these difficulties are described in Chapter 11 in which *fuzzy logic* is discussed.

One other point that has arisen is that the introduction of certain objectives such as life-cycle cost has proven to be computationally expensive. For this reason ways of reducing the cost using alternative approaches that simplify the problem have had to be introduced. These will also be described in this chapter and their use discussed in a practical example.

Before doing so, however, we will discuss *evolutionary algorithms*, and more specifically genetic algorithms. These are techniques that have proven successful in the search for Pareto front optimal solutions and have been applied in ship structure-related problems. Examples of their use in ship structural design are also given.

14.12 Evolutionary Algorithms, Genetic Algorithms and Applications

Genetic algorithms are one particular type of what are known as *evolutionary algorithms,* which are two-step processes of random variation and selection. They can be modelled mathematically as $x(t+1) = s(v(x(t)))$, where the population $x(t)$ at time t is operated upon by a random variable v and selection gives rise to a new population $x(t+1)$. The term *evolutionary algorithm* is used to denote in general computational algorithms that implement computational models of evolutionary processes that are key elements in their operation. The most important classes of evolutionary algorithms are (1) *genetic algorithms*, (2) *evolution strategies*, (3) *evolutionary programming* and (4) *genetic programming*. All these approaches are inspired by natural selection. They all share a common conceptual base of simulating the evolution of individual structures via selection, mutation and crossover. These terms will be explained in the next section that focuses on genetic algorithms, which are the type of evolutionary algorithm that has been used most extensively in ship structural design applications.

Evolutionary algorithms have found success because they can overcome some of the challenges that are due to multiple objectives, mixed design variables, irregular/noisy problem functions, implicit problem functions, expensive and/or unreliable function gradients and uncertainty in the model and the environment [6].

14.12.1 Genetics, Genetic Algorithms and the Fundamental Theorem

Genetic algorithms are so named because they have been inspired by, and function in ways that are similar to, the reproduction process of living organisms in the natural environment.

Therefore, before discussing the workings of genetic algorithms and their application in engineering, and in particular in structural optimisation problems, it is useful to briefly outline the process of evolution in nature as it is understood today, on the basis of the developments that have taken place since the middle of the 19th century. The modern theory of evolution of biological organisms is founded on two important principles: Darwin's theory of natural selection and genetics, which was founded by Mendel and has seen important advances during the 20th century (discovery of the structure of DNA, etc.).

The concept of *natural selection* is used to describe the gradual, non-random process by which biological traits become either more or less common in a population. This is achieved by the differential reproduction of their bearers.

The term *phenotype* is used to denote the composite of an organism‘s observable characteristics or traits: these include its morphology, development, biochemical or physiological properties, phenology, behaviour and products of behaviour (such as a bird's nest). Natural selection acts on an organism's phenotype, or physical characteristics. Phenotype is determined by an organism's genetic make-up (*genotype*) and the environment in which the organism lives. Often, natural selection acts on specific traits of an individual, and the terms phenotype and genotype are used narrowly to indicate these specific traits.

A *gene* is a molecular unit of *heredity* of a living organism. The total complement of genes in an organism or cell is known as its *genome*, which may be stored in one or more *chromosomes*. A chromosome consists of a single, very long DNA helix on which thousands of genes are encoded. When different organisms in a population possess different versions of a gene for a certain trait, each of these versions is known as an *allele*. It is this genetic variation that underlies phenotypic traits. A typical example is that certain combinations of genes for eye colour in humans that, for instance, give rise to the phenotype of blue eyes.

When some component of a trait is heritable, selection will alter the frequencies of the different alleles, or variants of the gene that produces the variants of the trait. Selection can be divided into three classes, on the basis of its effect on allele frequencies:

- *Directional selection* occurs when a certain allele has a greater fitness than others, resulting in an increase of its frequency.
- *Stabilising selection* lowers the frequency of alleles that have a deleterious effect on the phenotype – that is, produce organisms of lower fitness. This process can continue until the allele is eliminated from the population.
- *Balancing selection* does not result in fixation but maintains an allele at intermediate frequencies in a population.

Let us now consider *genetic algorithms*. These are search heuristics that mimic the process of natural evolution. They differ from classical optimisation techniques in that

1. The design variables are replaced by a set of coding variables.
2. The algorithm operates on a population of potential solutions rather than on individual solutions.

3. The objective function is used without any gradient information.
4. The transition scheme used is probabilistic and is not based on gradient information.

A genetic algorithm essentially consists of the following steps [19]:

1. Coding and decoding of design variables into *strings*. These are analogous to chromosomes encountered in nature. In general, they are expressed in binary form and contain the information describing their individual characteristics. Each individual bit position corresponds to a particular type of information that may be of the yes/no type or it may be quantitative. The number of bits required to represent the particular type of information will therefore vary. The position that encodes some feature is called a locus and, in an analogy to natural processes, each possible value at a locus is called an allele. Pseudo-random number generators are used to form an initial population of strings. Population size depends on the nature of the problem but may contain several hundreds or thousands of solutions.
2. Evaluation of the *fitness* of each string. Fitness is defined on the basis of the objective function and constraints and is always problem-dependent. Thus, in the case of structural engineering problems structural weight can be used to define the fitness function, giving $f = 1/W$, where f is the raw fitness and W is structural weight.
3. The selection of strings that will be used in the reproduction process will be the basis of strings for the next generation. Reproduction is analogous to Darwin's natural selection (survival of the fittest), and thus strings with a higher fitness will have a greater probability of being selected as 'parents'. There are several different selection methods used. These include *proportional* and *group selection*. In proportional selection, strings with higher fitness are allocated higher reproduction rates. In group selection, the whole population is divided into groups according to their fitness values. Every member of the group is assigned the same reproduction rate.
4. The application of genetic operators to generate the next solution of strings. *Crossover* is the mechanism whereby parts of two existing strings are combined to produce a string in a subsequent generation. It can take place in a number of ways. *Mutation* is a mechanism that involves the introduction of random changes to a population. This is beneficial in that it enhances diversity.
5. The algorithm terminates when either a maximum number of generations has been produced or a satisfactory fitness level has been reached for the population.

These steps are summarised in Figure 14.17.

It is necessary to describe in some more detail the crossover and mutation operators. As mentioned above, crossover involves the exchange of parts of strings. Consider the following two strings:

S1 010101010100011
S2 111111000111000

In *one-point crossover*, a single crossing site is selected at random. In the example illustrated below crossover takes place at the sixth bit.

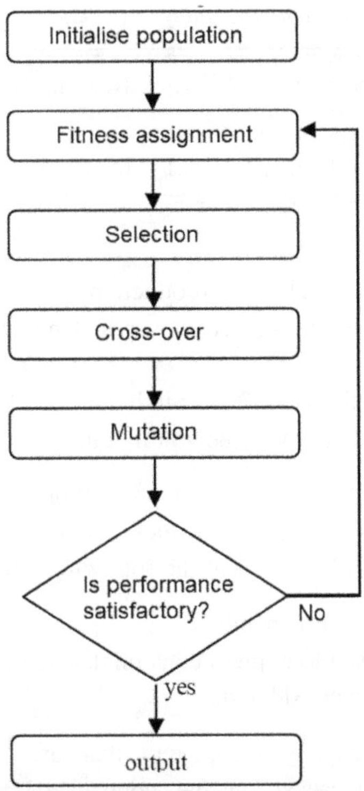

Figure 14.17 The basic structure of a genetic algorithm [18].

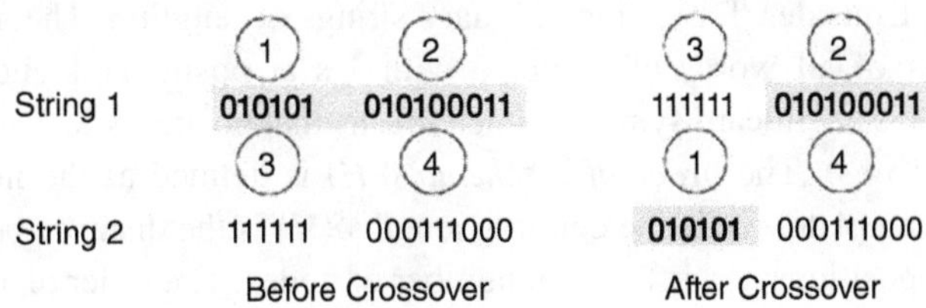

Other types of crossover are used as well (two-point, uniform, adapting).

Mutation involves the switching of individual bits (genes) provided a probability test is passed. The probability parameter used is generally low. Mutation acts as an insurance policy against premature loss of important information. Table 14.3 includes an example of mutation of two parent chromosomes of length 5. The mutation probability check is performed using randomly generated numbers.

Summarising, a simple generational genetic algorithm procedure consists of the following basic steps:

1. Choose the initial population of individuals.
2. Evaluate the fitness of each individual (chromosome) in that population.
3. Repeat on this generation until termination (time limit, sufficient fitness achieved, etc.):

Table 14.3 *Mutation example (mutation probability = 0.002)*

Old chromosomes	Random numbers				New chromosomes
1010	0.702	0.325	0.245	0.373	1010
0010	0.802	0.471	0.023	0.001	0011

(a) Select the best-fit individuals for reproduction.
(b) Breed new individuals through crossover and mutation operations to give birth to offspring.
(c) Evaluate the individual fitness of new individuals.
(d) Replace least-fit population with new individuals.

It is in fact the synergy of the processes of reproduction, mutation and crossover that can lead to significant improvements in quality (fitness). When compared to classical optimisation algorithms, genetic algorithms present the following advantages:

1. Discrete design variables can be handled.
2. An open format can be used to express constraint statements.
3. Multiple load cases can be considered.

The relationships between the parameters that are used in genetic algorithms are described by a model based on the concept of the *schema*, introduced by Holland [48].

A *schema* is a template that identifies a subset of strings with similarities at certain string positions. Consider, for example, binary strings of length 6. The schema 1**0*1 describes the set of all words of length 6 with 1's at positions 1 and 6 and a 0 at position 4. The * is a wildcard symbol, which means that positions 2, 3 and 5 can have a value of either 1 or 0. The *order of a schema* $0(H)$ is defined as the number of fixed positions in the template, while the defining length $\delta(H)$ is the distance between the first and last specific positions and is the total number of nodes. The order of 1**0*1 is 3 and its defining length is 5.

The *fitness of a schema* is the average fitness of all strings matching the schema. A *fitness function* is a particular type of objective function used to summarise, as a single figure of merit, how close a given design solution is to achieving the set aims. Fitness functions are designed by humans and, if specified wrongly, will either converge to an inappropriate solution or will have difficulty in converging. There are two types of fitness functions used: those that do not change and those that do, as when co-evolving the set of test cases. In certain cases the fitness function has to be defined using an iterative procedure.

In order to quantify the effect that the three co-existent processes of reproduction, crossover and mutation have on an existing schema within a population, we need first to consider the effect that each has individually. Since the probability of survival of a schema is proportional to its fitness, during subsequent generations the individuals with above-average fitness will survive at the expense of those with below-average fitness. That is, if *m(H,t)* is the number of schemata *H* in generation *t,* in generation *t+1*,

$$m(H,t+1) = m(H,t)\frac{f(H)}{f_{avg}} \tag{14.35}$$

where $f(H)$ is the average fitness of strings that include schema H and f_{avg} is the average fitness value of the whole population. We see that reproduction has the effect of increasing the number of schemata of above-average fitness exponentially with each generation.

We now consider the effect of crossover. This consists of the mutual replacement of strings of alleles in schemata that contain the same number of alleles. A cut divides the schema, and it is possible for more than one cut to occur. For present purposes we consider the following string S and schemata S_1, S_2.

S	1 0 0 1 1 | 0 1 0 1 0 0 1
S_1	* 0 * 1 * | 0 1 0 1 * * 1
S_2	* * * * * | 0 1 0 1 0 * *

If we assume that one-point crossover takes place and that the position of the cut is randomly chosen, the probability of it occurring at any one location depends on the length of the string and is thus equal to 1/*l*-1. Let us assume that the cut arises as indicated above. We observe that the effect of the cut on the two schemata differs. Schema S_1 is destroyed, since fixed alleles are replaced, whereas schema S_2 remains intact. It becomes clear then that the probability of survival p_s of a schema during crossover depends on its defining length. This probability is then

$$p_s = 1 - \frac{\delta(H)}{l-1}$$

and if p_c is the probability that crossover takes place,

$$p_s \geq 1 - p_c\left(\frac{\delta(H)}{l-1}\right) \tag{14.36}$$

We note that schemata that have a short defining length $\delta(H)$ have a greater probability of survival.

Finally we consider the effect of mutation. Mutation involves the random change of a single allele and has a probability of occurrence equal to p_m. A schema survives if all defined alleles survive. Since mutations are statistically independent, the probability that a schema survives is $(1 - p_m)^{O(H)}$, where $O(H)$ is the number of fixed positions. For small values of p_m this is approximately $1 - O(H)p_m$.

If the effects of reproduction, crossover and mutation are assumed to be independent, we can combine the above results in order to determine the effect they have on the number of schemata in a subsequent generation. The *schema theorem* or *fundamental theorem of genetic algorithms* is thus obtained and is expressed as follows:

$$m(H,t+1) \geq m(H,t)\frac{f(H)}{f_{avg}}\left[1 - p_c\frac{\delta(H)}{L-1} - O(H)p_m\right] \tag{14.37}$$

This theorem is due to Holland [21, 48] and is discussed in [19]. It states that short, low-order schemata are given exponentially increasing or decreasing numbers of samples, depending on a schema's average fitness.

14.12.2 Optimisation of a Catamaran Structure

In this section we shall describe the application of a genetic algorithm code to the optimisation of the hull structure of a catamaran, following [40]. Catamarans are a classic example of a marine vehicle that requires weight optimisation, since their TPCM immersion is particularly low. In this particular example optimisation is carried out with respect to this sole criterion. A profile view and a section of the vessel are shown in Figure 14.18. Figure 14.19 shows the portion of the hull structure that is being optimised. Ship structural designs have to comply with higher-level requirements such as payload distribution and subdivision requirements, in addition to classification society rules. As a result, the effort is directed at the topology and scantlings at a local level.

It is seen from Figure 14.19c that the hull girder is longitudinally stiffened throughout, with frames at fixed intervals. The general layout and distances between major structural items such as decks and bulkheads are treated as being fixed. Topology optimisation was enabled by introducing the number of transverse frames and the number of longitudinal stiffeners as design variables. Data on different stiffener cross-sections (bulbs and T bars) and their dimensions as well as plate thicknesses were provided, and this was by its nature a set of discrete variables. The material used throughout was aluminium alloy (series 5083 and 6082). The number of design variables was 37, as indicated in Figure 14.20.

The optimisation problem can then be formulated as the minimisation of a function $f(x)$ representing the structural weight, subject to equality and inequality constraints. The objective function takes the form

$$f(x) = \sum_{j=1}^{r} w_j W_j \qquad r = 9 \tag{14.38}$$

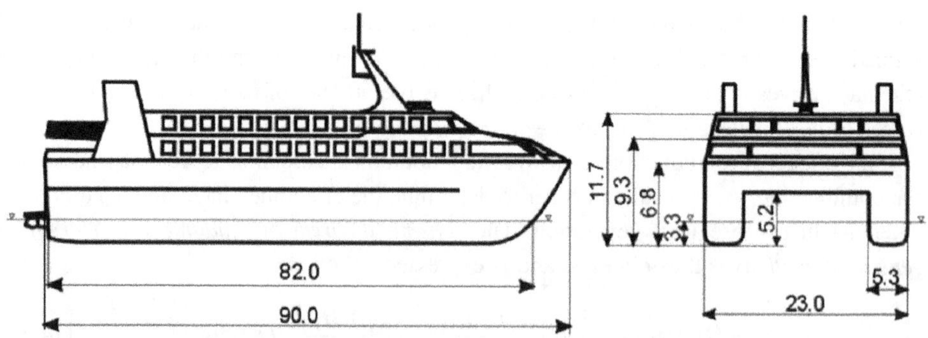

Figure 14.18 Overall views of vessel (Austral Auto Express 82) [40].

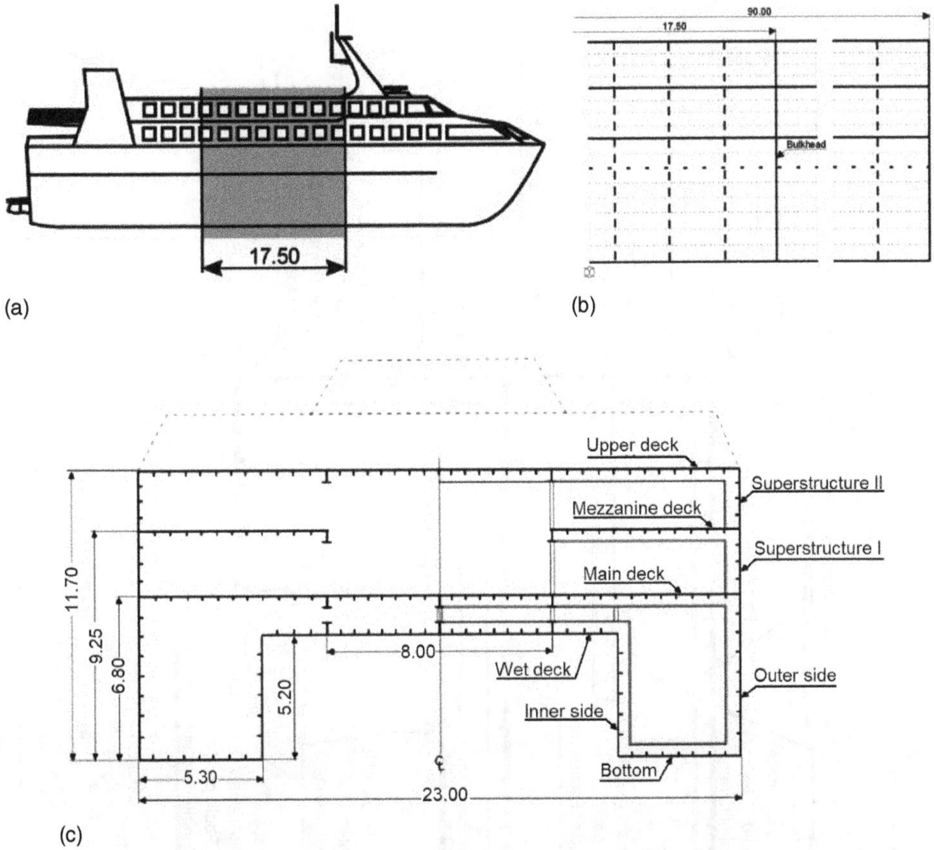

Figure 14.19 Catamaran structural sections and stiffening systems. (a) Section being optimised. (b) Plan view of section being optimised. (c) Transverse section.

where r is the number of structural regions, w = weight coefficient of particular region (in the region of 0.1) and W = weight of jth region. The constraints that are imposed concern safety (application of classification society rules) in the form of inequalities $\left(g_j(x) \geq 0\right)$, the range of available structural forms, termed side constraints $(h_k(x) = 0)$ and geometrical constraints that relate to good practice. Examples of the above are

$$\text{safety:} \qquad t_j - t_{j,rule} \geq 0 \tag{14.39a}$$

$$\text{geometry:} \qquad t_j - t_{f,w,j} \geq 0 \tag{14.39b}$$

In general terms, the genetic model described in Figure 14.21 that is used follows the description given earlier of genetic algorithms, with one exception. This concerns a new type of crossover operator that was introduced, which is random in nature. Thus the number of crossover points varied, and it was found that this led to improved performance of the algorithm.

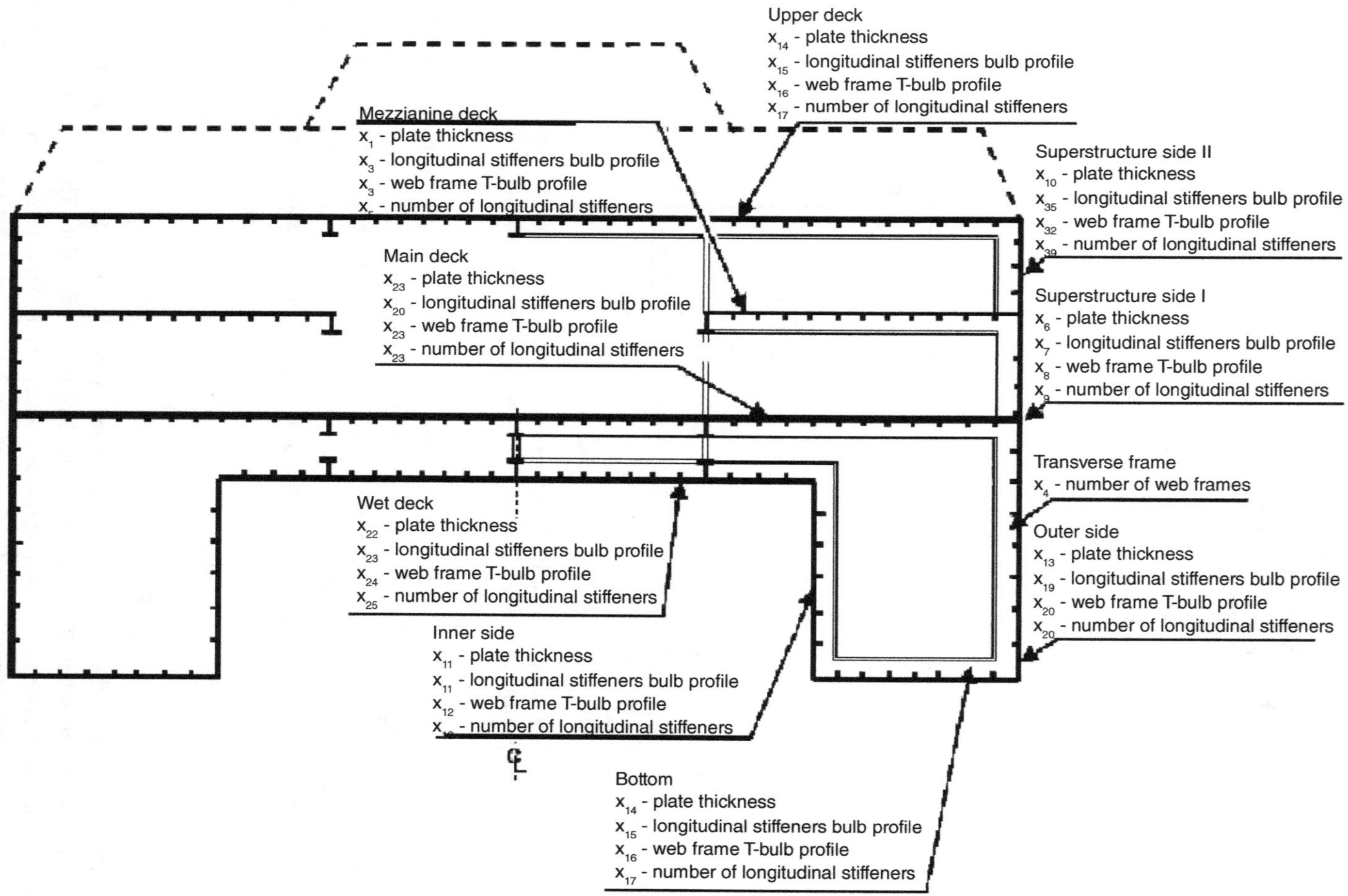

Figure 14.20 Design variables for catamaran hull transverse section [40].

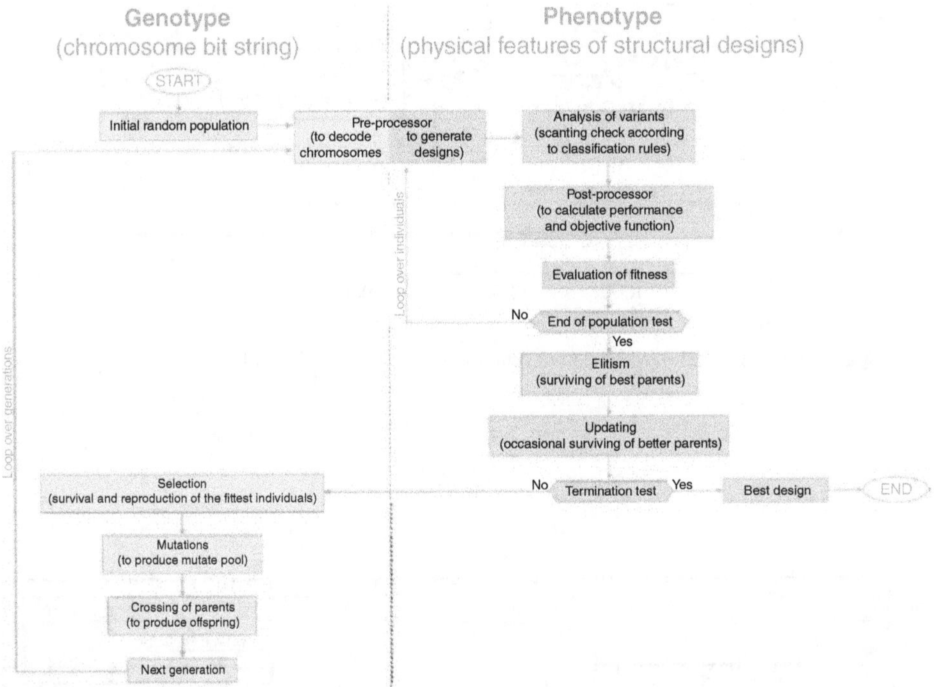

Figure 14.21 Genetic model used for the weight optimisation of catamaran structure.

The optimisation has been formulated as a constrained minimisation problem. This was transformed into an unconstrained problem by use of the penalty method. The *augmented objective function* $F(x)$ then becomes

$$F(x) = f(x) - \sum_{k=1}^{n_c} w_k P_k \tag{14.40}$$

where P_k is the penalty term corresponding to a violation of the kth constraint, w_k are the weight coefficients for the penalty terms (adjusted by trial) and n_c is the number of constraints.

The results of the optimisation are included in Figure 14.22. A total of 5,000 generations were produced, and the optimum arrangement was obtained in the 868th generation.

The corresponding structural arrangement is shown in Figure 14.23.

14.12.3 Multi-Objective Optimisation Using Genetic Algorithms

Earlier in this chapter the economic environment within which ship structural design takes place was described and some of the most important factors that affect decisions were outlined. It became apparent that it is necessary to resort to multi-objective, as opposed to single-objective, evolutionary algorithms.

We shall now reconsider multi-objective optimisation, and after an example that will serve to illustrate basic ideas, we will return to the catamaran structure optimisation

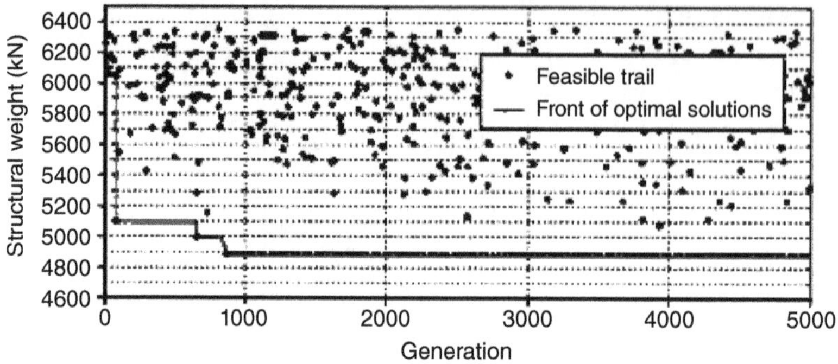

Figure 14.22 Evolution of objective function (weight of structure) [40].

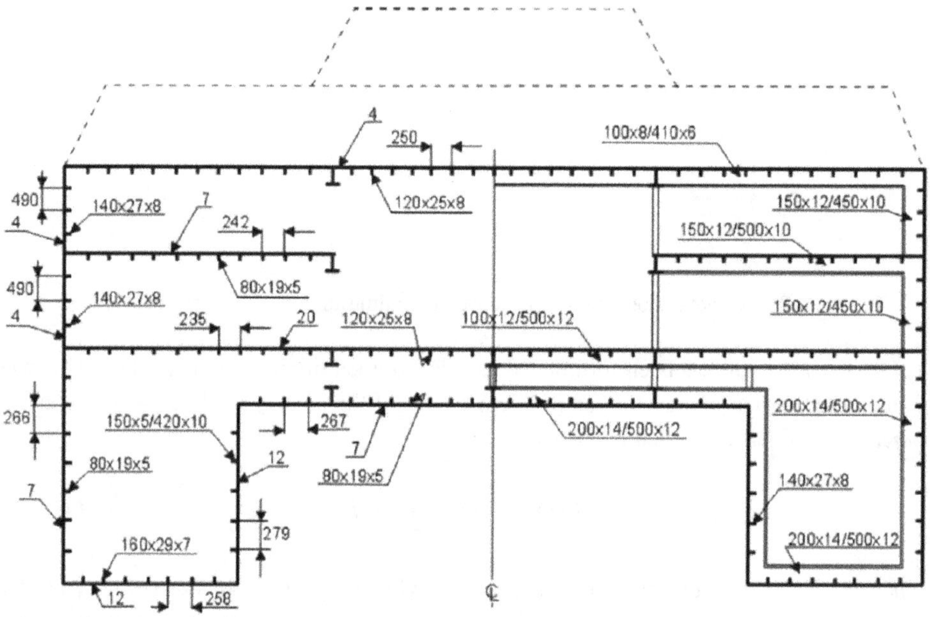

Figure 14.23 Minimum weight design of catamaran hull girder [40].

problem discussed in the previous section. Deb [11] gives the following example of a two-objective, two-variable unconstrained minimisation problem:

$$\text{Minimise} \qquad f_1(x) = x_1 \qquad f_2(x) = \frac{1 + x_2}{x_1} \tag{a}$$

$$\text{where} \qquad 0.1 \le x_1 \le 1, \qquad 0 \le x_2 \le 5$$

The solution of this problem consists of a set of Pareto-optimal solutions. By substituting, f_2 can be expressed in terms of f_1:

$$f_2 = \frac{1 + x_2}{f_1} \tag{b}$$

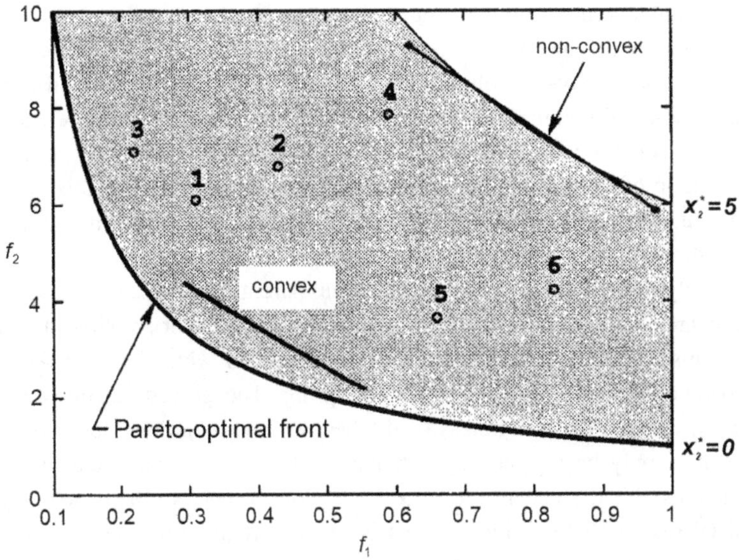

Figure 14.24 Solution of multi-objective optimisation minimisation problem (optimisation space) (adapted from [11]).

If our aim was to minimise f_2 alone, this would be achieved by minimising the numerator of the fraction on the right-hand side of (b) and by maximising the denominator. However, we also require to minimise f_1 that assumes values in the range 0.1 to 1. The search space, shown in Figure 14.24 as a shaded region, is then defined by the minimum and maximum values of the numerator ($1+0=1$ and $1+5=6$ corresponding to $x_2=0$ and $x_2=5$, respectively) and by the minimum and maximum values of the denominator. For $x_2=0$, all solutions are thus members of the Pareto optimal set. These are represented by the bold curve in Figure 14.24 and correspond to the ranges $10>f_2>1$, $1>f_1>0.1$. The solutions that correspond to points (1)–(6) are non-optimum, as these are dominated by the Pareto front solutions.

The first systematic solution obtained for a multi-objective optimisation problem, presented in [37], was termed the *vector-evaluated genetic algorithm* (VEGA). The name underlines the fact that the algorithm was used to provide a solution to a problem that assumes the form of a vector. Each element of this vector corresponds to a different objective function, in contradistinction to the single (scalar) objective function problem.

The basic thinking behind the functioning of VEGA is illustrated in Figure 94 in [11]. Let us assume that for a particular problem, five objective functions are stipulated. The population is then divided into five equal groups, each of which is assigned a fitness criterion based on a particular objective function. Selection is then performed using the proportionate operator within each group so that solutions are ranked within each group. In this way good solutions within the group are given prominence. In the next stage, reproduction is carried out by mating individuals from different groups.

The result of this approach is that solutions that rank highly with respect to specific objective functions are preferred. There do exist, however, other solutions that may not be near optimum but that may rank well with respect to a number of objective functions. Their contribution is required in order to ensure diversity. In order to detect these in VEGA, crossover takes place between solutions that belong to different groups. In this way it is hoped that good-quality compromise solutions will be generated. However, the solutions obtained are selected amongst the best ones for each objective function, and the sought-after contribution of diversity is not entirely successful. To overcome this problem, non-domination ranking of solutions and a diversity operator have been introduced in more recent algorithms.

In genetic algorithms described up to this point, parent solutions are always replaced by their descendants. If, however, it is desired to retain a particular solution in a subsequent generation, this has to be actively selected. One way in which this can be done is to compare parent solutions with their (two) offspring. The procedure followed is termed *elitism*, and it ensures that the fitness of the best existing solutions does not deteriorate, since these may only be replaced by better ones. It has been found that, in the case of certain functions, this has actually resulted in the convergence of the genetic algorithm to the global (Pareto) optimal solution. This is an important result and has pre-empted a number of approaches that introduce elitism in different ways.

In practical multi-objective optimisation problems constraints are imposed and need to be accounted for. This can be done by resorting to the penalty approach that has been applied to single objective optimisation problems, as described earlier.

Let us, however, return to the example treated in the case of the unconstrained problem. In this case constraints will be applied as indicated below:

$$\text{Minimise} \qquad f_1(x) = x_1 \qquad f_2(x) = \frac{1 + x_2}{x_1} \tag{c}$$

$$\text{where} \qquad 0.1 \leq x_1 \leq 1, \qquad 0 \leq x_2 \leq 5$$

subject to the following constraints:

$$g_1(x) \equiv x_2 + 9x_1 \geq 6 \qquad g_2(x) \equiv -x_2 + 9x_1 \geq 1 \tag{d}$$

We see that the Pareto optimal solution is modified and in the objective space is now indicated by the hatched line. The decision variable space is now defined by the constraints (figures 160 and 161 in ref [11]).

14.12.4 Multi-Objective Optimisation of a Catamaran Structure

In Section 14.12.2 the hull structure of a catamaran was treated as a single-objective optimisation problem and a minimum weight structure was obtained. In the present section a second objective function will be introduced, namely the area of the outer surface that requires to be maintained (cleaned and coated) [41].

In order to evaluate the Pareto optimal solutions, the concept of the *ideal point* was introduced (Figure 14.25a). This corresponds to the lowest values of all criteria analysed singly and not together. The optimal solutions can then be assessed by comparing their

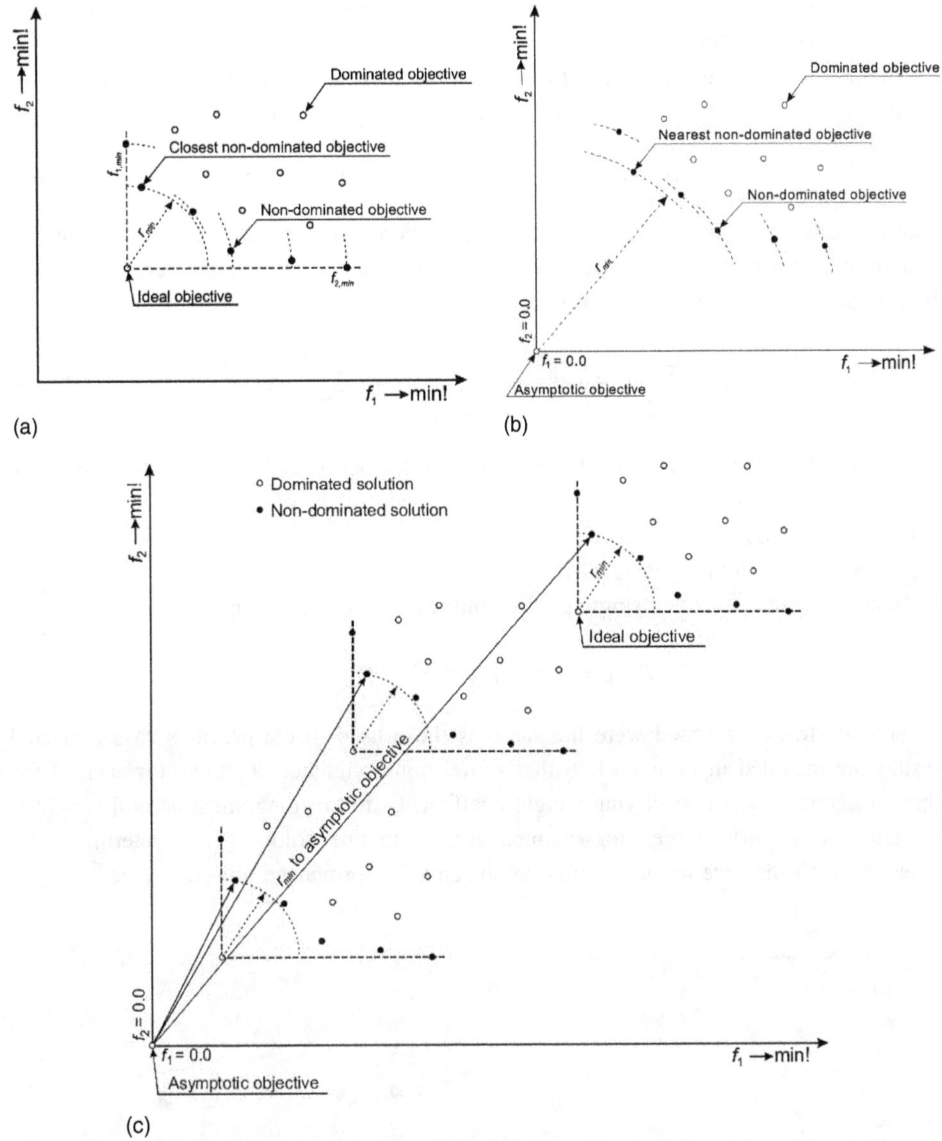

Figure 14.25 Non-dominated solutions, the ideal objective and the asymptotic objective [41]. (a) Ideal objective. (b) Asymptotic objective. (c) Evolution of non-dominated solutions and the asymptotic objective. Reprinted courtesy of Elsevier.

proximity to the ideal point. However, in multi-objective optimisation, the evolution of the set of non-dominated solutions may not be accompanied by a reduction in their distance from the ideal point. To circumvent this problem the concept of the *asymptotic objective* was introduced (Figure 14.25b), that in the case of the two-objective minimisation problem is specified by setting $f_1 = 0, f_2 = 0$ (Figure 14.25c). In the case of maximisation problems arbitrarily high (and practically unattainable) values are specified. In this way the ideal objective becomes stationary.

The previously used single-objective function (Equation (14.38)) was modified to allow for the additional objective by introducing the following scalar objective function:

$$F(x) = F(f_1(x), f_2(x)) = w_1 f_1(x) + w_2 f_2(x) \tag{14.41}$$

where f_1 is the structural weight of the midship section, f_2 is the area of the outer surface of structural members that requires maintenance, and w_1 and w_2 are weight coefficients. The fitness function then takes the following form:

$$f(x) = F(x) + \sum_{k=1}^{n_c} w_k P(x)_k^{r_k} = w_1 f_1(x) + w_2 f_2(x) \sum_{k=1}^{n_c} w_k P(x)_k^{r_k} \tag{14.42}$$

The weight coefficients used in both the objective function and the fitness function were specified in three different ways:

(a) $w_1 = w_2 = 0.5$ (user specified)

(b) w_1 and w_2 randomly generated in the range [0,1]

(c) using randomly selected single partial optimisation criteria (0 or 1)

$$F(x) = w_1 f_1(x) \text{ or } F(x) = w_2 f_2(x)$$

The constraints imposed were the same as those used in the previous case. Selected results are included in Figure 14.26 that shows non-dominated solutions for each of the three different ways of specifying weight coefficients. In this diagram structural weight is plotted against surface area for maintenance. A unique solution was determined by assessing their distance to the asymptotic objective in normalised objective space.

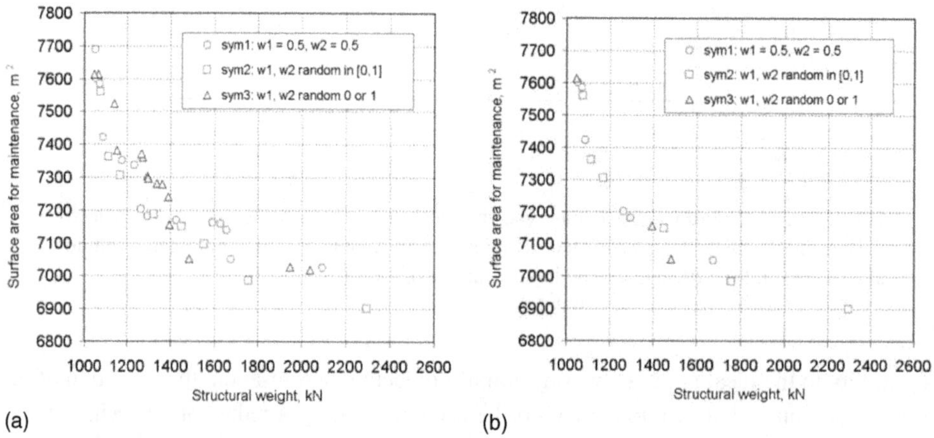

Figure 14.26 Multi-objective optimisation solutions obtained using multi-objective genetic algorithm [41]. (a) Pareto front solutions for three strategies. (b) Solution proximate to asymptotic objective. Reprinted courtesy of Elsevier.

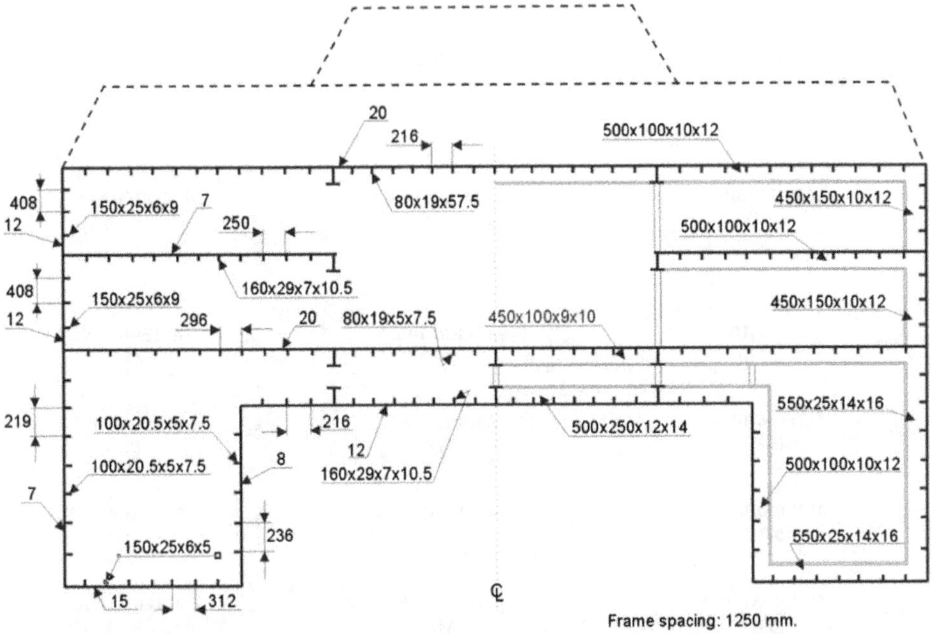

Figure 14.27 Catamaran structural arrangement based on multi-objective optimisation [41]. Reprinted courtesy of Elsevier.

The structural arrangement that corresponds to this solution is shown in Figure 14.27, and this may be compared with the design obtained previously using single-objective optimisation. The materials used in this design are 5083-H111 aluminium alloy for plates and 6082-T6 aluminium alloy for profile extrusions.

14.13 The Analytic Hierarchy Process

In multi-objective ship structure optimisation problems, as indeed in a large variety of optimisation problems, stakeholders have to (1) rank the different criteria in order of preference and (2) express the degree of preference for each of these criteria. The degree of preference for each criterion is expressed in linguistic terms and is most effectively treated using fuzzy logic, as discussed in Section 11.5.

With regard to the ranking of the different criteria, one of the most effective ways is the *analytic hierarchy process* (AHP). This is a technique with particularly broad applications, since it can be applied to any decision-making process that is based on subjective opinions [39]. In this, data consisting of practical alternatives or solutions are analysed and processed in a systematic manner in order to reach a management decision. The method is based on conducting pair-wise comparisons of selected criteria and possible alternatives in which subjective assessments of the relative importance of each are made. The method is used as a decision support tool and is versatile in the

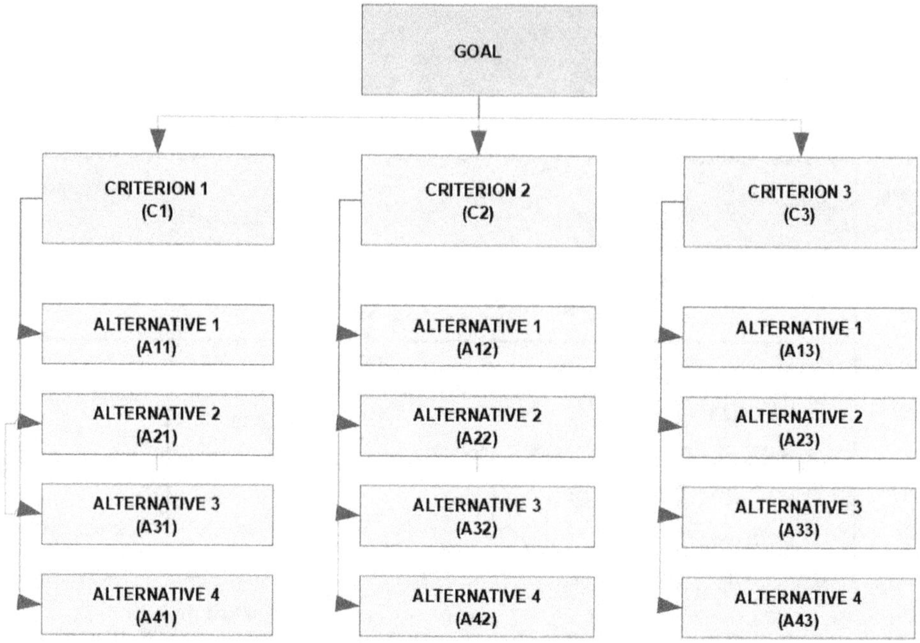

Figure 14.28 A simple AHP hierarchy.

sense that it is easily adaptable to any situation. There are three basic levels of data required:

1. Goal
2. Criteria
3. Alternatives

The decision problem is first decomposed into a *hierarchy* of more easily understood subproblems, each of which is treated independently. The hierarchy can extend to more than one level, depending on the complexity of the problem.

In Figure 14.28 the hierarchy involves three criteria for reaching the goal and four possible alternatives, one of which has to be selected. The design of any AHP hierarchy depends on the nature of the problem at hand as well as on the opinions of the participants that are involved in the decision-making process. Constructing a hierarchy may involve significant discussion, research and discovery by those involved. Even after its initial construction, it can be modified to accommodate newly thought-of criteria or criteria not originally considered to be important; alternatives can also be added, deleted or modified.

In the next step, the participants analyse the hierarchy by first comparing the criteria pair-wise against the goal and then comparing the alternatives pair-wise vis-à-vis each criterion. The purpose of these comparisons is to establish *weights* (*priorities*). A simple AHP hierarchy with default weights is shown in Figure 14.29.

In the hierarchy shown in Figure 14.29 default weights have been assigned and their values depend on the hierarchy structure. Different values would correspond to different

Table 14.4 *Preference scale and preference code p_{ij} (row* i*, column* j*) for attribute pair* (a_i, a_j)

$p_{ij}(j>1)$ (a_i preferred)	Pairwise preference	p_{ij} (a_j preferred)
1	Equally important	1
3	Slightly preferred	1/3
5	Strongly preferred	1/5
7	Demonstrably preferred	1/7
9	Absolutely preferred	1/9

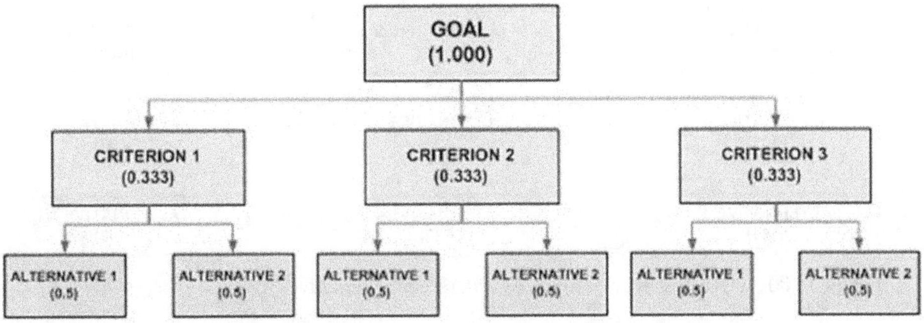

Figure 14.29 Simple AHP hierarchy with associated default weights.

hierarchical structures, although in all cases the sum of the weights for the criteria and for the alternatives has to equal unity (1.000).

The output of the method consists of a set of *ratio scales* and a *consistency index*. In order to obtain these, the first task is to perform a series of pair-wise comparisons of the different criteria. This leads to the formation of a *preference matrix*, an example of which is included in Section 14.14. In the AHP method the weights are assigned values in the range 1/9 to 9, as shown in Table 14.4.

The next task involves determining the normalised eigenvector of this matrix. The steps followed in this computation are listed below:

Computation of preference matrix eigenvector

1. Square the preference matrix.
2. Sum rows.
3. Normalise by dividing each sum by the sum of row total. This gives an estimate for the values of the elements of the eigenvector, and since these have normalized, their sum equals unity.
4. Iterate by repeating steps (1)–(3) above and in each case calculate the difference of the previous eigenvector from the current one. The matrix that is squared in step 1 is the squared matrix obtained in the previous iteration.
5. When the acceptable tolerance is no longer exceeded, iterations are ended.

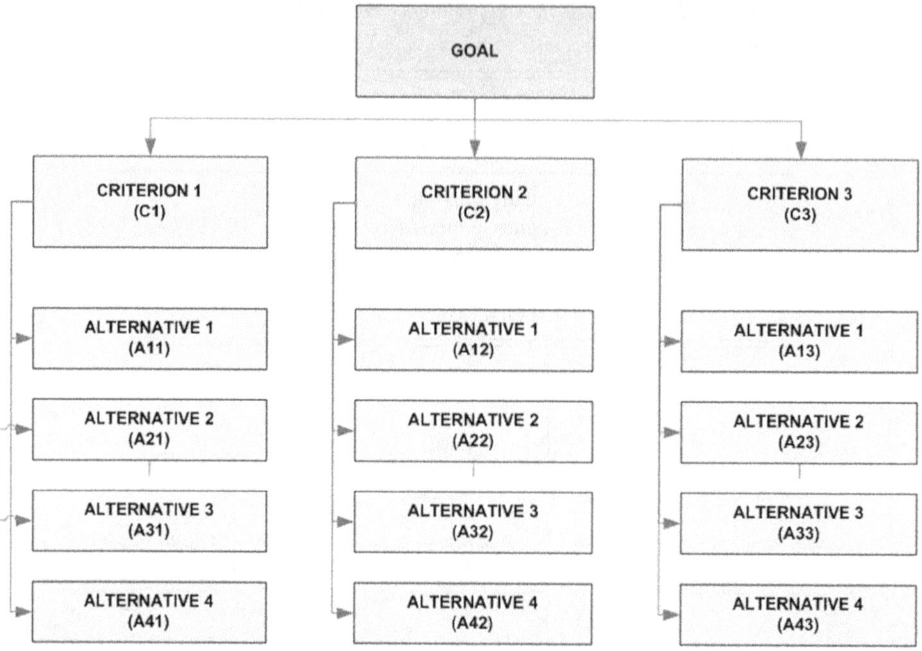

Figure 14.30 AHP tree with weights of criteria and alternatives.[12]

The values obtained for the elements of the eigenvector give the relative ranking (weights) of the criteria. The alternatives also need to be compared pair-wise with respect to each criterion and the results included in a preference matrix. The corresponding eigenvectors are calculated using the iterative procedure described above, which gives the ranking of the alternatives with respect to each criterion. This calculation may need to be extended in order to allow for additional branches of the hierarchy. The end-result is a tree that includes the weights for each of the criteria and alternatives, as shown in Figure 14.30.

The results are again arranged in a matrix, whose normalised eigenvector is calculated:

$$\mathbf{AC} = \begin{bmatrix} A_{11} & A_{12} & A_{13} \\ A_{21} & A_{22} & A_{23} \\ A_{31} & A_{32} & A_{33} \\ A_{41} & A_{42} & A_{43} \end{bmatrix} \begin{bmatrix} C_1 \\ C_2 \\ C_3 \end{bmatrix}$$

The complete result for the decision problem that includes the weights of the criteria (matrix **C**) and those of the alternatives (matrix **A**) is then obtained in tabular form by multiplying the individual weights of the alternatives by the weight of each criterion. This gives the ranking (or *benefit*) for each alternative.

If price is not included as a criterion, a cost-benefit analysis can be carried out following completion of the AHP process. This involves plotting benefit versus relative cost for each

[12] For example, *A23* denotes the weight of alternative 2 with respect to criterion 3.

alternative. Cost–benefit ratios can be calculated, and if cost needs to be included, the lowest cost–benefit ratio is the preferred solution.

14.14 Optimum Structural Design of a RoPax Ship

In this section the procedure followed in designing the structure of a RoPax ship (Figure 14.31) will be described based on [4]. The design methodology followed relates to both the concept and the preliminary design phases and consists of three major steps. Structural design is treated as a multi-criteria, multi-objective design problem, in which the stakeholders involved were the shipowner and the shipyard, with the support of the classification society. The novelty of this approach lies in that a series of studies were carried out in the concept design phase, thereby enabling important improvements and savings to be achieved with respect to the reference design. These are summarised in Table 14.8. The methodology involved extensive use of structural analysis software as well as a variety of optimisation software packages. Pertinent aspects of the method are discussed here, and further details are included in [4] and related publications.

14.14.1 Structural Design Procedure

The structural design procedure used had been formulated to cater especially for multi-deck ships such as the RoPax design described in this section. The procedure was used to solve questions of structural layout (topology), following which several optimisation techniques were applied in the case of the local geometry and scantlings. In the last step the proposed design was checked using classification society rules. The main steps were:

1. Topology and geometry optimisation
2. Scantlings and material optimisation of the variants obtained in (1)
3. Final analysis for classification society approval

A detailed flow chart of the procedure is included in Figure 14.32. The first two steps relate to the concept design phase (CDP) whilst the third corresponds to the preliminary design phase (PDP). The procedure is seen to be subdivided into a series of blocks. Blocks 1 and 2 constitute Step 1, blocks 3–6 constitute Step 2, and blocks 6–8 constitute Step 3. Table 14.5 details the layout of a representative block, included in Figure 14.32.

Figure 14.31 RoPax ferry.

Table 14.5 *Design block – description of format*

Block No.	Block name
Variables (**x**)	$[\mathbf{x}^G\mathbf{x}^T\mathbf{d}^S\mathbf{d}^M]$, (descriptors-obligatory)
Parameters (**z**)	(module name Φ, F, $\Phi^{\text{IACS Rule min}}$)
Attributes (**y**)	$\Omega^{(\max)\text{NAME}}$ [module name e.g. VCG]
Objectives (**o**)	for objective: specify direction (min-max)
Constraints (**g**)	$\alpha^{\text{SAFETY-NAME}}$: [module name e.g. IACS]
$\boldsymbol{\alpha}(\mathbf{g}) > g_{\min}$	$\alpha^{\text{SHIP-TEXT}}$: [text e.g. Global constr. or ShipGA or Deflection limits, etc.]
Analysis	$\varepsilon^{\text{LOADS TEXT}}$: [text e.g. Long. Strength – LC_1 or Racking BV – LC_2]
Modules (**AM**)	$\rho^{\text{RESPONSE SOLVER NAME}}$ [name e.g. OCTOPUS or MAESTRO]
Synthesis	$\Sigma^{\text{OPTIMIZ.SOLVER NAME}}$: [module name e.g. SLP, MOGA or MOPSO, etc.]
Modules (**SM**)	$\Gamma^{\text{SYNTHESIS TOOL NAME}}$: [module name e.g. DOE, ANOVA, AHP, etc]
Results	TEXT: [text e.g. Pareto designs, preferred designs, final design, etc.]

14.14.2 Use of Generic Models to Select a Topology of the Structure

The first step involves determining the preferred topology of the structure. This is a question that is governed by structural as well as functional (ship design) issues, and therefore collaboration between the designer and the owner is necessary.

In this particular procedure the geometry exploration is based on the development and use of *generic* 3D FEM models using commercially available FE software (MAESTRO in this case) (Figure 14.33). The purpose of a generic model is to enable the rapid analysis of the response of different variants of the proposed structure without having to model regions of the structure that are of lesser importance. To succeed in this the geometry of the generic model ought to follow the characteristics of the cross-section as well as the general layout of the intended design. The use of an extruded/tapered box model with transverse as well as longitudinal bulkheads leads to the rapid development of a series of generic 3D FE models. These are based on specific building blocks that in general consist of stiffened panels and pre-modelled sections of the large openings that arise in the side shell structure (Figure 14.33a). The use of 3D generic models represents a step forward in the overall structural optimisation of a multi-deck ship, since in current practice 2D models of transverse cross-sections are used.

When developing a generic model of a tapered multi-deck hull that has a low block coefficient, the basic requirement is for the model to generate realistic (expected) deformations that result from longitudinal bending and twisting (racking) loads. The generic model consists of *macro-elements*, *gross elements* and *super-elements*. Macro-elements are basic structural units such as stiffened panels, bracketed and locally reinforced girders and cellular units.[13]

Gross elements are sets of macro-elements that are assembled together and have a common natural boundary, such as a deck-at-side. They are used when a macro-element

[13] The macroelements listed above are described in detail in [22] as well as in [23] and [51].

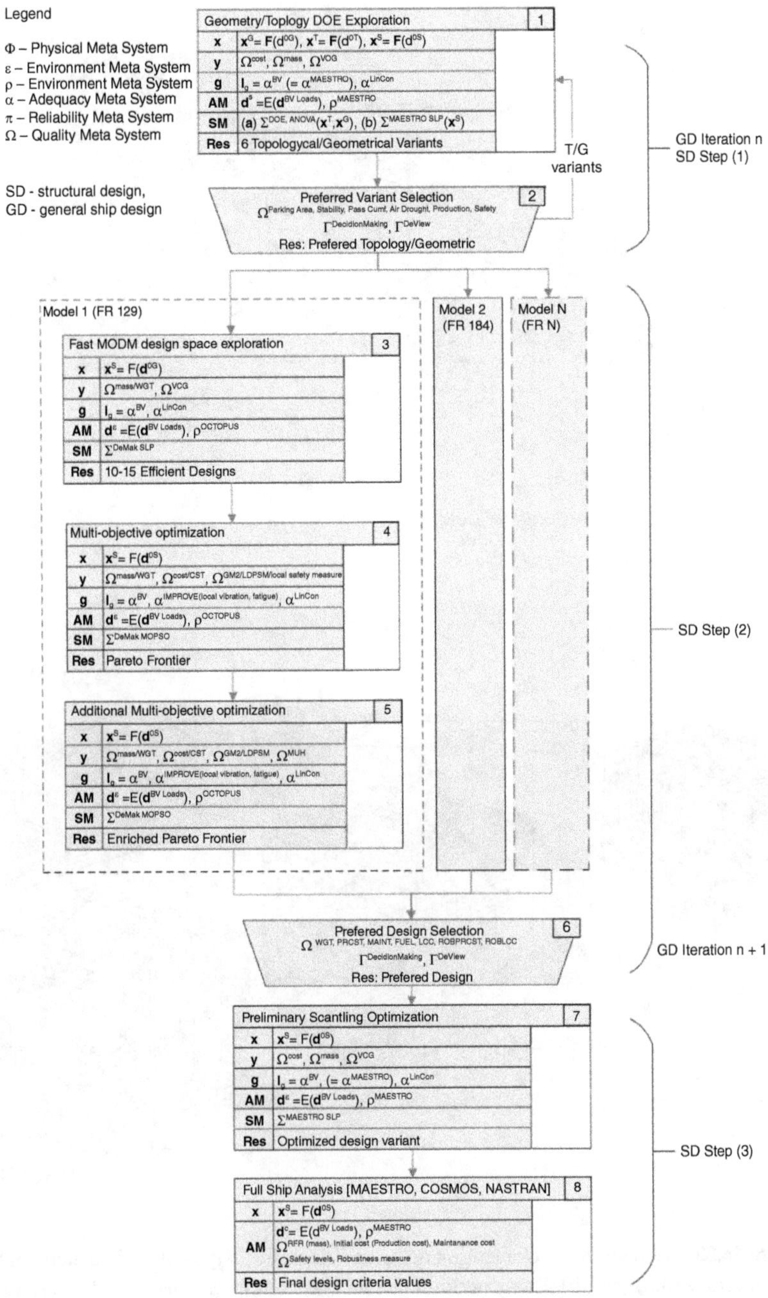

Figure 14.32 Structural design procedure for multi-deck ships [4]. Reprinted courtesy of Elsevier.

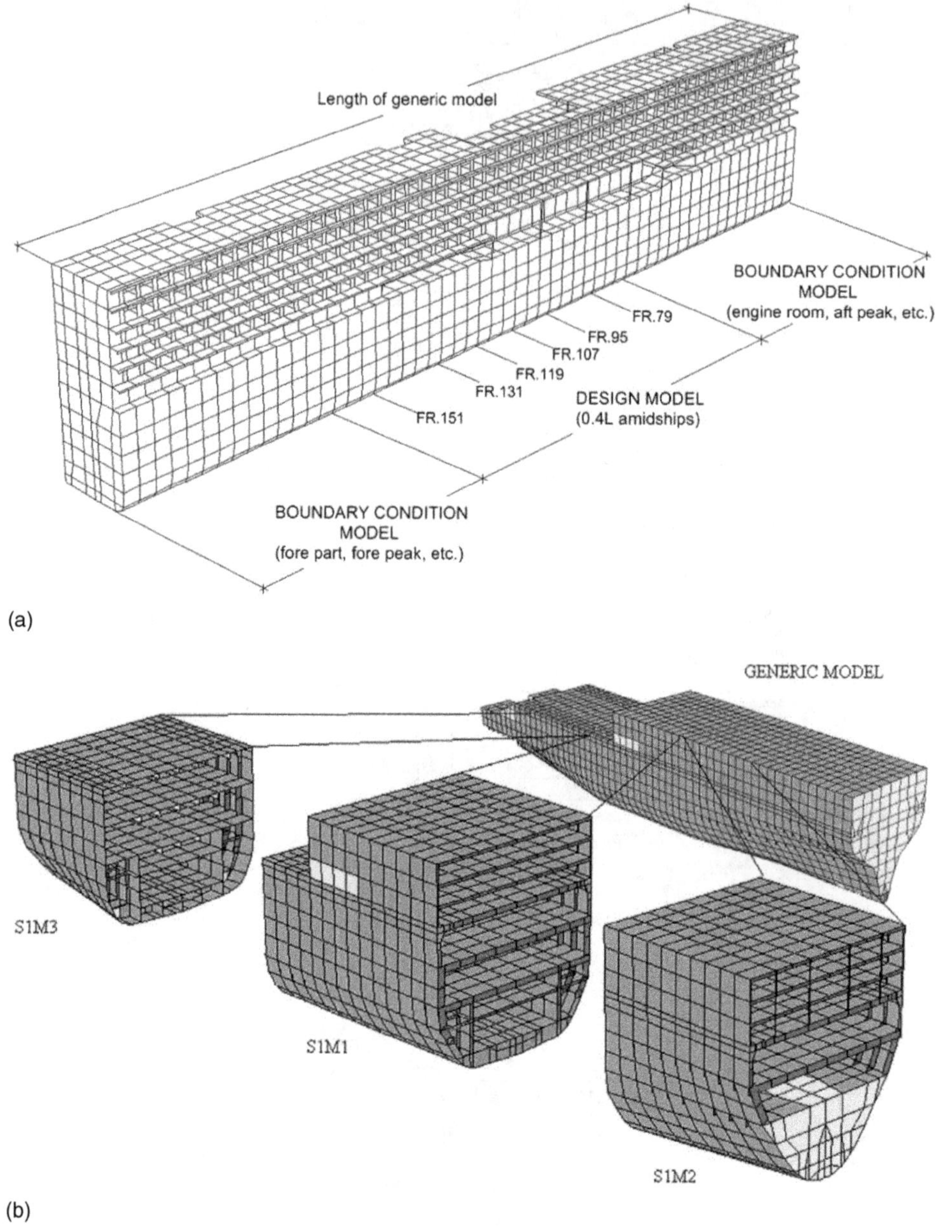

Figure 14.33 Generic finite element models. (a) Features of a typical basic generic 3D FE model (cruiseship) [5]. (b) The generic model and its components used in the case of the ROPAX ship [4]. Reprinted courtesy of Elsevier.

cannot sustain specific failure modes, such as overall buckling or vibration. The natural boundaries are stiff girders or other gross elements that provide boundary conditions for such failure modes.

The required features of macro- and gross elements used in the CDP of multi-deck ships are [5]:

- The displacement field should be of the third degree (cubic) at least. This will enable realistic modelling of response at locations such as the ship sides and decks, where shear lag effects arise. In transverse structural members the displacement fields should be of the third or fourth degree (cubic or quartic).
- The predicted normal and shear stress fields should vary linearly or parabolically, as expected.
- Membrane and bending effects should be modelled concurrently, although plate (tertiary) bending stresses can be excluded and transferred directly to the stress schedule[14] in the adequacy criteria.
- Each unsupported plate field between stiffeners must be accessible for adequacy control.
- Macro- and gross elements should permit direct connection with a supporting beam at a node or along its sides.

On multi-deck ships large side openings have an important influence on the effectiveness of the superstructure in resisting longitudinal hull girder bending loads. The reduced shear stiffness of the superstructure sides has a significant impact on the longitudinal hull girder stresses distributions of any transverse hull girder section. For this reason the accurate representation of large side openings is necessary in order to ensure accuracy of the primary stress calculations. Side panels with large openings and thus have to be modelled with an equivalent stiffness.

It has to be said that generic models can be examined, improved and verified because a chosen variant also acts as the starting point for the more detailed full ship 3D FE model that is developed in the final stage of the project.

14.14.3 Concept Design Phase (Steps 1 and 2)

In Step 1, topology, geometry and scantlings are considered, and thus it is necessary to specify relevant design variables. In this case, the one topological variable considered was the number of accommodation decks (either two or three tiers) with the constraint that the total deck area in both cases would be the same. The geometrical variable considered was the spacing between longitudinal bulkheads in the cargo spaces. Three possible values were considered, so that the combination of two topological and three geometrical solutions led to six possible variants (Figure 14.34). Details concerning the topological and geometrical design variables are included in Table 14.6 and are illustrated in Figure 14.34.

In this design problem three design objectives were used for the optimisation of each of the six variants shown in Figure 14.35. These were the mass of the structure, the cost of

[14] Section 12.5.4.

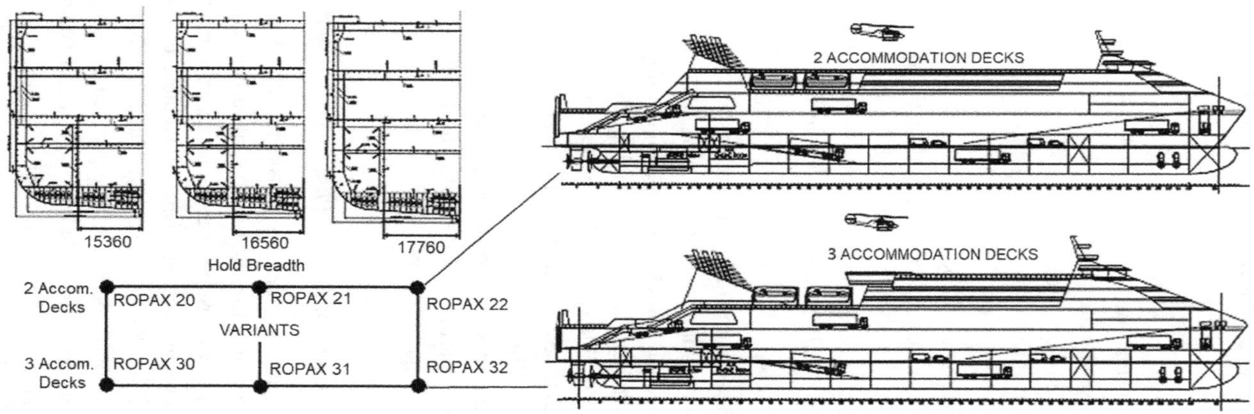

Figure 14.34 Topological and geometrical design variants for proposed RoPax design [4]. Reprinted courtesy of Elsevier.

Table 14.6 *Topological and geometrical design variables*

Item	Subsystem	Design value properties: Type	Name	Min	Max	Step	Comments
1	Cross-section	x^T	No of SS decks	2	3	1	Both versions have the same deck area
2	Cross-section	x^G	Lower hold breadth	1,5360	1,7760	1,200	One or more car lanes
3	Structure	x^S	Structural elements				Min/max values according to class rules, technology demands, experience, etc.

PARK ARR	STABILITY	AIR DRAUGHT	PRODUCTION	PASS COMF	SAFETY
1	3	1	3	1/3	3
1/3	1	1/3	1/3	0.1	1
1	3	1	3	1/3	3
1/3	3	1/3	1	0.1	1
3	9	3	9	1	9
1/3	1	1/3	1	0.1	1

(a)

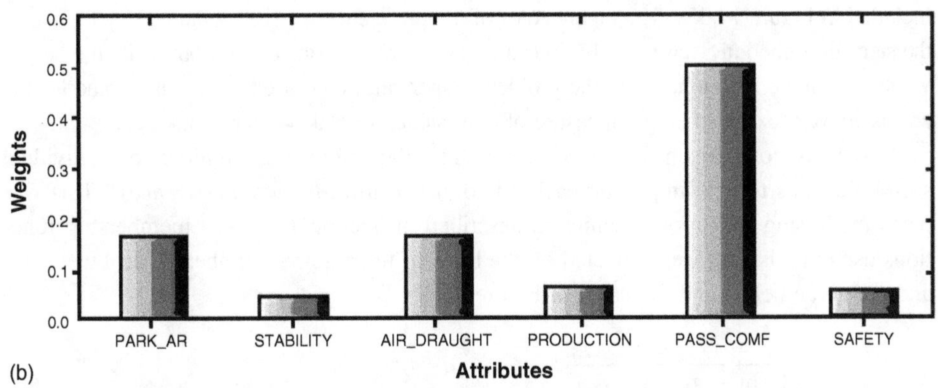

(b)

Figure 14.35 AHP results. (a) Preference matrix. (b) Attribute weights.

materials and the height of the vertical centre of gravity (VCG). The optimisation was performed using the sequential linear programming (SLP) algorithm described in Section 14.8, over 10 design cycles. The output of this step therefore involved six different variants optimised according to the above-mentioned three design objectives.

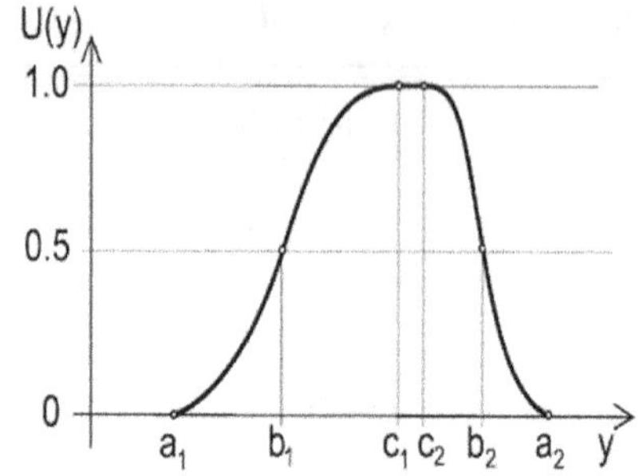

Figure 14.36 Novák membership grade functions used to model intra-attribute preferences [4]. Reprinted courtesy of Elsevier.

In the next step these designs were compared according to the following criteria proposed by the designer:

1. Ship stability
2. Air draught
3. Passenger comfort
4. Production cost
5. Parking area
6. Structural safety

Comparisons were carried out according to, first, stakeholders' *inter-attribute preferences* – that is, how each stakeholder ranks the above-listed criteria in order of importance. A systematic method of doing this is the analytic hierarchy method described in Section 14.13. The preference matrix for the selected criteria and the resulting weights obtained are included in Figure 14.35a. The eigenvector of the preference matrix is obtained by following the steps described in Section 14.13, in order to rank the criteria. The attribute with the largest value is then the best solution to the problem. Eigenvalues are used to obtain a value for the consistency index, this being a measure of consistency of stakeholders' rankings.

A second comparison involves interpreting the subjective opinions of individual stakeholders as to how important each criterion is (*intra-attribute preferences*). This was performed using fuzzy logic, which is described in Section 11.5. The membership functions used in this case were selected on the basis of the required number of attributes (six) and are given below and indicated in Figure 14.36.

$$U(y) = \begin{cases} \dfrac{(y-a_1)^2}{(b_1-a_1)(c_1-a_1)} & \Rightarrow \quad a_1 \le y < b_1 \\ 1 - \dfrac{(y-c_1)^2}{(c_1-b_1)(c_1-a_1)} & \Rightarrow \quad b_1 \le y < c_1 \\ 1 - \dfrac{(y-c_2)^2}{(b_2-c_2)(a_2-c_2)} & \Rightarrow \quad c_2 \le y < b_2 \\ \dfrac{(y-a_2)^2}{(a_2-b_2)(a_2-c_2)} & \Rightarrow \quad b_2 \le y < a_2 \end{cases} ; \quad \begin{cases} 1 \quad \Rightarrow \quad c_1 \le y \le c_2 \\ 0 \quad \Rightarrow \quad a_1 \ge y \ge a_2 \end{cases}$$

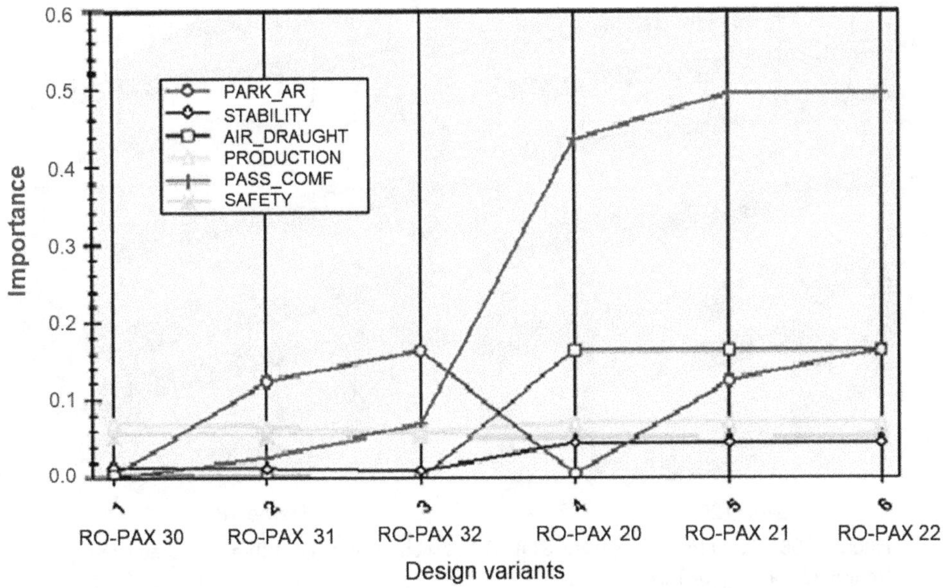

Figure 14.37 Parallel coordinates plot of design attributes [4]. Reprinted courtesy of Elsevier.

The selection of a preferred variant was performed using the parallel coordinates plot[15] of the design criteria (Figure 14.37), for which design variant 1 (ROPAX 30) was the starting point. On the basis of this diagram, design variant 6 (ROPAX 22) was chosen as the preferred solution.

This design is used in Step 2 of the concept design phase in order to develop a series of substructure designs using structural optimisation techniques. Initially two 2.5D FEM models were created, as shown in Figure 14.38. The model in way of frame 129 directly influences the scantlings of module S1M1, whilst that in way of frame 184 influences the scantlings of module S1M2 (Figure 14.33). For each of the structural models the respective synthesis models were created. The design variables included in Blocks 3–5 (Figure 14.32) are the longitudinal and transverse structure scantlings.

Following this (Block 3), fast MODM using a SLP optimisation algorithm was performed to produce 10 longitudinally optimised designs for each of the two 2.5D FE models. These were optimised for weight, construction cost and safety.

Multi-objective optimisation using multi-objective particle swarm optimisation (MOPSO)[16] was then performed (Block 4). The selected design objectives were production

[15] A parallel coordinates plot maps each row of a data table as a line. The aim of a parallel coordinate plot is to enable the comparison of different profiles; parallel coordinate plots are suitable for use with incommensurate data. The result is a series of discontinuous straight line segments, and by comparing profiles, similarities and differences become evident [43].

[16] Particle swarm optimisation is a nature-inspired optimisation method, like genetic algorithms, and mimics the behaviour of birds in a flock or of fish in a school. It has fewer algorithmic parameters than genetic algorithms and does not require binary encoding or decoding, and is therefore easier to use on a computer. For details, see, e.g. [6].

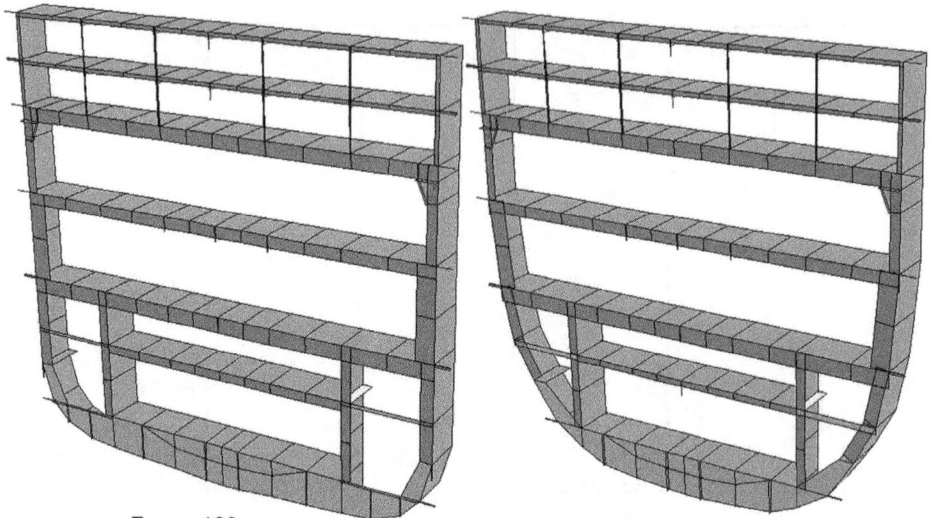

Figure 14.38 2.5D FEM models of ROPAX 22 sections in way of frames 129 and 184 [4]. Reprinted courtesy of Elsevier.

cost, mass of structure and structural safety, based on the local strength of stiffened panels. The resulting Pareto frontier for the frame 129 is included in Figure 14.39.

In Block 5 the hull girder ultimate strength calculation was performed for all Pareto solutions using the incremental-iterative Smith method. In addition, the weight of the structure was calculated and a safety evaluation was performed. The corresponding Pareto frontier is shown in Figure 14.40.

In Block 6 the preferred designs were chosen on the basis of the design attributes included in Table 14.7.

A study of Pareto designs led to the following conclusions of practical significance:

1. Designs with low mass have low production cost and low fuel cost whilst the maintenance cost is high.
2. The influence of maintenance cost on the life cycle cost is significantly smaller than that of fuel cost.
3. The preferred designs for the shipyard and the owner are similar to each other (point 1 above).

On this basis the shipyard and the owner agreed to select the design that is associated with the maximum profit margin, which in fact corresponds to that with the minimum fuel consumption.

14.14.4 Preliminary Design Phase (Step 3)

The calculations described up to this point bring to a conclusion the concept design phase (Steps 1 and 2, consisting of Blocks 1–5 included in Figure 14.32). The third and final step

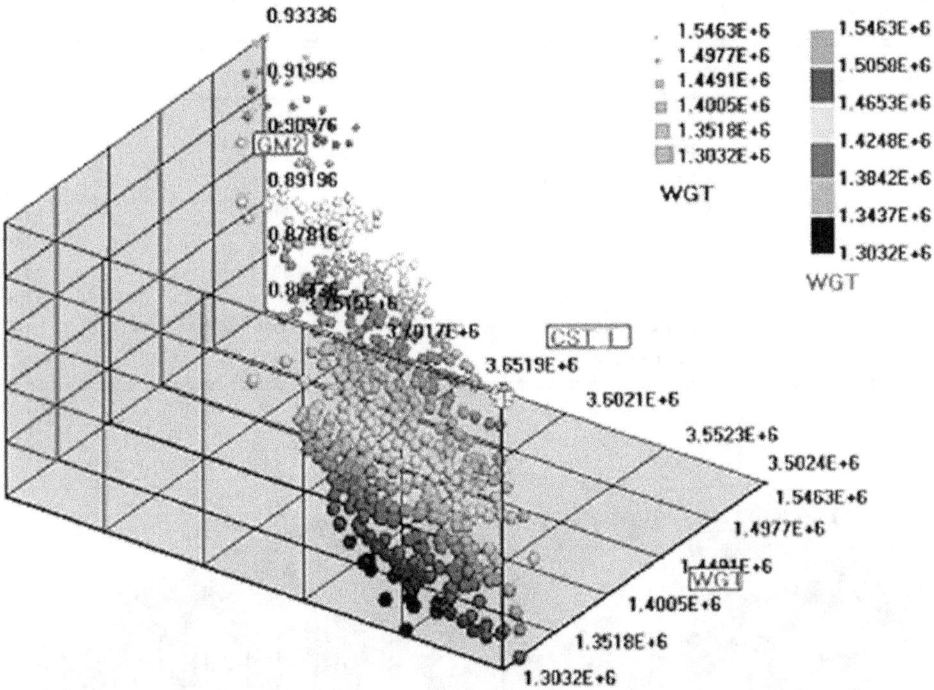

Figure 14.39 Pareto front obtained for model of frame 129 using MOPSO [4]. Reprinted courtesy of Elsevier.

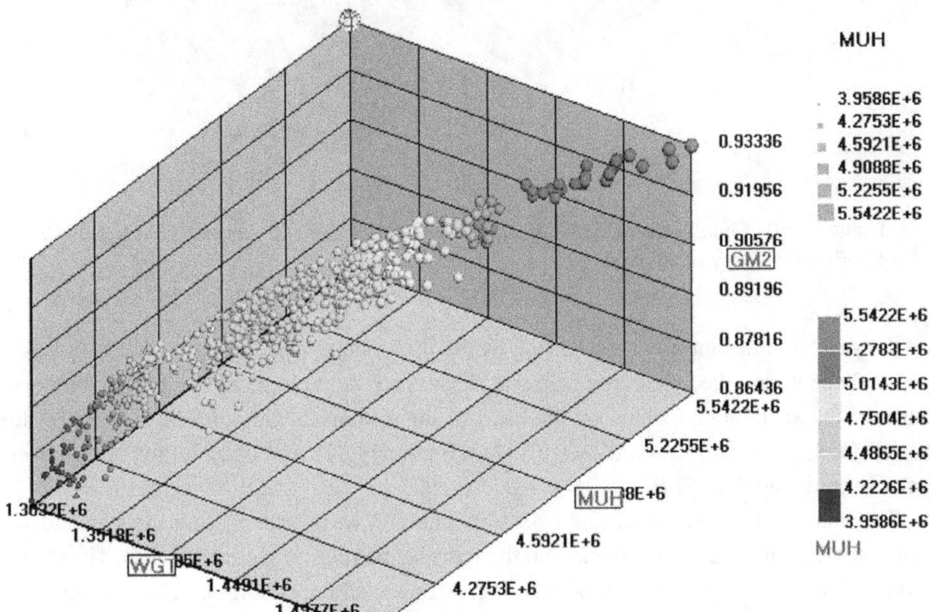

Figure 14.40 Pareto front for hull girder ultimate strength (MUH), weight (WGT) and safety (GM2). Reprinted courtesy of Elsevier.

Table 14.7 *Design attributes for final selection*

No	Description	Acronym	Type
1	Δ Lightship mass	D_LWT	Min
2	Δ Production cost	DC_PROD	Min
3	Δ Maintenance cost	DC_MAINT	Min
4	Δ Fuel cost	DC_FUEL	Min
5	Δ Earning	D_EARN	Max
6	Δ Dismantling	D_DISM	Max
7	Δ Life cycle cost	D_LCC	Max
8	Local safety	Gmean	Max
9	Global safety	SAF	max

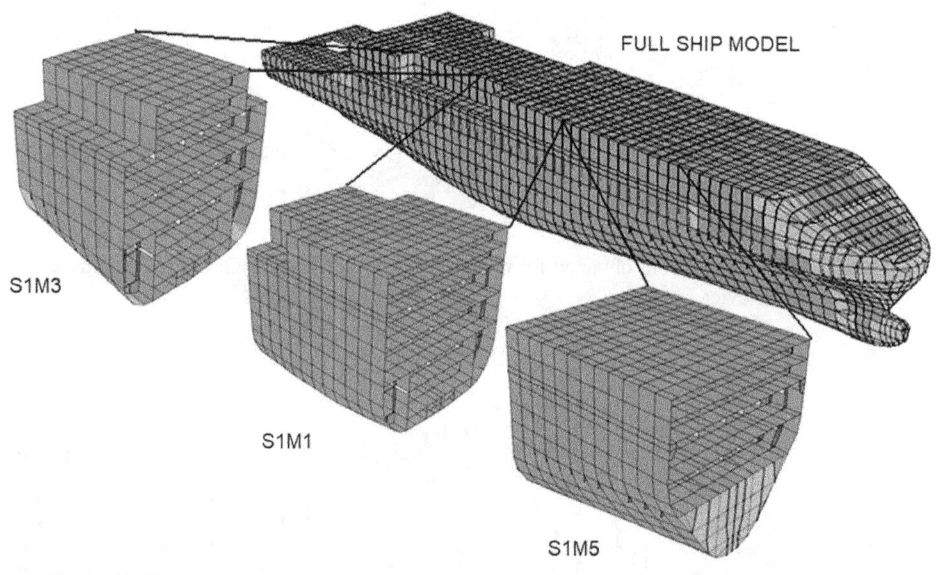

Figure 14.41 Finite element model of entire ship (PDP) (based on variant ROPAX 22) [4]. Reprinted courtesy of Elsevier.

of the structural design procedure concerns the preliminary design phase (PDP). In this, a finite element model of the entire ship was developed on the basis of the generic model of Step 1 with scantlings updated on the basis of the optimisation studies of Step 2. To this model fore and aft end structures were added (Figure 14.41), and optimisation studies were carried out using dual SLP.

The final 3D FEM response and feasibility analysis was performed for the standardised optimal solution obtained in Block 7 on the basis of the rules of Bureau Veritas. The ship is expected to spend 90 per cent of its time in one loading condition. This includes the lightship weight, 100 per cent of consumables and homogeneous loading of passengers on the accommodation decks and of cars and trailers on the car decks. The car decks should be

Table 14.8 *Summary of results*

Structural models	Mass (t)	Savings in mass (%)	Normalised cost (%)	VCG (mm)
ROPAX 30 (reference design)	3,283	-	1.00	13,695
ROPAX 22 (selected variant in 1st design loop)	3,313	-	1.01	13,378
Step 1 design	3,000	8.62	0.91	12,906
Step 2 design	2,768	7.73	0.84	13,290
Step 3 design	2,660	3.90	0.80	13,134
Final RO-PAX 22 design	2,690	−1.12	0.81	13,151
Total savings	593	18.1	19.0	544

designed to withstand the maximum homogeneous load specified as well as the wheel loads that arise in way of the wheel–deck contact patches.

In order to determine loads, the loads procedure included in the rules for the full ship FE analysis was implemented. Direct wave loads components were applied to the full ship model for the final evaluation of scantlings. The direct load analysis procedure consisted of two phases. The first phase concerned the estimation of load and/or motion RAOs whereas in the second phase these RAOs were combined with the idealised sea spectra of operation area to predict the design wave loads. Two procedures that are recommended in the rules were used and results compared:

1. The standard Bureau Veritas procedure for partial FEM models described in Part B chapter 7 App. 1-2 of the Rules [7].
2. The design wave method described in Part B chapter 7 App. 3 of the Rules [7]. This involves determining the short-term and long-term wave-induced pressure and inertia loads through direct sea-keeping analysis.

The success of the procedure can be partly gauged from the results included in Table 14.8. It is seen that compared to the reference design, the final design produces a saving in lightship mass of 18.1 per cent, which translates to a saving in fuel consumption, thereby increasing the owner's profit margin. Additionally, the final design has a larger parking area as well as improved life-cycle characteristics.

References

1. Adamchak J. (1975) Applications of Structural Synthesis and Optimization Techniques. In Evans J. H. (Ed) *Ship Structural Design Concepts*. Centreville, MD: Cornell Maritime Press 487-501.
2. Analytic Hierarchy Process AHP – business performance management. 19 April 2010. Available: www.youtube.com/watch?v=18GWVtVAAzs (accessed 19 December 2020).
3. André J. Beale R. Baptista A. M. (2015) New indices of structural robustness and structural fragility. *Struct. Eng. & Mech.* Vol. 56 1063–1093.

4. Andrić J. Prebeg P. Žanić V. (2019) Multi-level Pareto supported design methodology: application to RO-PAX structural design *Mar. Struct.* Vol. 67 102638.
5. Andrić J. Žanić V. (2010) The global structural response model for multi-deck ships in concept design phase. *Ocean Eng.* Vol. 37 688–704.
6. Arora J. S. (2012) *Introduction to Optimum Design*. 3rd ed. Amsterdam: Elsevier Academic Press.
7. Bureau Veritas (2003) *Rules for the Classification of Steel Ships.*
8. Burns S. A. (Ed) (2002) *Recent Advances in Optimal Structural Design*. ASCE Struct. Eng. Inst.
9. Caldwell J. B. Woodhead R. G. (1973) Ship structures: some possibilities for improvement. *Trans NECIES* Vol. 88 101–120.
10. *Calresco.org*. Available: www.calresco.org/lucas/pmo.html (accessed 19 December 2020).
11. Deb K. (2004) *Multi-objective Optimization Using Evolutionary Algorithms*. Chichester: John Wiley & Sons.
12. Det Norske Veritas (2005) *Hull Structure Course.*
13. Dextremel (1997) Design for Structural Safety under Extreme Loads State of the Art Review. BriteEuRam III contract BRPR-CT97–0513.
14. Evans J. H. Khoushy D. (1963) Optimized Design of Midship Section Structure. *SNAME Annual Meeting.*
15. Faulkner D. (1975) *Ship and Ocean Structures*. Lecture Notes Dept. Naval Arch & Ocean Eng. Univ. Glasgow Scotland.
16. Fleury C. (1993) *Mathematical programming methods for constrained optimization: dual Methods*. Structural optimization: Status and Promise, Progress in Astronautics and Aeronautics.
17. Gallagher R. H. (1973) Linear Programming in Structural Analysis and Design. In Gallagher R. H. Zienkiewicz O. C. *Optimum Structural Design. Theory and Applications*. New York: John Wiley & Sons 79–108.
18. Ghosh A. Dehuri S. (2004) Evolutionary algorithms for multi-criterion optimization: a survey. *Int. J. Comp. & Inform. Sc.* Vol. 2 38–57.
19. Goldberg D. E. (1989). *Genetic Algorithms in Search Optimization and Machine Learning*. Reading, MA: Addison Wesley Publishing Co.
20. van Hemmen H. F. (2014) Game Theory for the Maritime Professional. *Soc. Mar. Port Engrs Mtg.*
21. Holland J. H. (1992) *Adaptation in Natural and Artificial Systems: An Introductory Analysis with Applications to Biology Control and Artificial Intelligence*. Cambridge, MA: MIT Press.
22. Hughes O. F. (1983) *Ship Structural Design: A Rationally-Based Computer-Aided Optimization Approach*. New York: John Wiley & Sons.
23. Hughes O. F. Mistree F. Žanić V. (1980) A practical method for the rational design of ship structures. *J. Ship Res.* Vol. 24 101–113.
24. Hughes O. F. Paik J. K. (2010) *Ship Structural Analysis and Design*. Jersey City, NJ: SNAME.
25. Jendo S. (1990) Optimal Design of Elastic Frame Structures. In Save M. Prager W. (Eds) *Structural Optimization Vol. 2 Mathematical Programming* New York: Plenum Press 221–228.
26. Jendo S. (1990) Multi-objective Optimisation. In Save M. Prager W. (Eds) *Structural Optimisation. Vol. 2 Mathematical Programming*. New York: Plenum Press 311–341.
27. Karr D. et al. (2009) Design Methods. Committee IV.2 Rept. In Jang C. D. Hong S. Y. (Eds) *Proc. 17th Int. Ship & Offsh. Struct. Cong.* Vol. 1 689–756.

28. Kahraman C. Çevik Onar S. Oztaysi B. (2015) Fuzzy multicriteria decision-making: a literature review. *Int. J. Comp. Intell. Sys.* Vol. 8 637–666.
29. Kemene A. (2018) *Newbuildings and Yards*. Optima Shipping Services. MarineMoney week New York.
30. Klanac A. Varsta P. (2011) Design of marine structures with improved safety for environment. *Rel. Eng. & Sys. Saf.* Vol. 96 75–90.
31. Mistree F. Hughes O. F. Phuoc H. B. (1981) An optimization method for the design of large highly constrained complex systems. *Eng. Optim.* Vol. 5 179–197.
32. Moe J. (1974) Fundamentals of optimization. *Comp. & Struct.* Vol. 4 95–113.
33. Moe J. Lund S. (1968) Cost and weight minimization of structures with special emphasis on longitudinal strength members of tankers. *Trans RINA* Vol. 110 43–70.
34. Parsons M. G. (2009) Applications of optimization in early stage ship design. *Ship Sc. & Techn.* Vol. 5 9–31.
35. Pradillon J.-Y. et al. (2012) Design Methods. Committee IV.2 Rept. In Fricke W. Bronsart R. (Eds) *Proc. 18th Int. Ship & Offsh. Struct. Cong.* Vol. 1 Schiffarts-Verlag Hansa GmbH & Co KG Hamburg 507–576.
36. Rigo P. (2001) A module-oriented tool for optimum design of stiffened structures – part I. Mar. Struc. Vol. 14 611–629.
37. Schaffer J. D. (1984) Multiple Objective Optimization with Vector Evaluated Genetic Algorithms. *Proc. First Int. Conf. Gen. Alg.* 93–100.
38. Schipperen I. et al. (2012) Materials and Fabrication Technology. Committee V.3 Rept. In Fricke W. Bronsart R. (Eds) *Proc. 18th Int. Ship & Offsh. Struct. Cong.* Vol. 2 Schiffarts-Verlag Hansa GmbH & Co KG Hamburg 113–152.
39. Saaty T. L. (1980) *The Analytic Hierarchy Process: Planning Priority Setting Resource Allocation*. New York: McGraw-Hill.
40. Sekulski Z. (2009) Structural weight minimization of high speed vehicle-passenger catamaran by genetic algorithm. *Pol. Mar. Res.* Vol. 16 11–23.
41. Sekulski Z. (2010) Multi-objective topology and size optimization of high speed vehicle-passenger catamaran structure by genetic algorithm. *Mar. Struct.* Vol. 23 405–433.
42. Shanley F. H. (1957) *Strength of Materials* New York: McGraw-Hill.
43. *Tibco.com*. Available: https://docs.tibco.com/pub/spotfire/6.5.2/doc/html/para/para_what_is_a_parallel_coordinate_plot.htm (accessed 19 December 2020).
44. Triantaphyllou E. Shu B. Nieto Sanchez S. Ray T. (1998) Multi-Criteria Decision Making: An Operations Research Approach. In Webster J. G. (Ed) *Encyclopaedia of Electrical and Electronics Engineering* Vol. 15. Chichester: John Wiley & Sons 175–186.
45. Velasquez M. Hester P. T. (2013) An analysis of multi-criteria decision making-methods. *Int. J. Oper. Res.* Vol. 10 56–66.
46. Analytic hierarchy process. *Wikipedia The Free Encyclopedia* 5 December 2020. Available: https://en.wikipedia.org/w/index.php?title=Analytic_hierarchy_process&oldid=992518831 (accessed 19 December 2020).
47. Genetic algorithm. *Wikipedia The Free Encyclopedia* 1 December 2020. Available: https://en.wikipedia.org/w/index.php?title=Genetic_algorithm&oldid=991805472 (accessed 19 December 2020).
48. Holland's schema theorem. *Wikipedia The Free Encyclopedia* 1 December 2020. Available: https://en.wikipedia.org/w/index.php?title=Holland%27s_schema_theorem&oldid=991622109 (accessed 19 December 2020).
49. Multiple-criteria decision analysis. *Wikipedia The Free Encyclopedia* 25 October 2020. Available: https://en.wikipedia.org/w/index.php?title=Multiple-criteria_decision_analysis&oldid=985414997 (accessed 19 December 2020).

50. Schema (genetic algorithms). *Wikipedia The Free Encyclopedia* 21 September 2020. Available: https://en.wikipedia.org/w/index.php?title=Schema_(genetic_algorithms)&oldid=979504170 (accessed 19 December 2020).
51. Žanić V. Rogulj A. Jancijev T. Bralić, S. Hozmec, J. (2002) Methodology for Evaluation of Ship Structural Safety. *Proc. IMAM 2002 Cong.* Crete.
52. Zimmermann H.-J. (2001) *Fuzzy Set Theory – and Its Applications.* 4th ed. New York: Springer Science+Business Media.

Appendix A
Symbols, Coordinate System and Sign Convention

The symbols, coordinate system and sign convention used in the text follow those of the IACS Common Structural Rules and are summarised in this appendix.

A.1 Symbols

Tables A.1, A.2 and A.3 include symbols used in the text that follow those used in the IACS Common Structural Rules, with the exception of yield stress. Other symbols used are specified in the text as needed.

Table A.1 *Ship particulars*

Quantity	*Symbol*	*Units*
Length (rule)	L	m
Length between perpendiculars	L_{PP}	m
Breadth (moulded)	B	m
Depth (moulded)	D	m
Draught (moulded)	T	m
Draught (scantling)	T_{SC}	m
Displacement (moulded) at scantling draught	$\varDelta$	t
Block coefficient at scantling draught	C_B	-
Maximum service speed	V	knot

Table A.2 *Material particulars*

Quantity	*Symbol*	*Units*
Elastic modulus	E	kN/mm^2
Shear modulus	G	kN/mm^2
Yield stress	σ_0	N/mm^2
Poisson's ratio	ν	-
Material factor	k	-
Wave coefficient	C_w	m
Probability factor	f_p	-
Wavelength	λ	m

Table A.3 *Load resultants*

Quantity	*Symbol*	*Units*
Vertical still-water bending moment	M_{sw}	kNm
Vertical wave-induced bending moment	M_{wv}	kNm
Vertical still-water shear force	Q_{sw}	kN
Vertical wave-induced shear force	Q_{wv}	kN
Torsional wave moment	M_{wt}	kNm
Horizontal bending moment	M_{wh}	kNm

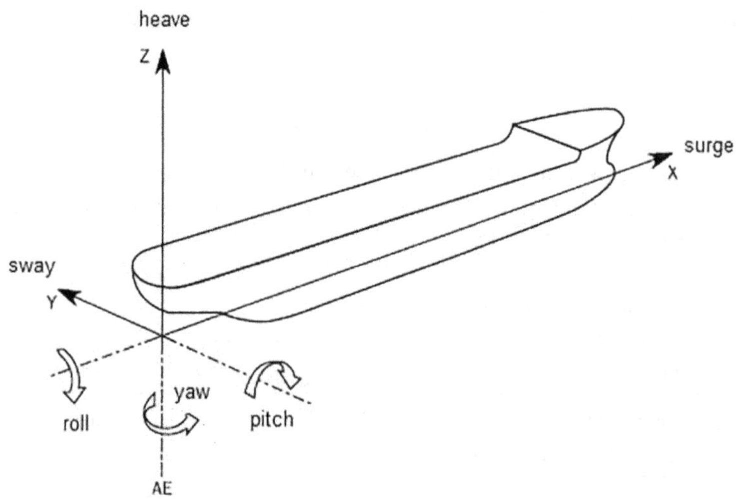

Figure A.1 Reference coordinate system and positive directions of ship motions (right-hand screw rule).

A.2 The Reference Coordinate System

The ship's geometry, motions, accelerations and loads are defined with respect to the right-hand system shown in Figure A1. The origin is specified to be at the intersection of the longitudinal vertical plane of symmetry, the aft end of the ship and the baseline.

The direction of incident waves is specified by the angle β between the x-axis and the propagating wave direction, as shown in Figure A.2. Thus, in a head sea, waves are propagating in the negative x-direction. In a beam sea, waves propagate in the positive or negative y-direction. In an oblique sea, waves propagate in a direction between a head (or following) sea and a beam sea.

A.3 Sign Convention

The sign convention for the load resultants is shown in Figure A.3.

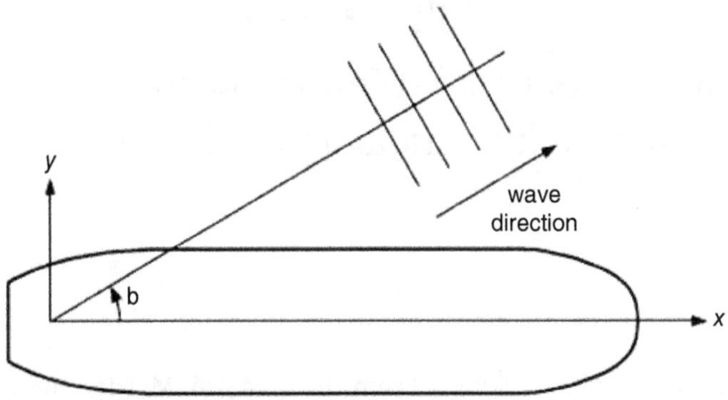

Figure A.2 Definition of wave heading.

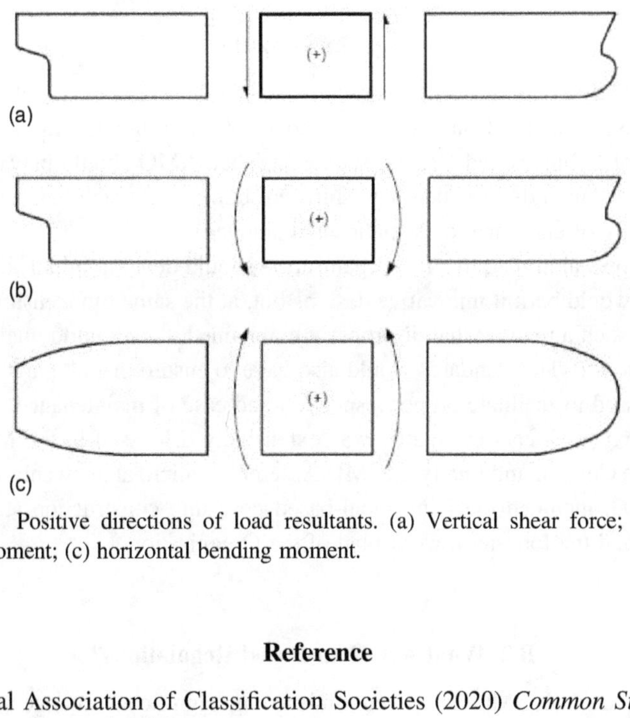

Figure A.3 Positive directions of load resultants. (a) Vertical shear force; (b) vertical bending moment; (c) horizontal bending moment.

Reference

1. International Association of Classification Societies (2020) *Common Structural Rules for Bulk Carriers and Oil Tankers.*

Appendix B

International Goal-Based Ship Construction Standards for Bulk Carriers and Oil Tankers

The following is based on texts extracted verbatim from two IMO documents and describes goal-based ship construction standards for bulk carriers and oil tankers.

B.1 Background

The concept of goal-based ship construction standards was introduced in IMO at the eighty-ninth session of the Council in November 2002 through a joint proposal from the Governments of Bahamas and Greece, suggesting that: 'IMO should play a larger role in determining the standards to which new ships are built . . .', which has been traditionally the responsibility of classification societies and shipyards.

The joint proposal argued that the Organisation should develop initial ship construction standards that would permit innovative designs but, at the same time, ensure that ships are constructed in such a manner that, if properly maintained, they could remain safe for their entire economic life. The standards would also have to ensure that all parts of a ship could be easily accessed to facilitate proper inspection and ease of maintenance.

Over the next two years the matter was extensively discussed in the Maritime Safety Committee, the Council and finally the IMO Assembly which, at its twenty-third session in November 2003, included the item 'Goal-based new ship construction standards' in the strategic plan and the long-term work plan of the Organisation.

B.2 What Are Goal-Based Regulations?

Over the past decade, IMO has been undergoing significant changes in the way it approaches the regulation of ship design. The ever-increasing speed of computers has opened a new world for designers and researchers, and as a result, IMO Member governments have started approaching safety from a completely new perspective – one that is goal and performance oriented, in lieu of the traditional prescriptive-based approach, taking into account the sophisticated nature of the industry.

Prescriptive regulations tend to be a representation of past experience and, as such, become less and less relevant over time and can hold back ship designers, who are technically innovative, from being able to properly address future design challenges.

As a result, safety regulations need to be frequently updated to keep pace with lessons learned and the latest technologies.

Example below demonstrates the difference between goal-based and prescriptive regulations:

- Goal-based: 'People shall be prevented from falling over the edge of a cliff.'
- Prescription: 'You shall install a 1-meter-high rail at the edge of the cliff.'

Going back to the 1990s, the Maritime Safety Committee (MSC) recognised that the prescriptive-based regulations were unable to cope with the new ship design challenges and already took action to incorporate the goal-based philosophy into the technical regulations of the Safety of Life at Sea Convention (SOLAS).

With the approval of the goal-based philosophy, the MSC, in effect, also agreed to a new way of viewing the regulatory development process. Traditionally, issues related to fire protection, marine engineering, naval architecture and other maritime disciplines would, in most circumstances, be considered in isolation of each other and, after deliberations, prescriptive regulations would be prepared on a piecemeal basis to address each specific area of safety. However, the new regulatory approach is holistic in nature and focuses on achieving goals such as 'a ship should be designed for improved survivability so that, in the event of a casualty, persons can stay safely on board (in a safe haven) as the ship proceeds to port'.

B.3 Basic Principles and Methodology

After in-depth discussions within the MSC and its ad hoc Working Group on Goal-Based Standards (GBS) over several years, the Committee, in May 2005, agreed the IMO goal-based standards are:

- Broad, overarching safety, environmental and/or security standards that ships are required to meet during their lifecycle;
- The required level to be achieved by the requirements applied by class societies and other recognised organisations, administrations and IMO;
- Clear, demonstrable, verifiable, long-standing, implementable and achievable, irrespective of ship design and technology; and
- Specific enough not to be open to differing interpretations.

It is understood that these basic principles were developed to be applicable to all goal-based standards developed by IMO and not only to ship construction standards, in recognition that, in the future, IMO may develop goal-based standards for other safety areas such as machinery, equipment, fire-protection and so forth, as well as security- and environment protection–related areas, and that all goal-based standards developed by the organisation should follow the same basic principles. It was agreed to proceed with the development of GBS using a deterministic approach, while, at the same time, the use of risk-based methodologies was to be further explored over the next few sessions of the Committee.

In May 2006, MSC 81 agreed to limit the scope of the GBS work initially to bulk carriers and oil tankers and consider expansion to other ship types and areas of safety at a later time. For GBS, in general, the following five-tier system was agreed:

Tier I – Goals

High-level objectives to be met.

Tier II – Functional requirements

Criteria to be satisfied in order to conform to the goals.

Tier III – Verification of conformity

Procedures for verifying that the rules and regulations for ship design and construction conform to the goals and functional requirements.

Tier IV – Rules and regulations for ship design and construction

Detailed requirements developed by IMO, national administrations and/or recognised organisations and applied by national administrations and/or recognised organisations acting on their behalf to the design and construction of a ship in order to conform to the goals and functional requirements.

Tier V – Industry practices and standards

Standards, codes of practice, and safety and quality systems for shipbuilding, ship operation, maintenance, training, manning and so forth which may be incorporated into, or referenced in, the rules and regulations for the design and construction of a ship.

B.4 Instruments Approved and Adopted

Goal-Based Standards (Tiers I to III) became mandatory on 1 January 2012 under the SOLAS Convention (new SOLAS regulation II-1/3-10), subsequent to the adoption of the following instruments at MSC 87 (May 2010):

- SOLAS regulation II-1/3-10 'Goal-based ship construction standards for bulk carriers and oil tankers' (resolution MSC.290(87));
- International goal-based ship construction standards for bulk carriers and oil tankers (resolution MSC.287(87)) (the GBS Standards); and
- Guidelines for the verification of conformity with goal-based ship construction standards for bulk carriers and oil tankers (resolution MSC.296(87)) (the GBS Verification Guidelines).
- The new SOLAS regulation II-1/3-10 made goal-based standards applicable to bulk carriers and oil tankers of 150 m in length and above, for which the building contract is placed on or after 1 July 2016;
- In the absence of a building contract, the keels of which are laid or which are at a similar stage of construction on or after 1 July 2017; or
- The delivery of which is on or after 1 July 2020.

The new SOLAS regulation also requires that a Ship Construction File be provided upon delivery of a new ship and kept on board the ship and/or ashore (see also the Guidelines for the information to be included in a Ship Construction File (MSC.1/Circ.1343)).

MSC 89 (May 2011), with a view to providing the process for the development, verification, implementation and monitoring of GBS to support regulatory development within IMO, approved the generic guidelines for developing IMO goal-based standards (MSC.1/Circ.1394).

B.4.1 Tier I – Goals

The Tier I goals are as defined in SOLAS regulation II-1/3-10 and are reproduced here for ease of reference, as follows:

Ships shall be designed and constructed for a specified design life to be safe and environmentally friendly, when properly operated and maintained under the specified operating and environmental conditions, in intact and specified damage conditions, throughout their life.

1. *Safe and environmentally friendly* means the ship shall have adequate strength, integrity and stability to minimise the risk of loss of the ship or pollution to the marine environment due to structural failure, including collapse, resulting in flooding or loss of watertight integrity.
2. *Environmentally friendly* also includes the ship being constructed of materials for environmentally acceptable recycling.
3. *Safety* also includes the ship's structure, fittings and arrangements providing for safe access, escape, inspection and proper maintenance and facilitating safe operation.
4. *Specified operating and environmental conditions* are defined by the intended operating area for the ship throughout its life and cover the conditions, including intermediate conditions, arising from cargo and ballasting operations in port, waterways and at sea.
5. *Specified design life* is the nominal period that the ship is assumed to be exposed to operating and/or environmental conditions and/or the corrosive environment and is used for selecting appropriate ship design parameters. However, the ship's actual service life may be longer or shorter depending on the actual operating conditions and maintenance of the ship throughout its life cycle.

B.4.2 Tier II – Functional Requirements

In Tier II a set of requirements relevant to the functions of ship structures to be complied with are specified, so as to meet the goals set in Tier I. The functional requirements have been structured to form three groups: design, construction and in-service considerations.

Design

II.1 Design Life

The specified design life is not to be less than 25 years.

II.2 Environmental Conditions

Ships should be designed in accordance with North Atlantic environmental conditions and relevant long-term sea state scatter diagrams.

II.3 Structural Strength

Ships should be designed with suitable safety margins:

1. To withstand, at *net scantlings*,[1] in the intact condition, the environmental conditions anticipated for the ship's design life and the loading conditions appropriate for them, which should include full homogeneous and alternate loads, partial loads, multi-port and ballast voyage, and ballast management condition loads and occasional overruns/overloads during loading/unloading operations, as applicable to the class designation
2. Appropriate for all design parameters whose calculation involves a degree of uncertainty, including loads, structural modelling, fatigue, corrosion, material imperfections, construction workmanship errors, buckling and residual strength. The structural strength should be assessed against excess deformation and failure modes, including but not limited to buckling, yielding and fatigue. Ultimate strength calculations should include ultimate hull girder capacity and ultimate strength of plates and stiffeners. The ship's structural members should be of a design that is compatible with the purpose of the space and ensures a degree of structural continuity. The structural members of ships should be designed to facilitate load/discharge for all contemplated cargoes to avoid damage by loading/discharging equipment which may compromise the safety of the structure.

II.4 Fatigue Life

The design fatigue life should not be less than the ship's design life and should be based on the environmental conditions in II.2.

II.5 Residual Strength

Ships should be designed to have sufficient strength to withstand the wave and internal loads in specified damaged conditions such as collision, grounding or flooding. Residual strength calculations should take into account the ultimate reserve capacity of the hull girder, including permanent deformation and post-buckling behaviour. Actual foreseeable scenarios should be investigated in this regard as far as is reasonably practicable.

II.6 Protection against Corrosion

Measures to protect against corrosion are to be applied to ensure that net scantlings that are required to meet structural strength provisions are maintained throughout the specified design life. Additional measures include, but are not limited to, coatings, cathodic protection, impressed current systems and so forth.

[1] *Net scantling*: a term introduced in the IACS Common Structural Rules signifying the scantlings required for purely strength requirements. This is discussed further in the next section (see also Section 6.2.6).

II.6.1 Coating Life Coatings should be applied and maintained in accordance with manufacturers' specifications concerning surface preparation, coating selection, application and maintenance. Where coating is required to be applied, the design coating life is to be specified. The actual coating life may be longer or shorter than the design coating life, depending on the actual conditions and maintenance of the ship. Coatings should be selected as a function of the intended use of the compartment, materials and application of other corrosion prevention systems, for example cathodic protection or other alternatives.

II.6.2 Corrosion Addition The corrosion addition should be added to the net scantling and should be adequate for the specified design life. The corrosion addition should be determined on the basis of exposure to corrosive agents such as water, cargo or corrosive atmosphere, and whether the structure is protected by corrosion prevention systems such as coating, cathodic protection or alternative means. The design corrosion rates (mm/year) should be evaluated in accordance with statistical information established from service experience and/or accelerated model tests. The actual corrosion rate may be greater or smaller than the design corrosion rate, depending on the actual conditions and maintenance of the ship.

II.7 Structural Redundancy

Ships should be of redundant design and construction so that localised damage of any one structural member will not lead to immediate consequential failure of other structural elements leading to loss of structural and watertight integrity of the ship.

II.8 Watertight and Weathertight Integrity

Ships should be designed to have adequate watertight and weathertight integrity for the intended service of the ship and adequate strength and redundancy of the associated securing devices of hull openings.

II.9 Design Transparency

Ships should be designed under a reliable, controlled and transparent process made accessible to the extent necessary to confirm the safety of the new as-built ship, with due consideration to intellectual property rights. Readily available documentation should include the main goal-based parameters and all relevant design parameters that may limit the operation of the ship.

Construction

II.10 Construction Quality Procedures

Ships should be built in accordance with controlled and transparent quality production standards with due regard to intellectual property rights. The ship construction quality procedures should include, but not be limited to, specifications for material, manufacturing, alignment, assembling, joining and welding procedures, surface preparation and coating.

II.11 Survey

A survey plan should be developed for the construction phase of the ship, taking into account the type and design. The survey plan should contain a set of requirements to ensure compliance of construction with classification rules and goal-based standards. The survey plan should also identify areas that need special attention during surveys throughout the ship's life.

In-service Considerations

II.12 Maintenance

Ships should be designed and constructed to facilitate ease of maintenance, in particular avoiding the creation of spaces too confined to allow for adequate maintenance activities.

II.13 Structural Accessibility

The ship should be designed, constructed and equipped to provide adequate means of access to all internal structures to facilitate overall and close-up inspections and thickness measurements.

B.4.3 Tier III – Verification of Compliance Criteria

In Tier III, instruments that are necessary to demonstrate that the detailed requirements in Tier IV comply with the Tier I goals and the Tier II functional requirements are defined. Verification should consist of the following four steps:

1. Verification that prescriptive rules by classification societies are in accordance with the goal-based standards;
2. Verification that the design of individual ships meets classification societies' rules;
3. Verification that the construction of ships meets classification societies' rules; and
4. Verification that the ship throughout its life meets applicable rules.

B.4.4 Tier IV – Technical Procedures and Guidelines, Classification Rules and Industry Standards

The detailed requirements developed by IMO, national administrations and/or classification societies and applied by national administrations and/or classification societies acting as recognised organisations to the design and construction of a ship in order to meet the Tier I goals and Tier II functional requirements.

In developing the bulk carrier rules JBP members mainly took into consideration the safety objectives (Tier I) and functional requirements (Tier II), considering that the common structural rules for bulk carriers themselves are part of Tier IV.

B.4.5 Tier V – Codes of Practice and Safety and Quality Systems for Shipbuilding, Safety, Ship Operation, Training, Manning etc.

Industry standards and shipbuilding design and building practices that are applied during the design and construction of a ship.

Appendix C
Rule Hull Girder Torsional Strength

It is useful to gain an idea as to how regulatory bodies stipulate the manner in which the effects of torsional loading are allowed for, especially in the case of large containerships. To this end this appendix includes relevant sections extracted from:

(a) The IACS CSR for bulk carriers and oil tankers relating to the determination of the wave torsional moment. (For these ship types the rules do not make provision for still-water torsional moment.)
(b) The Rules of the Classification Society Nippon Kaiji Kyokai for containerships of length up to 450 m.

C.1 IACS CSR for Bulk Carriers and Oil Tankers

The following have been excerpted from Part 1, Chapter 4, Section 4, Subsection 3.4 of the Rules (January 2020 edition).

The wave torsional moment at any longitudinal position with respect to the ship baseline, in kNm, is to be taken as:

$$M_{wt} = f_p(M_{wt1} + M_{wt2})$$

where

$$M_{wt1} = 0.4 f_{t1} C_W \sqrt{\frac{L}{T_{LC}}} B^2 D C_B$$

$$M_{wt2} = 0.22 f_{t2} C_w L B^2 C_B$$

f_{t1}, f_{t2} = distribution factors, taken as :

$$f_{t1} = \left| \sin \frac{2\pi x}{L} \right| \qquad 0 \leq x \leq L$$

$$f_{t2} = \sin^2\left(\frac{2\pi x}{L}\right) \qquad 0 \leq x \leq L$$

$$f_{t1}, f_{t2} = 0 \text{ for} \qquad x < 0, x > 0$$

$$f_p = \text{Coefficient to be taken as:}$$

$$= f_{ps} \text{ coefficient for strength assessment}$$

$$= 0.9\left[0.2 + (5f_T - 4.25)B\left(10^{-4}\right)\right] \text{ coefficient for fatigue assessment}$$

The coefficient for strength assessment, f_{ps}, is dependent on the applicable design load scenario specified in Chapter 4, Section 7, and to be taken as:

= 1.0 for extreme sea loads design load scenario
= 0.8 for the ballast water exchange design load scenario
= 0.8 for the accidental flooded design load scenario
= 0.4 for the harbour/sheltered water design load scenario

C_w = wave coefficient, to be taken as:

$$C_W = 10.75 - \left(\frac{300 - L}{100}\right)^{1.5} \qquad 90\text{ m} \le L \le 300\text{ m} \tag{a}$$

$$C_W = 10.75 \qquad 300\text{ m} < L \le 350\text{ m} \tag{b}$$

$$C_W = 10.75 - \left(\frac{L - 350}{150}\right)^{1.5} \qquad 350\text{ m} < L \le 500\text{ m} \tag{c}$$

In the above expressions B and D are, respectively, the moulded breadth and depth of the ship (m), and C_B is the block coefficient at the scantling draught, T_{SC}.

C.2 Nippon Kaiji Kyokai Rules for Strength Assessment Including Torsional Loading of Containerships

The following have been excerpted from the NKK Rules for the Survey and Construction of Steel Ships, Part C Hull Construction and Equipment, Chapter 32 Container carriers, Section 32.3.3–4 (2019 edition).

C.2.1 Evaluation Procedure

The procedure for calculating buckling and yield strength, allowing for vertical, horizontal and torsional moments in both still water and in waves is outlined in Figure C.1.

The steps to be followed are summarised below:

(1) Vertical still-water bending moments, vertical wave-induced bending moments, horizontal wave-induced bending moments, still-water torsional moments and wave-induced torsional moments are to be considered as applied loads;

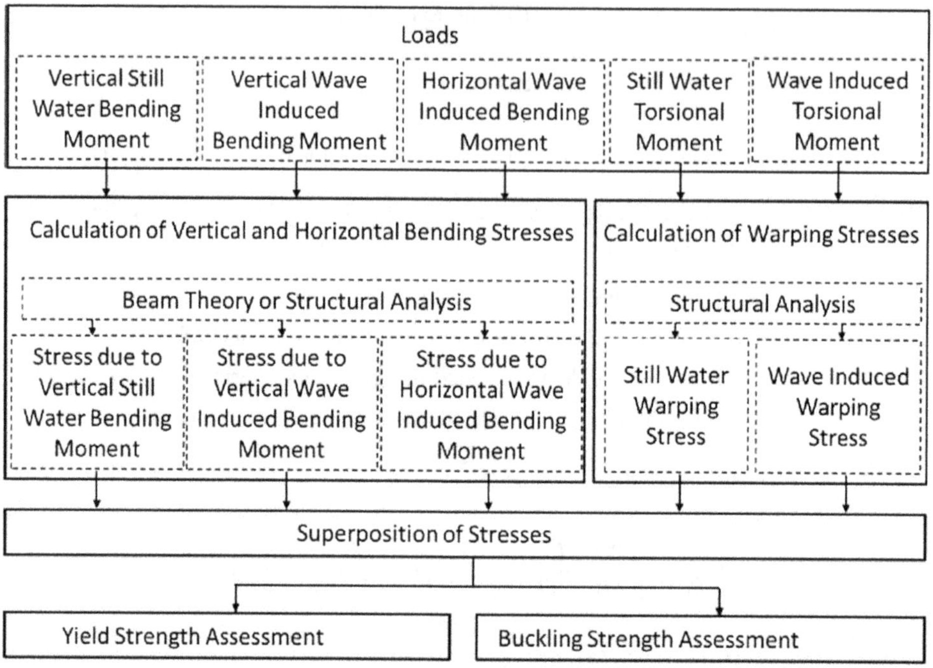

Figure C.1 Procedure for the determination of buckling and yield strength of containerships allowing for global vertical bending, horizontal bending and torsion.

(2) Stresses due to vertical still-water bending moments, vertical wave-induced bending moments and horizontal wave-induced bending moments are to be calculated using beam theory or structural analysis of a full ship FE model;

(3) Warping stresses due to still-water torsional moments and wave-induced torsional moments are to be calculated using structural analysis of a full ship FE model; and

(4) Yielding strength assessments and buckling strength assessments are to be carried out based upon evaluated stresses determined by combining the stress components.

C.2.2 Loads

1. The horizontal wave induced bending moments, M_{H1} and M_{H2}, are to be obtained from the following formulae:

$$M_{H1} = M_H \, C_{H1}$$
$$M_{H2} = M_H \, C_{H2}$$

M_H: As given by the following formula:

$$M_H = 0.32 C_1 C_2 L_1{}^2 d \sqrt{\frac{L_1 - 35}{L_1}} \qquad \text{(kNm)}$$

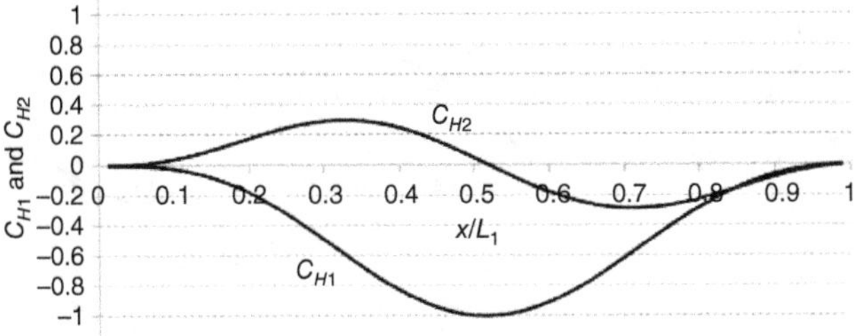

Figure C.2 Distribution of horizontal moments.

C_1 is the wave coefficient ($\equiv C_W$) to be taken as follows:

$$C_1 = 10.75 - \left(\frac{300 - L}{100}\right)^{1.5} \qquad 90 \le L \le 300 \tag{a}$$

$$C_1 = 10.75 \qquad 300 < L < 350 \tag{b}$$

$$C_1 = 10.75 - \left(\frac{L - 350}{150}\right)^{1.5} \qquad 350 \le L \tag{c}$$

C_2 is a coefficient to be taken equal to 0.9.

C_{H1}, C_{H2} are obtained from the following formulae (Figure C.2).

$$C_{H1} = -\cos\left(0.77\pi\left(\frac{x}{L_1} - 0.52\right)\right)\sin^2\left(\pi\frac{x}{L_1}\right)\left(\frac{1 - \exp(-6x/L_1)}{1 - \exp(-3)}\right)$$

$$C_{H2} = -\sin\left(0.77\pi\left(\frac{x}{L_1} - 0.52\right)\right)\sin^2\left(\pi\frac{x}{L_1}\right)\left(\frac{1 - \exp(-6x/L_1)}{1 - \exp(-3)}\right)$$

x is the distance from the AE (m)

2. Still-water torsional moments and wave-induced moments are to be in accordance with the following:

(1) The still-water torsional moments, M_{ST1} and M_{ST2}, are to be obtained from the following formulae:

$$M_{ST1} = M_{ST\_MAX} C_{T1}$$
$$M_{ST2} = M_{ST\_MAX} C_{T2}$$

M_{ST_MAX} is the maximum of values distributed along the longitudinal direction for the permissible maximum still-water torsional moment defined in the loading manual

C_{T1}, C_{T2} are given by the following formulae (Figure C.3):

$$C_{T1} = 1.0\left[\sin\left(2\pi\frac{x}{L_1}\right) + 0.1\sin^2\left(\pi\frac{x}{L_1}\right)\right]\exp\left(-0.35\frac{x}{L_1}\right)\exp\left(-8\left(\frac{x/L_1 - 0.5}{0.5}\right)^{10}\right)$$

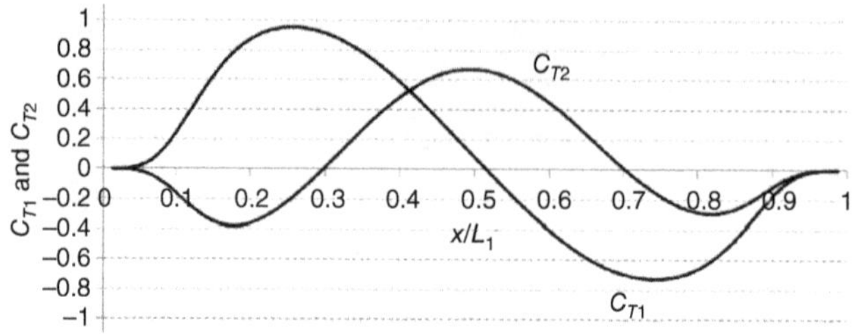

Figure C.3 Distribution of torsional moments along the ship length.

$$C_{T2} = 0.5\left[-\sin\left(3\pi\frac{x}{L_1}\right) + 0.65\sin^3\left(\pi\frac{x}{L_1}\right)\right]\exp\left(-0.4\frac{x}{L_1}\right)\exp\left(-8\left(\frac{x/L_1 - 0.5}{0.5}\right)^{10}\right)$$

x is the distance from the AE (m)

(2) The wave-induced torsional moments, M_{WT1} and M_{WT2}, are to be obtained from the following formulae:

$$M_{WT1} = M_{WT}C_{T1}$$
$$M_{WT2} = M_{WT}C_{T2}$$

M_{WT} as given by the following formula :

$$M_{WT} = 1.3C_1C_2L_1dC_b'(0.65d + e) + 0.2C_1C_2L_1B^2C_W$$

C_1, C_2 : as specified in (1) above

C_b' volume of displacement corresponding to the designed maximum load line divided by L_1Bd

e distance from baseline to centre of shear at the midship section (m)

C_W waterplane coefficient at the designed maximum load draught, to be taken as follows

$$C_W = \frac{A_W}{L_1B}$$

A_W waterplane area at the design maximum load draught (m^2)

C_{T1}, C_{T2} as specified in (1) above

C.2.3 Calculation of Stresses Due to Vertical Bending Moment and Horizontal Bending Moment

Stresses due to vertical still-water bending moment

$$\sigma_S = \frac{M_S}{1{,}000I}(z - z_n) \qquad (\text{N/mm}^2)$$

Stresses due to vertical wave-induced bending moment

$$\sigma_W = \frac{M_W}{1{,}000I}(z - z_n) \qquad (\mathrm{N/mm^2})$$

where

M_S, M_W permissible still-water and wave-induced bending moments specified in the rules at the section under consideration

I moment of inertia of the section (in m^4) about the horizontal neutral axis of bending based on gross scantlings

$z - z_n$ distance of member from horizontal neutral axis

Stresses due to the horizontal wave-induced bending moments σ_{H1}, σ_{H2}

$$\sigma_{H1} = -\frac{M_{H1}}{1{,}000I_H}y \qquad (\mathrm{N/mm^2})$$

$$\sigma_{H2} = -\frac{M_{H2}}{1{,}000I_H}y \qquad (\mathrm{N/mm^2})$$

where

M_{H1}, M_{H2} first and second components of horizontal bending moment (kN-m) at the cross-section under consideration

I_H moment of inertia around centreline (m^4) for the cross-section under consideration using the gross scantling approach

y y-coordinate at the location under consideration (horizontal distance from centreline) (m)

The warping stresses are to be calculated in accordance with (1) and (2) above by applying torsional moments to structural models satisfying the requirements of Subsection 32.3.5. The sign convention for tension stress is to be positive, while the sign convention for compressive stress is to be negative.

(1) The still-water warping stresses, $\sigma_{ST1}, \sigma_{ST2}$, are to be calculated by applying to the structural model the still-water torsional moments, M_{ST1}, M_{ST2}, specified in Subsection 32.3.4–2(1).

(2) The wave-induced warping stresses, $\sigma_{WT1}, \sigma_{WT2}$, are to be calculated by applying to the structural model the wave-induced torsional moments, M_{WT1}, M_{WT2}, specified in Subsection 32.3.4–2(1).

C.2.4 Superposition of Stresses

The evaluated stress, σ_T, is to be determined by combining each stress component in accordance with the following formula:

$$\sigma_T = C_3\left(\sqrt{\sigma_W{}^2 + (\sigma_{H1} + \sigma_{WT1})^2 + (\sigma_{H2} + \sigma_{WT2})^2} + |\sigma_S| + \sqrt{\sigma_{ST1}{}^2 + \sigma_{ST2}{}^2}\right)$$

C_3 coefficient to correct stress for net scantling approach, to be taken as1.05

σ_S, σ_W stresses due to the vertical still-water bending moments and vertical wave-induced bending moments (N/mm^2)

σ_{H1}, σ_{H2} stresses due to the horizontal wave-induced bending moments (N/mm^2)

σ_{WT1}, σ_{WT2} wave-induced warping stresses (N/mm^2)

σ_{ST1}, σ_{ST2} still-water warping stresses (N/mm^2)

Appendix D

IACS Approaches to Estimating Hull Girder Ultimate Strength

This appendix provides details of the methods included by IACS in the Common Structural Rules to calculate the ultimate strength of the hull girder.

D.1 Definition of Ultimate Strength Capacity

The ultimate strength under vertical bending is obtained from the bending moment – curvature diagram $(M - \chi_F)$, which is traced out for both the hogging and the sagging conditions, well into the post-collapse range (Figure D.1). It is thereby defined by the pair of maximum values of the response to hogging and sagging loads (M_{UH}, M_{US}).

These values are required to fulfil the following condition:

$$M \leq \frac{M_U}{\gamma_R} \tag{D.1}$$

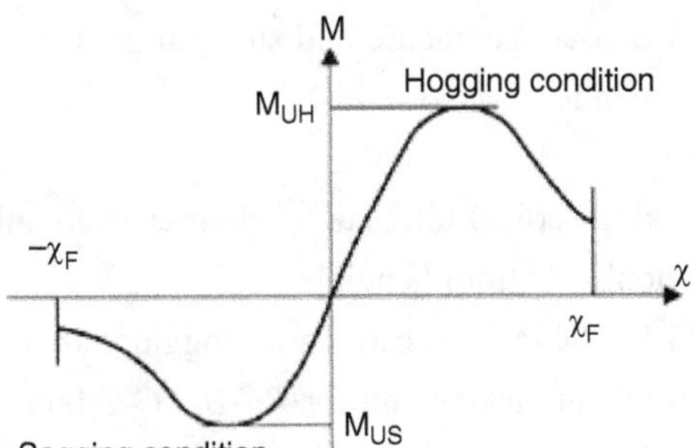

Figure D.1 General form of moment-curvature curve for hull girder in hogging and sagging conditions.

where

M_U = ultimate bending moment capacity at the section considered, calculated with net scantlings, based on a gross thickness reduced by one-half the corrosion allowance, t_C
= M_{UH} in hogging condition
= M_{US} in sagging condition

M = bending moment in sagging and hogging condition, calculated in the intact, flooded and harbour conditions
$= M_{sw} + \gamma_W M_{wv}$

M_{sw} = design still-water bending moment (hogging and sagging) for the ship in intact, flooded ($M_{SW,F}$) and harbour conditions ($M_{SW,P}$)

M_{wv} = design vertical wave-bending (hogging and sagging) for the ship in intact, flooded ($M_{sw,f}$) and harbour conditions ($M_{sw,p}$)

γ_R = partial safety factor for the vertical hull girder ultimate bending capacity allowing for the effect of double bottom bending
$= \gamma_M \gamma_{DB}$

γ_M = p.s.f. for vertical ultimate bending capacity allowing for material, geometric and strength prediction uncertainties
= 1.10

γ_{DB} = p.s.f. for vertical ultimate bending capacity allowing for double bottom bending
= 1.25 for *BC-A* bulk carriers in hogging condition
= 1.10 for oil tankers and for *BC-B*, *BC-C* bulk carriers in hogging condition
= 1.0 for vessel in a sagging condition

γ_W = safety factor on wave-bending moments
= 1.20

D.2 IACS Methods to Determine Ultimate Strength

A wide range of methods exist to calculate the ultimate bending moment capacity of the hull girder. These include:

(a) Approximate methods, which are based on simplifying assumptions concerning behaviour and the definition of collapse (an example is Caldwell's fully plastic collapse approach),
(b) The progressive collapse approach, whose principles are outlined in Chapter 6 and whose implementation is described in the present appendix, and
(c) Nonlinear FEM analysis. This is considered to be too time and resource consuming for everyday design work.

For use in design, IACS has proposed two ways of determining the ultimate strength of the hull girder. The first is a simple, easy-to-apply approach that provides an approximate value of the ultimate strength. The second is the incremental-iterative method.

We shall outline each of these in turn.

D.2.1 The Single-Step Ultimate Capacity Method

The single-step method is based on the assumption that the ultimate capacity of the entire hull girder is reached when the applied strain reaches the yield strain in the deck structure. This occurs in way of *hard corners*[2] in the strength deck. As a result, further plastic straining does not take place and permanent sets are avoided. The ultimate capacity thus obtained will underestimate the true value. At this strain, the deck structure will have buckled elasto-plastically. As a result, the capability of the hull girder comprises of two parts:

(a) The load carried by the deck when elasto-plastic failure occurs
(b) The load carried by the remaining portions of the hull girder, that is assumed to give rise to elastic stresses, apart from the deck portions that have not buckled (hard corners)

To determine the elasto-plastic collapse load of the deck structure, the method described in Appendix D of the CSR is used. In the single-step method, allowance is made for the reduction in stiffness due to deck buckling, by use of the modified position of the neutral axis.

The ultimate bending moment is found from:

$$M_U = EI_{red}\Delta\kappa_1 \qquad \text{(kNm)} \tag{D.2}$$

where

EI_{red} = reduced hull girder bending stiffness allowing for buckled deck

$\Delta\kappa_1$ = incremental curvature from zero load to yield strain, ε_0, in the deck

$$= \frac{\varepsilon_0}{z_{dk-mean} - z_{NA-red}} \tag{D.3}$$

[2] *Hard corners* are discussed in Section D.2.2

D.2.2 The Incremental-Iterative (Multistep) Method

The approach followed in this method is described in Chapter 6. The method is applied to a transverse section of the hull girder between two adjacent webs and is based on the following assumptions:

- Transverse sections of the hull girder remain plane throughout the loading sequence. (This assumption has been found to be correct for strains up to twice the yield strain.)
- The hull material behaves as an elastic–perfectly plastic material (no strain-hardening). The behaviour of a stiffened plate member will therefore be as shown in Figure D.2.
- The entire section is subdivided into several stiffened plate elements and hard corners as shown in Figure D.3. These are assumed to act independently, in the sense that the behaviour of one does not affect that of an adjacent member.

Hard corners are elements that consist of members that are attached to each other on either side of a structural boundary, such as a deck edge. The boundary ensures rotational clamping of the two members (the 90° angle is preserved) so that the entire element is assumed to fail when the squash load is reached (inelastic buckling does not arise). They generally refer to (Figure D.4):

- Plating area adjacent to intersecting plates
- Plating area adjacent to knuckles with an angle >30°
- Plating in rounded sections

Loading of the hull girder section is applied by increasing the curvature in a stepwise manner. This enables the full moment-curvature curve to be traced out. The response is thus displacement controlled, and the stress distribution across the section is the aggregate of the (average) stresses in each element.

The initial value of the curvature is set equal to zero, and the increment in curvature is set equal to 1 per cent of the curvature required to produce a stress equal to the yield stress in the deck.

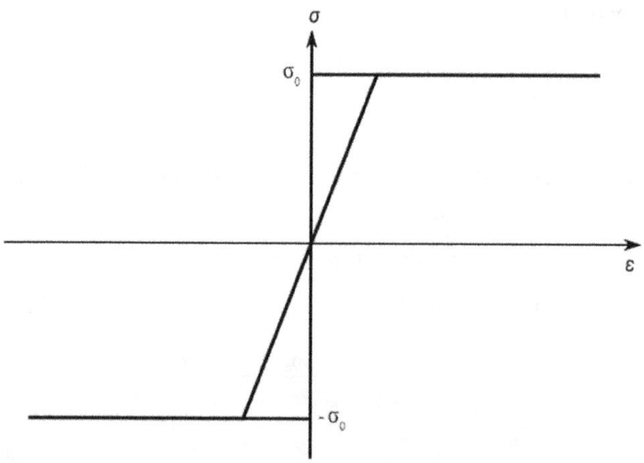

Figure D.2 Assumed behaviour of a stiffened plate element.

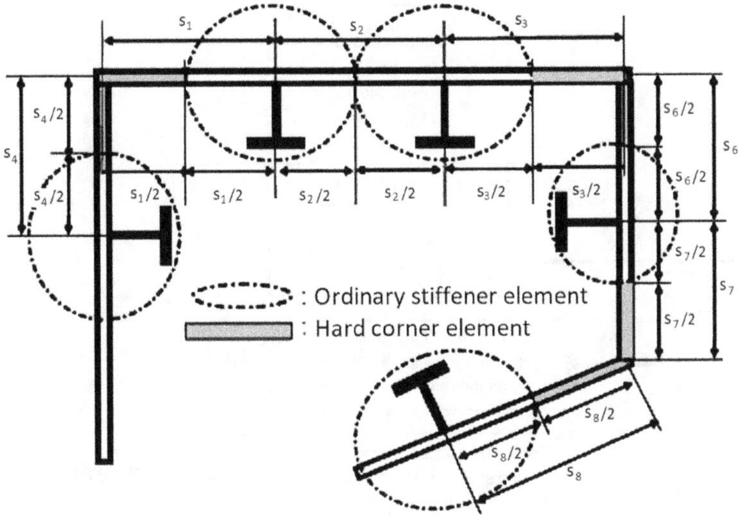

Figure D.3 Subdivision of structure into stiffened plate and hard corner elements.

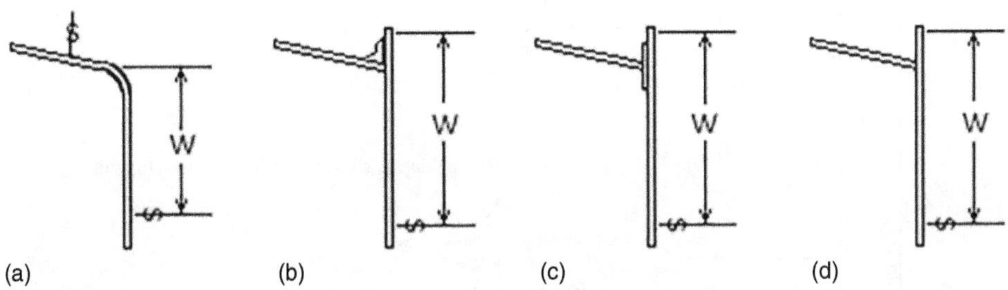

Figure D.4 Examples of hard corners. (a) Radiused gunwale; (b) Stringer angle; (c) Stringer bar welded to sheerstrake; (d) Stringer plate.

1. The curvature is increased by the specified increment.
2. For each value of curvature the strain that is applied to each element is found on the basis of linear theory.
3. Using the local strength curves, the average stress in each element is found.
4. The position of the neutral axis of the section is determined in order to ensure longitudinal equilibrium.
5. The bending moment about the current neutral axis is found by summation.
6. The resulting bending moment is compared with that of the previous load increment. If the slope in the moment-curvature curve is negative and exceeds a certain amount, the procedure is terminated. Otherwise, the procedure returns to step 1.

The steps of the method are included in the flowchart in Figure D.5. For practical implementation of the method, the MARS program prepared by Bureau Veritas can be used.

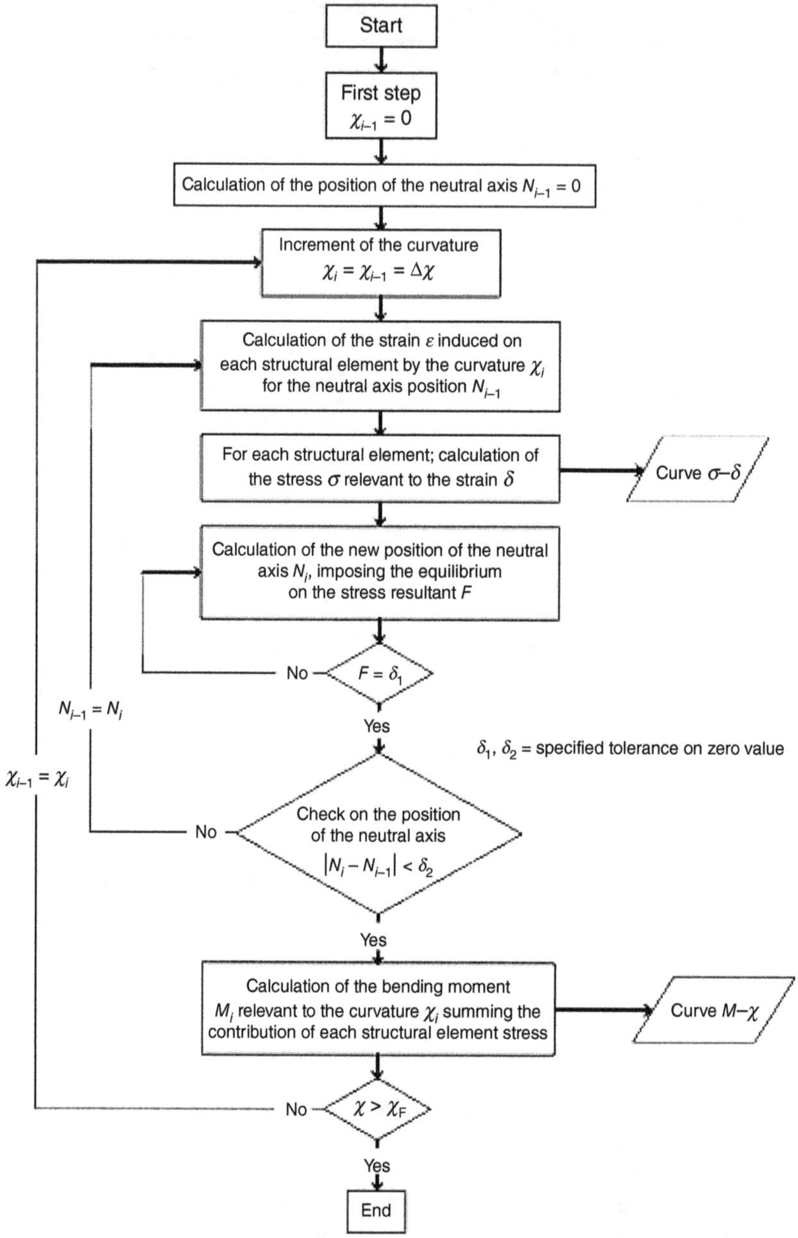

Figure D.5 Flow chart of the IACS method for the ultimate strength calculation [2].

D.2.3 Comparison of Results of IACS Method with FE Results

This section briefly describes the results obtained from a comparison of the IACS multi-step method with a NLFE analysis [3]. A study was carried out on several ship types (oil tankers, bulk carrier container ship), and it was found that differences do not exceed 5 per

Table D.1 *Principal particulars of chemical tanker*

Length overall	182.88 m	Young's modulus	210,000 GPa
Length (BP)	175.25 m	Poisson's ratio	0.3
Beam (moulded)	32.20 m	Yield stress (AH36)	355 MPa
Depth (to main deck)	15.00 m	Yield stress (Duplex SS)	455 MPa
Draught (scantling)	11.10 m		
Cargo tank capacity	44,000 m^3		

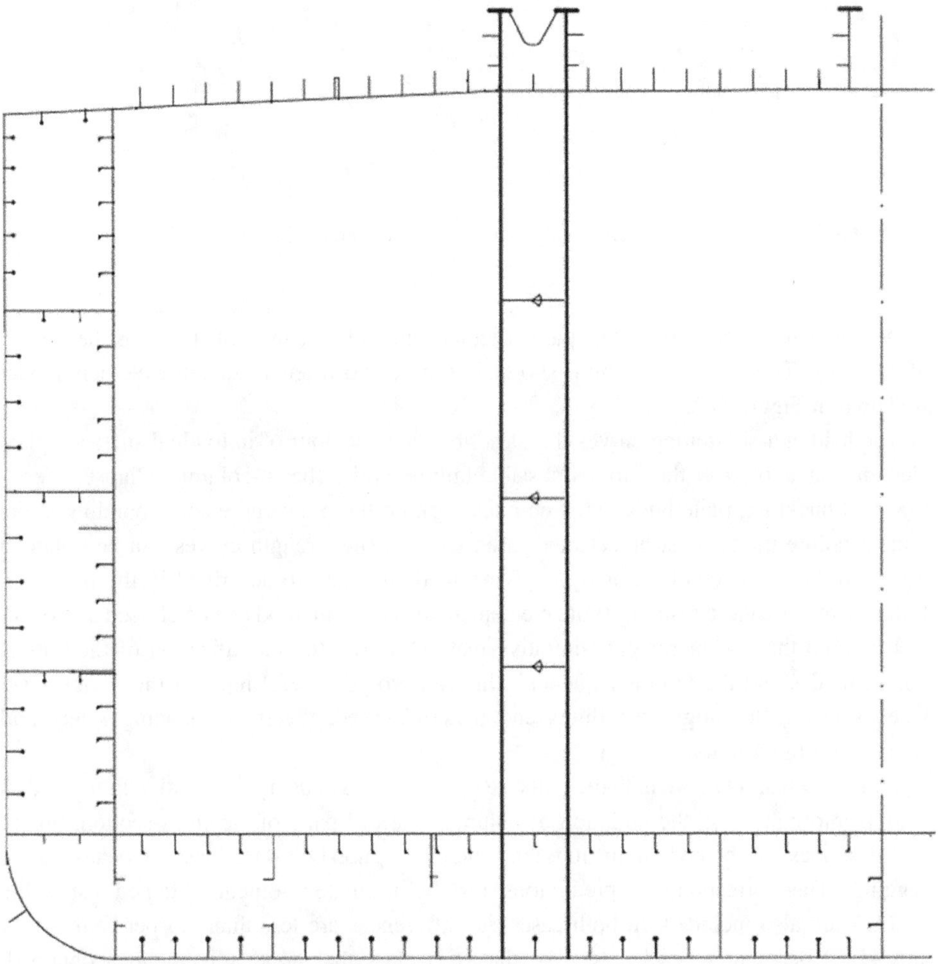

Figure D.6 Chemical tanker midship section.

cent. An interesting observation is that in both sagging and hogging conditions, failure was initiated in the strength deck. As expected, the ultimate bending moment in hogging is higher than that in sagging.

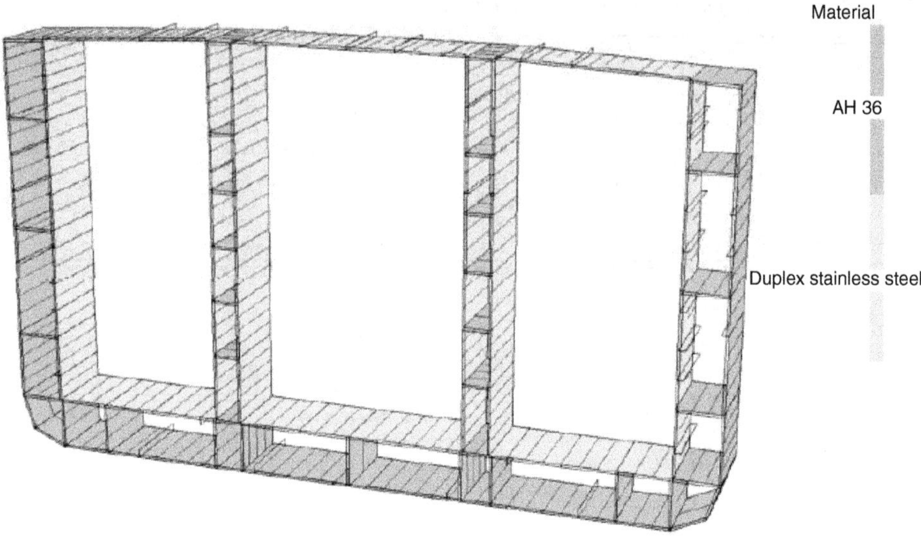

Figure D.7 FE model of a chemical tanker indicating materials used.

A chemical tanker whose principal particulars are included in Table D.1 was the subject of the study. The midship section is shown in Figure D.6 whereas the finite element model is shown in Figure D.7.

The load end-shortening curves that describe the behaviour of individual stiffened plate elements have to cover the various possible failure modes (beam-column collapse, lateral-torsional buckling, plate buckling, web buckling). Of these the curve corresponding to the critical failure mode must be selected in each case. The strength curves can be obtained using rule recommendations or by applying methods that are described in the literature. If data is obtained as a result of finite element analysis, the method can be classed as *hybrid*.

It is seen that the incremental-iterative method allows the identification of the critical failure modes and the failure sequence. This can provide useful input at the design level when selecting topology, scantlings and materials used, thereby enhancing safety and achieving a more rational design.

Figures D.8 and D.9 include the collapse curves obtained using the multi-step method as well as information on the initiation of failure in various parts of the cross-section. Figure D.8 describes the behaviour in hogging whereas Figure D.9 describes the behaviour in sagging. The corresponding predictions for the ultimate moment obtained using the NLFEA are also included. In both cases the differences are less than 2.5 per cent (–1.04 and +2.22 per cent in each case), so that it is clear that good agreement is obtained. A comparison of the failure modes at the ultimate bending moment showed that both methods predicted collapse in the same regions of the hull section, specifically the bottom and deck stiffened plating along with the upper portion of the double side shell structure and the longitudinal bulkheads. This was true in both the hogging and the sagging conditions.

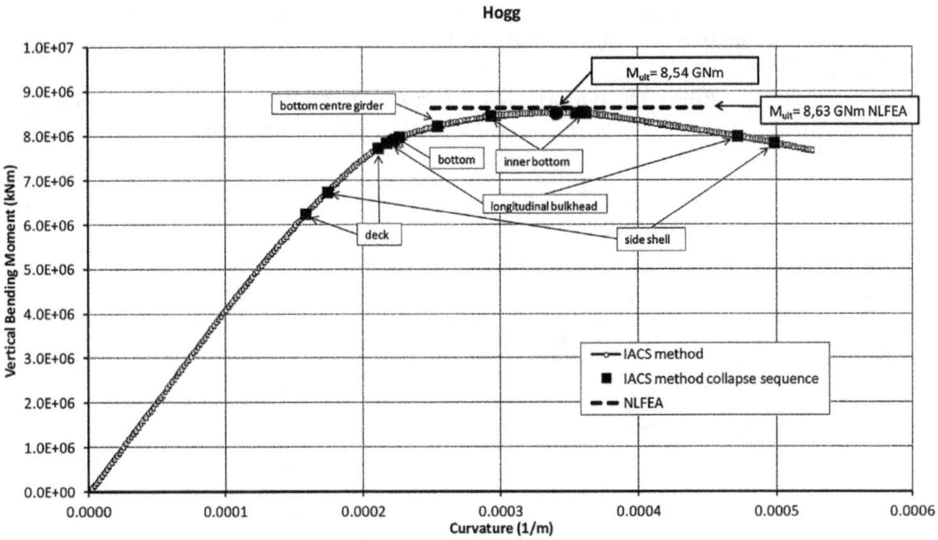

Figure D.8 Moment curvature curve for the chemical tanker in a hogging condition.

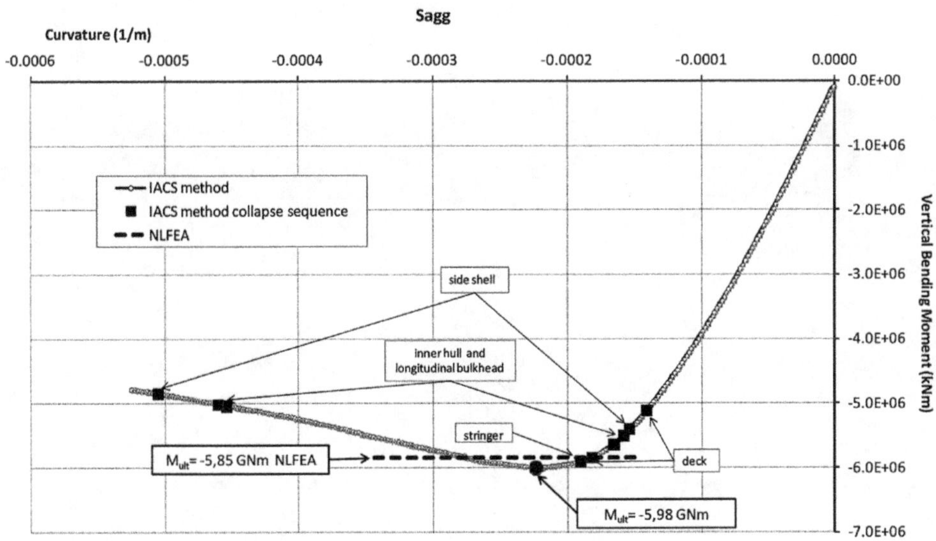

Figure D.9 Moment-curvature curve for the chemical tanker in a sagging condition.

References

1. International Association of Classification Societies (2012) Hull Girder Longitudinal Strength Report No. Pt 1 Chap. 5 Rev. 1.
2. International Association of Classification Societies (2006) Common Structural Rules for Double Hull Tankers. Background Document Appendix A - Hull Girder Ultimate Strength.

3. Andrić J. Kitarović S. Bičak M. (2014) IACS incremental-iterative method in progressive collapse analysis of various hull girder structures. *Brodogradnja (Shipbuilding)* Vol. 65 65–78.
4. Qi E. Cui W. (2006) Design-oriented Methods of Ultimate Hull Girder Strength, *Proc. 25th Int. Conf. Off. Mech. Arctic Eng. (OMAE2006)* Hamburg.

Appendix E

Orthogonality and the Modal Analysis of Structures

In this appendix the underlying concepts of the modal analysis method used in the vibration analysis of structures will be outlined. The method is used in numerical procedures and is therefore linked to the finite element method, whose application to the vibration analysis of the hull girder is discussed in Chapter 14. Before discussing the method itself, it is useful to clarify a concept from linear algebra which finds application in structural analysis, namely *orthogonality*. Orthogonality is an important concept that is employed in the solution of differential equations (both ordinary and partial) and in eigenvalue problems that are solved using finite element procedures. It is employed in modal analysis, which is used to solve problems such as the vibration of beams and, by extension, that of hull girder vibration.

Eigenvalue theory and the techniques used to extract eigenvectors are not covered here. Different bodies recommend and make use of different techniques. Thus, Lloyds Register uses static condensation whereas DNV-GL (formerly Germanischer Lloyd) uses the Ritz procedure coupled with the Lanczos method. There exist many texts describing eigenvector/eigenvalue extraction techniques that may be consulted.

E.1 Orthogonality

Definition:

Two non-zero vectors $\mathbf{q}_1, \mathbf{q}_2$ are termed *orthogonal* when their dot product is equal to zero. That is,

$$\mathbf{q}_1\mathbf{q}_2 = 0 \tag{E.1}$$

Hence, since

$$\mathbf{q}_1\mathbf{q}_2 = \|\mathbf{q}_1\|\|\mathbf{q}_2\| \cos\alpha = 0 \qquad \alpha = \pi/2 \tag{E.2}$$

That is, they are perpendicular to one another. Therefore, the magnitude of each vector can vary independently of the other. Orthogonality of vectors therefore implies that they are *independent*. A formal proof follows.

Independence of orthogonal vectors

Let $c_1, c_2, \ldots, c_k$ be constants such that nonzero orthogonal vectors $\boldsymbol{u}_1, \boldsymbol{u}_2, \ldots, \boldsymbol{u}_k$ satisfy the relation

$$c_1\mathbf{u}_1 + \ldots + c_k\mathbf{u}_k = 0 \tag{E.3}$$

Take the dot product of this equation with vector $\mathbf{u_j}$ to obtain the scalar relation

$$c_1\mathbf{u_1}\mathbf{u_j} + \ldots + c_k\mathbf{u_k}\mathbf{u_j} = 0 \tag{E.4}$$

Because all terms on the left are zero, except one ($j = k$), the relation reduces to the simpler equation

$$c_j\|\mathbf{u}_j\|^2 = 0 \tag{E.5}$$

This equation implies $c_j = 0$. Therefore, $c_1, c_2, \ldots, c_k = 0$, and the vectors are proved to be *independent.*

Furthermore, if the vectors are divided by their length, they become *orthonormal* vectors, and their length is equal to unity. These properties can be summarised as follows:

$$\phi_i^T\phi_j^T = \begin{cases} 0 & \text{if} \quad i \neq j \ \text{(orthogonality)} \\ 1 & \quad\quad i = j \ \text{(orthonormality)} \end{cases} \tag{E.6}$$

If a *matrix* **Q** contains *orthonormal* vectors **q**, the basis of that matrix is then said to be *orthonormal.* If we perform the multiplication $\mathbf{Q^TQ}$ for a square orthonormal matrix **Q**, we obtain:

$$\mathbf{Q^TQ} = \begin{bmatrix} q_1^T \\ q_2^T \\ \cdot \\ q_n^T \end{bmatrix} [q_1 \; q_2 \; q_n] = \begin{bmatrix} 1 & 0 & \ldots & 0 \\ 0 & 1 & \ldots & 0 \\ & & & \\ 0 & 0 & \ldots & 1 \end{bmatrix} = \mathbf{I} \tag{E.7}$$

That is,

$$\mathbf{Q}^T\mathbf{Q} = \mathbf{I} \tag{E.8}$$

This is an important property of orthonormal matrices. Orthonormal matrices are not necessarily square. If they are, however, they are easy to work with, and in this case, from $\mathbf{Q^TQ} = \mathbf{I}$ we have

$$\mathbf{Q^T} = \mathbf{Q}^{-1} \tag{E.9}$$

In this case **Q** is termed an *orthogonal* matrix. For an orthogonal matrix its transpose is therefore equal to its inverse. Since **Q** has orthonormal columns, the lengths are unchanged. That is,

$$\|\mathbf{Qx}\| = \|\mathbf{x}\| \qquad \text{for every vector } \mathbf{x} \tag{E.10}$$

If, however, the column vectors are orthogonal and not orthonormal, dot products still give a diagonal matrix, but not the identity matrix.

The fact that orthogonality implies independence is of fundamental importance in the solution of problems in which coupling arises, because if orthogonality is present or can be introduced, the problem can be simplified to a considerable extent.

In partial differential equations in which derivatives of a variable with respect to more than one independent variable exist, if an equivalent problem that is characterised by orthogonality can be derived, the original p.d.e. does not need to be solved but is replaced by a number of ordinary differential equations whose solution can be reached much more easily. An example that is included in this text is the Euler–Bernoulli equation of beam flexural vibration (Section 8.2.1).

Another example is that of the biharmonic equation, which represents equilibrium of a rectangular flat plate under generalised lateral loading. In this case the resulting deflections and the applied pressure can be represented by the double Fourier series, each of which involves products of sine and/or cosine terms. When the integral of a product of these terms over a specified interval is equal to zero, the terms are orthogonal with respect to one another – that is, they are independent. As a result, superposition is permitted, and the solution can be obtained by summation. Navier exploited this property in order to obtain his solution to this problem.

E.2 Basic Ideas of the Modal Analysis Method

Consider the 2DOF system shown in Figure E.1.

The two-storey frame with degrees of freedom u_1 and u_2 that is excited by the harmonic force $p_1(t) = p_0 sin\omega t$ can be idealised as a mass-spring system (Figure E.1b). The 2DOF system consists of two equations that may be written as follows:

$$\begin{bmatrix} m_1 & 0 \\ 0 & m_2 \end{bmatrix} \begin{Bmatrix} \ddot{u}_1 \\ \ddot{u}_2 \end{Bmatrix} + \begin{bmatrix} k_1 + k_2 & -k_2 \\ -k_2 & k_2 \end{bmatrix} \begin{Bmatrix} u_1 \\ u_2 \end{Bmatrix} = \begin{Bmatrix} p_0 \\ 0 \end{Bmatrix} \sin \omega t \qquad \text{(a)}$$

The stiffness matrix **K** includes off-diagonal terms, and therefore the equations are coupled. Hence the equations need to be solved simultaneously. This is entirely feasible, although, when dealing with large systems of equations such as in the case of a global ship model, the operation involves a large number of operations that render it a practical impossibility. It is possible to solve the system of the equations of motion much more efficiently by

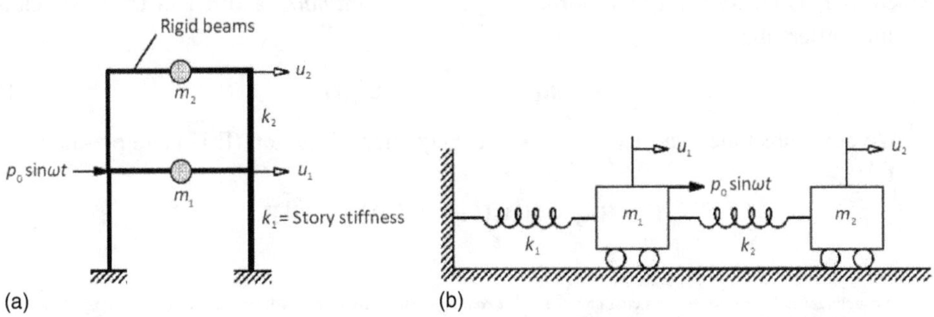

Figure E.1 Example of a 2DOF system [4]. (a) Two-storey frame excited by a harmonic force; (b) idealisation as a mass-spring system.

first uncoupling them.[3] When uncoupled, the off-diagonal terms of the stiffness matrix are all equal to zero, and the only non-zero terms are found along the leading diagonal. The simultaneous solution of the equations is avoided, and each equation is solved independently of others. In this manner the problem becomes equivalent to solving a set of single-degree-of-freedom equations, a process that involves far fewer operations.

The uncoupling of the equations of motion is carried out using the *mode superposition* method. The basic concept is that we can model any vibration as the sum of individual components or modes. This applies to both continuous and discrete systems. In the mode superposition method, the equations of motion are transformed into another coordinate system, termed the *normal* or *modal* coordinate system, in which the structural matrices (mass, damping and stiffness) are in diagonal form. In finite element analysis, this results in a reduction in the bandwidth of these matrices.

Consider the following equation of motion[4]

$$\mathbf{M}\ddot{\mathbf{x}} + \mathbf{C}\dot{\mathbf{x}} + \mathbf{k}\mathbf{x} = \mathbf{F}(t) \tag{E.11}$$

where the unknown components of the response vector $\boldsymbol{x}$ are termed *generalised coordinates* that include translations or rotations for each degree of freedom.

The basic premise of modal analysis is the *modal expansion theorem*, which is the assertion that any set of vectors can be represented as the superposition of each natural mode multiplied by its time-dependent behaviour. That is,

$$\mathbf{x}(t) = \{\mathbf{u}\}^{(1)}\mathbf{q}_1(t) + \{\mathbf{u}\}^{(2)}\mathbf{q}_2(t) + \ldots\{\mathbf{u}\}^{(n)}\mathbf{q}_n(t) = \left[\{\mathbf{u}\}^{(1)}\ldots\{\mathbf{u}\}^{(n)}\right]\begin{Bmatrix} \mathbf{q}_1(t) \\ \mathbf{q}_2(t) \\ \cdot \\ \mathbf{q}_n(t) \end{Bmatrix} \tag{E.12}$$

where $\{\mathbf{u}\}^{(n)}$ is the nth mode shape (e.g. one half-sine wave) and $\mathbf{q}_n(t)$ is the time-dependent contribution (amplitude).

The modal expansion theorem states that the generalised coordinates, $\boldsymbol{x}$, in which the differential equation has been derived, can be written as

$$\mathbf{x} = \mathbf{u}\mathbf{q}(t) \tag{E.13}$$

where $\mathbf{q}(t)$ is the *vector of the normal* or *modal co-ordinates* and $\mathbf{u}$ is a transformation vector. Differentiating,

$$\dot{\mathbf{x}} = \mathbf{u}\dot{\mathbf{q}}(t) \qquad \ddot{\mathbf{x}} = \mathbf{u}\ddot{\mathbf{q}}(t) \tag{E.14}$$

We now substitute these expressions in the equation of motion (E.11) and pre-multiply by $\mathbf{U}^{\mathrm{T}}$:

$$\mathbf{U}^{\mathrm{T}}\mathbf{M}\mathbf{U}\mathbf{q} + \mathbf{U}^{\mathrm{T}}\mathbf{C}\mathbf{U}\,\dot{\mathbf{q}} + \mathbf{U}^{\mathrm{T}}\mathbf{k}\mathbf{U}\mathbf{q} = \mathbf{U}^{\mathrm{T}}\mathbf{F}$$

[3] The coupling may be due to stiffness coupling or *inertial coupling*. The mass matrix will be uncoupled for a lumped mass system and coupled for a consistent mass system.

[4] The presentation that follows is based on a lecture entitled *Modal Analysis* delivered by K. Vandiver as part of a course in Engineering Dynamics at the Massachussetts Institute of Technology, USA [3]. The example used is available in [4].

When performing the multiplication $\mathbf{U}^{\mathrm{T}}\mathbf{MU}$ it is found that the product is equal to zero, unless the multiplication involves the element in the same, say the rth, position. That is,

$$\mathbf{U}^{\mathrm{T}}\mathbf{U}^{(s)T}\mathbf{MU}^{(r)} = 0 \qquad \text{for } r \neq s \tag{E.15}$$

Hence,

$$\mathbf{U}^{\mathrm{T}}\mathbf{MU} = \begin{bmatrix} M_1 & 0 & 0 & 0 \\ 0 & M_2 & 0 & 0 \\ . & . & . & . \\ 0 & 0 & 0 & M_n \end{bmatrix} = \overline{\mathbf{M}} \tag{E.16a}$$

(diagonal modal mass matrix)

$$\mathbf{U}^{\mathrm{T}}\mathbf{CU} = \begin{bmatrix} C_1 & 0 & 0 & 0 \\ 0 & C_2 & 0 & 0 \\ . & . & . & . \\ 0 & 0 & 0 & C_n \end{bmatrix} = \overline{\mathbf{C}} \tag{E.16b}$$

(diagonal modal damping matrix)

The damping matrix **C** is of diagonal form only for ideal conditions of damping. Orthogonality does not always apply to damping.

also

$$\mathbf{U}^{\mathrm{T}}\mathbf{KU} = \begin{bmatrix} K_1 & 0 & 0 & 0 \\ 0 & K_2 & 0 & 0 \\ . & . & . & . \\ 0 & 0 & 0 & K_n \end{bmatrix} = \overline{\mathbf{K}} \tag{E.16c}$$

(diagonal modal stiffness matrix)

It is seen that the operation involving the pre-multiplication by $\mathbf{U}^{\mathrm{T}}$ and the post-multiplication by **U** transforms the mass, damping and stiffness matrices into corresponding diagonal matrices.

The natural modes of a system form a complete and independent set of vectors. Any kinematically admissible displacement of a system can be described by a linear combination (weighted sum) of the mode shapes of the system. This arises because the mode shapes are orthogonal to one another.

Substituting (D.6a)–(D.6c) in (D.11), the equation of motion becomes:

$$\begin{bmatrix} M_1 & 0 & 0 & 0 \\ 0 & M_2 & 0 & 0 \\ . & . & . & . \\ 0 & 0 & 0 & M_n \end{bmatrix}\ddot{\mathbf{q}} + \begin{bmatrix} C_1 & 0 & 0 & 0 \\ 0 & C_2 & 0 & 0 \\ . & . & . & . \\ 0 & 0 & 0 & C_n \end{bmatrix}\dot{\mathbf{q}} + \begin{bmatrix} K_1 & 0 & 0 & 0 \\ 0 & K_2 & 0 & 0 \\ . & . & . & . \\ 0 & 0 & 0 & K_n \end{bmatrix}\mathbf{q} = \ = \mathbf{U}^{\mathrm{T}}\mathbf{F} = \mathbf{Q}(t) \tag{E.17}$$

where $\mathbf{Q}(t)$ is the *modal force*. This is the equation of motion expressed in modal form and coordinates (*modal equation*). From (E.17) select any mode, (rth). The equation of motion for that mode is then:

$$M_r\ddot{q}_r + C_r\dot{q}_r + K_r q_r = Q_r(t) \tag{E.18}$$

where M_r, C_r, and K_r are the *modal mass, damping* and *stiffness* (scalar) terms, respectively, for mode r. It is seen that (E.18) is the equation of motion of a single-degree-of-freedom oscillator. Therefore, Eq. (E.18) represents N *uncoupled* single-degree-of-freedom systems that can be solved individually. This is much easier than solving the original, coupled equations that were expressed in generalised coordinates $\mathbf{x}$.

However, the solution of the modal equation (E.18) must revert to the generalised coordinate system for it to be comprehensible. This is done by performing the following summation, where only the modes that matter need to be included:

$$\mathbf{x} = \sum_i \{\mathbf{u}\}^i \mathbf{q}_i(t) \tag{E.19}$$

To determine the natural frequencies of vibration ω_n and mode shapes u_n, the equation of motion in free vibration (zero forcing function) needs to be transformed into an algebraic problem. The equation of motion of an undamped system undergoing free vibration that corresponds to (E.11) is:

$$\mathbf{M}\ddot{\mathbf{x}} + \mathbf{k}\mathbf{x} = 0 \tag{E.20}$$

whose solution has the following form:

$$\mathbf{x} = \{\mathbf{u}\}e^{i\omega t}$$

Substituting in (D.20), the solution obtains the following form

$$\left[-\omega^2\mathbf{M} + \mathbf{K}\right]\mathbf{u}e^{i\omega t} = 0 \tag{E.21}$$

Since for a non-trivial solution $\mathbf{u}e^{i\omega t} \neq 0$,

$$\text{we have } -\omega^2\mathbf{M} + \mathbf{K} = \mathbf{0} \tag{E.22}$$

Hence,

$$\det\left[-\omega^2\mathbf{M} + \mathbf{K}\right] = 0 \tag{E.23}$$

This is an eigenvalue equation, where $\omega^2 = \lambda$. Solving this equation using standard techniques, the natural frequencies of the problem, ω_n, and from these the mode shapes (eigenvectors) $\boldsymbol{u}$ are subsequently obtained. The mode shapes tell us, for each case, the relative motion of each degree of freedom with respect to a reference d.o.f.

References

1. Bathe K.-J. (1996) *Finite element procedures*. Upper Saddle River, NJ: Prentice-Hall.
2. Strang W. G. (2006) *Linear algebra and its applications*. 4th ed. Cambridge: Cambridge University Press.

3. Modal analysis: Orthogonality, mass stiffness, damping matrix (3 September 2013). Retrieved 15 February 2021 from www.youtube.com/watch?v=OxcCPTc_bXw
4. (N.d.). LinkedIn.com. Retrieved 15 February 2021 from www.linkedin.com/pulse/revisiting-mode-superposition-analysis-aka-modal-ajay-taneja
5. (N.d.-b). Utah.edu. Retrieved 15 February 2021 from www.math.utah.edu/~gustafso/s2009/orthogonality.pdf

Appendix F

Statistics and Probability Concepts with Applications

F.1 Basic Concepts and Short-Term Loading

The purpose of this appendix is to provide several definitions of basic probability concepts that are of particular relevance to ship structural analysis as well as the probabilistic analysis of ship structural response to random wave loading.

F.1.1 Conditional Probability

Given an event M, such that $P[M] \neq 0$, we form the ratio $P[A \cap M]/P[M]$ where A is an event of S. This ratio is denoted by $P[A|M]$ and is called the *conditional probability of A given M*. Thus,

$$P[A|M] = \frac{P[A \cap M]}{P[M]} \tag{F.1}$$

Conditional probability: given a short-term sea state described by m_0, such that $P[m_0] \neq 0$, we form the ratio $P[x \cap m_0]/P[m_0]$ where x is the wave amplitude, which lies in a range S. This ratio is denoted by $P[x|m_0]$ and is the conditional probability of x given m_0. Thus,

$$P[x|m_0] = \frac{P[x \cap m_0]}{P[m_0]} \tag{F.2}$$

F.1.2 Joint Probability Distribution Function

The *joint cumulative distribution function* or joint distribution function of random variables X, Y is the probability of the event

$$[x \leq X, y \leq Y)] = [x \leq X] \cap [y \leq Y]$$

that consists of all outcomes ζ such that $x(\zeta) \leq X$ and $y(\zeta) \leq Y$. Thus,

$$F_{xy}(x, y) = P[x \leq X, y \leq Y] \tag{F.3}$$

F1.3 Central Limit Theorem

The summed random variable

$$X = X_1 + X_2 + \ldots + X_n = \sum_{j=1}^{n} X_j$$

will be normally distributed as $n \to \infty$ with an expected value μ_x and variance $\sigma_x{}^2$ given by

$$\mu_x = \sum_{j=1}^{n} \mu_j \qquad \sigma_x{}^2 = \sum_{j=1}^{n} \sigma_j{}^2 \tag{F.4}$$

For a proof of the theorem see [7] or [2].

F.1.4 Correlation Coefficients

Pearson's *population correlation coefficient*, ρ, for two variables X, Y is given by:

$$\rho_{X,Y} = \frac{\text{cov}(X, Y)}{\sigma_X \sigma_Y} = \frac{E[(X - \mu_X)(Y - \mu_Y)]}{\sigma_X \sigma_Y} \tag{F.5}$$

The *sample correlation coefficient* r can be obtained by substituting estimates of the covariances and variances based on a sample into the formula above. The formula for r is:

$$r = \frac{\sum_{i=1}^{n} (X_i - \overline{X})(Y_i - \overline{Y})}{\sqrt{\sum_{i=1}^{n} (X_i - \overline{X})^2} \sqrt{\sum_{i=1}^{n} (Y_i - \overline{Y})^2}} \tag{F.6}$$

F.2 Hull Girder Loading

F.2.1 Short-Term Analysis of Ocean Waves

In the long-term analysis of wave loading, M is the characteristic parameter of a particular short-term sea spectrum (Rayleigh) such as m_0 $(= \sigma^2)$ and A is the peak to mean amplitude of the loading (such as vertical bending moment). The value of m_0 in turn depends on significant wave height, $H_{1/3}$, spectral shape S and heading ψ that are assumed to be mutually independent.

The long-term probability of A can then be expressed as a joint probability as follows:

$$q(x, \sigma) = P[X|\sigma]P[\sigma] \tag{F.7}$$

where $P[X|\sigma]$ is the probability of X with respect to σ. Thus, for a Rayleigh distribution (Figure F.1),

$$P[X|\sigma] = \left(\frac{X}{\sigma}\right) e^{-\frac{x^2}{2\sigma^2}} \tag{F.8}$$

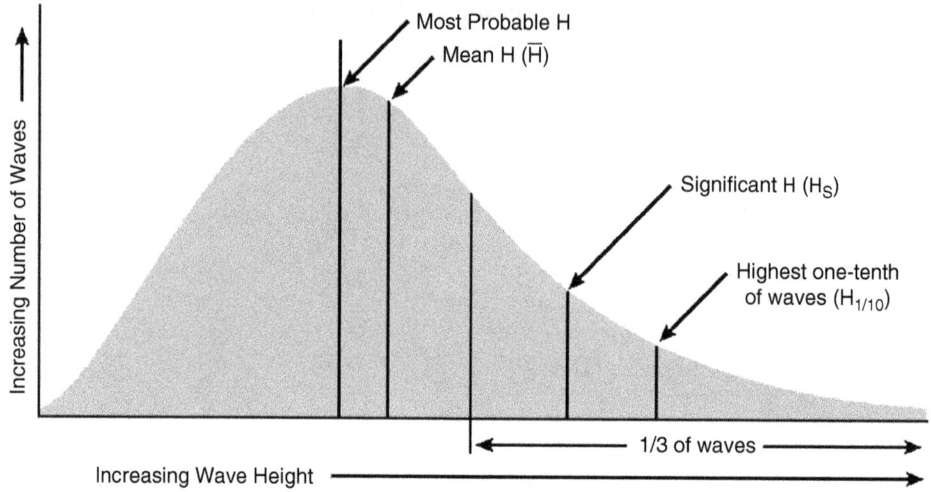

Figure F.1 Probability density function of the Rayleigh distribution.

and $P[\sigma]$ is a group of functions that define the probabilities of σ on the basis of the values of the different variables (significant wave height, $H_{1/3}$, spectral shape S and heading ψ).

F.2.2 Long-Term Loading of the Hull Girder: Order Statistics

$X_1, X_2, \ldots, X_n$ are n continuous random variables with a common density f, a distribution function F. Define:

$$X_{(1)} = \min(X_1, X_2, \ldots, X_n)$$

$$X_{(2)} = \min\{\{(X_1, X_2, \ldots, X_n)\} - \{X_{(1)}\}\}$$

$$X_{(n)} = \max(X_1, X_2, \ldots, X_n)$$

The ordered values of the sample are known as *order statistics*. The question we will try to answer is: What is the formula for the (probability) density (function) of the kth order statistic? We are interested in the extremes:

$$\begin{aligned} F_{\max}(x) &= F_{(n)}(x) = P[X_1 \le x, X_2 \le x, \ldots, X_n \le x] \\ &= F_1(x)F_2(x)\ldots F_n(x) \\ &= F^n(x) \end{aligned} \tag{F.9}$$

$$f_{\max}(x) = \frac{d}{dx}(F(x))^n = n(F(x))^{n-1}f(x) \tag{F.10}$$

$$\begin{aligned} F_{\min}(x) = F_{(1)}(x) &= 1 - P[X_{\min} > x] \\ &= 1 - P[X_1 > x, X_2 > x, \ldots, X_n > x] \\ &= 1 - (1 - F_1(x))(1 - F_2(X)) \ldots (1 - F_n(x)) \end{aligned} \tag{F.11}$$

$$f_{\min}(x) = -\frac{d}{dx}(1 - F(x))^n = n(1 - F(x))^{n-1} f(x) \tag{F.12}$$

F.2.3 Maximum-Likelihood Estimation

The *maximum-likelihood estimation method* is used to obtain values for the parameters of a statistical model when these are not explicitly available and the data set available is insufficient to perform direct calculations. The method is based on the use of a set of limited samples of data in conjunction with knowledge of the appropriate statistical model. If the number of data points is extremely small, then it is not possible to obtain parameter estimates in which case a confidence interval is defined, within which the true value of the parameter, say θ_0, lies. The basis of the method is to determine values of the unknown parameters θ that make the observed results the most probable within the framework of the model. This is achieved by determining an estimator $\hat{\theta}$ that is as close to the true value θ_0 as possible.

Consider a sample $x_1, x_2, \ldots, x_n$ of n independent and identically distributed data points[5] that correspond to a probability density function $f_0(.)$. In the case of ocean wave heights, the p.d.f. of the Rayleigh distribution is appropriate.

We require first to specify the joint density function of the available data points. For an independent and identically distributed sample, the joint density function is

$$f(x_1, x_2, \ldots, x_n|\theta) = f(x_1|\theta) f(x_2|\theta \ldots f(x_n|\theta) \tag{F.13}$$

where θ can be a vector of parameters of the distribution.

The observed values $x_1, x_2, \ldots, x_n$ are treated as fixed 'parameters' of this function, whereas θ will be the function's variable that is allowed to vary freely and will be called the *likelihood*:

$$\mathcal{L}(\theta|x_1, \ldots, x_n) = f(x_1, x_2, \ldots, x_n|\theta) = \prod_{i=1}^{n} f(x_i|\theta) \tag{F.14}$$

It is convenient to work with the logarithm of the likelihood function thereby converting the product of multiple terms into an addition. The logarithm of the likelihood function is called the *log-likelihood*:

$$\ln \mathcal{L}(\theta|x_1, \ldots, x_n) = \sum_{i=1}^{n} \ln f(x_i|\theta) \tag{F.15}$$

[5] A sequence or other set of random variables is *independent and identically distributed* (*i.i.d.*) if each random variable has the same probability distribution as the others and all are mutually independent.

As an example, consider the case of the Rayleigh distribution. Substituting in (F.14) from (F.8), the likelihood is [5]:

$$\mathcal{L} = \prod_{i=1}^{n} \frac{2x_i}{R} \exp\left(-\frac{x_i^2}{R}\right) = \left(\frac{2}{R}\right)^n \left(\prod_{i=1}^{n} x_i\right) \exp\left\{-\frac{\sum_{i=1}^{n} x_i^2}{R}\right\} \tag{F.16}$$

The maximum-likelihood estimator of R, $\hat{R}$, is then defined so that the likelihood $\mathcal{L}$ is larger than any other likelihood function evaluated using other possible estimator, R'. That is,

$$\mathcal{L}(x_1, x_2, \ldots, x_n | \hat{R}) > \mathcal{L}(x_1, x_2, \ldots, x_n | R') \tag{F.17}$$

Therefore, a solution to the following equation is sought

$$\frac{d\mathcal{L}}{dR} = 0$$

In the case of log-likelihood,

$$\frac{d}{dR}(\ln \mathcal{L}) = -\frac{n}{R} + \sum_{i=1}^{n} \frac{x_i^2}{R^2} = 0$$

Hence,

$$\hat{R} = \frac{1}{n} \sum_{i=1}^{n} x_i^2 \tag{F.18}$$

where x_i are random samples of wave height.

It is seen that the value of the estimator depends on the sample size, n. A minimum of 120–150 readings are required to enable a sufficiently accurate calculation of $\mathcal{L}$ for a Rayleigh distribution of wave heights [5].

F.2.4 Modes of Convergence of Random Variables

In this section we shall give some results from mathematical statistics that are used in Section F.2.5.

We say that X_n *converges in probability* to X:

$$X_n \xrightarrow{P} X$$

if for any $\varepsilon > 0$

$$P(|X_n - X| > \varepsilon) \to 0 \qquad \text{as } n \to \infty \tag{F.19}$$

We say that the sequence $X_1, X_2, \ldots, X_n$ *converges in distribution* to X:

$$X_n \xrightarrow{D} X$$

If

$$P(X_n \le x) \to P(X \le x) \qquad \text{as } n \to \infty \tag{F.20}$$

Weak Law of Large Numbers

If $X_1, X_2, \ldots, X_n$ is a sequence of independent, identically distributed random variables each with finite μ, and if $\overline{X}$ is the average of $X_1, X_2, \ldots, X_n$, then

$$\bar{X} \xrightarrow{P} \mu \tag{F.21}$$

Slutsky's Lemma

If $X_n \xrightarrow{D} X$ and $Y_n \xrightarrow{D} y_0$(a constant), then

$$\text{(a)} \quad X_n + Y_n \xrightarrow{D} X + y_0 \tag{F.22}$$

$$\text{(b)} \quad X_n Y_n \xrightarrow{D} y_0 X \tag{F.22}$$

A proof of the lemma is included in [1].

F.2.5 Proof of the Stationarity of the Correlation Coefficient, R [9]

The proof given in this section makes use of the results included in Section F.2.4. The correlation coefficient between loads X and Y calculated from time series of two voyages is given by

$$R_{2voy} = \frac{\sum_{i=1}^{2n} (x_i - \bar{x})(y_i - \bar{y})}{\sqrt{\sum_{j=1}^{2n} (x_j - \bar{x})^2 \sum_{k=1}^{2n} (y_k - \bar{y})^2}} = \frac{\frac{1}{n}\sum_{i=1}^{n} (x_i y_i - \overline{xy}) + \frac{1}{n}\sum_{i=n+1}^{2n} (x_i y_i - \overline{xy})}{\sqrt{\frac{1}{n}\left[\sum_{i=1}^{n} (x_j - \bar{x})^2 + \sum_{j=n+1}^{2n} (x_j - \bar{x})^2\right] \frac{1}{n}\left[\sum_{k=1}^{n} (y_k - \bar{y})^2 + \sum_{k=n+1}^{2n} (y_k - \bar{y})^2\right]}} \tag{F.23}$$

where n is the number of load values in one voyage. It is assumed that each simulated voyage time series contains all known randomness including seasonal effects.

If the randomisation of both portions (i = 1, 2, . . ., n and $i = n$ + 1, n + 2, . . . , 2n) of time series for each load is performed (preserving the value of R as explained in Section 9.3.4), the load values from both voyages can be thought of as realisations of 2n independent and identically distributed random variables $X_1, X_2, \ldots, X_{2n}$ and $Y_1, Y_2, \ldots, Y_n$ with each having its own distribution. Changing 2n to n, 3n, 4n, etc. will not alter these distributions. Therefore, the weak law of large numbers (*WLLN*) can be applied to each term of (F.23). The *WLLN* together with Slutsky's lemma and the convergence in probability and convergence in distribution rules yield, as $n \to \infty$, the following:

$$\bar{x} \xrightarrow{P} E[X] \Rightarrow \bar{x} \xrightarrow{D} E[X] = \mu_x \tag{F.24a}$$

$$\bar{y} \xrightarrow{P} E[Y] \Rightarrow \bar{y} \xrightarrow{D} E[Y] = \mu_y \tag{F.24b}$$

$$\bar{x}^2 \xrightarrow{P} E[X]^2 \Rightarrow \bar{x}^2 \xrightarrow{D} E[X]^2 = \mu_x{}^2 \tag{F.24c}$$

$$\bar{y}^2 \xrightarrow{P} E[Y]^2 \Rightarrow \bar{y}^2 \xrightarrow{D} E[Y]^2 = \mu_y{}^2 \tag{F.24d}$$

$$\frac{1}{n}\sum_{i=1}^{n}(x_i y_i - \bar{x}\bar{y}) \xrightarrow{D,P} E[XY] - E[X]E[Y] \tag{F.24e}$$

$$\frac{1}{n}\sum_{i=n+1}^{2n}(x_i y_i - \bar{x}\bar{y}) \xrightarrow{D,P} E[XY] - E[X]E[Y] \tag{F.24f}$$

$$\frac{1}{n}\sum_{j=1}^{n}(x_j - \bar{x})^2 \xrightarrow{D,P} Var[X] \tag{F.24g}$$

$$\frac{1}{n}\sum_{j=n+1}^{2n}(x_j - \bar{x})^2 \xrightarrow{D,P} Var[X] \tag{F.24h}$$

$$\frac{1}{n}\sum_{k=1}^{n}(y_j - \bar{y})^2 \xrightarrow{D,P} Var[Y] \tag{F.24i}$$

$$\frac{1}{n}\sum_{k=n+1}^{2n}(y_k - \bar{y})^2 \xrightarrow{D,P} Var[Y] \tag{F.24j}$$

Finally using Slutsky's lemma again,

$$\begin{aligned} R_{2voy} &\xrightarrow{D} \frac{E[XY] - E[X]E[Y] + E[XY] - E[X]E[Y]}{\sqrt{(2Var[X])(2Var[Y])}} \\ &= \frac{E[XY] - E[X]E[Y]}{\sqrt{(Var[X]Var[Y])}} = \rho \end{aligned} \tag{F.25}$$

But since ρ is a constant, it is also true that

$$R_{2voy} \xrightarrow{P} \rho \tag{F.26}$$

This can also be extended to $R_{3voy}, R_{4voy}, \ldots.$ and of course R_{1voy}. The stationarity of the correlation coefficient estimate in the case of multiple voyages is thereby proven.

References

1. Bickel P. J. Doksum K. A. (2001) *Mathematical Statistics. Basic Ideas and Selected Topics*. Vol. 1, 2nd ed. Upper Saddle River, NJ: Prentice Hall.
2. Cramér H. (1946) *Mathematical Methods of Statistics.* Princeton, NJ: Princeton University Press.
3. Mansour A. E. (1995) Extreme loads and load combinations. *J. Ship Res.* Vol. 39 53–61.
4. Mansour A. Liu D. (2008) *Strength of Ships and Ocean Structures*, Paulling J. R. (Ed.) Principles of Naval Architecture Series, Jersey City, NJ: SNAME.
5. Ochi M. E. (1999) *Ocean Waves: The Stochastic Approach*. Cambridge: Cambridge University Press.
6. Papoulis A. (1990) *Probabiliy & Statistics* Englewood Cliffs, NJ: Prentice Hall International Inc.
7. Price W. G. Bishop R. E. D. (1974) *Probabilistic Theory of Ship Dynamics*. London: Chapman Hall.
8. Petricić M. Mansour A. E. (2011) Long-term correlation structure of wave loads using simulation. *Mar. Struct.* Vol. 24 97–116.
9. Petricić M. Mansour A. E. (2010) Estimation of the long-term correlation coefficients by simulation. *Proc. ASME 2010 29th Int. Conf. Ocean Offshore & Arctic Eng. OMAE2010 6–11 June 2010 Shanghai*. 1–11.
10. Wikipedia. (11 February 2021). Maximum likelihood estimation. Wikipedia. Retrieved 14 February 2021 from https://en.wikipedia.org/w/index.php?title=Maximum_likelihood_estimation&oldid=1006144516
11. (N.d.). Stanford.edu. Retrieved 15 February 2021 from www-stat.stanford.edu/~susan/courses/s116/node79.html

Appendix G

Fourier Series, Fourier Transforms and Convolution Integrals

G.1 Introduction

The use of infinite series has wide applications in physical sciences in general. In naval architecture, Fourier series and related concepts have been used since early on, in both hydrodynamics and structural mechanics. A brief introduction to relevant theory will be given, illustrated with an example from hydrodynamics (ocean wave representation).

G.2 Ocean Waves

The sea environment consists of a large number of waves all interacting with one another, each having different amplitude, period, speed and direction. For a moment we will assume a situation in which a disturbance consists of waves all travelling in one direction. The waveform consists of a wave of fundamental frequency and waves whose frequencies are multiples of the fundamental frequency and whose phases differ. The resulting profile may have the form shown in Figure G.1a.

It is possible to analyse the resultant wave profile in order to determine the amplitude of the individual wave components and thus the amplitudes and phases of each component can be listed. Results can also be plotted in the form of an amplitude-frequency graph. In this case the graph will look like Figure G.1b.

If we now consider a situation in which waves of different periods, amplitudes and speeds are reaching a particular position, the outcome will be different to that obtained previously, and will include components over a broad range of frequencies (Figure G.2a). The amplitude-frequency plot (spectrum) will then look like Figure G.2b.

In the first case the description can be given by the fundamental frequency and the corresponding amplitude and those of its multiples. This would have the form

$$F(t) = a_0 + a_1 \cos 2\pi f_0 t + b_1 \sin 2\pi f_0 t + a_2 \cos 4\pi f_0 t + b_2 \sin 4\pi f_0 t + \dots \qquad \text{(G.1)}$$

where f_0 is the fundamental wave frequency. This can also be written as

$$F(t) = \sum_{n=-\infty}^{\infty} a_n \cos(2\pi n f_0 t) + b_n \sin(2\pi n f_0 t) \qquad \text{(G.2)}$$

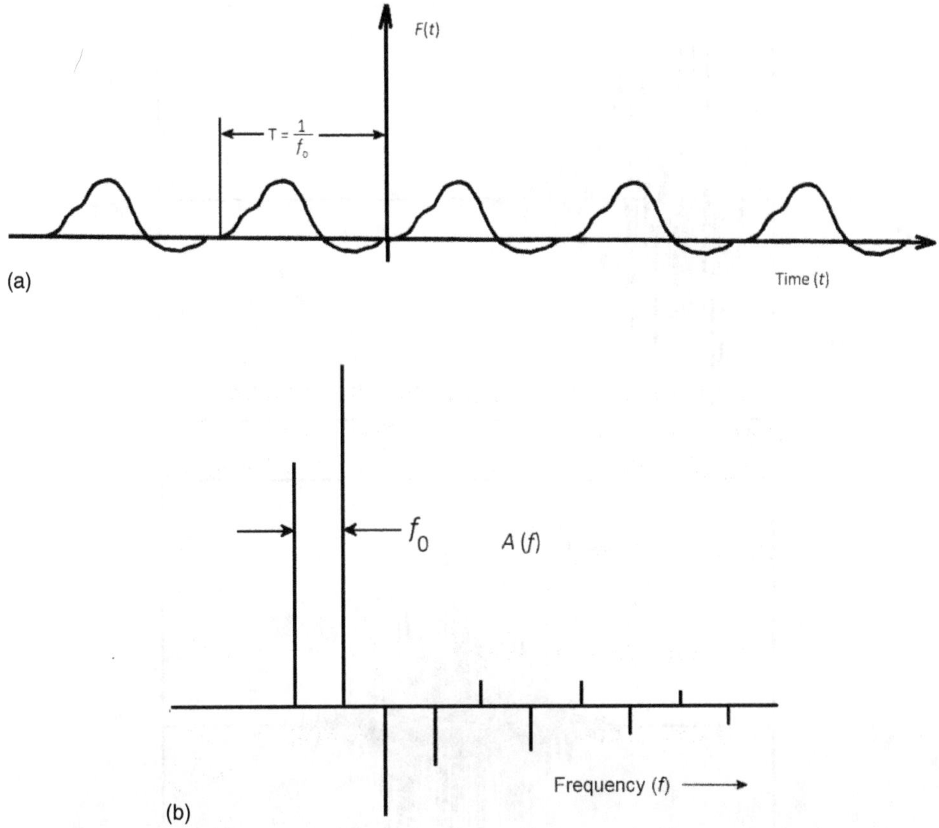

Figure G.1 Spectrum of waves travelling in one direction. (a) Time domain; (b) frequency domain.

The semi-infinite series ($n = 1$ to $+\infty$) can be written as

$$F(t) = \frac{A_0}{2} + \sum_{n=1}^{\infty} A_n \cos(2\pi n f_0 t) + B_n \sin(2\pi n f_0 t) \tag{G.3}$$

since $\cos x = \cos(-x)$ and $\sin x = -\sin(-x)$. The coefficients A_n and B_n are chosen such that

$$A_n = a_{-n} + a_n \tag{G.4a}$$

$$B_n = b_{-n} + b_n \tag{G.4b}$$

The term A_0 is found from the formula used to determine the A_n terms and is halved in order to avoid counting it twice.

If stress measurements have been obtained on board a ship at sea, it is necessary to find the properties of the component stresses, that is, their amplitudes and frequencies. This process is termed *Fourier analysis* and in the case of ship response leads to the identification of the low-frequency overall bending component and high-frequency springing and

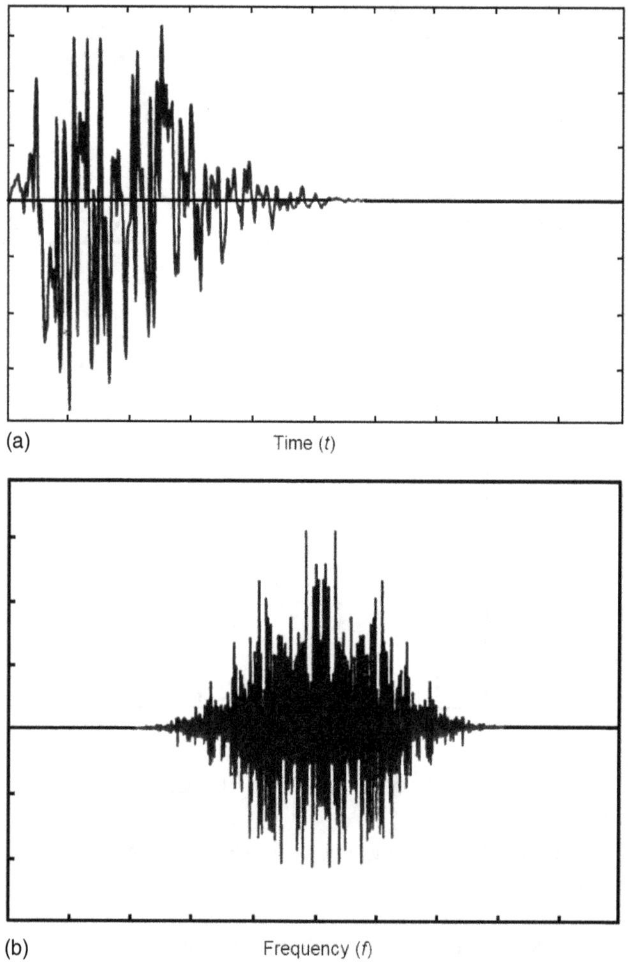

Figure G.2 Wave spectrum of a confused sea. (a) Time domain; (b) frequency domain.

slamming components, if one or more of these are present. In order to do so it is necessary to identify as many values of the Fourier coefficients A_n, B_n as are necessary.

To do this, advantage is taken of the *orthogonality property* of sines and cosines. Two functions are orthogonal over an interval [*0-b*] if the integral of their product over this interval is equal to zero.[6] That is,

$$\int_0^b f(x)g(x)dx = 0 \tag{G.5}$$

[6] Appendix E.

In the case of sines and cosines this property holds except for certain special cases. If $P = 1/f_0$ is one period,

$$\int_0^P \cos(2\pi n f_0 t)\cos(2\pi m f_0 t)dt = 0 \tag{G.6a}$$

$$\int_0^P \sin(2\pi n f_0 t)\sin(2\pi m f_0 t)dt = 0 \tag{G.6b}$$

if $m \neq n$. If $m = n$, these integrals are equal to $1/2f_0$. If (G.3) is multiplied by $\sin(2\pi m f_0 t)$ and the product is integrated over one period, we obtain

$$\begin{aligned}\int_0^P F(t)\sin(2\pi m f_0 t)dt &= \frac{A_0}{2}\int_0^P \sin(2\pi m f_0 t)dt + \\ &+ \int_0^P \left[\sum_{n=1}^{\infty} A_n\cos(2\pi n f_0 t) + B_n \sin(2\pi n f_0 t)\right]\sin(2\pi m f_0 t)dt\end{aligned} \tag{G.7}$$

All terms of the sum vanish except for

$$\int_0^P B_m \sin^2(2\pi m f_0 t)dt = B_m \int_0^P \sin^2(2\pi m f_0 t)dt = \frac{B_m}{2f_0} = B_m\left(\frac{P}{2}\right) \tag{G.8}$$

Hence,

$$B_m = 2f_0 \int_0^{1/f_0} F(t)\sin(2\pi m f_0 t)dt \tag{G.9a}$$

so that, provided $F(t)$ is known over one period P, the coefficient B_m can be found. The corresponding formula for A_m is

$$A_m = 2f_0 \int_0^{1/f_0} F(t)\cos(2\pi m f_0 t)dt \tag{G.9b}$$

Equations (G.9a) and (G.9b) are termed *inversion formulae*. If $F(t)$ has a simple form, direct integration is possible. Otherwise, numerical integration is required.

Up to this point it has been assumed that the Fourier series $F(t)$ is used to represent a time-varying physical process. The process can be described by trigonometric functions (sines and cosines) regardless of whether or not it is periodic [2]. If it is not periodic, all frequencies are necessary for it to be represented. A non-periodic function may be treated as the limiting case of a periodic one when the period tends to infinity. In this case the fundamental frequency tends to zero. All frequencies are required, and the sum of terms is replaced by an infinite series of components each having infinitesimal amplitude $a(f)$. The Fourier representation then becomes

$$F(t) = \int_{-\infty}^{+\infty} a(f)df \cos(2\pi ft) + \int_{-\infty}^{+\infty} b(f)df \sin(2\pi ft) \tag{G.10}$$

or

$$F(t) = \int_{-\infty}^{+\infty} a(f) \cos(2\pi ft + \varphi(f))df \tag{G.11}$$

where $\varphi(f)$ is the *phase angle* at frequency *f*. It is, however, possible and useful to represent the process in frequency terms so that in this case a Fourier series $\Phi(f)$ is used.

G.2.1 The Dirichlet Conditions

The Dirichlet conditions are based on harmonic analysis and are sufficient but not necessary for convergence of a Fourier series. These are:

- *The function has to be single-valued over the period (apart possibly from a finite number of points).*
- *The function has to be piecewise continuous and bounded over the same range.*

When the Dirichlet conditions hold and the function is continuous, the Fourier series converges to the corresponding value of the function (Figure G.3). At points of discontinuity of the first kind it converges to the mean value of the two limits $p(x_0 + 0)$ and $p(x_0 - 0)$, that is,

$$p(x_0) = \frac{p(x_0 + 0) + p(x_0 + 0)}{2} \tag{G.12}$$

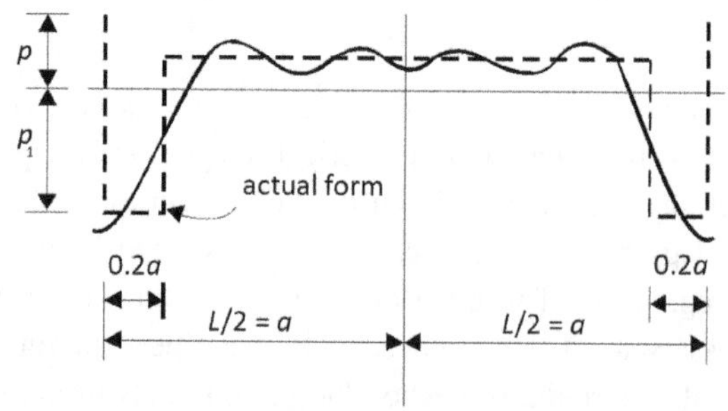

Figure G.3 Approximation using a waveform.

As a general rule, only a few terms of any Fourier series are determined. In this way only an approximate form of the true shape of the function is obtained following the superposition of the trigonometric (sinusoidal) curves that are included in the analysis.

It can be shown that the approximation of a function using Fourier series using n terms is the best of all approximations that is obtained using any kind of trigonometric polynomial having n terms, since it has the least mean square error.

G.3 Fourier Transform

When manipulating Fourier series it is preferable to use complex exponentials. In order to convert an expression such as (G.2) to complex exponential form, Euler's identity, which states that

$$e^{im\theta} = \cos m\theta + i \sin m\theta \tag{G.13}$$

is used. Applying (G.13) to (G.2) we obtain

$$F(t) = \sum_{m=-\infty}^{m=+\infty} \frac{a_m}{2}\left(e^{im\theta} + e^{-im\theta}\right) + \frac{b}{2i}\left(e^{im\theta} - e^{-im\theta}\right) =$$

$$= \sum_{m=-\infty}^{+\infty} \frac{a_m - ib_m}{2} e^{im\theta} + \sum_{m=-\infty}^{+\infty} \frac{a_m + ib_m}{2} e^{-im\theta}$$

Let now $n = -m$. Then,

$$F(t) = \sum_{m=-\infty}^{+\infty} \frac{a_m - ib_m}{2} e^{im\theta} + \sum_{n=-\infty}^{+\infty} \frac{a_{-n} + ib_{-n}}{2} e^{in\theta}$$

Since $a_m = a_{-m}$ and $b_m = -b_{-m}$ and m and n are summed from -∞ to +∞, we have

$$F(t) = \sum_{m=-\infty}^{+\infty} e^{im\theta}\left\{\frac{a_m - ib_m}{2} + \frac{a_{-m} + ib_{-m}}{2}\right\} =$$

$$= \sum_{m=-\infty}^{+\infty} e^{im\theta}\left\{\frac{A_m - iB_m}{2}\right\} =$$

$$= \frac{A_0}{2} + \sum_{m=1}^{\infty} e^{im\theta}(A_m - iB_m)$$

Substituting θ $(= 2\pi f_0 t)$ we obtain

$$\begin{aligned} F(t) &= \frac{A_0}{2} + \sum_{m=1}^{\infty} e^{2\pi i m f_0 t}(A_m - iB_m) = \\ &= \frac{A_0}{2} + \sum_{m=1}^{\infty} e^{2\pi i m f_0 t} C_m \end{aligned} \tag{G.14}$$

The inversion formula for (A7.14) is then analogous to (G.9a) and (G.9b):

$$C_m = 2f_0 \int_0^{1/\nu_0} F(t)e^{-2\pi imf_0t}dt \tag{G.15}$$

From (G.11) and (G.15) we have

$$F(t) = \int_{-\infty}^{+\infty} \Phi(f)e^{2\pi ift}dt \tag{G.16}$$

It can be shown that[7]

$$\Phi(f) = \int_{-\infty}^{+\infty} F(t)e^{-2\pi ift}dt \tag{G.17}$$

where the expression $\Phi(f)$ in (G.16) is termed the *Fourier transform* of $F(t)$ and $F(t)$ in (G.17) is the Fourier transform of $\Phi(f)$. The two together constitute a *Fourier pair*. It is important to note that f and t have inverse dimensions (t^{-1} and t) so that the product ft is a dimensionless number.

G.3.1 Spectral Density

In the case of wave loading of a ship, energy is transmitted and the output O varies with time as $O(t)$. $\Phi(f)$, the Fourier transform of $O(t)$, may well be complex. However, the energy transmitted per unit frequency is proportional to $\Phi(f)\Phi^*(f)$, where the constant of proportionality is given by the response amplitude operator (RAO). The function $S(f) = \Phi(f)\Phi^*(f) = |\Phi(f)|^2$ is called the *energy spectrum* or *energy spectral density* of $F(t)$.

G.3.2 Gaussian Function

Let us consider the Gaussian function

$$G(x) = e^{-x^2/a^2} \tag{G.18}$$

where a is the *width parameter* of the function. Its Fourier transform, $g(p)$ is given by

$$g(p) = \int_{-\infty}^{+\infty} e^{-x^2/a^2}e^{2\pi ipx}dx \tag{G.19}$$

[7] See any advanced textbook on Fourier analysis.

The exponent can be rewritten as $-(x/a - \pi ipa)^2 - (\pi pa)^2$ so that

$$g(p) = e^{-(\pi pa)^2} \int_{-\infty}^{+\infty} e^{-(x/a-\pi ipa)^2} dx$$

Substituting $x/a - \pi ipa = z$ so that $dx = adz$ we obtain

$$\begin{aligned} g(p) &= ae^{-(\pi pa)^2} \int_{-\infty}^{+\infty} e^{-z^2} dz = \\ &= a\sqrt{\pi} e^{-(\pi pa)^2} \end{aligned} \tag{G.20}$$

We see that $g(p)$ is another Gaussian function with width parameter $1/\pi a$.

G.3.3 Dirichlet Conditions for Fourier Transformability

In bringing this brief exposition of Fourier transforms to a close it has to be underscored that not all functions can be Fourier-transformed. The conditions they need to fulfil are termed the *Dirichlet conditions for Fourier transforms*:

1. The functions $F(t)$ and $\Phi(f)$ need to be square integrable, that is $\int_{-\infty}^{+\infty} |F(x)|^2 dx$ is finite. This implies that $F(x) \to 0$ as $|x| \to \infty$.
2. The functions $F(t)$ and $\Phi(f)$ are single-valued.
3. The functions $F(t)$ and $\Phi(f)$ are piecewise continuous. To obtain the transform, it is necessary to consider continuous portions that are connected at discontinuities.
4. The functions $F(t)$ and $\Phi(f)$ have upper and lower bounds (sufficient but not necessary condition).

G.4 The Convolution Integral

Convolution is an operation that is used to obtain the response of a linear system that is at zero initial conditions. The response of a system may be governed by an equation of the following type [1]:

$$\frac{d^2y}{dt^2} + \frac{dy}{dt} + y(t) = f(t) \tag{G.21}$$

where $y(t)$ is the output and $f(t)$ is the input. If $f(t)$ is well defined, it is possible to determine the output $y(t)$ using direct methods. The *convolution integral* enables the output to be determined for any kind of input provided the unit response is known. It is therefore a particularly powerful technique. To gain an intuitive understanding of the procedure it is necessary to treat the response of a linear system to a particular input as the sum of a sequence of unit impulse responses.

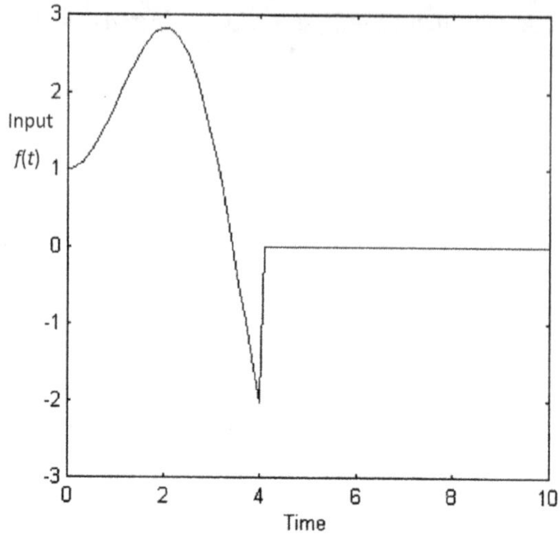

Figure G.4 Input function $f(t)$.

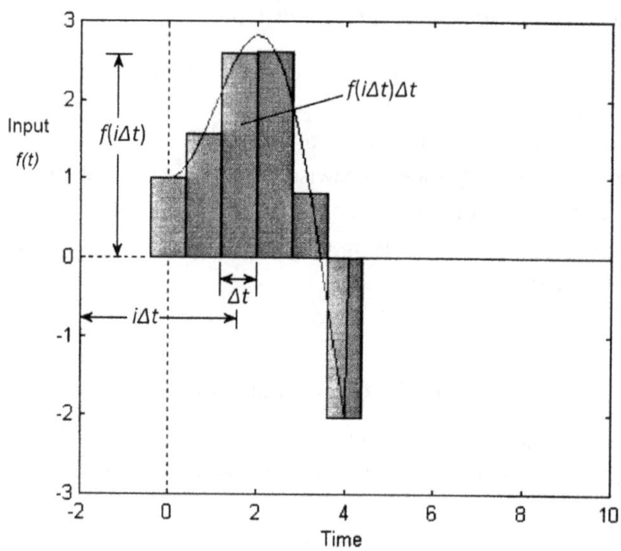

Figure G.5 Discretised input function.

To illustrate this, we consider (G.21) for which the unit impulse response, $h(t)$ can be determined using Laplace transform tables to be

$$h(t) = \frac{2}{\sqrt{3}} e^{-t/2} \sin\left(\frac{\sqrt{3}}{2} t\right) \quad (t > 0) \tag{G.22}$$

The input function $f(t)$ is shown in Figure G.4.

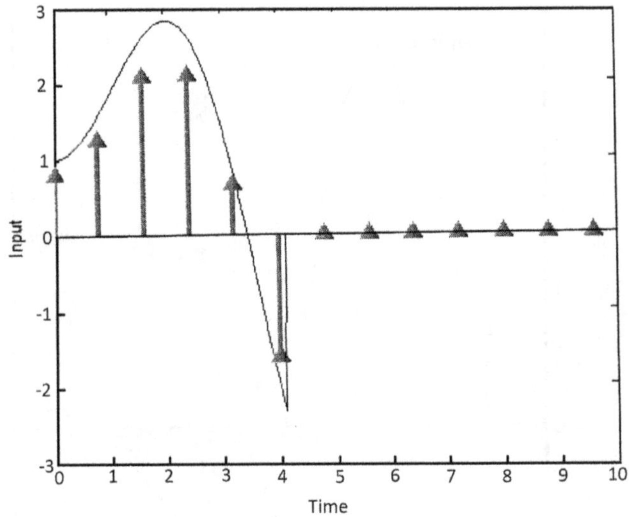

Figure G.6 Input treated as a finite sum of impulses.

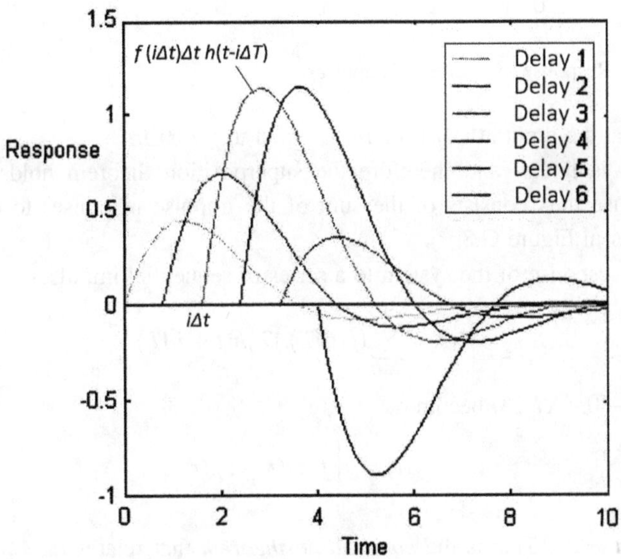

Figure G.7 System response to a finite sum of impulses.

If the continuous input function $f(t)$ is treated as a series of discrete functions with a constant time interval, it may assume the following form (Figure G.5). It is seen that the input at any time $i\Delta t$ is equal to $f(i\Delta t)\Delta t$.

The discretised input can then be considered as a series of impulses acting at regular intervals (Figure G.6).

The responses to these individual inputs are shown in Figure G.7. Since for unit input at time t the response is $h(t)$, for an input of magnitude $f(i\Delta t)\Delta t$ at time $t - i\Delta t$ will be $f(i\Delta T)(\Delta T)(h(t - i\Delta T))$.

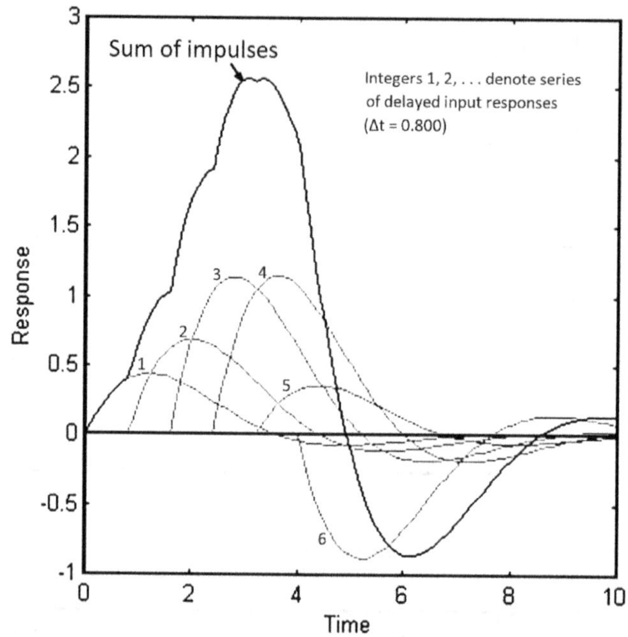

Figure G.8 Response to a series of impulses.

It is seen that the input at any time $i\Delta t$ is equal to $f(i\Delta t)\Delta t$.

The system is linear, and therefore the superposition theorem holds. The resultant impulse response thus consists of the sum of the impulse responses to each individual input, as shown in Figure G.8.

The overall response of the system to a series of sequential impulses is therefore

$$y(t) = \sum_{i=0}^{\infty} (f(i\Delta T)\Delta T)h(t - i\Delta T) \tag{G.23}$$

Letting $\Delta t \to 0$, (A7.23) becomes

$$y(t) = \int_{0}^{\infty} f(\tau)(t - \tau)d\tau \tag{G.24a}$$

and $i\Delta t = \tau$, $\Delta t = d\tau$. This is the *convolution theorem* that relates the output of a linear system to any type of input. It is also expressed as

$$y(t) = f(t)^{*}h(t) = h(t)^{*}f(t) \tag{G.24b}$$

References

1. Cheever E. Swarthmore College. (n.d.). *The Convolution Integral*. Swarthmore.edu. Retrieved 16 February 2021 from http://lpsa.swarthmore.edu/Convolution/Convolution.html
2. James J. F. (1995) *A Student's Guide to Fourier Transforms* 3rd ed. Cambridge: Cambridge University Press.

Index

Note: The italicized page numbers refer to figures.